Springer Monographs in Mathematics

Springer
Berlin
Heidelberg
New York
Hong Kong
London
Milan
Paris
Tokyo

Georges Gras

Class Field Theory

From Theory to Practice

Springer

Georges Gras
University of Franche-Comté
Faculty of Sciences
Laboratory of Mathematics and CNRS
16, route de Gray
25030 Besançon Cedex, France
e-mail: gras@math.univ-fcomte.fr *or* g.mn.gras@wanadoo.fr

Translator of the original French manuscript
Henri Cohen
University of Bordeaux I
Mathematics and Computer Sciences
351, Cours de la Libération
33405 Talence Cedex, France

Library of Congress Cataloging-in-Publication Data applied for

Bibliographic information published by Die Deutsche Bibliothek
Die Deutsche Bibliothek lists this publication in the DeutscheNationalbibliografie;
detailed bibliographic data is available in the Internet at <http://dnb.ddb.de>.

Mathematics Subject Classification (2000): 11RXX, 12R37, 11R29, 11R70, 11R34, 11S31, 11Y40

ISSN 1439-7382
ISBN 3-540-44133-6 Springer-Verlag Berlin Heidelberg New York

Springer-Verlag Berlin Heidelberg New York
a member of BertelsmannSpringer Science+Business Media GmbH

http://www.springer.de

Printed in Germany

Cover design: *Erich Kirchner*, Heidelberg
Typesetting: by the author
Printed on acid-free paper SPIN 10883751 41/3142ck-5 4 3 2 1 0

À Marie-Nicole
À mes parents

Preface

This book is intended for students, for researchers, and for all those who are familiar with classical algebraic number theory. Its aim is to help in the practical use and understanding of the principles of global class field theory for number fields, without any attempt to give proofs of the foundations or their chronological appearance. It does give some historical landmarks, however, for a theory which began at the beginning of the twentieth century and involved many first-rank mathematicians.

More precisely, we will assume the existence and functorial properties of local and global reciprocity maps, as well as the existence of the Galois correspondence of class field theory (between class groups of K and abelian extensions of K, called class fields). These results are essentially given in Theorems II.1.4 and II.3.3, as well as Theorems II.1.5 and II.3.5 of Chapter II, and their proofs can easily be found in the literature.

Even though the proofs of the four basic results are omitted, we give detailed justifications for the consequences that we deduce from these results. This is more effective since the proofs of the foundations rely on a logical chain of reasoning which is different from that of their use. The space that we gain in proceeding in this manner is used to give examples, remarks, to insist on certain technical aspects, and to give practical applications, and finally to give some new or little-known results (such as the precise description of the Galois group of the abelian closure of a number field).

This is done at an elementary level, but with a sufficient degree of generality. For example, for a fixed number field K, the generalized ideal class groups $\mathcal{C}\ell_{\mathfrak{m}}^S$ (indexed by an integral ideal $\mathfrak{m} \neq 0$ of support T corresponding to ramification, and a set of places S corresponding to decomposition (i.e., splitting)), which will play a central role, include both the ordinary class group $\mathcal{C}\ell^{\mathrm{ord}}$ and the restricted (or narrow) class group $\mathcal{C}\ell^{\mathrm{res}}$ which do not come from two different theories, but a single one thanks to a suitable choice of $\mathfrak{m}$ and S. The presence of the decomposition parameter S is also essential because of the symmetry theorems between ramification and decomposition occurring in the theory.

In addition, so as not to create a methodological bias, we will often compare the two classical formalisms: on the one hand that of idèle class groups, and on the other hand that of generalized ideal class groups which are Galois groups of ray class fields. Concerning this, we note that a number of expositions of class field theory begin with the formalism of congruence groups introduced by Weber, which is essentially that of generalized ideal class groups as above after lifting to the group of fractional ideals of K. However, once we know the existence of ray class fields containing a given abelian number field (assumption which we can make *from the beginning*, contrary to most textbooks), we obtain a much nicer and more efficient Galois-theoretic description of the classification of abelian extensions: this is the generalization of the Kronecker–Weber theorem, where ray class fields (over $\mathbb{Q}$) are cyclotomic fields, and introduces the conductor. At this finite level, the idelic viewpoint differs little from the generalized ideal class viewpoint. We will, however, use both techniques since some theoretical computations are much simpler in idelic terms, and some arithmetical interpretations are more natural in terms of ideal classes. The situation is different for limiting processes.

We assume known the classical theory of algebraic number fields (finite extensions of $\mathbb{Q}$), and that of local fields obtained as completions of the above, thus giving finite extensions of the fields $\mathbb{Q}_\ell$ and $\mathbb{R}$. We also assume known the elementary properties of profinite groups. Since we will use idelic and profinite objects, we will naturally have to use topological arguments. We will also use elementary results of the representation theory of finite groups.

We give a number of exercises, together with their solutions. They are usually results which are used elsewhere. Thus, we will consider them as supplementary results with proofs. Sometimes several solutions are proposed, as well as a number of comments.

In addition to giving the main references concerning the topics we will be treating, we will almost systematically quote the book of Koch [e, Ko3] which for each subject gives a “main reference”. This will be more useful for the reader who does not intend to read a large number of papers. In addition, Koch's book contains material going beyond class field theory; this is also the case for the book of Manin–Panchishkin [e, MP], which deals with more geometric aspects of number theory.

For the foundations of class field theory we have not tried to cite all the original papers, which are very numerous and often not easily accessible. This is perhaps unfair, and cannot do justice to the remarkable work of the pioneers and their descendants, which has led to the modern multifaceted approach to the theory, as well as its extensions (local class field theory in arbitrary dimensions, the Langlands programme for nonabelian class field theory, and higher class field theory in the case of arithmetic geometry). Among the main pioneers, from the very end of the nineteenth century to the first third of the twentieth, we must at least mention D. Hilbert, L. Kronecker, H. Weber, P. Furtwängler, T. Takagi, E. Artin. They have been followed by H. Hasse, F.K. Schmidt, C. Chevalley, J. Herbrand, A. Scholz, O. Taussky, S. Iyanaga, whose contributions precede the major references used today (Artin–Tate [d, AT], Cassels–Fröhlich [d, CF], Iyanaga [d, Iy1], Lang [d, Lang1], Serre [d, Se2], Weil [d, Wel]), all published in the sixties, which contain (except Lang's book, more in accordance with our point of view) the cohomological setting of class field theory developed by Hochschild–Nakayama, Tate, Šafarevič, Weil, Serre, ... The reader may reconstruct the history of class field theory by using the references of Section (i) of the bibliography, Chapter XI of [d, CF], Appendix 2 of [d, Iy1]; in [h, Iy2], in addition to interesting biographical information (in particular on Takagi), one can find a more personal and detailed description of the scientific and human relationships between several mathematicians, around class field theory, in France, Germany, and Japan, during a large part of the twentieth century. At the 1998 international conference held in Tokyo, many talks where devoted to the history of class field theory (see in [i, Miy0] the articles by Iyanaga, Frei, Koch, Roquette).

More recent books extend beyond the subject of class field theory alone, which is then considered as a basic tool, and deal with Galois cohomology (Section (g) of the bibliography) or higher K-theory. They are at a much higher technical level than that of the present book.

Finally, we would like to stress that we have used notations which seem to be as much as possible considered the most common. They are taken from several "cultures", and are sufficiently precise so as to be understandable without reference to their definition, using some natural general principles. However, we have dropped a number of bad notations which have unfortunately become classical, such as $K_S^\times$ and J_S for the S-units (global and idelic respectively), since for $S = \emptyset$ they do not represent $K^\times$ and J, but the unit groups (global and idelic respectively) E and U!

I would like to express my warmest thanks to all the people who have looked at the manuscript of the book, at different stages of writing, and/or who have helped in its publication; in particular, in a rough chronological order, to Paul Rey for his pertinent remarks, for his persistence in finding incorrect statements, Michel Waldschmidt for his editorial action and his bibliographic comments, Éva Bayer-Fluckiger for her friendly help, Thong Nguyen Quang Do for fruitful technical exchanges, Jean Cougnard, Chazad Movahhedi, Franz Lemmermeyer for a number of simplifications and improvements, Christian Maire for many interesting discussions on pro-p-groups and restricted ramification, Jean-François Jaulent about the logarithmic class group, also to Henri Lombardi, Detlev Hoffmann for valuable help, anonymous colleagues for their suggestions, then especially to Henri Cohen for the work of translating the first draft of the book into English and for his influence through his books on computational number theory.

La Bérarde, August 2002 *Georges Gras*

Table of Contents

Preface . . . vii

Introduction to Global Class Field Theory . . . 1

I. Basic Tools and Notations . . . 7
§1 Places of K . . . 9
§2 Embeddings of a Number Field in its Completions . . . 12
§3 Number and Ideal Groups . . . 21
a) The Local Case: The Group $K_v^\times$ . . . 21
b) The Global Case: Numbers, Ideals, and Units . . . 23
§4 Idèle Groups — Generalized Class Groups . . . 27
a) Idèle Groups — Topology . . . 28
b) Generalized Class Groups — Rank Formulas . . . 37
§5 Reduced Idèles — Topological Aspects . . . 45
a) The Fundamental Exact Sequence . . . 45
b) Topological Lemmas . . . 49
c) Characters of Profinite Groups . . . 53
§6 Kummer Extensions . . . 54
a) Algebraic Kummer Theory . . . 54
b) Arithmetic Aspects of Kummer Theory . . . 59

II. Reciprocity Maps — Existence Theorems . . . 65
§1 The Local Reciprocity Map — Local Class Field Theory . . . 65
a) Decomposition of Places: Local and Global Cases . . . 66
b) Local Class Field Theory Correspondence . . . 74
c) Local Conductors and Norm Groups . . . 80
d) Infinite Local Class Field Theory . . . 86
§2 Idèle Groups in an Extension L/K . . . 91
a) Canonical Injection of C_K in C_L . . . 91
b) Relations Between Local and Global Norms . . . 92
c) Galois Structure of J_L: Semi-Local Theory . . . 94
d) Local Norm Groups — The Non-Galois Case . . . 98
§3 Global Class Field Theory: Idelic Version . . . 104
a) Global Reciprocity Map — The Product Formula — Global Class Field Theory Correspondence . . . 104

b) Global Class Field Theory in $\overline{K}^{\mathrm{ab}}/K$ 121
§4 Global Class Field Theory: Class Group Version 125
a) Global Norm Conductor — Properties 125
b) Artin's Reciprocity Map — Reciprocity Law — Global Computation of Hasse Symbols — Decomposition Law 130
§5 Ray Class Fields — Hilbert Class Fields 143
a) Elementary Properties — Decomposition Law 144
b) Rank Formulas — The Reflection Theorem 152
c) Class Field Theory Over $\mathbb{Q}$ 161
d) Congruence Groups 164
e) Norm Action on Generalized Class Groups 164
f) The Principal Ideal Theorem — Hilbert Towers 168
§6 The Hasse Principle — For Norms — For Powers 176
§7 Symbols Over Number Fields — Hilbert and Regular Kernels 195

III. Abelian Extensions with Restricted Ramification — Abelian Closure 221
§1 Generalities on H_T^S/H^S and its Subextensions 221
a) Description of $\mathrm{Gal}(K(\mathfrak{m})^S/H^S)$ 221
b) The Case of p-Extensions 226
c) The Structure of $\mathrm{Gal}(H_T^S/H^S)$ — p-Adic Ranks 233
§2 Computation of $\mathcal{A}_T^S := \mathrm{Gal}(H_T^S(p)/K)$ and $\mathcal{T}_T^S := \mathrm{tor}_{\mathbb{Z}_p}(\mathcal{A}_T^S)$. 240
a) $\mathbb{Z}_p$-Free-Extensions — Logarithms 240
b) $\mathcal{A}_T^S$ as an Infinitesimal Ray Class Group 243
c) Computation of $\mathcal{T}_T^S$ 250
d) Class Field Theory Correspondence in $H_T^{\mathrm{res}}(p)/K$ 256
§3 Compositum of the S-Split $\mathbb{Z}_p$-Extensions — The p-Adic Conjecture 258
a) p-Adic Ranks: The Leopoldt–Jaulent–Roy Conjecture . 258
b) The Galois Case 264
c) The Monogeneous Case 268
§4 Structure Theorems for the Abelian Closure of K 274
a) Deployment of $\mathrm{Gal}(\overline{K}^{\mathrm{ab}}(p)/H_p^{\mathrm{ord}}(p))$ 275
b) Triviality Criterion for $\mathcal{T}_T^{\mathrm{ord}}$: When is $\mathcal{G}_T^{\mathrm{ord}}$ Pro-p-Free? 282
c) The Schmidt–Chevalley Theorem — Inertia Groups in $\overline{K}^{\mathrm{ab}}(p)/K$ 287
d) Galois Diagram for $\overline{K}^{\mathrm{ab}}(p)/K$ — Structure of the Connected Component D_0 — The Fundamental Equality: $\overline{K}_v^{\mathrm{ab}} = (\overline{K}^{\mathrm{ab}})_v$ 291
e) Decomposition Law of Wild Places in $\overline{K}^{\mathrm{ab}}(p)/H_p^{\mathrm{ord}}(p)$. 300
f) The Strong p-Adic Conjecture — Other p-Adic Aspects 305
g) Structural Properties of $\overline{G}^{\mathrm{ab}}$ — Divisibility of the Connected Component — Cyclic Embedding Criterion 323

h) The Grunwald–Wang Theorem — Weak Deployment Theorem for Decomposition Groups 330
§5 Explicit Computations in Incomplete p-Ramification 342
§6 Initial Radical of the $\mathbb{Z}_p$-Extensions 348
§7 The Logarithmic Class Group 354

IV. Invariant Class Groups in p-Ramification — Genus Theory 361
§1 Reduction to the Case of p-Ramification 362
§2 Injectivity of the Transfer Map $\mathcal{A}_K^{\mathrm{ord}} \longrightarrow \mathcal{A}_L^{\mathrm{ord}}$ 363
§3 Determination of $(\mathcal{A}_L^{\mathrm{ord}})^G$ and $(\mathcal{T}_L^{\mathrm{ord}})^G$ — p-Rational Fields .. 365
a) Invariant Classes Formulas 366
b) p-Primitive Ramification — p-Rationality 371
§4 Genus Theory with Ramification and Decomposition 375
a) Computation of the Number of Genera — Examples .. 377
b) The Genus Exact Sequence 390
c) Central Classes — Knot Groups 398

V. Cyclic Extensions with Prescribed Ramification 407
§1 Study of an Example 408
§2 Construction of a Governing Field 410
a) Solution to the Cyclic Case of Degree p 412
b) Minimal Ramification Sets 421
c) Approach to the Cyclic Case of Degree p^e 423
d) Solution to the Weak Form 432
§3 Conclusion and Perspectives 434

Appendix: Arithmetical Interpretation of $H^2(\mathcal{G}_T^S, \mathbb{Z}/p^e\mathbb{Z})$. 441
§1 A General Approach by Class Field Theory 442
a) Study of $\mathrm{Ker}(H^2(G, \mathbb{Z}/p^e\mathbb{Z}) \xrightarrow{\mathrm{Inf}} H^2(\mathcal{G}_T^S, \mathbb{Z}/p^e\mathbb{Z}))$.... 443
b) Study of $H^2(G, \mathbb{Z}/p^e\mathbb{Z})$ — The Schur Multiplier 444
c) A Class Field Theory Formula for $|\mathrm{Inf}(H^2(G, \mathbb{Z}/p^e\mathbb{Z}))|$ 449
§2 Complete p-Ramification Without Finite Decomposition 450
§3 The General Case — Infinitesimal Knot Groups 453
a) Infinitesimal Computations 454
b) Infinitesimal Knot Groups — The Number of Relations — A Generalization of Šafarevič's Results 456
c) Finite Generalized p-Class Fields Towers 460
d) A Lower Bound for $\mathrm{rk}_p(H^2(\mathcal{G}_T^S, \mathbb{Z}/p\mathbb{Z}))$ — Conclusion . 463

Bibliography 467

Index of Notations 481

General Index 487

Introduction to Global Class Field Theory

Global class field theory for number fields contains, roughly speaking, two classes of largely interdependent results:

(A) Results showing that the classical invariants of the base field K alone (generalized class groups, unit groups) are sufficient to describe the Galois group of the abelian closure $\overline{K}^{\mathrm{ab}}$ of K and the decomposition law of places of K in its subextensions.

(B) Results which illustrate the local-global principle, coming from the existence of the global reciprocity law which is (correctly) considered as a monumental generalization of Gauss's quadratic reciprocity law [1], and which rely on normic aspects of class field theory, developed essentially by Takagi, Artin, and Hasse.

The complexity of the use of class field theory comes not only from the multiplicity of the above viewpoints, but also from the fact that it is often necessary to use nonalgebraic arguments (which are, in fact, independent from class field theory since, as Chevalley has shown in [h, Che2, § 9], it is possible to avoid any analytical argument in the proofs). There are two aspects of this, as follows.

(C) The construction of abelian number fields, whose existence is assured by (A); this is the starting point for "Kronecker's Jugendtraum" (for an imaginary quadratic base field), or more generally Hilbert's twelfth problem whose aim is to extend to an arbitrary base field K the theory of cyclotomic fields over $\mathbb{Q}$. This aspect can be studied using three fundamentally different paths:

(α) Stark's conjectures, described in [TBS]: these are analytical conjectures linked to classical Artin L-functions of number fields. They allow a numerical approach to the construction of ray class fields; [2]

(β) arithmetic geometry: it shows that certain algebraic numbers coming from torsion points of curves or algebraic varieties also give constructions of abelian extensions; this quite old aspect has its origins in complex multiplication, and has had numerous theoretical and numerical developments; [3]

---

[1] and a few others, the history of which is described in [f, Lem].

[2] See [e, Ko3, Ch. 5, § 1], then [Ro], [j, Coh2, Ch. 6, §§ 1, 2], and [CohR].

[3] See [Deu], [d, CF, Ch. XIII], then [e, Ko3, Ch. 2, § 2], [j, Coh2, Ch. 6, § 3].

(γ) classical Kummer theory: this theory cannot be separated from class field theory since it is the tool used to prove the existence theorems of (A). It allows a direct numerical approach, however. [4] Indeed, in the extension K' of K obtained by adjoining suitable roots of unity, we must determine a suitable radical which presupposes the knowledge of the class and unit group of K', which is possible only in terms of classical geometry of numbers. We can then come back down to K and give explicitly a polynomial defining the desired abelian extension of K which had been characterized in terms of class field.

(D) Finally, density theorems such as the Čebotarev density theorem on Frobenius' automorphisms, which for $K = \mathbb{Q}$ is simply the Dirichlet theorem on the distribution of prime numbers in arithmetic progressions. [5]

In addition to the above four fundamental aspects of algebraic number theory, we can also mention a number of remarkable results linked to class field theory, such as Iwasawa theory [6] and the Main Theorem on class groups of abelian number fields showing that the analytic aspect (in the present case p-adic L-functions) gives part of the Galois structure of these class groups: indeed, for this result we have a geometric approach initiated by Ribet [MaW], and a number field approach initiated by Thaine, with the theory of Euler systems ([Koly], [f, Lang2, Appendix by Rubin], [Ru1, Ru2]). We could also mention results in Galois cohomology dealing with certain nonabelian aspects, and higher K-theory which has roots in the notion of symbol attached to class field theory.

All of this leads us far from our initial goal which is rather classical class field theory (in other words essentially points (A) and (B)), and we come back to it with the following two remarks:

— The parts of class field theory that we will describe in detail, in particular when Kummer theory is involved, either implicitly or explicitly as in Chapter V, hinge around a *symmetry* principle which, historically, has been limited to inequalities, giving the famous "Spiegelungssatz" of Scholz–Leopoldt–Kuroda on p-ranks of class groups; it will however be seen that it governs much more general and precise aspects, as we have explained in [Gr10]. We will prove the general reflection theorem with characters in Chapter II, Section 5.

— Algebraic number theory having potentially many applications, a large body of algorithmic results has recently emerged in class field theory. For this reason, in this book we have always given computable versions of the different mathematical entities we encounter, including the deepest cohomological invariants which, as we will see, measure the ultimate p-adic aspects of the theory. In particular in the Appendix, we have given the translation of certain

[4] [j, Coh2, Ch. 5], [CoDO], [DaP], [Fi].

[5] [e, Ko3, Ch. 1, § 6], [d, Lang1, Ch. VIII, § 4; CF, Ch. VIII, § 3], [c, Nar1, Ch. 7].

[6] [Iw2; Iw3; Iw4], [c, Wa], [e, Ko3, Ch. 4, § 3], [J], [g, NSW, Ch. XI], and a survey by Greenberg in [i, Miy0].

cohomology groups in terms of more familiar arithmetic invariants to which we will shortly return in this Introduction.

For an up-to-date description of the algorithmic aspects of class field theory, one can read the commented bibliography of [j, Coh1], but the reader who wants to systematically use the algorithmic point of view can use [j, Coh2], which regroups the most modern techniques.

* * *

The classical formalism of class field theory deals essentially with *finite* abelian extensions of the base field K, which is equivalent to the study of ray class fields, which we will do in Chapter II. It is however essential to obtain results on some canonical infinite extensions such as the maximal p-ramified abelian extension of K (i.e., unramified outside the places dividing the prime number p), or more generally with restricted ramification. Indeed, they are key objects in the theory since they lead to a good description of $\overline{K}^{\text{ab}}/K$, which is one of the main goals of class field theory. It is of course possible to use a limiting process with the classical formalism; this is the viewpoint taken in Chevalley [h, Che2], Weil [h, We2], Artin–Tate [d, AT]. More precisely, Chevalley studies the dual group of $\text{Gal}(\overline{K}^{\text{ab}}/K)$ and defines a special topology on the idèle class group C of the field K giving a direct result. In the other cases, using the standard topology, the problem is equivalent to the study of the connected component D of the unit element of C. When appropriate, we will use the quotient C/D, with the simplification which consists in removing from the start the archimedean part of the idèle group J, in other words its connected component $(\mathbb{R}^{\times +})^{r_1} \times (\mathbb{C}^{\times})^{r_2}$.

These objects do not enable us to give a direct description of the p-Sylow subgroups of the profinite groups occurring in class field theory; this localization is however essential in practice since the study of $\overline{K}^{\text{ab}}(p)$, the maximal pro-$p$-extension of K contained in $\overline{K}^{\text{ab}}$, relies on delicate p-adic techniques (and on p-adic independence conjectures such as the Leopoldt conjecture) which also require a distinction between tame and wild ramifications. By a suitable localization-completion of the idèle group, it is possible to obtain a class field theory which is directly adapted to the pro-p-abelian extensions of a number field K: this is the so-called p-adic class field theory, initiated by Jaulent in his thesis [Ja2], and which has been rewritten in [Ja7] together with many applications. But this approach supposes the knowledge of the whole classical theory.

For the study of these abelian pro-p-groups, we have chosen an intermediate approach (which will be developed in Chapter III), which consists in working in the $\mathbb{Z}_p$-module $I \otimes_{\mathbb{Z}} \mathbb{Z}_p$. If A is a $\mathbb{Z}$-module, we will call $A \otimes_{\mathbb{Z}} \mathbb{Z}_p$ the p-completion of A, not to be confused with its profinite p-completion

$\varprojlim_U (A/U)$ for the subgroups of finite index U of A; the two notions coincide when A is of finite type.

The topological setting for these two techniques is similar to the one considered by Chevalley in [h, Che2, § 2]. The interest for changing the topology, regarding the natural one, will become clear later (this occurs already for the local infinite class field theory, thus a fortiori for the global one).

To obtain the structure of the abelian closure $\overline{K}^{\text{ab}}$ of K, it is thus sufficient to obtain, for all prime p, that of $\overline{K}^{\text{ab}}(p)$; for this, it is sufficient to know the structure of the maximal abelian T-ramified pro-p-extensions of K (i.e., unramified outside T), where T is any finite set of finite places of K containing the places above p.[7] It represents the abelian aspect of the analogous (local or global) Galois problem[8], to which it corresponds, which will also study at least by giving the fundamental results coming from a general study given in the Appendix.

One of the most striking results (assuming the p-adic Leopoldt conjecture) is then stated in Chapter III in the form of an exact sequence of $\mathbb{Z}_p$-modules:

$$1 \to \prod_{v \nmid p} (F_v^\times)_p \longrightarrow \mathrm{Gal}(\overline{K}^{\text{ab}}(p)/K) \longrightarrow \mathrm{Gal}(H_p^{\text{ord}}(p)/K) \simeq \mathcal{T}_p^{\text{ord}} \times \mathbb{Z}_p^{r_2+1} \to 1,$$

in which enters the noncomplexified maximal abelian p-ramified pro-p-extension $H_p^{\text{ord}}(p)$ of K, the number r_2 of complex places, as well as the isomorphism of profinite groups:

$$\mathrm{Gal}(\overline{K}^{\text{ab}}(p)/H_p^{\text{ord}}(p)) \simeq \prod_{v \nmid p} (F_v^\times)_p.$$

In this direct product, the $(F_v^\times)_p$ (corresponding to the inertia groups of the tame finite places and of the decomposition groups of the real infinite places) are the p-Sylow subgroups of the multiplicative groups of the residue fields of K at all its noncomplex places v, with $(F_v^\times)_p \simeq (\mathbb{Z}/2\mathbb{Z})_p$ for the real places at infinity. We call this isomorphism the deployment theorem. We shall use on several occasions the (perhaps unusual) word of "deployment" to suggest that the Galois group under consideration is the direct product of a canonical family of subgroups (like inertia and/or decomposition groups), which has an interesting arithmetical meaning.

This exact sequence shows that an important arithmetic invariant attached to $\mathrm{Gal}(\overline{K}^{\text{ab}}(p)/K)$ is a finite p-group, $\mathcal{T}_p^{\text{ord}}$, thus equal to the p-torsion subgroup of $\mathrm{Gal}(H_p^{\text{ord}}(p)/K)$, which is not the p-class group $(\mathcal{C}\ell^{\text{ord}})_p$, contrary

[7] This well-studied problem has its origins in Kubota [Kub1] (who, after Chevalley, studies the dual of the group $\mathrm{Gal}(\overline{K}^{\text{ab}}/K)$), then [Ša], [Mi]. It has been developed in [Gr1; Gr2], [Ja2; Ja7], and has led to many applications, such as those given in [Gr4; Gr5], [GrJ], [JaN], [Mo], [MoNg], [JaS].

[8] [g, Ko4, § 10; Hab; NSW, Ch. X; Se3, Ch. II, § 5.6], [Neum2], [Schn].

to what one could expect, but a group which can in a certain sense be put in Kummer duality with a class group through the reflection theorem, and whose order is, in analytic terms, related to the residue at $s = 1$ of the p-adic zeta function of the field K, when it is totally real ([Coa], [Col], [Se6]).

In a cohomological approach to class field theory, the group $\mathcal{T}_p^{\rm ord}$ occurs as the dual of $H^2(\mathcal{G}_p^{\rm ord}, \mathbb{Z}_p)$, where $\mathcal{G}_p^{\rm ord}$ is the Galois group of the maximal noncomplexified p-ramified pro-p-extension of K in $\overline{K}$ (see the Section 2 of the Appendix). It thus gives essential information on the group $\mathcal{G}_p^{\rm ord}$ defined by generators and relations, since its minimal number of generators is well known (again as a consequence of class field theory since, by the Burnside basis theorem, this number is that of the abelianization of $\mathcal{G}_p^{\rm ord}$). This fundamental invariant is made computable thanks to the use of a p-adic logarithm defined on $I \otimes_{\mathbb{Z}} \mathbb{Z}_p$, introduced in [Gr1; Gr2], [GrJ], and which we give here in a slightly more general form. From a numerical point of view, these logarithmic computations assume known only the class and unit group of K, which is indeed the philosophy of class field theory. This interpretation of $H^2(\mathcal{G}_p^{\rm ord}, \mathbb{Z}_p)$ (and its generalizations) in terms of a computable arithmetic invariant, directly linked to class field theory, is one of the themes of the present book.

We will show that we have the following exact sequence (independent of the Leopoldt conjecture) [Ja7, § 2.2]:

$$1 \longrightarrow \bigoplus_{v|p} U_v^1 \longrightarrow \mathrm{Gal}(\overline{K}^{\rm ab}(p)/K) \longrightarrow \mathrm{Gal}(H_{\rm ta}^{\rm res}(p)/K) \longrightarrow 1$$

which involves the maximal abelian tamely ramified pro-p-extension $H_{\rm ta}^{\rm res}(p)$ of K and the group of principal local units $\bigoplus_{v|p} U_v^1$ of the v-completions of K above p, each $U_v^1 \simeq \mu_p(K_v) \times \mathbb{Z}_p^{[K_v \,:\, \mathbb{Q}_p]}$ being interpreted as the inertia group of $v|p$ in $\overline{K}^{\rm ab}(p)/K$.

We will also prove results on "weak deployment" of the inertia and decomposition groups (i.e., for any *finite* number of places of K), in $\overline{K}^{\rm ab}(p)/K$ and $H_{\rm ta}^{\rm res}(p)/K$, which are independent of the Leopoldt conjecture, and which come, because of the Schmidt–Chevalley theorem, from the local-global principle for powers (a fundamental tool for the study of $\mathrm{Gal}(\overline{K}^{\rm ab}/K)$). The weak deployment theorem for decomposition groups (of finite and/or infinite places) is, roughly speaking, equivalent to the Grunwald–Wang theorem. Then we will give a detailed expression for the inertia and decomposition groups of the places of K in the extensions $\overline{K}^{\rm ab}(p)/H_p^{\rm ord}(p)$ and $H_p^{\rm ord}(p)/K$. This will lead us to formulate a new p-adic conjecture on units, which is more precise than the Leopoldt conjecture (Ch. III, § 4, (f)).

Refer to III.4.4.1 for a complete view of the structure of $\mathrm{Gal}(\overline{K}^{\rm ab}(p)/K)$ which can also be partly expressed by the following exact sequence (assuming the Leopoldt conjecture for p) [Ja7, § 2.2]:

$$1 \longrightarrow E^{\text{ord}} \otimes \mathbb{Z}_p \longrightarrow \prod_{v \nmid p} (F_v^{\times})_p \longrightarrow \text{Gal}(H_{\text{ta}}^{\text{res}}(p)/K) \longrightarrow (\mathcal{C}\ell^{\text{ord}})_p \longrightarrow 1,$$

where E^{ord} and $\mathcal{C}\ell^{\text{ord}}$ are the usual unit and class group of K. This exact sequence should not be mistaken for the (better-known) relatively trivial p-ramified class field exact sequence:

$$1 \longrightarrow E^{\text{ord}} \otimes \mathbb{Z}_p \longrightarrow \bigoplus_{v|p} U_v^1 \longrightarrow \text{Gal}(H_p^{\text{ord}}(p)/K) \longrightarrow (\mathcal{C}\ell^{\text{ord}})_p \longrightarrow 1.$$

Apart from these aspects that we have wanted to describe in detail since they are less known or new, the reader will find, starting with Chapter II, the study of notions classically attached to class field theory: local norms, Hasse principles, conductors, Artin maps, symbols, regular and wild kernels, reciprocity laws, the Grunwald–Wang theorem, etc ... The rest of the book, in addition to the concrete study of the properties of the group $\mathcal{T}_p^{\text{ord}}$, of abelian pro-$p$-extensions with restricted ramification, and of the decomposition of places in the compositum of $\mathbb{Z}_p$-extensions of K (Chapter III), deals with a number of practical applications, such as: logarithmic class groups in connection with Gross's conjecture (Chapter III), p-rational and p-regular fields, genus theory with ramification and decomposition (Chapter IV) which is used for the p-class fields towers problem, as well as an effective and original approach to the problem of the existence of a global abelian extension L/K satisfying a number of ramification conditions, and which will enable us to have some idea of the complicated structure of $H_{\text{ta}}^{\text{res}}/K$ (Chapter V).

I. Basic Tools and Notations

This chapter gives the definitions of the objects which will be used throughout this book. We are thus led to give the main general notations.

Let K be a number field (i.e., a finite extension of the field $\mathbb{Q}$ of rational numbers), of signature (r_1, r_2). We will denote by Pl_0 (resp. Pl_∞^{r}) the set of finite (resp. infinite real) places of K (see § 1). Classically, the general framework uses two disjoint finite sets T and S of places of K, associated respectively to the ramification and to the decomposition (i.e., splitting), implicitly understood in the abelian extensions of K which are the subject of the correspondence of class field theory.

The r_1 real places at infinity of K play a special role, that of a possible *complexification*[1] in some abelian extension L of K, which classical books consider, after Hasse, as a ramification. In a completely equivalent manner, it is possible to consider that a real place at infinity of K, complexified in L, has a ramification index equal to 1, a residue degree equal to 2, and defines a Frobenius automorphism of order 2 (the corresponding extension of residue fields being $\mathbb{C}/\mathbb{R}$). This point of view allows us to take for T a set of *finite* places (in other words of nonzero *prime ideals*) and for the moduli $\mathfrak{m}$ (or the conductors $\mathfrak{f}$) nonzero *integral ideals* built from T, in other words elements of the multiplicative monoid $\langle T \rangle_{\mathbb{N}}$, the set S being equal to the union of a finite set S_0 of finite places of K together with a set S_∞ of infinite real places of K.

There are many arguments in favor of the above approach, and we will come back to them later. Let us for now give a single example: in the simplest case ($K = \mathbb{Q}$), to define the cyclotomic fields $\mathbb{Q}(\mu_m)$, $m > 0$, which are clearly the universal ray class fields for class field theory over $\mathbb{Q}$ both for arithmetical and for analytical reasons, we would constantly have to prescribe the ramification of the place at infinity of $\mathbb{Q}$, which is illogical since it is not

[1] We use this neologism at least for two reasons: the first one will be enlarged on below, the second one is that for an arbitrary extension L/K, the kernel of the transfer map $\overline{G}_K^{\mathrm{ab}} \longrightarrow \overline{G}_L^{\mathrm{ab}}$, for the Galois groups of the corresponding abelian closures, is isomorphic to $(\mathbb{Z}/2\mathbb{Z})^{r_1^{\mathrm{c}}}$, where r_1^{c} is the number of real places of K becoming totally complex in L. Generally speaking, in the reasonings of class field theory, complex places never enter, so that the above places "desappear" for the class field theory of L. Therefore, it seems essential to bring out this distinctive nature.

necessary for any other place. From our point of view, the ray class field modulo $m\mathbb{Z}$ is equal to $\mathbb{Q}(\mu_m)$ and its maximal real subfield is obtained by specifying the total splitting of ∞, in other words by choosing $S = \{\infty\}$. Note that in a 1951 paper of Weil [h, We2, § 3], one can already find the above principle which consists in considering only finite moduli, in defining the corresponding groups of unit idèles and, consequently, to have as basic (global) units the totally positive units. This viewpoint, used by Jaulent in [Ja2] and his subsequent work, has also been used by Neukirch in [c, Neu1, Ch. VI], but without the use of the parameter S, which reduces its efficiency (see his remark pp. 400–401).

It follows that the natural basic framework for generalized ideal class groups (resp. unit groups) is the restricted (or narrow) sense ($S_\infty = \emptyset$) (resp. that of totally positive units), which corresponds to abelian extensions of K in which we do not specify the behavior of the places at infinity (as is the case for all the finite places not belonging to T), the ordinary sense ($S_\infty = Pl_\infty^{\rm r}$) corresponding to the total splitting of the places at infinity (in other words, to their noncomplexification).

Important remark. We will see that these two senses are involved in *all* the invariants of class field theory (class groups, unit groups, torsion groups, K-theory groups, zeta functions, ...). To *unify* the vocabulary *independently of any tradition*, in the notations we will use the superscript "res" (as restricted) instead of narrow (for class groups), totally positive (for units), etc ... In the opposite situation, we will use the superscript "ord" (as ordinary) instead of wide, absolute, etc ...

The situation is in fact even more subtle: indeed, the reflection theorem mentioned in the Introduction *exchanges* among other things the sets S_∞ and $Pl_\infty^{\rm r} \backslash S_\infty$, so that the best notion depends on the invariant which is considered, and in certain situations one would like to invert the above principles, which we will indeed do when necessary. Briefly stated, we can for instance say that the "best" torsion groups, given in the Introduction as fundamental invariants, are the $\mathcal{T}_p^{\rm ord}$ and not the $\mathcal{T}_p^{\rm res}$, contrary to the case of class groups.

All the above is quite premature but completely justifies the use of the sets T and S, and informs the reader that he will constantly have to keep in mind these dualities.

Convention. To avoid many repetitions, in the entire book we will say that the number field K is given "*together with sets of places T and S*" if we fix $T \subseteq Pl_0$ (a finite set of finite places of K) and $S = S_0 \cup S_\infty \subseteq Pl_0 \cup Pl_\infty^{\rm r}$ (a finite set of noncomplex places of K) such that $T \cap S_0 = \emptyset$.

We will now give the main results concerning places, embeddings in completions, distinguished subgroups, generalized class groups, unit groups, idèle groups, and idèle class groups of K.

§1 Places of K

Recall, following Cassels in [d, CF, Ch. II, § 1] after Artin, the basic definition.

1.1 Definition (absolute value). An absolute value $|\bullet|$ on K is a map from K to the set of nonnegative real numbers, satisfying $|1+x| \leq c$ whenever $|x| \leq 1$ (for a suitable constant c), whose restriction to $K^\times$ is a group homomorphism from $K^\times$ to $\mathbb{R}^{\times +}$, and such that $|0| = 0$. The trivial absolute value on K is the map which sends $x \in K$ to 1 (resp. 0) if $x \neq 0$ (resp. $x = 0$). □

1.1.1 Remark. Any absolute value $|\bullet|$ on K induces a field topology in which a fundamental system of neighbourhoods of 0 is given by the sets $\{x \in K,\ |x| < \varepsilon\}$, $\varepsilon \in \mathbb{R}^{\times +}$. We have actually $c \geq 1$ and there exists $\lambda \in \mathbb{R}^{\times +}$ such that $|\bullet|_0 := |\bullet|^\lambda$ gives the same topology and satisfies the triangle inequality: $|x+y|_0 \leq |x|_0 + |y|_0$ for all $x,\ y \in K$ (this is clear if $c = 1$; if $c > 1$, chose λ such that $c_0 = 2$ and see the footnote in the above reference for the proof). Thus this topology is that of a metric space, but we will not put emphasis on this aspect which is useless for class field theory. □

We say that two absolute values are equivalent if the topological spaces that they define are homeomorphic. We have: [2]

1.1.2 Proposition. *The absolute values $|\bullet|_1$ and $|\bullet|_2$ are equivalent if and only if there exists $\lambda \in \mathbb{R}^{\times +}$ such that $|\bullet|_2 = |\bullet|_1^\lambda$.* □

As the local-global principle is indeed fundamentally related to a certain topological approximation idea with optimal (i.e., independent) informations, we put: [3]

1.1.3 Definition. An equivalence class of nontrivial absolute values of the number field K is called a place of K. □

The set Pl_K of places v of K is then described by what we will call the generalized Ostrowski theorem which we recall: [4]

1.2 Theorem (Ostrowski). *The set of places of K is represented by the following elements:*

(i) the finite or $\mathfrak{p}$-adic (ultrametric or nonarchimedean) [5] *absolute values:*

---

[2] [d, CF, Ch. II, §§ 4, 5, 6], [b, Rob, Ch. 2, § 1.7].

[3] [a, D; Lg], [c, Ca; Ko2; Neu1; WsE], [d, Iy1; Lang1].

[4] [a, BŠa; D], [b, Gou; K], [d, CF, Ch. II], [c, Ha1; Ca; Nar1, Ch. 3, § 1].

[5] i.e., such that $|x+y| \leq \max(|x|, |y|)$ for all $x,\ y \in K$, and which here verify the additional property $|x+y| = |x|$ when $|y| < |x|$; they correspond to "$c = 1$".

$$|\bullet|_{\mathfrak{p}},\ \text{where } |x|_{\mathfrak{p}} := \left(\frac{1}{\mathrm{N}\mathfrak{p}}\right)^{\mathrm{v}_{\mathfrak{p}}(x)} \quad \text{for all } x \in K^\times,$$

for prime ideals $\mathfrak{p} \neq 0$ *of the ring of integers of* K*, where* $\mathrm{N}\mathfrak{p}$ *is the absolute norm of* $\mathfrak{p}$ *and* $\mathrm{v}_{\mathfrak{p}}$ *is the normalized* $\mathfrak{p}$*-adic valuation (i.e., with image equal to* $\mathbb{Z}$*);*

(ii) the infinite (archimedean) absolute values, which are of two kinds:

(ii$_1$*) the real ones:*

$$|\bullet|_\sigma,\ \text{where } |x|_\sigma := |\sigma(x)| \text{ for all } x \in K^\times,$$

for the r_1 *real* $\mathbb{Q}$*-isomorphisms* σ *of* K *in* $\mathbb{C}$*, where* $|\bullet|$ *is the usual absolute value on* $\mathbb{R}$*;*

(ii$_2$*) the complex ones:*

$$|\bullet|_\sigma = |\bullet|_{c\sigma},\ \text{where } |x|_\sigma := |\sigma(x)|^2 \text{ for all } x \in K^\times,$$

for the r_2 *pairs of nonreal* $\mathbb{Q}$*-isomorphisms* $\{\sigma, c\sigma\}$ *of* K *in* $\mathbb{C}$*, where* c *denotes complex conjugation, and* $|\bullet|$ *is the usual modulus on* $\mathbb{C}$*.* □

The complex absolute values $|\bullet|_\sigma$ do not satisfy the triangle inequality since here the best constant c is equal to 4.

The above classification of places of K is inhomogeneous in the sense that the second type (that of archimedean places) uses the embeddings of K in $\mathbb{C}$ (complete and algebraically closed) built beforehand, while the first type seems to be of a different nature. We will in fact see in Section 2 that this is not the case and that there does exist a unified formulation.

From now on, we will denote by $|\bullet|_v$ these representatives of the places v of K.

1.2.1 Remark (places in an extension L/K). The theory of the decomposition of the prime ideals in a finite extension L/K and that of the extensions of the $\mathbb{Q}$-isomorphisms of K to L will allow us to define the notions of decomposition of finite and infinite places in L/K, respectively; thus, we will obtain the description of Pl_L as the disjoint union of the $Pl_{L,v}$, $v \in Pl_K$, where $Pl_{L,v}$ is the set of places w of L above v (which we will denote by $w|v$).

More precisely, if $\mathfrak{p}$ is a prime ideal of K and if the extension of $\mathfrak{p}$ in L is, as usual:

$$(\mathfrak{p}) = \prod_{i=1}^{n} \mathfrak{P}_i^{e_i},$$

then this yields the places w_i of L represented by the absolute values $|\bullet|_{\mathfrak{P}_i}$, $i = 1, \ldots, n$.

Now let σ be a $\mathbb{Q}$-isomorphism of K in $\mathbb{C}$. If σ is real, denote by $\sigma_1, \ldots, \sigma_{\rho_1}$ the real extensions of σ to L, and by $\sigma_{\rho_1+1}, c\sigma_{\rho_1+1}, \ldots, \sigma_{\rho_1+\rho_2}, c\sigma_{\rho_1+\rho_2}$, the complex ones (with $\rho_1 + 2\rho_2 = [L : K]$). Then this yields the real absolute values $|\bullet|_{\sigma_i}$, $i = 1, \ldots, \rho_1$, and the complex ones $|\bullet|_{\sigma_{\rho_1+j}} = |\bullet|_{c\,\sigma_{\rho_1+j}}$, $j = 1, \ldots, \rho_2$. If σ is complex, the extensions σ_k, $k = 1, \ldots, [L : K]$, are complex and define $[L : K]$ different complex absolute values $|\bullet|_{\sigma_k}$. But $c\,\sigma$ gives rise to the same absolute values $|\bullet|_{c\,\sigma_k} = |\bullet|_{\sigma_k}$, $k = 1, \ldots, [L : K]$.

In other words, complex places give $[L : K]$ complex places above, and real ones give ρ_1 real places and ρ_2 complex ones. It is clear (using the Ostrowski theorem in L) that all the places of L are obtained in this way.

Finally, one sees that the set of places of a number field is canonically descriebed by that of $\mathbb{Q}$ (i.e., the prime numbers and ∞).

We will not dwell further on these questions, nor on the corresponding properties (residue characteristics and degrees, ramification indices, ...). □

1.2.2 Remark (product formula). The normalizations given in the Ostrowski theorem for the finite and infinite absolute values are necessary if we want to have the product formula $\prod_v |x|_v = 1$ for all $x \in K^\times$ (the residue degree of the place always occurs in these normalizations, and this is an additional reason for considering that the complex places are unramified). □

1.3 Remark. If we want the result to be independent of the field (containing x) in which we perform the computation, we must consider on K the expression:

$$\|\bullet\|_v := |\bullet|_v^{\frac{1}{e_v\, f_v}},$$

where e_v and f_v are the ramification index and the residue degree of v in $K/\mathbb{Q}$ (with $e_v = 1$ if v is an infinite place). More precisely, if $K' \supseteq K$ and if v' is a place of K' above v, we have the identity:

$$\|x\|_v = \|x\|_{v'} \ \text{ for all } x \in K.$$

It allows us to define the absolute values $\|\bullet\|_{\overline{v}}$ of an algebraic closure $\overline{\mathbb{Q}}$ of $\mathbb{Q}$, for its places $\overline{v} = \varinjlim_K v$, from the fact that $\overline{\mathbb{Q}} = \varinjlim_K K$ (for finite $K/\mathbb{Q}$). The field $\overline{\mathbb{Q}}$ does not have discrete (ultrametric) valuations $\overline{\mathrm{v}}$; ramification implies that $\overline{\mathrm{v}}(\overline{\mathbb{Q}}^\times) = \mathbb{Q}$. □

1.4 Notations. (i) We denote by:

- Pl, the set of places v of K, union of:
- Pl_0, the set of finite places (see 1.1, (i)), and of:
- $Pl_\infty := Pl_\infty^{\mathrm{r}} \cup Pl_\infty^{\mathrm{c}}$, the set of real and complex places at infinity (see 1.1, (ii)), and we denote by:
- $Pl^{\mathrm{nc}} := Pl_0 \cup Pl_\infty^{\mathrm{r}}$, the set of noncomplex places of K.

(ii) We use the following specific notations:

- $\mathfrak{p}_v$, the prime ideal corresponding to $v \in Pl_0$,

- i_v, the real or complex $\mathbb{Q}$-isomorphism corresponding to $v \in Pl_\infty$. If v is real, $i_v = \sigma$ is unique; if v is complex, i_v represents an arbitrary choice of one of the two $\mathbb{Q}$-isomorphisms σ or $c\,\sigma$. □

1.5 Definitions (valuations, signatures). (i) For every place $v \in Pl$, we define the corresponding valuation $\mathrm{v}_v =: \mathrm{v}$ on $K^\times$ as follows.

- $v \in Pl_0$: $\mathrm{v} := \mathrm{v}_{\mathfrak{p}_v}$ is the normalized $\mathfrak{p}_v$-adic valuation (i.e., with image equal to $\mathbb{Z}$);
- $v \in Pl_\infty^{\mathrm{r}}$: v, with values in $\mathbb{Z}/2\mathbb{Z}$, is such that:

$$\mathrm{v}(x) = 0 \text{ (resp. 1) if } i_v(x) > 0 \text{ (resp. } i_v(x) < 0) \ ;$$

- $v \in Pl_\infty^{\mathrm{c}}$: $\mathrm{v} := 0$.

(ii) For the real places at infinity, we use the corresponding sign functions, with values in $\mathbb{R}^\times/\mathbb{R}^{\times +} \simeq \{\pm 1\}$, and defined on $K^\times$ by:

$$\mathrm{sgn}_v(x) := \mathrm{sgn}(i_v(x)) = (-1)^{\mathrm{v}(x)} \ ;$$

with this notation, the signature homomorphism on $K^\times$ is the map:

$$\mathrm{sgn}_\infty := (\mathrm{sgn}_v)_{v \in Pl_\infty^{\mathrm{r}}} \ : \ K^\times \longrightarrow \bigoplus_{v \in Pl_\infty^{\mathrm{r}}} \{\pm 1\},$$

sending $x \in K^\times$ to $(\mathrm{sgn}_v(x))_{v \in Pl_\infty^{\mathrm{r}}}$. We say that x is totally positive if x belongs to the kernel of sgn_∞. If δ_∞ is only a subset of Pl_∞^{r}, we denote by $\mathrm{sgn}_{\delta_\infty}$ the corresponding partial signature homomorphism:

$$\mathrm{sgn}_{\delta_\infty} := (\mathrm{sgn}_v)_{v \in \delta_\infty} \ : \ K^\times \longrightarrow \bigoplus_{v \in \delta_\infty} \{\pm 1\}. \qquad \square$$

For instance, if K is a totally complex field (i.e., $r_1 = 0$), every element of $K^\times$ is totally positive.

§2 Embeddings of a Number Field in its Completions

2.1 ABSOLUTE CASE. The notion of a $\mathbb{Q}$-embedding of K in $\mathbb{C} := \mathbb{C}_\infty$ (as the algebraic closure of the completion $\mathbb{R} := \mathbb{Q}_\infty$ of $\mathbb{Q}$ at the place ∞) is natural from the point of view of the geometry of numbers. Here, we must extend it to the case of the fields $\mathbb{C}_p$ for the prime numbers p ($\mathbb{C}_p$ being the completion of an algebraic closure of the completion $\mathbb{Q}_p$ of $\mathbb{Q}$ for the p-adic absolute value). In addition, as mentioned in Section 1 after the Ostrowski theorem, we will explain the relationship between $\mathbb{Q}$-embeddings and places of K (finite or not).

So as to see that we are dealing with the same situation, we will consider once again the general problem, denoting by p an arbitrary place of $\mathbb{Q}$, identified either with a prime number or with ∞. This use of the letter p as a place of $\mathbb{Q}$ is temporary and will not be used in the sequel since places of K will always be denoted v, including in the case where $K = \mathbb{Q}$, and the residue characteristic of a finite place will be denoted ℓ.

Thus, let K be a number field; to begin with, we can write:

$$K := \mathbb{Q}[X]/(Q),$$

where Q is a suitable irreducible polynomial of $\mathbb{Q}[X]$, and we denote by θ the class of X modulo (Q). Let Pl_p be the set of places v of K above p. The completions K_v of K, for $v \in Pl_p$, can be constructed in the following way: we assume $\mathbb{Q}_p$ known, and we consider the $\mathbb{Q}_p$-algebra $K \otimes_{\mathbb{Q}} \mathbb{Q}_p$, whose decomposition as a product of fields yields:

$$K \otimes_{\mathbb{Q}} \mathbb{Q}_p =: \bigoplus_{v|p} K_v.$$ [6]

More precisely, if we use the writing $K = \mathbb{Q}[X]/(Q)$, the decomposition of Q into irreducible polynomials of $\mathbb{Q}_p[X]$ yields, by uniqueness of the decomposition of $K \otimes_{\mathbb{Q}} \mathbb{Q}_p$ as a product of fields:

$$\begin{aligned} K \otimes_{\mathbb{Q}} \mathbb{Q}_p &= (\mathbb{Q}[X]/(Q)) \otimes_{\mathbb{Q}} \mathbb{Q}_p \simeq \mathbb{Q}_p[X]/(Q) \\ &\simeq \mathbb{Q}_p[X]\Big/\Big(\prod_{v|p} Q_v\Big) \simeq \bigoplus_{v|p}(\mathbb{Q}_p[X]/(Q_v)) \simeq \bigoplus_{v|p} K_v, \end{aligned}$$

for a suitable indexing of the $\mathbb{Q}_p$-irreducible factors of Q.

Since it is very useful to see all the K_v for $v|p$ in the same algebraic closure of $\mathbb{Q}_p$ (in particular for local norm problems and genus theory, when L/K is not Galois), we can proceed in the following way (still with p a finite or infinite place of $\mathbb{Q}$). Let $\mathbb{C}_p$ be a completion of an algebraic closure of $\mathbb{Q}_p$.

2.1.1 Definitions (p-completions of K — embeddings of K). (i) For each $v \in Pl_p$ we fix some root θ_v of Q_v in $\mathbb{C}_p$ and we set $K_v := \mathbb{Q}_p(\theta_v) \subset \mathbb{C}_p$. This defines all the completions of K above p, uniquely up to $\mathbb{Q}_p$-conjugation.

(ii) We define the embeddings of K in the K_v by:

$$\begin{aligned} i_v \,:\, K &\longrightarrow K_v. \\ \theta &\longmapsto \theta_v \end{aligned}$$

□

We immediately obtain:

2.1.2 Proposition. *The field K_v is the closure of $i_v(K)$ for the topology of $\mathbb{C}_p$. By construction, we thus have:*

$$K_v = i_v(K)\,\mathbb{Q}_p.$$

[6] A way used by Chevalley in [h, Che2, § 2]; see [d, Se2, Ch. II, § 3], [b, A], [c, Ca], [d, CF, Ch. II].

For a finite place v and for the ideal $\mathfrak{p}_v$ of the valuation v *of K, $i_v(\mathfrak{p}_v)$ is contained in the maximal ideal of $\mathbb{C}_p$.* □

One of the embeddings, i_{v_1} say, can be considered as the identity map if we have realized K as an extension of $\mathbb{Q}$ in $\mathbb{C}_p$, but we can do this only for a single place p (often for $p = \infty$), and not for the idelic embedding that we will have to use.

2.1.3 Remark. A priori, the absolute value on K_v is obtained by extending by continuity the absolute value $|\bullet|_v$ of K, by setting:[7]

$$|\alpha|_{K_v} := \lim_{\substack{i_v(x) \to \alpha \\ x \in K}} |x|_v \quad \text{for all } \alpha \in K_v.$$

However, we also know that, up to normalization, the absolute value on K_v is unique and given by the well-known (and more practical) formula:

$$|\alpha|_{K_v} = |\mathrm{N}_{K_v/\mathbb{Q}_p}(\alpha)|_p,$$

coming from the absolute value $|\bullet|_p$ of $\mathbb{Q}_p$ (p a prime number or ∞). The *standard* absolute value of $\mathbb{C}_p$ is then given by extending by continuity the absolute value:

$$\|z\| := \Big|\mathrm{N}_{\mathbb{Q}_p(z)/\mathbb{Q}_p}(z)\Big|_p^{\frac{1}{[\mathbb{Q}_p(z)\,:\,\mathbb{Q}_p]}},$$

for every $z \in \overline{\mathbb{Q}}_p$ (algebraic closure of $\mathbb{Q}_p$ in $\mathbb{C}_p$) (see 1.3). □

2.2 Another Definition of Places. For each v, the choice of a conjugate of K_v (i.e., the choice of i_v) is not canonical (choice of a root of Q_v) hence, for the converse aspect, we introduce the following equivalence relation in the set of $\mathbb{Q}$-embeddings of K into $\mathbb{C}_p$.

2.2.1 Definition. We say that two $\mathbb{Q}$-embeddings (or $\mathbb{Q}$-isomorphisms) σ and σ' of $K = \mathbb{Q}(\theta)$ into $\mathbb{C}_p$ are equivalent (or define the same place $v \in Pl_p$) if $\sigma(\theta)$ and $\sigma'(\theta)$ are $\mathbb{Q}_p$-conjugate (i.e., if $\sigma(\theta)$ and $\sigma'(\theta)$ are roots of the same $\mathbb{Q}_p$-irreducible factor of $\mathrm{Irr}(\theta, \mathbb{Q})$). □

It is easy (but essential) to check that all this (i.e., 2.1.1 and 2.2.1) does not depend on the choice of the primitive element θ (indeed, each element of K is a rational polynomial in θ), so that it is equivalent to say that $\mathbb{Q}_p\,\sigma(K)$ and $\mathbb{Q}_p\,\sigma'(K)$ are $\mathbb{Q}_p$-conjugate fields.

2.2.2 Remarks. (i) The arbitrarily fixed $\mathbb{Q}$-embeddings i_v form a complete system of representatives of classes for the equivalence relation given above.

[7] The notation $|\bullet|_v$ makes sense on K but not on $i_v(K) \subset \mathbb{C}_p$.

Thus, a class C_v is of the form $C_v = \{\sigma,\ \sigma = \tau \circ i_v\}$, where τ ranges over all the $\mathbb{Q}_p$-isomorphisms of K_v in $\mathbb{C}_p$.

The field $i_v(K) \cap \mathbb{Q}_p$ is independent of the choice of $i_v \in C_v$, and is in general different from $\mathbb{Q}$. We have $K_v = i_v(K)\,\mathbb{Q}_p$ but this compositum is not necessarily a direct compositum over the intersection.

(ii) This way of considering the v-completions of K is not the naïve one which would give some abstract fields K'_v containing K as a dense subfield; it corresponds instead exactly to what is usually done for $p = \infty$ where each completion is equal to $\mathbb{R}$ or to $\mathbb{C}$ with the counterpart that it is necessary to describe precisely the $\mathbb{Q}$-embeddings of K in these completions. For example, if $K = \mathbb{Q}(\sqrt[3]{2})$ is considered as a real field, for the first place at infinity v we have $i_v(K) = K \subset \mathbb{R}$, but for the other place at infinity v' we have $i_{v'}(K) = \mathbb{Q}(j\sqrt[3]{2}) \subset \mathbb{C}$ and $c \circ i_{v'}(K) = \mathbb{Q}(j^2\sqrt[3]{2}) \subset \mathbb{C}$ which are $\mathbb{R}$-conjugate. Thus, this interpretation of Pl_p as the set of equivalence classes of $\mathbb{Q}$-embeddings of K into $\mathbb{C}_p$ indeed generalizes what is well known in the case $p = \infty$.

(iii) It is also related to the fact, mentioned in 2.1.3, that the theorem about the extensions of the absolute value $|\bullet|_p$ in an algebraic extension of $\mathbb{Q}_p$ (p a prime number or ∞) states uniqueness and yields the following formula, when we compute the different absolute values (above p) of a given $x \in K$ *in the corresponding completions*:

$$|x|_v = |\mathrm{N}_{K_v/\mathbb{Q}_p}(i_v(x))|_p \text{ for all } x \in K\ ;$$

this expression is invariant under $\mathbb{Q}_p$-conjugation of the image $i_v(x)$, and corresponds to the same equivalence relation. Although it is typical of the p-adic case (p a prime number), it is also valid in the case $p = \infty$: if $K_v/\mathbb{Q}_p = \mathbb{C}/\mathbb{R}$, $\mathrm{N}_{K_v/\mathbb{Q}_p}(i_v(x)) = \mathrm{N}_{\mathbb{C}/\mathbb{R}}(i_v(x)) = i_v(x)c(i_v(x))$, and $|x|_v = |i_v(x)c(i_v(x))|_p$ defines a complex absolute value on K; since $|\bullet|_p$ denotes the usual absolute value on $\mathbb{R}$, we indeed obtain the *square* of the complex modulus. □

2.2.3 Example. Let $K := \mathbb{Q}[X]/(X^6 - 2) =: \mathbb{Q}(\theta)$, and fix a place p of $\mathbb{Q}$. Denote by $\sqrt[6]{2}$ a root (possibly specified in some way) of $X^6 - 2$ in $\mathbb{C}_p$, the roots of $X^6 - 2$ being the $\zeta^i\sqrt[6]{2}$ for $0 \le i \le 5$, where ζ is a fixed primitive 6th root of unity in $\mathbb{C}_p$; set also $\xi := \zeta^2$ (a primitive cube root of unity), so that $\zeta = -\xi^2$. This defines the six $\mathbb{Q}$-embeddings σ_i of K into $\mathbb{C}_p$ by:

$$\sigma_i(\theta) := \zeta^i\sqrt[6]{2},\ 0 \le i \le 5.$$

(i) For $p = 2$, the polynomial $X^6 - 2$ is irreducible in $\mathbb{Q}_2[X]$; we can thus fix the unique completion K_v of K above 2, up to $\mathbb{Q}_2$-conjugation, as being $\mathbb{Q}_2(\sqrt[6]{2})$ (i.e., $\theta_v := \sqrt[6]{2}$). The six $\mathbb{Q}$-embeddings of K are equivalent and correspond to the six $\mathbb{Q}_2$-isomorphisms τ_i of K_v, by means of the relation $\tau_i(\sqrt[6]{2}) := \zeta^i\sqrt[6]{2}$ for $0 \le i \le 5$.

We thus have $i_v(\theta) = \sqrt[6]{2}$.

This unique class corresponds to the unique prime ideal $\mathfrak{p}_v := (\sqrt[6]{2})$ above (2) in K.

(ii) For $p = 7$, we have the following factorization as a product of irreducibles of $\mathbb{Q}_7[X]$:

$$X^6 - 2 = (X^3 - \sqrt[6]{2}^{\,3})(X^3 + \sqrt[6]{2}^{\,3}),$$

with $\sqrt[6]{2}^{\,3} \equiv 4 \bmod (7)$, which defines the completions (of degree 3, equal to $\mathbb{Q}_7(\sqrt[6]{2}) = \mathbb{Q}_7(\sqrt[3]{2})$):

$$\begin{aligned} &K_v, \ \text{corresponding to the root } \theta_v := \sqrt[6]{2}, \\ &K_{v'}, \ \text{corresponding to the root } \theta_{v'} := -\sqrt[6]{2}\ ; \end{aligned}$$

the two classes of $\mathbb{Q}$-embeddings are then:

$$C_v := \{\sigma_0, \sigma_2, \sigma_4\}, \ C_{v'} := \{\sigma_1, \sigma_3, \sigma_5\}.$$

We have $i_v(\theta) = \sqrt[6]{2}$ and $i_{v'}(\theta) = -\sqrt[6]{2}$.

These classes correspond to the two prime ideals of K above 7 which are:

$$\mathfrak{p}_v := (7, \theta^3 - 4), \ \mathfrak{p}_{v'} := (7, \theta^3 + 4).$$

Note that the completions K_v and $K_{v'}$ are abelian over $\mathbb{Q}_7$ since $\zeta \in \mathbb{Q}_7$.

(iii) For $p = \infty$, we have the following factorization as a product of irreducibles of $\mathbb{Q}_\infty[X]$:

$$X^6 - 2 = (X - \sqrt[6]{2})(X + \sqrt[6]{2})(X^2 - \sqrt[6]{2}X + \sqrt[6]{2}^{\,2})(X^2 + \sqrt[6]{2}X + \sqrt[6]{2}^{\,2}),$$

with $\sqrt[6]{2} \in \mathbb{R}$, giving the expected completions:

$$K_{v_1} = \mathbb{R}, \ \ K_{v_2} = \mathbb{R}, \ \ K_{v'_1} = \mathbb{C}, \ \ K_{v'_2} = \mathbb{C},$$

corresponding to the following roots:

$$\theta_{v_1} := \sqrt[6]{2} \in \mathbb{R}, \ \ \theta_{v_2} := -\sqrt[6]{2}, \ \ \theta_{v'_1} := -\xi\sqrt[6]{2}, \ \ \theta_{v'_2} := \xi\sqrt[6]{2},$$

and to the four classes of $\mathbb{Q}$-embeddings:

$$\begin{aligned} C_{v_1} &:= \{\sigma_0\}, \quad & C_{v_2} &:= \{\sigma_3\}, \\ C_{v'_1} &:= \{\sigma_1, \ \sigma_5 = c\,\sigma_1\}, \quad & C_{v'_2} &:= \{\sigma_2, \ \sigma_4 = c\,\sigma_2\}. \end{aligned}$$

(iv) For $p = 5$, we have the following factorization as a product of irreducibles of $\mathbb{Q}_5[X]$:

$$\begin{aligned} X^6 - 2 = (X^2 - \sqrt[6]{2}^{\,2}) \times \\ (X^2 - \sqrt{-3}\,\sqrt[6]{2}\,X - \sqrt[6]{2}^{\,2})(X^2 + \sqrt{-3}\,\sqrt[6]{2}\,X - \sqrt[6]{2}^{\,2}), \end{aligned}$$

with $\sqrt[6]{2}^{\,2} \equiv 3 \bmod (5)$ and where $\sqrt{-3}\,\sqrt[6]{2} \equiv 1 \bmod (5)$ in $\mathbb{Q}_5$, which defines the completions (equal to $\mathbb{Q}_5(\sqrt{3}) = \mathbb{Q}_5(\xi)$ since -1 is a square in $\mathbb{Q}_5$):

K_v, defined by the root $\theta_v := \sqrt[6]{2} = \sqrt{3}\,u$,
$K_{v'}$, defined by the root $\theta_{v'} := \xi\sqrt[6]{2} = \xi\sqrt{3}\,u$,
$K_{v''}$, defined by the root $\theta_{v''} := \xi^2\sqrt[6]{2} = \xi^2\sqrt{3}\,u$,

where $u \in \mathbb{Z}_5^\times$, $u \equiv 1 \bmod (25)$; the three classes of $\mathbb{Q}$-embeddings are:

$$C_v := \{\sigma_0, \sigma_3\},\ C_{v'} := \{\sigma_1, \sigma_2\},\ C_{v''} := \{\sigma_4, \sigma_5\},$$

and the corresponding prime ideals above 5 are:

$$\mathfrak{p}_v := (5, \theta^2 - 3),\quad \mathfrak{p}_{v'} := (5, \theta^2 - \theta - 3),\quad \mathfrak{p}_{v''} := (5, \theta^2 + \theta - 3).$$

(v) For $p = 31$, $X^6 - 2$ splits as a product of six polynomials of degree 1 in $\mathbb{Q}_{31}[X]$, giving six places corresponding to six completions equal to $\mathbb{Q}_{31}$, to six inequivalent $\mathbb{Q}$-embeddings i_v, and to six prime ideals of the form $\mathfrak{p}_i := (31, \theta - 19(-5)^i)$ for $0 \le i \le 5$. □

2.3 Absolute Galois Case. Let $K/\mathbb{Q}$ be Galois with Galois group G. In this case, the study given in Paragraph 2.2 becomes simpler.

2.3.1 Decomposition Group of a Place — Decomposition Field. We first define a group action of G on Pl by setting, for all $s \in G$:

$$|s(x)|_{sv} := |x|_v \text{ for all } x \in K,$$

or, equivalently, $|x|_{sv} := |s^{-1}(x)|_v$ for all $x \in K$. If p is the place of $\mathbb{Q}$ below v, the group G acts transitively on the set Pl_p. We set:

$$D_v := \{s \in G,\ sv = v\},$$

which is called the decomposition group of v. The subfield K^{D_v} of K fixed under D_v is called the decomposition field of v in $K/\mathbb{Q}$, and one shows that $i_v(K^{D_v}) = i_v(K) \cap \mathbb{Q}_p$; thus, it is dense in $\mathbb{Q}_p$. If $v \in Pl_0$ and if $\mathfrak{p}_v$ is the corresponding prime ideal of K, D_v is also the decomposition group of $\mathfrak{p}_v$ in $K/\mathbb{Q}$.

If θ' is a $\mathbb{Q}_p$-conjugate of $i_v(\theta)$ in $\mathbb{C}_p$, it is a $\mathbb{Q}$-conjugate of $i_v(\theta)$, and by assumption $\mathbb{Q}(\theta') = \mathbb{Q}(i_v(\theta))$, hence $\mathbb{Q}_p(\theta') = \mathbb{Q}_p(i_v(\theta)) = K_v$ which is then a Galois extension of $\mathbb{Q}_p$; moreover, it is independent of $v|p$ since all the $i_{v'}(\theta)$, $v'|p$, are $\mathbb{Q}$-conjugate (see also 2.5.1). It follows that if we set:

$$G_v := \mathrm{Gal}(K_v/\mathbb{Q}_p),$$

the map which sends $\tau \in G_v$ to the restriction to K of $i_v^{-1} \circ \tau \circ i_v$ defines, by linear disjunction of $i_v(K)$ and $\mathbb{Q}_p$ over their intersection, an isomorphism between G_v and D_v. This can be summarized by the following diagram:

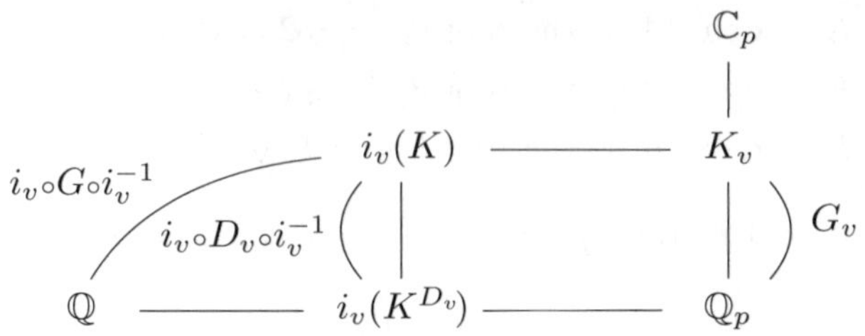

Fig. 2.1

in which $i_v(K^{D_v}) = i_v(K) \cap \mathbb{Q}_p$ is independent of the choice of i_v in C_v, and $K_v = i_v(K)\,\mathbb{Q}_p$. Thus $\sigma \in C_v$ if and only if there exists $\tau \in G_v$ such that $\sigma = \tau \circ i_v$, hence if and only if $i_v^{-1} \circ \sigma = i_v^{-1} \circ \tau \circ i_v \in D_v$.

To summarize, we thus have the following results in the Galois case.

2.3.2 Proposition. *If $K/\mathbb{Q}$ is Galois with Galois group G, the equivalence class C_v of elements σ equivalent to i_v is characterized by the condition:*

$$i_v^{-1} \circ \sigma \in D_v := D_v(K/\mathbb{Q}),$$

where $D_v(K/\mathbb{Q}) \subseteq G$ is the decomposition group of the place v in $K/\mathbb{Q}$. Furthermore, the extension $K_v/\mathbb{Q}_p$ is Galois with Galois group G_v canonically isomorphic to D_v. □

2.3.3 Proposition. *We have the formula:*

$$D_{sv} = s\,D_v\,s^{-1} \text{for all } s \in G. \qquad \square$$

2.4 Relative Case. It we are careful, it is easy to extend these embedding questions to the case of relative extensions.

Let L/K be a finite extension of number fields (we will set $L := K(\theta)$) and, for $v \in Pl_K$, fix some conjugate of the completion K_v of K in $\mathbb{C}_p$ and the corresponding embedding i_v of K.

If $Pl_{L,v}$ denotes the set of places w of L above v, we can fix the representatives i_w of the classes of $\mathbb{Q}$-embeddings of L in $\mathbb{C}_p$ so that ${i_w}_{|_K} = i_v$, the w-completions $L_w := i_w(L)\,\mathbb{Q}_p = K_v(\theta_w)$ (where $\theta_w := i_w(\theta)$) then being extensions of K_v in $\mathbb{C}_p$. Thus, we have distributed the $[L : K]$ extensions σ of i_v in classes C_w corresponding to the elements of $Pl_{L,v}$, for the equivalence relation: $\sigma \sim \sigma'$ if and only if $\sigma(\theta)$ and $\sigma'(\theta)$ are K_v-conjugate; these classes are represented by the i_w, and we have $C_w = \{\sigma,\ \sigma = \tau \circ i_w\}$ where τ ranges in the set of K_v-isomorphisms of L_w in $\mathbb{C}_p$.

The field $i_w(L) \cap K_v$ is independent of the choice of $i_w \in C_w$ and, by construction, $L_w = i_w(L)\,K_v$. Strictly speaking, these extensions i_w are not K-embeddings of L since i_v is not the identity anymore, contrary to the absolute case where $K = \mathbb{Q}$. If we introduce the classes which are relative to $L/\mathbb{Q}$, it is easy to connect them to the C_w for $w|v|p$.

2.5 RELATIVE GALOIS CASE. If, in addition, L/K is Galois with Galois group G, the field $i_w(L) \cap K_v$ is equal to the image under i_w of the decomposition field L^{D_w} of w in L/K, where:

$$D_w := D_w(L/K) := \{s \in G,\ sw = w\}$$

is the decomposition group of w in L/K, and the equivalence class C_w is still given by the condition $\sigma \in C_w$ if and only if:

$$i_w^{-1} \circ \sigma \in D_w.$$

Once again, we have:

$$D_{sw} = s\, D_w\, s^{-1} \text{ for all } s \in G.$$

Finally, if $G_w := \mathrm{Gal}(L_w/K_v)$, the canonical isomorphism $G_w \simeq D_w$ is given by the map which sends $\tau \in G_w$ to the restriction to L of $i_w^{-1} \circ \tau \circ i_w$. We thus have the following diagram, analogous to that of Figure 2.1.

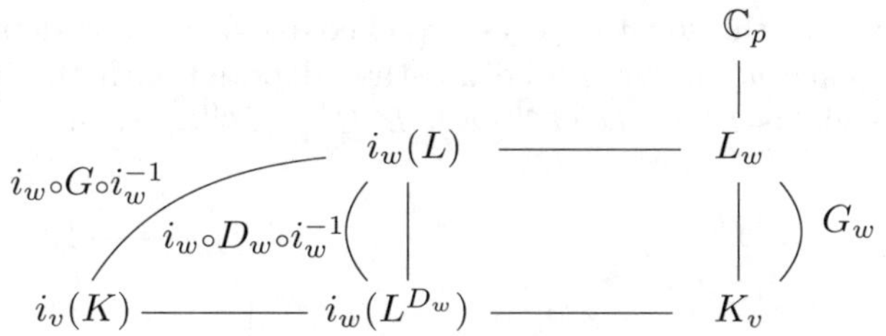

Fig. 2.2

where $i_w(L^{D_w}) = i_w(L) \cap K_v$ and $L_w = i_w(L)K_v$ (direct compositum over the intersection).

2.5.1 Remarks. (i) In the Galois case the fields L_w for $w|v$ are all equal: indeed, consider temporarily $L = K(\theta)$ as an algebraic extension in $\mathbb{C}_p/\mathbb{Q}$; then, if $L_{w_0} := K_v(\theta)$, L_{w_0} contains $K(\theta) = L$, hence contains $K(\theta') = L$ for every K-conjugate θ' of θ, so L_{w_0} contains all the $L_w := K_v(\theta')$, proving our claim. Beware that although the G_w for $w|v$ are all equal, the isomorphisms $G_w \longrightarrow D_w$ are not.

(ii) As the case $\mathbb{Q}(\sqrt[6]{2})/\mathbb{Q}$ shows (see 2.2.3), even in the non-Galois case the completions often coincide, and are often Galois and even abelian (the reason for this is that there are less algebraic extensions of $\mathbb{Q}_p$ than algebraic extensions of $\mathbb{Q}$). □

2.6 COMPLETION OF A SUBEXTENSION. If L'/K is a subextension of the finite extension L/K, for $w|v$ we define the w-completion of L'/K to be the subextension of L_w/K_v equal to the closure of $i_w(L')$ in L_w, which is equal to the compositum $i_w(L')K_v$ (see 2.4). It is also equal to $L'_{w'}$ for the unique place w' of L' below w and for $i_{w'} := i_w{}_{|L'}$.

2.6.1 Remark. The compositum $i_w(L)K_v$ is not closed when the algebraic extension L/K is infinite[8]. In this case, we will still denote by L_w the compositum $L_w := i_w(L)K_v$, which we will call the local extension at $w|v$ corresponding to L/K. Indeed, the completion of L_w is not interesting for the arithmetical study of L/K. For instance, if $K = \mathbb{Q}$, $L = \overline{\mathbb{Q}}$, and $v = \ell$ is finite, then $L_w = i_w(\overline{\mathbb{Q}})\mathbb{Q}_\ell$ is an algebraic closure of $\mathbb{Q}_\ell$ (Krasner's lemma) whose topological closure is equal to $\mathbb{C}_\ell$ (see [Rob, Ch. 3, § 1.4]).

In the general case, the nice notion is thus that of *local extensions corresponding to an algebraic extension* L/K; however, we will often use the above abuse of language. □

2.7 MAXIMAL ABELIAN SUBEXTENSIONS. We give the following diagram which, in the case where L/K is a finite Galois extension, details the correspondence between global fields and their completions when one introduces the maximal abelian subextensions. Concerning this, note that when $L' \subseteq L$ is abelian over K, the local objects attached to L' do not depend on the choice of the place w' above v, and are thus denoted with the index v (for $L' = L^{\rm ab}$, this yields $D_v := D_v(L^{\rm ab}/K)$, $L^{{\rm ab}\,D_v}$, $(L^{\rm ab})_v, \ldots$):

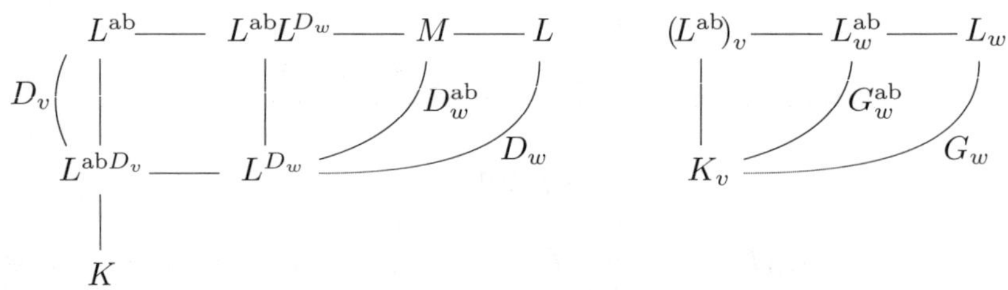

Fig. 2.3

In this diagram, under completion the fields L^{D_w} and L give respectively K_v and L_w, the field $L^{\rm ab}L^{D_w}$ gives what we have denoted $(L^{\rm ab})_v$ (completion of $L^{\rm ab}$ above v); the field M fixed under the commutator subgroup $[D_w, D_w]$ gives $L_w^{\rm ab}$, which is in general different from $(L^{\rm ab})_v$ (thus $[D_w, D_w]$ has an image equal to ${\rm Gal}(L_w/L_w^{\rm ab})$ under the isomorphism $D_w \simeq G_w$). □

[8] Since $L = \varinjlim_{L'} L'$ for finite extensions L'/K, $L' \subset L$, the places w of L are thus of the form $w = \varinjlim_{L'} w'$ with evident definition.

§3 Number and Ideal Groups

Global class field theory involves either global invariants (i.e., attached to the arithmetic properties of the field K alone), or *families* of local invariants attached to the set of all completions of K, so that whatever aspect is considered, to begin with we must always have a fixed base number field K.

a) The Local Case: The Group $K_v^\times$

3.1 RESULTS ON THE STRUCTURE OF $K_v^\times$.[9] For $v \in Pl$, let K_v be the v-completion of K; we have three different cases.

3.1.1 FINITE PLACES. Let $v \in Pl_0$. We denote respectively by:

$$\pi_v, \ \ U_v, \ \ U_v^i \, (i \geq 1), \ \ F_v, \ \ q_v,$$

a uniformizer, the unit group, the subgroups $\{u \in U_v, \ u \equiv 1 \bmod (\pi_v^i)\}$, the residue field (which is of nonzero characteristic ℓ), the number of elements of F_v, for the local field K_v; we then have:

$$K_v^\times = \pi_v^{\mathbb{Z}} \bigoplus \mu_{q_v - 1} \bigoplus U_v^1,$$

where μ_{q_v-1} is the group of $(q_v - 1)$th roots of unity. For convenience, we define U_v^i for $i \geq 0$ by setting $U_v^0 := U_v$.

We have:

$$U_v/U_v^1 \simeq \mu_{q_v-1} \simeq F_v^\times \ \text{ and } \ U_v^i/U_v^{i+1} \simeq F_v \ \text{ for all } \ i \geq 1.$$

Recall that U_v^1 is a $\mathbb{Z}_\ell$-module of finite type whose torsion sub-$\mathbb{Z}_\ell$-module is the group $\mu_\ell(K_v)$ of roots of unity of order a power of ℓ and belonging to K_v, and that:

$$U_v^1/\mu_\ell(K_v) \simeq \mathbb{Z}_\ell^{[K_v \,:\, \mathbb{Q}_\ell]}.$$

The group U_v is compact and the group $K_v^\times$ is locally compact. The groups U_v^i for $i \geq 0$ form a canonical fundamental system of neighbourhoods of 1 in $K_v^\times$; since they are closed subgroups of finite index of U_v, it follows that U_v is a profinite group and that (algebraically and topologically):

$$U_v \simeq \varprojlim_{i \geq 0} U_v/U_v^i.$$

3.1.2 REAL INFINITE PLACES. Let $v \in Pl_\infty^{\mathrm{r}}$. In this case, $K_v^\times = \mathbb{R}^\times$ and we set:

$$U_v^i := \mathbb{R}^{\times +} \text{ for all } i \geq 0,$$

[9] [a, Ma; Lg], [c, Nar1, Ch. 5, § 1], [d, Se2], [e, Ko3, Ch. 1, § 4.4], [g, Ko4].

and a uniformizer will be any element $\pi_v \in -\mathbb{R}^{\times +}$ (in practice, we will choose $\pi_v := -1$); we evidently have:

$$K_v^\times = \{\pm 1\} \bigoplus \mathbb{R}^{\times +} = \pi_v{}^{\mathbb{Z}/2\mathbb{Z}} \bigoplus U_v.$$

By definition, the residue field at v is $F_v := \mathbb{R}$, so that $F_v^\times = \mathbb{R}^\times$. We will see later that the multiplicative groups of residue fields are used in class field theory only through their p-Sylow subgroups $(F_v^\times)_p$, for every prime number p. If by convention we set:

$$(\mathbb{R}^\times)_p \simeq (\{\pm 1\})_p$$

so that, for $v \in Pl_\infty^{\mathrm{r}}$ we have:

$$(F_v^\times)_p = 1 \text{ if } p \neq 2, \ (F_v^\times)_2 \simeq \{\pm 1\},$$

then this is exactly what we will need for a unified way of writing the formulas. In other words, if v is a finite place, $\bigoplus_p (F_v^\times)_p = F_v^\times$, while if v is a real place at infinity, $\bigoplus_p (F_v^\times)_p = \{\pm 1\}$ only.

3.1.3 COMPLEX INFINITE PLACES. Let $v \in Pl_\infty^{\mathrm{c}}$. In this case, $K_v^\times = \mathbb{C}^\times$ and we set:

$$U_v^i := \mathbb{C}^\times \text{ for all } i \geq 0.$$

There would not be any notion of p-Sylow subgroup for the multiplicative group of the residue field $F_v := \mathbb{C}$, but the latter does not enter the theory.

Finally, recall a convenient way of stating the Hensel lemma for the rare cases where we will use it in this book (such as the characterization of nth powers in $K_v^\times$).

3.2 Lemma (Hensel). *Let P be a nonzero polynomial in $K_v[X]$ (v finite place of K) and let P' be its derivative; assume that all the coefficients of P have nonnegative valuation (i.e., are integers of K_v). If there exists an integer α_0 of K_v such that $\lambda := \mathrm{v}(P(\alpha_0)) - 2\,\mathrm{v}(P'(\alpha_0)) > 0$, then there exists an integer α of K_v such that $P(\alpha) = 0$ and such that $\mathrm{v}(\alpha - \alpha_0) \geq \lambda$.* □

If $\mathrm{v}(P'(\alpha_0)) = 0$, the condition of the lemma means simply that $P(\alpha_0) \equiv 0 \bmod \mathfrak{p}_v$ (i.e., $\overline{P(\alpha_0)} = \overline{0}$ in F_v). In this case, we speak of the weak Hensel lemma.

3.2.1 Example 1. Let $n \geq 1$ be an integer. If $\mathrm{v}(n) = 0$ then $u \in U_v$ is a nth power in $K_v^\times$ if and only if its residual image is a nth power in $F_v^\times$. □

3.2.2 Example 2. The fact that $\mu_{q_v - 1} \subset K_v^\times$ (used in 3.1.1) comes from the weak Hensel lemma applied to $P = X^{q_v - 1} - 1$. □

b) The Global Case: Numbers, Ideals, and Units

We start by defining number and ideal groups attached to K.

3.3 RAY GROUPS ATTACHED TO K. Let T be a finite set of finite places of K, and let $\mathfrak{m} \in \langle T \rangle_{\mathbb{N}}$ be a modulus of K built from T, i.e., an integral ideal of the form $\mathfrak{m} = \prod_{v \in T} \mathfrak{p}_v^{m_v}$, $m_v \geq 0$. Let Δ_∞ be a set of real places at infinity of K.

3.3.1 NUMBER GROUPS. We define:

- $K^\times_{T,\Delta_\infty} := \{x \in K^\times,\ \mathrm{v}(x) = 0\ \ \forall v \in T \cup \Delta_\infty\}$, the subgroup of $K^\times$ of elements x prime to T (i.e., such that $\mathrm{v}(x) = 0$ for all $v \in T$) and positive on Δ_∞ (i.e., such that $i_v(x) > 0$ for all $v \in \Delta_\infty$);
- $K^\times_{T,\mathfrak{m},\Delta_\infty} := \{x \in K^\times_{T,\Delta_\infty},\ x \equiv 1 \bmod \mathfrak{m}\}$ (the congruence $x \equiv 1 \bmod \mathfrak{m}$ should be understood as $\mathrm{v}(x-1) \geq m_v$ for all $v \in T$; for $x \in K^\times_T$, it is equivalent to $i_v(x) \in U_v^{m_v}$ for all $v \in T$).

In particular, if $\Delta_\infty = Pl^{\mathrm{r}}_\infty$:

$$K^\times_{T,\mathfrak{m},Pl^{\mathrm{r}}_\infty} =: K^\times_{T,\mathfrak{m},\mathrm{pos}}$$

is the subgroup of totally positive elements of $K^\times_{T,\mathfrak{m}}$, and if $\Delta_\infty = \emptyset$, we simply have:

$$K^\times_{T,\mathfrak{m},\emptyset} =: K^\times_{T,\mathfrak{m}}.$$

3.3.2 IDEAL GROUPS. We define:

- I_T, the subgroup of the group I of nonzero fractional ideals of K which are prime to T;
- $P_T := \{(x),\ x \in K^\times_T\}$, the subgroup of principal ideals of I_T;
- $P_{T,\Delta_\infty} := \{(x),\ x \in K^\times_{T,\Delta_\infty}\}$;
- $P_{T,\mathfrak{m},\Delta_\infty} := \{(x),\ x \in K^\times_{T,\mathfrak{m},\Delta_\infty}\}$.

If $\Delta_\infty = Pl^{\mathrm{r}}_\infty$:

$$P_{T,\mathfrak{m},Pl^{\mathrm{r}}_\infty} =: P_{T,\mathfrak{m},\mathrm{pos}}$$

is the ray group modulo $\mathfrak{m}$ which will correspond to the restricted sense, and if $\Delta_\infty = \emptyset$:

$$P_{T,\mathfrak{m},\emptyset} =: P_{T,\mathfrak{m}}$$

is the ray group modulo $\mathfrak{m}$ which will correspond to the ordinary sense.

We now give definitions relative to the unit groups of K given together with sets of places T and S.

3.4 UNIT GROUPS. Let $\mathfrak{m} \in \langle T \rangle_{\mathbb{N}}$. We set:

$$E_{\mathfrak{m}}^{S} := \{x \in K_{T,\mathfrak{m}}^{\times},\ \mathrm{v}(x) = 0\ \ \forall v \notin S\},$$

which is the group of S-units congruent to 1 modulo $\mathfrak{m}$.

For instance, for $\mathfrak{m} = 1$ and $S = \emptyset$ (resp. $S = Pl_\infty^{\mathrm{r}}$) this defines:

$$E_1^{\emptyset} =: E =: E^{\mathrm{res}}\ (\text{resp. } E_1^{Pl_\infty^{\mathrm{r}}} =: E^{Pl_\infty^{\mathrm{r}}} =: E^{\mathrm{ord}}),$$

which is the group of units of K in the restricted sense (i.e., totally positive) (resp. the group of ordinary units of K). More generally, if $S_\infty := \emptyset$ (resp. Pl_∞^{r}) we will also write (for mnemonic reasons):

$$E^{S_0} =: E^{S_0\,\mathrm{res}}\quad (\text{resp. } E^{S_0 \cup Pl_\infty^{\mathrm{r}}} =: E^{S_0\,\mathrm{ord}}),$$

for the group of S_0-units in the restricted (resp. ordinary) sense. We will also use a similar notation for the subgroup corresponding to units congruent to 1 modulo $\mathfrak{m}$:

$$E_{\mathfrak{m}}^{S_0\,\mathrm{res}}\quad (\text{resp. } E_{\mathfrak{m}}^{S_0\,\mathrm{ord}}).$$

3.5 Remark. Note that the rules of notation for the sets of places at infinity used in 3.3 for certain subgroups of $K^\times$ are not the same as those used in 3.4 for units. In the first case, we want to define the subgroups of $K^\times$ corresponding to the given ray groups, and which describe the complexification (of the places of Δ_∞) through a signature condition; in the second case, the generalized unit groups correspond to the decomposition (of the places of $S = S_0 \cup S_\infty$). It is thus natural for S_∞-decomposition to be equivalent to the Δ_∞-complexification for $\Delta_\infty := Pl_\infty^{\mathrm{r}} \backslash S_\infty$. □

We now prove the Dirichlet–Herbrand theorem on S-units of K (i.e., the Dirichlet theorem seen in the context of representation theory).[10]

Let g be a group of automorphisms of K, and $k := K^g$. We denote by $Pl_{k,\infty}^{\mathrm{rc}}$ (resp. $Pl_{k,\infty}^{\mathrm{rnc}}$) the set of real infinite places of k which are complexified (resp. noncomplexified) in K. We denote by $Pl_{k,\infty}^{\mathrm{s}}$ the set of infinite places of k which are totally split in K; thus $Pl_{k,\infty}^{\mathrm{s}} = Pl_{k,\infty}^{\mathrm{c}} \cup Pl_{k,\infty}^{\mathrm{rnc}}$ (see 1.2.1).

For an infinite place u of k and an arbitrary place v of K above u, we put $c_u := i_v^{-1} \circ c \circ i_v$, where c is complex conjugation restricted to K_v (so that c_u is a Frobenius of u in K/k defined up to conjugation, in other words a generator of the decomposition group d_u of a place $v|u$).

3.6 Lemma. *Let E be a sub-g-module of finite index of E_K^{ord}. Then we have the following isomorphisms of g-modules:*

$$\mathbb{Q} \oplus (E \otimes_{\mathbb{Z}} \mathbb{Q}) \simeq \bigoplus_{u \in Pl_{k,\infty}^{\mathrm{rc}}} \mathbb{Q}[g](1 + c_u) \bigoplus_{u \in Pl_{k,\infty}^{\mathrm{s}}} \mathbb{Q}[g],$$

$$\mathbb{F}_p \oplus (E \otimes_{\mathbb{Z}} \mathbb{F}_p) \simeq \mathrm{tor}_{\mathbb{Z}}(E) \otimes_{\mathbb{Z}} \mathbb{F}_p \bigoplus_{u \in Pl_{k,\infty}^{\mathrm{rc}}} \mathbb{F}_p[g](1 + c_u) \bigoplus_{u \in Pl_{k,\infty}^{\mathrm{s}}} \mathbb{F}_p[g],$$

[10] [c, WsE, Ch. 5, § 3], [d, Lang1, Ch. 5, § 1]; see [Se4] for representation theory, especially [Se4, §§ 3.3, 15.5].

for any prime number p not dividing $|g|$.

Proof. Let

$$\log_\infty : E \longrightarrow \bigoplus_{u \in Pl^{\rm rc}_{k,\infty}} \mathbb{R}_u^{|g|/2} \bigoplus_{u \in Pl^{\rm s}_{k,\infty}} \mathbb{R}_u^{|g|},$$

with $\mathbb{R}_u = \mathbb{R}$, be the logarithmic embedding of E defined for all $\varepsilon \in E$ in the following way: for $u \in Pl^{\rm rc}_{k,\infty}$ (resp. $Pl^{\rm s}_{k,\infty}$), the component of $\log_\infty(\varepsilon)$ on the summand $\mathbb{R}_u^{|g|/2}$ (resp. $\mathbb{R}_u^{|g|}$) is given by:

$$(\log |i_v \circ s^{-1}(\varepsilon)|)_{s \in g/\langle c_u \rangle} \text{ (resp. } (\log |i_v \circ s^{-1}(\varepsilon)|)_{s \in g}),$$

where i_v is a fixed embedding $K \longrightarrow K_v = \mathbb{R}$ or $\mathbb{C}$ extending i_u on k. Each vector space $\mathbb{R}_u^{|g|/2}$ (resp. $\mathbb{R}_u^{|g|}$) is considered as a permutation representation, in other words, as the g-module defined by:

$$t((x_{u,s})_s) = (x_{u,t^{-1}s})_s \ \text{ for all } t \in g,$$

$s \in g/\langle\, c_u \,\rangle$ (resp. $s \in g$). We verify that $\log_\infty$ is then a g-module homomorphism.

Let $r_1^{\rm c}(k) = |Pl^{\rm rc}_{k,\infty}|$, $r_1^{\rm nc}(k) = |Pl^{\rm rnc}_{k,\infty}|$; thus $r_1(K) = r_1^{\rm nc}(k)|g|$, $r_2(K) = r_1^{\rm c}(k)|g|/2 + r_2(k)|g|$. Finally, we put:

$$r := r_1 + r_2 = r_1^{\rm c}(k)|g|/2 + (r_1^{\rm nc}(k) + r_2(k))|g|,$$

and we identify the spaces $\bigoplus_{u \in Pl^{\rm rc}_{k,\infty}} \mathbb{R}_u^{|g|/2} \bigoplus_{u \in Pl^{\rm s}_{k,\infty}} \mathbb{R}_u^{|g|}$ with $\mathbb{R}^r$. Then the classical Dirichlet theorem [a, Sam, Ch. IV] states that $\log_\infty(E)$ is a free $\mathbb{Z}$-module of rank $r-1$ contained in the hyperplane of $\mathbb{R}^r$ defined by:

$$\sum_{\substack{u \in Pl^{\rm rnc}_{k,\infty} \\ s \in g}} x_{u,s} \; + \; 2 \sum_{\substack{u \in Pl^{\rm rc}_{k,\infty} \\ s \in g/\langle c_u \rangle}} x_{u,s} \; + \; 2 \sum_{\substack{u \in Pl^{\rm c}_{k,\infty} \\ s \in g}} x_{u,s} = 0.$$ [11]

We still denote by $\mathbb{Z}$ the diagonal embedding of $\mathbb{Z}$ in $\mathbb{R}^r$. The $\mathbb{Z}[g]$-module $\mathbb{Z} \oplus \log_\infty(E)$, which is a $\mathbb{Z}$-lattice of $\mathbb{R}^r$, leads, for any field Q of characteristic zero, to representations $V_Q = (\mathbb{Z} \oplus \log_\infty(E)) \otimes Q$ with the same character; it is then possible to work with the representation $V_\mathbb{R}$ which is isomorphic to the above representation $\mathbb{R}^r$. Hence, $V_\mathbb{Q}$ is the sum, taken over $u \in Pl_{k,\infty}$, of the corresponding permutation representations. Point (i) of the lemma follows as a way of writing for the permutation representations modulo d_u, $u \in Pl_{k,\infty}$.

For point (ii), we have the exact sequence of g-modules:

[11] This relation comes from $\mathrm{N}_{K/\mathbb{Q}}(\varepsilon) = \mathrm{N}_{k/\mathbb{Q}}(\mathrm{N}_{K/k}(\varepsilon)) = \pm 1$, using the identity $\mathrm{N}_{k/\mathbb{Q}}(x) = \prod_{u|\infty} \mathrm{N}_{k_u/\mathbb{R}}(i_u(x))$ that we will prove in II.2.2, and the fact that $i_u(\mathrm{N}_{K/k}(\varepsilon)) = \prod_{s \in g} i_v \circ s^{-1}(\varepsilon)$.

$$1 \longrightarrow \mathrm{tor}(E) \longrightarrow E \xrightarrow{\log_\infty} \log_\infty(E) \longrightarrow 0,$$

which, because of the equality $\mathrm{tor}(E) \cap E^p = (\mathrm{tor}(E))^p$, yields:

$$1 \longrightarrow \mathrm{tor}(E) \otimes \mathbb{F}_p \longrightarrow E \otimes \mathbb{F}_p \longrightarrow \log_\infty(E) \otimes \mathbb{F}_p \longrightarrow 0.$$

The g-module isomorphism:

$$\mathbb{F}_p \oplus (E \otimes \mathbb{F}_p)/(\mathrm{tor}(E) \otimes \mathbb{F}_p) \simeq \bigoplus_{u \in Pl_{k,\infty}^{\mathrm{rc}}} \mathbb{F}_p[g](1 + c_u) \bigoplus_{u \in Pl_{k,\infty}^{\mathrm{s}}} \mathbb{F}_p[g],$$

comes from the representation theory in characteristic p when p does not divide the order of g. □

We now consider the case of S-units, where S is a finite set of noncomplex places invariant under g. We denote by $\langle S_0 \rangle$ the subgroup of I generated by S_0.

Let F be a sub-g-module of finite index of E_K^S and $P(S) := \{(x),\ x \in E_K^S\}$. We have the exact sequence:

$$1 \longrightarrow E \longrightarrow F \xrightarrow{(\,\bullet\,)} P(S),$$

where E is a submodule of E_K^{ord} of finite index of E_K^{ord}. Since $\langle S_0 \rangle$ is a free $\mathbb{Z}$-module of finite dimension, $P(S)$ is a submodule of finite index (the ordinary class group is finite), and this also holds for the image (F) of F in $P(S)$, which is thus a free $\mathbb{Z}$-module of finite dimension equal to that of $\langle S_0 \rangle$.

If p does not divide the order of g, the isomorphism of the corresponding representations (in characteristic zero) gives an isomorphism of $\mathbb{F}_p$-representations:

$$(F) \otimes \mathbb{F}_p \simeq \langle S_0 \rangle \otimes \mathbb{F}_p.$$

Put $S_0 = \bigcup_u S_u$, with $S_u = \{v \in Pl_K,\ v|u\}$, $u \in S_{0,k}$, where $S_{0,k}$ is the set of places of k below those of S_0; then:

$$\langle S_0 \rangle \otimes \mathbb{F}_p \simeq \bigoplus_u (\langle S_u \rangle \otimes \mathbb{F}_p),$$

where each representation $\langle S_u \rangle \otimes \mathbb{F}_p$ is the permutation representation modulo the decomposition subgroup of a place of K above u.

Since $E \cap F^p = E^p$ and by the above, we have the exact sequence:

$$1 \longrightarrow E \otimes \mathbb{F}_p \longrightarrow F \otimes \mathbb{F}_p \longrightarrow (F) \otimes \mathbb{F}_p \longrightarrow 1,$$

giving the following result for which we recall that g is a group of automorphisms of K, k is the subfield of K fixed under g, d_u denotes the decomposition group in K/k of a place v of K above u for each $u \in Pl_k$, and $\mathrm{Ind}_{d_u}^g(1_{d_u})$ is the character of the representation of g induced by the unit representation of d_u (this characterizes the corresponding permutation representation).

3.7 Theorem (generalized Dirichlet–Herbrand theorem on S-units). *Let S be a finite set of noncomplex places of K invariant under g, and let F be a sub-g-module of finite index of E_K^S. Then the character of the representation $F \otimes \mathbb{Q}$ of g is given by:*

$$\sum_{u \in Pl_{k,\infty}} \mathrm{Ind}_{d_u}^g(1_{d_u}) + \sum_{u \in S_{0,k}} \mathrm{Ind}_{d_u}^g(1_{d_u}) - 1_g.$$

If p does not divide the order of g, the character of $F \otimes \mathbb{F}_p$ is given by:

$$\sum_{u \in Pl_{k,\infty}} \mathrm{Ind}_{d_u}^g(1_{d_u}) + \sum_{u \in S_{0,k}} \mathrm{Ind}_{d_u}^g(1_{d_u}) + \delta_p(F)\omega - 1_g,$$

where $\delta_p(F) = 1$ or 0 according as $\mu_p \subset F$ or not, and where ω is the Teichmüller character. □

3.7.1 Corollary (classical Dirichlet theorem on S-units). *The case $g = 1$ yields (for any p) the formula:*

$$\mathrm{rk}_p(F) := \dim_{\mathbb{F}_p}(F/F^p) = r_1 + r_2 + |S_0| + \delta_p(F) - 1. \qquad \square$$

3.7.2 Corollary (Dirichlet–Herbrand theorem on units [12]). *If $K/\mathbb{Q}$ is Galois and if $g = G := \mathrm{Gal}(K/\mathbb{Q})$, the character of the representation $E_K^{\mathrm{ord}} \otimes \mathbb{Q}$ is given by $\mathrm{Ind}_{d_\infty}^G(1_{d_\infty}) - 1_G$; in other words, $\mathbb{Q} \oplus (E_K^{\mathrm{ord}} \otimes \mathbb{Q})$ is the permutation representation modulo $d_\infty = \langle\, c_\infty \,\rangle$, which is a subrepresentation of the regular one $\mathbb{Q}[G]$.* □

3.7.3 Remark. Because of the above property, we will say that E_K^{ord} is a monogeneous $\mathbb{Q}[G]$-module. It is also common to say that E_K^{ord} admits a Minkowski unit (i.e., a unit ε such that the conjugates of ε generate a subgroup of finite index of E_K^{ord}; this index may be assumed prime to any prime number p not dividing $|G|$). □

§4 Idèle Groups — Generalized Class Groups

The notion of idèle was introduced by Chevalley as early as 1936 in [Che4], so as to be able to describe the correspondence of class field theory for infinite abelian extensions. In that paper, he uses the terminology of "ideal elements", quoting Prüfer (1925), and already proves the main properties of idèles (topological properties, properties of the norm, definition of the reciprocity map from that of Artin, ...), and he mentions the possibility of a complete formalism of class field theory in this setting, which will indeed be developed a few

[12] [d, Lang1, Ch. IX, § 4], [WsA].

years later. This leads to the contributions of Chevalley (1940), Artin–Tate (1951/1952), Weil (1951; 1961/1962), which were published respectively in [h, Che2], [d, AT], [h, We2], [d, We1]. Recall that in his book, Weil uses central simple algebras as basic tool for class field theory, and that the Chapters XII and XIII (local and global class field theories) rely on an unpublished work of Chevalley.

As we will see, it is the tool which best exhibits the local-global principle which underlies class field theory.

a) Idèle Groups — Topology

We keep the local and global notations of Section 3 above, relative to the number field K.

4.1 Idèle Group — Idèle Class Group. (i) We define the idèle group of K by:

$$J := J_K := \prod_{v \in Pl}{}' K_v^\times ,$$

where $\prod'$ denotes the restricted product to local unit subgroups; in other words, an idèle $\boldsymbol{x} \in J$ is a family:

$$\boldsymbol{x} =: (x_v)_{v \in Pl} \in \prod_{v \in Pl} K_v^\times$$

for which $x_v \in U_v$ for almost all $v \in Pl$ (i.e., all v except a finite number).

(ii) We define the idèle class group of K by:

$$C := C_K := J/i(K^\times),$$

where $i(K^\times)$ is the image of $K^\times$ under the injective diagonal embedding:

$$\begin{aligned} i := i_{Pl} : K^\times &\longrightarrow \prod_{v \in Pl} K_v^\times . \\ x &\longmapsto (i_v(x))_v \end{aligned}$$

The inclusion $i(K^\times) \subset J$ comes from the fact that only a finite number of prime ideals enter into the decomposition of (x), so that $\mathrm{v}(x) = 0$ for almost all $v \in Pl_0$, hence for almost all $v \in Pl$. The group $i(K^\times)$ is called the group of principal idèles of K.

(iii) Finally, we denote by $\mathcal{c\ell}$ the canonical map $J \longrightarrow C = J/i(K^\times)$.

Note. In the literature one sees the notation $K^\times$ instead of $i(K^\times)$ (and thus $J/K^\times$ for C), the embedding i being implicitly understood. We will, however, use this abuse of notation with caution since it may be quite incorrect (for instance, compute explicitly the embedding of $x = 1 + \sqrt[6]{2}$ in J for $K = \mathbb{Q}(\sqrt[6]{2})$, at least for a few places, using 2.2.3). More precisely, we will use this simplification, for $K^\times$ and its subgroups, only in general formulas when the idelic context is clear and when no precise computations are involved.

The idèle class group is the most sophisticated object, in that it mixes local and global properties in a rather subtle way; for us, however, it will be considered as a functional tool which will simplify certain types of computations, for instance by avoiding the explicit use of moduli $\mathfrak{m}$, as we will see in 4.3.3. Nonetheless, Weil [h, We2] has looked for an interpretation of C and has written:

"La recherche d'une interprétation pour C $(\dots)$ me semble constituer l'un des problèmes fondamentaux de la théorie des nombres à l'heure actuelle; il se peut qu'une telle interprétation renferme la clef de l'hypothèse de Riemann." [13]

Idèle groups are used to construct some C^*-dynamical systems whose partition function is the Riemann zeta function, and number theory has, indeed, some links with quantum statistical mechanic (Julia, Connes, ...). For more information, see for instance the paper of Paula Cohen (Journal de Théorie des Nombres de Bordeaux, 11 (1999), 15–30).

4.1.1 Ray Subgroups of J. Let T and Δ_∞ be two finite sets of finite and real infinite places of K, and let $\mathfrak{m} = \prod_{v\in T} \mathfrak{p}_v^{m_v}$, $m_v \geq 0$. We define:

$$\begin{aligned} J_{T,\mathfrak{m},\Delta_\infty} &:= \{\boldsymbol{x} =: (x_v)_v \in J, \\ &\qquad \mathrm{v}(x_v) = 0,\ x_v \equiv 1 \bmod (\pi_v^{m_v})\ \ \forall\, v \in T,\ x_v > 0\ \ \forall\, v \in \Delta_\infty\} \\ &= \prod_{v \notin T \cup \Delta_\infty}{}' K_v^\times \prod_{v\in T} U_v^{m_v} \prod_{v\in\Delta_\infty} \mathbb{R}^{\times +}, \end{aligned}$$

which is the subgroup of idèles prime to T, congruent to 1 modulo $\mathfrak{m}$, and Δ_∞-positive (this is the idelic analog of $K^\times_{T,\mathfrak{m},\Delta_\infty}$).

If $\Delta_\infty = Pl_\infty^{\mathrm{r}}$:

$$J_{T,\mathfrak{m},Pl_\infty^{\mathrm{r}}} =: J_{T,\mathfrak{m},\mathrm{pos}}$$

is the group of idèles prime to T, congruent to 1 modulo $\mathfrak{m}$, and totally positive, while if $\Delta_\infty = \emptyset$:

$$J_{T,\mathfrak{m},\emptyset} =: J_{T,\mathfrak{m}}$$

is the group of idèles prime to T and congruent to 1 modulo $\mathfrak{m}$.

We will now give a number of definitions about unit idèles. Let S be a set of places disjoint from T.

[13] "The search for an interpretation of C $(\dots)$ seems to me to be one of the most basic problems in number theory today; it is possible that such an interpretation holds the key to the Riemann hypothesis."

4.1.2 Unit Idèle Groups. We define the following subgroups of J:

(i) The group of S-unit idèles of K congruent to 1 modulo $\mathfrak{m}$ (which is the analog of $E^S_{\mathfrak{m}}$):

$$\begin{aligned} U^S_{\mathfrak{m}} &:= \{\boldsymbol{x} =: (x_v)_v \in J,\ x_v \in U_v \ \ \forall\, v \notin S,\ x_v \equiv 1 \bmod (\pi_v^{m_v}) \ \ \forall\, v \in T\} \\ &= \prod_{v \notin T \cup S} U_v \prod_{v \in T} U_v^{m_v} \prod_{v \in S} K_v^\times . \end{aligned}$$

(ii) The group of idèles with support equal to S:

$$\langle\, S\,\rangle := \prod_{v \in S} K_v^\times \prod_{v \in Pl \backslash S} \{1\},$$

which is a subgroup of U^S, also denoted $\bigoplus_{v \in S} K_v^\times$; one easily checks that:

$$U^{\mathrm{res}}_{\mathfrak{m}} \,.\, \langle\, S\,\rangle = U^S_{\mathfrak{m}}.$$

4.1.3 Example. For $T = S_0 = \emptyset$ we obtain:

$$U^{\mathrm{res}} := U = \prod_{v \in Pl_0} U_v \prod_{v \in Pl^{\mathrm{r}}_\infty} \mathbb{R}^{\times +} \prod_{v \in Pl^{\mathrm{c}}_\infty} \mathbb{C}^\times ,$$

$$U^{\mathrm{ord}} := U^{Pl^{\mathrm{r}}_\infty} = \prod_{v \in Pl_0} U_v \prod_{v \in Pl^{\mathrm{r}}_\infty} \mathbb{R}^{\times} \prod_{v \in Pl^{\mathrm{c}}_\infty} \mathbb{C}^\times ,$$

which are the unit idèle groups (in the restricted sense and the ordinary sense). □

4.1.4 Notation. When $S_\infty = \emptyset$ and Pl^{r}_∞, we denote the corresponding unit idèle groups by $U^{S_0\,\mathrm{res}}_{\mathfrak{m}}$ and $U^{S_0\,\mathrm{ord}}_{\mathfrak{m}}$. □

4.1.5 Remark. As mentioned in 3.5, the principles of notation used for the idèle groups $J_{T,\mathfrak{m},\Delta_\infty}$ and that for the groups of unit idèles $U^S_{\mathfrak{m}}$ are not the same; the former, analogous to $K^\times_{T,\mathfrak{m},\Delta_\infty}$, involves the Δ_∞-complexification, while the latter, where $U^S_{\mathfrak{m}}$ corresponds to $E^S_{\mathfrak{m}}$, involves the S-decomposition. The analogy between the two situations is strengthened by the following identities:

$$K^\times \cap J_{T,\mathfrak{m},\Delta_\infty} = K^\times_{T,\mathfrak{m},\Delta_\infty} \text{ and } K^\times \cap U^S_{\mathfrak{m}} = E^S_{\mathfrak{m}}. \qquad \square$$

4.2 Topological Aspects. We begin with the following classical result (see Bourbaki or Pontrjagin [Pon]).

4.2.1 Lemma. *Let A be a locally compact topological group. Then the connected component of the unit element of A is equal to the intersection of the open subgroups of A.*

If B is a normal closed subgroup of A, then the connected component of the unit element of A/B is equal to the closure of the image in A/B of the connected component of the unit element of A.[14] □

4.2.2 TOPOLOGY OF THE IDÈLE GROUP. We define a restricted product topology on the group J by asking that a fundamental system of neighbourhoods of the unit element is given by the $W := \prod_{v \notin \Sigma} U_v \prod_{v \in \Sigma} V_v$, for finite subsets Σ of Pl, where the V_v are neighbourhoods of 1 in $K_v^\times$ for their natural topology (ultrametric or archimedean).[15] Thus, the restriction of the topology of J to $U^{\rm res}$ yields the usual product topology of the U_v (these being compact for $v \in Pl_0$, locally compact otherwise). In the same way, for S fixed, it induces the product topology on U^S and on $\langle\, S\,\rangle$.

For finite places v, the $U_v^{m_v}$, $m_v \geq 0$, form a fundamental system of open neighbourhoods of 1 in $K_v^\times$.

4.2.3 Proposition. *Every open subgroup N of J containing the diagonal embedding of $K^\times$ is of finite index in J.*

Proof. Indeed, for each infinite place v, N contains a neighbourhood V_v of 1 in $K_v^\times$ ($K_v^\times = \mathbb{R}^\times$ or $\mathbb{C}^\times$), hence N contains $U_\infty := \prod_{v \in Pl_\infty^{\rm r}} \mathbb{R}^{\times +} \prod_{v \in Pl_\infty^{\rm c}} \mathbb{C}^\times$ (if $x \in U_\infty$, by choosing n large enough, we can find an nth root u of x in U_∞ which belongs to $\prod_{v \mid \infty} V_v$, hence to N; since N is a group, $u^n = x \in N$). It follows that N contains a subgroup of the form:

$$U_{\mathfrak m}^{\rm res} := \prod_{v \in Pl_0 \backslash T} U_v \prod_{v \in T} U_v^{m_v} \bigoplus U_\infty,$$

for a suitable finite set $T \subset Pl_0$ and $\mathfrak m = \prod_{v \in T} \mathfrak p_v^{m_v}$, $m_v \geq 0$, and we will see that $J/K^\times U_{\mathfrak m}^{\rm res} \simeq \mathcal{C}\ell_{\mathfrak m}^{\rm res}$ (a generalized class group; see Subsection (b)) is finite. □

4.2.4 Definitions. An element of the abelian group A is divisible if it is a nth power in A for all $n \geq 1$. A subgroup B of A is divisible if each element of B is divisible in B. One speaks of p-divisible element, group (p prime), when the n are limited to the p-powers. □

The set of divisible elements of A is then a subgroup (not necessarily divisible!) which contains the maximal divisible subgroup of A.

[14] Warning: if B and U are subgroups of a topological group A with B a normal subgroup, the group isomorphism $BU/B \simeq U/B \cap U$ is not always a homeomorphism (we will meet this situation). It is one if U is open, or if B is compact and U is closed; otherwise, we improvise.

[15] [c, Neu1, Ch. VI; WsE, Ch. 5, § 5], [d, CF, Ch. II; Iy1, Ch. III, § 4; Lang1, Ch. VII; Wel, Ch. IV, § 3].

Note that U_∞ is equal to the connected component of the unit element of J, and also to its maximal divisible subgroup.

4.2.5 Connected Component. The groups J and C are Hausdorff and locally compact (see in 4.2.8 that the image of $K^\times$ is closed in J). By the above lemma 4.2.1, the connected component D of the unit element of C is equal to the closure of the image of U_∞; this can be written:

$$D = \mathrm{adh}(c\ell(U_\infty)) = \bigcap_W (c\ell(W)c\ell(U_\infty)) = c\ell\Big(\bigcap_W (K^\times W U_\infty)\Big) ;$$

hence, since the WU_∞ can be taken to be the $U_{\mathfrak{m}}^{\mathrm{res}}$:

$$D = c\ell\Big(\bigcap_{\mathfrak{m}} (K^\times U_{\mathfrak{m}}^{\mathrm{res}})\Big) = \bigcap_{\mathfrak{m}} (K^\times U_{\mathfrak{m}}^{\mathrm{res}}) \Big/ K^\times .$$

4.2.6 Remark. It is easily checked that $\bigcap_{\mathfrak{m}} (K^\times U_{\mathfrak{m}}^{\mathrm{res}}) = \mathrm{adh}(K^\times U_\infty)$, so that we can also write $D = \mathrm{adh}(K^\times U_\infty)/K^\times$. □

We will show in III.4.15.1 that D is also a divisible subgroup of C and in II.3.7.3, II.3.7.4, that $C/D \simeq J \Big/ \bigcap_{\mathfrak{m}} (K^\times U_{\mathfrak{m}}^{\mathrm{res}})$ is a profinite group (hence compact) which class field theory interprets as the Galois group of the abelian closure of K.[16] This does not at all help to know the precise structure of this group; this requires a lot of work, which will be the object of Chapter III, Section 4.

In fact, the group C/D is more complicated than necessary, as we will see by introducing reduced idèles; for historical reasons, however, we will explain the relationships between these different variants, even though it may sometimes give quite redundant presentations.

We will come back in detail to all these questions in II.3.7. For some easy additional results on the topology of idèle groups, see 4.2.7 and 4.2.8 below, then 5.3 to 5.5.

4.2.7 Proposition. *Let $\boldsymbol{x} =: (x_v)_v \in J$ be an idèle of K; we define the volume of $\boldsymbol{x}$ to be the positive real number $|\boldsymbol{x}| := \prod_v |x_v|_{K_v}$ (see 2.1.3). Set $J^0 := \{\boldsymbol{x} \in J,\ \ |\boldsymbol{x}| = 1\}$; then J^0 contains $K^\times$ and $J^0/K^\times$ is compact.*

Proof. See [d, Lang1, Ch. VII, § 3; Neu1, Ch. VI, § 1], [e, Ko3, Ch. 1, § 5.4], or [e, MP, Ch. 2, § 3.8]. It is clear that the inclusion $K^\times \subset J^0$ comes from the product formula 1.2.2, hence enables us to define the volume of an idèle class. This function volume is continuous. □

This compactness result is the idelic version of theorems in the geometry of numbers (in the archimedean sense), more precisely the Dirichlet theorem

[16] [d, CF, Ch. VII, § 5; AT, Ch. 9, Ch. 14, § 6; Iy1, Ch. 3, § 7.2], [e, Ko3, Ch. II, § 7.2].

on units and on the finiteness of the class group; it deals with the archimedean part of the idèle class group which is *suppressed* in the correspondence of class field theory (this explains that we will not have to use this result, which must not be confused with the compactness of C/D).

4.2.8 Exercise. (i) Show that the diagonal embedding of $K^\times$ in J is closed.

(ii) Show that $D := \mathcal{cl}\Big(\bigcap_{\mathfrak{m}}(K^\times U_{\mathfrak{m}}^{\mathrm{res}})\Big)$ is equal to $\mathcal{cl}\Big(\bigcap_{\mathfrak{m}}(E^{\mathrm{ord}} U_{\mathfrak{m}}^{\mathrm{res}})\Big)$, where $\mathcal{cl}$ is the canonical map $J \longrightarrow J/K^\times$ (see 4.2.5).

(iii) (reduced idèles). Let $J_0 := \prod'_{v \in Pl_0} K_v^\times \prod_{v \in Pl_\infty^{\mathrm{r}}} \{\pm 1\}$ and $i_0 := (i_v)_{v \in Pl_0 \cup Pl_\infty^{\mathrm{r}}}$ where, for $v \in Pl_\infty^{\mathrm{r}}$, we identify i_v with the function sgn_v. We use the same rules of notation as in 4.1 for the diagonal embedding of $K^\times$ and its subgroups in J_0 (using i_0 instead of i). The quotient $C_0 := J_0/K^\times$ is called the reduced idèle class group and the canonical map is denoted $\mathcal{cl}_0$.

Show that $K^\times$ is closed in J_0 if and only if E^{ord} is finite.

(iv) Assume $K = \mathbb{Q}$, consider $A := \mathbb{Q}_2^\times$ as a subgroup of J_0 for its topology (defined in 4.2), and let A' be the image of A in C_0 with the quotient topology.

Show that A and A' are isomorphic but not homeomorphic.[17]

Answer. (i) The product formula $\prod_v |x-1|_v = 1$ for all $x \in K^\times$, $x \neq 1$, shows that an element of $i(K^\times)$ different from 1 cannot be close to 1 in J: consider a neighbourhood $W := \prod_{v \notin \Sigma} U_v \prod_{v \in \Sigma} V_v$, Σ containing Pl_∞, for any sufficiently small V_v, and let $i(x) \in W$, $x \neq 1$; this means that for $v \notin \Sigma$, $|x-1|_v \leq 1$ (since x, hence $1-x$, is v-integral) and that for $v \in \Sigma$, $|x-1|_v < 1$ (for example), contradiction, hence $W \cap i(K^\times) = 1$. In other words, $i(K^\times)$ is a discrete subgroup of J, proving (i).

(ii) Let $\boldsymbol{x} \in \bigcap_{\mathfrak{m}}(K^\times U_{\mathfrak{m}}^{\mathrm{res}})$ and let $\mathfrak{n}$ be *fixed*; then $\boldsymbol{x} =: i(y)\,\boldsymbol{v}$, $y \in K^\times$, $\boldsymbol{v} \in U_{\mathfrak{n}}^{\mathrm{res}}$, and for all $\mathfrak{m}'$ multiple of $\mathfrak{n}$ we have in the same way $\boldsymbol{x} =: i(x)\,\boldsymbol{u}$, $x \in K^\times$, $\boldsymbol{u} \in U_{\mathfrak{m}'}^{\mathrm{res}}$. It follows that $\boldsymbol{x} = i(x)\,\boldsymbol{u} = i(y)\,\boldsymbol{v}$, which implies $i(x)\,i(y)^{-1} \in U_{\mathfrak{n}}^{\mathrm{res}}$, so that:

$$x =: y\,\varepsilon, \quad \varepsilon \in E_{\mathfrak{n}}^{\mathrm{res}}.$$

Finally, $\boldsymbol{x}\,i(y)^{-1} = i(\varepsilon)\,\boldsymbol{u}$ is an element of $E^{\mathrm{ord}} U_{\mathfrak{m}'}^{\mathrm{res}}$ where the left hand side of the equality is constant; thus:

$$\boldsymbol{x}\,i(y)^{-1} \in \bigcap_{\mathfrak{m}}(E^{\mathrm{ord}} U_{\mathfrak{m}}^{\mathrm{res}})$$

since the multiples $\mathfrak{m}'$ of $\mathfrak{n}$ form a cofinal family, giving the result.

The above proof shows that we can even replace E^{ord} by $E_{\mathfrak{n}}^{\mathrm{res}}$ for an arbitrary $\mathfrak{n}$.

(iii) In this case, we cannot use the product formula; consider:

[17] To simplify, homeomorphism will mean topological group isomorphism, the group isomorphism being always clear.

$$U_0^{\mathrm{ord}} := \prod_{v \in Pl_0} U_v \prod_{v \in Pl_\infty^{\mathrm{r}}} \{\pm 1\}$$

and denote by $\mathrm{adh}_0(E^{\mathrm{ord}})$ the closure of the image of E^{ord} in J_0 (hence in U_0^{ord}), and by $\mathrm{adh}_0(K^\times)$ that of $K^\times$ in J_0. The group U_0^{ord} is a compact subgroup of J_0 (on U_0^{ord} we get the product topology and the noncompact factors have been removed). Since the projections $\mathrm{pr}_v : U_0^{\mathrm{ord}} \longrightarrow U_v$, $v \in Pl_0$, are continuous, if $i_0(E^{\mathrm{ord}})$ is closed (hence compact) the corresponding images $i_v(E^{\mathrm{ord}})$ are compact. However this is impossible if there exists an element ε of infinite order in E^{ord}: fix such a place v, and use the fact that $U_v \simeq \mathbb{Z}/m\mathbb{Z} \times \mathbb{Z}_\ell^{[K_v\,:\,\mathbb{Q}_\ell]}$, where m is a nonzero integer and ℓ is the residue characteristic of v, does not have any closed subgroup isomorphic to $\mathbb{Z}/n\mathbb{Z} \times \mathbb{Z}^r$, for some $n \geq 1$ and $r \geq 1$. The reader who uses the fact that U_v^{ord} is a profinite group can deduce that $i_v(E^{\mathrm{ord}})$ (closed) is also profinite, which can happen only if E^{ord} is torsion.

This is sufficient to show that if E^{ord} is infinite, then $i_0(K^\times)$ is not closed. Indeed, otherwise since $E^{\mathrm{ord}} \subset K^\times$, we would have $\mathrm{adh}_0(E^{\mathrm{ord}}) \subset i_0(K^\times)$, a contradiction since:

$$\mathrm{adh}_0(E^{\mathrm{ord}}) = \mathrm{adh}_0(E^{\mathrm{ord}}) \cap i_0(K^\times) \subseteq U_0^{\mathrm{ord}} \cap i_0(K^\times) = i_0(E^{\mathrm{ord}}).$$

Using a proof identical to that of (ii), we have:

$$\bigcap_{\mathfrak{m}} (K^\times U_{0,\mathfrak{m}}^{\mathrm{res}}) = K^\times \,.\, \bigcap_{\mathfrak{m}} (E^{\mathrm{ord}} U_{0,\mathfrak{m}}^{\mathrm{res}}),$$

where $U_{0,\mathfrak{m}}^{\mathrm{res}} := U_{\mathfrak{m}}^{\mathrm{res}} \cap J_0 = \prod_{v \in Pl_0 \backslash T} U_v \prod_{v \in T} U_v^{m_v}$, except that in the reduced case, for any subset A of $K^\times$ we can write:

$$\bigcap_{\mathfrak{m}} (A\, U_{0,\mathfrak{m}}^{\mathrm{res}}) = \mathrm{adh}_0(A),$$

contrary to the usual case above since the $U_{\mathfrak{m}}^{\mathrm{res}}$ do not form a fundamental system of neighbourhoods of 1 in J. This yields the identity:

$$\mathrm{adh}_0(K^\times) = i_0(K^\times) \,.\, \mathrm{adh}_0(E^{\mathrm{ord}}),$$

where we can even replace E^{ord} by $E_{\mathfrak{n}}^{\mathrm{res}}$ for an arbitrary $\mathfrak{n}$. We can then obtain the converse, in other words that if E^{ord} is finite, then $i_0(E^{\mathrm{ord}})$ is closed, hence $\mathrm{adh}_0(K^\times) = i_0(K^\times)$, so $i_0(K^\times)$ is closed.

Thus, to compute the connected component of the unit element of C_0 we cannot use the Lemma 4.2.1; it is however very simple (see III.1.5.1, III.4.4.6).

(iv) We have $A' \simeq A/A \cap \mathbb{Q}^\times = A$. Let n be an integer taking infinitely many values. The following idèle (whose first component corresponds to $v = 2$ and the last one to $v = \infty$):

$$\boldsymbol{x}_n := (2^{-n}; \ldots, 1, \ldots; +1)$$

is unbounded (if $n \in \mathbb{N}$) or without any limit in A (if $n \in \mathbb{Z}$), but $\mathcal{cl}_0(\boldsymbol{x}_n)$ is represented by:

$$\boldsymbol{x}'_n := (1; \ldots, 2^n, \ldots; +1),$$

which we can make arbitrarily close to 1 in J_0: for every finite subset Σ of odd places of $\mathbb{Q}$, choose n of the form $n := \prod_{\ell \in \Sigma} (\ell - 1)\ell^m, \quad m \in \mathbb{N}$. Hence, we have $\mathcal{cl}_0(\boldsymbol{x}_n) \to 1$ in A'.

We invite the reader to replace the place 2 by the place ∞, starting from the idèle $(1; \ldots, 1, \ldots; -1)$, and to see that the above trick is impossible (indeed, in this case we have a homeomorphism). The situation is the same for $A := \bigoplus_{v \in S_0} K_v^\times \bigoplus_{v \in S_\infty} \{\pm 1\} \subset J_0$ with a finite set S of places of any number field K (we can prove that the map $A' := AK^\times/K^\times \longrightarrow A$ is continuous if and only if $S_0 = \emptyset$).

If we take $A := \bigoplus_{v \in S} K_v^\times$ in J and $A' := \mathcal{cl}(A)$ in C, the map $A \longrightarrow A'$ is always a continuous isomorphism, and it is a homeomorphism if and only if $|S| = 1$. Indeed, if $\boldsymbol{x} =: (x_v)_{v \in S}$ is such that $\mathcal{cl}(\boldsymbol{x})$ is close to 1 in C, then $|\mathcal{cl}(\boldsymbol{x})| = |\boldsymbol{x}|$ (with the meaning given by 4.2.7) is close to 1, so we only have that $\prod_{v \in S} |x_v|_{K_v}$ is close to 1, and not necessarily all the $|x_v|_{K_v}$ individually close to 1 (necessary condition to have $\boldsymbol{x}$ close to 1). This tells us how to construct counterexamples: for $K = \mathbb{Q}$ take an idèle with support $S = \{2, 3\}$ of the form $(2^{-n}, 3^{-m}; \ldots, 1, \ldots; 1)$, and represent it by $(3^m, 2^n; \ldots, 2^n \times 3^m, \ldots; 2^n \times 3^m)$, which can be made arbitrarily close to 1 (including at the place ∞!).

If $S = \{v_0\} \subset Pl_0$ and if the representative $(i_{v_0}(x)x_{v_0}; \ldots, i_v(x), \ldots)$ of $\boldsymbol{x}$ is close to 1, we deduce that its volume, equal to $|x_{v_0}|_{K_{v_0}}$, is close to 1, hence equal to 1, so $x_{v_0} \in U_{v_0}$. Then we may suppose that $i_{v_0}(x) \in U_{v_0}$ which implies $|x - 1|_{v_0} \leq 1$; by reproducing the proof of (i), we easily deduce that $x = 1$ and that x_{v_0} is close to 1. Thus in this case, $\boldsymbol{x}$ is close to 1 in J. The case $v_0 \in Pl_\infty$ is similar, but using in this case that $|i_{v_0}(x)|_{K_{v_0}}$ is only close to 1 and thus that $|x - 1|_{v_0}$ is bounded.

We thus see here that reduced idèles enable us to suppress an awkward aspect since, as we will see, the fact that the above isomorphism is not a homeomorphism has a precise meaning at the level of reciprocity maps and decomposition groups of finite places. We will come back to reduced idèles in 5.2. □

An immediate property of the idèle class group (which will be stated in 4.3.3) is that it expresses in a convenient and natural way the classical generalized chinese remainder theorem, which we recall in its multiplicative form.

4.3 Theorem (multiplicative chinese remainder). *Let T and Δ_∞ be two finite sets of finite and real infinite places of K. For each $v \in T$, let $m_v \geq 0$*

be a given integer. Then, for any family $(a_v)_{v\in T\cup\Delta_\infty}$ of elements $a_v \in K_v^\times$, there exists $x \in K^\times$ which satisfies the following properties:

$$\begin{aligned} &i_v(x)\,a_v^{-1} \in U_v^{m_v} \text{ for all } v \in T,\\ &i_v(x)\,a_v^{-1} > 0 \quad\ \text{ for all } v \in \Delta_\infty. \end{aligned} \qquad \square$$

The usual approximation theorem is additive and says that the image of K under the diagonal map $(i_v)_{v\in T\cup\Delta_\infty}$ is dense in $\prod_{v\in T\cup\Delta_\infty} K_v$ [d, CF, Ch. II, § 6]. However class field theory does not need the geometric aspect represented by the density result in the archimedean factors where only the signature enters.

Finally, we mention the strong approximation theorem, of a very arithmetical nature, and which is of practical interest, but which is also not required for class field theory ([d, CF, Ch. II, § 15], [j, Coh2, Ch. 1, § 1.2]).

4.3.1 Remarks. (i) If for some $v \in T$ we have $m_v = 0$, recall that the relation $i_v(x)a_v^{-1} \in U_v^{m_v}$ means that $i_v(x)a_v^{-1} \in U_v$. If necessary, this allows us to enlarge T and Δ_∞ so as to obtain a solution x prime to a given finite set of places Σ, disjoint from $T \cup \Delta_\infty$: for each $v \in \Sigma$ we choose $a_v := 1$ and $m_v := 0$ for each $v \in \Sigma_0$. The solution x corresponding to $T \cup \Sigma_0$ and $\Delta_\infty \cup \Sigma_\infty$ is then prime to Σ_0 (in the usual sense) and positive on Σ_∞.

(ii) If $m_v > 0$, then $i_v(x)a_v^{-1} \in U_v^{m_v}$ is equivalent to $i_v(x)a_v^{-1} \equiv 1 \bmod (\pi_v^{m_v})$, but not to $i_v(x) \equiv a_v \bmod (\pi_v^{m_v})$ since we are dealing here with the equivalence relation on $K_v^\times$ defined by its subgroup $U_v^{m_v}$. Historically, the multiplicative congruences of Hasse were written $i_v(x) \equiv a_v \bmod^\times \pi_v^{m_v}$; we will not use this ambiguous and useless notation.

This formulation of the chinese remainder theorem can be justified by the fact that class field theory is exclusively multiplicative. $\square$

4.3.2 Corollary. *Let $\mathfrak{m} = \prod_{v\in T} \mathfrak{p}_v^{m_v}$. Let $\mathfrak{n} = \prod_{v\in T} \mathfrak{p}_v^{n_v}$, with $0 \le n_v \le m_v$, be a divisor of $\mathfrak{m}$, and let $\delta_\infty \subseteq \Delta_\infty$. Then the diagonal map:*

$$K^\times_{T,\mathfrak{n},\Delta_\infty\backslash\delta_\infty} \longrightarrow \bigoplus_{v\in T} U_v^{n_v}/U_v^{m_v} \bigoplus_{v\in\delta_\infty} \{\pm 1\}$$

gives rise to the exact sequence:

$$1 \longrightarrow K^\times_{T,\mathfrak{m},\Delta_\infty} \longrightarrow K^\times_{T,\mathfrak{n},\Delta_\infty\backslash\delta_\infty} \longrightarrow \bigoplus_{v\in T} U_v^{n_v}/U_v^{m_v} \bigoplus_{v\in\delta_\infty} \{\pm 1\} \longrightarrow 1.$$

Proof. Let us show that the map is surjective. Let:

$$(a_v)_{v\in T\cup\delta_\infty} \in \bigoplus_{v\in T} U_v^{n_v} \bigoplus_{v\in\delta_\infty} \mathbb{R}^\times\,;$$

for $v \in \Delta_\infty\backslash\delta_\infty$, we choose $a_v := 1$. Then there exists $x \in K^\times$ such that:

$$\begin{aligned} &i_v(x)\, a_v^{-1} \in U_v^{m_v} \text{ for all } v \in T,\\ &i_v(x)\, a_v^{-1} > 0 \quad\ \text{ for all } v \in \Delta_\infty \ ; \end{aligned}$$

such an x belongs to $K^\times_{T,\mathfrak{n},\Delta_\infty \setminus \delta_\infty}$ and is a suitable preimage. □

The most important consequence is the following fundamental theorem.

4.3.3 Theorem. *Let T and Δ_∞ be two finite sets of finite and real infinite places of K, and let $\mathfrak{m} = \prod_{v \in T} \mathfrak{p}_v^{m_v}$, $m_v \geq 0$. Then the canonical map from $J_{T,\mathfrak{m},\Delta_\infty}$ to $J/K^\times$ induces the canonical isomorphism:*

$$j_{T,\mathfrak{m},\Delta_\infty} \ : \ J_{T,\mathfrak{m},\Delta_\infty} / K^\times_{T,\mathfrak{m},\Delta_\infty} \longrightarrow J/K^\times .$$

Proof. The injectivity comes from the fact that $K^\times \cap J_{T,\mathfrak{m},\Delta_\infty} = K^\times_{T,\mathfrak{m},\Delta_\infty}$. Now let $\boldsymbol{x} =: (x_v)_v \in J$. By the chinese remainder theorem, we can find $x \in K^\times$ such that:

$$\begin{aligned} &i_v(x)\, x_v^{-1} \in U_v^{m_v} \text{ for all } v \in T,\\ &i_v(x)\, x_v^{-1} > 0 \quad\ \text{ for all } v \in \Delta_\infty \ ; \end{aligned}$$

the idèle $\boldsymbol{x}' := i(x)^{-1}\, \boldsymbol{x}$ is an element of $J_{T,\mathfrak{m},\Delta_\infty}$ representing $\boldsymbol{x}$ modulo $K^\times$, proving the result. □

b) Generalized Class Groups — Rank Formulas

We will now give a definition which will allow us to describe the most general correspondence of class field theory for finite extensions; it will of course contain the usual definition of the ordinary class group, as well as the restricted (or narrow) class group. Let K be a number field together with sets of places T and S, and let $\mathfrak{m} \in \langle\, T \,\rangle_{\mathbb{N}}$.

4.4 Definition (generalized classes). Using the definitions of 3.3, we set:

$$\mathcal{C}\ell^S_{\mathfrak{m}} := I_T / P_{T,\mathfrak{m},\mathrm{pos}} \langle\, S \,\rangle,$$

where, by definition, $P_{T,\mathfrak{m},\mathrm{pos}}\langle\, S \,\rangle := P_{T,\mathfrak{m},\Delta_\infty} . \langle\, S_0 \,\rangle$, with $\Delta_\infty := Pl^{\mathrm{r}}_\infty \setminus S_\infty$, and where $\langle\, S_0 \,\rangle := \langle\, S_0 \,\rangle_{\mathbb{Z}}$ is the subgroup of I_T generated by the $\mathfrak{p}_v$ for $v \in S_0$.[18] The group $\mathcal{C}\ell^S_{\mathfrak{m}}$ is called the S-ray class group modulo $\mathfrak{m}$ of K. □

These groups are also called generalized class groups or S-class groups depending on whether one wishes to mention especially the modulus $\mathfrak{m}$ or the set of places S.

[18] The notation $\langle \bullet \rangle$ has been used in 4.1.2, (ii) for an idèle group, and here an ideal group; this is convenient, corresponds to the same use, and should not create any confusion.

We will see in 5.1.2 that the groups $\mathcal{C}\ell_{\mathfrak{m}}^{S}$ do not depend on the choice of the set T containing the support of $\mathfrak{m}$.

4.4.1 Remarks. (i) If we take $T = \emptyset$ and $S = Pl_{\infty}^{\mathrm{r}}$, we recover the definition of the ordinary class group, which we denote more simply:

$$\mathcal{C}\ell^{\mathrm{ord}} := I/P\ ;$$

if we take $S = \emptyset$, we recover the definition of the restricted class group:

$$\mathcal{C}\ell^{\mathrm{res}} := I/P_{\mathrm{pos}}.$$

(ii) The group $\mathcal{C}\ell_{\mathfrak{m}}^{S}$ (for fixed $\mathfrak{m}$) should be considered as a quotient of the basic group $\mathcal{C}\ell_{\mathfrak{m}}^{\mathrm{res}} := I_T/P_{T,\mathfrak{m},\mathrm{pos}}$. More precisely:

$$\mathcal{C}\ell_{\mathfrak{m}}^{S} =: \mathcal{C}\ell_{\mathfrak{m}}^{\mathrm{res}}/\langle\, c\ell_{\mathfrak{m}}^{\mathrm{res}}(S)\,\rangle,$$

$c\ell_{\mathfrak{m}}^{\mathrm{res}}$ being the canonical map $c\ell_{\mathfrak{m}}^{\mathrm{res}} : I_T \longrightarrow I_T/P_{T,\mathfrak{m},\mathrm{pos}}$, and where $\langle\, c\ell_{\mathfrak{m}}^{\mathrm{res}}(S)\,\rangle$, the subgroup of $\mathcal{C}\ell_{\mathfrak{m}}^{\mathrm{res}}$ generated by the classes of elements of S can be described as a group of ideal classes in the following way:

$$\langle\, c\ell_{\mathfrak{m}}^{\mathrm{res}}(S)\,\rangle := c\ell_{\mathfrak{m}}^{\mathrm{res}}(\langle\, S_0\,\rangle)\,.\,c\ell_{\mathfrak{m}}^{\mathrm{res}}(\langle\, (u_{\mathfrak{m},v})\,\rangle_{v\in S_\infty}),$$

where, for each $v \in S_\infty$, $u_{\mathfrak{m},v}$ is any element of $K_{T,\mathfrak{m}}^{\times}$ of signature given by:

$$\begin{aligned} &i_{v'}(u_{\mathfrak{m},v}) > 0 \text{ for all } v' \in Pl_{\infty}^{\mathrm{r}},\ v' \neq v,\\ &i_v(u_{\mathfrak{m},v}) < 0. \end{aligned}$$

It is easily checked that we indeed have $P_{T,\mathfrak{m},\Delta_\infty} = P_{T,\mathfrak{m},\mathrm{pos}}\,.\,\langle\, (u_{\mathfrak{m},v})\,\rangle_{v\in S_\infty}$, so that the explicit use of the elements $u_{\mathfrak{m},v}$ is not necessary. □

This point of view, which by definition sets:

$$\mathcal{C}\ell_{\mathfrak{m}}^{S} := I_T/P_{T,\mathfrak{m},\mathrm{pos}}\langle\, S\,\rangle := I_T/P_{T,\mathfrak{m},\Delta_\infty}\,.\,\langle\, S_0\,\rangle,$$

avoids dealing with classes of divisors as in [Ja2], or with signed ideals as in [AJ], hence uses only ideal classes. In other words, at the level of ideal groups, we give up the possibility of giving any meaning to $\langle\, S_\infty\,\rangle$, limiting ourselves to the grouping:

$$P_{T,\mathfrak{m},\mathrm{pos}}\langle\, S_\infty\,\rangle := P_{T,\mathfrak{m},\Delta_\infty}.$$

From a conceptual point of view, the use of divisors is perhaps right, but having divisors denoted $\mathfrak{a}$ as for ideals (although some are of order 2) is in a dual manner similar to the problem of cycles $\mathfrak{m} = \mathfrak{m}_0\mathfrak{m}_\infty$ classically used in class field theory to define ramification; thus, we have simply transfered the problem. Finally, this choice is in accordance with the classical Artin map which is defined on a group of ideals (prime to T), and not on a group of divisors. The advantage of our point of view is that there are only two clear and natural questions (see 3.5):

• congruence questions (modulo an integral ideal) and signatures of number groups ($K^{\times}_{T,\mathfrak{m},\Delta_\infty}$, $E^{S_0\cup S_\infty}_{\mathfrak{m}}$), which enter anyway whatever the choice of point of view;

• Frobenius' questions, for which the only new aspect (due to the fact that we consider infinite places to be unramified) will consist in considering complex conjugations (in other words the generators of the decomposition groups of the infinite places) as Frobenius' of the infinite places, which the analytic aspects of the theory of number fields also seem to suggest ([Se6, § 2], [Gr7, § 3]), as well as the K-theory of the ring of integers of number fields (this will be seen in more detail in II.7.6.3 and III.2.6.5).

Thus, from our point of view, class field theory assigns to $\mathcal{C}\!\ell^S_{\mathfrak{m}}$ a T-ramification condition (limited by the modulus $\mathfrak{m}$) and of S-decomposition (for S_0 as well as for S_∞).

We can even take the inverse limit on those groups, considered as Galois groups.

4.4.2 Definition. Let K be a number field together with sets of places T and S. We set:

$$\mathcal{C}\!\ell^S_T := \varprojlim_{\mathfrak{m}\in\langle T\rangle_{\mathbb{N}}} \mathcal{C}\!\ell^S_{\mathfrak{m}},$$

where the ordering on the multiplicative monoid $\langle T\rangle_{\mathbb{N}}$ generated by T is divisibility, the transition homomorphisms being the canonical maps:

$$I_T/P_{T,\mathfrak{m},\mathrm{pos}}\langle S\rangle \longrightarrow I_T/P_{T,\mathfrak{n},\mathrm{pos}}\langle S\rangle, \text{ if } \mathfrak{n}\,|\,\mathfrak{m}. \qquad \square$$

Chapters III and IV will be devoted to the study, for any prime p, of the p-Sylow subgroup of this group $\mathcal{C}\!\ell^S_T$ which class field theory interprets as the Galois group $\mathcal{A}^S_T$ of the maximal abelian T-ramified and S-split pro-p-extension of K (i.e., unramified outside T and totally split at the places belonging to S).

The groups $\mathcal{C}\!\ell^S_{\mathfrak{m}}$ can be studied in different ways, the most classical being to use the basic group $\mathcal{C}\!\ell^S = \mathcal{C}\!\ell^{\mathrm{res}}/\langle c\!\ell^{\mathrm{res}}(S)\rangle$. We can, however, prove the following more general result which we be useful later, and which not only yields classical relations between the orders of these groups, but also the corresponding relations between their p-ranks. For this recall the:

4.4.3 Definition. For a $\mathbb{Z}$-module A of finite type, we denote by $\mathrm{rk}_p(A)$ the $\mathbb{F}_p$-dimension of the $\mathbb{F}_p$-vector space A/A^p. □

4.5 Theorem. *Let* $\mathfrak{m} = \prod_{v\in T}\mathfrak{p}_v^{m_v}$, $m_v \geq 0$. *Let* $\mathfrak{n} = \prod_{v\in T}\mathfrak{p}_v^{n_v}$, $0 \leq n_v \leq m_v$, *be a divisor of* $\mathfrak{m}$, *and* $\delta_\infty \subseteq \Delta_\infty := Pl^{\mathrm{r}}_\infty\backslash S_\infty$.

(i) We have the following exact sequence:

$$1 \longrightarrow E_{\mathfrak{n}}^{S\cup\delta_\infty}/E_{\mathfrak{m}}^{S} \xrightarrow{i_{T,\delta_\infty}} \bigoplus_{v\in T} U_v^{n_v}/U_v^{m_v} \bigoplus_{v\in\delta_\infty} \{\pm 1\} \xrightarrow{h_{\mathfrak{m}}^S} P_{T,\mathfrak{n},\mathrm{pos}}\langle\, S\cup\delta_\infty\,\rangle/P_{T,\mathfrak{m},\mathrm{pos}}\langle\, S\,\rangle \longrightarrow 1.$$

(ii) For every prime number p, we have:

$$\mathrm{rk}_p(\mathcal{C}\ell_{\mathfrak{m}}^S) = \mathrm{rk}_p(\mathcal{C}\ell_{\mathfrak{n}}^{S\cup\delta_\infty}) + \sum_{v\in T} \mathrm{rk}_p((U_v)^p U_v^{n_v}/(U_v)^p U_v^{m_v}) + \delta_{2,p}|\delta_\infty| - (\mathrm{rk}_p(Y_{T,\mathfrak{n}}^{S\cup\delta_\infty}/K_T^{\times p}) - \mathrm{rk}_p(Y_{T,\mathfrak{m}}^S/K_T^{\times p})),$$

with $\delta_{2,p} := 1$ (resp. 0) if $p = 2$ (resp. $p \neq 2$), where we have set:

$$Y_{T,\mathfrak{n}}^{S\cup\delta_\infty} := \{\alpha \in K_T^{\times p} K_{T,\mathfrak{n},\Delta_\infty\backslash\delta_\infty}^{\times},\ (\alpha) = \mathfrak{a}^p\mathfrak{a}_{S_0},\ \mathfrak{a}\in I_T,\ \mathfrak{a}_{S_0}\in\langle\, S_0\,\rangle\},$$
$$Y_{T,\mathfrak{m}}^{S} := \{\alpha \in K_T^{\times p} K_{T,\mathfrak{m},\Delta_\infty}^{\times},\ (\alpha) = \mathfrak{a}^p\mathfrak{a}_{S_0},\ \mathfrak{a}\in I_T,\ \mathfrak{a}_{S_0}\in\langle\, S_0\,\rangle\}.$$

4.5.1 Corollary. *We have:*

$$|\mathcal{C}\ell_{\mathfrak{m}}^S| = |\mathcal{C}\ell_{\mathfrak{n}}^{S\cup\delta_\infty}|\ \frac{\varphi(\mathfrak{m})\,\varphi(\mathfrak{n})^{-1}\,2^{|\delta_\infty|}}{(E_{\mathfrak{n}}^{S\cup\delta_\infty} : E_{\mathfrak{m}}^S)},$$

where φ denotes the generalized Euler function defined by:

$$\varphi(\mathfrak{m}) := \prod_{v\in T}(U_v : U_v^{m_v}) = \prod_{\substack{v\in T\\ m_v\geq 1}} \mathrm{N}\mathfrak{p}_v^{m_v-1}(\mathrm{N}\mathfrak{p}_v - 1).$$

Proof of the theorem. By definition, we have:

$$\mathcal{C}\ell_{\mathfrak{m}}^S = I_T/P_{T,\mathfrak{m},\Delta_\infty}.\langle\, S_0\,\rangle,\quad \mathcal{C}\ell_{\mathfrak{n}}^{S\cup\delta_\infty} = I_T/P_{T,\mathfrak{n},\Delta_\infty\backslash\delta_\infty}.\langle\, S_0\,\rangle,$$

from which we deduce the exact sequence:

$$1 \longrightarrow P_{T,\mathfrak{n},\mathrm{pos}}\langle\, S\cup\delta_\infty\,\rangle/P_{T,\mathfrak{m},\mathrm{pos}}\langle\, S\,\rangle \longrightarrow \mathcal{C}\ell_{\mathfrak{m}}^S \longrightarrow \mathcal{C}\ell_{\mathfrak{n}}^{S\cup\delta_\infty} \longrightarrow 1$$

with the usual meaning of the notation $P_{T,\bullet,\mathrm{pos}}\langle\,\bullet\,\rangle$.

Now consider the canonical (surjective) map:

$$K_{T,\mathfrak{n},\Delta_\infty\backslash\delta_\infty}^{\times}/K_{T,\mathfrak{m},\Delta_\infty}^{\times} \longrightarrow P_{T,\mathfrak{n},\Delta_\infty\backslash\delta_\infty}.\langle\, S_0\,\rangle/P_{T,\mathfrak{m},\Delta_\infty}.\langle\, S_0\,\rangle,$$

whose kernel is equal to $E_{\mathfrak{n}}^{S\cup\delta_\infty}/E_{\mathfrak{m}}^S$. Using 4.3.2, we have:

$$K_{T,\mathfrak{n},\Delta_\infty\backslash\delta_\infty}^{\times}/K_{T,\mathfrak{m},\Delta_\infty}^{\times} \simeq \bigoplus_{v\in T} U_v^{n_v}/U_v^{m_v} \bigoplus_{v\in\delta_\infty}\{\pm 1\},$$

and the exact sequence of the theorem follows.

In Exercise 5.1.3 we will see an idelic proof of this exact sequence.

We now prove statement (ii) relating the p-ranks of the given groups: the above exact sequence yields:

$$1 \longrightarrow P_{T,\mathfrak{n},\Delta_\infty \backslash \delta_\infty} . \langle S_0 \rangle I_T^p / P_{T,\mathfrak{m},\Delta_\infty} . \langle S_0 \rangle I_T^p \longrightarrow \\ \mathcal{C}\ell_{\mathfrak{m}}^S / (\mathcal{C}\ell_{\mathfrak{m}}^S)^p \longrightarrow \mathcal{C}\ell_{\mathfrak{n}}^{S \cup \delta_\infty} / (\mathcal{C}\ell_{\mathfrak{n}}^{S \cup \delta_\infty})^p \longrightarrow 1$$

whose kernel denoted X can be interpreted in terms of:

$$1 \longrightarrow Y_{T,\mathfrak{n}}^{S \cup \delta_\infty} / Y_{T,\mathfrak{m}}^S \longrightarrow K_T^{\times p} K_{T,\mathfrak{n},\Delta_\infty \backslash \delta_\infty}^\times / K_T^{\times p} K_{T,\mathfrak{m},\Delta_\infty}^\times \longrightarrow X \longrightarrow 1.$$

By the chinese remainder theorem, the middle term of this exact sequence is isomorphic to:

$$\bigoplus_{v \in T} (U_v)^p U_v^{n_v} / (U_v)^p U_v^{m_v} \bigoplus_{v \in \delta_\infty} (\{\pm 1\})_p,$$

finishing the proof of the rank formula. □

Note. For $v \in T \backslash T_p$ we have $\mathrm{rk}_p((U_v)^p U_v^{n_v} / (U_v)^p U_v^{m_v}) = 1$ if and only if $\mu_p \subset K_v$ (i.e., $q_v \equiv 1 \bmod (p)$), $n_v = 0$, and $m_v \geq 1$; otherwise this rank is equal to zero.

4.5.2 Remark. In the exact sequence that we have obtained:

$$1 \longrightarrow E_{\mathfrak{n}}^{S \cup \delta_\infty} / E_{\mathfrak{m}}^S \xrightarrow{i_{T,\delta_\infty}} \bigoplus_{v \in T} U_v^{n_v} / U_v^{m_v} \bigoplus_{v \in \delta_\infty} \{\pm 1\} \xrightarrow{h_{\mathfrak{m}}^S} \\ P_{T,\mathfrak{n},\mathrm{pos}} \langle S \cup \delta_\infty \rangle / P_{T,\mathfrak{m},\mathrm{pos}} \langle S \rangle \longrightarrow 1,$$

i_{T,δ_∞} comes from the diagonal embedding in $\bigoplus_{v \in T} U_v^{n_v} \bigoplus_{v \in \delta_\infty} \mathbb{R}^\times$ and the map $h_{\mathfrak{m}}^S$ can be described in the following way. If:

$$\boldsymbol{a} =: (a_v)_{v \in T \cup \delta_\infty} \in \bigoplus_{v \in T} U_v^{n_v} \bigoplus_{v \in \delta_\infty} \mathbb{R}^\times,$$

let $x \in K_{T,\mathfrak{n},\Delta_\infty \backslash \delta_\infty}^\times$ a preimage of $\boldsymbol{a}$ modulo $\bigoplus_{v \in T} U_v^{m_v} \bigoplus_{v \in \delta_\infty} \mathbb{R}^{\times +}$, under the surjective map of Corollary 4.3.2; then $h_{\mathfrak{m}}^S(\boldsymbol{a})$ is the class of the ideal (x) modulo $P_{T,\mathfrak{m},\mathrm{pos}} \langle S \rangle$.

This map $h_{\mathfrak{m}}^S$ is closely related to the map $\gamma_{\mathfrak{m}}^S$ that we will define in 5.1 to relate the idèle class group with the generalized class groups. Note that these definitions are incompatible. Indeed, if we *represent* the idèle $\boldsymbol{a}$ with support $T \cup \delta_\infty$ by an idèle $\boldsymbol{a}' =: (a'_v)_v$ of $J_{T,\mathfrak{m},\Delta_\infty}$ (so that we have $\boldsymbol{a}' = i(y)\, \boldsymbol{a}$, $y \in K^\times$), the ideal corresponding to $\boldsymbol{a}'$ is equal to $\prod_{v \in Pl_0} \mathfrak{p}_v^{\mathrm{v}(a'_v)} = (y)$ since $\mathrm{v}(a_v) = 0$ for all $v \in Pl_0$. Hence, we obtain the relation $(x\, y) \in P_{T,\mathfrak{m},\Delta_\infty}$, which shows that $h_{\mathfrak{m}}^S$ and $\gamma_{\mathfrak{m}}^S$ differ by the inversion automorphism ι on the class group, in other words:

$$\gamma_{\mathfrak{m}}^S = \iota \circ h_{\mathfrak{m}}^S.$$

This is unavoidable, except if we change the definition of $\gamma_{\mathfrak{m}}^S$ or that of the natural isomorphism deduced from Corollary 4.3.2 whose inverse is used here. □

4.5.3 Remark. If Z_K denotes the ring of integers of K, we have:

$$\bigoplus_{v\in T} U_v/U_v^{m_v} \simeq (Z_K/\mathfrak{m})^\times,$$

where $(Z_K/\mathfrak{m})^\times$ is the group of invertible elements of the finite ring $Z_K/\mathfrak{m}$.

This point of view is often used in the literature but does not reflect the underlying idelic aspect, and we will not have any use for it (moreover, in all these questions we do not need the knowledge of the ring of integers Z_K). The abelian group structure of $(Z_K/\mathfrak{m})^\times$ (in other words, of the $U_v/U_v^{m_v}$) is not simple in a computational point of view (see details concerning this in [j, Coh2, Ch. 4, § 4.2]). □

Since we have the inclusions $K_T^{\times p} \subseteq Y_{T,\mathfrak{m}}^S \subseteq Y_{T,\mathfrak{n}}^{S\cup\delta_\infty}$, the quantity:

$$\mathrm{rk}_p(Y_{T,\mathfrak{n}}^{S\cup\delta_\infty}/K_T^{\times p}) - \mathrm{rk}_p(Y_{T,\mathfrak{m}}^S/K_T^{\times p}) = \mathrm{rk}_p(Y_{T,\mathfrak{n}}^{S\cup\delta_\infty}/Y_{T,\mathfrak{m}}^S)$$

is positive or zero, which yields:

4.5.4 Corollary. *We have the upper bound:*

$$\mathrm{rk}_p(\mathcal{C}\ell_{\mathfrak{m}}^S) - \mathrm{rk}_p(\mathcal{C}\ell_{\mathfrak{n}}^{S\cup\delta_\infty}) \le \sum_{v\in T} \mathrm{rk}_p((U_v)^p U_v^{n_v}/(U_v)^p U_v^{m_v}) + \delta_{2,p}|\delta_\infty|.$$

In particular, if we assume that T is equal to the support of $\mathfrak{m}$, we obtain:

$$\mathrm{rk}_p(\mathcal{C}\ell_{\mathfrak{m}}^{\mathrm{res}}) \le \mathrm{rk}_p(\mathcal{C}\ell^{\mathrm{res}}) + \sum_{v\in T_p} [K_v : \mathbb{Q}_p] + \sum_{v\in T} \delta_v,$$

where $\delta_v = 1$ or 0 according as $\mu_p \subset K_v$ or not. □

4.5.5 Corollary. *(i) We have:*

$$|\mathcal{C}\ell_{\mathfrak{m}}^S| = |\mathcal{C}\ell^S| \, \frac{\varphi(\mathfrak{m})}{(E^S : E_{\mathfrak{m}}^S)}.$$

(ii) For every prime number p, we have:

$$\mathrm{rk}_p(\mathcal{C}\ell_{\mathfrak{m}}^S) = \mathrm{rk}_p(\mathcal{C}\ell^S) + \sum_{v\in T} \mathrm{rk}_p(U_v/U_v^{m_v}) - \mathrm{rk}_p(Y_T^S/Y_{T,\mathfrak{m}}^S),$$

with:

$$Y_{T,\mathfrak{m}}^S = \{\alpha \in K_T^{\times p} K_{T,\mathfrak{m},\Delta_\infty}^\times,\ (\alpha) = \mathfrak{a}^p \mathfrak{a}_{S_0},\ \mathfrak{a} \in I_T,\ \mathfrak{a}_{S_0} \in \langle\, S_0 \,\rangle\},$$
$$Y_T^S = \{\alpha \in K_T^{\times p} K_{T,\Delta_\infty}^\times,\ (\alpha) = \mathfrak{a}^p \mathfrak{a}_{S_0},\ \mathfrak{a} \in I_T,\ \mathfrak{a}_{S_0} \in \langle\, S_0 \,\rangle\}.$$ □

4.5.6 Corollary. *(i) When $S = \emptyset$, we obtain:*

$$|\mathcal{C}\ell_{\mathfrak{m}}^{\mathrm{res}}| = |\mathcal{C}\ell^{\mathrm{res}}| \, \frac{\varphi(\mathfrak{m})}{(E^{\mathrm{res}} : E_{\mathfrak{m}}^{\mathrm{res}})},$$
$$\mathrm{rk}_p(\mathcal{C}\ell_{\mathfrak{m}}^{\mathrm{res}}) = \mathrm{rk}_p(\mathcal{C}\ell^{\mathrm{res}}) + \sum_{v\in T} \mathrm{rk}_p(U_v/U_v^{m_v}) - \mathrm{rk}_p(Y_T^{\mathrm{res}}/Y_{T,\mathfrak{m}}^{\mathrm{res}}),$$

with:

$$Y^{\rm res}_{T,\mathfrak m} = \{\alpha \in K_T^{\times p} K^{\times}_{T,\mathfrak m,\rm pos},\ (\alpha) = \mathfrak a^p,\ \mathfrak a \in I_T\},$$
$$Y^{\rm res}_{T} = \{\alpha \in K_T^{\times p} K^{\times}_{T,\rm pos},\ (\alpha) = \mathfrak a^p,\ \mathfrak a \in I_T\}.$$

(ii) When $S = Pl^{\rm r}_{\infty}$, *we obtain:*

$$|\mathcal{C}\ell^{\rm ord}_{\mathfrak m}| = |\mathcal{C}\ell^{\rm ord}|\ \frac{\varphi(\mathfrak m)}{(E^{\rm ord} : E^{\rm ord}_{\mathfrak m})},$$
$$\mathrm{rk}_p(\mathcal{C}\ell^{\rm ord}_{\mathfrak m}) = \mathrm{rk}_p(\mathcal{C}\ell^{\rm ord}) + \sum_{v\in T} \mathrm{rk}_p(U_v/U_v^{m_v}) - \mathrm{rk}_p(Y^{\rm ord}_T/Y^{\rm ord}_{T,\mathfrak m}),$$

with:

$$Y^{\rm ord}_{T,\mathfrak m} = \{\alpha \in K_T^{\times p} K^{\times}_{T,\mathfrak m},\ (\alpha) = \mathfrak a^p,\ \mathfrak a \in I_T\},$$
$$Y^{\rm ord}_{T} = \{\alpha \in K^{\times}_{T},\ (\alpha) = \mathfrak a^p,\ \mathfrak a \in I_T\}. \qquad \square$$

By specifying δ_∞, the theorem gives us a convenient way to go from the restricted sense to the ordinary sense.

4.5.7 Corollary. *We have:*

$$|\mathcal{C}\ell^{\rm res}_{\mathfrak m}| = |\mathcal{C}\ell^{\rm ord}_{\mathfrak m}|\ \frac{2^{r_1}}{(E^{\rm ord}_{\mathfrak m} : E^{\rm res}_{\mathfrak m})},$$
$$|\mathcal{C}\ell^{\rm res}| = |\mathcal{C}\ell^{\rm ord}|\ \frac{2^{r_1}}{(E^{\rm ord} : E^{\rm res})},$$
$$|\mathcal{C}\ell^{\rm res}_{\mathfrak m}| = |\mathcal{C}\ell^{\rm ord}|\ \frac{\varphi(\mathfrak m)\, 2^{r_1}}{(E^{\rm ord} : E^{\rm res}_{\mathfrak m})},$$

and the corresponding formulas for the 2-ranks:

$$\mathrm{rk}_2(\mathcal{C}\ell^{\rm res}_{\mathfrak m}) = \mathrm{rk}_2(\mathcal{C}\ell^{\rm ord}_{\mathfrak m}) + r_1 - \mathrm{rk}_2(Y^{\rm ord}_{T,\mathfrak m}/Y^{\rm res}_{T,\mathfrak m}),$$
$$\mathrm{rk}_2(\mathcal{C}\ell^{\rm res}) = \mathrm{rk}_2(\mathcal{C}\ell^{\rm ord}) + r_1 - \mathrm{rk}_2(Y^{\rm ord}/Y^{\rm res}),$$
$$\mathrm{rk}_2(\mathcal{C}\ell^{\rm res}_{\mathfrak m}) = \mathrm{rk}_2(\mathcal{C}\ell^{\rm ord}) + \sum_{v\in T} \mathrm{rk}_2(U_v/U_v^{m_v}) + r_1 - \mathrm{rk}_2(Y^{\rm ord}_T/Y^{\rm res}_{T,\mathfrak m}). \qquad \square$$

Note. Because of the above formulas, the finiteness of $\mathcal{C}\ell^{\rm ord}$, which is a result coming from the geometry of numbers, implies the finiteness of the groups $\mathcal{C}\ell^{\rm res}_{\mathfrak m}$ which has been used in 4.2.3 for an important topological property of the idèle group.

4.5.8 Remarks. (i) It follows from the exact sequence 4.5, (i) that, for $\mathfrak n | \mathfrak m$, we have $\mathcal{C}\ell^S_{\mathfrak n} \simeq \mathcal{C}\ell^S_{\mathfrak m}$ (i.e., $P_{T,\mathfrak m,\rm pos}\langle\, S\,\rangle = P_{T,\mathfrak n,\rm pos}\langle\, S\,\rangle$) if and only if:

$$i_T(E^S_{\mathfrak n})\,.\, \bigoplus_{v\in T} U_v^{m_v} = \bigoplus_{v\in T} U_v^{n_v}$$

(i.e., the image of $E^S_{\mathfrak{n}}$ generates $\bigoplus_{v\in T} U_v^{n_v}/U_v^{m_v}$), which is equivalent to:

$$(E^S_{\mathfrak{n}} : E^S_{\mathfrak{m}}) = \prod_{v\in T} (U_v^{n_v} : U_v^{m_v}) = \varphi(\mathfrak{m})\varphi(\mathfrak{n})^{-1}.$$

We will find again later this characterization for the computation of the conductor of an S-split ray class field. Similarly, we have $\mathcal{C}\ell^{S\cup\delta_\infty}_{\mathfrak{m}} = \mathcal{C}\ell^S_{\mathfrak{m}}$ if and only if:

$$\mathrm{sgn}_{\delta_\infty}(E^{S\cup\delta_\infty}_{\mathfrak{m}}) \simeq (\mathbb{Z}/2\mathbb{Z})^{|\delta_\infty|}$$

(i.e., the signature group of $E^{S\cup\delta_\infty}_{\mathfrak{m}}$ on δ_∞ is as large as possible, in other words has order $2^{|\delta_\infty|}$).

(ii) The group $Y^{S\cup\delta_\infty}_{T,\mathfrak{n}}$ contains $E^{S\cup\delta_\infty}_{\mathfrak{n}}$, and $Y^S_{T,\mathfrak{m}}$ contains $E^S_{\mathfrak{m}}$. By definition, they contain $K_T^{\times p}$. □

4.5.9 Remark. Let p be a fixed prime number. If g is a group of automorphisms of K of order not divisible by p, we can decompose the p-class groups in χ-components, according to the $\mathbb{Q}_p$-irreducible characters χ of g, thus giving χ-rank formulas analogous to those of 4.5, (ii) which correspond to the case $g = 1$. These formulas lead to the generalization of a formula of Šafarevič [Ša] which will be given in II.5.4.7 and, in the Kummer case, to the reflection theorem proved in II.5.4.5. □

If we take $g = 1$, there are no representation problems, and we can already give the following formulas valid for all p (see their proofs in Exercise II.5.4.1). Let K be a number field together with sets of places T and S; set $\Delta_\infty := Pl^{\mathrm{r}}_\infty \backslash S_\infty$, $T_p := T \cap Pl_p$, $S_p := S \cap Pl_p$, where Pl_p is the set of places of K above p.

4.6 Theorem (Šafarevič's formula (1964) — reflection formula (1998)). *Consider the group* $\mathcal{C}\ell^S_T := \varprojlim_{\mathfrak{m}\in\langle T\rangle_{\mathbb{N}}} \mathcal{C}\ell^S_{\mathfrak{m}}$ *(see 4.4.2).*

(i) We have:

$$\mathrm{rk}_p(\mathcal{C}\ell^S_T) = \mathrm{rk}_p(V^S_T/K_T^{\times p}) + \sum_{v\in T}\delta_v - \delta + \sum_{v\in T_p}[K_v : \mathbb{Q}_p]$$
$$+1 - r_1 - r_2 - |S_0| + \delta_{2,p}|\Delta_\infty|,$$

where:

$$V^S_T := \{\alpha \in K_T^{\times p}K^{\times}_{T,\Delta_\infty},$$
$$\mathrm{v}(\alpha) \equiv 0 \bmod (p)\ \ \forall v \in Pl_0\backslash S_0,\ i_v(\alpha) \in K_v^{\times p}\ \ \forall v \in T\}$$
$$= \{\alpha \in Y^S_T,\ i_v(\alpha) \in K_v^{\times p}\ \ \forall v \in T\},$$

$\delta_v := 1$ or 0 according as K_v contains μ_p or not, $\delta := 1$ or 0 according as K contains μ_p or not, $\delta_{2,p} := 1$ or 0 according as $p = 2$ or not.

(ii) In addition, if K contains μ_p, we have the following reflection formula:

$$\mathrm{rk}_p(\mathcal{C}\ell_T^{S_0 \cup S_\infty}) - \mathrm{rk}_p(\mathcal{C}\ell_{\mathfrak{m}^*}^{T \cup \Delta_\infty}) = |T| + \sum_{v \in T_p} [K_v : \mathbb{Q}_p] - r_1 - r_2 - |S_0| + \delta_{2,p}|\Delta_\infty|,$$

where $\mathfrak{m}^* := \prod_{v \in S_0 \backslash S_p} \mathfrak{p}_v \prod_{v \in S_p} \mathfrak{p}_v^{pe_v+1} \prod_{v \in Pl_p \backslash (T_p \cup S_p)} \mathfrak{p}_v^{pe_v}$, *where e_v denotes the ramification index of v in $K/\mathbb{Q}(\mu_p)$.*

If $T_p \cup S_p = Pl_p$, we have $\mathcal{C}\ell_{\mathfrak{m}^*}^{T \cup \Delta_\infty} = \mathcal{C}\ell_{S_0}^{T \cup \Delta_\infty}$ and we obtain:

$$\mathrm{rk}_p(\mathcal{C}\ell_T^{S_0 \cup S_\infty}) - \mathrm{rk}_p(\mathcal{C}\ell_{S_0}^{T \cup \Delta_\infty}) = |T| + \sum_{v \in T_p} [K_v : \mathbb{Q}_p] - r_1 - r_2 - |S_0| + \delta_{2,p}|\Delta_\infty|. \qquad \square$$

Šafarevič's formula (i) is often used, but one should not think that the situation is simple since, even though the group V_T^S seems more computable, its practical computation is a problem of units and ideal class groups which needs the algorithmic methods of [j, Coh2]; for example, in the simplest case ($T = \emptyset$, $S = Pl_\infty^{\mathrm{r}}$) we obtain the formula:

$$\mathrm{rk}_p(\mathcal{C}\ell^{\mathrm{ord}}) = \mathrm{rk}_p(V^{\mathrm{ord}}/K^{\times p}) - \delta + 1 - r_1 - r_2,$$

which can be proved in one's head from the definition of V^{ord}. On the contrary, the reflection formula (ii) is less immediate.

As mentioned in [j, Coh2], the groups V_T^S may be considered as good analogs of the Selmer groups of the theory of elliptic curves. They are also closely related to the $\mathcal{T}_T^S$ whose formal properties are more canonical and whose meaning is deeper (see III.4.2 and the results which follow).

§5 Reduced Idèles — Topological Aspects

a) The Fundamental Exact Sequence

The fact that it is always possible to compare the formalism of idèle class groups with that of generalized ideal class groups comes essentially from the exact sequence which we give below.

We refer to Definitions 3.3, 3.4, for number groups, then 4.1.1, 4.1.2, for idèle groups, and 4.4 for class groups of K given together with sets of places T and S.

5.1 Theorem (fundamental exact sequence). *For any modulus $\mathfrak{m} \in \langle T \rangle_{\mathbb{N}}$, we have the following exact sequence:*

$$1 \longrightarrow K^\times U_{\mathfrak{m}}^S / K^\times \simeq U_{\mathfrak{m}}^S / E_{\mathfrak{m}}^S \longrightarrow C = J/K^\times \xrightarrow{\gamma_{\mathfrak{m}}^S} \mathcal{C}\ell_{\mathfrak{m}}^S \longrightarrow 1,$$

where $\mathcal{C}\ell_{\mathfrak{m}}^{S} := I_T/P_{T,\mathfrak{m},\Delta_\infty} \cdot \langle S_0 \rangle$, *with* $\Delta_\infty := Pl_\infty^{\mathrm{r}} \setminus S_\infty$, *where* $\gamma_{\mathfrak{m}}^{S}$ *is the map which sends the class* $c\ell(\boldsymbol{x}) \in J/K^\times$, *represented by* $\boldsymbol{x}_{\mathfrak{m},\Delta_\infty} =: (x'_v)_v \in J_{T,\mathfrak{m},\Delta_\infty}$ *(see 4.3.3), to* $c\ell_{\mathfrak{m}}^{S}(\mathfrak{a})$, *with* $\mathfrak{a} := \prod_{v \in Pl_0} \mathfrak{p}_v^{\mathrm{v}(x'_v)}$. *In particular, we have:*

$$\mathcal{C}\ell_{\mathfrak{m}}^{S} \simeq J/K^\times U_{\mathfrak{m}}^{S}.$$

Proof. Put $\mathfrak{m} = \prod_{v \in T} \mathfrak{p}_v^{m_v}$, $m_v \geq 0$. Let $x \in K^\times$ be such that $i(x)\,\boldsymbol{x} = \boldsymbol{x}_{\mathfrak{m},\Delta_\infty} =: (x'_v)_v$, with:

$$\begin{aligned} &x'_v \in U_v^{m_v} \text{ for all } v \in T, \\ &x'_v > 0 \qquad \text{for all } v \in \Delta_\infty\ ; \end{aligned}$$

it is clear that the ideal $\mathfrak{a}$ belongs to I_T. If we change the representative of the class, we change $\boldsymbol{x}_{\mathfrak{m},\Delta_\infty}$ by an element $i(x_{\mathfrak{m},\Delta_\infty})$ of $i(K_{T,\mathfrak{m},\Delta_\infty}^\times)$ whose image under $\gamma_{\mathfrak{m}}^{S}$ is trivial (since if $i(x)$ is any principal idèle, then $\prod_{v \in Pl_0} \mathfrak{p}_v^{\mathrm{v}(x)}$ yields the principal ideal (x)), so that $\gamma_{\mathfrak{m}}^{S}$ is well defined. The fact that it is surjective is clear (the class of $\mathfrak{a} \in I_T$ is the image of the idèle $\boldsymbol{x} := (x_v)_v$ such that $x_v = \pi_v^{\mathrm{v}(\mathfrak{a})}$ for $v \in Pl_0 \backslash T$, and $x_v = 1$ for $v \in T \cup Pl_\infty$).

Hence, let us assume that $\boldsymbol{x}_{\mathfrak{m},\Delta_\infty} = (x'_v)_v$ yields:

$$\mathfrak{a} \in P_{T,\mathfrak{m},\Delta_\infty} \cdot \langle S_0 \rangle\ ;$$

we thus have $\mathfrak{a} =: (x_{\mathfrak{m},\Delta_\infty})\,\mathfrak{a}_{S_0}$, $x_{\mathfrak{m},\Delta_\infty} \in K_{T,\mathfrak{m},\Delta_\infty}^\times$, $\mathfrak{a}_{S_0} \in \langle S_0 \rangle$. Since $\boldsymbol{x}_{\mathfrak{m},\Delta_\infty} = i(x)\boldsymbol{x}$, consider the idèle:

$$(u_v)_v := \boldsymbol{u} := i(x_{\mathfrak{m},\Delta_\infty})^{-1}\boldsymbol{x}_{\mathfrak{m},\Delta_\infty} = i(x)\, i(x_{\mathfrak{m},\Delta_\infty})^{-1}\boldsymbol{x}$$

which defines the same idèle class as $\boldsymbol{x}$; we note the following facts:

- if $v \in Pl_0$, $\mathrm{v}(u_v) = \mathrm{v}(\boldsymbol{x}_{\mathfrak{m},\Delta_\infty}) - \mathrm{v}(x_{\mathfrak{m},\Delta_\infty}) = \mathrm{v}(\mathfrak{a}_{S_0})$,
- if $v \in \Delta_\infty$, $u_v > 0$,
- if $v \in T$, $u_v \in U_v^{m_v}$.

Thus, we obtain $\boldsymbol{u} \in U_{\mathfrak{m}}^{S}$, so finally:

$$\boldsymbol{x} \in K^\times U_{\mathfrak{m}}^{S}.$$

Conversely, if $\boldsymbol{x} := i(a)\boldsymbol{u}$, $a \in K^\times$, $\boldsymbol{u} =: (u_v)_v \in U_{\mathfrak{m}}^{S}$, we can represent $\boldsymbol{x}$ by $\boldsymbol{x}_{\mathfrak{m},\Delta_\infty} =: i(x)\,\boldsymbol{x}$, $x \in K^\times$, hence we obtain $\boldsymbol{x}_{\mathfrak{m},\Delta_\infty} = i(xa)\,\boldsymbol{u} =: i(b)\,\boldsymbol{u} =: (x'_v)_v$, for which we want to show that:

$$\mathfrak{a} := \prod_{v \in Pl_0} \mathfrak{p}_v^{\mathrm{v}(x'_v)} \in P_{T,\mathfrak{m},\Delta_\infty} \cdot \langle S_0 \rangle\ ;$$

we have:

$$\mathrm{v}(x'_v) = \mathrm{v}(b) + \mathrm{v}(\boldsymbol{u}) \text{ for all } v \in Pl,$$

hence the ideal $\mathfrak{b} := \mathfrak{a}\,(b)^{-1}$ and the element $b \in K^\times$ are such that:

- if $v \in Pl_0$, $\mathrm{v}(\mathfrak{b}) = \mathrm{v}(\boldsymbol{u})$ (hence $\mathfrak{b} \in \langle\, S_0 \,\rangle$),
- if $v \in \Delta_\infty$, $i_v(b) = x'_v\, u_v^{-1} > 0$ (hence $b \in K^\times_{\Delta_\infty}$),
- if $v \in T$, $i_v(b) = x'_v\, u_v^{-1} \in U_v^{m_v}$ (hence $b \in K^\times_{T,\mathfrak{m}}$).

We have thus obtained $\mathfrak{b} \in \langle\, S_0 \,\rangle$ and $(b) \in P_{T,\mathfrak{m},\Delta_\infty}$, so it follows that:

$$\mathfrak{a} = (b)\,\mathfrak{b} \in P_{T,\mathfrak{m},\Delta_\infty} \cdot \langle\, S_0 \,\rangle.$$

We have thus shown that the kernel of $\gamma^S_\mathfrak{m}$ is equal to the image of $U^S_\mathfrak{m}$ in C, in other words $K^\times U^S_\mathfrak{m}/K^\times \simeq U^S_\mathfrak{m}/K^\times \cap U^S_\mathfrak{m} = U^S_\mathfrak{m}/E^S_\mathfrak{m}$. □

This proves the exact sequence of the theorem, which links idèle classes with generalized ideal classes, and which can also be written in the following less canonical forms:

$$1 \longrightarrow K^\times_{T,\mathfrak{m},\Delta_\infty} U^S_\mathfrak{m} \longrightarrow J_{T,\mathfrak{m},\Delta_\infty} \longrightarrow \mathcal{C}\ell^S_\mathfrak{m} \longrightarrow 1,$$
$$1 \longrightarrow K^\times U^S_\mathfrak{m} \longrightarrow J \longrightarrow \mathcal{C}\ell^S_\mathfrak{m} \longrightarrow 1.$$

Note that in the second exact sequence above, the map $J \longrightarrow \mathcal{C}\ell^S_\mathfrak{m}$ is the one which sends the idèle $\boldsymbol{x} =: (x_v)_v$ to the class of $\mathfrak{a} := \prod_{v \in Pl_0} \mathfrak{p}_v^{\mathrm{v}(x_v)}$ only when $T = \emptyset$, $S_\infty = Pl^{\mathrm{r}}_\infty$ (the ordinary sense).

5.1.1 Corollary. *We have the following exact sequences:*

$$1 \longrightarrow U^{\mathrm{ord}}/E^{\mathrm{ord}} \longrightarrow C \longrightarrow \mathcal{C}\ell^{\mathrm{ord}} \longrightarrow 1,$$
$$1 \longrightarrow U^{\mathrm{res}}/E^{\mathrm{res}} \longrightarrow C \longrightarrow \mathcal{C}\ell^{\mathrm{res}} \longrightarrow 1.$$

□

In the second case, the class of $\boldsymbol{x}$ in C must be represented by a totally positive idèle, of which we take the corresponding ideal.

5.1.2 Remark. It is often useful to represent the class $c \in \mathcal{C}\ell^S_\mathfrak{m}$, corresponding to the idèle $\boldsymbol{x}$, by an ideal $\mathfrak{a}$ prime to a finite set Σ_0 of finite places of K (disjoint from T). This is always possible: in the computation of $c := \gamma^S_\mathfrak{m}(c\ell(\boldsymbol{x}))$, we represent the idèle $\boldsymbol{x}$ by $\boldsymbol{x}'_{\mathfrak{m},\Delta_\infty} \in J_{T\cup\Sigma_0,\mathfrak{m},\Delta_\infty}$, where we have chosen $m_v = 0$ for each $v \in \Sigma_0$; it is then clear that the ideal $\mathfrak{a}$ thus obtained is prime to Σ_0. In particular, we can write:

$$I_{T\cup\Sigma_0}/P_{T\cup\Sigma_0,\mathfrak{m},\mathrm{pos}} \simeq I_T/P_{T,\mathfrak{m},\mathrm{pos}},$$

whence the notation $\mathcal{C}\ell^S_\mathfrak{m}$ where we suppress any explicit mention of the set T containing the support of $\mathfrak{m}$. □

5.1.3 Exercise. Give an idelic proof of 4.5, (i).

Answer. Theorem 5.1 yields:

$$\mathcal{C}\ell_{\mathfrak{m}}^{S} \simeq J/K^{\times}U_{\mathfrak{m}}^{S} \ \text{ and } \ \mathcal{C}\ell_{\mathfrak{n}}^{S\cup\delta_\infty} \simeq J/K^{\times}U_{\mathfrak{n}}^{S\cup\delta_\infty},$$

hence the following exact sequence:

$$1 \longrightarrow K^{\times}U_{\mathfrak{n}}^{S\cup\delta_\infty}/K^{\times}U_{\mathfrak{m}}^{S} \longrightarrow \mathcal{C}\ell_{\mathfrak{m}}^{S} \longrightarrow \mathcal{C}\ell_{\mathfrak{n}}^{S\cup\delta_\infty} \longrightarrow 1,$$

where the kernel can be studied by means of the exact sequence:

$$1 \longrightarrow E_{\mathfrak{n}}^{S\cup\delta_\infty}/E_{\mathfrak{m}}^{S} \longrightarrow U_{\mathfrak{n}}^{S\cup\delta_\infty}/U_{\mathfrak{m}}^{S} \longrightarrow K^{\times}U_{\mathfrak{n}}^{S\cup\delta_\infty}/K^{\times}U_{\mathfrak{m}}^{S} \longrightarrow 1.$$

It is then immediate to prove that:

$$U_{\mathfrak{n}}^{S\cup\delta_\infty}/U_{\mathfrak{m}}^{S} \simeq \bigoplus_{v\in T} U_v^{n_v}/U_v^{m_v} \bigoplus_{v\in\delta_\infty} \{\pm1\},$$

which again leads to 4.5 and its corollaries. Note that the map:

$$U_{\mathfrak{n}}^{S\cup\delta_\infty}/U_{\mathfrak{m}}^{S} \longrightarrow P_{T,\mathfrak{n},\mathrm{pos}}\,\langle\, S\cup\delta_\infty \,\rangle/P_{T,\mathfrak{m},\mathrm{pos}}\,\langle\, S \,\rangle,$$

coming from it, is equal to the composite of the following maps:

$$U_{\mathfrak{n}}^{S\cup\delta_\infty}/U_{\mathfrak{m}}^{S} \xrightarrow{\ \mathrm{can}\ } J/K^{\times}U_{\mathfrak{m}}^{S} \xrightarrow{\ \gamma_{\mathfrak{m}}^{S}\ } \mathcal{C}\ell_{\mathfrak{m}}^{S},$$

which can be identified, not with $h_{\mathfrak{m}}^{S}$, but with $\iota\circ h_{\mathfrak{m}}^{S}$ (see 4.5.2). □

5.2 Definition (reduced idèles). We put, using the data T, S, $\mathfrak{m}\in\langle\, T\,\rangle_{\mathbb{N}}$:

$$\begin{aligned} J_0 &:= {\prod_{v\in Pl_0}}' K_v^{\times} \prod_{v\in Pl_\infty^{\mathrm{r}}} \{\pm1\},\\ U_{0,\mathfrak{m}}^{S} &:= \prod_{v\in Pl_0\backslash(T\cup S_0)} U_v \prod_{v\in T} U_v^{m_v} \prod_{v\in S_0} K_v^{\times} \prod_{v\in S_\infty} \{\pm1\},\\ U_\infty &:= \bigoplus_{v\in Pl_\infty} U_v = \bigoplus_{v\in Pl_\infty^{\mathrm{r}}} \mathbb{R}^{\times+} \bigoplus_{v\in Pl_\infty^{\mathrm{c}}} \mathbb{C}^{\times}. \end{aligned}$$

Then, we have:

$$J = J_0 \bigoplus U_\infty \ \text{ and } \ U_{\mathfrak{m}}^{S} = U_{0,\mathfrak{m}}^{S} \bigoplus U_\infty.$$

The elements of J_0 are called reduced idèles (see 4.1.2, 4.2.8, (iii)).

Let $i_0 := (i_v)_{v\in Pl_0\cup Pl_\infty^{\mathrm{r}}}$ (where only enter the embeddings at finite places and the signature homomorphism of the field, the i_v for $v\in Pl_\infty^{\mathrm{r}}$ here being identified with the functions sgn_v). □

Since $\mathcal{C}\ell(U_\infty)$ is in the kernel of all the $\gamma_{\mathfrak{m}}^{S}$, we have:

5.2.1 Proposition. *The fundamental exact sequence can be written:*

$$1 \longrightarrow U^S_{0,\mathfrak{m}}/i_0(E^S_{\mathfrak{m}}) \longrightarrow C_0 := J_0/i_0(K^\times) \xrightarrow{\gamma^S_{\mathfrak{m}}} \mathcal{C}\ell^S_{\mathfrak{m}} \longrightarrow 1.$$

Proof. Indeed, we can write:

$$1 \longrightarrow U^S_{\mathfrak{m}}/i(E^S_{\mathfrak{m}})U_\infty \longrightarrow J/i(K^\times)U_\infty \xrightarrow{\gamma^S_{\mathfrak{m}}} \mathcal{C}\ell^S_{\mathfrak{m}} \longrightarrow 1\ ;$$

but (using in J and J_0 the rules of notation stated in 4.1, Note), the canonical projection $J \longrightarrow J_0$ yields the isomorphism $J/K^\times U_\infty \simeq J_0/K^\times$, and under this isomorphism, the image of the group $U^S_{\mathfrak{m}}/E^S_{\mathfrak{m}}U_\infty$ is equal to $U^S_{0,\mathfrak{m}}/E^S_{\mathfrak{m}}$. These isomorphisms are also homeomorphisms. □

5.2.2 Remark. We even have the following exact sequence (with evident notations):

$$1 \longrightarrow K^\times_{T,\mathfrak{m},\Delta_\infty}U^S_{0,\mathfrak{m}} \longrightarrow J_{0,T,\mathfrak{m},\Delta_\infty} \longrightarrow \mathcal{C}\ell^S_{\mathfrak{m}} \longrightarrow 1,$$

which yields, for $T = \emptyset$ and $S = Pl^{\mathrm{r}}_\infty$:

$$1 \longrightarrow K^\times U^{\mathrm{ord}}_0 \longrightarrow J_0 \longrightarrow \mathcal{C}\ell^{\mathrm{ord}} \longrightarrow 1.$$ □

b) Topological Lemmas

We begin by the following exercise, whose purpose is to justify the usual hypothesis concerning norm groups in J or in J_0.

5.3 Exercise. Show that it is not true that every subgroup N of finite index of J, containing the image of $K^\times$, is closed (or open) in J.

Answer. Since U_∞ is a divisible group and N is of finite index, we already have $U_\infty \subset N$ and N is of the form $N_0 \oplus U_\infty$, with $N_0 := N \cap J_0$ of finite index in J_0 and containing the image of $K^\times$ (see 5.2). Set $V_0 := N_0 \cap U^{\mathrm{res}}_0$. We have the exact sequence:

$$1 \longrightarrow U^{\mathrm{res}}_0/V_0 \longrightarrow J_0/N_0 = C_0/c\ell_0(N_0) \longrightarrow \mathcal{C}\ell' \longrightarrow 1,$$

where $\mathcal{C}\ell' := \mathcal{C}\ell^{\mathrm{res}}/\gamma^{\mathrm{res}}(c\ell_0(N_0))$, showing that V_0 is of finite index in U^{res}_0. If V_0 is open in U^{res}_0, then $K^\times V_0$ and N_0 are open in J_0, so that N is open in J. Hence, it is necessary (and in fact sufficient) to build a V_0 of finite index which is not open (or not closed) in $A := U^{\mathrm{res}}_0$.

Consider the $\mathbb{F}_2$-vector space $A/A^2 \simeq (\mathbb{Z}/2\mathbb{Z})^I$, where I is an enumerable set of indices (each odd finite place v yields a $\mathbb{Z}/2\mathbb{Z}$ component and the even finite places yield $(\mathbb{Z}/2\mathbb{Z})^{[K_v\,:\,\mathbb{Q}_2]+1}$); for each $i \in I$, let $\overline{e}_i$ be the nontrivial element of A/A^2 with support $\{i\}$, and let $\overline{e}_\infty$ be the element (with support I) whose components are all nontrivial. Since I is infinite, the set of the $\overline{e}_i$

for $i \in I \cup \{\infty\}$ is free. Let $(\overline{e}_i)_{i\in\overline{I}}$, $\overline{I} \supset I \cup \{\infty\}$, be an $\mathbb{F}_2$-basis of A/A^2 containing $\{\overline{e}_i,\ i \in I \cup \{\infty\}\}$; we have:

$$A/A^2 \simeq \langle\, \overline{e}_i,\ i \in \overline{I}\backslash\{\infty\}\,\rangle \oplus \langle\, \overline{e}_\infty\,\rangle.$$

Denote by e_i a lift to A of $\overline{e}_i$, and consider:

$$V_0 := \langle\, e_i,\ i \in \overline{I}\backslash\{\infty\}\,\rangle_{\mathbb{Z}}\, A^2,$$

so that:

$$A/A^2 = V_0/A^2 \oplus \langle\, \overline{e}_\infty\,\rangle$$

and:

$$A/V_0 \simeq \langle\, \overline{e}_\infty\,\rangle \simeq \mathbb{Z}/2\mathbb{Z}.$$

But V_0 cannot be closed in A since V_0/A^2 is not closed in A/A^2 ($\overline{e}_\infty$ belongs to the closure of $\langle\, \overline{e}_i,\ i \in I\,\rangle \subset V_0/A^2$, hence to the closure of V_0/A^2, but it does not belong to V_0/A^2). It follows that the closure of V_0 is equal to $A = U_0^{\mathrm{res}}$.

Let $N_0 := K^\times V_0$ for which $N_0 \cap U_0^{\mathrm{res}} = E^{\mathrm{res}}\, V_0$. Since the construction of V_0 is not always convenient, we will assume that $K = \mathbb{Q}$ (i.e., $E^{\mathrm{res}} = 1$) so that it will be suitable for our needs (i.e., to have $V_0 = N_0 \cap U_0^{\mathrm{res}}$). If N_0 was closed, it would contain the closure of V_0 and would be equal to $K^\times U_0^{\mathrm{res}}$, which is absurd since it would yield $V_0 = U_0^{\mathrm{res}}$. This gives our counterexample.

In the case where the connected component of the unit element is nontrivial, such groups clearly exist, but their explicit construction is more complicated. □

Still in a topological context (which for us will only be for *abelian* groups), the following argument is often used.

5.4 Topological Lemma 1. *Let A be a Hausdorff topological group, and let $(A_i)_{i\in I}$ be a family of closed normal subgroups of A, indexed by the directed (or inductive)*[19] *set I, and such that $i \leq j$ is equivalent to $A_j \subseteq A_i$. Let C be a compact subset of A, and for any $A' \subseteq A$ set $CA' := \{ca',\ c \in C,\ a' \in A'\}$. Then:*

$$\bigcap_{i\in I} (C\, A_i) = C\Big(\bigcap_{i\in I} A_i\Big).$$

In particular, if $A = C\, A_i$ for all $i \in I$, this yields $A = C\Big(\bigcap\limits_{i\in I} A_i\Big)$.

Proof. One inclusion ($\supseteq$) is obvious. Thus, let $a \in \bigcap\limits_{i\in I} (C\, A_i)$; we must show that there exists $c \in C$ such that $ac^{-1} \in \bigcap\limits_{i\in I} A_i$. If $a\, C^{-1} \cap \bigcap\limits_{i\in I} A_i = \emptyset$, the complements of the A_i form an open cover of the compact set $a\, C^{-1}$, hence there exists a finite set $J \subset I$ such that $a\, C^{-1} \cap \bigcap\limits_{j\in J} A_j = \emptyset$. By choosing

[19] i.e., for all $i,\ j \in I$, there exists $k \in I$ such that $k \geq i$ and $k \geq j$.

some upper bound k of J, we get $a\,C^{-1} \cap A_k = \emptyset$, a contradiction, proving the result. □

In applications, the A_i will often be open.

5.4.1 Remark. If the A_i form a fundamental system of neighbourhoods of the unit element of A and if C is any subset, we know that:

$$\bigcap_{i\in I} (C\,A_i) = \mathrm{adh}(C)\ ;$$

in this case, since $\bigcap_{i\in I} A_i = 1$, the above result is true if we only assume that C is closed. □

5.4.2 Exercise. By an idelic counterexample, show that if the (open) subsets A_i do not form such a fundamental system of neighbourhoods, the assumption that C is closed is not sufficient to imply $\bigcap_{i\in I} (C\,A_i) = C\Big(\bigcap_{i\in I} A_i\Big)$.

Answer. Let us consider $A := J$ together with the family of open subsets $U_{\mathfrak{m}}^{\mathrm{res}} = U_{0,\mathfrak{m}}^{\mathrm{res}} \oplus U_\infty$ (see 5.2) and $C := i(K^\times)$. By 4.2.8, (i), C is closed in J. If $\bigcap_{\mathfrak{m}} (K^\times U_{\mathfrak{m}}^{\mathrm{res}}) = K^\times\Big(\bigcap_{\mathfrak{m}} U_{\mathfrak{m}}^{\mathrm{res}}\Big)$, it follows that $\bigcap_{\mathfrak{m}} (K^\times U_{\mathfrak{m}}^{\mathrm{res}}) = K^\times U_\infty$ since $\bigcap_{\mathfrak{m}} (U_{0,\mathfrak{m}}^{\mathrm{res}} \oplus U_\infty) = U_\infty$. Thus, we would have:

$$D := \mathcal{C}\!\ell\Big(\bigcap_{\mathfrak{m}} (K^\times U_{\mathfrak{m}}^{\mathrm{res}})\Big) = K^\times U_\infty / K^\times\ ;$$ [20]

however, this is the case only if the $\mathbb{Z}$-rank of E^{ord} is trivial. Indeed:

$$\bigcap_{\mathfrak{m}} (K^\times U_{\mathfrak{m}}^{\mathrm{res}}) =: \mathrm{adh}(K^\times U_\infty) = K^\times U_\infty$$

is equivalent to the fact that $i(K^\times) U_\infty = i_0(K^\times) \oplus U_\infty$ is closed in J, hence to the fact that $i_0(K^\times)$ is closed in J_0; the result then follows from 4.2.8, (iii). □

The following lemma states an important classical property of inverse limits which will also often be used.

5.5 Topological Lemma 2. *Let A be a Hausdorff topological group, and let $(A_i)_{i\in I}$ be a family of closed normal subgroups of A, indexed by the directed set I and such that $i \le j$ is equivalent to $A_j \subseteq A_i$. Assume that there exists a subgroup B of A contained in $\bigcap_{i\in I} A_i$, such that $A = BC := \{b\,c,\ b \in B,\ c \in$*

[20] Note that we can here find an example where $K^\times U_\infty / K^\times$ is isomorphic, but not homeomorphic to U_∞: choose a real quadratic field and, in the spirit of 4.2.8, (iv), use the powers of its fundamental unit ε in the idèle $(\ldots, 1, \ldots\ ;\ \varepsilon^{-n}, \varepsilon^{-n})$.

$C\}$, where C is a compact subset of A. Consider the inverse system of the A/A_i, $i \in I$, for the canonical transition homomorphisms:

$$h_{i,j} : A/A_j \longrightarrow A/A_i, \text{ for each } (i,j) \text{ such that } j \geq i.$$

Then the map:

$$f : A \longrightarrow \varprojlim_{i \in I} A/A_i$$

sending $a \in A$ to $(a \bmod A_i)_i$ is surjective and yields the following homeomorphism of compact groups:

$$A \Big/ \bigcap_{i \in I} A_i \simeq \varprojlim_{i \in I} A/A_i.$$

Proof. The A_i (hence their intersection) being closed, the groups $A \Big/ \bigcap_{i \in I} A_i$ and A/A_i, $i \in I$, are Hausdorff, hence compact as continuous images of the compact set C, therefore $\varprojlim_{i \in I} A/A_i$ is compact. Since $f(A) = f(C)$, it is thus sufficient to show that f is continuous with a dense image.

A fundamental system of neighbourhoods of the unit element of $\varprojlim_{i \in I} A/A_i$ can be taken to be the set of the:

$$W := \Big(\prod_{i \in I \setminus J} A/A_i \times \prod_{j \in J} V_j A_j / A_j \Big) \bigcap \varprojlim_{i \in I} A/A_i,$$

for finite $J \subset I$, and for neighbourhoods V_j of the unit element of A.

(i) f is continuous: for any W, consider $V := \bigcap_{j \in J} V_j$, which is a neighbourhood in A such that $f(V) \subseteq W$.

(ii) $f(C) = f(A)$ is dense: let $(a_i \bmod A_i)_i \in \varprojlim_{i \in I} A/A_i$ and W as above. Let $k \in I$ be an upper bound for J, and let $a := a_k$; by coherence, $aa_j^{-1} \in A_j$ for all $j \in J$ and $(aa_i^{-1} \bmod A_i)_i \in W$.

This shows that f indeed induces a homeomorphism between the given compact groups. □

Note. We could also have found a preimage from a limit point of $(c_i)_i$ representing $(a_i)_i$ modulo B in C. We will use limit points in some proofs; in this case, the filter basis corresponding to I is formed by the $I_i := \{k \in I, \ k \geq i\}$.

This lemma is often used in the case where the A_i are closed (or open) of finite index; we then obtain an homeomorphism of profinite groups.

Note that this situation is typical of the subject since, for instance, in the study of the group of classes of reduced idèles $J_0/K^\times$, already considered in 4.2.8 ($A = J_0$, $B = i_0(K^\times)$, the A_i being here the closed subgroups of finite index $K^\times U_{0,\mathrm{m}}^{\mathrm{res}}$), then neither A nor A/B are compact, but A can be written as BC for a compact subset C (which is not a subgroup). We will see in II.3.7 that $A/\cap A_i \simeq \mathrm{Gal}(\overline{K}^{\mathrm{ab}}/K)$.

It would of course be sufficient to state the lemma for $B = \bigcap_{i \in I} A_i$.

For other (simpler) practical applications, A will be for example a profinite group such as:

$$\bigoplus_{v \in T} U_v \simeq \varprojlim_{(m_v)_v} \bigoplus_{v \in T} U_v / U_v^{m_v} \ (T \text{ fixed}) \quad \text{or} \quad \prod_{v \in Pl_0} F_v^\times \simeq \varprojlim_{T} \bigoplus_{v \in T} F_v^\times,$$

and the assumptions on the set I and on the A_i will be trivially satisfied so as to be able to apply the lemma with $B = 1$.

c) Characters of Profinite Groups

5.6 PROFINITE GROUPS. Recall that a group A is profinite (i.e., an inverse limit of discrete finite groups) if and only if it is compact and totally discontinuous. Since infinite Galois theory will be our main source of profinite groups, the following characterization is the most suitable: the group A is profinite if and only if it is compact and if the set of its (normal) closed subgroups of finite index forms a fundamental system of neighbourhoods of the unit element (note that, by compactness, closed subgroup of finite index is equivalent to open subgroup). We then have:

$$A \simeq \varprojlim_{U} A/U,$$

for the family of (normal) open subgroups of A. Finally, any closed subgroup and any quotient by a (normal) closed subgroup of a profinite group is also profinite.

5.7 DUALITY. Let A be an *abelian* topological group. The character group of A is by definition the topological group (for the topology of pointwise convergence):

$$\widehat{A} := \mathrm{Hom}_{\mathrm{cont}}(A, \mathbb{C}_1^\times), \quad \text{where } \mathbb{C}_1^\times := \{z \in \mathbb{C}^\times,\ |z| = 1\} \simeq \mathbb{R}/\mathbb{Z}.$$

If, in addition, A is profinite, it is immediately checked that $\widehat{A}$ coincides with the group A^* of continuous characters χ of *finite order* of A: indeed, the $\chi(U)$ are then subgroups of $\mathbb{C}_1^\times$, and such a subgroup is contained in a neighbourhood of 1, different from $\mathbb{C}_1^\times$, if and only if it is trivial; thus, we have $\chi(U) = 1$ for a suitable U, so χ is a character of the finite group A/U. If A is of the form $\mathrm{Gal}(L^{\mathrm{ab}}/K)$, an element of A^* can be identified with a character of a finite extensions of K contained in L^{ab} according to the usual point of view of Dirichlet type characters of the case $K = \mathbb{Q}$ [c, Wa, Ch. 3].

The duality * on profinite abelian groups has identical algebraic and topological properties to those of the finite case, but only for those which are functorial. In other words, we do not have necessarily $A^* \simeq A$; for example $\mathbb{Z}_p^* \simeq \mathbb{Q}_p/\mathbb{Z}_p$, and more generally if A is profinite, then A^* is discrete.

When the abelian group A is only *locally compact* (for example an idèle group), $\widehat{A}$ is called the Pontrjagin dual of A;[21] it is in general larger than:

$$A^* := \{\chi \in \widehat{A},\ \chi \text{ of finite order}\}$$

(for example, in $A = \mathbb{R}$ or $A = \mathbb{R}/\mathbb{Z}$ (compact but not profinite), the map sending x to $\exp(2i\pi x)$ is an element of $\widehat{A}\backslash A^*$). We will not use this more general situation except for the proof of the Grunwald–Wang theorem where J^* is needed.

In the next paragraph we prove the main results of the theory of Kummer extensions, which is used in an essential way in class field theory, in particular in the proof of the existence theorem. These results are very classical, but not always easy to find in the literature in full generality.[22] We will work in the case where K will be given together with an automorphism group. This will enable the reader to understand the origin of the reflection theorem that we will state in complete generality in II.5.4.5.

§6 Kummer Extensions

Let K be a number field, let g be an automorphism group of K, and denote by k the subfield of K fixed under g; we assume that K contains the group μ_n of nth roots of unity for some integer $n \geq 2$. We consider finite abelian extensions L/K with exponent a divisor of n and Galois over k. We recover classical Kummer theory (i.e., without a group action) by taking $g = 1$, in other words $k = K$.

a) Algebraic Kummer Theory

We fix such an extension L of K. Consider the radical of L/K, which is the subgroup of $K^\times/K^{\times n}$ defined as follows.

$$W := \mathrm{Rad}(L/K) := \{\overline{\alpha} := \alpha K^{\times n} \in K^\times/K^{\times n},\ \sqrt[n]{\alpha} \in L\}.$$

This makes sense since if $\alpha' \in \alpha K^{\times n}$, $\sqrt[n]{\alpha'} \in \sqrt[n]{\alpha}\,\mu_n K^\times$, but $\mu_n \subset K$; thus, by abuse of notation we will denote by $\sqrt[n]{\overline{\alpha}}$ an nth root of any representative of $\overline{\alpha}$. Since L/k is Galois, it is clear that g acts on W (but we do not yet know that $L = K(\sqrt[n]{W})$).

[21] [Pon], [e, Ko3, App. 3; MP, Part II, Ch. 2, § 3.7], [g, NSW, Ch. I, § 1].
[22] [d, CF, Ch. III; Iy1, Ch. IV, § 2], [e, Ko3, Ch. 1, § 4.7], [j, Coh2, App. A, § 2]).

We denote by:

$$G := \mathrm{Gal}(L/K)$$

the Galois group of L/K; for the same reason, g acts by conjugation on G, and so for all $s \in g$ and all $\sigma \in G$ we set:

$$\sigma^s := s'\sigma s'^{-1},$$

for an arbitrary $s' \in \mathrm{Gal}(L/k)$ extending s.

6.1 Proposition. *We have the following canonical group isomorphism:*

$$W \simeq \mathrm{Hom}(G, \mu_n) \text{ denoted } G^*.$$

Proof. For a given $\overline{\alpha} \in W$, consider the following map $\varphi_{\overline{\alpha}}$ (in which the root of unity $(\sqrt[n]{\overline{\alpha}})^{\sigma-1}$ does not depend on the choice of $\sqrt[n]{\alpha}$ defined modulo $K^\times$):

$$\begin{aligned} \varphi_{\overline{\alpha}} : G &\longrightarrow \mu_n \,; \\ \sigma &\longmapsto (\sqrt[n]{\overline{\alpha}})^{\sigma-1} \end{aligned}$$

then $\varphi_{\overline{\alpha}} \in G^*$ since (using the crucial fact that G acts trivially on μ_n):

$$\begin{aligned} \varphi_{\overline{\alpha}}(\sigma\tau) &= (\sqrt[n]{\overline{\alpha}})^{\sigma\tau-1} = (\sqrt[n]{\overline{\alpha}})^{\tau(\sigma-1)+\tau-1} \\ &= (\sqrt[n]{\overline{\alpha}})^{\sigma-1}(\sqrt[n]{\overline{\alpha}})^{\tau-1} = \varphi_{\overline{\alpha}}(\sigma)\varphi_{\overline{\alpha}}(\tau) \end{aligned}$$

for all σ, $\tau \in G$, which we can write as a map f:

$$\begin{aligned} f : W &\longrightarrow G^* \\ \overline{\alpha} &\longmapsto \varphi_{\overline{\alpha}} = ((\sqrt[n]{\overline{\alpha}})^{\sigma-1})_\sigma \end{aligned}$$

which is easily checked to be a group homomorphism.

If, for a representative α of $\overline{\alpha}$ we have $(\sqrt[n]{\alpha})^{\sigma-1} = 1$ for all $\sigma \in G$, then $\sqrt[n]{\alpha} \in K^\times$, hence $\overline{\alpha} = K^{\times n}$, proving the injectivity of f.

The surjectivity of f comes from the Hilbert–Speiser–Noether Theorem 90 [d, Se2, Ch. X]:

$$H^1(G, L^\times) = 1.$$

Indeed, any element $\varphi =: (\zeta_\sigma)_\sigma \in G^*$ defines a 1-cocycle; hence there exists $u \in L^\times$ such that $\zeta_\sigma = u^{\sigma-1}$ for all $\sigma \in G$, and $u^n K^{\times n} \in W$ is a suitable preimage. □

6.1.1 Corollary. *If* $W = \mathrm{Rad}(L/K)$, *then* $L = K(\sqrt[n]{W})$.

Proof. If we set $L' := K(\sqrt[n]{W})$, we trivially have the inclusion $L' \subseteq L$.

(i) (first proof). Let W' be the radical of L' and let $G' := \mathrm{Gal}(L'/K)$; it is clear that $W = W'$, and because of 6.1 this yields $G^* = G'^*$, so that $L = L'$.

(ii) (second proof). The dual isomorphism $f^* : G \longrightarrow W^*$ of f (see 6.2.2) is such that $\sigma \in \mathrm{Gal}(L/L')$ is equivalent to $\sigma \in \mathrm{Ker}(f^*) = 1$; it follows that $L = L'$. □

6.1.2 Definition. We define a canonical g-module structure on G^* by setting (for any $\varphi \in G^*$ and $s \in g$):

$$\varphi^s(\sigma) := (\varphi(\sigma^{s^{-1}}))^s \text{ for all } \sigma \in G. \qquad \square$$

Note. This is the definition of the g-module structure on the groups $\mathrm{Hom}(A, B)$ for g-modules A, B used in more general situations since we need that φ is fixed under g if and only if it is a g-module homomorphism.

The problem is then the following: how does the map f behave under the actions of g on G, G^*, and W? For this, it is necessary to introduce the cyclotomic character θ.

6.1.3 Definition. We denote by $\theta : g \longrightarrow (\mathbb{Z}/n\mathbb{Z})^\times$ the character of the action of g on $\mu_n \subset K$ (i.e., such that $s(\zeta) =: \zeta^{\theta(s)}$ for all $\zeta \in \mu_n$ and all $s \in g$). □

The main result, which is Kummer duality in the Galois sense of the term, is the following (from 6.1 and using the g-module structure defined in 6.1.2).

6.2 Theorem (Galois action). *The isomorphism $W \simeq G^*$ is a g-module isomorphism; in other words, for all $\overline{\alpha} \in W := \mathrm{Rad}(L/K)$ and for all $s \in g$, we have:*

$$(\sqrt[n]{\overline{\alpha}^s})^{\sigma - 1} = (\sqrt[n]{\overline{\alpha}})^{\sigma^{\theta(s)s^{-1}} - 1}$$

for all $\sigma \in G$.

Proof. Let s' be an extension of s in $\mathrm{Gal}(L/k)$; since radicals are defined modulo $K^\times$, we may always assume that $\sqrt[n]{\overline{\alpha}^s} = (\sqrt[n]{\overline{\alpha}})^{s'}$, in which case we have:

$$\begin{aligned}\varphi_{\overline{\alpha}^s}(\sigma) &= (\sqrt[n]{\overline{\alpha}^s})^{\sigma-1} = (\sqrt[n]{\overline{\alpha}})^{(\sigma-1)s'} = (\sqrt[n]{\overline{\alpha}})^{\sigma s' - s'} \\ &= (\sqrt[n]{\overline{\alpha}})^{s'(\sigma^{s^{-1}} - 1)} = ((\sqrt[n]{\overline{\alpha}})^{\sigma^{s^{-1}} - 1})^s\end{aligned}$$

since $(\sqrt[n]{\overline{\alpha}})^{\sigma^{s^{-1}} - 1} \in \mu_n$, so that by definition $\varphi_{\overline{\alpha}^s}(\sigma) = (\varphi_{\overline{\alpha}}(\sigma^{s^{-1}}))^s = \varphi^s_{\overline{\alpha}}(\sigma)$, proving the first claim; furthermore, we have:

$$\varphi^s_{\overline{\alpha}}(\sigma) = (\varphi_{\overline{\alpha}}(\sigma^{s^{-1}}))^s = (\varphi_{\overline{\alpha}}(\sigma^{s^{-1}}))^{\theta(s)} = \varphi_{\overline{\alpha}}(\sigma^{\theta(s)s^{-1}}),$$

since $\varphi_{\overline{\alpha}} \in G^*$, hence $\varphi^s_{\overline{\alpha}}(\sigma) = (\sqrt[n]{\overline{\alpha}})^{\sigma^{\theta(s)s^{-1}} - 1}$. □

6.2.1 Corollary (Kummer duality). *The map:*

$$\begin{aligned} \lambda : W \times G &\longrightarrow \mu_n \\ (\overline{\alpha}, \sigma) &\longmapsto (\sqrt[n]{\overline{\alpha}})^{\sigma-1} = \varphi_{\overline{\alpha}}(\sigma) \end{aligned}$$

is a nondegenerate bilinear form for which, for all $s \in g$ we have:

$$\lambda(\overline{\alpha}^s, \sigma) = \lambda(\overline{\alpha}, \sigma^{\theta(s)s^{-1}}).$$

If for any subgroup V of W and any subgroup H of G, we set:

- $V^\perp := \{\sigma \in G, \ \lambda(\overline{\alpha}, \sigma) = 1, \ \ \forall\, \overline{\alpha} \in V\}$,
- $H^\perp := \{\overline{\alpha} \in W, \ \lambda(\overline{\alpha}, \sigma) = 1, \ \ \forall\, \sigma \in H\}$,

we have the following relations:

(i) $V^\perp = \mathrm{Gal}(L/K(\sqrt[n]{V}))$,
(ii) $H^\perp = \mathrm{Rad}(L^H/K)$.

Proof. The bilinear form λ is nondegenerate: if $(\sqrt[n]{\overline{\alpha}})^{\sigma-1} = 1$ for all $\sigma \in G$, then $\sqrt[n]{\overline{\alpha}} \in K$ (i.e., $\overline{\alpha} = K^{\times n}$). If $(\sqrt[n]{\overline{\alpha}})^{\sigma-1} = 1$ for all $\overline{\alpha} \in W$, then σ leaves $K(\sqrt[n]{W}) = L$ fixed, hence $\sigma = 1$.

This action of g is another way of writing the relation given by Theorem 6.2.

(i) We have $\sigma \in V^\perp$ if and only if $(\sqrt[n]{\overline{\alpha}})^{\sigma-1} = 1$ for all $\overline{\alpha} \in V$, which is equivalent to σ fixing $\sqrt[n]{V}$, hence to $\sigma \in \mathrm{Gal}(L/K(\sqrt[n]{V}))$.

(ii) We have $\overline{\alpha} \in H^\perp$ if and only if $(\sqrt[n]{\overline{\alpha}})^{\sigma-1} = 1$ for all $\sigma \in H$, which is equivalent to $\sqrt[n]{\overline{\alpha}}$ fixed under H, hence to $\sqrt[n]{\overline{\alpha}} \in L^H$, and finally to $\overline{\alpha} \in \mathrm{Rad}(L^H/K)$. □

6.2.2 Exercise. The g-module isomorphism $W \simeq G^*$ yields by duality a dual isomorphism $G \simeq W^*$. Write explicitly this isomorphism, its inverse, and the g-module structure on W^*.

Answer. The dual isomorphism of: $\overline{\alpha} \in W \longmapsto \varphi_{\overline{\alpha}} \in G^*$ (see 6.1) is the isomorphism: $\sigma \in G \longmapsto \chi_\sigma \in W^*$, where $\chi_\sigma(\overline{\alpha}) = (\sqrt[n]{\overline{\alpha}})^{\sigma-1}$ for all $\overline{\alpha} \in W$.

Let $\chi \in W^*$. Since the group $W/\mathrm{Ker}(\chi)$ is cyclic there exists $\overline{\beta} \in W$ such that:

$$W = \langle\, \overline{\beta}\,\rangle \mathrm{Ker}(\chi).$$

Let us introduce the field $M := K(\sqrt[n]{\mathrm{Ker}(\chi)}\,) \subseteq L$; then $L = M(\sqrt[n]{\overline{\beta}})$, and if we apply 6.1 to L/K and M/K we obtain:

$$[L:M] = \frac{[L:K]}{[M:K]} = \frac{|W|}{|\mathrm{Ker}(\chi)|} = |\langle\, \chi(\overline{\beta})\,\rangle|.$$

It follows that $\mathrm{Gal}(L/M) \simeq \langle\, \chi(\overline{\beta})\,\rangle$, hence there exists a unique generator σ_χ of $\mathrm{Gal}(L/M)$ such that $(\sqrt[n]{\overline{\beta}})^{\sigma_\chi - 1} = \chi(\overline{\beta})$; it is clear that σ_χ does not depend on the choice of $\overline{\beta}$. Consider now χ_{σ_χ}. It is such that:

$$\chi_{\sigma_\chi}(\overline{\alpha}) = (\sqrt[n]{\overline{\alpha}})^{\sigma_\chi - 1} \text{ for all } \overline{\alpha} \in W \ ;$$

but $\overline{\alpha} = \overline{\beta}^{\,i}\overline{\gamma}$ with $\overline{\gamma} \in \mathrm{Ker}(\chi)$, so that:

$$(\sqrt[n]{\overline{\alpha}})^{\sigma_\chi - 1} = \left(\sqrt[n]{\overline{\beta}^{\,i}}\right)^{\sigma_\chi - 1} = \chi(\overline{\beta}^{\,i}) = \chi(\overline{\alpha}),$$

which implies $\chi_{\sigma_\chi} = \chi$. This is enough to show that the map $\chi \longmapsto \sigma_\chi$ is the required inverse.

In the same way, we can show that the isomorphism $G^* \longrightarrow W$ sends $\psi \in G^*$ to the unique generator $\overline{\alpha}_\psi$ of the radical of the subfield N of L fixed under $\mathrm{Ker}(\psi)$ and such that $(\sqrt[n]{\overline{\alpha}_\psi})^{\sigma - 1} = \psi(\sigma)$ for all $\sigma \in G$; the dual isomorphism thus coincides with the one that we have just obtained.

There remains to define χ^s for all $\chi \in W^*$ and $s \in g$. Since every such χ is of the form χ_σ and that we must have $\chi_\sigma^s = \chi_{\sigma^s}$, it follows that:

$$\chi_\sigma^s(\overline{\alpha}) = (\sqrt[n]{\overline{\alpha}})^{\sigma^s - 1} \text{ for all } \overline{\alpha} \in W,$$

and the computations made in the proof of 6.2 easily yield the g-module structure defined for $\chi \in W^*$ and $s \in g$ by:

$$\chi^s(\overline{\alpha}) = \chi^{\theta(s)}(\overline{\alpha}^{\,s^{-1}}) \text{ for all } \overline{\alpha} \in W.$$

As expected, it coincides with that of Definition 6.1.2. □

This Kummer duality, which allows us to replace W^* by G, will be used in Chapter V.

6.2.3 Exercise (Albert's theorem). Let p be a prime number and let K be a field of characteristic different from p, containing $\mu_{p^e} =: \langle\, \zeta_e \,\rangle$ for some $e \geq 1$. Let L/K be an arbitrary cyclic extension of degree p^r for some $r \geq 0$. Show that there exists a cyclic extension M of K of degree p^{r+e} containing L if and only if there exists $y \in L^\times$ such that $\mathrm{N}_{L/K}(y) = \zeta_e$. If this condition is satisfied, check that there exists $z \in L^\times$ such that $M = L(\sqrt[p^e]{z})$ and $(\sqrt[p^e]{z})^{\tau - 1} = y$ for a suitable generator τ of $\mathrm{Gal}(M/K)$.

Answer. The case $r = 0$ being nothing else than Kummer theory, we assume $r \geq 1$. Fix a generator σ of $\mathrm{Gal}(L/K)$.

(i) Assume that there exists $y \in L^\times$ such that $\mathrm{N}_{L/K}(y) = \zeta_e$. Since $\mathrm{N}_{L/K}(y^{p^e}) = 1$, there exists $z \in L^\times$ such that $y^{p^e} = z^{\sigma - 1}$. Let $\sqrt[p^e]{z}$ be a fixed p^eth root of z, and let $M := L(\sqrt[p^e]{z})$; we can then check the following facts:

• M/L has degree equal to p^e (if $z \in L^{\times p}$, we can write $y^{p^{e-1}} =: \zeta_1\, t^{\sigma - 1}$, $\zeta_1 \in \mu_p$, $t \in L^\times$, and since L/K is nontrivial, we obtain $\mathrm{N}_{L/K}(y^{p^{e-1}}) = 1$, a contradiction);

• M/K is Galois and there exists a K-isomorphism τ of M extending σ and such that $\tau(\sqrt[p^e]{z}) = \sqrt[p^e]{z}\, y$;

• τ is of order p^{r+e} (we take $i = p^r$ in the identity $\tau^i(\sqrt[p^e]{z}) = \sqrt[p^e]{z}\, y^{1+\sigma+\cdots+\sigma^{i-1}}$).

(ii) Assume that the extension $M/L/K$ exists, and let τ be a generator of $\mathrm{Gal}(M/K)$ whose restriction to L is σ. There exists $z \in L^\times$ such that $M = L(\sqrt[p^e]{z})$, and since M/K and L/K are Galois, we easily deduce that:

$$\tau(\sqrt[p^e]{z}) =: \sqrt[p^e]{z^\lambda}\, y,\ \lambda \not\equiv 0 \bmod (p),\ y \in L^\times .$$

If we write that τ and τ^{p^r} commute and that $\tau^{p^r}(\sqrt[p^e]{z}) = \zeta_e \sqrt[p^e]{z}$, we obtain $\lambda \equiv 1 \bmod (p^e)$. Changing y if necessary, we may assume that $\lambda = 1$, which again leads to $\tau^i(\sqrt[p^e]{z}) = \sqrt[p^e]{z}\, y^{1+\sigma+\cdots+\sigma^{i-1}}$. Since τ^{p^r} is a generator of $\mathrm{Gal}(M/L)$, we obtain $\mathrm{N}_{L/K}(y) = \zeta_e$ (once again we use the above identity for $i = p^r$). □

b) Arithmetic Aspects of Kummer Theory

We will now study the arithmetic aspects of Kummer theory. In the case of general n, it is difficult to give explicitly the decomposition law of the wild places in an arbitrary Kummer extension with exponent a divisor of n (this can be considered as a justification of class field theory, which does not have this problem); thus, in general the statement is only given for prime degree p (for a general numerical approach, use [j, Coh2, Ch. 5]). We then have (after [a, He, Ch. V, § 39] and [h, Ha2]), using notations and results given in 3.1:

6.3 Theorem (decomposition of places in a Kummer extension of prime degree p). *Let K be a number field containing the group μ_p of pth roots of unity, and let $L = K(\sqrt[p]{\alpha})$ for some $\alpha \in K^\times \backslash K^{\times p}$. Let $v \in Pl_0$.*

(i) (case $v \nmid p$). The place v is ramified in L/K if and only if $\mathrm{v}(\alpha) \not\equiv 0 \bmod (p)$. If v is unramified in L/K, it is split if and only if $i_v(\alpha) \in K_v^{\times p}$, hence if and only if the residual image of $i_v(\alpha)\pi_v^{-\mathrm{v}(\alpha)}$ belongs to $F_v^{\times p}$.

(ii) (case $v|p$). Let e_v be the ramification index of v in $K/\mathbb{Q}(\mu_p)$. Then v is ramified in L/K if and only if the congruence:

$$\frac{\alpha}{x_v^p} \equiv 1 \bmod \mathfrak{p}_v^{pe_v},\ x_v \in K^\times$$

(which means $i_v(\alpha/x_v^p) \in U_v^{pe_v}$) does not have a solution in x_v (this contains the case $\mathrm{v}(\alpha) \not\equiv 0 \bmod (p)$).

If v is unramified in L/K, it is split if and only if $i_v(\alpha) \in K_v^{\times p}$, which can be characterized in the following way (noting that if ζ is a generator of μ_p, then $\mathrm{v}(p(1-\zeta)) = pe_v$). Write:

$$\frac{\alpha}{x_v^p} =: 1 + p(1-\zeta)y_v,\ \ y_v \in K^\times\ (y_v \text{ is } v\text{-integral})\ ;$$

then $i_v(\alpha) \in K_v^{\times p}$ if and only if the absolute trace $\mathrm{tr}_{F_v/\mathbb{F}_p}(\overline{i_v(y_v)})$ of the residual image of $i_v(y_v)$ is equal to zero.

Proof. The first result of (i) is clear and the second follows from the Hensel Lemma 3.2.

(ii) Assume that $v|p$. Recall that K_v has a unique unramified cyclic extension of degree p, defined as a cyclotomic extension (see II.1.1.5), which we must define here in a Kummer manner. If, to ease notation, we still denote by $1-\zeta$ the image of $1-\zeta$ in K_v, we have the following result.

Lemma. *Let $\kappa := 1+p(1-\zeta)\eta$, where η is an integer of K_v. Then $K_v(\sqrt[p]{\kappa})/K_v$ is unramified; it is of degree p (i.e., $\kappa \notin K_v^{\times p}$) if and only if $\mathrm{tr}_{F_v/\mathbb{F}_p}(\overline{\eta}) \neq \overline{0}$.*

Proof. Note that $x := \sqrt[p]{\kappa} - 1$ is a root of the following polynomial of $K_v[X]$:

$$X^p + pX^{p-1} + \cdots + pX - p(1-\zeta)\eta.$$

It follows that $y := \dfrac{x}{1-\zeta}$ is a root of the polynomial:

$$P := Y^p + \sum_{i\in[2,p-1]} \pi_i\, Y^i + \frac{p}{(1-\zeta)^{p-1}}\, Y - \frac{p}{(1-\zeta)^{p-1}}\, \eta,$$

where the π_i are multiples of $1-\zeta$ in $\mathbb{Z}[\zeta]$. Since we have $\prod_{i\in[1,p[} (1-\zeta^i) = p$, we obtain:

$$\frac{p}{(1-\zeta)^{p-1}} = \prod_{i\in[1,p[} \frac{1-\zeta^i}{1-\zeta} = \prod_{i\in[1,p[} (1+\zeta+\cdots+\zeta^{i-1})$$
$$\equiv (p-1)! \equiv -1 \bmod (1-\zeta).$$

Hence the residual image of P in $F_v[Y]$ is the Artin–Schreier polynomial:

$$\overline{P} = Y^p - Y + \overline{\eta}\ ;$$

this is a separable polynomial over F_v which has 0 or p roots in F_v since, if $P(\overline{y}) = \overline{0}$ for some $\overline{y} \in F_v$, then $\overline{P}(\overline{y}+\overline{a}) = \overline{0}$ for all $\overline{a} \in \mathbb{F}_p$. We also know that it is irreducible over F_v if and only if it does not have any roots: indeed, a nontrivial F_v-isomorphism σ of $F_v(\overline{y})$ ($P(\overline{y}) = \overline{0}$, $\overline{y} \notin F_v$) is such that $\sigma(\overline{y}) = \overline{y} + \overline{a}$, $\overline{a} \in \mathbb{F}_p^{\times}$; but this defines an F_v-automorphism of order p of $F_v(\overline{y})$. The Hensel lemma then shows that if $\overline{P}$ has a root in F_v, then P has a root in K_v, hence $\kappa \in K_v^{\times p}$; otherwise (i.e., if $\overline{P}$ is irreducible over F_v), P is irreducible in $K_v[Y]$ and since the residue extension is of degree p, it follows that $K_v(\sqrt[p]{\kappa})$ is an unramified extension of degree p of K_v. Since $\mathrm{Ker}(\mathrm{tr}_{F_v/\mathbb{F}_p}) = \{\overline{t} - \overline{t}^p,\ \overline{t} \in F_v\}$, the lemma follows. □

Thus, knowing the residue extension $F_v/\mathbb{F}_p$, to define the unique unramified Kummer extension of degree p of K_v, we can set:

$$\kappa_0 := 1 + p(1-\zeta)\eta_v^0, \text{ with } \mathrm{tr}_{F_v/\mathbb{F}_p}(\overline{\eta_v^0}) = \overline{1}\ ;$$

if $[F_v : \mathbb{F}_p]$ is prime to p, we can choose:

$$\kappa_0 = 1 + p(1 - \zeta).$$

Let us now come back to the extension $K_v(\sqrt[p]{i_v(\alpha)}\,)/K_v$. By uniqueness, if this extension is unramified this means that:

$$i_v(\alpha) = \xi_v^p \, \kappa_0^\lambda, \ \xi_v \in K_v^\times, \ \lambda \in \mathbb{Z},$$

and we obtain the congruence:

$$i_v(\alpha)\xi_v^{-p} \equiv 1 \bmod (p(1 - \zeta)) \text{ in } K_v.$$

Conversely, if $i_v(\alpha)\xi_v^{-p} =: 1 + p(1 - \zeta)\eta$ with η an integer of K_v, the lemma implies that the extension is unramified.

Thus the statements are indeed equivalent, proving part (ii) of the theorem. □

6.3.1 Remark (local p-power criterion). For $v|p$ it is clear that $i_v(\alpha) \in K_v^{\times p}$ if and only if there exists $x'_v \in K^\times$ such that $\alpha x_v'^{-p} \equiv 1 \bmod \mathfrak{p}_v^{pe_v+1}$; but once we have obtained $\alpha =: x_v^p(1 + p(1 - \zeta)y_v)$, y_v being v-integral, the solubility of this congruence is characterized by the (easier) condition that the residual trace of $i_v(y_v)$ vanishes. □

The case $v \in Pl_\infty^{\mathrm{r}}$ has not been mentioned since it cannot contribute to ramification, but v is split in L/K if and only if $i_v(\alpha) \in K_v^{\times p} = \mathbb{R}^{\times p}$ (i.e., $p > 2$, or $p = 2$ and $i_v(\alpha) > 0$).

6.3.2 Remarks. (i) Assume that we are in the case where $v|p$ is unramified, and choose α prime to v (which is always possible since $\mathrm{v}(\alpha) \equiv 0 \bmod (p)$). Since $\mathrm{v}(p(1 - \zeta)) = pe_v$, the relation:

$$\alpha = x_v^p(1 + p(1 - \zeta)y_v), \ x_v \in K_{\{v\}}^\times, \ y_v \ \ v\text{-integral}$$

is equivalent to the congruence:

$$\alpha \equiv x_v^p \bmod \mathfrak{p}_v^{pe_v}, \ x_v \in K_{\{v\}}^\times,$$

but from this formula it is not easy to determine the decomposition type of v in $K(\sqrt[p]{\alpha})/K$. Finally, to obtain an approximation to y_v in the formula:

$$\alpha = x_v^p(1 + p(1 - \zeta)y_v)$$

(in view of the computation of its residual trace), we simply compute:

$$\frac{\alpha x_v^{-p} - 1}{p(1 - \zeta)} \bmod \mathfrak{p}_v.$$

(ii) The trace map in $F_v/\mathbb{F}_p$ is surjective; hence, we can compute once and for all $y_v^0 \in K_{\{v\}}^\times$ such that for instance $\mathrm{tr}_{F_v/\mathbb{F}_p}(\overline{i_v(y_v^0)}) = \overline{1}$, and then the place v will be inert in $K(\sqrt[p]{\alpha})/K$ (i.e., $i_v(\alpha) \notin K_v^{\times p}$) if and only if there exist $\lambda \in \mathbb{Z}\backslash p\mathbb{Z}$ and $x_v \in K_{\{v\}}^\times$ such that:

$$\frac{\alpha}{1 + p(1-\zeta)\,\lambda\, y_v^0} \equiv x_v^p \bmod \mathfrak{p}_v^{pe_v+1}.$$

When the residue degree of v in $K/\mathbb{Q}$ is prime to p, we can choose $y_v^0 = 1$, in which case v is inert in $K(\sqrt[p]{\alpha})/K$ if and only if there exist $\lambda \in \mathbb{Z}\backslash p\mathbb{Z}$ and $x_v \in K_{\{v\}}^\times$ such that:

$$\frac{\alpha}{1 + p(1-\zeta)\,\lambda} \equiv x_v^p \bmod \mathfrak{p}_v^{pe_v+1}. \qquad \square$$

6.3.3 Example. For $K = \mathbb{Q}(\sqrt{-23}, \zeta)$, ζ a primitive cube root of unity, consider $K(\sqrt[3]{\varepsilon})$ for $\varepsilon := \frac{1}{2}(25 + 3\sqrt{69})$ (the fundamental unit of $\mathbb{Q}(\sqrt{69}) \subset K$). By noting that:

$$\varepsilon \equiv -(1 + 3\sqrt{69}) \bmod (9),$$

we thus have:

$$\varepsilon \equiv (-1)^3(1 + 3\sqrt{-3}\sqrt{-23}) \equiv (-1)^3(1 + 3(1-\zeta)\zeta\sqrt{-23}) \bmod (9)$$

since $\sqrt{-3} = (1-\zeta)\zeta$, hence $\varepsilon =: (-1)^3(1 + 3(1-\zeta)\eta)$ with $\eta = \zeta\sqrt{-23} \equiv \sqrt{-23} \bmod \mathfrak{p}_v$ (for every place $v|3$ in K). Since 3 is split in $\mathbb{Q}(\sqrt{-23})$, $\sqrt{-23} \equiv \lambda \bmod \mathfrak{p}_v$ (with $\lambda = \pm 1$), and this shows that v is inert in $K(\sqrt[3]{\varepsilon})/K$. It is clear that $K(\sqrt[3]{\varepsilon})/K$ is everywhere unramified, and finally we can check that $K(\sqrt[3]{\varepsilon})/K$ is equal to the compositum $H\mathbb{Q}(\zeta)$, where H is the Hilbert class field of $\mathbb{Q}(\sqrt{-23})$. Indeed, we have:

$$H = \mathbb{Q}(\sqrt{-23}\,)(\theta), \quad \theta := \frac{3}{\psi - 1},$$

where $\psi := \sqrt[3]{\varepsilon} + \sqrt[3]{\varepsilon^{-1}}$. Then $\mathrm{Irr}(\theta, \mathbb{Q}(\sqrt{-23}\,)) = X^3 - X - 1$ (this example of class field will be used to illustrate the general law of reciprocity in II.5.2.1). $\square$

6.3.4 Remarks. (i) By using the chinese remainder theorem, we can say that $K(\sqrt[p]{\alpha})/K$ (where α is assumed prime to p) is unramified at p if and only if there exists $x \in K^\times$ prime to p such that in K we have:

$$\alpha \equiv x^p \bmod (p(1-\zeta)).$$

If this is the case, the decomposition of p in $K(\sqrt[p]{\alpha})/K$ can be read on the residual traces for $v|p$ of the p-integral element:

$$\frac{\alpha x^{-p} - 1}{p(1-\zeta)}.$$

(ii) The case $p = 2$ is especially simple since for any number field K and for $\alpha \in K^\times$ prime to 2, the extension $K(\sqrt{\alpha})/K$ is unramified at 2 if and only if there exists $x \in K^\times$ prime to 2 such that:

$$\alpha \equiv x^2 \bmod (4).$$

If this is the case, from the 2-integral expressions:

$$\frac{\alpha x^{-2} - 1}{4} \bmod \mathfrak{p}_v,$$

we deduce the decomposition of the places $v|2$ in $K(\sqrt{\alpha})/K$.

(iii) The lemma in 6.3 shows that, for $v|p$, the smallest modulus $\mathfrak{p}_v^h$ of K such that:

$$U_v^h = 1 + (\pi_v^h) \subset K_v^{\times p},$$

is given by:

$$\mathfrak{p}_v^h = \mathfrak{p}_v^{pe_v+1}. \qquad \square$$

The above can be summarized in the following way.

6.3.5 Proposition. *Assume that K contains $\mu_p =: \langle \zeta \rangle$. For any place v of K above p we have the following canonical exact sequence:*

$$1 \longrightarrow U_v^{pe_v} \cap (U_v^1)^p \longrightarrow U_v^{pe_v} \xrightarrow{\ \tau\ } \mu_p \longrightarrow 1,$$

where τ is the map which sends $1 + p(1-\zeta)\eta$ (for η an integer of K_v) to ζ^t, where $t := \mathrm{tr}_{F_v/\mathbb{F}_p}(\overline{\eta})$. $\square$

II. Reciprocity Maps Existence Theorems

The fundamental results given in this chapter do not necessarily form a sequence of logical steps for a proof of class field theory, but are written and commented so as to be used. This is so true that, as we will see several times, a classical proof consists in *deducing* local class field theory from global class field theory, as was initiated by Hasse and Schmidt in 1930, and in particular to base some local computations on global arguments (a typical example being the global computation of a local Hilbert symbol in 7.5); however here, in the description of the results, we will go from local to global, which seems more natural.

Chapter I contains tools coming directly from elementary considerations on number fields and local fields; on the contrary the present chapter relies on the (nontrivial) existence, for each place of the number field K, of the local reciprocity map (or local norm residue symbol). [1] The existence of the local reciprocity map, and thus of the global one as we will see in 3.2, is in fact of a cohomological nature, even though other approaches are possible, such as the Lubin–Tate theory of formal groups. [2]

§1 The Local Reciprocity Map — Local Class Field Theory

Let K be a number field and let $v \in Pl$ fixed. Since the case $v \in Pl_\infty$ is immediate (see 1.4.6), we will usually assume that we are dealing with a finite place, but everything which is detailed below can also be applied to the local extension $\mathbb{C}/\mathbb{R}$, an unramified abelian extension whose Frobenius is equal to complex conjugation c.

As usual, if v is finite *all* the corresponding local fields are taken in the completion $\mathbb{C}_\ell$ of an algebraic closure $\overline{\mathbb{Q}}_\ell$ of $\mathbb{Q}_\ell$, where ℓ is the residue characteristic of v. Let K_v be the v-completion of K; let $\overline{K}_v = \overline{\mathbb{Q}}_\ell$ (resp. $\overline{K}_v^{\,\mathrm{ab}}$) be the algebraic (resp. abelian) closure of K_v in $\mathbb{C}_\ell$, and $\overline{G}_v$ (resp. $\overline{G}_v^{\,\mathrm{ab}}$) the profi-

[1] Its proof can be found in [d, CF, Ch. VI; Se2, Ch. XI, XIII], [f, Art1; Haz], [c, Neu1, Ch. IV]; the first proof is actually due to Hasse–Chevalley [h, Che1].

[2] [f, Lang2, Ch. 8], [d, CF, Ch. VI, § 3], [c, Neu1, Ch. V, § 4].

nite group $\mathrm{Gal}(\overline{K}_v/K_v)$ (resp. $\mathrm{Gal}(\overline{K}_v^{\mathrm{ab}}/K_v) \simeq \overline{G}_v/[\overline{G}_v, \overline{G}_v]$), where $[\overline{G}_v, \overline{G}_v]$ denotes the topological closure of the commutator subgroup of $\overline{G}_v$.

Now let L be a finite extension of K; for each $w \in Pl_{L,v}$, L_w is the w-completion of L. We fix L_w/K_v and the embeddings i_v and i_w for $w|v$, as explained in I.2.4.

Note. Since any finite extension of local fields can be written (in infinitely many ways) as L_w/K_v, we are indeed studying an arbitrary local extension; here, the use of this global point of view will not matter since it will in any case be the natural setting for the definition of the global reciprocity map corresponding to L/K (the only one which interests us here and for which we must consider simultaneously all the completions of L/K and all the corresponding local reciprocity maps).

a) Decomposition of Places: Local and Global Cases

In this subsection, we recall the main classical properties of the places in an arbitrary extension L/K, both in the local and global cases. As explained in the above Note, we work in the setting of a global extension L/K.

1.1 LOCAL GALOIS GROUPS, INERTIA GROUPS, FROBENIUS'. In this paragraph, we will constantly refer to Section 2 of Chapter I.

1.1.1 GALOIS CASE. If L/K is Galois with Galois group G, the decomposition group of $w \in Pl_{L,v}$ in L/K, denoted D_w, can be canonically identified with $G_w := \mathrm{Gal}(L_w/K_v)$, and under this isomorphism the inertia group I_w of w in L/K, which is a normal subgroup of D_w, corresponds to the inertia group G_w^0 of L_w/K_v. We will sometimes use the notation D_w^0 instead of I_w when the higher ramification groups are needed in a global situation, and then more generally $D_w^i \simeq G_w^i$ or $D_{w,i} \simeq G_{w,i}$, $i \geq 0$, with the definitions recalled in 1.3.

Since the field L^{D_w} is the decomposition field of w in L/K, we know that $i_w(L^{D_w})$ is dense in K_v. The inertia field is L^{I_w}; similarly $i_w(L^{I_w})$ is dense in the subfield L_w^{nr} of L_w fixed under G_w^0 (the largest subfield of L_w unramified over K_v). [3]

1.1.2 NON-GALOIS CASE. Even when the extension L/K is not Galois the extension L_w/K_v may still be Galois (see Example I.2.2.3, (ii)), so that G_w (hence G_w^0 which is still a normal subgroup of G_w) can exist even when D_w does not make sense; this explains the independent choice of notations between the local and global cases.

[3] We use the superscript "nr" (as "non ramifié" from the french), thus following most authors.

1.1.3 ABELIAN CASE. If L/K is abelian, the groups D_w, I_w, do not depend on the choice of $w \in Pl_{L,v}$ and, by abuse of notation, are simply denoted D_v, I_v; similarly for the fixed subfields L^{D_v}, L^{I_v}. The notations G_v, G_v^0, L_v, L_v^{nr} are also legitimate since L_w does not depend on the choice of $w|v$, and neither does the isomorphism $G_w \simeq D_w$ which sends $\tau \in G_w$ to $i_w^{-1} \circ \tau \circ i_w$ on L.

In the Galois case the G_w for $w|v$ are equal (since the L_w are equal), but the canonical isomorphism $G_w \to D_w$ depends on the choice of $w|v$, which explains that in this case G_w is not denoted G_v.

1.1.4 MAXIMAL ABELIAN SUBEXTENSIONS. We now assume that L/K is any finite extension. In the sequel, we will refer to the following diagram, in which $L'_{w'}/K_v$ comes from a subextension L'/K of L/K, w' is the place of L' below w, L_w^{ab} (resp. $L'^{\mathrm{ab}}_{w'} = L_w^{\mathrm{ab}} \cap L'_{w'}$) is the maximal abelian extension of K_v in L_w (resp. in $L'_{w'}$), and where $L_w^{\mathrm{ab}'}$ is the maximal abelian extension of $L'_{w'}$ in L_w ($L_w^{\mathrm{ab}'}$ contains $L'_{w'}L_w^{\mathrm{ab}}$, but is not necessarily equal to it).

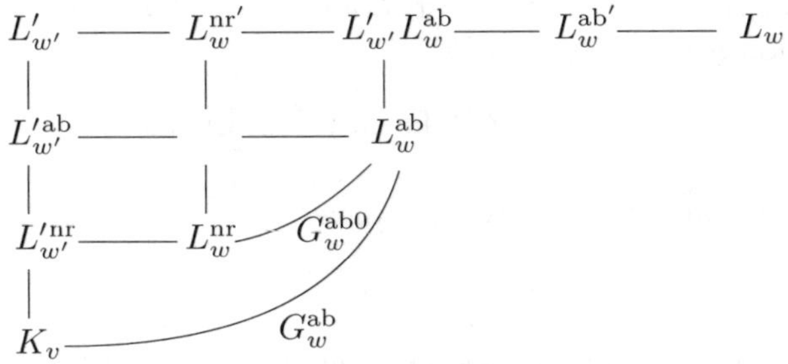

Fig. 1.1

By abuse of notation we set $G_w^{\mathrm{ab}} := \mathrm{Gal}(L_w^{\mathrm{ab}}/K_v)$; if L_w/K_v is Galois then G_w exists and we indeed have $G_w^{\mathrm{ab}} \simeq G_w/[G_w, G_w]$. We denote by $G_w^{\mathrm{ab}\,0}$ the inertia group of L_w^{ab}/K_v. If L_w/K_v is Galois (with Galois group G_w), the inertia group of L_w^{ab}/K_v is equal to $G_w^{\mathrm{ab}\,0} \simeq G_w^0/[G_w, G_w]$ which is not the abelianization of G_w^0 but the image of G_w^0 in G_w^{ab}, in accordance with the general property of higher ramification groups (in upper numbering) which is that for any normal subgroup H of G_w, we have $(G_w/H)^i = G_w^i H/H$ [d, Se2, Ch. IV, § 3].

We proceed in an analogous manner to define the groups D_w^{ab} and I_w^{ab} which lift G_w^{ab} and $G_w^{\mathrm{ab}\,0}$ respectively, when L/K is Galois; we have $D_w^{\mathrm{ab}} = D_w/[D_w, D_w]$ and $I_w^{\mathrm{ab}} = I_w/[D_w, D_w]$.

Note that, even in the Galois case, G_w^{ab} is not necessarily isomorphic to the decomposition group $D_w(L^{\mathrm{ab}}/K)$ of w in L^{ab}/K, the latter corresponding to the quotient of G_w^{ab} which gives $\mathrm{Gal}((L^{\mathrm{ab}})_v/K_v)$ (see I.2.7).

1.1.5 Maximal Unramified Subextensions — Frobenius'. We denote by $\overline{K}_v^{\,\mathrm{nr}}$ the maximal unramified extension of K_v in $\overline{K}_v$; it will of course be obtained by the local infinite class field theory correspondence 1.7, but its direct construction is classical and elementary: recall that, when v is finite, $\overline{K}_v^{\,\mathrm{nr}}$ is the lift, via the Hensel Lemma I.3.2, of the algebraic closure $\overline{F}_v$ of the residue field $F_v \simeq \mathbb{F}_{q_v}$ of K_v, since for each degree $n \geq 1$, the unique unramified extension of degree n of K_v is equal to $K_v(\mu_{q_v^n-1})$ [d, Se2, Ch. III, § 5]; it is a cyclic extension. It follows that $\overline{K}_v^{\,\mathrm{nr}}$ is contained in $\overline{K}_v^{\,\mathrm{ab}}$, that it is procyclic with Galois group:

$$\mathrm{Gal}(\overline{K}_v^{\,\mathrm{nr}}/K_v) \simeq \varprojlim_{n\geq 1} \mathbb{Z}/n\mathbb{Z} =: \widehat{\mathbb{Z}} \simeq \prod_{p\,\mathrm{prime}} \mathbb{Z}_p,$$

the profinite completion of $\mathbb{Z}$. For any subextension M/K_v (finite or not) of $\overline{K}_v^{\,\mathrm{nr}}/K_v$, we denote by:

$$(M/K_v)$$

the Frobenius automorphism of M/K_v which is a topological generator of the group $\mathrm{Gal}(M/K_v)$; it is the restriction of $(\overline{K}_v^{\,\mathrm{nr}}/K_v)$ to M, where the Frobenius action is characterized by:

$$(\overline{K}_v^{\,\mathrm{nr}}/K_v)\,\zeta = \zeta^{q_v} \ \text{ for any } \zeta \in \bigcup_{n\geq 1} \mu_{q_v^n-1}.$$

In the above context 1.1.4, we then have:

$$L_w^{\mathrm{nr}} = L_w \cap \overline{K}_v^{\,\mathrm{nr}}\ ;$$

this is the maximal unramified subextension of L_w/K_v. As we have just mentioned, it is cyclic, unique, contained in L_w^{ab}, and L_w/L_w^{nr} is totally ramified [d, Se2, Ch. III, Cor. 3]; L_w^{nr} exists even when L_w/K_v is not Galois; in the Galois case, it is the subfield fixed under G_w^0 (see 1.1.1).

We immediately check that $\overline{L}_w^{\,\mathrm{nr}} = L_w\overline{K}_v^{\,\mathrm{nr}}$ or, more canonically:

$$\overline{k}^{\,\mathrm{nr}} = k\,\overline{\mathbb{Q}}_\ell^{\,\mathrm{nr}} = \bigcup_{n\geq 1} k(\mu_{\ell^n-1}),$$

for any algebraic extension k of $\mathbb{Q}_\ell$, where ℓ is the corresponding residue characteristic (for this, we must check that $k(\mu_{\ell^n-1})/k$ is unramified, even if ℓ^n is not a power of q_v, which is immediate).

In (Fig. 1.1) above we have given the various maximal unramified extensions using a principle of notation identical to the one used for maximal abelian extensions (noting that since L_w/L_w^{nr} is totally ramified, we indeed have here that $L_w^{\mathrm{nr}'} = L'_{w'}L_w^{\mathrm{nr}}$). The diagram can be justified by the very nature of $\overline{K}_v^{\,\mathrm{nr}}$. This gives the following result.

1.1.6 Exact Sequence of Inertia Groups. Let L'/K be a subextension of L/K. If L_w/K_v and $L'_{w'}/K_v$ are Galois with respective Galois groups G_w

and $G_{w'}$, we have the following exact sequence of inertia groups (which is still valid if these local extensions are infinite as we will see in 1.2.3):

$$1 \longrightarrow G'^0_w \longrightarrow G^0_w \longrightarrow G^0_{w'} \longrightarrow 1,$$

where $G'^0_w := \mathrm{Gal}(L_w/L_w^{\mathrm{nr}'})$, $G^0_{w'} := \mathrm{Gal}(L'_{w'}/L'^{\mathrm{nr}}_{w'})$.

The cyclicity of L_w^{nr}/K_v implies the following property of the Frobenius automorphism.

1.1.6.1 Proposition. *We have:*

$$(L_w^{\mathrm{nr}}/K_v)^{f'_{w'}} = (L_w^{\mathrm{nr}}/L'^{\mathrm{nr}}_{w'}) = (L_w^{\mathrm{nr}'}/L'_{w'})_{|L_w^{\mathrm{nr}}},$$

where $f'_{w'} := [L'^{\mathrm{nr}}_{w'} : K_v]$ is the residue degree of $L'_{w'}/K_v$. □

1.1.6.2 Notations. We denote by $e_w := [L_w : L_w^{\mathrm{nr}}]$ and $e_w^{\mathrm{ab}} := [L_w^{\mathrm{ab}} : L_w^{\mathrm{nr}}]$ the ramification index of L_w/K_v and of L_w^{ab}/K_v, respectively; since the extension L_w/L_w^{ab} is totally ramified, the residue degree $f_w := [L_w^{\mathrm{nr}} : K_v]$ of L_w/K_v is equal to f_w^{ab}, that of L_w^{ab}/K_v. □

1.2 GLOBAL DECOMPOSITION AND INERTIA GROUPS. We still consider a finite extension L/K with a subextension L'/K.

1.2.1 EXACT SEQUENCES OF DECOMPOSITION AND INERTIA GROUPS. When L/K and L'/K are Galois, recall how the groups $D_{w'}$ and $I_{w'}$ (in L'/K), $D'_w := D_w(L/L')$ and $I'_w := I_w(L/L')$, are related to the groups D_w and I_w in L/K (with $w \mid w' \mid v$ in $L \supseteq L' \supseteq K$).

We have the following diagram, where L^{D_w} and $L'^{D_{w'}}$ are respectively the decomposition fields of w and w' in L/K and L'/K, where L^{I_w} and $L'^{I_{w'}}$ are the corresponding inertia fields, and where $L^{D'_w}$ and $L^{I'_w}$ are the decomposition and inertia fields of w in the extension L/L'.

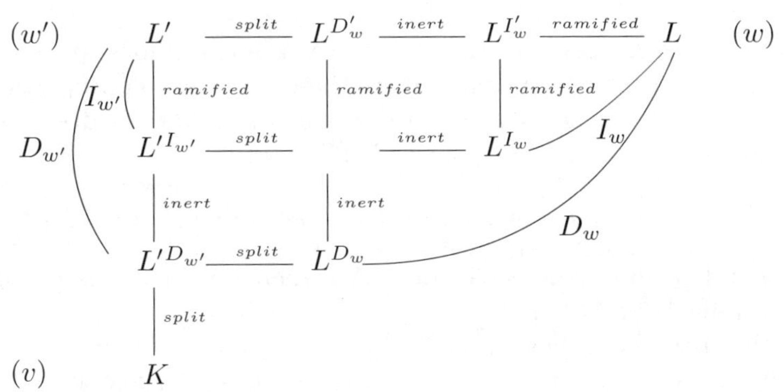

Fig. 1.2

In this diagram, all the field compositums are direct, the linear disjunction (on their intersection) coming from the fact that in each case at least one of the extensions is Galois. In particular, note that the decomposition and inertia properties propagate under ground field extension. For L/K Galois, all of this comes from existence and uniqueness of the fields L_1, L_2, $K \subseteq L_1 \subseteq L_2 \subseteq L$ for which w is totally ramified in L/L_2, totally inert in L_2/L_1 (which is a cyclic extension), and totally split in L_1/K. This means that w is of residue degree and ramification index equal to 1 in L_1/K, of residue degree $[L_2 : L_1]$ in L_2/L_1, and of ramification index $[L : L_2]$ in L/L_2 ([a, Sam, Ch. VI, § 2], [d, Lang1, Ch. I]). We can summarize the above with the following.

1.2.1.1 Proposition. *When L/K and L'/K are Galois, we have the exact sequences:*

$$1 \longrightarrow D'_w = D_w(L/L') \longrightarrow D_w \longrightarrow D_{w'} \longrightarrow 1,$$
$$1 \longrightarrow I'_w = I_w(L/L') \longrightarrow I_w \longrightarrow I_{w'} \longrightarrow 1,$$

which come from the restriction map $\mathrm{Gal}(L/K) \longrightarrow \mathrm{Gal}(L'/K)$. □

1.2.1.2 Definition (global Frobenius'). Let L/K be Galois. Let v be a place of K and $w|v$ in L. When w is unramified in L/K, we denote by $\left(\frac{L/K}{w}\right)$ the global Frobenius of w in L/K, i.e., the image of the local Frobenius (L_w/K_v) in $\mathrm{Gal}(L/K)$ under the canonical isomorphism $G_w \simeq D_w$. □

Then, in an analogous way as for 1.1.6.1:

1.2.1.3 Proposition. *When L/K and L'/K are Galois, and if w is unramified in L/K, we have:*

$$\left(\frac{L/K}{w}\right)^{|D_{w'}(L'/K)|} = \left(\frac{L/L'}{w}\right). \qquad \square$$

1.2.2 Non-Galois Case. In the case where the extensions are not necessarily Galois, see 1.2.5 which again proves these propagation properties using local arguments. In the non-Galois case, we can still define the decomposition and inertia fields L_1 and L_2 of w in L/K if we set:

$$L_1 := \{x \in L,\ i_w(x) \in K_v\},\ \ L_2 := \{x \in L,\ i_w(x) \in L_w^{\mathrm{nr}}\}.$$

These fields depend on the choice of w. We then say that w is totally split (resp. unramified) in L/K if $L_1 = L$ (resp. $L_2 = L$).

Note that now the extensions $i_w(L)/i_v(K)$ and $K_v/i_v(K)$ are not anymore necessarily linearly disjoint over their intersection $i_w(L_1)$: look at the case $K = \mathbb{Q}$, $L = \mathbb{Q}(\sqrt[3]{2})$, where w is the place with residue characteristic equal to $\ell = 5$ for which $i_w(\sqrt[3]{2})$ is not contained in $\mathbb{Q}_5$.

1.2.3 Infinite Galois Case. When L/K is an infinite Galois extension with Galois group G, we define the decomposition field (resp. the inertia field) of w in L/K as the union of the finite extensions which are totally split (resp. unramified) at w, i.e., whose image under i_w is contained in K_v (resp. $\overline{K}_v^{\rm nr}$); thus, this defines the groups D_w and I_w in a way which is compatible with the finite case. It is also possible to define these groups as in the finite case [c, Wa, App., § 2].

1.2.3.1 Proposition. *We have the following homeomorphisms:*

$$D_w \simeq \varprojlim_{L'} D_{w'}(L'/K), \quad I_w \simeq \varprojlim_{L'} I_{w'}(L'/K),$$

where L' ranges in the set of all finite Galois extensions of K in L ordered by inclusion, and where, for each L', w' is the place of L' below w.[4]

Proof. Since D_w is a closed subgroup of the profinite group $G \simeq \varprojlim_{H} G/H$, where H ranges in the set of all closed normal subgroups of finite index of G, a general result gives:

$$D_w \simeq \varprojlim_{H} D_w H/H \ ;$$

if L' denotes the subfield of L fixed under H, by the above we thus have:

$$D_w H/H \simeq \mathrm{Gal}(L'/L' \cap L^{D_w}) = D_{w'}(L'/K),$$

proving the result. The proof is the same for the inertia group. □

In applications, L/K will usually be abelian (infinite class field theory) and so the groups $D_w =: D_v$ and $I_w =: I_v$ will thus be independent of the choice of $w|v$.

1.2.4 Infinite Non-Galois Case. When L/K is an (arbitrary) infinite algebraic extension, we use the definition given in 1.2.2 to define the decomposition and inertia fields of w; this means that by definition an infinite extension is totally split (resp. unramified) at w if and only if any finite subextension is totally split (resp. unramified) at w.

The following exercise justifies by local arguments the existence and basic properties of the decomposition and inertia fields in the most general situation.

---

[4] The place w is defined by means of a choice of coherent extensions w' of v, and denoted $w = \varinjlim_{L'} w'$.

1.2.5 Exercise (propagation of decomposition and nonramification). Let L/K be a fixed extension and let M be another extension of K.

(i) Let $v' \in Pl_M$ be totally split in M/K. Show that every place $w' \in Pl_{LM,v'}$ is totally split in LM/L.

(ii) Let $v' \in Pl_M$ be a finite place unramified in M/K. Show that every place $w' \in Pl_{LM,v'}$ is unramified in LM/L.

Answer. We first show the evident relation (where v is the place of K below v', and w that of L below w', which is thus above v):

$$(LM)_{w'} = L_w M_{v'},$$

by writing that $(LM)_{w'} := i_{w'}(LM)K_v$ is the compositum of $i_w(L)i_{v'}(M)$ with K_v (see I.2.6, I.2.6.1).

(i) By assumption, we have $M_{v'} = K_v$ so that $(LM)_{w'} = L_w$.

(ii) By existence and uniqueness of the maximal unramified subextensions of local extensions (or using the formula $F^{\rm nr} = F \cap \overline{K}_v^{\rm nr}$ for any extension F of K_v), we have:

$$L_w \cap (LM)^{\rm nr}_{w'} = L^{\rm nr}_w,$$

and since by assumption $M_{v'} \subseteq (LM)^{\rm nr}_{w'}$, we have:

$$(LM)_{w'} = L_w M_{v'} \subseteq L_w (LM)^{\rm nr}_{w'},$$

so $(LM)_{w'}$ is equal to the direct compositum of L_w with $(LM)^{\rm nr}_{w'}$ over $L^{\rm nr}_w$. Since $L_w/L^{\rm nr}_w$ is totally ramified, when these extensions are finite, the multiplicativity property of ramification indices immediately shows that $(LM)_{w'}/L_w$ is unramified. For infinite extensions, simply note that $M_{v'} =: K_v(\mu)$, where μ is of the form $\bigcup_n \mu_{q_v^n - 1}$ for suitable integers n, and that $L_w M_{v'} = L_w(\mu)$ is contained in $\overline{L}^{\rm nr}_w$.

If M/K is abelian and if $v \in Pl_K$, then in case (i) (resp. (ii)), every place $w \in Pl_{L,v}$ is totally split (resp. unramified) in LM/L. □

1.3 Higher Ramification. It is also useful to keep in mind a number of results on higher ramification (which will of course be in parallel with the fundamental results of class field theory when the extension is abelian), in particular what follows.[5]

Let v be a finite place of K and denote by ℓ the residue characteristic of v. Assume that L_w/K_v is a finite Galois extension with Galois group G_w.

1.3.1 Higer Ramification Groups. For each $i \geq 0$, we define the higher ramification group (in lower numbering) by:

$$G_{w,i} := \{s \in G_w,\ \mathrm{w}(s(x) - x) \geq i + 1 \ \ \forall x \text{ integer of } L_w\},$$

[5] After [d, Se2, Ch. IV, §§ 1, 2], [e, Ko3, Ch. 1, § 3.7].

which is a normal subgroup of G_w. We have $G_{w,0} = G_w^0$, the inertia group.

We also have for all $i \geq 1$:

$$G_{w,i} = \{s \in G_{w,0},\ \mathrm{w}(\pi_w^{s-1} - 1) \geq i\},$$

where π_w is a uniformizer of L_w.

1.3.2 INERTIA GROUP. Recall that the group $G_{w,1}$ is an ℓ-group and the quotient $G_{w,0}/G_{w,1}$ is a cyclic group isomorphic to a subgroup of $F_w^\times$ (hence whose order is prime to ℓ): indeed, the kernel of the map sending $s \in G_{w,0}$ to the residue class $\overline{u_s}$ of $u_s := \pi_w^{s-1}$ is by definition equal to $G_{w,1}$. This result shows that for a global finite p-extension[6] L/K and for a finite place v of K with residue characteristic $\ell \neq p$, the order of $I_w \simeq G_{w,0}$ divides $|F_w^\times| = q_w - 1$.

1.3.3 INERTIA GROUP IN THE ABELIAN CASE. When L_w/K_v is *abelian*, we can write for $s \in G_{w,0}$ and $u_s := \pi_w^{s-1}$:

$$u_s^{t-1} = (\pi_w^{s-1})^{t-1} = (\pi_w^{t-1})^{s-1} \ \text{ for all } t \in G_w\ ;$$

since $\pi_w^{t-1} =: u(t) \in U_w$, we obtain:

$$\overline{u_s}^{\,t-1} = \overline{u(t)}^{\,s-1} = \overline{1} \ \text{ for all } t \in G_w$$

since F_w (equal to the residue field of L_w^{nr}) is fixed under $G_{w,0}$. It follows that $\overline{u_s} \in F_v^\times$; in this case, we have an injection of the form:

$$G_{w,0}/G_{w,1} \longrightarrow F_v^\times .$$

We have obtained:

1.3.3.1 Proposition. *In any abelian extension L/K (finite or not), the group $D_{v,0}(L/K)/D_{v,1}(L/K)$ (which measures the tame ramification)[7] is isomorphic to a subgroup of $F_v^\times$.* □

For example, if $\mathrm{Gal}(L/K) \simeq \mathbb{Z}_p^r$ (i.e., if L/K is $\mathbb{Z}_p$-free of finite type), then L/K is unramified outside places dividing p (same result if $\mathrm{Gal}(L/K)$ is an arbitrary free pro-p-group [Y]).

For the definition of the ramification groups in upper numbering G_w^i, see [d, Se2, Ch. IV, § 3].

We will come back to higher ramification groups in 1.6.2 when we will perform conductor computations.

---

[6] Recall that "p-extension" or "pro-p-extension" always means Galois extension whose Galois group is a p-group or a pro-p-group.

[7] As explained in 1.1.1, for $i \geq 0$, $D_{w,i}(L/K)$ denotes the ramification groups in a global Galois extension L/K; in particular, $D_{w,0}(L/K) = I_w(L/K)$ and $D_{w,i}(L/K) \simeq G_{w,i}$.

b) Local Class Field Theory Correspondence

Let L/K be a finite extension of number fields, L'/K a subextension, and let $v \in Pl$, $w \in Pl_{L,v}$, and w' below w (i.e., above v) in L'. We denote by U_v, $U'_{w'}$, and U_w the unit groups of the fields K_v, $L'_{w'}$, and L_w. The first fundamental result in the local case is the following (use Fig. 1.1).

1.4 Theorem (local reciprocity map, norm residue symbol). *There exists a canonical homomorphism:*

$$\begin{aligned}(\bullet\,,\,L_w/K_v)\,:\,K_v^\times &\longrightarrow G_w^{\mathrm{ab}} := \mathrm{Gal}(L_w^{\mathrm{ab}}/K_v)\\ x &\longmapsto (x\,,\,L_w/K_v)\end{aligned}$$

having the following properties:

(i) We have the exact sequence:

$$1 \longrightarrow \mathrm{N}_{L_w/K_v}(L_w^\times) \longrightarrow K_v^\times \xrightarrow{(\bullet\,,L_w/K_v)} G_w^{\mathrm{ab}} \longrightarrow 1\ ;$$

(ii) the composition of $(\bullet\,,\,L_w/K_v)$ and of the projection $G_w^{\mathrm{ab}} \longrightarrow \mathrm{Gal}(L'^{\mathrm{ab}}_{w'}/K_v)$ is equal to $(\bullet\,,\,L'_{w'}/K_v)$;

(iii) the image of U_v (resp. of U_v^i, $i \geq 1$) under $(\bullet\,,\,L_w/K_v)$ is equal to the inertia group $G_w^{\mathrm{ab}\,0}$ (resp. to the ith higher ramification group in upper numbering $G_w^{\mathrm{ab}\,i}$) in L_w^{ab}/K_v, and in particular we have the exact sequence:

$$1 \longrightarrow \mathrm{N}_{L_w/K_v}(U_w) \longrightarrow U_v \longrightarrow G_w^{\mathrm{ab}\,0} \longrightarrow 1\ ;$$

(iv) for all $x' \in L'^\times_{w'}$, the image of $(x'\,,\,L_w/L'_{w'})$ in G_w^{ab} is equal to $(\mathrm{N}_{L'_{w'}/K_v}(x')\,,\,L_w/K_v)$; in particular, we have:

$$\mathrm{Gal}(L_w^{\mathrm{ab}}/L'^{\mathrm{ab}}_{w'}) = (\mathrm{N}_{L'_{w'}/K_v}(L'^\times_{w'})\,,\,L_w/K_v),$$

and the inertia group of $L_w^{\mathrm{ab}}/L'^{\mathrm{ab}}_{w'}$ is equal to $(\mathrm{N}_{L'_{w'}/K_v}(U'_{w'})\,,\,L_w/K_v)$;

(v) for all $x \in K_v^\times$, the image of $(x\,,\,L_w/K_v)$ under the transfer map[8] *(from G_w^{ab} to $\mathrm{Gal}(L_w^{\mathrm{ab}'}/L'_{w'})$), is equal to $(x\,,\,L_w/L'_{w'})$;*

(vi) for any isomorphism τ of L_w and all $x \in K_v^\times$, we have:

$$(\tau x\,,\,\tau L_w/\tau K_v) = \tau \circ (x\,,\,L_w/K_v) \circ \tau^{-1} \ \text{on}\ \tau L_w^{\mathrm{ab}}\ ;$$

(vii) if L_w^{ab}/K_v is unramified, then for all $x \in K_v^\times$ we have:

$$(x\,,\,L_w/K_v) = (L_w^{\mathrm{ab}}/K_v)^{\mathrm{v}(x)},$$

where (L_w^{ab}/K_v) denotes the Frobenius[9] *of L_w^{ab}/K_v; in other words:*

[8] See Remark 1.

[9] See Remark 2.

$$(\pi_v\,,\,L_w/K_v) = (L_w^{\mathrm{ab}}/K_v),$$

for any uniformizer π_v of K_v. □

The symbol $(\bullet, L_w/K_v)$ is called the local norm residue symbol or the local reciprocity map.

1.4.1 Remark 1 (transfer map). For the cohomological definition and the properties of the transfer map, see [d, Se2, Ch. VII, § 8] or [f, Neu2, th. 8.8]. Here we do not assume that L_w/K_v is Galois, and the result is obtained (by restriction) from the analogous computation in the Galois closure of L_w over K_v. Recall how to compute $\mathrm{Ver} : G/[G,G] \longrightarrow H/[H,H]$ for any subgroup H of finite index of a group G: let $(s_i)_{i=1,\dots,(G:H)}$ be a system of representatives of the elements of G/H; for any fixed $s \in G$ and for each i put $s\,s_i =: s_j\,t_i$, $t_i \in H$, then we have:

$$\mathrm{Ver}(s \bmod [G,G]) = \prod_{i=1}^{(G:H)} t_i \bmod [H,H].$$ □

1.4.2 Remark 2 (local Frobenius). Recall that if L_w^{ab}/K_v is unramified (i.e., $L_w^{\mathrm{ab}} = L_w^{\mathrm{nr}}$), it is cyclic and its Frobenius (L_w^{ab}/K_v) is the unique generator σ of G_w^{ab} such that $\sigma(x) \equiv x^{q_v} \bmod (\pi_v)$ for all integers x of L_w^{ab}, where $q_v := |F_v|$, or such that $\sigma(\zeta) = \zeta^{q_v}$ for a root of unity ζ of order $q_v^{f_w^{\mathrm{ab}}} - 1$ generating L_w^{ab} over K_v. □

Note. If we denote respectively by N' and j' the norm map in $L'_{w'}/K_v$ and the canonical injection $K_v \longrightarrow L'_{w'}$, we have:

$$\mathrm{N}' \circ j' = [L'_{w'} : K_v] \text{ on } K_v^\times,\ \ j' \circ \mathrm{N}' = \sum_i \sigma'_i \text{ on } L'^{\times}_{w'},$$

where the σ'_i are the $[L'_{w'} : K_v]$ K_v-isomorphisms of $L'_{w'}$ in $\mathbb{C}_\ell$ (in the Galois case, $\sum_i \sigma'_i =: \nu'$ is the algebraic norm in $L'_{w'}/K_v$). This applies to (iv) and (v).

1.4.3 Corollary. *We have (with Notations 1.1.6.2):*

(i) $K_v^\times/\mathrm{N}_{L_w/K_v}(L_w^\times) \simeq G_w^{\mathrm{ab}}$ has order equal to $e_w^{\mathrm{ab}}\, f_w^{\mathrm{ab}}$;

(ii) $U_v/\mathrm{N}_{L_w/K_v}(U_w) \simeq G_w^{\mathrm{ab}\,0}$ has order equal to e_w^{ab}. If L_w^{ab}/K_v is unramified, we have:

$$U_v = \mathrm{N}_{L_w/K_v}(U_w)$$

(i.e., $\mathrm{N}_{L_w/K_v}(L_w^\times) = \pi_v^{f_w^{\mathrm{ab}}\mathbb{Z}} \oplus U_v$), and $x \in K_v^\times$ is a norm in L_w/K_v if and only if:

$$\mathrm{v}(x) \equiv 0 \bmod (f_w^{\mathrm{ab}}).$$ □

By 1.4, (iii), and the fact that for a finite place v with residue characteristic equal to ℓ the group U_v^1 is the ℓ-Sylow subgroup of U_v, we deduce that $G_w^{\mathrm{ab}\,1} = G_{w,1}^{\mathrm{ab}}$ (the ℓ-Sylow subgroup of $G_{w,0}^{\mathrm{ab}}$) (see 1.3.2).

1.4.4 Proposition. *The map* $\mathrm{N}_{L_w/K_v} : L_w^\times \longrightarrow K_v^\times$ *is an open map.*

Proof. It is enough to show that for all $j \geq 0$ there exists $i \geq 0$ such that $U_v^i \subseteq \mathrm{N}_{L_w/K_v}(U_w^j)$. The properties of the logarithm and exponential in K_v [c, Wa, Ch. 5, § 1] imply that, for i sufficiently large we have:

$$\log(U_v^i) = (\pi_v^i) =: [L_w : K_v](\pi_v^{i-h}),$$

where $h := \mathrm{v}([L_w : K_v])$, and for $i \geq j + h$ sufficiently large:

$$U_v^i = (U_v^{i-h})^{[L_w:K_v]} = \mathrm{N}_{L_w/K_v}(U_v^{i-h}) \subseteq \mathrm{N}_{L_w/K_v}(U_w^j). \qquad \square$$

It is clear that 1.4.3, (ii) is a deep result which is not simply elementary v-adic analysis, but it implies that $\mathrm{N}_{L/K}$ is an open map as a map from J_L to J_K (this will be clear in Section 2); see also 1.6.4.

1.4.5 Remark. By its very nature, in a certain sense the norm residue symbol:

$$(\bullet\,,\, L_w/K_v),$$

does not depend on the extension L_w/K_v, but only on its maximal abelian subextension; for instance, this allows us to write:

$$(\bullet\,,\, L_w/K_v) = (\bullet\,,\, L_w^{\mathrm{ab}}/K_v) \ \text{ and } \ \mathrm{N}_{L_w/K_v}(L_w^\times) = \mathrm{N}_{L_w^{\mathrm{ab}}/K_v}(L_w^{\mathrm{ab}\times}),$$

showing that the definition of this symbol for an arbitrary extension is useful in practice and gives more precise information. This proves for example that any element of $\mathbb{Q}_2^\times$ is the norm of an element of $\mathbb{Q}_2(\sqrt[3]{2})$. In particular, we deduce the following equality:

$$L_w^{\mathrm{ab}\times} = \mathrm{N}_{L_w/L_w^{\mathrm{ab}}}(L_w^\times)\,.\,{}_{\mathrm{N}}L_w^{\mathrm{ab}\times},$$

where ${}_{\mathrm{N}}L_w^{\mathrm{ab}\times}$ is the kernel of $\mathrm{N}_{L_w^{\mathrm{ab}}/K_v}$. $\square$

1.4.6 Remark. Let us explicitly describe the case $v \in Pl_\infty^{\mathrm{r}}$ (the reciprocity map is the trivial map when v is complex). In this case $K_v = \mathbb{R}$ hence, since the only nontrivial algebraic extension of $\mathbb{R}$ is $\mathbb{C}$ (which is in addition abelian and unramified over $\mathbb{R}$), the reciprocity map $(\bullet\,,\,\mathbb{C}/\mathbb{R})$ is given by:

$$(x\,,\,\mathbb{C}/\mathbb{R}) := c^{\mathrm{v}(x)} \text{ for all } x \in \mathbb{R}^\times,$$

where c is complex conjugation, and where $\mathrm{v}(x) = 0$ (resp. 1) if $x > 0$ (resp. $x < 0$). This is the only way to have the exact sequence in 1.4, (i). It is also the formula of statement (vii) since $c = (\mathbb{C}/\mathbb{R})$ (the Frobenius of v).

In this case, when $L_w = \mathbb{C}$ (i.e., $i_w(L)$ is a nonreal extension of $i_v(K) \subset \mathbb{R}$), we have $G_w = G_w^{\mathrm{ab}} = \langle c \rangle$, $G_w^0 = 1$, $f_w = 2$, $e_w = 1$. If L/K is Galois, the decomposition group D_w is generated by the global Frobenius $c_w := i_w^{-1} \circ c \circ i_w$; c_w is called "a" complex conjugation of L/K. □

Let K_v be the v-completion of the number field K.

1.5 Theorem (local existence). *For any subgroup of finite index N of $K_v^\times$ there exists*[10] *a unique finite abelian extension M of K_v such that $\mathrm{N}_{M/K_v}(M^\times) = N$; the norm residue symbol in M/K_v yields the exact sequence:*

$$1 \longrightarrow N \longrightarrow K_v^\times \longrightarrow \mathrm{Gal}(M/K_v) \longrightarrow 1.$$

In addition, the bijection from the set of subgroups of finite index of $K_v^\times$ to the set of finite abelian extensions of K_v is a Galois correspondence; in other words we have the following properties (where M_1 and M_2 are abelian over K_v and correspond respectively to N_1 and $N_2 \subseteq K_v^\times$):

(i) we have $M_1 \subseteq M_2$ if and only if $N_2 \subseteq N_1$;

(ii) $M_1 M_2$ corresponds to $N_1 \cap N_2$;

(iii) $M_1 \cap M_2$ corresponds to $N_1 N_2$;

(iv) if $M_1 \subseteq M_2$, we have $\mathrm{Gal}(M_2/M_1) \simeq N_1/N_2$, where the isomorphism is obtained from the restriction of $(\,\bullet\,,\, M_2/K_v)$ to N_1. □

Note. The subgroups of finite index of $K_v^\times$ are open, but the converse is false (look for example at the case of U_v). However, when we take limits to describe $\mathrm{Gal}(\overline{K}_v^{\mathrm{ab}}/K_v)$, it is not $K_v^\times$ and its topology which occur (see 1.7).

1.5.1 Remarks. (i) Properties (i) to (iv) logically follow from the existence of this bijection, because of 1.4, (i), (ii) (the equality $\mathrm{Gal}(M_2/M_1) = (N_1, M_2/K_v)$ is a particular case of 1.4, (iv)).

(ii) By 1.4, (iii), the subgroup of $K_v^\times$ corresponding to the inertia subfield M^{nr} of M is $U_v\, N$, where N corresponds to M; similarly, its maximal tamely ramified subextension corresponds to $U_v^1\, N$. It is clear that $U_v N = \pi_v^{f\mathbb{Z}} \oplus U_v$, where f is the residue degree of M/K_v. We recover the existence and the uniqueness of the unramified extension of degree n of K_v: it corresponds to $\pi_v^{n\mathbb{Z}} \oplus U_v$.

(iii) The group N corresponding to M/K_v is called the norm group of the extension M/K_v. □

1.5.2 Exercise. Prove the above Remark (i).

Answer. Suppose $M_1 \subseteq M_2$; since $(N_2,\, M_2/K_v) = 1$ by definition, one gets $(N_2,\, M_1/K_v) = 1$, proving that $N_2 \subseteq N_1$. We now put:

[10] in some fixed algebraic closure of K_v; here it is convenient to use $\overline{K}_v = \overline{\mathbb{Q}}_\ell \subset \mathbb{C}_\ell$, where ℓ is the residue characteristic or ∞ (see the introduction to Section 1).

$$H := (N_1,\, M_2/K_v) \subseteq \mathrm{Gal}(M_2/M_1)\ ;$$

then $|H| = (N_1 : N_2) = (K_v^\times : N_2)(K_v^\times : N_1)^{-1} = [M_2 : M_1]$, proving (iv):

$$(N_1,\, M_2/K_v) = \mathrm{Gal}(M_2/M_1),$$

which we will now use systematically.

Let M_1, M_2 be arbitrary, and let N and N' be the norm groups of $M := M_1M_2$ and $M' := M_1 \cap M_2$; we put:

$$H_i := \mathrm{Gal}(M/M_i) = (N_i,\, M/K_v),\ i = 1,\, 2\ ;$$

we have the inclusions:

$$N \subseteq N_1 \cap N_2 \subseteq N_1N_2 \subseteq N'.$$

We have $H_1H_2 = \mathrm{Gal}(M/M')$, hence, since $H_i = (N_i,\, M/K_v)$, we have $(N_1N_2,\, M/K_v) = (N',\, M/K_v)$, and finally $N' = N_1N_2$ since these groups contain the kernel N of $(\,\bullet\,,\, M/K_v)$. In the same way, $H_1 \cap H_2 = 1$ yields:

$$(N_1,\, M/K_v) \cap (N_2,\, M/K_v) = (N_1 \cap N_2,\, M/K_v) = 1,$$

thus $N_1 \cap N_2 = N$.

If $N_2 \subseteq N_1$ then $N_1N_2 = N_1$ yields (by uniqueness) $M_1 \cap M_2 = M_1$ (or $N_1 \cap N_2 = N_2$ and $M_1M_2 = M_2$), which finishes the proof. □

To illustrate the local class field theory correspondence, let us look at the following situation.

1.5.3 Example (local extensions coming from non-Galois extensions). Let L/K be a finite extension of number fields and, for $v \in Pl$, let L_w for $w \in Pl_{L,v}$ be the completions of L above v; recall that the L_w are defined only up to K_v-conjugation. Let L_v^{ab} be the maximal abelian subextension of $L_v := \bigcap_{w|v} L_w$; it is independent of the choice of the K_v-conjugates of the L_w since we have $L_v^{\mathrm{ab}} = \bigcap_{w|v} L_w^{\mathrm{ab}}$, while L_v does depend on them, but we will see that L_v will not really be used as such. We set:

$$G_v^{\mathrm{ab}} := \mathrm{Gal}(L_v^{\mathrm{ab}}/K_v).$$

Then the subgroup of $K_v^\times$ corresponding to L_v^{ab} is the subgroup generated by the $\mathrm{N}_{L_w/K_v}(L_w^\times)$ for $w|v$; in particular we have the equality:

$$\mathrm{N}_{L_v/K_v}(L_v^\times) = \langle\, \mathrm{N}_{L_w/K_v}(L_w^\times)\,\rangle_{w|v}$$

and the exact sequence:

$$1 \longrightarrow \mathrm{N}_{L_v/K_v}(L_v^\times) \longrightarrow K_v^\times \longrightarrow G_v^{\mathrm{ab}} \longrightarrow 1,$$

which can also be written using the corresponding abelianizations:

$$\mathrm{N}_{L_v^{\mathrm{ab}}/K_v}(L_v^{\mathrm{ab}\times}) = \langle\, \mathrm{N}_{L_w^{\mathrm{ab}}/K_v}(L_w^{\mathrm{ab}\times})\,\rangle_{w|v},$$

$$1 \longrightarrow \mathrm{N}_{L_v^{\mathrm{ab}}/K_v}(L_v^{\mathrm{ab}\times}) \longrightarrow K_v^\times \longrightarrow G_v^{\mathrm{ab}} \longrightarrow 1.$$

We will also encounter the field compositum $\hat{L}_v^{\mathrm{ab}} := \langle\, L_w^{\mathrm{ab}}\,\rangle_{w|v}$; by local class field theory, the field $\hat{L}_v^{\mathrm{ab}}$ corresponds to the subgroup:

$$\bigcap_{w|v} \mathrm{N}_{L_w/K_v}(L_w^\times) = \bigcap_{w|v} \mathrm{N}_{L_w^{\mathrm{ab}}/K_v}(L_w^{\mathrm{ab}\times}). \qquad \square$$

Let us return to the general situation; we then have the following additional property which can easily be deduced from 1.4 and which we state in a slightly different setting.

1.5.4 Corollary (norm lifting theorem). *Let L/K be a number field extension and for $v \in Pl$, let M/K_v be a finite abelian extension. If N is the subgroup of $K_v^\times$ corresponding to M over K_v, then for $w|v$ the subgroup of $L_w^\times$ corresponding to L_wM over L_w is:*

$$N' = \{y \in L_w^\times,\ \mathrm{N}_{L_w/K_v}(y) \in N\} =: \mathrm{N}_{L_w/K_v}^{-1}(N).$$

Proof. We give the proof using the following diagram:

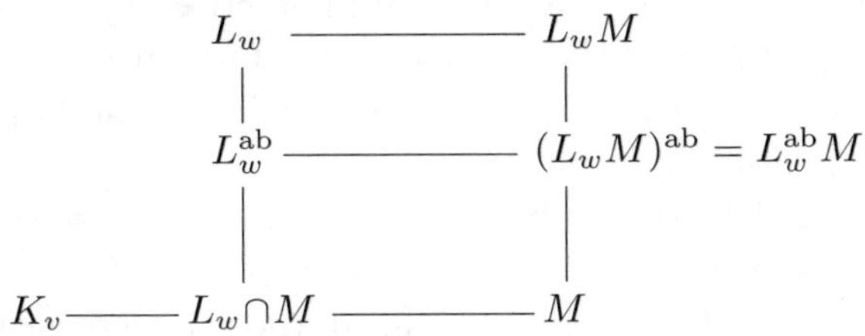

We have $N' = \mathrm{Ker}((\,\bullet\,,\, L_wM/L_w))$. Since $(L_wM)^{\mathrm{ab}} = L_w^{\mathrm{ab}}M$, we have the isomorphisms:

$$\mathrm{Gal}(L_wM/L_w) \simeq \mathrm{Gal}(L_w^{\mathrm{ab}}M/L_w^{\mathrm{ab}}) \simeq \mathrm{Gal}(M/L_w \cap M)\ ;$$

it follows that $y \in N'$ if and only if the image of $(y\,,\, L_wM/L_w)$ in $\mathrm{Gal}(M/L_w \cap M)$ is trivial. Using 1.4, (iv) applied to L_wM/K_v, the image of $(y\,,\, L_wM/L_w)$ in $\mathrm{Gal}(L_w^{\mathrm{ab}}M/K_v)$, which is an element of $\mathrm{Gal}(L_w^{\mathrm{ab}}M/L_w^{\mathrm{ab}})$, is equal to $(\mathrm{N}_{L_w/K_v}(y)\,,\, L_w^{\mathrm{ab}}M/K_v)$ whose image in $\mathrm{Gal}(M/L_w \cap M)$ is obtained by restriction to M, giving $(\mathrm{N}_{L_w/K_v}(y)\,,\, M/K_v)$ by 1.4, (ii). Since by definition $\mathrm{Ker}((\,\bullet\,,\, M/K_v)) = N$, we obtain the given formula for N'. $\square$

c) Local Conductors and Norm Groups

Let L/K be an extension of number fields, and let $v \in Pl_0$ and $w \in Pl_{L,v}$.

1.6 Definitions (local conductors). (i) The smallest power $\mathfrak{p}_v^{m_w}$, $m_w \geq 0$, such that:

$$U_v^{m_w} \subseteq \mathrm{N}_{L_w/K_v}(L_w^\times)$$

(or, equivalently, $U_v^{m_w} \subseteq \mathrm{N}_{L_w/K_v}(U_w)$) is called the norm conductor or conductor of L_w/K_v and is denoted:

$$\mathfrak{f}_{L_w/K_v}.$$

(ii) The conductor of $(L^{\mathrm{ab}})_v/K_v$, the completion of L^{ab}/K at v, is called the norm v-conductor or v-conductor of L/K and denoted:

$$\mathfrak{f}_v := \mathfrak{f}_v(L/K).$$ □

1.6.1 Remarks. (i) By 1.4, (iii), m_w is the smallest integer m for which we have (using upper numbering):

$$G_w^{\mathrm{ab}\, m} = 1.$$

Since $\mathrm{N}_{L_w/K_v}(L_w^\times) = \mathrm{N}_{L_w^{\mathrm{ab}}/K_v}(L_w^{\mathrm{ab}\times})$, we have equality of the conductors of the extensions L_w/K_v and L_w^{ab}/K_v (so that in practice we always are reduced to compute $\mathfrak{f}_{L_w^{\mathrm{ab}}/K_v}$ by using the formula that we will give in 1.6.2).

(ii) By definition, we have $\mathfrak{f}_v(L/K) = \mathfrak{f}_v(L^{\mathrm{ab}}/K)$; in addition $\mathfrak{f}_v(L^{\mathrm{ab}}/K)$ divides the $\mathfrak{f}_{L_w^{\mathrm{ab}}/K_v}$ for $w|v$.

(iii) Local class field theory implies the local conductor theorem which says that v is ramified in L_w^{ab}/K_v if and only if $\mathfrak{f}_{L_w^{\mathrm{ab}}/K_v} \neq 1$ (use 1.4.3, (ii)). Note however that L_w/L_w^{ab} is totally ramified; the conductor is thus equal to 1 if and only if $L_w^{\mathrm{ab}} = L_w^{\mathrm{nr}}$. □

It seems that it would be useful to define a generalized v-conductor $\mathfrak{f}_v[L/K]$ when L/K is any extension; it should be the conductor of L_v^{ab} because of the normic properties that we will see in 2.6 and of Definition 3.1.4 of the generalized norm residue symbol.

In the Galois case, $L_v^{\mathrm{ab}} = L_{w_0}^{\mathrm{ab}}$ for any place $w_0|v$, and $\mathfrak{f}_v[L/K]$ is given by $\mathfrak{f}_{L_{w_0}/K_v}$ (which is then independent of the choice of w_0).

Concerning $\hat{L}_v^{\mathrm{ab}}$, we easily check that its conductor is equal to the l.c.m. of the $\mathfrak{f}_{L_w/K_v}$, $w|v$.

We can summarize the above by the following diagram where the corresponding conductors divide each other in the given order:

$$\begin{array}{ccccccccc} K_v & \text{———} & (L^{\mathrm{ab}})_v & \text{———} & L_v^{\mathrm{ab}} & \text{———} & L_w^{\mathrm{ab}} & \text{———} & \hat{L}_v^{\mathrm{ab}} \\ (1) & & \mathfrak{f}_v(L/K) & & \mathfrak{f}_v[L/K] & & \mathfrak{f}_{L_w/K_v} & & \text{l.c.m.}(\mathfrak{f}_{L_w/K_v}) \end{array}$$

However, we will not have any use for the generalized v-conductor and in global class field theory, only the $\mathfrak{f}_v$ (the v-conductors for L^{ab}/K) will enter, whose product will define a global conductor.

1.6.2 Remark (conductor computation). For the explicit computation of the conductor $\mathfrak{f}_{L_w/K_v}$, we refer to [d, Se2, Ch. XV, § 2, Cor. 2 to Th. 1, Ch. IV, § 3] from which we deduce the following formula:

$$\mathfrak{f}_{L_w/K_v} = \mathfrak{f}_{L_w^{\mathrm{ab}}/K_v} =: \mathfrak{p}_v^{m_w}, \text{ with } m_w := \frac{1}{g_0^{\mathrm{ab}}} \sum_{\substack{i \geq 0 \\ g_i^{\mathrm{ab}} > 1}} g_i^{\mathrm{ab}},$$

where g_i^{ab} is the order of the higher ramification group $G_{w,i}^{\mathrm{ab}}$ (in lower numbering) in L_w^{ab}/K_v; for each $i \geq 1$, $G_{w,i}^{\mathrm{ab}}$ is defined from $G_{w,0}^{\mathrm{ab}}$ (of order $g_0^{\mathrm{ab}} = e_w^{\mathrm{ab}}$) by:

$$G_{w,i}^{\mathrm{ab}} = \{s \in G_{w,0}^{\mathrm{ab}},\ \mathrm{w}((\pi_w^{\mathrm{ab}})^{s-1} - 1) \geq i\},$$

where π_w^{ab} is a uniformizer of L_w^{ab} (see 1.3.1).

If v is tamely ramified in L_w^{ab}/K_v (i.e., if the residue characteristic ℓ of v does not divide e_w^{ab}), we have $G_{w,1}^{\mathrm{ab}} = 1$, hence $m_w = 1$. □

We will assume known this conductor formula since it can be obtained by a direct study of the norm on the groups U_v, as is done in [d, Se2, Ch. V and XV] following Hasse, study which reduces to proving the property of the local reciprocity map assumed in 1.4, (iii). It is nothing but the translation of the lower numbering to the upper numbering for ramification groups when we look for the first trivial $G_w^{\mathrm{ab}\,m}$ (see 1.6, (i)). This is a great advantage since the computation of the higher ramification groups *in lower numbering* is always effective and easy in practice (see the example given below).

We use the same method to compute the v-conductor $\mathfrak{f}_v$ from the groups $(G^{\mathrm{ab}})_{v,i}$, where $(G^{\mathrm{ab}})_v := \mathrm{Gal}((L^{\mathrm{ab}})_v/K_v)$.

As an application, we give the following result for the Kummer case, which illustrates the computation of local conductors from the classical results on higher ramification groups mentioned above.

1.6.3 Proposition (v-conductors of a Kummer extension of prime degree p). *Let K be a number field containing the group μ_p of pth roots of unity, and let $L = K(\sqrt[p]{\alpha})$ with $\alpha \in K^\times \setminus K^{\times p}$. Let v be a (finite) place of K ramified in L/K and let e_v be the ramification index of v in $K/\mathbb{Q}(\mu_p)$. The norm v-conductor of L/K is equal to $\mathfrak{p}_v$ if $v \nmid p$, and to $\mathfrak{p}_v^{pe_v+1-r}$ if $v|p$, where r is the largest integer for which the congruence:*

$$\frac{\alpha}{x^p} \equiv 1 \bmod \mathfrak{p}_v^r,\ x \in K^\times,$$

has a solution (the case $\mathrm{v}(\alpha) \not\equiv 0 \bmod (p)$ *meaning* $r = 0$*).*

Proof. Let $\alpha_v := i_v(\alpha) \in K_v^\times$; then $L_v := K_v(\sqrt[p]{\alpha_v})$ is the completion L_w of L at some place $w|v$, and by definition the conductor of L_v/K_v is equal to $\mathfrak{f}_v$. We set $G_v := \mathrm{Gal}(L_v/K_v)$.

If $v \nmid p$ is ramified (i.e., $\mathrm{v}(\alpha) \not\equiv 0 \bmod (p)$), we have $\mathfrak{f}_v = \mathfrak{p}_v$ (tame ramification). This case follows in fact trivially, directly from Definition 1.6.

Assume now that $v|p$. In this case the formula of 1.6.2 yields $\mathfrak{f}_v = \mathfrak{p}_v^{t+1}$, where t is the largest integer such that $g_t \neq 1$, and we have:

$$t = \mathrm{w}(\pi_w^{\sigma-1} - 1),$$

where π_w is a uniformizer of L_v and σ a generator of G_v.

(i) If $\mathrm{v}(\alpha) \not\equiv 0 \bmod (p)$, we can always assume that $\mathrm{v}(\alpha) = 1$, hence that $\pi_w = \sqrt[p]{\alpha_v}$. We then have $\pi_w^{\sigma-1} - 1 =: \zeta - 1$, where ζ is a generator of μ_p, giving $t = pe_v$. But in this case $r = 0$, proving the result.

(ii) If $\mathrm{v}(\alpha) \equiv 0 \bmod (p)$, we can reduce to the case where $\alpha_v \in U_v$. By I.6.3, (ii), the integer r satisfies:

$$1 \leq r \leq pe_v - 1 \ ;$$

changing $\alpha_v \bmod (U_v)^p$ if necessary, we can assume that $\alpha_v \in U_v^r$; then, by definition of r, $\alpha_v (U_v^1)^p$ is disjoint from U_v^{r+1}. Let us write:

$$\sqrt[p]{\alpha_v} := 1 + \pi_w^\rho \, u_w, \ \rho \geq 1, \ u_w \in U_w \ ;$$

this yields:

$$\alpha_v = 1 + \pi_w^{pe_v(p-1)+\rho} \, u'_w + \pi_w^{p\rho} \, u_w^p, \ u'_w \in U_w$$

since $\mathrm{v}(p) = e_v(p-1)$ and v is ramified in L_v/K_v. We must have $\rho < pe_v$, otherwise we would get $r \geq pe_v$, a contradiction. Thus we must have $p\rho < pe_v(p-1) + \rho$, hence $\rho = r$. Writing that $\sigma(\sqrt[p]{\alpha_v}) = \zeta \sqrt[p]{\alpha_v}$ and that $\pi_w^{\sigma-1} = \xi \in U_w^t \backslash U_w^{t+1}$, we obtain:

$$1 + \pi_w^r \, \xi^r \, u_w^\sigma = \zeta(1 + \pi_w^r \, u_w),$$

hence, since $\mathrm{w}(1 - \zeta) = pe_v > r$:

$$\mathrm{w}(\xi^r \, u_w^{\sigma-1} - 1) = pe_v - r.$$

Lemma. *We have* $r \not\equiv 0 \bmod (p)$.

Proof. Assume that $r = \lambda p$ and set $\alpha_v =: 1 + \pi_v^{\lambda p} \, \eta_v, \ \eta_v \in U_v$; since $r = \lambda p < pe_v$ we have $\lambda < e_v$. Since F_v is a finite field, there exists $\eta'_v \in U_v$ such that $\eta_v \equiv \eta_v'^p \bmod (\pi_v)$; it is then immediately checked that:

$$\frac{\alpha_v}{(1 + \pi_v^\lambda \, \eta'_v)^p} \in U_v^{r+1},$$

a contradiction. □

It follows that $\xi^r \in U_w^t \backslash U_w^{t+1}$. But it is clear that $u_w^{\sigma-1} \in U_w^{t+1}$, which yields:

$$\mathrm{w}(\xi^r - 1) = pe_v - r.$$

It follows that $\mathrm{w}(\xi^r - 1) = \mathrm{w}(\xi - 1) = t = pe_v - r$, finishing the computation of the v-conductor in the wild case. □

Note. If $v|p$ and if $\mathrm{v}(\alpha) \not\equiv 0 \bmod (p)$, then $r = 0$ and the v-conductor is maximal; if $\mathrm{v}(\alpha) \equiv 0 \bmod (p)$, we have $1 \leq r \leq pe_v - 1$, so that $2 \leq pe_v + 1 - r \leq pe_v$, where the lower bound is in agreement with statement III.1.3.2.

The following result on norm actions can be useful in practice.

1.6.4 Proposition. *Let L_w/K_v be a completion of an arbitrary finite extension L/K of number fields. There exists a function ψ_w from $\mathbb{N}$ to $\mathbb{N}$ such that $\mathrm{N}_{L_w/K_v}(U_w^{\psi_w(n)}) = U_v^n$ for all sufficiently large n. If in addition L_w/K_v is Galois with Galois group G_w, the above relation holds as soon as $G_{w,\psi_w(n)}$ (the $\psi_w(n)$th higher ramification group) is trivial.*

Proof. Referring to [d, Se2], we can sketch the following proof: we use [IV, 3, Rem., 2] which allows us to define ψ_w in complete generality from the Galois case, and using [V, 6, Cor. 3] for norm aspects. It is then clear that if L_w/K_v is unramified, $\mathrm{N}_{L_w/K_v}(U_w^n) = U_v^n$ for all $n \geq 0$ ([V, 2, Prop. 3] or 1.4.3, (ii)); if L_w/K_v is tamely ramified, the above equality holds for all $n \geq 1$ ([XV, 2, Cor. 1 to Th. 1] or once again 1.4.3, (ii)). □

1.6.5 Exercise (norm groups and conductors of quadratic extensions of $\mathbb{Q}_\ell$). Let ℓ be a prime number. If $\ell \neq 2$, since:

$$\mathbb{Q}_\ell^\times = \langle\, \ell \,\rangle \oplus \langle\, \zeta \,\rangle \oplus \langle\, 1 + \ell \,\rangle_{\mathbb{Z}_\ell}, \text{ where } \langle\, \zeta \,\rangle = \mu_{\ell-1},$$

Kummer theory shows, through the study of $\mathbb{Q}_\ell^\times/\mathbb{Q}_\ell^{\times 2}$, that quadratic extensions of $\mathbb{Q}_\ell$ are:

$$\mathbb{Q}_\ell(\sqrt{\zeta}\,),\ \mathbb{Q}_\ell(\sqrt{\ell}\,),\ \mathbb{Q}_\ell(\sqrt{\ell\zeta}\,).$$

If $\ell = 2$, knowing that in this case:

$$\mathbb{Q}_2^\times = \langle\, 2 \,\rangle \oplus \langle\, -1 \,\rangle \oplus \langle\, 5 \,\rangle_{\mathbb{Z}_2},$$

we obtain the following list of quadratic extensions of $\mathbb{Q}_2$:

$$\mathbb{Q}_2(\sqrt{5}\,),\ \mathbb{Q}_2(\sqrt{-1}\,),\ \mathbb{Q}_2(\sqrt{-5}\,),\ \mathbb{Q}_2(\sqrt{2}\,),\ \mathbb{Q}_2(\sqrt{-2}\,),\ \mathbb{Q}_2(\sqrt{10}\,),\ \mathbb{Q}_2(\sqrt{-10}\,).$$

Compute all the norm groups and conductors.

Answer. For $\ell \neq 2$, the subgroups of index 2 of $\mathbb{Q}_\ell^\times$ are the following:

$$\begin{aligned} N_1 &:= \langle \ell^2 \rangle \oplus \langle \zeta \rangle \oplus \langle 1+\ell \rangle_{\mathbb{Z}_\ell}, \\ N_2 &:= \langle \ell \rangle \oplus \langle \zeta^2 \rangle \oplus \langle 1+\ell \rangle_{\mathbb{Z}_\ell}, \\ N_3 &:= \langle \ell\zeta \rangle \oplus \langle \zeta^2 \rangle \oplus \langle 1+\ell \rangle_{\mathbb{Z}_\ell}. \end{aligned}$$

The unique unramified extension $\mathbb{Q}_\ell(\sqrt{\zeta}\,)$ corresponds to N_1 since the residue degree is 2 or the ramification index is 1 using 1.4.3; then, if we denote by N the norm in these quadratic extensions, we have:

$$\mathrm{N}(\sqrt{-\ell}\,) = \ell, \ \mathrm{N}(\sqrt{-\ell\zeta}\,) = \ell\zeta.$$

Thus, it is more natural to write the three quadratic extensions of $\mathbb{Q}_\ell$ in the form:

$$\mathbb{Q}_\ell(\sqrt{\zeta}\,), \ \mathbb{Q}_\ell(\sqrt{-\ell}\,), \ \mathbb{Q}_\ell(\sqrt{-\ell\zeta}\,),$$

in which case they correspond respectively to N_1, N_2, N_3 (conductors (1), (ℓ), and (ℓ) using 1.6). We then have:

$$\mathbb{Q}_\ell(\sqrt{-\ell}\,) = \mathbb{Q}_\ell(\sqrt{\ell}\,) \ \text{ and } \ \mathbb{Q}_\ell(\sqrt{-\ell\zeta}\,) = \mathbb{Q}_\ell(\sqrt{\ell\zeta}\,),$$

if and only if $\ell \equiv 1 \bmod (4)$ (otherwise the extensions on the right hand sides are permuted).

For $\ell = 2$, the norm groups are the following:

$$\begin{aligned} N_1 &:= \langle 4 \rangle \oplus \langle -1 \rangle \oplus \langle 5 \rangle_{\mathbb{Z}_2}, \\ N_2 &:= \langle 2 \rangle \oplus \langle 5 \rangle_{\mathbb{Z}_2}, \\ N_3 &:= \langle -2 \rangle \oplus \langle 5 \rangle_{\mathbb{Z}_2}, \\ N_4 &:= \langle 2 \rangle \oplus \langle -1 \rangle \oplus \langle 5^2 \rangle_{\mathbb{Z}_2}, \\ N_5 &:= \langle 2 \rangle \oplus \langle -5 \rangle_{\mathbb{Z}_2}, \\ N_6 &:= \langle 2 \times 5 \rangle \oplus \langle -1 \rangle \oplus \langle 5^2 \rangle_{\mathbb{Z}_2}, \\ N_7 &:= \langle -2 \rangle \oplus \langle -5 \rangle_{\mathbb{Z}_2}. \end{aligned}$$

The unramified extension is $\mathbb{Q}_2(\sqrt{5}\,)$ and corresponds to N_1 (we can also note that $\mathrm{N}(2+\sqrt{5}\,) = -1$ and $\mathrm{N}(5+2\sqrt{5}\,) = 5$). We then have the following computations:

$$\begin{aligned} &\mathrm{N}(1+\sqrt{-1}\,) = 2, && \mathrm{N}(2+\sqrt{-1}\,) = 5, \\ &\mathrm{N}(\sqrt{-5}\,) = 5, && \mathrm{N}(3+\sqrt{-5}\,) \in -2\mathbb{Q}_2^{\times 2}, \\ &\mathrm{N}(\sqrt{2}\,) = -2, && \mathrm{N}(1+\sqrt{2}\,) = -1, \\ &\mathrm{N}(\sqrt{-2}\,) = 2, && \mathrm{N}(1+\sqrt{-2}\,) \in -5\mathbb{Q}_2^{\times 2}, \\ &\mathrm{N}(\sqrt{10}\,) = -10, && \mathrm{N}(3+\sqrt{10}\,) = -1, \\ &\mathrm{N}(\sqrt{-10}\,) = 10, && \mathrm{N}(2+\sqrt{-10}\,) \in -2\mathbb{Q}_2^{\times 2}; \end{aligned}$$

they show that the (ramified) extensions $\mathbb{Q}_2(\sqrt{-1})$, $\mathbb{Q}_2(\sqrt{-5})$, $\mathbb{Q}_2(\sqrt{2})$, $\mathbb{Q}_2(\sqrt{-2})$, $\mathbb{Q}_2(\sqrt{10})$, and $\mathbb{Q}_2(\sqrt{-10})$ correspond respectively to N_2, N_3, N_4, N_5, N_6, N_7 and have as conductors (4), (4), (8), (8), (8), (8). □

1.6.6 Exercise. Let v be a finite place of K and let n be a nonzero integer. Show that $K_v^\times/K_v^{\times n}$ is finite.

Assume that K_v contains the group μ_n of nth roots of unity; compute the norm group N corresponding to $M := K_v(\sqrt[n]{K_v^\times})$.

Answer. Let p be a prime number and let p^e be the largest power of p dividing n; it is sufficient to show that the p-torsion group $K_v^\times/K_v^{\times p^e}$ is finite. By I.3.1.1, we have $K_v^\times \simeq \mathbb{Z} \oplus \mu_{q_v-1} \oplus \mu_\ell(K_v) \oplus \mathbb{Z}_\ell^{[K_v\,:\,\mathbb{Q}_\ell]}$, where ℓ is the residue characteristic of v, and the result follows.

Classical Kummer theory says that M is the maximal abelian extension of exponent n of K_v; but the quotient $K_v^\times/N$ is maximal of exponent n if and only if $N = K_v^{\times n}$.

When $\mu_n \subset K_v^\times$, for each p dividing n we have more precisely:

$$K_v^\times/K_v^{\times p^e} \simeq (\mathbb{Z}/p^e\mathbb{Z})^2 \quad (\text{resp. } (\mathbb{Z}/p^e\mathbb{Z})^{[K_v\,:\,\mathbb{Q}_\ell]+2})$$

if $\ell \neq p$ (resp. $\ell = p$). Without the Kummer hypothesis the norm group $K_v^{\times n}$ still corresponds to the maximal abelian extension of exponent n of K_v (which is not a Kummer extension and cannot be generated by radicals) and the structure of its Galois group is modified (in an explicit way). □

1.6.7 Remark (local Hilbert symbols). One might think that in the Kummer case ($\mu_n \subset K_v$, $M := K_v(\sqrt[n]{K_v^\times})$) the symbol $(\bullet\,,\, M/K_v)$ is "easy"; as the long search for explicit formulas shows, this is not the case. If we set for all $x,\ y \in K_v^\times$:

$$(y\,,\, M/K_v)(\sqrt[n]{x}) = (y\,,\, K_v(\sqrt[n]{x})/K_v)(\sqrt[n]{x}) =: \ (x\,,\, y)_v\, \sqrt[n]{x},$$

we thus define the local Hilbert symbol of order n:[11]

$$(\bullet\,,\,\bullet)_v : K_v^\times \times K_v^\times \longrightarrow \mu_n$$

whose knowledge is equivalent to that of the norm residue symbol (we will study it in Section 7). □

Note. Since the 1928 original papers of Artin–Hasse, a very large number of contributions (Mills, Hasse, Kneser, Šafarevič, Shiratani, Brückner, Iwasawa, Vostokov, Wiles, Henniart, Sen, Kolyvagin, Coleman, de Shalit, Miki, Jaulent, ...) have given explicit formulas for the local Hilbert symbol and reciprocity laws; these techniques, closely related to the theory of formal groups that we have already mentioned

[11] [a, Se1; D, Ch. IV], [d, AT, Ch. 12; Se2, Ch. XIV], [e, Ko3, Ch. 2, § 1].

(Lubin–Tate (1965)) [f, Lang2, Ch. 9] are outside the setting studied here. In fact, from a theoretical point of view, all these laws can be expressed in the unified setting of p-adic Galois representations, developed in particular by Fontaine, Messing, ... In this setting, one can say that all the known results on the local Hilbert symbol are contained in the reciprocity law of Bloch and Kato, conjecturally generalized by Perrin-Riou, and independently proved by Benois, Colmez, Kato–Kurihara–Tsuji, ...

1.6.8 Remarks. (i) However, in the regular case, also called by abuse of language the tame case (i.e., when the residue characteristic ℓ of v does not divide n), n is then a divisor of $q_v - 1$, and we have the following simple formula (proved in 7.1.5) for the Hilbert symbol of order n:

$$(x\,,\,y)_v \equiv \left((-1)^{\mathrm{v}(x)\mathrm{v}(y)}\,\frac{x^{\mathrm{v}(y)}}{y^{\mathrm{v}(x)}}\right)^{\frac{q_v-1}{n}} \mod (\pi_v)\ ;$$

this indeed pinpoints $(x\,,\,y)_v \in \mu_n(K_v)$ since the residue map:

$$\mu_n(K_v) \longrightarrow \mu_n(F_v) = (F_v^{\times})^{\frac{q_v-1}{n}}$$

is an isomorphism. When v is a real place at infinity ($K_v = \mathbb{R}$ and $n = 2$), $(x\,,\,y)_v$ is given by the sign of the analogous expression:

$$(x\,,\,y)_v = \mathrm{sgn}((-1)^{\mathrm{v}(x)\mathrm{v}(y)} x^{\mathrm{v}(y)} y^{-\mathrm{v}(x)}) = (-1)^{\mathrm{v}(x)\mathrm{v}(y)}.$$

(ii) In the absolute quadratic case, Exercise 1.6.5 gives the answer in complete generality.

(iii) Finally, we will see, perhaps surprisingly, that to compute explicitly a local Hilbert symbol in the irregular (or wild) case, we can always proceed globally, without knowing any explicit formula; this will be explained together with the statements of global class field theory (see 7.5). □

We will devote Section 7 of this chapter to the more general notion of symbol and their properties; we will see that Hilbert symbols play an important role.

d) Infinite Local Class Field Theory

We will conclude by showing that finite local class field theory contains all information concerning the structure of the abelian closure $\overline{K}_v^{\mathrm{ab}}$ of K_v, for $v \in Pl_0$, and the class field theory correspondence.

1.7 Limiting Procedure. By infinite Galois theory and the local class field theory correspondence, we can relate the topological groups:

$$\mathrm{Gal}(\overline{K}_v^{\mathrm{ab}}/K_v) \simeq \varprojlim_M \mathrm{Gal}(M/K_v),$$

for the set of finite abelian extensions M of K_v, to:

$$\varprojlim_{N} (K_v^\times/N),$$

where N ranges in the set of subgroups of finite index of $K_v^\times$, and by definition we obtain the profinite completion $\widehat{K_v^\times} := \varprojlim_{N} (K_v^\times/N)$ of $K_v^\times$. It is easily checked (see 1.6.6) that the subgroups $K_v^{\times n}$ for $n > 0$ form a cofinal subset of the set of subgroups N of finite index, hence that:

$$\widehat{K_v^\times} = \varprojlim_{n\geq 1} (K_v^\times/K_v^{\times n})\ ;$$

since U_v is a profinite group, we immediately obtain (choosing a uniformizer π_v):

$$\widehat{K_v^\times} = \pi_v^{\widehat{\mathbb{Z}}} \bigoplus U_v,$$

where by abuse of notation we have set:

$$\pi_v^{\widehat{\mathbb{Z}}} := \varprojlim_{n\geq 1} (\langle\, \pi_v \,\rangle/\langle\, \pi_v \,\rangle^n) \simeq \varprojlim_{n\geq 1} (\mathbb{Z}/n\mathbb{Z}) = \widehat{\mathbb{Z}},$$

which is legitimate since $\langle\, \pi_v \,\rangle$ has no $\mathbb{Z}$-torsion. Recall that if ℓ is the residue characteristic and q_v the order of the residue field of v, we have:

$$U_v \simeq \mu_{q_v-1} \bigoplus \mu_\ell(K_v) \bigoplus \mathbb{Z}_\ell^{[K_v\,:\,\mathbb{Q}_\ell]}.$$

A fundamental system of neighbourhoods of 1 in the profinite group $\widehat{K_v^\times}$ is given by the $(\widehat{K_v^\times})^n$ for $n > 0$, or by the $\pi_v^{n\widehat{\mathbb{Z}}} \oplus U_v^i,\ \ n > 0,\ i \geq 0$.

More precisely, properties 1.4, (i), (ii) of the norm residue symbol imply the existence of an isomorphism of inverse systems giving the homeomorphism $\rho_v : \widehat{K_v^\times} \longrightarrow \overline{G}_v^{\,\mathrm{ab}}$ and showing that there exists an analog to Theorem 1.5 on the correspondence of infinite local class field theory, replacing $K_v^\times$ by $\widehat{K_v^\times}$ and the notion of subgroup of finite index (of $K_v^\times$) by that of *closed* subgroup of $\widehat{K_v^\times}$.

Let us describe this correspondence in a little more detail. Let M be a finite abelian extension of K_v with norm group $N := \mathrm{N}_{M/K_v}(M^\times)$, and consider the local reciprocity exact sequence:

$$1 \longrightarrow N \xrightarrow{\qquad\quad} K_v^\times \xrightarrow{(\,\bullet\,, M/K_v)} \mathrm{Gal}(M/K_v) \longrightarrow 1.$$

The norm residue symbol $(\,\bullet\,,\ M/K_v)$ is still continuous for the topology of $K_v^\times$, diagonally embedded in $\widehat{K_v^\times}$, induced by that of $\widehat{K_v^\times}$ as a profinite group (neighbourhoods in $K_v^\times$: the $K_v^{\times n}$, $n > 0$); thus extending by continuity we obtain the exact sequence:

$$1 \longrightarrow \mathrm{adh}(N) \longrightarrow \widehat{K_v^\times} \xrightarrow{(\bullet\,, M/K_v)} \mathrm{Gal}(M/K_v) \longrightarrow 1,$$

where adh denotes closure in $\widehat{K_v^\times}$ for its topology, and the norm group which now corresponds to M is:

$$\mathrm{adh}(N) := \bigcap_{n>0} (N\,.\,(\widehat{K_v^\times})^n).$$

This defines in an evident way the local reciprocity map (or norm residue symbol):

$$(\bullet\,,\, M/K_v) : \widehat{K_v^\times} \longrightarrow \mathrm{Gal}(M/K_v),$$

for any abelian extension M (finite or not); it is also the composition of ρ_v and of the projection $\overline{G}_v^{\mathrm{ab}} \longrightarrow \mathrm{Gal}(M/K_v)$.

One checks, from the finite case, that the image of U_v (compact) under $(\bullet\,,\, M/K_v)$ is the inertia group. To summarize:

1.7.1 Theorem. *There exists a homeomorphism of profinite groups (the infinite local reciprocity map):*

$$\rho_v =: (\bullet\,,\, \overline{K}_v/K_v) \;:\; \widehat{K_v^\times} \longrightarrow \overline{G}_v^{\mathrm{ab}} := \mathrm{Gal}(\overline{K}_v^{\mathrm{ab}}/K_v),$$

whose composition with the projection $\overline{G}_v^{\mathrm{ab}} \longrightarrow \mathrm{Gal}(M/K_v)$ *is equal to* $(\bullet\,,\, M/K_v)$ *for any abelian extension* M/K_v.

The inertia group $\mathrm{Gal}(\overline{K}_v^{\mathrm{ab}}/\overline{K}_v^{\mathrm{nr}})$ *is the image of* U_v *under* ρ_v, *and the higher ramification groups (in upper numbering) correspond to the* U_v^i, $i \geq 1$.

The image of π_v *under* ρ_v *is a (noncanonical) extension of the Frobenius automorphism of* $\overline{K}_v^{\mathrm{nr}}/K_v$.

Finally, there exists a bijective correspondence, between the set of abelian extensions of K_v *and the set of closed subgroups of* $\widehat{K_v^\times}$, *which satisfies the Galois properties (i) to (iv) of 1.5.* □

1.8 Norm Groups in Infinite Local Class Field Theory. Note that if M/K_v is infinite, the notation $\mathrm{N}_{M/K_v}(M^\times)$ does not make any sense directly, but since:

$$\mathrm{Gal}(M/K_v) = \varprojlim_{M'} \mathrm{Gal}(M'/K_v),$$

for $K_v \subseteq M' \subseteq M$, M'/K_v finite with norm group N', we have:

$$\mathrm{Gal}(M/K_v) = \varprojlim_{M'} \widehat{K_v^\times}/\mathrm{adh}(N') \simeq \widehat{K_v^\times} \Big/ \bigcap_{M'} \mathrm{adh}(N')$$

(by I.5.5, applied to $A = \widehat{K_v^\times}$ compact and $B = 1$), so that the norm group corresponding to M in $\widehat{K_v^\times}$ can be written:

$$\bigcap_{\substack{M' \subseteq M \\ M'/K_v \text{ finite}}} \mathrm{adh}(\mathrm{N}_{M'/K_v}(M'^\times)).$$

Note also that the usual (locally compact) topology of $K_v^\times$ is absolutely not used here, and is not induced by that of $\widehat{K_v^\times}$ (which is compact); in particular, U_v is not open in $\widehat{K_v^\times}$ since it is not of finite index.

If M (finite or not) corresponds to the norm group N, then M^{nr} still corresponds to the group $U_v\, N$, and its maximal tamely ramified subextension corresponds to $U_v^1\, N$.

We thus easily obtain the structure of the group $\overline{G}_v^{\mathrm{ab}}$ since that of $\widehat{K_v^\times} \simeq \widehat{\mathbb{Z}} \oplus U_v$ is known; we deduce a number of consequences, such as the following result.

1.8.1 Proposition. *Let $v \in Pl_0$. The extension $\overline{K}_v^{\mathrm{ab}}$ is the direct compositum over K_v of $\overline{K}_v^{\mathrm{nr}}$ and of a (nonunique) maximal totally ramified abelian extension of K_v, the extension $\overline{K}_v^{\mathrm{nr}}$ being fixed by the image of U_v under the local reciprocity map, while the maximal totally ramified extension is fixed by that of the subgroup $\pi_v^{\widehat{\mathbb{Z}}}$.* □

1.8.2 Remark. If we want to limit ourselves to the maximal pro-p-subex tension $\overline{K}_v^{\mathrm{ab}}(p)$ of $\overline{K}_v^{\mathrm{ab}}$, p prime, $v \in Pl_0$ of residue characteristic equal to ℓ, we simply note that in terms of p-Sylow subgroups we have:

(i) for $p \neq \ell$, $(\widehat{K_v^\times})_p \simeq \mathbb{Z}_p \oplus (\mu_{q_v - 1})_p$, where $(\mu_{q_v-1})_p \simeq (F_v^\times)_p$ corresponds to the inertia group, giving the following diagram:

$$\begin{array}{ccc} \overline{K}_v^{\mathrm{nr}}(p) & \overset{(\mu_{q_v-1})_p}{\text{―――――}} & \overline{K}_v^{\mathrm{ab}}(p) \\ \Big| & & \Big|\,\mathbb{Z}_p \\ K_v & \text{―――――} & M \end{array}$$

(ii) for $p = \ell$, $(\widehat{K_v^\times})_p \simeq \mathbb{Z}_p \oplus U_v^1$, which corresponds to the following analogous diagram with an inertia group which is here isomorphic to $U_v^1 \simeq \mu_p(K_v) \oplus \mathbb{Z}_p^{[K_v\,:\,\mathbb{Q}_p]}$:

$$\begin{array}{ccc} \overline{K}_v^{\mathrm{nr}}(p) & \overset{U_v^1}{\text{―――――}} & \overline{K}_v^{\mathrm{ab}}(p) \\ \Big| & & \Big|\,\mathbb{Z}_p \\ K_v & \text{―――――} & M \end{array}$$

In these two diagrams, the (nonunique) field M defines a maximal totally ramified abelian pro-p-extension of K_v, finite in the case $p \neq \ell$, containing $[K_v : \mathbb{Q}_p]$ independent totally ramified $\mathbb{Z}_p$-extensions in the case $p = \ell$.

In case (i), if p^h is the p-part of $q_v - 1$, K_v contains the group μ_{p^h}, in which case Kummer theory shows that we can choose $M = K_v(\sqrt[p^h]{-\pi_v})$. □

After treating the global case (Ch. III, § 4, (c), (d)), it will be useful to compare the structures of $\overline{K}_v^{\,\mathrm{ab}}/K_v$ and of $\overline{K}^{\mathrm{ab}}/K$, for instance by checking that for each place v, the decomposition group of v in $\overline{K}^{\mathrm{ab}}/K$ does give a quotient of the Galois group of the abelian closure of K_v. In fact we will obtain the much stronger result that the trivial inclusion $(\overline{K}^{\mathrm{ab}})_v \subseteq \overline{K}_v^{\,\mathrm{ab}}$ is an equality (Theorem III.4.5, in the direction of the Grunwald–Wang theorem).

1.8.3 Exercise (the case of $\overline{\mathbb{Q}}_\ell^{\,\mathrm{ab}}$). Assume that $K = \mathbb{Q}$ and that v is finite; we have $K_v = \mathbb{Q}_\ell$, where ℓ is the corresponding residue characteristic. We thus have $\mathrm{Gal}(\overline{\mathbb{Q}}_\ell^{\,\mathrm{ab}}/\mathbb{Q}_\ell) \simeq \ell^{\widehat{\mathbb{Z}}} \oplus \mathbb{Z}_\ell^\times$.

(i) Show that $\overline{\mathbb{Q}}_\ell^{\,\mathrm{nr}} = \mathbb{Q}_\ell(\mu')$, where μ' is the group of roots of unity of order prime to ℓ, and that the field M fixed under the image of $\ell^{\widehat{\mathbb{Z}}}$ is equal to $\mathbb{Q}_\ell(\mu_{\ell^\infty})$.

(ii) Assume that $\ell \neq 2$. Since $\mathbb{Q}_\ell$ contains a primitive $(\ell - 1)$th root of unity, it is clear that the extension $\mathbb{Q}_\ell(\sqrt[\ell-1]{-\ell})/\mathbb{Q}_\ell$ is abelian. Show that it is equal to $\mathbb{Q}_\ell(\mu_\ell)$, and deduce that there exists in $\mathbb{Q}_\ell(\mu_\ell)$ a uniformizer π (called Dwork's uniformizer) such that:

$$\pi^{\ell-1} = -\ell.$$

Answer. (i) The elementary theory of cyclotomic fields over $\mathbb{Q}$ shows that $\mathbb{Q}_\ell(\mu')/\mathbb{Q}_\ell$ is unramified and that $\mathbb{Q}_\ell(\mu_{\ell^\infty})/\mathbb{Q}_\ell$ is totally ramified. Hence we already have that $\mathbb{Q}_\ell(\mu') \subseteq \overline{\mathbb{Q}}_\ell^{\,\mathrm{nr}}$. If $n \geq 1$ is some integer, we know that the field $\mathbb{Q}_\ell(\mu_{\ell^n-1})$ has degree n (the Frobenius is of order n), which defines the unique unramified extension of degree n of $\mathbb{Q}_\ell$, proving the first result of (i).

The norm group of the field M is $\ell^{\widehat{\mathbb{Z}}}$ and we have $\mathrm{Gal}(M/\mathbb{Q}_\ell) \simeq U_v = \mathbb{Z}_\ell^\times$. Using the cyclotomic polynomials Φ_m, we see that for all $t \geq 1$, $\ell = \Phi_{\ell^t}(1)$ is the norm of $1 - \zeta_t$ in $\mathbb{Q}_\ell(\mu_{\ell^t})/\mathbb{Q}_\ell$, where ζ_t generates μ_{ℓ^t}. Thus $\mathbb{Q}_\ell(\mu_{\ell^t}) \subset M$. Let N_t be the norm group of $\mathbb{Q}_\ell(\mu_{\ell^t})$. Since $\mathbb{Q}_\ell(\mu_{\ell^t})/\mathbb{Q}_\ell$ is totally ramified of degree $\ell^{t-1}(\ell - 1)$, we have $N_t = \ell^{\mathbb{Z}} \oplus V$ with V of index $\ell^{t-1}(\ell - 1)$ in U_ℓ. If $\ell \neq 2$, the only possibility is $V = 1 + \ell^t\mathbb{Z}_\ell$; if $\ell = 2$ and $t \geq 2$, we have $\mathrm{Gal}(L_\ell/\mathbb{Q}_\ell) \simeq \mathbb{Z}/2\mathbb{Z} \times \mathbb{Z}/2^{t-2}\mathbb{Z}$ and the only possibility is $V = 1 + 4.2^{t-2}\mathbb{Z}_2$ since V must be contained in the norm group of $\mathbb{Q}(\mu_4)$ which is equal to $2^{\mathbb{Z}} \oplus (1 + 4\mathbb{Z}_2)$ (using 1.6.5). Thus in all cases we have $N_t = \ell^{\mathbb{Z}} \oplus U_v^t$ and, by 1.8, the norm group of $\mathbb{Q}_\ell(\mu_{\ell^\infty})$ is equal to $\bigcap_t \mathrm{adh}(\ell^{\mathbb{Z}} \oplus U_v^t) = \ell^{\widehat{\mathbb{Z}}}$ proving that $M = \mathbb{Q}_\ell(\mu_{\ell^\infty})$.

Thus, here we have $\overline{\mathbb{Q}}_\ell^{\,ab} = \mathbb{Q}_\ell(\mu)$, the field generated by all the roots of unity.

(ii) The norm of $\sqrt[\ell-1]{-\ell}$ in $\mathbb{Q}_\ell(\sqrt[\ell-1]{-\ell})/\mathbb{Q}_\ell$ is equal to ℓ since:

$$\mathrm{Irr}(\sqrt[\ell-1]{-\ell}, \mathbb{Q}_\ell) = X^{\ell-1} + \ell \; ;$$

the norm group of $\mathbb{Q}_\ell(\sqrt[\ell-1]{-\ell})$ thus contains that of M, hence $\mathbb{Q}_\ell(\sqrt[\ell-1]{-\ell}) \subset M$, whence the equality $\mathbb{Q}_\ell(\sqrt[\ell-1]{-\ell}) = \mathbb{Q}_\ell(\mu_\ell)$ (note that $\mathbb{Q}_\ell(\sqrt[\ell-1]{\ell})$ is also totally ramified and abelian over $\mathbb{Q}_\ell$, but is not contained in M). The conclusion is clear.

Note that $1 - \zeta_1$ is also a uniformizer, hence $\frac{(1-\zeta_1)^{\ell-1}}{-\ell}$ is the $(\ell - 1)$th power of a unit of $\mathbb{Q}_\ell(\mu_\ell)$. □

In the case where $K = \mathbb{Q}$, we will be able to compute by global means the local norm residue symbol for abelian extensions of the completions of $\mathbb{Q}$ (see Exercise 3.4.3).

1.9 Exercise (Abhyankar's lemma). Let M_1 and M_2 be finite extensions of a nonarchimedean local field k. Assume that M_2/k is tamely ramified (i.e., $e(M_2/k)$ is not divisible by the residue characteristic of k) and that $e(M_2/k)$ divides $e(M_1/k)$. Show that $M_1 M_2/M_1$ is unramified.

Answer. See [Cor1, Th. 3] for the use of this result, and more generally [d, Lang1, Ch. II, § 5] for the study of not necessarily Galois tamely ramified extensions. □

§2 Idèle Groups in an Extension L/K

Let L/K be a finite extension of number fields. We use the local notations of Section 1 (in particular of 1.1); if L/K (resp. L_w/K_v for $w \in Pl_{L,v}$) is Galois, we set $G := \mathrm{Gal}(L/K)$ (resp. $G_w := \mathrm{Gal}(L_w/K_v)$) and we introduce the decomposition group D_w of w in L/K.

a) Canonical Injection of C_K in C_L

Let J_K and J_L be the respective idèle groups of K and L. For $v \in Pl_K$, recall the relations between the different embeddings of K and L in the corresponding components $K_v^\times$ and $\bigoplus_{w|v} L_w^\times$ of J_K and J_L. The embedding:

$$i_v \; : \; K \longrightarrow K_v$$

comes from the choice of a conjugate K_v of the v-completion of K; for all $w \,|\, v$, L_w is defined in a similar way as an extension of K_v; the embedding:

$$i_w : L \longrightarrow L_w$$

is then an extension of i_v, such that the family $(i_w)_{w|v}$ is a complete set of representatives of the classes of $\mathbb{Q}$-embeddings of L in $\overline{K}_v$ extending i_v.

It is convenient to consider J_K as a subgroup of J_L using the diagonal embedding $j_{L/K} : J_K \longrightarrow J_L$ for which the image of $\boldsymbol{x} =: (x_v)_v \in J_K$ is given by $(x_w)_w$, with $x_w = x_v$ for all $w \,|\, v$. This map is injective. Similarly:

2.1 Proposition. *The canonical map:*

$$j_{L/K} : C_K \longrightarrow C_L,$$

induced by $J_K \longrightarrow J_L$, is injective.

Proof. Let $\boldsymbol{x} =: (x_v)_v \in J_K$ be an idèle such that $j_{L/K}(\boldsymbol{x}) = i_L(y)$ for $y \in L^\times$, and let v be a fixed place of K; we thus have:

$$i_w(y) = x_v \ \text{ for all } w|v.$$

This implies that all the K-conjugates of y are equal (seen in $\overline{K} \subset \overline{K}_v$, these conjugates are the $\tau \circ i_w(y) = \tau(x_v)$ for the $w|v$ and the K_v-isomorphisms τ of L_w); thus there is only one, so $y \in K^\times$, proving the result. □

2.1.1 Remarks. (i) This property leads to a simple definition of the idèle class group of an infinite algebraic extension L/K by taking the direct limit of the $C_{L'}$ for $L' \subset L$, L'/K finite.

(ii) We will see later that the corresponding map $C_K/D_K \longrightarrow C_L/D_L$, which class field theory identifies with the transfer map $\overline{G}_K^{\text{ab}} \longrightarrow \overline{G}_L^{\text{ab}}$ (for L/K finite), has a kernel isomorphic to $(\mathbb{Z}/2\mathbb{Z})^{r_1^{\text{c}}}$, where r_1^{c} is the number of real places of K totally complexified in L/K. □

b) Relations Between Local and Global Norms

Let L/K be an arbitrary finite extension, $\mathrm{N}_{L/K}$ the norm in L/K, and fix a place v of K. For $y \in L^\times$, we have, giving in detail the computations:

$$i_v(\mathrm{N}_{L/K}(y)) = \prod_{\sigma} \sigma(y)$$

(where σ ranges in the set of $\mathbb{Q}$-embeddings of L which extend i_v)

$$= \prod_{w|v} \prod_{\tau} \tau \circ i_w(y)$$

(where τ, which depends on w, ranges in the set of K_v-isomorphisms of L_w)

$$= \prod_{w|v} \mathrm{N}_{L_w/K_v}(i_w(y)),$$

which can be summarized as follows.

2.2 Proposition. *For any place v of K we have:*

$$i_v(\mathrm{N}_{L/K}(y)) = \prod_{w|v} \mathrm{N}_{L_w/K_v}(i_w(y)) \quad \text{for all } y \in L^\times. \qquad \square$$

By abuse of notation, this formula is in general written:

$$\mathrm{N}_{L/K}(y) = \prod_{w|v} \mathrm{N}_{L_w/K_v}(y),$$

by saying that, for each place v of K, "the global norm of y is equal to the product of its local norms above v". This can be reinterpreted as the commutativity of the following diagram.

$$\begin{array}{ccc} L^\times & \xrightarrow{\;\bigoplus_{w|v} i_w\;} & \bigoplus_{w|v} L_w^\times \\ \Big\downarrow{\scriptstyle \mathrm{N}_{L/K}} & & \Big\downarrow{\scriptstyle \prod_{w|v} \mathrm{N}_{L_w/K_v}} \\ K^\times & \xrightarrow{\;i_v\;} & K_v^\times \end{array}$$

Fig. 2.1

From this we obtain a canonical definition of the norm in L/K of an idèle $\boldsymbol{y} \in J_L$.

2.2.1 Definition. Let $\boldsymbol{y} =: (y_w)_w \in J_L$, we set:

$$\mathrm{N}_{L/K}(\boldsymbol{y}) := \Big(\prod_{w|v} \mathrm{N}_{L_w/K_v}(y_w)\Big)_v. \qquad \square$$

This norm map indeed extends that defined on $L^\times$ thanks to the above commutative diagram. Taking quotients, we also define:

$$\mathrm{N}_{L/K} \;:\; C_L = J_L/L^\times \longrightarrow C_K = J_K/K^\times.$$

c) Galois Structure of J_L: Semi-Local Theory

When L/K is Galois with Galois group G, it is necessary to put on J_L a G-module structure compatible (algebraically and topologically) with that of the diagonal embedding of $L^\times$ in J_L. For this, it is sufficient to define explicitly the operation of G on the semi-local factor (seen as a K_v-algebra):

$$\bigoplus_{w|v} L_w, \quad v \in Pl_K,$$

operation which we will then restrict to $\bigoplus_{w|v} L_w^\times$. Thus, it must be such that the diagonal embedding:

$$(i_w)_{w|v} : L \longrightarrow \bigoplus_{w|v} L_w$$

is a G-module homomorphism, is continuous for the v-topology of a K-algebra on L, i.e., $L \simeq K^{[L:K]}$ with the product of the topologies induced by $|\bullet|_v$ on K. Thus, by density of $((i_w)_{w|v})(L)$ in $\bigoplus_{w|v} L_w$ (chinese remainder Theorem I.4.3), this defines it uniquely. From this remark we can give the following more precise algorithmic proof. For another direct proof, see 2.3.4, (i).

2.3 Existence and Definition of the Galois Action. Let $w_0 \in Pl_{L,v}$ fixed. Sometimes, by abuse of notation, we consider L_{w_0} as a subspace of $\bigoplus_{w|v} L_w$.[12] Therefore, it will be sufficient to know the action of G on such a subspace L_{w_0}. Let V_{w_0} be a neighbourhood of 0 in L for w_0; for each $s \in G$, sV_{w_0} is an analogous neighbourhood for sw_0, which we can denote V_{sw_0}; this defines V_w for each $w|v$ since G acts transitively on $Pl_{L,v}$. The approximation theorem means that $i_{w_0}\Big(\bigcap_{w \neq w_0} V_w\Big)$ is dense in the field L_{w_0} for every V_{w_0}, and that the closure of $((i_w)_{w|v})\Big(\bigcap_{w \neq w_0} V_w\Big)$ in $\bigoplus_{w|v} L_w$ is of the form $L_{w_0} \oplus V$, where V is a neighbourhood of 0 in $\bigoplus_{w \neq w_0} L_w$.

Note. When v is finite we can for instance take $\bigcap_{w \neq w_0} V_w = \prod_{w \neq w_0} \mathfrak{p}_w^m$ for m as large as we like, hence $V = \bigoplus_{w \neq w_0} (\pi_w)^m$ since $i_w(\mathfrak{p}_w) = (\pi_w)$, where $\pi_w = \pi$ is a suitable element (independent of $w|v$) of the maximal ideal of C_ℓ.

Furthermore, if $s \in G$ we have $s\Big(\bigcap_{w \neq w_0} V_w\Big) = \bigcap_{w \neq w_0} (sV_w) = \bigcap_{w \neq sw_0} V_w$, which, by going to the limit, easily gives the definition of the action of G which in particular is such that (in terms of subspaces):

[12] It is important to distinguish between the two sets since an approximation of $y \in L_{w_0}$ by an element of $L^\times$ may be very different from an approximation of $(y, 0, \dots, 0)$, but the context will be clear.

$$s.L_{w_0} = L_{sw_0} \text{ for all } s \in G.$$

Hence, for all $s \in G$ we obtain a continuous K_v-isomorphism s_{w_0} depending on w_0, still denoted s by abuse of notation:

$$s : L_{w_0} \longrightarrow L_{sw_0},$$

which defines an element of $G_{w_0} = \mathrm{Gal}(L_{w_0}/K_v)$ if and only if $s \in D_{w_0}$ (in this case, we recover the canonical isomorphism $D_{w_0} \simeq G_{w_0}$ of I.2.5).

2.3.1 Exercise. (i) Check that for all $y_0 \in L_{w_0}$:

$$s\,(y_0, 0, \ldots, 0) = (0, \ldots, \tau(y_0), \ldots, 0)$$

(as element of the subspace L_{sw_0}), where $\tau \in \mathrm{Gal}(L_{sw_0}/K_v)$ is the extension by continuity of:

$$i_{sw_0} \circ s \circ i_{w_0}^{-1} \text{ on } i_{w_0}(L)$$

(i.e., if $z \in L$ is such that $i_{w_0}(z)$ is an approximation of y_0 in the field L_{w_0}, then $\tau(i_{w_0}(z)) = i_{sw_0}(s(z))$ is an approximation of $s(y_0)$ in L_{sw_0}).

(ii) Apply this (for $K = \mathbb{Q}$) to the fields $L = \mathbb{Q}(\sqrt{2}, \sqrt{5})$ and $L = \mathbb{Q}(\sqrt{2+\sqrt{2}})$ for the residue characteristic $\ell = 7$, and compute the action of G on $(\sqrt{3}, 0)$ and $(\sqrt{2}, 0)$ in each case.

(iii) Check the formula $s_{s'w_0} \circ s'_{w_0} = (ss')_{w_0}$, for any $s,\ s' \in G$.

Answer. Let $y \in L^{\times}$ such that:

$$y \equiv z \bmod \mathfrak{p}_{w_0}^n,$$
$$y \equiv 0 \bmod \mathfrak{p}_w^n \ \ \forall w \neq w_0,$$

where $i_{w_0}(z) \equiv y_0 \bmod \pi_{w_0}^n$ in L_{w_0}; thus we have:

$$s(y) \equiv s(z) \bmod \mathfrak{p}_{sw_0}^n,$$
$$s(y) \equiv 0 \bmod \mathfrak{p}_{w'}^n \ \ \forall w' \neq sw_0.$$

Since the embedding of y is an approximation of $(y_0, 0, \ldots, 0)$, an approximation of $s(y_0, 0, \ldots, 0)$ is given by the embedding of $s(y)$, which clearly is close to $(0, \ldots, i_{sw_0}(s(z)), \ldots, 0)$.

Points (ii) and (iii) are left to the reader. □

We deduce from all the above the following explicit result (semi-local theory) stated in terms of representations.

2.3.2 Theorem. *Let L/K be Galois with Galois group G. For any place v of K, the K_v-representation $\bigoplus_{w|v} L_w$ of G is induced by the representation of* *the decomposition group* D_{w_0} *of* $w_0|v$ *defined by* L_{w_0}*. Thus it is the regular representation of* G*.*

Proof. Since G acts transitively on $Pl_{L,v}$, we have $\bigoplus_{w|v} L_w = \sum_{s\in G} L_{sw_0} = \bigoplus_{\overline{s}\in G/D_{w_0}} L_{\overline{s}w_0} = \bigoplus_{\overline{s}\in G/D_{w_0}} \overline{s}.L_{w_0}$, giving $\bigoplus_{w|v} L_w$ as induced representation. The representation L_{w_0}, of $D_{w_0} \simeq G_{w_0}$, is the regular one (normal basis theorem); the uniqueness of the induced representation yields the result by [Se4, § 3.3 or § 7.1]. □

2.3.3 Corollary. *The action of* G *on an* $\boldsymbol{y} =: (y_w)_{w|v} \in \bigoplus_{w|v} L_w$, *is such that* $(s.\boldsymbol{y})_{sw} = s(y_w)$ *for all* $w|v$, *where* $s : L_w \longrightarrow L_{sw}$ *is the* K_v*-isomorphism defined above.* □

2.3.4 Remarks. (i) Since we also have $\bigoplus_{w|v} L_w \simeq L \otimes_K K_v$, in this context the G-module action is defined by $s\,.\,(x \otimes a) = (s\,.\,x) \otimes a$ for all $s \in G$, $x \in L$, and $a \in K_v$, giving again 2.3.2; writing this explicitly as in (Ch. I, § 2), we would recover the above results.

(ii) Finally, if we introduce the algebraic norm $\nu_{L/K} := \sum_{s\in G} s$, we have on J_L the relation $j_{L/K} \circ \mathrm{N}_{L/K} = \nu_{L/K}$. □

Note. In the non-Galois case, we would have, on J_L, $j_{L/K} \circ \mathrm{N}_{L/K} = \sum_i \tau_i$, where the τ_i are the isomorphisms $J_L \longrightarrow J_{\sigma_i L}$ corresponding to the $[L : K]$ K-isomorphisms σ_i of L in $\mathbb{C}_\ell$ by density. In other words, on the local factor L_w, τ_i is the extension by continuity of $i_{\sigma_i w} \circ \sigma_i \circ i_w^{-1}$ on $i_w(L)$.

2.4 Proposition. *Let* L/K *be a finite Galois extension, and put* $G = \mathrm{Gal}(L/K)$*. Then* $J_L^G = j_{L/K}(J_K)$, $H^1(G, J_L) = 1$, *and* $C_L^G = j_{L/K}(C_K)$*.*

Proof. Consider the following more general situation. Let G be a *finite* group and H a subgroup of G. Let A be a G-module and B a sub-H-module of A (considered as a H-module). For $\overline{s} \in G/H$, $B_{\overline{s}} := \overline{s}.B := s.B$ does not depend on the choice of the representative $s \in \overline{s}$. Suppose that $A = \bigoplus_{\overline{s}\in G/H} B_{\overline{s}}$ (in other words, the G-module A is induced by the H-module B); then, for the usual cohomology H^r, $r \geq 0$, as well as for Tate's modified cohomology $\widehat{H}^r$, $r \in \mathbb{Z}$, we have (Shapiro's lemma):

$$H^r(G, A) \overset{\text{can}}{\simeq} H^r(H, B).$$ [13]

[13] [d, CF, Ch. VII, § 7, Prop. 7.2]; for the most general situation of Shapiro's lemma concerning the links between cohomology and representation theory, see [g, NSW, Ch. I, § 6, Prop. 1.6.3 and Rem.].

For example, this is the case for the regular representation $A = \bigoplus_{w|v} L_w$ of $G = \mathrm{Gal}(L/K)$, with $B = L_{w_0}$, $H = D_{w_0}$ for any $w_0|v$, hence for the modules $A = \bigoplus_{w|v} L_w^\times$ (or $\bigoplus_{w|v} U_w$), with $B = L_{w_0}^\times$ (or U_{w_0}) (review the definitions to see that the action of D_{w_0} on B becomes the natural one of G_{w_0} on B under the isomorphism $D_{w_0} \simeq G_{w_0}$).

We have $J_L^G = j_{L/K}(J_K)$ because of 2.3.1, 2.3.2.

Using the fact that $J_L = \varinjlim_{\Sigma} U_L^{\Sigma'}$, $\Sigma \subset Pl^{\mathrm{nc}}$ finite containing the ramified and complexified places, and the fact that the cohomology of finite groups commutes with direct limits, the proof of $H^1(G, J_L) = 1$ follows from the above, the identity $H^r\Big(G, \prod_{i\in I} A_i\Big) \simeq \prod_{i \in I} H^r(G, A_i)$ for all $r \geq 0$ (here with $A_i = \bigoplus_{w|v} L_w^\times$ or $\bigoplus_{w|v} U_w$, and $r = 1$), then from the Theorem 90 and Corollary 1.4.3 (see more details in [d, CF, Ch. VII, Prop. 7.3]).

The proof of $C_L^G = j_{L/K}(C_K)$ then uses the cohomology exact sequence $1 \to L^{\times G} \xrightarrow{\ i\ } J_L^G \longrightarrow C_L^G \longrightarrow H^1(G, L^\times) = 1.$ □

2.4.1 Remarks. Let G be a finite group, and A a G-module.

(i) We recall that $\widehat{H}^{-r-1} := \widehat{H}_r$, for $r \geq 0$, in the context of Tate's modified cohomology, and that we have:

$$\begin{aligned} H^0(G,A) &= A^G, & \widehat{H}^0(G,A) &= A^G/\nu A, \\ H_0(G,A) &= A/I_G A, & \widehat{H}_0(G,A) &= {}_\nu A/I_G A, \end{aligned}$$

where $\nu := \nu_G := \sum_{s\in G} s$, and where I_G is the augmentation ideal of G.

(ii) Recall also that we have the canonical isomorphisms:

$$\begin{aligned} \widehat{H}^r(G,A)^* &\simeq \widehat{H}^{-r-1}(G,A^*), \quad r \in \mathbb{Z}, \\ H_0(G,A)^* &\simeq H^0(G,A^*), \\ \widehat{H}^1(G,\mathbb{Q}/\mathbb{Z}) &\simeq \mathrm{Hom}\,(G,\mathbb{Q}/\mathbb{Z}) \simeq G^{\mathrm{ab}\,*}, \\ \widehat{H}_1(G,\mathbb{Z}) &\simeq I_G/I_G^2, \end{aligned}$$

where * (see I.5.7) is the usual duality (i.e., $X^* := \mathrm{Hom}(X, \mathbb{Q}/\mathbb{Z})$) [g, NSW, Ch. III, § 1, Prop. 3.1.1], and where $G^{\mathrm{ab}} := G/[G, G]$.

(iii) For instance, the case $r = 1$, $A = \mathbb{Q}/\mathbb{Z}$, gives the canonical isomorphism $I_G/I_G^2 \simeq G^{\mathrm{ab}}$. □

We would thus have all the necessary tools to start the computation of the cohomology of idèle groups and of idèle class groups, as developed by Hochschild, Nakayama, and Weil, then by Tate[14], which leads to the cohomological statement of class field theory, which is probably its most intrinsic

[14] [h, HN; We2; Che3], [d, CF, Ch. VII; Iy1, Ch. IV; Se2, Ch. XI], [e, Ko3, Ch. 2], [g, NSW, Ch. VIII, § 1]; see also the formalism developed in [f, Neu2].

form (hence the most generalizable), but which does not allow the explicit description of the arithmetic invariants which are involved (see 3.2 for some insights about these cohomological aspects).

d) Local Norm Groups — The Non-Galois Case

We come back to the situation of an arbitrary finite extension L/K, and we will lay the groundwork for a fundamental local to global principle, that which corresponds to the norm in L/K.

2.5 LOCAL NORM GROUPS — GENERAL DEFINITIONS. (i) We say that $x \in K^\times$ is a local norm at $v \in Pl$ for L/K if:

$$i_v(x) \in \mathrm{N}_{L/K}\Big(\bigoplus_{w|v} L_w^\times\Big),$$

which is equivalent to the existence of elements $y_w \in L_w^\times$ such that:

$$i_v(x) = \prod_{w|v} \mathrm{N}_{L_w/K_v}(y_w).$$ [15]

(ii) We say that $x \in K^\times$ is a local norm everywhere for L/K if x is a local norm at v for L/K for every place v.

2.5.1 Remark. It follows from 1.5.3 that x is a local norm at v for L/K if and only if:

$$i_v(x) \in \mathrm{N}_{L_v^{\mathrm{ab}}/K_v}(L_v^{\mathrm{ab}\times}),$$

with $L_v^{\mathrm{ab}} = \bigcap_{w|v} L_w^{\mathrm{ab}}$. In practice the field L_v^{ab} has in general a small degree and we can search directly whether or not $i_v(x) \in \mathrm{N}_{L_v^{\mathrm{ab}}/K_v}(L_v^{\mathrm{ab}\times})$. Of course a sufficient condition is that $i_v(x)$ must be a norm in a local extension L_{w_0}/K_v for some $w_0|v$ (which is the case if for example $L_{w_0}^{\mathrm{ab}} = K_v$).

As for the notion of v-conductor, we will also have to distinguish between the local norm group at v for L/K and the local norm group at v for L^{ab}/K, the former being the group $i_v^{-1}(\mathrm{N}_{L_v^{\mathrm{ab}}/K_v}(L_v^{\mathrm{ab}\times}))$, contained in the latter $i_v^{-1}(\mathrm{N}_{(L^{\mathrm{ab}})_v/K_v}((L^{\mathrm{ab}})_v^\times))$. □

2.5.2 Proposition. *The group of elements of $K^\times$ which are local norms everywhere for L/K is equal to:*

$$\{x \in K^\times,\ i(x) \in \mathrm{N}_{L/K}(J_L)\}.$$

Proof. One inclusion is trivial and the other comes from the fact that, apart from the places v which are ramified in all the extensions L_w/K_v for $w|v$,

[15] This formula shows, by approximation in L, that x is a local norm at v if and only if it is arbitrary close, at v, to a global norm.

and those for which $\mathrm{v}(x) \neq 0$, we have $i_v(x) \in \mathrm{N}_{L_{w_0}/K_v}(U_{w_0})$, where $w_0|v$ is unramified (see 1.4.3, (ii)), hence $i_v(x) = \mathrm{N}_{L_{w_0}/K_v}(u_0) \in \mathrm{N}_{L/K}\Big(\bigoplus_{w|v} U_w\Big)$, completing $u_0 \in U_{w_0}$ outside w_0 by components equal to 1. We can thus obtain $i(x)$ as the norm of an idèle of L. □

Because of this fact it is not necessary to give specific notations for the local norm groups and in particular the subgroups of elements of $K^\times$ which are local norms everywhere for L/K is denoted *by abuse of notation*:

$$K^\times \cap \mathrm{N}_{L/K}(J_L)$$

(instead of $i^{-1}(i(K^\times) \cap \mathrm{N}_{L/K}(J_L)))$. In the same way:

$$K^\times \cap \mathrm{N}_{L/K}\Big(\bigoplus_{w|v} L_w^\times\Big),$$

denotes the local norm group at v for L/K.

2.5.3 Remarks. (i) It is clear that any $x \in K^\times$ is a local norm almost everywhere for L/K.

(ii) More generally, we could say that x is a local norm at $w|v$ for L/K when $i_v(x) \in \mathrm{N}_{L_w/K_v}(L_w^\times)$, but in the non-Galois case this depends on the choice of w and does not have the desired meaning (it is the same problem as that of local conductors defined in 1.6 since we want to define local notions attached only to the places v of the base field K). For instance, if $K = \mathbb{Q}$ and $L = \mathbb{Q}(\sqrt[3]{2})$, -1 is a local norm at the real place w above $v = \infty$ (trivial since $L_w = K_v = \mathbb{R}$) but not at the complex place w' ($L_{w'} = \mathbb{C}$, $K_v = \mathbb{R}$); however -1 is a local norm at v, and *must be* since -1 is here a global norm:

$$-1 = \mathrm{N}_{L/\mathbb{Q}}(-1) = \mathrm{N}_{L/\mathbb{Q}}(1 - \sqrt[3]{2}) = \cdots$$

Still in $\mathbb{Q}(\sqrt[3]{2})/\mathbb{Q}$, we have a similar example with $5 = \mathrm{N}_{L/\mathbb{Q}}\Big(\frac{5}{3-\sqrt[3]{2}}\Big)$ which is a local norm at $w|5$ such that $L_w = \mathbb{Q}_5$ but not at the place w' such that $L_{w'} = \mathbb{Q}_5(j)$. In these two examples we have $L_v^{\mathrm{ab}} = K_v$.

In other words, the idea of a local norm is attached to the formula of Subsection (b):

$$\text{“}\, i_v(\mathrm{N}_{L/K}(y)) = \prod_{w|v} \mathrm{N}_{L_w/K_v}(i_w(y))\, \text{”},$$

which suggests a necessary condition to have $x = \mathrm{N}_{L/K}(y)$. Indeed, the local-global principle attached to the norm for the extension L/K is the fact (true or not) that $x \in K^\times$ is a norm for L/K (i.e., $x =: \mathrm{N}_{L/K}(y)$ for $y \in L^\times$) if and only if x is a local norm everywhere for L/K; the least one can ask is that the trivial direction be true. □

2.5.4 Corollary (Galois case). *Assume that L/K is Galois and, for $v \in Pl$, consider the semi-local factor $\bigoplus_{w|v} L_w^\times$. Since the L_w for $w|v$ are equal, we have $L_v^{\mathrm{ab}} = L_{w_0}^{\mathrm{ab}}$ for $w_0|v$ arbitrarily fixed, and we obtain:*

$$\mathrm{N}_{L/K}\Big(\bigoplus_{w|v} L_w^\times\Big) = \mathrm{N}_{L_{w_0}/K_v}(L_{w_0}^\times) = \mathrm{N}_{L_{w_0}^{\mathrm{ab}}/K_v}(L_{w_0}^{\mathrm{ab}\times}),$$

$$\mathrm{N}_{L/K}\Big(\bigoplus_{w|v} U_w\Big) = \mathrm{N}_{L_{w_0}/K_v}(U_{w_0}) = \mathrm{N}_{L_{w_0}^{\mathrm{ab}}/K_v}(U_{L_{w_0}^{\mathrm{ab}}}).$$

Hence, $x \in K^\times$ is a local norm at v for L/K if and only if, for some arbitrary $w_0|v$, there exists $y_{w_0} \in L_{w_0}^{\mathrm{ab}\times}$ such that:

$$i_v(x) = \mathrm{N}_{L_{w_0}^{\mathrm{ab}}/K_v}(y_{w_0}).$$

If $\mathrm{v}(x) = 0$*, then x is a local norm at v if and only if $i_v(x)$ is a local norm of local units, in other words:*

$$i_v(x) \in \mathrm{N}_{L_{w_0}^{\mathrm{ab}}/K_v}(U_{L_{w_0}^{\mathrm{ab}}}).$$ □

Once again, $L_{w_0}^{\mathrm{ab}}$ can strictly contain the completion $(L^{\mathrm{ab}})_v$ of L^{ab}.

For questions concerning local norms of local units in the *non-Galois* case, the intersection and the compositum of the fields L_w^{ab} (for $w|v$) play a fundamental role, and the above discussion is not valid; hence it is necessary to study the following subsection whose results (apparently not in the literature) will be used later in genus theory (Ch. IV, § 4).

2.6 Local Norm Invariants for Non-Galois Extensions. Let L/K be an arbitrary finite extension. For $v \in Pl$, let L_w be the completions of L for $w|v$; denote by L_v^{ab} (resp. by $\hat{L}_v^{\mathrm{ab}}$) the intersection (resp. the compositum) of the L_w^{ab} for $w|v$. To ease notations, we set:

$$N_w := \mathrm{N}_{L_w/K_v}(L_w^\times),\ \ V_w := \mathrm{N}_{L_w/K_v}(U_w)\ ;$$

then we get:

$$\mathrm{N}_{L/K}\Big(\bigoplus_{w|v} L_w^\times\Big) = \langle\, N_w \,\rangle_{w|v},\ \ \mathrm{N}_{L/K}\Big(\bigoplus_{w|v} U_w\Big) = \langle\, V_w \,\rangle_{w|v}.$$

2.6.1 Lemma 1. *We have $(K_v^\times : \langle\, N_w \,\rangle_{w|v}) = [L_v^{\mathrm{ab}} : K_v] = e_v^{\mathrm{ab}} f_v^{\mathrm{ab}}$, where e_v^{ab} and f_v^{ab} are the ramification index and the residue degree of the extension L_v^{ab}/K_v, repectively, and we have $(U_v : U_v \cap \langle\, N_w \,\rangle_{w|v}) = e_v^{\mathrm{ab}}$.* □

Denote by $G^0(M/M')$ the inertia group of v in M/M', where M and M' are abelian extensions of K_v such that $M' \subseteq M$.

2.6.2 Lemma 2. *We have the following canonical isomorphisms:*

$$G^0(L_{w_0}^{\mathrm{ab}}/K_v) \simeq U_v/V_{w_0} \text{ for any } w_0|v,$$
$$G^0(\hat{L}_v^{\mathrm{ab}}/K_v) \simeq U_v\Big/\bigcap_{w|v} V_w,$$
$$G^0(\hat{L}_v^{\mathrm{ab}}/L_{w_0}^{\mathrm{ab}}) \simeq V_{w_0}\Big/\bigcap_{w|v} V_w \text{ for any } w_0|v. \qquad \square$$

2.6.3 Proposition. *Let $\check{L}_v^{\mathrm{ab}}$ be the subfield of $\hat{L}_v^{\mathrm{ab}}$ fixed under the subgroup of $\mathrm{Gal}(\hat{L}_v^{\mathrm{ab}}/K_v)$ generated by the $G^0(\hat{L}_v^{\mathrm{ab}}/L_w^{\mathrm{ab}})$ for $w|v$, and let $\check{e}_v^{\mathrm{ab}}$ be the ramification index of $\check{L}_v^{\mathrm{ab}}/K_v$. We have the canonical isomorphism:*

$$G^0(\check{L}_v^{\mathrm{ab}}/K_v) \simeq U_v\Big/\mathrm{N}_{L/K}\Big(\bigoplus_{w|v} U_w\Big),$$

and therefore the formula $\Big(U_v : \mathrm{N}_{L/K}\Big(\bigoplus_{w|v} U_w\Big)\Big) = \check{e}_v^{\mathrm{ab}}$. In other words:

$$(U_K^{\mathrm{res}} : \mathrm{N}_{L/K}(U_L^{\mathrm{res}})) = \prod_{v \in Pl_0} \check{e}_v^{\mathrm{ab}}. \qquad \square$$

2.6.4 Proposition. *The number $\check{e}_v^{\mathrm{ab}}$ is a multiple of e_v^{ab} and these indices are equal when the L_w^{ab} for $w|v$ are all equal (for instance in the Galois case).* $\square$

Proof of the statements. The results 2.6.1 and 2.6.2 are proved by giving systematically the norm groups corresponding to the abelian extensions under study and their inertia subfields, and by using properties 1.5 of the correspondence of local class field theory (if N is the norm group corresponding to the abelian extension M of K_v, by 1.4, (iii), the group corresponding to the inertia field of M is equal to U_vN). If, to simplify notations we set:

$$N_v := \langle N_w \rangle_{w|v},\ \hat{N}_v := \bigcap_{w|v} N_w,\ V_v := \langle V_w \rangle_{w|v},\ \hat{V}_v := \bigcap_{w|v} V_w,$$

we obtain more precisely the following list:

FIELDS	CORRESPONDING	NORM GROUPS
K_v	$K_v^\times$	$K_v^\times$
L_w^{ab}	N_w	U_vN_w
L_v^{ab}	N_v	U_vN_v
$\hat{L}_v^{\mathrm{ab}}$	$\hat{N}_v$	$U_v\hat{N}_v$

Hence:

$$\mathrm{Gal}(L_v^{\mathrm{ab}}/K_v) \simeq K_v^\times/N_v, \text{ of order } e_v^{\mathrm{ab}} f_v^{\mathrm{ab}},$$

and the inertia group of L_v^{ab}/K_v is given by:

$$U_v N_v/N_v \simeq U_v/U_v \cap N_v, \text{ of order } e_v^{\mathrm{ab}},$$

proving Lemma 1.

Furthermore:

$$G^0(L_{w_0}^{\mathrm{ab}}/K_v) \simeq U_v N_{w_0}/N_{w_0} \simeq U_v/U_v \cap N_{w_0} \simeq U_v/V_{w_0},$$
$$G^0(\hat{L}_v^{\mathrm{ab}}/K_v) \simeq U_v \hat{N}_v/\hat{N}_v \simeq U_v/U_v \cap \hat{N}_v \simeq U_v\Big/ \bigcap_{w|v} (U_v \cap N_w) = U_v/\hat{V}_v.$$

From the exact sequence of inertia groups 1.1.6:

$$1 \longrightarrow G^0(\hat{L}_v^{\mathrm{ab}}/L_{w_0}^{\mathrm{ab}}) \longrightarrow G^0(\hat{L}_v^{\mathrm{ab}}/K_v) \longrightarrow G^0(L_{w_0}^{\mathrm{ab}}/K_v) \longrightarrow 1,$$

we deduce that the kernel of the map:

$$U_v \hat{N}_v/\hat{N}_v \longrightarrow U_v N_{w_0}/N_{w_0}$$

is equal to:

$$G^0(\hat{L}_v^{\mathrm{ab}}/L_{w_0}^{\mathrm{ab}}) \simeq N_{w_0} \cap (U_v \hat{N}_v)/\hat{N}_v = V_{w_0}\hat{N}_v/\hat{N}_v \simeq V_{w_0}/\hat{V}_v,$$

finishing the proof of Lemma 2.

The subgroup of $\mathrm{Gal}(\hat{L}_v^{\mathrm{ab}}/K_v)$ generated by the $G^0(\hat{L}_v^{\mathrm{ab}}/L_w^{\mathrm{ab}})$ is thus:

$$\mathrm{Gal}(\hat{L}_v^{\mathrm{ab}}/\check{L}_v^{\mathrm{ab}}) \simeq V_v \hat{N}_v/\hat{N}_v \simeq V_v/\hat{V}_v,$$

showing that the field $\check{L}_v^{\mathrm{ab}}$ and its inertia subfield have respective norm groups equal to:

$$V_v \hat{N}_v \text{ and } U_v V_v \hat{N}_v = U_v \hat{N}_v,$$

so that:

$$G^0(\check{L}_v^{\mathrm{ab}}/K_v) \simeq U_v \hat{N}_v/V_v \hat{N}_v \simeq U_v/V_v,$$

since $U_v \cap (V_v \hat{N}_v) = V_v(U_v \cap \hat{N}_v) = V_v \hat{V}_v = V_v$, proving 2.6.3.

Since L_v^{ab} is a subfield of $\check{L}_v^{\mathrm{ab}}$, we have $e_v^{\mathrm{ab}} | \check{e}_v^{\mathrm{ab}}$. Since $e_v^{\mathrm{ab}} = (U_v : U_v \cap N_v)$ and $V_v \subseteq U_v \cap N_v$, we have more precisely:

$$\frac{\check{e}_v^{\mathrm{ab}}}{e_v^{\mathrm{ab}}} = (U_v \cap N_v : V_v) \ ;$$

if the L_w^{ab} are all equal, $L_v^{\mathrm{ab}} = \hat{L}_v^{\mathrm{ab}} = \check{L}_v^{\mathrm{ab}}$, hence $e_v^{\mathrm{ab}} = \check{e}_v^{\mathrm{ab}}$, proving 2.6.4. □

2.6.5 Remark. In the Galois case, we thus have the formula:

$$(U_K^{\mathrm{res}} : \mathrm{N}_{L/K}(U_L^{\mathrm{res}})) = \prod_{v \in Pl_0} e_v^{\mathrm{ab}},$$

which must not be mistaken for the corresponding formula for the maximal abelian subextension of L/K:

$$(U_K^{\mathrm{res}} : \mathrm{N}_{L^{\mathrm{ab}}/K}(U_{L^{\mathrm{ab}}}^{\mathrm{res}})) = \prod_{v \in Pl_0} e_v(L^{\mathrm{ab}}/K). \qquad \square$$

2.6.6 Exercise. Consider the following irreducible polynomial in $\mathbb{Q}[X]$:

$$P := X^4 + 14X^2 - 19,$$

and take $K = \mathbb{Q}$, $L = K(\theta)$ with $\mathrm{Irr}(\theta, \mathbb{Q}) = P$, and $v = 2$.

(i) Show that $Pl_{L,v} = \{w_1, w_2\}$ with:

$$L_{w_1} = \mathbb{Q}_2(\sqrt{-1}),\ L_{w_2} = \mathbb{Q}_2(\sqrt{3}).$$

Compute the indices in U_v of $\mathrm{N}_{L/K}\Big(\bigoplus_{w|v} U_w\Big)$ and of $U_v \cap \mathrm{N}_{L/K}\Big(\bigoplus_{w|v} L_w^\times\Big)$.

(ii) Give also a direct numerical check by showing that the norm groups N_{w_1} and N_{w_2} are respectively equal to:

$$\langle\, 2 \,\rangle \oplus (1 + 4\mathbb{Z}_2) \ \text{ and } \ \langle\, -2 \,\rangle \oplus (1 + 4\mathbb{Z}_2).$$

Answer. We can check (by computing the roots) that we have the following factorization into irreducibles of $\mathbb{Q}_2[X]$:

$$P = (X^2 + a^2)(X^2 - 3b^2),\ a,\ b \in \mathbb{Q}_2^\times,$$

giving the two completions $\mathbb{Q}_2(\sqrt{-1})$ and $\mathbb{Q}_2(\sqrt{3})$. We easily obtain:

$$\hat{L}_v^{\mathrm{ab}} = \mathbb{Q}_2(\sqrt{-1}, \sqrt{3}),$$

hence $\check{L}_v^{\mathrm{ab}} = \hat{L}_v^{\mathrm{ab}}$ and $\check{e}_v^{\mathrm{ab}} = 2$ since $\hat{L}_v^{\mathrm{ab}}/\mathbb{Q}_2(\sqrt{-1})$ and $\hat{L}_v^{\mathrm{ab}}/\mathbb{Q}_2(\sqrt{3})$ are unramified. Since $L_v^{\mathrm{ab}} = \mathbb{Q}_2$, we have $e_v^{\mathrm{ab}} = 1$, which gives an example for which:

$$\Big(U_v : \mathrm{N}_{L/K}\Big(\bigoplus_{w|v} U_w\Big)\Big) \neq \Big(U_v : U_v \cap \mathrm{N}_{L/K}\Big(\bigoplus_{w|v} L_w^\times\Big)\Big).$$

For (ii), Exercise 1.6.5 then yields the norm groups. It follows that $V_{w_1} = V_{w_2} = 1 + 4\mathbb{Z}_2$, hence finally $\mathrm{N}_{L/K}\Big(\bigoplus_{w|v} U_w\Big) = 1 + 4\mathbb{Z}_2$, which is of index 2 in $U_v = \langle -1 \rangle \oplus (1 + 4\mathbb{Z}_2)$.

One checks that -1 can be written:

$$\mathrm{N}_{\mathbb{Q}_2(\sqrt{3})/\mathbb{Q}_2}(1 + \sqrt{3}) \,.\, \mathrm{N}_{\mathbb{Q}_2(\sqrt{-1})/\mathbb{Q}_2}(1 + \sqrt{-1})^{-1},$$

but is not the norm of local units. $\square$

2.6.7 Conclusion. (i) We can keep in mind for later use (in particular for genus theory) that when L/K is not Galois, local norm problems involve the following diagrams of local fields (for a finite number of finite places v):

$$K_v \longrightarrow L_v^{\mathrm{ab}} := \bigcap_{w|v} L_w^{\mathrm{ab}} \longrightarrow \check{L}_v^{\mathrm{ab}} \longrightarrow \hat{L}_v^{\mathrm{ab}} := \langle\, L_w^{\mathrm{ab}} \,\rangle_{w|v}$$

and the inertia groups of $\check{L}_v^{\mathrm{ab}}/K_v$, where $\mathrm{Gal}(\hat{L}_v^{\mathrm{ab}}/\check{L}_v^{\mathrm{ab}})$ is generated by the inertia groups of the $\hat{L}_v^{\mathrm{ab}}/L_w^{\mathrm{ab}}$ for $w|v$.

(ii) The norm group of L_v^{ab}/K_v is equal to:

$$\langle\, \mathrm{N}_{L_w/K_v}(L_w^{\times})\,\rangle_{w|v},$$

and the norm group of $\check{L}_v^{\mathrm{ab}}/K_v$ is equal to:

$$\langle\, \mathrm{N}_{L_w/K_v}(U_w)\,\rangle_{w|v} \cdot \bigcap_{w|v} \mathrm{N}_{L_w/K_v}(L_w^{\times}).$$

In particular $u \in U_v$ is a *norm of local elements* (i.e., u is an element of $\langle\, \mathrm{N}_{L_w/K_v}(L_w^{\times})\,\rangle_{w|v}$) if and only if:

$$u \in \mathrm{N}_{L_v^{\mathrm{ab}}/K_v}(L_v^{\mathrm{ab}\,\times}),$$

and u is a *norm of local units* (i.e., u is an element of $\langle\, \mathrm{N}_{L_w/K_v}(U_w)\,\rangle_{w|v}$, which is more difficult to characterize), if and only if:

$$u \in \mathrm{N}_{\check{L}_v^{\mathrm{ab}}/K_v}(\check{L}_v^{\mathrm{ab}\,\times}),$$

which can be expressed in terms of the corresponding norm residue symbol *without referring to units.*

(iii) The case of the places at infinity is trivial for unit groups; we can simply note that for $v \in Pl_\infty$ the field $L_v^{\mathrm{ab}} = L_v$, equal to $\mathbb{R}$ or to $\mathbb{C}$, is different from K_v if and only if v is real and *all* the places $w|v$ are complex (i.e., $f_v^{\mathrm{ab}} = 2$).

§3 Global Class Field Theory: Idelic Version

We now start the study of the fundamental step in global class field theory; it consists in giving the properties of the global reciprocity map (whose existence, from the point of view that we have adopted, only relies on the existence of the local reciprocity maps which has been assumed). We will state in parallel the existence theorem (understood to mean of abelian extensions corresponding to norm groups) whose proof uses independent direct techniques of Kummer extensions and which, because of this, is generally proved at the end of the exposition.

a) Global Reciprocity Map — The Product Formula — Global Class Field Theory Correspondence

Let L/K be a finite extension of number fields and let L^{ab}/K be its maximal abelian subextension whose Galois group will be denoted $G^{\mathrm{ab}} := G^{\mathrm{ab}}(L/K)$

by abuse of notation (we have $G^{\rm ab} \simeq G/[G,G]$ when the extension L/K is Galois with Galois group G). We give here the crucial definition of the book.

3.1 Global Reciprocity Map. Let $J := J_K$ be the idèle group of K. We define the global reciprocity map as being:

$$\rho_{L/K} \ : \ J \longrightarrow G^{\rm ab}$$

sending $\boldsymbol{x} =: (x_v)_v \in J$ to:

$$\rho_{L/K}(\boldsymbol{x}) := \prod_{v \in Pl} \left(\frac{x_v \, , \, L^{\rm ab}/K}{v}\right),$$

where $\left(\frac{x_v \, , \, L^{\rm ab}/K}{v}\right) \in G^{\rm ab}$ is the image of $(x_v \, , \, (L^{\rm ab})_v/K_v) \in {\rm Gal}((L^{\rm ab})_v/K_v)$ under the canonical isomorphism:

$${\rm Gal}((L^{\rm ab})_v/K_v) \simeq D_v(L^{\rm ab}/K) \subseteq G^{\rm ab},$$

where $D_v(L^{\rm ab}/K)$ is the decomposition group of v in the extension $L^{\rm ab}/K$ (see I.2.7, Fig. 2.3).

3.1.1 Remarks. (i) Since the x_v are almost all units, property 1.4, (vii) of the local norm residue symbol shows that the $\left(\frac{x_v \, , \, L^{\rm ab}/K}{v}\right)$ are almost all equal to 1, so the product makes sense.

(ii) The definition of $\rho_{L/K}(\boldsymbol{x})$ shows that $\rho_{L/K} = \rho_{L^{\rm ab}/K}$ but, as in the local case, we must also define $\rho_{L/K}$ for an arbitrary extension. In keeping with the notations that we have used, we can also by definition denote by $\left(\frac{\bullet \, , \, L/K}{v}\right)$ the symbol $\left(\frac{\bullet \, , \, L^{\rm ab}/K}{v}\right)$, all the more so that a little later we will introduce a generalized symbol denoted $\left[\frac{\bullet \, , \, L/K}{v}\right]$ to avoid any confusion.

(iii) The definition of $\rho_{L/K}$ does not use all the local information relative to L/K: indeed, the local symbols $(\bullet, \, (L^{\rm ab})_v/K_v)$ are the restrictions of the $(\bullet, \, L_v^{\rm ab}/K_v)$, themselves restrictions of the symbols $(\bullet, \, L_w^{\rm ab}/K_v)$ of local class field theory. This can be explained by the fact that we globalize and that, for each v, $(L^{\rm ab})_v/K_v$ is the largest local extension whose Galois group can be interpreted as a subgroup of $G^{\rm ab}$. □

3.1.2 Definition (Hasse symbols). Restricting the symbols $\left(\frac{\bullet \, , \, L^{\rm ab}/K}{v}\right)$ to $i_v(K^\times) \subset K_v^\times$, by composition with i_v we define symbols on $K^\times$, called Hasse symbols, and denoted in an analogous manner:

$$\left(\frac{\bullet \, , \, L/K}{v}\right) := \left(\frac{\bullet \, , \, L^{\rm ab}/K}{v}\right) \ : \ \begin{array}{ccc} K^\times & \longrightarrow & G^{\rm ab} \\ x & \longmapsto & \left(\frac{i_v(x) \, , \, L^{\rm ab}/K}{v}\right). \end{array}$$

□

They are not essentially different from the preceding ones since the image of $K^\times$ is dense in each $K_v^\times$, but the point is that we will see that on $K^\times$ these symbols are not anymore independent. More precisely, for $x \in K^\times$ the notation $\left(\frac{x\,,\,L^{\mathrm{ab}}/K}{v}\right)$, allows us to distinguish between the Hasse symbol (defined on $K^\times$) and its analog $\left(\frac{x_v\,,\,L^{\mathrm{ab}}/K}{v}\right)$ (defined on $K_v^\times$) used to define $\rho_{L/K}$.

The Hasse symbols only depend on L^{ab}/K. Their properties follow of course from those of the symbols $(\bullet, (L^{\mathrm{ab}})_v/K_v)$, except that the global context modifies certain statements (compare with 1.4 whose notations we again use), such as 3.1.3, (iv) below which uses the fact that the global norm in L'/K is the product of the local norms.

Let L/K be a finite extension of number fields, L'/K a subextension of L/K, and let $v \in Pl$.

3.1.3 Theorem (properties of the Hasse symbol). *(i) We have the exact sequence:*

$$1 \longrightarrow K^\times \cap \mathrm{N}_{(L^{\mathrm{ab}})_v/K_v}((L^{\mathrm{ab}})_v^\times) \longrightarrow K^\times \xrightarrow{\left(\frac{\bullet\,,\,L^{\mathrm{ab}}/K}{v}\right)} D_v(L^{\mathrm{ab}}/K) \longrightarrow 1,$$

where the kernel is the local norm group at v for L^{ab}/K (see 2.5.1);

(ii) the composition of $\left(\frac{\bullet\,,\,L^{\mathrm{ab}}/K}{v}\right)$ and of the projection $G^{\mathrm{ab}} \longrightarrow \mathrm{Gal}(L'^{\mathrm{ab}}/K)$, is equal to $\left(\frac{\bullet\,,\,L'^{\mathrm{ab}}/K}{v}\right)$;

(iii) the image of $K_{\{v\}}^\times$ (the subgroup of $K^\times$ of elements prime to v) under $\left(\frac{\bullet\,,\,L^{\mathrm{ab}}/K}{v}\right)$ is the group $I_v(L^{\mathrm{ab}}/K)$;

(iv) for all $x' \in L'^\times$, the image of $\prod_{w'|v} \left(\frac{x'\,,\,L^{\mathrm{ab}\prime}/L'}{w'}\right)$ in G^{ab} is equal to $\left(\frac{\mathrm{N}_{L'/K}(x')\,,\,L^{\mathrm{ab}}/K}{v}\right)$;

(v) for all $x \in K^\times$, the image of $\left(\frac{x\,,\,L^{\mathrm{ab}}/K}{v}\right)$ under the transfer map (from G^{ab} to $\mathrm{Gal}(L^{\mathrm{ab}\prime}/L')$), is equal to $\prod_{w'|v} \left(\frac{x\,,\,L^{\mathrm{ab}\prime}/L'}{w'}\right)$;

(vi) for any $\mathbb{Q}$-isomorphism τ of L^{ab} in $\overline{\mathbb{Q}}$ and all $x \in K^\times$, we have:

$$\left(\frac{\tau x\,,\,\tau L^{\mathrm{ab}}/\tau K}{\tau v}\right) = \tau \circ \left(\frac{x\,,\,L^{\mathrm{ab}}/K}{v}\right) \circ \tau^{-1} \text{ on } \tau L^{\mathrm{ab}}\,;$$

(vii) if v is unramified in L^{ab}/K then we have, for all $x \in K^\times$:

$$\left(\frac{x\,,\,L^{\mathrm{ab}}/K}{v}\right) = \left(\frac{L^{\mathrm{ab}}/K}{v}\right)^{\mathrm{v}(x)},$$

where $\left(\frac{L^{\mathrm{ab}}/K}{v}\right)$ denotes the Frobenius of v for L^{ab}/K. □

3.1.3.1 Remark (global Frobenius). Recall that for a place v of K, unramified in $L^{\rm ab}/K$, the Frobenius of v for $L^{\rm ab}/K$ is the canonical image in $G^{\rm ab}$ of the local Frobenius $((L^{\rm ab})_v/K_v)$ (i.e., $\left(\frac{L^{\rm ab}/K}{v}\right) = i_{w_0}^{-1} \circ ((L^{\rm ab})_v/K_v) \circ i_{w_0}$ for any $w_0|v$ in $L^{\rm ab}$). In particular, if $v \in Pl_\infty^{\rm r}$, then $\left(\frac{L^{\rm ab}/K}{v}\right) = i_{w_0}^{-1} \circ c \circ i_{w_0}$, is the image of the restriction to $(L^{\rm ab})_v$ of complex conjugation c.

If v is finite, the Frobenius of v in $L^{\rm ab}/K$ is thus the unique generator σ of the decomposition group of $\mathfrak{p}_v$ such that:

$$\sigma(x) \equiv x^{q_v} \bmod \mathfrak{p}_v \text{ for all integers } x \text{ of } L^{\rm ab},$$

where $q_v := |F_v| = {\rm N}\mathfrak{p}_v$. □

3.1.3.2 Examples. (i) For $K = \mathbb{Q}$, $L = \mathbb{Q}(\sqrt[6]{2})$, $v = (7)$, $x = 7$, we have:

$$\left(\frac{x,\, L/K}{v}\right) = \left(\frac{7,\, \mathbb{Q}(\sqrt{2})/\mathbb{Q}}{(7)}\right) = 1$$

since $L^{\rm ab} = \mathbb{Q}(\sqrt{2})$ and 2 is a square in $\mathbb{Q}_7^\times$, but:

$$(x,\, L_v/K_v) = (7,\, \mathbb{Q}_7(\sqrt[6]{2})/\mathbb{Q}_7) = (\mathbb{Q}_7(\sqrt[6]{2})/\mathbb{Q}_7),$$

the Frobenius (of order 3) since $\mathbb{Q}_7(\sqrt[6]{2})$ is the unramified extension of degree 3 of $\mathbb{Q}_7$ (see 1.4, (vii)). This means that 7 is a local norm at (7) in $L^{\rm ab}/K$ but not in L/K.

(ii) For $K = \mathbb{Q}$, $L = \mathbb{Q}(\sqrt[6]{2})$, $v = (43)$, $L' = \mathbb{Q}(\sqrt[3]{2})$ (for which $L^{\rm ab'} = L$), $x' = -1 + 15\sqrt[3]{2} - 10\sqrt[3]{4}$ (for which ${\rm N}_{L'/K}(x') = 43^2$), we have:

$$\left(\frac{{\rm N}_{L'/K}(x'),\, L^{\rm ab}/K}{v}\right) = \left(\frac{43^2,\, \mathbb{Q}(\sqrt{2})/\mathbb{Q}}{(43)}\right) = \left(\frac{\mathbb{Q}(\sqrt{2})/\mathbb{Q}}{(43)}\right)^2 = 1,$$

the square of the Frobenius (of order 2) since 43 is inert in $\mathbb{Q}(\sqrt{2})/\mathbb{Q}$ (see 3.1.3, (vii)). Since $v = (43)$ is totally split in L'/K, we check that:

$$\prod_{w'|v} \left(\frac{x',\, L/L'}{w'}\right) = \left(\frac{L/L'}{w'_1}\right)\left(\frac{L/L'}{w'_2}\right),$$

the product of two of the three (nontrivial) relative Frobenius', giving the result by restriction in $L^{\rm ab}/K$ and illustrating 3.1.3, (iv). □

3.1.4 Remark (generalized norm residue symbol for L/K). As already noted, we have $\left(\frac{x\,,\, L^{\rm ab}/K}{v}\right) = 1$ if and only if x is a local norm at v for $L^{\rm ab}/K$. If we want to characterize the subgroup of elements of $K^\times$ which are local norms at v for L/K, we need to define a generalized norm residue symbol which must thus be intermediate between the symbol $(\bullet\,,\, L_w/K_v)$, which characterizes the elements which are local norms in L_w/K_v, which is not suitable (see 2.5.3, (ii)), and the Hasse symbol which does not deal with L/K but with L^{ab}/K. Since this subgroup is $\{x \in K^\times,\ i_v(x) \in \mathrm{N}_{L_v^{\mathrm{ab}}/K_v}(L_v^{\mathrm{ab}\times})\}$, where $L_v^{\mathrm{ab}} := \bigcap_{w|v} L_w^{\mathrm{ab}}$, local class field theory tells us that we must use $(\,\bullet\,,\, L_v^{\mathrm{ab}}/K_v)$ restricted to $i_v(K^\times)$; by composition with i_v, this defines the symbol:

$$\left[\frac{\bullet\,,\, L/K}{v}\right] : K^\times \longrightarrow G_v^{\mathrm{ab}} := \mathrm{Gal}(L_v^{\mathrm{ab}}/K_v)$$

called the generalized norm residue symbol for L/K. This symbol cannot be interpreted in G^{ab} since G_v^{ab} can be strictly larger than $\mathrm{Gal}((L^{\mathrm{ab}})_v/K_v) \simeq D_v(L^{\mathrm{ab}}/K)$ but, for all $x \in K^\times$,

$$\left[\frac{x\,,\, L/K}{v}\right]\Big|_{(L^{\mathrm{ab}})_v} \text{ can be identified with } \left(\frac{x\,,\, L^{\mathrm{ab}}/K}{v}\right). \qquad \square$$

Note that if L/K is Galois, the generalized norm residue symbol at v can be identified, for any $w_0|v$, with the local norm residue symbol $(\,\bullet\,,\, L_{w_0}^{\mathrm{ab}}/K_v)$ restricted to $i_v(K^\times)$.

3.2 Cohomological Statement of Class Field Theory (1951/1952). Let L be a finite Galois extension of the number field K, and let $G := \mathrm{Gal}(L/K)$. Assuming that the cohomological version of class field theory can be (in part) summarized by the magical formulas:

$$\widehat{H}^r(G,\mathbb{Z}) \overset{\text{can}}{\simeq} \widehat{H}^{r+2}(G,C_L), \quad \text{for the global case,}$$
$$\widehat{H}^r(G_w,\mathbb{Z}) \overset{\text{can}}{\simeq} \widehat{H}^{r+2}(G_w,L_w^\times),\ w \in Pl_L, \quad \text{for the local case,}$$

where the $\widehat{H}^r$ for $r \in \mathbb{Z}$ are Tate's modified cohomology groups, we deduce once again the existence of "a" global reciprocity map in the following way.

3.2.1 Global Reciprocity Map. Take $r = -2$, which in the global case yields:

$$\widehat{H}^{-2}(G,\mathbb{Z}) \overset{\text{can}}{\simeq} \widehat{H}^0(G,C_L)\ ;$$

but classically, we have:

$$\widehat{H}^0(G,C_L) = C_L^G/\nu_{L/K}(C_L) \simeq C_K/\mathrm{N}_{L/K}(C_L) \simeq J_K/K^\times \mathrm{N}_{L/K}(J_L),$$

the fact that:

$$C_L^G = j_{L/K}(C_K) \simeq C_K \ \text{ and } \ \nu_{L/K}(C_L) = j_{L/K} \circ \mathrm{N}_{L/K}(C_L) \simeq \mathrm{N}_{L/K}(C_L),$$

with the usual definitions of ν, j, N, being elementary (see (§ 2, (a), (b), (c))).

On the other hand, we have:

$$\widehat{H}^{-2}(G,\mathbb{Z}) := \widehat{H}_1(G,\mathbb{Z}) \overset{\text{can}}{\simeq} G^{\text{ab}}$$

(see 2.4.1, (iii), or [d, CF, Ch. IV, § 3, Prop. 1]), giving the result, except that we must identify the map $J_K/K^\times \mathrm{N}_{L/K}(J_L) \longrightarrow G^{\text{ab}}$. But the surjection:

$$J_K/\mathrm{N}_{L/K}(J_L) \longrightarrow G^{\text{ab}},$$

which we obtain from it, and the fact (which is immediate by 1.4.3, (ii)) that:

$$J_K/\mathrm{N}_{L/K}(J_L) \simeq \bigoplus_v (K_v^\times/\mathrm{N}_{L_{w_0}/K_v}(L_{w_0}^\times)),$$

indeed suggests that it is the map defined in 3.1 from the local reciprocity maps:

$$\widehat{H}^0(G_{w_0}, L_{w_0}^\times) \simeq K_v^\times/\mathrm{N}_{L_{w_0}/K_v}(L_{w_0}^\times) \longrightarrow \widehat{H}^{-2}(G_{w_0},\mathbb{Z}) \simeq G_{w_0}^{\text{ab}},$$

where for each place v of K we have chosen a place w_0 of L above v.

This does not make it any easier to obtain the kernel of this surjection, which is the very heart of global class field theory.

3.2.2 Fundamental Class. Note that for $r = 0$ we obtain:

$$\widehat{H}^2(G, C_L) \overset{\text{can}}{\simeq} \widehat{H}^0(G,\mathbb{Z}) = \mathbb{Z}/[L:K]\mathbb{Z} \simeq [L:K]^{-1}\mathbb{Z}/\mathbb{Z},$$

since $\nu_{L/K}$ acts on $\mathbb{Z}$ as multiplication by $[L:K]$. The element $u_{L/K} \in \widehat{H}^2(G, C_L)$, which is the inverse image of the class $[L:K]^{-1} + \mathbb{Z} \in \mathbb{Q}/\mathbb{Z}$, is called the fundamental class of L/K.

Hence, the isomorphisms of global class field theory:

$$\widehat{H}^r(G,\mathbb{Z}) \overset{\text{can}}{\simeq} \widehat{H}^{r+2}(G, C_L),\ \ r \in \mathbb{Z},$$

are given by the cup product $x \mapsto x \smile u_{L/K}$ for all $x \in \widehat{H}^r(G,\mathbb{Z})$.

For a general view of the cohomological approach, we refer the reader to [d, CF, Ch. VII; Iy1, Ch. IV; Se2, Ch. XI] or to [e, Ko3, Ch. 2], [g, NSW, Ch. III, § 1], as well as to the concrete explanations of [i, Gar], and to Koch's lecture in [i, Miy0] for the history of the concept of class formation for which the fundamental class plays a basic role.

We will now introduce a finite set S of places, which will be a parameter allowing us to specify the decomposition (i.e., splitting) of the elements of S in the correspondence of class field theory.

Notations. (i) Let $S = S_0 \cup S_\infty$ be a finite set of noncomplex places of K. For any finite extension L/K, we denote by $L^{\text{ab}\,S}/K$ the maximal S-split subextension of L^{ab}/K (i.e., in which every place of S is totally split).

(ii) We denote by $\rho_{L/K}^S$ the composite:

$$J \xrightarrow{\rho_{L/K}} G^{\mathrm{ab}} := \mathrm{Gal}(L^{\mathrm{ab}}/K) \longrightarrow G^{\mathrm{ab}\,S} := \mathrm{Gal}(L^{\mathrm{ab}\,S}/K).$$ [16]

(iii) We set $\langle S \rangle := \prod_{v\in S} K_v^\times \prod_{v \in Pl\setminus S}\{1\} =: \bigoplus_{v\in S} K_v^\times$, considered as a subgroup of J (see I.4.1.2, (ii)). □

The fundamental Theorem 1.4 (for the local reciprocity maps) has a global analog in which the multiplicative group $K_v^\times$ is replaced by the multiplicative group J. Let L/K be a finite extension of number fields and L'/K a subextension of L/K. We denote by S' the set of places of L' above those of S.

3.3 Theorem (properties of the global reciprocity map). *The global reciprocity map $\rho^S_{L/K}$ has the following properties:*

(i) We have the exact sequence:

$$1 \longrightarrow K^\times \langle S \rangle \mathrm{N}_{L/K}(J_L) \longrightarrow J_K \xrightarrow{\rho^S_{L/K}} G^{\mathrm{ab}\,S} \longrightarrow 1\ ;$$

(ii) the composition of $\rho^S_{L/K}$ and of the projection $G^{\mathrm{ab}\,S} \longrightarrow \mathrm{Gal}(L'^{\mathrm{ab}\,S}/K)$ is equal to $\rho^S_{L'/K}$;

(iii) for any $v \in Pl$, the image of $K_v^\times$ (resp. of U_v, resp. of U_v^i for $i \geq 1$)[17] *under $\rho^S_{L/K}$ is the decomposition group (resp. the inertia group, resp. the ith higher ramification group in upper numbering) of v for $L^{\mathrm{ab}\,S}/K$;*

(iv) for all $\boldsymbol{x}' \in J_{L'}$, the image of $\rho^{S'}_{L/L'}(\boldsymbol{x}')$ in $G^{\mathrm{ab}\,S}$ is equal to $\rho^S_{L/K}(\mathrm{N}_{L'/K}(\boldsymbol{x}'))$; in particular, we have:

$$\mathrm{Gal}(L^{\mathrm{ab}\,S}/L'^{\mathrm{ab}\,S}) = \rho^S_{L/K}(\mathrm{N}_{L'/K}(J_{L'}))\ ;$$

(v) for all $\boldsymbol{x} \in J_K$, the image of $\rho^S_{L/K}(\boldsymbol{x})$ under the transfer map, from $G^{\mathrm{ab}\,S}$ to $\mathrm{Gal}(L^{\mathrm{ab}'\,S'}/L')$, is equal to $\rho^{S'}_{L/L'}(\boldsymbol{x}')$, where $\boldsymbol{x}'$ is the image of $\boldsymbol{x}$ under the canonical injection $J_K \longrightarrow J_{L'}$;

(vi) for any $\mathbb{Q}$-isomorphism τ of L in $\overline{\mathbb{Q}}$ and all $\boldsymbol{x} \in J_K$, we have:

$$\rho^{\tau S}_{\tau L/\tau K}(\tau \boldsymbol{x}) = \tau \circ \rho^S_{L/K}(\boldsymbol{x}) \circ \tau^{-1} \text{ on } \tau L^{\mathrm{ab}\,S},$$

noting that $\tau L^{\mathrm{ab}\,S} = (\tau L^{\mathrm{ab}})^{\tau S} = (\tau L)^{\mathrm{ab}\,\tau S}$ (abelianized over τK);

(vii) if the support of $\boldsymbol{x} =: (x_v)_v \in J_K$ is prime to the ramification of $L^{\mathrm{ab}\,S}/K$ (i.e., $x_v = 1$ if v is ramified), we have:

$$\rho^S_{L/K}(\boldsymbol{x}) = \prod_{v \in Pl} \left(\frac{L^{\mathrm{ab}\,S}/K}{v}\right)^{\mathrm{v}(x_v)},$$

[16] where $\rho_{L/K}$ will also be denoted $\rho^{\mathrm{res}}_{L/K}$, in accordance with the principles of notation given in Sections 3 and 4 of Chapter I.

[17] where $K_v^\times$, U_v, and U_v^i are considered as subgroups of J_K.

where $\left(\frac{L^{\mathrm{ab}\,S}/K}{v}\right)$ *denotes the Frobenius of* v *for* $L^{\mathrm{ab}\,S}/K$*. If* π_v *is a uniformizer of* K_v *(seen as an idèle of support* $\{v\}$*) and if* $L^{\mathrm{ab}\,S}/K$ *is unramified at* v*, then* $\rho^S_{L/K}(\pi_v) = \left(\frac{L^{\mathrm{ab}\,S}/K}{v}\right)$. □

Note. In (vii), we can replace the assumption on ramification by the weaker condition: x_v sufficiently close to 1 if v is ramified.

At this point we can note that the norm group N_v corresponding to $(L^{\mathrm{ab}})_v/K_v$ is $N \cap K_v^\times$, where $N := K^\times \mathrm{N}_{L/K}(J_L)$: indeed, by 3.1, the restriction to $K_v^\times \subset J_K$ of $\rho_{L^{\mathrm{ab}}/K}$ can be identified with the norm residue symbol $(\,\bullet\,, (L^{\mathrm{ab}})_v/K_v)$, proving our claim (considering N_v as canonically embedded in J_K).

This proves the following important relationship between local and global class field theories for L^{ab}/K.

3.3.1 Corollary. *For any place* v *of* K*, we have the identity:*

$$(K^\times \mathrm{N}_{L/K}(J_L)) \cap K_v^\times = \mathrm{N}_{(L^{\mathrm{ab}})_v/K_v}((L^{\mathrm{ab}})_v^\times).$$ □

3.3.2 Remarks. (i) There exists an infinity of finite sets Σ of places of K such that, by restricting $\rho_{L/K}$ to $\bigoplus_{v \in \Sigma} K_v^\times$, we obtain the exact sequence:

$$1 \longrightarrow N \cap \Big(\bigoplus_{v\in\Sigma} K_v^\times\Big) \longrightarrow \bigoplus_{v\in\Sigma} K_v^\times \xrightarrow{\rho_{L/K}} G^{\mathrm{ab}} \longrightarrow 1,$$

where $N := K^\times \mathrm{N}_{L/K}(J_L)$ (for this, by 3.3, (iii), it suffices that the decomposition groups of the places $v \in \Sigma$ for L^{ab}/K generate G^{ab}, which uses the density theorem which we will recall in 4.6).

(ii) In terms of reduced idèles, since $U_\infty \subset \mathrm{N}_{L/K}(J_L)$, we systematically replace the exact sequence of 3.3, (i) by:

$$1 \longrightarrow K^\times . \bigoplus_{v\in S_0} K_v^\times \bigoplus_{v\in S_\infty} \{\pm 1\} . \mathrm{N}_{L/K}(J_{L,0}) \longrightarrow J_{K,0} \xrightarrow{\rho^S_{L/K}} G^{\mathrm{ab}\,S} \longrightarrow 1.$$

We will do this only if it is technically necessary. □

3.3.3 Corollary. *For any finite extension* L/K *of number fields, we have:*

$$K^\times \mathrm{N}_{L/K}(J_L) = K^\times \mathrm{N}_{L^{\mathrm{ab}}/K}(J_{L^{\mathrm{ab}}}).$$ □

By giving a numerical example, it is easy to show that the equality:

$$\mathrm{N}_{L/K}(J_L) = \mathrm{N}_{L^{\mathrm{ab}}/K}(J_{L^{\mathrm{ab}}})$$

is in general false.

3.3.4 Corollary. *We obtain the exact sequences:*

$$1 \longrightarrow K^\times \mathrm{N}_{L/K}(J_L) \longrightarrow J_K \xrightarrow{\rho_{L/K}} G^{\mathrm{ab}} \longrightarrow 1,$$

$$1 \longrightarrow K^\times \langle\, Pl_\infty^{\mathrm{r}} \,\rangle \mathrm{N}_{L/K}(J_L) \longrightarrow J_K \xrightarrow{\rho_{L/K}^{\mathrm{ord}}} \mathrm{Gal}(L^{\mathrm{ab\,nc}}/K) \longrightarrow 1,$$

where $L^{\mathrm{ab\,nc}}$ is the maximal noncomplexified (i.e., Pl_∞^{r}-split) abelian subextension of L or, equivalently, the maximal abelian subextension of L which stays real under all the real embeddings of K. □

Let L/K be a finite extension, and let $N := K^\times \mathrm{N}_{L/K}(J_L)$, so that we have the exact sequence:

$$1 \longrightarrow N \longrightarrow J_K \xrightarrow{\rho_{L/K}} G^{\mathrm{ab}} \longrightarrow 1.$$

Theorem 3.3 gives then the important result:

3.3.5 Corollary (decomposition law of places in L^{ab}/K). *For each place $v \in Pl$, we have the isomorphisms:*

$$K_v^\times N/N \simeq D_v(L^{\mathrm{ab}}/K), \quad U_v N/N \simeq I_v(L^{\mathrm{ab}}/K),$$

where $K_v^\times$ and U_v are considered as subgroups of J_K. In particular, v is unramified in L^{ab}/K if and only if $U_v \subset N$. Hence, if v is unramified in L^{ab}/K, we have $K_v^\times N/N = \langle\, \pi_v \,\rangle N/N$ since $U_v \subset N$, and the residue degree f_v of v for L^{ab}/K is equal to the order in J_K/N of any uniformizer π_v (seen as an idèle with support $\{v\}$). □

Recall that for $v \in Pl_\infty^{\mathrm{r}}$, $K_v^\times = \mathbb{R}^\times$, $\pi_v = -1$, and $U_v = \mathbb{R}^{\times +}$, so that in this case we always have $U_v \subseteq N$ (i.e., nonramification of the infinite places). Indeed, $v \in Pl_\infty^{\mathrm{r}}$ does not become complex in L^{ab}/K if and only if -1 (seen as an idèle with support $\{v\}$ and not diagonally embedded!) belongs to N.

It also follows that, for all $S \subset Pl^{\mathrm{nc}}$, we have the exact sequence:

$$1 \longrightarrow \langle\, S \,\rangle N \longrightarrow J_K \longrightarrow G^{\mathrm{ab}\,S} \longrightarrow 1.$$

3.4 Product Formula — Classical Applications. The fact that the subgroup $\mathrm{N}_{L/K}(J_L)$ is in the kernel of $\rho_{L/K}$ is clear since for each v, $\langle\, \mathrm{N}_{L_w/K_v}(L_w^\times) \,\rangle_{w|v}$ which corresponds to $L_v^{\mathrm{ab}} := \bigcap_{w|v} L_w^{\mathrm{ab}}$ by local class field theory (see 1.5.3), is in the kernel of $(\,\bullet\,, L_v^{\mathrm{ab}}/K_v)$, hence a fortiori in that of $(\,\bullet\,, (L^{\mathrm{ab}})_v/K_v)$, hence of $\left(\frac{\bullet\,,\, L^{\mathrm{ab}}/K}{v}\right)$ since L_v^{ab} contains $(L^{\mathrm{ab}})_v$.

On the contrary, the fact that the kernel of $\rho_{L/K}$ contains the diagonal embedding of $K^\times$ is the *most remarkable* fact (and the least trivial) of global class field theory. We can consider this fact as the idelic version of Artin's reciprocity law (1924/1927), that we will give in 4.3.2 and 4.4; it is also called the product formula since it can be stated in the following way in terms of Hasse symbols (see 3.1.2).

3.4.1 Theorem. *Let L/K be a finite extension. For any $x \in K^\times$ we have:*

$$\prod_{v \in Pl} \left(\frac{x \, , \, L^{\mathrm{ab}}/K}{v} \right) = 1. \qquad \square$$

This property allows us to define the reciprocity map on the idèle class group:

$$\rho_{L/K} \; : \; C_K \longrightarrow G^{\mathrm{ab}}.$$

This product formula, which says that the Hasse symbols are not independent on $K^\times$, can also be considered as the general reciprocity law, since it generalizes (among other results) the quadratic reciprocity law of Gauss. To illustrate this, we are going to show that we can deduce the quadratic reciprocity law without using any additional deep arguments (we will give in 7.4 the nth power reciprocity law analogous to the quadratic reciprocity law when K contains μ_n; see also [f, Lem] and [Wy] for further examples and the history of the subject).

3.4.2 Example (quadratic reciprocity law). Take $K = \mathbb{Q}$ and consider $L = \mathbb{Q}\Big(\sqrt{(-1)^{\frac{p-1}{2}} p}\Big)$ for a positive odd prime p; thus the extension $L/\mathbb{Q}$ is only ramified at p. For convenience, we identify $\mathrm{Gal}(L/\mathbb{Q})$ with the multiplicative group $\{\pm 1\}$. The places of $\mathbb{Q}$ will be denoted either v, or else ℓ and ∞.

The computation of the Hasse symbols $\left(\frac{x \, , \, L/\mathbb{Q}}{v}\right)$, $x \in \mathbb{Q}^\times$, can be reduced successively, by multiplicativity, to that of the:

$$\left(\frac{-1 \, , \, L/\mathbb{Q}}{v} \right),$$

and of the:

$$\left(\frac{q \, , \, L/\mathbb{Q}}{v} \right),$$

for any positive prime q. Recall that, by 3.1.3, (vii), $\left(\frac{x \, , \, L/\mathbb{Q}}{v}\right) = 1$ except perhaps for (finite or infinite) places v such that $\mathrm{v}(x) \neq 0$ and the ramified places v (hence here p).

(i) $\left(\frac{-1 \, , \, L/\mathbb{Q}}{v}\right)$ is equal to 1 except perhaps for $v \in \{\infty, p\}$; but we have:

$$\left(\frac{-1 \, , \, L/\mathbb{Q}}{\infty} \right) = (-1)^{\frac{p-1}{2}}$$

(indeed, $L_\infty/\mathbb{Q}_\infty = \mathbb{C}/\mathbb{R}$ or $\mathbb{R}/\mathbb{R}$ if $(-1)^{\frac{p-1}{2}} = -1$ or 1 respectively, and -1 is a local norm only in the second case, or (by 3.1.3, (vii)) $\left(\frac{-1\,,\,L/\mathbb{Q}}{\infty}\right)$ is the Frobenius of ∞ for $L/\mathbb{Q}$); using the product formula, we obtain:

$$\left(\frac{-1\,,\,L/\mathbb{Q}}{p}\right) = \left(-1\right)^{\frac{p-1}{2}}$$

(even though a direct computation is easy).

(ii) $\left(\frac{q\,,\,L/\mathbb{Q}}{v}\right)$ is equal to 1 except perhaps for $v \in \{p, q\}$:
If $q = p$, the product formula (reduced to a single term) yields:

$$\left(\frac{p\,,\,L/\mathbb{Q}}{p}\right) = 1.$$

Assume now that $q \neq p$:

- If $v = p$, then $\left(\frac{q\,,\,L/\mathbb{Q}}{p}\right) = 1$ if and only if q (which belongs to U_p) is a norm for the extension $L_p/\mathbb{Q}_p$ (since this extension is ramified, we have $(U_p : U_p \cap \mathrm{N}_{L_p/\mathbb{Q}_p}(L_p^\times)) = 2$ by local theory); but the only subgroup of index 2 of $U_p = \mathbb{Z}_p^\times = \mu_{p-1} \oplus (1 + p\mathbb{Z}_p)$ is $\mu_{p-1}^2 \oplus (1 + p\mathbb{Z}_p) = (U_p)^2$, hence q is a norm for $L_p/\mathbb{Q}_p$ if and only if $q \in (U_p)^2$, hence if and only if $\overline{q} \in \mathbb{F}_p^{\times 2}$, so that:

$$\left(\frac{q\,,\,L/\mathbb{Q}}{p}\right) = \left(\frac{q}{p}\right)$$

(the usual quadratic residue symbol). We can of course use 1.6.5.

- If $v = q$, to compute $\left(\frac{q\,,\,L/\mathbb{Q}}{q}\right)$ we see that $L_q/\mathbb{Q}_q$ is unramified and we have, by 3.1.3, (vii), $\left(\frac{q\,,\,L/\mathbb{Q}}{q}\right) = \left(\frac{L/\mathbb{Q}}{q}\right)$ which is equal to 1 if and only if q is split in $L/\mathbb{Q}$, hence if and only if $L_q = \mathbb{Q}_q\Big(\sqrt{(-1)^{\frac{p-1}{2}}p}\Big) = \mathbb{Q}_q$, hence:

$$\left(\frac{q\,,\,L/\mathbb{Q}}{q}\right) = \left(\frac{(-1)^{\frac{p-1}{2}}p}{q}\right),$$

where we use here the Kronecker symbol, equal to the quadratic residue symbol if $q \neq 2$, and otherwise defined by $(\frac{a}{2}) = 1$ or -1 according as $a \equiv 1 \bmod (8)$ or not.[18]

Hence, using the product formula we have $\left(\frac{q\,,\,L/\mathbb{Q}}{p}\right)\left(\frac{q\,,\,L/\mathbb{Q}}{q}\right) = 1$, which can be interpreted as follows.

- For $q \notin \{2, p\}$ we get $\left(\frac{q}{p}\right)\left(\frac{(-1)^{\frac{p-1}{2}}p}{q}\right) = 1$, hence (using (i)):

[18] This symbol at 2 is not multiplicative: $(\frac{3}{2}) = (\frac{5}{2}) = (\frac{15}{2}) = -1$; it is multiplicative however on $1 + 4\mathbb{Z}_2$, which is the present context.

$$\left(\frac{q}{p}\right)\left(\frac{p}{q}\right) = (-1)^{\frac{p-1}{2}\frac{q-1}{2}} \;;$$

• for $q = 2$ this yields $\left(\frac{2}{p}\right) = \left(\frac{(-1)^{\frac{p-1}{2}} p}{2}\right)$; we check, by choosing $p \equiv 1, 3, 5, 7 \bmod (8)$, that this can be written:

$$\left(\frac{2}{p}\right) = (-1)^{\frac{p^2-1}{8}} .$$

Note that the above computations never went any further than the use of the Hensel lemma in the $\mathbb{Q}_\ell$ (to characterize the elements of $\mathbb{Q}_\ell^{\times 2}$) and ramification theory in a quadratic field. □

The product formula enables us to make convenient explicit computations, even for the local case, as is shown in the following exercise which gives the local reciprocity map for abelian extensions of the completions of $\mathbb{Q}$ (by Exercise 1.8.3 any abelian extension of $\mathbb{Q}_\ell$ is contained in the maximal cyclotomic extension $\mathbb{Q}_\ell(\mu)$). This procedure will be systematized in 4.4.3 and illustrated in 7.5 for the computation of local Hilbert symbols.

3.4.3 Exercise (local reciprocity map in $\overline{\mathbb{Q}_\ell}^{\mathrm{ab}}/\mathbb{Q}_\ell$). Assume that $K = \mathbb{Q}$ and consider for a fixed prime ℓ the abelian extension $L = \mathbb{Q}(\mu_{\ell^n})$ with $n \geq 1$; we will denote by $L_q/\mathbb{Q}_q = \mathbb{Q}_q(\mu_{\ell^n})/\mathbb{Q}_q$ the completion of $L/\mathbb{Q}$ at $v = q$ finite, and we will consider the embedding $i : \mathbb{Q}^\times \longrightarrow J_\mathbb{Q}$ as being the identity.

(i) Find the norm group N corresponding to L_ℓ.

(ii) Check that for all prime q, $q > 0$, and $q \neq \ell$, the local norm residue symbol $(q\,,\, L_q/\mathbb{Q}_q)$ is the Frobenius automorphism σ_q defined by $\zeta \to \zeta^q$ for all $\zeta \in \mu_{\ell^n}$.

(iii) Show that $(q\,,\, L_\ell/\mathbb{Q}_\ell) = \sigma_q^{-1}$.

(iv) Let $x \in \mathbb{Q}_\ell^\times$, and write $x =: \ell^{\mathrm{v}_\ell(x)} u$. Deduce from the above that:

$$(x\,,\, L_\ell/\mathbb{Q}_\ell) = \sigma_u^{-1} = \sigma_{u^{-1}},$$

defined by $\zeta \to \zeta^{u^{-1}}$ for all $\zeta \in \mu_{\ell^n}$.

Show that for $L = \mathbb{Q}(\mu_{\ell^\infty})$, the local reciprocity map:

$$\widehat{\mathbb{Q}_\ell^\times} = \ell^{\widehat{\mathbb{Z}}} \oplus U_\ell \longrightarrow \mathrm{Gal}(L_\ell/\mathbb{Q}_\ell)$$

induces the isomorphism $U_\ell \simeq \mathrm{Gal}(L_\ell/\mathbb{Q}_\ell)$ which sends $u \in U_\ell$ to $\sigma_u^{-1} = \sigma_{u^{-1}}$ defined by $\zeta \to \zeta^{u^{-1}}$ for all $\zeta \in \mu_{\ell^\infty}$.

Answer. (i) We have $\mathbb{Q}_\ell^\times = \ell^{\mathbb{Z}} \oplus U_\ell$ with $U_\ell = \mu_{\ell-1} \oplus (1 + \ell\mathbb{Z}_\ell)$ if $\ell \neq 2$, and $U_2 = \{\pm 1\} \oplus (1 + 4\mathbb{Z}_2)$. We know that if ζ_n is a generator of μ_{ℓ^n} we have $\mathrm{N}_{L/\mathbb{Q}}(1 - \zeta_n) = \Phi_{\ell^n}(1) = \ell \in N$; since in addition $L_\ell/\mathbb{Q}_\ell$ is totally ramified of degree $\ell^{n-1}(\ell - 1)$, we have $N = \ell^{\mathbb{Z}} \oplus V$ with V of index $\ell^{n-1}(\ell - 1)$ in U_ℓ.

If $\ell \neq 2$, the only possibility is $V = 1 + \ell^n\mathbb{Z}_\ell$; if $\ell = 2$ and $n \geq 2$, we have $\mathrm{Gal}(L_\ell/\mathbb{Q}_\ell) \simeq \mathbb{Z}/2\mathbb{Z} \times \mathbb{Z}/2^{n-2}\mathbb{Z}$ and the only possibility is $V = 1 + 4.2^{n-2}\mathbb{Z}_2$ since V must be contained in the norm group of $\mathbb{Q}(\mu_4)$ which is equal to $2^{\mathbb{Z}} \oplus (1 + 4\mathbb{Z}_2)$ (using 1.6.5). Thus in all cases we have:

$$N = \ell^{\mathbb{Z}} \oplus (1 + \ell^n\mathbb{Z}_\ell).$$

(ii) By 1.4, (vii), we have $(q\,,\,L_q/\mathbb{Q}_q) = (L_q/\mathbb{Q}_q)$, the Frobenius of q, equal to σ_q for a cyclotomic field.

(iii) We have the product formula:

$$\prod_v \left(\frac{q\,,\,L/\mathbb{Q}}{v}\right) = \prod_v (q\,,\,L_v/\mathbb{Q}_v) = 1,$$

and we know that $(q\,,\,L_v/\mathbb{Q}_v) = 1$ except perhaps if $L_v/\mathbb{Q}_v$ is ramified (which occurs only for $v = \ell$) or if q is not a unit at v (hence only for $v = q$ since we chose $q > 0$); since $(q\,,\,L_q/\mathbb{Q}_q) = \sigma_q$ by (ii), $(q\,,\,L_\ell/\mathbb{Q}_\ell)$ is its inverse.

(iv) We deduce that, for any rational number $a > 0$ prime to ℓ, we have $(a\,,\,L_\ell/\mathbb{Q}_\ell) = \sigma_a^{-1}$ hence, by density, that $(a\,,\,L_\ell/\mathbb{Q}_\ell) = \sigma_a^{-1}$ for all $a \in U_\ell$; in particular it follows that $(x\,,\,L_\ell/\mathbb{Q}_\ell) = \sigma_u^{-1}$ since $(\ell\,,\,L_\ell/\mathbb{Q}_\ell) = 1$. It is then immediate to obtain the local reciprocity map for $\mathbb{Q}_\ell(\mu_{\ell^\infty})/\mathbb{Q}_\ell$ by taking inverse limits.

Note that if μ' is the group of roots of unity of order prime to ℓ, then $\mathbb{Q}_\ell(\mu')$ is the maximal unramified extension of $\mathbb{Q}_\ell$ and its norm group is equal to U_ℓ (see 1.8.3); the isomorphism $\ell^{\widehat{\mathbb{Z}}} \longrightarrow \mathrm{Gal}(\mathbb{Q}_\ell(\mu')/\mathbb{Q}_\ell)$ is given by $\ell \to \sigma_\ell$ (Frobenius of ℓ). Since $\overline{\mathbb{Q}}_\ell^{\mathrm{ab}}$ is the direct compositum of $\mathbb{Q}_\ell(\mu')$ with $\mathbb{Q}_\ell(\mu_{\ell^\infty})$ over $\mathbb{Q}_\ell$, the case of abelian extensions of $\mathbb{Q}_\ell$ is completely explicit. □

See in 4.4.3.3 a slightly more global version of this exercise.

The product formula has the following converse which gives more precise information on the dependence of the Hasse symbols.

3.4.4 Theorem (converse of the product formula). *Let $(s_v)_{v\in Pl}$ be a family of elements $s_v \in \mathrm{Gal}(L^{\mathrm{ab}}/K)$ satisfying the following conditions:*

(i) $s_v \in D_v(L^{\mathrm{ab}}/K)$ for all v,

(ii) $s_v = 1$ for almost all v, and $\prod_v s_v = 1$.

Then there exists $x \in K^\times$ such that $\left(\frac{x\,,\,L^{\mathrm{ab}}/K}{v}\right) = s_v$ for all $v \in Pl$.

Proof. Let Σ be the (finite) support of $(s_v)_v$. Since the image of $K^\times$ under the Hasse symbol $\left(\frac{\bullet\,,\,L^{\mathrm{ab}}/K}{v}\right)$ is equal to $D_v(L^{\mathrm{ab}}/K)$, for each $v \in \Sigma$ there exists $x(v) \in K^\times$ such that $\left(\frac{x(v)\,,\,L^{\mathrm{ab}}/K}{v}\right) = s_v$; consider the idèle $\boldsymbol{x} := (x_v)_v$, whose components outside Σ are equal to 1, and where we have chosen $x_v := i_v(x(v))$ for $v \in \Sigma$. We then have:

$$\rho_{L/K}(\boldsymbol{x}) = \prod_v \left(\frac{x_v \, , \, L^{\mathrm{ab}}/K}{v} \right) = \prod_v s_v = 1,$$

so that there exist $x \in K^\times$ and $\boldsymbol{y} =: (y_w)_w \in J_L$ such that:

$$\boldsymbol{x} = i(x) \mathrm{N}_{L/K}(\boldsymbol{y}) \; ;$$

but $\Big(\prod_{w|v} \mathrm{N}_{L_w/K_v}(y_w) \, , \, (L^{\mathrm{ab}})_v/K_v \Big) = 1$ for all v since by definition we already have $(\mathrm{N}_{L_w/K_v}(y_w) \, , \, L_w^{\mathrm{ab}}/K_v) = 1$ for all $w|v$. Hence x is a solution to our problem. □

This result can in fact be expressed in terms of generalized norm residue symbols (see 3.1.4). For all $v \in Pl$, let $L_v^{\mathrm{ab}} := \bigcap_{w|v} L_w^{\mathrm{ab}}$, $G_v^{\mathrm{ab}} := \mathrm{Gal}(L_v^{\mathrm{ab}}/K_v)$ (see 1.5.3).

3.4.4′ Theorem. *For any family $(\sigma_v)_{v \in Pl} \in \bigoplus_{v \in Pl} G_v^{\mathrm{ab}}$ such that the product of the images of the $\sigma_v|_{(L^{\mathrm{ab}})_v}$ in G^{ab} is equal to the identity, there exists $x \in K^\times$ such that $(i_v(x), L_v^{\mathrm{ab}}/K_v) =: \left[\frac{x \, , \, L/K}{v} \right] = \sigma_v$ for all $v \in Pl$.*

Proof. We use here the fact that the local symbol:

$$(\bullet, L_v^{\mathrm{ab}}/K_v) : K_v^\times \longrightarrow G_v^{\mathrm{ab}}$$

is surjective to construct an idèle $\boldsymbol{x} := (x_v)_v$ such that $(x_v, L_v^{\mathrm{ab}}/K_v) = \sigma_v$ for each $v \in \Sigma$, $x_v = 1$ outside Σ (where Σ is the support of $(\sigma_v)_v$). Since $\left(\frac{x_v \, , \, L^{\mathrm{ab}}/K}{v} \right)$ is the canonical image in G^{ab} of $\sigma_v|_{(L^{\mathrm{ab}})_v}$ (see 3.1.4), we still have $\rho_{L/K}(\boldsymbol{x}) = 1$, $\boldsymbol{x} = i(x)\mathrm{N}_{L/K}(\boldsymbol{y})$, and x is still a solution, but note that we now use the (more precise) fact that for all v, $\Big(\prod_{w|v} \mathrm{N}_{L_w/K_v}(y_w) \, , \, L_v^{\mathrm{ab}}/K_v \Big) = 1$ for the same reasons as in the preceding case, or note that by 1.5.3 we have directly:

$$\langle \mathrm{N}_{L_w/K_v}(L_w^\times) \rangle_{w|v} = \mathrm{N}_{L_v^{\mathrm{ab}}/K_v}(L_v^{\mathrm{ab}\times}).$$

□

We will see in IV.4.5.5 that genus theory gives some additional information on this converse aspect of the product formula and shows that x can be chosen in a suitable S-unit group.

This finishes the first applications of the product formula.

We now come to the global existence theorem (i.e., the existence of abelian extensions of the global field K). By opposition to the local case, all the abelian extensions of K will be taken in a fixed algebraic closure $\overline{K}$ of K which may be independant of our various complex fields $\mathbb{C}_\ell$ or $\mathbb{C}_\infty$.

The analog of the local existence theorem can be obtained from the idèle class group C_K of K in the following way (coming back to J_K for convenience).

3.5 Theorem (global existence). *For any closed subgroup N of finite index of J_K containing $K^\times$, there exists a unique abelian extension M of K such that $K^\times \mathrm{N}_{M/K}(J_M) = N$; the reciprocity map yields the exact sequence:*

$$1 \longrightarrow N \longrightarrow J_K \longrightarrow \mathrm{Gal}(M/K) \longrightarrow 1.$$

In addition, the bijection between the closed subgroups of finite index of J_K containing $K^\times$ and the finite abelian extensions of K is a Galois correspondence which has the following properties (where M_1 and M_2 are abelian over K and correspond respectively to N_1 and N_2):

(i) we have $M_1 \subseteq M_2$ if and only if $N_2 \subseteq N_1$;

(ii) $M_1 M_2$ corresponds to $N_1 \cap N_2$;

(iii) $M_1 \cap M_2$ corresponds to $N_1 N_2$;

(iv) if $M_1 \subseteq M_2$, we have $\mathrm{Gal}(M_2/M_1) \simeq N_1/N_2$. □

3.5.1 Remarks. (i) As in the local case, the above Galois properties come from the existence of the correspondence and from 3.3, (i), (ii) on the global reciprocity map.

(ii) Similarly, if M corresponds to N, the decomposition subfield (resp. the inertia subfield) of a place v in M/K corresponds to $K_v^\times N$ (resp. to $U_v N$). The subfield of M fixed under $\rho_{M/K}(U_v^1) =: D_v^1(M/K) = D_{v,1}(M/K)$ is the maximal v-tamely ramified subextension of K in M. Hence the maximal tamely ramified extension of K in M corresponds to the idèle group:

$$\prod_v U_v^1 \, . \, N.$$

In the statement it is not necessary to refer to a set S; if we want that the places in such a set split completely, it is necessary and sufficient to include $\langle S \rangle$ in the subgroup N under consideration (see 3.3.5).

(iii) By abuse of notation, we will say that N is the norm group corresponding to the extension M/K.

(iv) Finally, recall that an open subgroup of J_K containing $K^\times$ is of finite index and necessarily contains a subgroup of the form $U_{\mathfrak{m}}^{\mathrm{res}}$ (see I.4.2.3). Thus there is an *equivalence* between a closed subgroup of finite index of J_K containing $K^\times$ and an open subgroup of J_K containing $K^\times$ (which corresponds to an open subgroup of C_K). Hence this contains the assertion about the existence of a conductor which will be studied in Section 4.

(v) The situation is the same if we express the correspondence in terms of subgroups of $J_{K,0}$ containing the diagonal embedding of $K^\times$ since we can go from one point of view to the other thanks to the identity $N = N_0 \oplus U_\infty$, which is self-explanatory. □

Note that in the correspondence of class field theory, the group $\mathrm{N}_{L/K}(J_L)$ does not characterize the extension L^{ab} (in other words, although the equality $\mathrm{N}_{L'/K}(J_{L'}) = \mathrm{N}_{L''/K}(J_{L''})$ clearly implies $L'^{\mathrm{ab}} = L''^{\mathrm{ab}}$, the converse is false). More precisely, Stern has given the following result (for the proof and the study of some consequences for norms, see [St]).

3.5.2 Proposition. *Let L' and L'' be two finite extensions of K, and let L be a Galois extension of K containing L' and L''. Set:*

$$G := \mathrm{Gal}(L/K), \quad H' := \mathrm{Gal}(L/L'), \quad H'' := \mathrm{Gal}(L/L'').$$

Denote by $H_{[*]}$ the set of primary elements (i.e., of order a prime power) of a group H. Then the following conditions are equivalent:
(i) $\mathrm{N}_{L'/K}(J_{L'}) \subseteq \mathrm{N}_{L''/K}(J_{L''})$,
(ii) $K^\times \cap \mathrm{N}_{L'/K}(J_{L'}) \subseteq K^\times \cap \mathrm{N}_{L''/K}(J_{L''})$,
(iii) $\mathrm{N}_{L'/K}(L'^\times) \cap \mathrm{N}_{L''/K}(L''^\times)$ *is of finite index in* $\mathrm{N}_{L''/K}(L''^\times)$,
(iv) $\bigcup_{s \in G} s\, H'_{[*]}\, s^{-1} \subseteq \bigcup_{s \in G} s\, H''_{[*]}\, s^{-1}$. □

Finally, as in the local case (same proof), we have the following consequence of 3.3.

3.5.3 Corollary (norm lifting theorem). *Let L/K be a finite extension of number fields and let M/K be an abelian extension. If N is the subgroup of J_K corresponding to M, then the subgroup N' of J_L corresponding to LM over L is given by:*

$$N' = \{\boldsymbol{y} \in J_L,\ \ \mathrm{N}_{L/K}(\boldsymbol{y}) \in N\} =: \mathrm{N}_{L/K}^{-1}(N).$$ □

3.5.4 Proposition (relative decomposition and inertia groups). *Let L/K be a finite extension and let L'/K be a subextension. Denote by N and N' the subgroups of J corresponding to L^{ab} and L'^{ab} (so that $N \subseteq N'$). Then, under the isomorphisms $D_v(L^{\mathrm{ab}}/K) \simeq K_v^\times N/N$ and $I_v(L^{\mathrm{ab}}/K) \simeq U_v N/N$ (see 3.3.5), we have:*

$$D_v(L^{\mathrm{ab}}/L'^{\mathrm{ab}}) \simeq N' \cap K_v^\times / N \cap K_v^\times,$$
$$I_v(L^{\mathrm{ab}}/L'^{\mathrm{ab}}) \simeq N' \cap U_v / N \cap U_v.$$

Proof. Indeed, we have the general exact sequence (see 1.2):

$$1 \longrightarrow D_v(L^{\mathrm{ab}}/L'^{\mathrm{ab}}) \longrightarrow D_v(L^{\mathrm{ab}}/K) \longrightarrow D_v(L'^{\mathrm{ab}}/K) \longrightarrow 1,$$

which can be written:

$$1 \longrightarrow N' \cap (K_v^\times N)/N \longrightarrow K_v^\times N/N \longrightarrow K_v^\times N'/N' \longrightarrow 1\,;$$

it is then immediate to check that $N' \cap (K_v^\times N)/N \simeq N' \cap K_v^\times / N \cap K_v^\times$. The case of inertia groups is completely similar. □

3.6 Theorem (Galois action). *Let M/K be a finite abelian extension of number fields and let g be an automorphism group of K with fixed subfield k. Let $N := K^\times \mathrm{N}_{M/K}(J_M)$ be the norm group corresponding to M/K. We have the following facts:*

(i) M/k is Galois if and only if g acts on N;

(ii) M/k is abelian if and only if g is commutative and there exists a subgroup n of J_k, containing the diagonal embedding of $k^\times$, such that $N = \mathrm{N}_{K/k}^{-1}(n)$, in which case M is the compositum of K with the abelian extension of k corresponding to n.

Proof. (i) Let τ be a k-isomorphism of M extending $t \in g$. By 3.3, (vi), the group corresponding to τM over $\tau K = K$ is equal to $\tau N = tN$ (since $\rho_{\tau M/K}(\tau \boldsymbol{x}) = 1$ is equivalent to $\rho_{M/K}(\boldsymbol{x}) = 1$); thus the uniqueness theorem indeed implies that $\tau M = M$ if and only if $tN = N$ (i.e., g acts on N).

(ii) If M/k is abelian, g is commutative and there exists n in J_k containing $k^\times$, corresponding to M/k. By 3.5.3, we have $N = \mathrm{N}_{K/k}^{-1}(n)$.

Conversely, assume that g is commutative and that N is of the form $\mathrm{N}_{K/k}^{-1}(n)$. Since by 1.4.3 and 1.4.4, for all $\mathfrak{m}$ (built on the ramified places in K/k), $\mathrm{N}_{K/k}(U_{K,\mathfrak{m}}^{\mathrm{res}})$ contains $U_{k,\mathfrak{n}}^{\mathrm{res}}$ for a suitable $\mathfrak{n}$ in k, it follows that $\mathrm{N}_{K/k} : J_K \longrightarrow J_k$ is an open map and so n, which contains $\mathrm{N}_{K/k}(N)$, is an open subgroup of J_k, hence of finite index since it contains $k^\times$.

Let k' be the abelian extension corresponding to n over k; since g is commutative, the field Kk' is the compositum of two abelian extensions of k and corresponds, over K, to $\mathrm{N}_{K/k}^{-1}(n) = N$. Hence, by uniqueness we have $Kk' = M$. □

This theorem is the starting point for a more general Galois study; for instance, if M/k is Galois and $[M : K]$ is prime to $|g|$, the action of g on N or, equivalently, that of g on $\mathrm{Gal}(M/K)$, characterizes the semidirect product $\mathrm{Gal}(M/k) = \mathrm{Gal}(M/K) \rtimes g$.

3.6.1 Example. Let $K/\mathbb{Q}$ be Galois with Galois group $G =: g$, and let H (resp. $\mathcal{C}\ell$) be the restricted or the ordinary Hilbert class field (resp. class group) of K. If $|G|$ and $|\mathcal{C}\ell|$ are coprime, $\mathrm{Gal}(H/\mathbb{Q}) \simeq \mathcal{C}\ell \rtimes G$ is characterized by the relations:

$$s' \left(\frac{H/K}{c\ell(\mathfrak{a})}\right) s'^{-1} = \left(\frac{H/K}{c\ell(s\,\mathfrak{a})}\right),$$

for any s' extending $s \in G$ and any ideal $\mathfrak{a}$. Thus the Galois structure of $\mathcal{C}\ell$ gives that of $\mathrm{Gal}(H/\mathbb{Q})$. Note that if $\mathcal{C}\ell \neq 1$, $\mathcal{C}\ell \rtimes G$ is never a direct product since this is equivalent to $\mathcal{Cl} = \mathcal{Cl}^G$, therefore to $\mathcal{Cl} = 1$ because of the assumption on the orders (hint: if the class h is fixed under G, then, since $\mathbb{Q}$ is principal, $1 = \nu_{L/K}(h) := \prod_{s\in G} h^s = h^{|G|}$; or use the fact that one can write $H = KM$ with $M/\mathbb{Q}$ abelian and unramified). □

The most complete achievement is then the Šafarevič–Weil theorem,[19] which characterizes the group extension:

$$1 \longrightarrow \mathrm{Gal}(M/K) \longrightarrow \mathrm{Gal}(M/k) \longrightarrow g \longrightarrow 1,$$

which is of a cohomological nature, in terms of the fundamental class briefly mentioned in 3.2.2. More precisely, the element of $H^2(g, \mathrm{Gal}(M/K))$ associated to this group extension is the image of the fundamental class under the composite of canonical maps:

$$H^2(g, C_K) \longrightarrow H^2(g, C_K/\mathcal{cl}_K(N)) \xrightarrow{\rho_{M/K}} H^2(g, \mathrm{Gal}(M/K)),$$

where $N \subset J_K$ is the norm group corresponding to M/K.

b) Global Class Field Theory in $\overline{K}^{\mathrm{ab}}/K$

To conclude, we want to show how the global reciprocity map behaves when we take the inverse limit of the J/N (from the correspondence of 3.5), hoping that this will not create some new and dreadful object; we will see that this is not the case.

3.7 Reciprocity Map in $\overline{K}^{\mathrm{ab}}/K$. By the general principles, we can go to the limit as in the local case by writing that:

$$\overline{G}^{\mathrm{ab}} := \mathrm{Gal}(\overline{K}^{\mathrm{ab}}/K) \simeq \varprojlim_{N} J/N,$$

where N ranges in the set of open (or closed of finite index) subgroups of J containing $K^\times$. As already explained, these subgroups N must necessarily contain a subgroup of the form $U_{\mathfrak{m}}^{\mathrm{res}} = U_{0,\mathfrak{m}}^{\mathrm{res}} \oplus U_\infty$, where (see I.5.2):

$$U_{0,\mathfrak{m}}^{\mathrm{res}} := \prod_{v\in Pl_0\backslash T} U_v \prod_{v\in T} U_v^{m_v}$$

if $\mathfrak{m} = \prod_{v\in T} \mathfrak{p}_v^{m_v}$, and where $U_\infty := \bigoplus_{v|\infty} U_v \simeq (\mathbb{R}^{\times +})^{r_1} \times (\mathbb{C}^\times)^{r_2}$ is the connected component of the unit element of J. Thus the group $\overline{G}^{\mathrm{ab}}$ is also of the form (using reduced idèles):

---

[19] See [e, Ko3, Ch. 2, § 7.1] and [i, Miy0, Koch] for the history of this result whose name would aptly be "Šafarevič–Hochschild–Nakayama–Jehne theorem", as explained by Koch.

$$\varprojlim_{N_0} J_0/N_0,$$

where N_0 ranges in the set of open subgroups of J_0 containing $K^\times$, of which a cofinal subset is formed by the $K^\times U_{0,\mathfrak{m}}^{\mathrm{res}}$.

We are going to see however that these inverse limits can easily be written in terms of quotients of J or J_0, which in practice avoids working in C or C_0.

3.7.1 Definition. Let ρ be the limit reciprocity map: $\rho : J \longrightarrow \overline{G}^{\mathrm{ab}}$, defined by:

$$\rho(\boldsymbol{x}) := (\rho_{M/K}(\boldsymbol{x}))_M \in \varprojlim_M \mathrm{Gal}(M/K) \simeq \varprojlim_N J/N,$$

for the finite abelian extensions M/K, N denoting the norm group of M (i.e., the kernel of the reciprocity map $\rho_{M/K}$ defined in 3.1). □

The fundamental exact sequence (in terms of reduced idèles):

$$1 \longrightarrow K^\times U_0^{\mathrm{ord}} \longrightarrow J_0 \longrightarrow \mathcal{C}\ell^{\mathrm{ord}} \longrightarrow 1,$$

shows that there exists a finite number of representative idèles $\boldsymbol{x}_0^i \in J_0$, $1 \leq i \leq \gamma$, such that $J_0 = \{\boldsymbol{x}_0^i,\ 1 \leq i \leq \gamma\}\, K^\times\, U_0^{\mathrm{ord}}$.

We have:

$$J/K^\times U_\infty \simeq J_0/K^\times = \{\boldsymbol{x}_0^i,\ 1 \leq i \leq \gamma\}\, K^\times\, U_0^{\mathrm{ord}}/K^\times,$$

which is represented by the set $\{\boldsymbol{x}_0^i,\ 1 \leq i \leq \gamma\}\, U_0^{\mathrm{ord}}$; we can then apply I.5.5 to $A = J$, the subgroups $N \subset J$ corresponding to the finite abelian extensions M/K, $B = K^\times U_\infty$, and the compact set $C = \{\boldsymbol{x}_0^i,\ 1 \leq i \leq \gamma\}\, U_0^{\mathrm{ord}}$, and so we deduce that ρ is *surjective* (its kernel being trivially equal to $\bigcap_N N$).

We have of course the analogous surjective map $\rho : J_0 \longrightarrow \overline{G}^{\mathrm{ab}}$, defined by $\rho(\boldsymbol{x}_0) := (\rho_{M/K}(\boldsymbol{x}_0))_M$, whose kernel is $\bigcap_{N_0} N_0$.

We thus have the following homeomorphisms:

$$\begin{aligned}\overline{G}^{\mathrm{ab}} &\simeq \varprojlim_N J/N \simeq J\Big/\bigcap_N N \simeq J\Big/\bigcap_{\mathfrak{m}}(K^\times U_{\mathfrak{m}}^{\mathrm{res}}) \simeq C/D \\ &\simeq \varprojlim_{N_0} J_0/N_0 \simeq J_0\Big/\bigcap_{N_0} N_0 \simeq J_0\Big/\bigcap_{\mathfrak{m}}(K^\times U_{0,\mathfrak{m}}^{\mathrm{res}}) \simeq C_0/D_0,\end{aligned}$$

where $D := \mathcal{C}\ell\Big(\bigcap_{\mathfrak{m}}(K^\times U_{\mathfrak{m}}^{\mathrm{res}})\Big)$, $D_0 := \mathcal{C}\ell_0\Big(\bigcap_{\mathfrak{m}}(K^\times U_{0,\mathfrak{m}}^{\mathrm{res}})\Big)$, are the connected components of the unit element of C and C_0 respectively, and where we recall that (see I.4.2.5, I.4.2.8, (ii)):

$$D = \mathcal{C}\ell\Big(\bigcap_{\mathfrak{m}}(E^{\mathrm{ord}} U_{\mathfrak{m}}^{\mathrm{res}})\Big),\quad D_0 = \mathcal{C}\ell_0\Big(\bigcap_{\mathfrak{m}}(E^{\mathrm{ord}} U_{0,\mathfrak{m}}^{\mathrm{res}})\Big).$$

3.7.2 Remark. Note that in general the $U^{\mathrm{ord}}_{0,\mathfrak{m}}$ do not form a fundamental system of neighbourhoods of 1 in J_0; in a similar way, although the $U^{\mathrm{res}}_{0,\mathfrak{m}}$ form such a fundamental system, this is not the case for the $U^{\mathrm{res}}_{\mathfrak{m}}$ in J (because of the archimedean factors), and only $\bigcap_{\mathfrak{m}}(K^\times U^{\mathrm{res}}_{0,\mathfrak{m}})$ represents the closure of the image of $K^\times$ in J_0, so that we have:

$$D_0 = \mathrm{adh}_0(K^\times)/K^\times = \mathrm{adh}_0(E^{\mathrm{ord}})/E^{\mathrm{ord}}.$$

However, we can write $D = \mathrm{adh}(K^\times U_\infty)/K^\times$. □

Summarizing these results, we obtain the following description of $\overline{G}^{\mathrm{ab}} := \mathrm{Gal}(\overline{K}^{\mathrm{ab}}/K)$ by means of the usual reciprocity maps $\rho_{M/K}$ for finite abelian extensions M/K.

3.7.3 Theorem. *The infinite reciprocity map* $\rho : J \longrightarrow \overline{G}^{\mathrm{ab}}$ *(resp.* $\rho : J_0 \longrightarrow \overline{G}^{\mathrm{ab}}$*), which associates with* $\boldsymbol{x} \in J$ *(resp.* $\boldsymbol{x}_0 \in J_0$*),* $(\rho_{M/K}(\boldsymbol{x}))_M$ *(resp.* $(\rho_{M/K}(\boldsymbol{x}_0))_M$*, is surjective.*

Thus ρ *induces the canonical homeomorphisms:*

$$\overline{G}^{\mathrm{ab}} \simeq J/\mathrm{adh}(K^\times U_\infty) \simeq J_0/\mathrm{adh}_0(K^\times).$$ □

Of course, the composition of ρ and the restriction $\overline{G}^{\mathrm{ab}} \longrightarrow \mathrm{Gal}(M/K)$ yields the reciprocity map $\rho_{M/K}$ for any abelian extension M/K (finite or not).

3.7.4 Corollary. *Taking quotients by the diagonal embeddings of* $K^\times$*, we can write:*

$$\overline{G}^{\mathrm{ab}} \simeq C/D \simeq C_0/D_0,$$

where $D = c\ell\Big(\bigcap_{\mathfrak{m}}(K^\times U^{\mathrm{res}}_{\mathfrak{m}})\Big) = \mathrm{adh}(K^\times U_\infty)/K^\times$, $D_0 = c\ell_0\Big(\bigcap_{\mathfrak{m}}(K^\times U^{\mathrm{res}}_{0,\mathfrak{m}})\Big) = \mathrm{adh}_0(K^\times)/K^\times$. □

3.8 INFINITE GLOBAL CLASS FIELD THEORY CORRESPONDENCE. The correspondence for *infinite* idelic class field theory can be expressed in terms of either:

- closed subgroups of J containing $K^\times U_\infty$,
- closed subgroups of C containing D,
- closed subgroups of J_0 containing $K^\times$,
- closed subgroups of C_0 containing D_0.

In addition, the bijection between the set of abelian extensions of K and the set of closed subgroups of J_0 containing $K^\times$ (for instance) is a Galois correspondence having the properties (i) to (iv) of 3.5.

3.8.1 Remarks. (i) Under this correspondence, the decomposition and inertia groups are still related to the images under ρ of the groups $K_v^\times$ and U_v but this is not enough to identify them; in other words the computation of $K_v^\times \mathrm{adh}_0(K^\times)/\mathrm{adh}_0(K^\times)$ or of $K_v^\times/K_v^\times \cap \mathrm{adh}_0(K^\times)$ is neither sufficient nor a priori easy. It is however easy to see that we have $\rho(U_v) = I_v(\overline{K}^{\mathrm{ab}}/K)$ and that $\rho(K_v^\times)$ is dense in $D_v(\overline{K}^{\mathrm{ab}}/K)$. In addition there is a topological problem since J induces on $K_v^\times$ its usual topology (with neighbourhoods U_v^m), while it is that induced by $\widehat{K_v^\times}$ (with neighbourhoods $\pi_v^{n\mathbb{Z}} \oplus U_v^m$) which is suitable since $D_v(\overline{K}^{\mathrm{ab}}/K)$ is obtained by an inverse limiting process (which we will give in III.4.12.5 following III.4.5); recall also the problem that we have met in I.4.2.8, (iv). All this needs Theorem III.4.3 of Schmidt–Chevalley, which uses the local-global principle 6.3.3 on powers. It is thus natural to delay the study of all questions dealing with the global structure of $\overline{K}^{\mathrm{ab}}/K$ which are logically equivalent to the study of the properties of D and D_0, and will be the object of the next chapter.

(ii) The group D is also called the universal norm group simply because it corresponds to $\overline{K}^{\mathrm{ab}}$ by infinite class field theory, and because it is contained in the images in C of all the norm groups $K^\times \mathrm{N}_{L/K} J_L$ of finite extensions L of K. As we have already mentioned, D is the connected component of the unit element of C and also its maximal divisible subgroup. We will show this last point in III.4.15.1. Exactly the same things can be said about D_0 in J_0. □

In terms of classes of reduced idèles, the above yields:

$$C_0 = \{\mathcal{cl}_0(\boldsymbol{x}_0^i),\ 1 \le i \le \gamma\} \,.\, \mathcal{cl}_0(U_0^{\mathrm{ord}}),$$

and the exact sequence:

$$1 \longrightarrow \mathcal{cl}_0(U_0^{\mathrm{ord}})D_0/D_0 \longrightarrow C_0/D_0 \longrightarrow \mathcal{Cl}^{\mathrm{ord}} \longrightarrow 1,$$

shows that the study of C_0/D_0 can be reduced to that of:

$$\mathcal{cl}_0(U_0^{\mathrm{ord}})D_0/D_0 \simeq K^\times U_0^{\mathrm{ord}} \Big/ \bigcap_{\mathfrak{m}} (K^\times U_{0,\mathfrak{m}}^{\mathrm{res}}) \simeq U_0^{\mathrm{ord}} \Big/ \bigcap_{\mathfrak{m}} (E^{\mathrm{ord}} U_{0,\mathfrak{m}}^{\mathrm{res}}),$$

which is $U_0^{\mathrm{ord}}/\mathrm{adh}_0(E^{\mathrm{ord}})$ or, equivalently, the quotient of $U_0^{\mathrm{ord}}/E^{\mathrm{ord}}$ by the connected component D_0.

We obtain the following result which will be found again in III.1.6.7.

3.8.2 Theorem (global exact sequence of class field theory). *We have the exact sequence:*

$$1 \longrightarrow U_0^{\mathrm{ord}}/\mathrm{adh}_0(E^{\mathrm{ord}}) \xrightarrow{\ \rho\ } \overline{G}^{\mathrm{ab}} \longrightarrow \mathcal{Cl}^{\mathrm{ord}} \longrightarrow 1,$$

in which $U_0^{\mathrm{ord}}/\mathrm{adh}_0(E^{\mathrm{ord}}) \simeq \mathrm{Gal}(\overline{K}^{\mathrm{ab}}/H^{\mathrm{ord}})$ *and* $\mathcal{Cl}^{\mathrm{ord}} \simeq \mathrm{Gal}(H^{\mathrm{ord}}/K)$. □

The determination of the structure of D_0 is the object of Theorem III.4.4.6.

The point of view of the notes of Artin–Tate [d, AT, Ch. 9, § 1], and before of the book of Weil [h, We2, III], is to give explicitly the structure of D (in particular for the computation of the cohomology of C and of C/D). Our point of view consists also in looking at the formulas:

$$\overline{G}^{\mathrm{ab}} \simeq \varprojlim_{N} J/N \simeq \varprojlim_{N_0} J_0/N_0 \simeq \varprojlim_{\mathfrak{m}} \mathcal{C}\ell_{\mathfrak{m}}^{\mathrm{res}},$$

which show how the finite case, which is amenable to numerical computations, regularizes when one takes the limit, but there is no difficulty in expressing and in proving certain results thanks to the properties of C/D or of C_0/D_0, and we will do so when needed (for instance in Chapter III, Section 4, and in the Appendix).

The structure of the inverse limit $\varprojlim_{\mathfrak{m}} \mathcal{C}\ell_{\mathfrak{m}}^{\mathrm{res}}$ and especially those of its p-Sylow subgroups are quite complex, and their arithmetic computation will be the object of Chapter III.

3.8.3 Remark. It is interesting to note a difference between the local and global cases. In the global case, the map:

$$\rho : J \longrightarrow \mathrm{Gal}(\overline{K}^{\mathrm{ab}}/K)$$

is surjective, while in the local case, the analogous map:

$$\rho_v := (\,\bullet\,, \overline{K}_v^{\mathrm{ab}}/K_v) : K_v^\times \longrightarrow \mathrm{Gal}(\overline{K}_v^{\mathrm{ab}}/K_v)$$

only has a dense image. This comes from the fact that in the local case $\overline{K}_v^{\mathrm{nr}}/K_v$ is infinite, contrary to the global case where H^{ord}/K is finite: indeed, the relative Galois groups $\mathrm{Gal}(\overline{K}_v^{\mathrm{ab}}/\overline{K}_v^{\mathrm{nr}})$ and $\mathrm{Gal}(\overline{K}^{\mathrm{ab}}/H^{\mathrm{ord}})$ are the images under the continuous maps ρ_v and ρ of the compact groups U_v and U_0^{ord}, but $K_v^\times/U_v \simeq \mathbb{Z}$ is not compact, contrary to $J_0/K^\times U_0^{\mathrm{ord}}$ which is finite. Note also that for any $x_v \in K_v^\times$ (seen as an idèle), $\rho(x_v)$ corresponds to $\rho_v(x_v)|_{(\overline{K}^{\mathrm{ab}})_v}$ under the identification of $D_v(\overline{K}^{\mathrm{ab}}/K)$ with $\mathrm{Gal}((\overline{K}^{\mathrm{ab}})_v/K_v)$. But we will prove in III.4.5 that $(\overline{K}^{\mathrm{ab}})_v = \overline{K}_v^{\mathrm{ab}}$; thus we indeed have that $\rho|_{K_v^\times}$ corresponds to ρ_v, which confirms the topological problems we have mentionned at several occasions. □

§4 Global Class Field Theory: Class Group Version

a) Global Norm Conductor — Properties

Let L/K be a finite extension of number fields; the use of generalized ideal class groups implies that we must introduce the fundamental notion of a global norm conductor. To begin with, recall that from our point of view, the existence of a modulus $\mathfrak{m}$ satisfying the condition:

$$U_{\mathfrak{m}}^{\mathrm{res}} \subseteq \mathrm{N}_{L/K}(J_L),$$

which says that the group $\mathrm{N}_{L/K}(J_L)$ is open, hence that the group:

$$K^\times \mathrm{N}_{L/K}(J_L),$$

is open with finite index in J_K, comes essentially from local class field theory since, for every place v unramified in L/K we have, for the places w of L above v:

$$U_v = \mathrm{N}_{L_w/K_v}(U_w)$$

(see 1.4.3, (ii)), and that for every place v ramified in L/K, there exists a sufficiently large i such that for all $w|v$ we have (see 1.4.4):

$$U_v^i \subseteq (U_v)^{[L_w:K_v]} \subseteq \mathrm{N}_{L_w/K_v}(U_w).$$

The existence of $\mathfrak{m}$ (also called an admissible modulus) trivially implies that of a *smallest* admissible modulus since we have $U_v^{m_1} U_v^{m_2} = U_v^{\min(m_1, m_2)}$, for any $v \in Pl$. The support of this modulus is contained in the set of places ramified in L/K.[20] We can thus state the following.

4.1 Theorem and Definition (global norm conductor). *Let L/K be a finite extension of number fields. There exists a smallest modulus $\mathfrak{f}_{L/K} =: \mathfrak{f}$ of K, such that:*

$$U_{\mathfrak{f}}^{\mathrm{res}} \subseteq K^\times \mathrm{N}_{L/K}(J_L).$$

This modulus is called the global norm conductor or the conductor of L/K. It only depends on the maximal abelian subextension L^{ab}/K of L/K and hence is also equal to $\mathfrak{f}_{L^{\mathrm{ab}}/K}$, the global norm conductor of L^{ab}/K (see 3.3.3). □

Note. According to our point of view, $\mathfrak{f}$ is a nonzero integral ideal of K and in particular does not involve any infinite places. Moreover, its support is contained in the set of places ramified in L/K (even with the meaning of the above footnote) since it is an admissible module for $K^\times \mathrm{N}_{L/K}(J_L)$ which contains $\mathrm{N}_{L/K}(J_L)$. See the more precise result 4.2.

4.1.1 Proposition (conductor of a compositum of fields). *If L^{ab} is the compositum, over K, of the extensions $M_1, \ldots, M_n$, then its conductor is equal to the l.c.m. of the conductors of the M_i for $1 \leq i \leq n$.*

[20] More precisely, in the non-Galois case we have $U_v \subset \mathrm{N}_{L/K}(J_L)$ if and only if L_v^{ab}/K_v is unramified, so that this support is contained in the set of places v such that *all* $w|v$ ramify in L/K. For the sequel (abelian case), this information is useless.

Proof. Immediate from Definition 4.1 and the fact that we have:

$$K^\times \mathrm{N}_{L/K}(J_L) = \bigcap_{i=1}^{n} (K^\times \mathrm{N}_{M_i/K}(J_{M_i})),$$

by 3.5, (ii). □

For example, if L/K is a p-elementary extension (i.e., $\mathrm{Gal}(L/K) \simeq (\mathbb{Z}/p\mathbb{Z})^r$, p prime), its norm group and its conductor can be computed from the case of a cyclic extension of degree p; if $\mu_p \subset K$, the result immediately follows from 1.6.3, otherwise we can be reduced to the Kummer case because of norm lifting Theorem 3.5.3.

4.1.2 Remark. If M is the intersection of the M_i, we can only say that the conductor of M divides the g.c.d. of the conductors of the M_i; for example, for $K = \mathbb{Q}$, $M_1 = \mathbb{Q}(\sqrt{-1})$, $M_2 = \mathbb{Q}(\sqrt{2})$, we have $\mathfrak{f}_1 = (4)$, $\mathfrak{f}_2 = (8)$ (see 1.6.5), g.c.d.$(\mathfrak{f}_1, \mathfrak{f}_2) = (4)$, but $M_1 \cap M_2 = \mathbb{Q}$. □

The global norm conductor has the following property (which comes from the local case, as is shown by the proof of Lemma 4.2.1).

4.2 Theorem (of the conductor). *Let L/K be a finite extension of number fields and let $\mathfrak{f}$ be its global norm conductor. Then a (finite) place v of K is ramified in L^{ab}/K if and only if $\mathfrak{p}_v$ divides $\mathfrak{f}$ (i.e., the support of $\mathfrak{f}$ is equal to the set R of places which are ramified in L^{ab}/K).* □

4.2.1 Lemma (computation of a global norm conductor). *Let L/K be a finite extension. Then $\mathfrak{f} := \mathfrak{f}_{L/K}$ is equal to the product of the local v-conductors of L^{ab}/K, in other words $\mathfrak{f} = \prod_{v \in Pl_0} \mathfrak{f}_v(L^{\mathrm{ab}}/K)$ (see 1.6, (ii)).*

Proof of the statements. In the fundamental Corollary 3.3.1 we have observed that the norm groups N_v (corresponding to $(L^{\mathrm{ab}})_v/K_v$) are the $N \cap K_v^\times$, where $N := K^\times \mathrm{N}_{L/K}(J_L)$; thus, using 1.4.3, (ii), it is clear that $U_v^{m_v} \subseteq N_v$ for all v (taking $m_v = 0$ if $v \notin R$), is equivalent to $U_{\mathfrak{m}}^{\mathrm{res}} \subseteq N$, for $\mathfrak{m} = \prod_{v \in R} \mathfrak{p}_v^{m_v}$. This proves the result as well as the theorem of the conductor. □

The above lemma gives a result which is essential for the practical computation of a global conductor: indeed, in general we know a multiple of the discriminant of L^{ab}/K, so that we are reduced to a finite (explicit) number of computations of local v-conductors of cyclic extensions by 4.1.1 (for this, we use Formula 1.6.2).

We illustrate the above on an example showing that the local information coming from the L_v^{ab}/K_v should not be mistaken for that coming from the subextension L^{ab}/K, even when L/K is Galois.

Example. Consider the extension $L = \mathbb{Q}(\sqrt[3]{7}, j)$ of $K = \mathbb{Q}$, in which the ramified places are 3 and 7. For the place $v = 7$ of K, we obtain $L_v^{\mathrm{ab}} = \mathbb{Q}_7(\sqrt[3]{7})$ which is a cyclic extension of degree 3 of $\mathbb{Q}_7$; it follows that we have $\mathfrak{f}_{L_v^{\mathrm{ab}}/K_v} = (7)$ (7 is tamely ramified in L_v^{ab}/K_v) while the global conductor $\mathfrak{f} = \prod_v \mathfrak{f}_v(L^{\mathrm{ab}}/K)$, which is the conductor of $L^{\mathrm{ab}}/K = \mathbb{Q}(j)/\mathbb{Q}$, is equal to (3) (i.e., for $v = 7$ the local v-conductor of L^{ab}/K is equal to 1, or equivalently, we have $L_v^{\mathrm{ab}} = \mathbb{Q}_7(\sqrt[3]{7})$, but $(L^{\mathrm{ab}})_v = \mathbb{Q}_7$). □

4.2.2 Exercise. Deduce from the proof of Lemma 4.2.1 the equivalence of the following conditions:

$$U_{\mathfrak{m}}^{\mathrm{res}} \subseteq N := K^\times \mathrm{N}_{L/K}(J_L) \ \text{ and } \ U_{\mathfrak{m}}^{\mathrm{res}} \subseteq \mathrm{N}_{L^{\mathrm{ab}}/K}(U_{L^{\mathrm{ab}}}^{\mathrm{res}}).$$

Answer. One direction is clear since $N = K^\times \mathrm{N}_{L^{\mathrm{ab}}/K}(J_{L^{\mathrm{ab}}})$; for the other, it suffices to check that $N \cap U_v = N_v \cap U_v = \mathrm{N}_{(L^{\mathrm{ab}})_v/K_v}(U_{(L^{\mathrm{ab}})_v})$, where $U_{(L^{\mathrm{ab}})_v}$ is the unit group of $(L^{\mathrm{ab}})_v$.

Beware that the results of 2.6 show that $U_{\mathfrak{m}}^{\mathrm{res}} \subseteq K^\times \mathrm{N}_{L/K}(J_L)$ is in general not equivalent to $U_{\mathfrak{m}}^{\mathrm{res}} \subseteq \mathrm{N}_{L/K}(U_L^{\mathrm{res}}) = \prod_v \mathrm{N}_{\check{L}_v^{\mathrm{ab}}/K_v}(U_{\check{L}_v^{\mathrm{ab}}})$ since we have:

$$\Big(\mathrm{N}_{L^{\mathrm{ab}}/K}(U_{L^{\mathrm{ab}}}^{\mathrm{res}}) : \prod_v \mathrm{N}_{\check{L}_v^{\mathrm{ab}}/K_v}(U_{\check{L}_v^{\mathrm{ab}}})\Big) = \prod_v \frac{\check{e}_v^{\mathrm{ab}}}{e_v(L^{\mathrm{ab}}/K)}\ ;$$

neither is it equivalent to $U_{\mathfrak{m}}^{\mathrm{res}} \subset \mathrm{N}_{L/K}(J_L)$, which means that for all v, $U_v^{m_v} \subseteq \mathrm{N}_{L_v^{\mathrm{ab}}/K_v}(U_{L_v^{\mathrm{ab}}})$, since the index:

$$\Big(\mathrm{N}_{L^{\mathrm{ab}}/K}(U_{L^{\mathrm{ab}}}^{\mathrm{res}}) : \prod_v \mathrm{N}_{L_v^{\mathrm{ab}}/K_v}(U_{L_v^{\mathrm{ab}}})\Big) = \prod_v \frac{e_v^{\mathrm{ab}}}{e_v(L^{\mathrm{ab}}/K)},$$

can also be different from 1 (recall that $\check{e}_v^{\mathrm{ab}} := e(\check{L}_v^{\mathrm{ab}}/K_v)$ and $e_v^{\mathrm{ab}} := e(L_v^{\mathrm{ab}}/K_v)$). This remark remains in the case where L/K is only Galois (use the example given above). □

We also give the following classical property which is summarized under the name “Führerdiskriminantenproduktformel”.

4.2.3 Proposition. *Let L/K be a finite extension of number fields. The relative discriminant of the subextension L^{ab}/K is:*

$$\mathfrak{d}_{L^{\mathrm{ab}}/K} = \prod_\chi \mathfrak{f}_\chi,$$

where χ ranges in the dual of G^{ab} and where $\mathfrak{f}_\chi$ is the global norm conductor of the subfield of L^{ab} fixed under the kernel of χ. □

4.2.4 Nonabelian Artin Conductors. Recall, without any justification, that these (abelian) conductors have the following Galois generalization which comes from higher ramification theory ([d, Se2, Ch. VI, § 2; CF, Ch. VI, § 4], [c, Neu1, Ch. VII, § 11]).

Denote by $\Psi(\Gamma)$ the set of absolutely irreducible characters of a finite group Γ. Let L/K be a finite Galois extension with Galois group G, and let L_{w_0}/K_v, with Galois group G_{w_0}, be a completion of L/K at $v \in Pl_0$ and a fixed $w_0|v$; for any $\psi \in \Psi(G_{w_0})$ we set:

$$\mathfrak{f}_v^{\mathrm{art}}(\psi) := \mathfrak{p}_v^{m_{v,\psi}},$$

with:

$$m_{v,\psi} := \frac{1}{g_0} \sum_{i \geq 0} g_i \Big(\psi(1) - \frac{1}{g_i} \sum_{s \in G_{w_0,i}} \psi(s)\Big),$$

where $G_{w_0,i}$ is the ith higher ramification group of L_{w_0}/K_v (in lower numbering) and $g_i := |G_{w_0,i}|$. For an arbitrary character χ we define $\mathfrak{f}_v^{\mathrm{art}}(\chi)$ by linearity, and this modulus is called the local v-conductor of the character χ. If ψ is of degree 1, the factor:

$$\psi(1) - \frac{1}{g_i} \sum_{s \in G_{w_0,i}} \psi(s),$$

is equal to 0 or 1 depending on whether or not the restriction of ψ to $G_{w_0,i}$ is the unit character, and we recover the norm conductor of the cyclic extension fixed under the kernel of ψ [d, Se2, Ch. VI, § 2, Prop. 5, Cor.].

This gives the local Artin v-conductors for the extension L/K. We then define the global Artin conductors, for any $\psi \in \Psi(G)$, by:

$$\mathfrak{f}^{\mathrm{art}}(\psi) := \prod_{v \in Pl_0} \mathfrak{f}_v^{\mathrm{art}}(\mathrm{Res}_v(\psi)),$$

where $\mathrm{Res}_v(\psi)$ is the restriction of ψ to $D_{w_0}(L/K) \simeq G_{w_0}$ (it does not depend on the choice of w_0). We thus have the corresponding formula for the relative discriminant of L/K (Artin–Hasse):

$$\mathfrak{d}_{L/K} = \prod_{\psi \in \Psi(G)} (\mathfrak{f}^{\mathrm{art}}(\psi))^{\psi(1)}.$$

An important property of the Artin conductor is that it *characterizes* the ramification for the extension L/K (and not only for the extension L^{ab}/K), but this is not anymore part of class field theory.

The reader can refer to [e, Ko3, Ch. 5, § 1] to have an overview on questions dealing with Artin L-functions, whose study at $s = 0$ is the object of Stark's conjectures.

b) Artin's Reciprocity Map — Reciprocity Law — Global Computation of Hasse Symbols — Decomposition Law

To go from the idelic to the generalized class group point of view, we have at our disposal the fundamental exact sequence of Theorem I.5.1, relative to the usual data T, $\mathfrak{m}$, and S prime to T:

$$1 \longrightarrow K^\times U_{\mathfrak{m}}^S/K^\times \simeq U_{\mathfrak{m}}^S/E_{\mathfrak{m}}^S \longrightarrow C \xrightarrow{\gamma_{\mathfrak{m}}^S} \mathcal{C}\ell_{\mathfrak{m}}^S \longrightarrow 1.$$

Furthermore, the above fundamental results (in idelic terms) for a finite extension L/K and a finite set S of noncomplex places of K unramified in L^{ab}/K, are:

(α) the exact sequence:

$$1 \longrightarrow K^\times \langle S \rangle \mathrm{N}_{L/K}(J_L)/K^\times \longrightarrow C \xrightarrow{\rho_{L/K}^S} G^{\mathrm{ab}\,S} \longrightarrow 1,$$

with $\langle S \rangle := \bigoplus_{v \in S} K_v^\times$, where $K^\times \mathrm{N}_{L/K}(J_L)$ is an open subgroup of J, and $G^{\mathrm{ab}\,S} := \mathrm{Gal}(L^{\mathrm{ab}\,S}/K)$;

(β) the existence theorem which says that, conversely, for any open subgroup N of J containing the image of $K^\times$, there exists L/K such that we indeed have:

$$1 \longrightarrow N\langle S \rangle/K^\times \longrightarrow C \xrightarrow{\rho_{L/K}^S} G^{\mathrm{ab}\,S} \longrightarrow 1.$$

We then see that we may successively:

(α') factor $\rho_{L/K}^S$ as a map from $\mathcal{C}\ell_{\mathfrak{m}}^S$ to $G^{\mathrm{ab}\,S}$, for $\mathfrak{m}$ multiple of the conductor of L/K;

(β') express the existence theorem in terms of subgroups of $\mathcal{C}\ell_{\mathfrak{m}}^S$, which will be equivalent to classifying abelian extensions of K by their conductor.

4.3 The Fundamental Diagram for Artin and Reciprocity Maps. The translation in terms of generalized class groups of the properties of the global reciprocity map relies on the following commutative diagram, in which L/K is a finite extension of number fields, $\mathfrak{m}$ is any multiple of the norm conductor $\mathfrak{f}$ of L/K, and S is a finite set of noncomplex places of K, disjoint from the set T containing the support of $\mathfrak{m}$. Recall that $U_{\mathfrak{m}}^S = U_{\mathfrak{m}}^{\mathrm{res}}\langle S \rangle$ and, by our assumption on $\mathfrak{m}$, that we have:

$$U_{\mathfrak{m}}^{\mathrm{res}} \subseteq K^\times \mathrm{N}_{L/K}(J_L)\ ;$$

we also recall that the map $\gamma_{\mathfrak{m}}^S$ defines the fundamental exact sequence of I.5.1 and that $c\ell_{\mathfrak{m}}^S$ is the canonical map:

$$I_T \longrightarrow \mathcal{C}\ell_{\mathfrak{m}}^S.$$

This commutative diagram has the following form (where N denotes $\mathrm{N}_{L/K}$):

$$\begin{array}{ccccccc}
 & 1 & & 1 & & & \\
 & \downarrow & & \downarrow & & & \\
 & K^\times U^S_{\mathfrak{m}}/K^\times & = & K^\times U^S_{\mathfrak{m}}/K^\times & & & \\
 & \downarrow & & \downarrow & & & \\
1 \longrightarrow & K^\times\langle S\rangle \mathrm{N}(J_L)/K^\times & \longrightarrow & C & \xrightarrow{\rho^S_{L/K}} & \mathrm{Gal}(L^{\mathrm{ab}\,S}/K) & \longrightarrow 1 \\
 & \downarrow & & \downarrow \gamma^S_{\mathfrak{m}} & & \| & \\
1 \longrightarrow & \mathcal{C}\ell^S_{\mathfrak{m}}(\mathrm{N}(I_{L,T})) & \longrightarrow & \mathcal{C}\ell^S_{\mathfrak{m}} & \xrightarrow{\alpha^S_{L/K}} & \mathrm{Gal}(L^{\mathrm{ab}\,S}/K) & \longrightarrow 1 \\
 & \downarrow & & \downarrow & & & \\
 & 1 & & 1 & & &
\end{array}$$

To show its validity, it is sufficient to define $\alpha_{L/K}$ since, as for $\rho^S_{L/K}$, $\alpha^S_{L/K}$ will be the composition of $\alpha_{L/K}$ with the canonical projection $G^{\mathrm{ab}} \longrightarrow G^{\mathrm{ab}\,S}$.[21] The map $\alpha_{L/K}$ must thus be such that $\alpha_{L/K} \circ \gamma_{\mathfrak{m}} = \rho_{L/K}$.

Recall that if $\boldsymbol{x} =: (x_v)_v \in J$, $\rho_{L/K}(\boldsymbol{x}) = \prod_v \left(\frac{x_v\,,\,L^{\mathrm{ab}}/K}{v}\right)$, and that $K^\times \subseteq \mathrm{Ker}(\rho_{L/K})$; it follows that we can replace $\boldsymbol{x}$ modulo $K^\times$ by $\boldsymbol{x}_{\mathfrak{m},\mathrm{pos}} =: (x'_v)_v \in J_{T,\mathfrak{m},\mathrm{pos}}$ (see I.4.3.3) so that we now have:

$$\rho_{L/K}(\boldsymbol{x}) = \rho_{L/K}(\boldsymbol{x}_{\mathfrak{m},\mathrm{pos}}) = \prod_v \left(\frac{x'_v\,,\,L^{\mathrm{ab}}/K}{v}\right) = \prod_{v \in Pl_0 \backslash T} \left(\frac{x'_v\,,\,L^{\mathrm{ab}}/K}{v}\right),$$

the symbols on $T \cup Pl_\infty$ being trivial since $U^{\mathrm{res}}_{\mathfrak{m}} \subseteq \mathrm{N}_{L^{\mathrm{ab}}/K}(U^{\mathrm{res}}_{L^{\mathrm{ab}}})$ (see 4.2.2); furthermore $\gamma_{\mathfrak{m}}(\boldsymbol{x}_{\mathfrak{m},\mathrm{pos}})$ is of the form $\mathcal{C}\ell^{\mathrm{res}}_{\mathfrak{m}}(\mathfrak{a})$, where $\mathfrak{a} := \prod_{v \in Pl_0} \mathfrak{p}_v^{\mathrm{v}(x'_v)}$ is an ideal prime to T. For $v \in Pl_0 \backslash T$, v is unramified in L^{ab}/K and we have $\left(\frac{x'_v\,,\,L^{\mathrm{ab}}/K}{v}\right) = \left(\frac{L^{\mathrm{ab}}/K}{v}\right)^{\mathrm{v}(x'_v)}$ (see 1.4, (vii), or 3.1.3, (vii) by density), so that we must set:

$$\alpha_{L/K}(\mathcal{C}\ell^{\mathrm{res}}_{\mathfrak{m}}(\mathfrak{a})) := \prod_{v \in Pl_0} \left(\frac{L^{\mathrm{ab}}/K}{v}\right)^{\mathrm{v}(\mathfrak{a})}.$$

The diagram for $S = \emptyset$ follows by computing $\gamma_{\mathfrak{m}}(K^\times \mathrm{N}(J_L))$. The general case is immediate by taking quotients with $\langle\, S \,\rangle$.

It is classical to lift $\alpha_{L/K}$ to I_T, denoting also the Frobenius $\left(\frac{L^{\mathrm{ab}}/K}{v}\right)$ by $\left(\frac{L^{\mathrm{ab}}/K}{\mathfrak{p}_v}\right)$ for any finite place v unramified in L^{ab}/K.

4.3.1 Definitions (Artin map and Artin group). (i) Let L/K be a finite extension of number fields and let T be a finite set of finite places containing

[21] In accordance with our general principles, we have $\alpha_{L/K} =: \alpha^{\mathrm{res}}_{L/K}$, and similarly for $\gamma_{\mathfrak{m}}$ and for $\rho_{L/K}$.

the set R of places ramified in L^{ab}/K. The Artin map (or Artin symbol) on I_T is the map:

$$\alpha_{L/K} \ : \ I_T \longrightarrow G^{\mathrm{ab}} := \mathrm{Gal}(L^{\mathrm{ab}}/K)$$

which sends $\mathfrak{a} \in I_T$ to:

$$\left(\frac{L^{\mathrm{ab}}/K}{\mathfrak{a}}\right) := \prod_{\mathfrak{p}} \left(\frac{L^{\mathrm{ab}}/K}{\mathfrak{p}}\right)^{\mathrm{v}_{\mathfrak{p}}(\mathfrak{a})}.$$

(ii) Its kernel $A_{L/K,T}$ is called the Artin group of L^{ab}/K in $I_T \subseteq I_R$. □

The diagram shows, by lifting $\mathcal{C}\!\ell^{\mathrm{res}}_{\mathfrak{m}}(\mathrm{N}(I_{L,T}))$ to I_T, the following.

4.3.2 Theorem. *The kernel of the Artin map $\alpha_{L/K}$ on I_T, $T \supseteq R$, is equal to the group:*

$$P_{T,\mathfrak{m},\mathrm{pos}}\mathrm{N}_{L/K}(I_{L,T}),$$

for any $\mathfrak{m}$ which is a multiple of the norm conductor $\mathfrak{f}$ of L^{ab}/K. □

This shows that $P_{T,\mathfrak{m},\mathrm{pos}}\mathrm{N}_{L/K}(I_{L,T})$ is independent of $\mathfrak{m}$, as long as this modulus is a multiple of $\mathfrak{f}$, a result which is not a priori clear.

As for the composite map $\alpha^S_{L/K} \ : \ I_T \longrightarrow G^{\mathrm{ab}\,S}$, its kernel is equal to:

$$A^S_{L/K,T} := P_{T,\mathfrak{m},\mathrm{pos}}\langle\, S \,\rangle\mathrm{N}_{L/K}(I_{L,T}) := P_{T,\mathfrak{m},\Delta_\infty}\cdot\langle\, S_0 \,\rangle\mathrm{N}_{L/K}(I_{L,T}),$$

for any $\mathfrak{m}$ multiple of $\mathfrak{f}$, where $\Delta_\infty := Pl^{\mathrm{r}}_\infty \backslash S_\infty$ (see I.4.4).

4.4 ARTIN'S RECIPROCITY LAW. The canonical isomorphism:

$$I_T/P_{T,\mathfrak{m},\mathrm{pos}}\mathrm{N}_{L/K}(I_{L,T}) \xrightarrow{\ \alpha_{L/K}\ } G^{\mathrm{ab}}$$

defines the Artin reciprocity law. It is the ideal version of the idelic version of the global reciprocity law asserting that $K^\times$ is in the kernel of $\rho_{L/K}$.

4.4.1 TAKAGI GROUPS — ARTIN AND NORM CONDUCTORS. The groups:

$$T_{L/K,T,\mathfrak{m}} := P_{T,\mathfrak{m},\mathrm{pos}}\mathrm{N}_{L/K}(I_{L,T}),$$

which were introduced by Takagi to make explicit the congruence groups of Weber (see Subsection (d)), are thus independent of the choice of $\mathfrak{m}$ (multiple of the norm conductor $\mathfrak{f}$ of L/K); hence the canonical choice is that of:

$$T_{L/K,R,\mathfrak{f}} := P_{R,\mathfrak{f},\mathrm{pos}}\mathrm{N}_{L/K}(I_{L,R}),$$

where R is the support of $\mathfrak{f}$. This group is called simply the Takagi group of L/K and "the groups" $T_{L/K,T,\mathfrak{m}}$, the Takagi groups modulo $\mathfrak{m}$; they only depend on $L^{\rm ab}/K$, but the possibility of choosing $\mathfrak{m}$ (multiple of $\mathfrak{f}$) may have some practical importance. In particular, we have the equality (for any $\mathfrak{m}$ multiple of $\mathfrak{f}$):

$$P_{T,\mathfrak{m},\mathrm{pos}}\mathrm{N}_{L/K}(I_{L,T}) = P_{T,\mathfrak{m},\mathrm{pos}}\mathrm{N}_{L^{\rm ab}/K}(I_{L^{\rm ab},T}).$$

The identity $A_{L/K,T} = T_{L/K,T,\mathfrak{m}} = P_{T,\mathfrak{m},\mathrm{pos}}\mathrm{N}_{L/K}(I_{L,T})$ is classically stated by saying that:

"The Artin group is equal to the Takagi group".

If we say that the (abelian!) Artin conductor of L/K is by definition the smallest modulus $\mathfrak{f}_A$ of K, with support equal to R, such that $P_{R,\mathfrak{f}_A,\mathrm{pos}}$ is in the kernel of the Artin map $\alpha_{L/K}$, the above results show that $\mathfrak{f} = \mathfrak{f}_A$. We can thus speak of the conductor of L/K (or of $L^{\rm ab}/K$) without being specific. The fact that $P_{T,\mathfrak{m}_1,\mathrm{pos}}\, P_{T,\mathfrak{m}_2,\mathrm{pos}} = P_{T,\mathrm{g.c.d.}(\mathfrak{m}_1,\mathfrak{m}_2),\mathrm{pos}}$, where $\mathfrak{m}_1$ and $\mathfrak{m}_2$ have supports contained in T, can easily be checked directly thanks to the chinese remainder theorem, but the idelic formulation in the introduction to Subsection (a) is much more immediate; we thus obtain the existence of the abelian Artin conductor (by choosing $T = R$).

4.4.2 History. These statements in terms of ideal groups (the equality of the Artin and Takagi groups, and the isomorphism $I_T/T_{L/K,T,\mathfrak{m}} \simeq G^{\rm ab}$) form the historical approach to the fundamental results of class field theory. In particular, the only proof of the equality, valid for any number field:

$$(I_T : P_{T,\mathfrak{m},\mathrm{pos}}\mathrm{N}_{L/K}(I_{L,T})) = [L^{\rm ab} : K],$$

was split in the difficult proofs of the first inequality of class field theory ("$\geq$", Takagi) and of the second inequality or universal equality ("$\leq$", Weber (1897) using analytic methods, and Hasse–Scholz (1929) in the general case). It is only later (1920/1924) that Artin introduced the map $\alpha_{L/K}$, constructed with the Frobenius symbols, and showed (1927), using ideas of Čebotarev (the crossing with a cyclic cyclotomic field), that $\alpha_{L/K}$ gave the exact sequence:

$$1 \longrightarrow P_{T,\mathfrak{m},\mathrm{pos}}\mathrm{N}_{L/K}(I_{L,T}) \longrightarrow I_T \longrightarrow G^{\rm ab} \longrightarrow 1,$$

thus giving for the first time the general notion of a global reciprocity map; the idelic version of Section 3 (Chevalley (1936/1940)) representing only the translation in the other direction, showing that it is possible (although apparently illogical but very useful) to go from a global approach of class field theory to a local approach (after Hasse–Schmidt (1930)).

The direct proof of the existence of an Artin conductor for an abelian extenion L/K, i.e., the existence of $\mathfrak{m}$ such that $\alpha_{L/K}$ is trivial on $P_{T,\mathfrak{m},\mathrm{pos}}$ (or such that $K^\times U_{\mathfrak{m}}^{\rm res} \subseteq \mathrm{Ker}(\rho_{L/K})$ in idelic terms) uses very strongly the properties of cyclotomic fields (in other words, essentially class field theory for $\mathbb{Q}$

with which we must begin); although it is only a series of elementary exercises [d, Lang1, Ch. X, § 2], this proof is still considered as deep since it involves the construction of abelian extensions of K satisfying certain local conditions and giving already enough information on $\mathrm{Gal}(\overline{K}^{\mathrm{ab}}/K)$. For instance, one of the key arguments is Lemma 1 of [d, Lang1, Ch. X, § 2] which originates in Birkhoff–Vandiver (1907), of which several proofs have been given by Chevalley [h, Che1], Iyanaga [h, Iy2], van der Waerden (1934), Takagi (1948); this lemma states that if $a > 1$ and $e \geq 1$ are integers and p a prime number, there exists a prime number q such that a is of order equal to p^e modulo (q).

At this step, we should mention (among other interesting studies of Kubota [Kub2, Kub3]) the paper of Kubota–Oka [KO] (2000) proving that Artin's reciprocity law can be deduced from the case of cyclotomic extensions and Kummer extensions. This paper is based on the Schmidt–Chevalley theorem.

The necessity of performing such constructions shows that it seems impossible to give a naïve proof of the fact that (to give a minimal example in the case of conductor 1):

$$\text{“}\mathfrak{p} \text{ is a principal prime ideal of } K = \mathbb{Q}(\sqrt{10}\,),$$

if and only if:

$$\text{the Frobenius of } \mathfrak{p} \text{ in } K(\sqrt{5}\,)/K \text{ is trivial”}$$

(since $K(\sqrt{5}\,)$ is the Hilbert class field of K). This example may not be completely convincing since it can be solved using genus theory (here Gauss's genus theory of quadratic forms, see IV.4.2.10), which can be considered as intermediate between naïve and highly nontrivial. On the contrary, the analogous result:

$$\text{“}\mathfrak{p} \text{ is a principal prime ideal of } K = \mathbb{Q}(\sqrt{-23}),$$

if and only if:

$$\text{the Frobenius of } \mathfrak{p} \text{ in } K(\theta)/K \text{ is trivial”}$$

(where $\mathrm{Irr}(\theta, \mathbb{Q}) := X^3 - X - 1$), seems faultless (see 5.2.1 for more details), except that $\mathbb{Q}(\sqrt{-23})$ has no nontrivial units, and to stress even more the origin of the difficulties for an arbitrary base field, we can cite Tate [d, CF, Ch. VII, § 6] who asserts: “It may well be that it is the connected component that prevents a simple proof of the reciprocity law in the general case”. Indeed, we will see that $D_0 = 1$ if and only if the $\mathbb{Z}$-rank of E^{ord} is equal to zero (i.e., K is equal to $\mathbb{Q}$ or to an imaginary quadratic field, fields for which questions having to do with reciprocity laws are indeed simpler).

Finally, we can replace $\mathbb{Q}(\sqrt{-23})$ by $\mathbb{Q}(\sqrt{79})$, whose Hilbert class field is also of degree 3, and obtain the same conclusion.

The advantage of Artin's formulation above is that in general we know how to compute the Frobenius' (in particular numerically). Thus, we are going to give a global method for the computation of the symbols:

$$\left(\frac{x\,,\,L^{\mathrm{ab}}/K}{v}\right),\ x \in K^\times,\ v \in Pl.$$

4.4.3 COMPUTATION OF HASSE SYMBOLS BY GLOBAL MEANS. Call R the set of (finite) places ramified in L^{ab}/K, and let $\mathfrak{m}$ be a multiple of the conductor $\mathfrak{f}$ (it does not matter if the support T of $\mathfrak{m}$ strictly contains R, which will be the case if the conductor and its support are not precisely known). Finally, set $\mathfrak{m} =: \prod_{v\in T} \mathfrak{p}_v^{m_v}$ with $m_v \geq 0$.

Let $x \in K^\times$; fix a place v of K, and let us consider several cases:

(α) $v \in Pl_\infty^{\mathrm{r}}$. By 3.1.3, (vii), we have:

$$\left(\frac{x\,,\,L^{\mathrm{ab}}/K}{v}\right) = \left(\frac{L^{\mathrm{ab}}/K}{v}\right)^{\mathrm{v}(x)},$$

where $\mathrm{v}(x) = 0$ (resp. 1) if $i_v(x) > 0$ (resp. $i_v(x) < 0$).

(β) $v \in Pl_0 \backslash T$. Similarly, since v is unramified, we have:

$$\left(\frac{x\,,\,L^{\mathrm{ab}}/K}{v}\right) = \left(\frac{L^{\mathrm{ab}}/K}{v}\right)^{\mathrm{v}(x)}.$$

(γ) $v \in T$. Let $x' \in K^\times$ be such that (chinese remainder theorem):

(i) $i_v(x'x^{-1}) \in U_v^{m_v}$,

(ii) $i_{v'}(x') \in U_{v'}^{m_{v'}}$, for each place $v' \in T,\ v' \neq v$,

(iii) $i_{v'}(x') > 0$ for each place $v' \in Pl_\infty^{\mathrm{r}}$ (or only each place v' complexified in L^{ab}/K).

Then, by the product formula we have:

$$\left(\frac{x'\,,\,L^{\mathrm{ab}}/K}{v}\right) = \prod_{v' \in Pl, v' \neq v} \left(\frac{x'\,,\,L^{\mathrm{ab}}/K}{v'}\right)^{-1},$$

and since $\left(\frac{x\,,\,L^{\mathrm{ab}}/K}{v}\right) = \left(\frac{x'\,,\,L^{\mathrm{ab}}/K}{v}\right)$, by (i) and the definition of the local v-conductor of L^{ab}/K, we have:

$$\left(\frac{x\,,\,L^{\mathrm{ab}}/K}{v}\right) = \prod_{v' \in Pl, v' \neq v} \left(\frac{x'\,,\,L^{\mathrm{ab}}/K}{v'}\right)^{-1};$$

let us compute the symbols occurring in the right hand side:

- if $v' \in T\backslash\{v\}$, $i_{v'}(x') \in U_{v'}^{m_{v'}}$ (by (ii)) and we have $\left(\frac{x'\,,\,L^{\mathrm{ab}}/K}{v'}\right) = 1$,

• if $v' \in Pl_\infty$, $\left(\frac{x'\,,\,L^{\mathrm{ab}}/K}{v'}\right) = 1$ since either $\left(\frac{L^{\mathrm{ab}}/K}{v'}\right) = 1$ if v' is complex or noncomplexified real, or $\mathrm{v}'(x') = 0$ for v' complexified real (by (iii)),

• if $v' \in Pl_0 \backslash T$, v' is unramified and we can write (by 3.1.3, (vii)):

$$\left(\frac{x'\,,\,L^{\mathrm{ab}}/K}{v'}\right)^{-1} = \left(\frac{L^{\mathrm{ab}}/K}{v'}\right)^{-\mathrm{v}'(x')} ;$$

finally, we have obtained:

$$\left(\frac{x\,,\,L^{\mathrm{ab}}/K}{v}\right) = \prod_{v' \in Pl_0 \backslash T} \left(\frac{L^{\mathrm{ab}}/K}{v'}\right)^{-\mathrm{v}'(x')} .$$

It follows that if we write $(x') =: \mathfrak{p}_v^{\mathrm{v}(x')}\mathfrak{a} = \mathfrak{p}_v^{\mathrm{v}(x)}\mathfrak{a}$ (we have $\mathrm{v}(x') = \mathrm{v}(x)$ by (i) even when $m_v = 0$), then $\mathfrak{a}$ is prime to T by (ii) and we obtain, since $Pl_0 \backslash T$ does not contain v:

$$\left(\frac{x\,,\,L^{\mathrm{ab}}/K}{v}\right) = \left(\frac{L^{\mathrm{ab}}/K}{\mathfrak{a}}\right)^{-1} = \alpha_{L/K}(\mathfrak{a})^{-1},$$

which finishes the hand computation of the Hasse symbol of an $x \in K^\times$ which is not necessarily prime to the place v under consideration.

We will come back to this procedure in 7.5 for the practical computation of Hilbert symbols.

4.4.3.1 Remarks. (i) The auxiliary element x' is called a v-associate (or a $\mathfrak{p}_v$-associate) of x.

(ii) In the case where $v \in T$ is unramified (i.e., $v \notin R$), the above computation still yields $(x') = \mathfrak{p}_v^{\mathrm{v}(x)}\mathfrak{a}$, $\mathfrak{a}$ prime to T, but the v-associate x' is then such that $\alpha_{L/K}((x')) = 1$ (the Artin map being defined since here (x') is prime to R, so we have $\rho_{L/K}(i(x')) = 1$), and we find once again that by 3.1.3, (vii):

$$\left(\frac{x\,,\,L^{\mathrm{ab}}/K}{v}\right) = \alpha_{L/K}(\mathfrak{a})^{-1} = \left(\frac{L^{\mathrm{ab}}/K}{\mathfrak{p}_v}\right)^{\mathrm{v}(x)} .$$

(iii) If $\mathfrak{f}$ is primary (i.e., a power of $\mathfrak{p}_v$), then any $x \in K^\times$ (positive at the complexified real places) is equal to its own v-associate, and we have:

$$\left(\frac{x\,,\,L^{\mathrm{ab}}/K}{v}\right) = \alpha_{L/K}\Big((x)\mathfrak{p}_v^{-\mathrm{v}(x)}\Big)^{-1} . \qquad \square$$

4.4.3.2 Example. Let $K = \mathbb{Q}$ and let $L = L^{\mathrm{ab}} = \mathbb{Q}(\sqrt{5}, \sqrt{-3}\,)$. Let us compute the Hasse symbol $\left(\frac{15\,,\,L/\mathbb{Q}}{(3)}\right)$. The conductor of L is equal to (15); we must find $x' \in \mathbb{Q}^\times$ such that:

$$\frac{x'}{15} \equiv 1 \bmod (3),$$
$$x' \equiv 1 \bmod (5),$$
$$x' > 0\ ;$$

$x' = 6$ is suitable, so that $\mathfrak{a} = (2)$ and:

$$\left(\frac{15\,,\, L/\mathbb{Q}}{(3)}\right) = \left(\frac{L/\mathbb{Q}}{(2)}\right)^{-1}.$$

But it is easy to see that (2) is inert in $\mathbb{Q}(\sqrt{5}\,)$ and in $\mathbb{Q}(\sqrt{-3}\,)$; the Frobenius of (2) is thus a generator of $\mathrm{Gal}(L/\mathbb{Q}(\sqrt{-15}\,))$. It follows that 15 is not a local norm at (3).

Since the product formula is reduced here to:

$$\left(\frac{15\,,\, L/\mathbb{Q}}{(3)}\right)\left(\frac{15\,,\, L/\mathbb{Q}}{(5)}\right) = 1,$$

the symbol at (5) is the same, but the (3)-associate x' is not suitable anymore for the direct computation of $\left(\frac{15\,,\, L/\mathbb{Q}}{(5)}\right)$; a (5)-associate is for example 40 which indeed gives the expected result.

Finally, if we omit the condition $x' > 0$, for instance for a (3)-associate we can try:

$$x'' = -39,$$

which yields $\mathfrak{a} = (13)$; but the Frobenius of (13) being the generator of $\mathrm{Gal}(L/\mathbb{Q}(\sqrt{-3}\,))$, the result is false! □

4.4.3.3 Exercise (the case of cyclotomic fields). Let $K = \mathbb{Q}$ and let $L = \mathbb{Q}(\mu_m)$; we assume that m is odd or divisible by 4. Describe the method for the computation of $\left(\frac{x\,,\, L/\mathbb{Q}}{(\ell)}\right)$, $x \in \mathbb{Q}^\times$, for a prime divisor ℓ of m.

Deduce the values of $\left(\frac{\ell\,,\, L/\mathbb{Q}}{(\ell)}\right)$ and of $\left(\frac{y\,,\, L/\mathbb{Q}}{(\ell)}\right)$ for y prime to ℓ.

Characterize the x which are local norms at (ℓ) for $L/\mathbb{Q}$.

Answer. The conductor of $L/\mathbb{Q}$ is equal to $m\mathbb{Z}$ (see 5.5); set $m =: \ell^a n$ and $x =: \ell^b y$ with n and y prime to ℓ. We must find $x' = \ell^b y'$, with $y' \in \mathbb{Q}^\times$ such that:

$$y' \equiv y \bmod (\ell^a),$$
$$\ell^b y' \equiv 1 \bmod (n),$$
$$y' > 0,$$

which can be achieved thanks to suitable extended Euclid relations. The result is the Artin symbol:

$$\left(\frac{L/\mathbb{Q}}{(y')}\right)^{-1}$$

corresponding to the inverse of $\overline{y'} \in (\mathbb{Z}/m\mathbb{Z})^\times$ under the usual canonical isomorphism $\mathrm{Gal}(L/\mathbb{Q}) \simeq (\mathbb{Z}/m\mathbb{Z})^\times$.

It follows that $\left(\frac{\ell\,,\,L/\mathbb{Q}}{(\ell)}\right)$ (take $b=1$, $y=1$) is the lift of $\left(\frac{\mathbb{Q}(\mu_n)/\mathbb{Q}}{(\ell)}\right)$ to $\mathrm{Gal}(\mathbb{Q}(\mu_m)/\mathbb{Q}(\mu_{\ell^a}))$, and that for all $y>0$ prime to ℓ (take $b=0$), $\left(\frac{y\,,\,L/\mathbb{Q}}{(\ell)}\right)$ is the lift of $\left(\frac{\mathbb{Q}(\mu_{\ell^a})/\mathbb{Q}}{(y)}\right)^{-1}$ to $\mathrm{Gal}(\mathbb{Q}(\mu_m)/\mathbb{Q}(\mu_n))$. This should be compared with the results of Exercise 3.4.3.

The rational number $x=\ell^b y$ is a local norm at ℓ in $L/\mathbb{Q}$ if and only if $y' \equiv 1 \bmod (m)$; but this is equivalent to:

$$y \equiv 1 \bmod (\ell^a),$$
$$\ell^b \equiv 1 \bmod (n).$$

We see that $N_1 := \ell^{\mathbb{Z}} \oplus (1+\ell^a\mathbb{Z}_\ell)$ and $N_2 := \ell^{f\mathbb{Z}} \oplus \mathbb{Z}_\ell^\times$ are the norm groups of $\mathbb{Q}_\ell(\mu_{\ell^a})$ and $\mathbb{Q}_\ell(\mu_n)$, where f is the residue degree of ℓ in $\mathbb{Q}(\mu_n)/\mathbb{Q}$ (smallest integer such that $\ell^f \equiv 1 \bmod (n)$) (use respectively 3.4.3, (i), and the fact that $\mathbb{Q}_\ell(\mu_n)/\mathbb{Q}_\ell$ is unramified); the norm group of $\mathbb{Q}_\ell(\mu_m)$ is then $N_1 \cap N_2 = \ell^{f\mathbb{Z}} \oplus (1+\ell^a\mathbb{Z}_\ell)$ giving again the result. □

Let L/K be a finite extension of number fields and L'/K a subextension of L/K. Let S be a set of noncomplex places of K, disjoint from $T \supseteq R$; we denote by S' the set of places of L' above those of S, and put $G^{\mathrm{ab}\,S} := \mathrm{Gal}(L^{\mathrm{ab}\,S}/K)$. Let us state the functorial properties of the Artin map on:

$$\mathcal{C}\ell_{\mathfrak{m}}^S := I_T/P_{T,\mathfrak{m},\mathrm{pos}}\langle\, S\,\rangle := I_T/P_{T,\mathfrak{m},\Delta_\infty}\cdot\langle\, S_0\,\rangle,$$

with $\Delta_\infty := Pl_\infty^{\mathrm{r}} \setminus S_\infty$, which follow from those of $\rho_{L/K}^S$ (this is a simple transcription of Theorem 3.3).

4.5 Theorem (properties of the Artin map). *Let $\mathfrak{m}$ with support contained in T be a multiple of the norm conductor of L/K (i.e., of L^{ab}/K). We have the following properties:*

(i) We have the exact sequence:

$$1 \longrightarrow c\ell_{\mathfrak{m}}^S(\mathrm{N}_{L/K}(I_{L,T})) \longrightarrow \mathcal{C}\ell_{\mathfrak{m}}^S \xrightarrow{\ \alpha_{L/K}^S\ } G^{\mathrm{ab}\,S} \longrightarrow 1,$$

where $c\ell_{\mathfrak{m}}^S$ is the map $I_T \longrightarrow \mathcal{C}\ell_{\mathfrak{m}}^S$;

(ii) the composition of $\alpha_{L/K}^S$ and of the projection $G^{\mathrm{ab}\,S} \longrightarrow \mathrm{Gal}(L'^{\mathrm{ab}\,S}/K)$ is equal to $\alpha_{L'/K}^S$;

(iii) for each place $v \in T$, set $\mathfrak{m}_v := \mathfrak{p}_v^{\mathrm{v}(\mathfrak{m})}$; then the decomposition group (resp. the inertia group, resp. the higher ramification group with

upper index $i \geq 1$) of $v \in T$ in $L^{\mathrm{ab}\,S}/K$ is the image under $\alpha^S_{L/K}$ of $(P_{T\setminus\{v\}, \frac{\mathfrak{m}}{\mathfrak{m}_v}, \mathrm{pos}} \cdot \langle\, \mathfrak{p}_v \,\rangle) \cap I_T$ [22] *(resp. of $P_{T, \frac{\mathfrak{m}}{\mathfrak{m}_v}, \mathrm{pos}}$, resp. of $P_{T, \frac{\mathfrak{m}}{\mathfrak{m}_v}\mathfrak{p}_v^i, \mathrm{pos}}$); if $v \notin T$ is finite, the decomposition group of v is the image of $\langle\, \mathfrak{p}_v \,\rangle$; if $v \in Pl^{\mathrm{r}}_\infty$, the decomposition group of v for $L^{\mathrm{ab}\,S}/K$ is the image under $\alpha^S_{L/K}$ of $P_{T, \mathfrak{m}, Pl^{\mathrm{r}}_\infty \setminus\{v\}}$;*

(iv) for all $\mathfrak{a}' \in I_{L',T}$, prime to the norm conductor of L/L', the image of $\left(\frac{L^{\mathrm{ab}'\,S'}/L'}{\mathfrak{a}'}\right)$ in $G^{\mathrm{ab}\,S}$ is $\left(\frac{L^{\mathrm{ab}\,S}/K}{\mathrm{N}_{L'/K}(\mathfrak{a}')}\right)$; in particular, we have:

$$\mathrm{Gal}(L^{\mathrm{ab}\,S}/L'^{\mathrm{ab}\,S}) = \alpha^S_{L/K}(\mathrm{N}_{L'/K}(I_{L',T}))\ ;$$

(v) for all $\mathfrak{a} \in I_T$, prime to the norm conductor of L/L', the image of $\left(\frac{L^{\mathrm{ab}\,S}/K}{\mathfrak{a}}\right)$ under the transfer map (from $G^{\mathrm{ab}\,S}$ to $\mathrm{Gal}(L^{\mathrm{ab}'\,S'}/L')$) is $\left(\frac{L^{\mathrm{ab}'\,S'}/L'}{\mathfrak{a}'}\right)$, where $\mathfrak{a}'$ is obtained by extending $\mathfrak{a}$ to L';

(vi) for any $\mathbb{Q}$-isomorphism τ of L in $\overline{\mathbb{Q}}$, we have for all $\mathfrak{a} \in I_T$:

$$\left(\frac{\tau\, L^{\mathrm{ab}\,S}/\tau K}{\tau\mathfrak{a}}\right) = \tau \circ \left(\frac{L^{\mathrm{ab}\,S}/K}{\mathfrak{a}}\right) \circ \tau^{-1} \text{ on } \tau L^{\mathrm{ab}\,S} = (\tau L)^{\mathrm{ab}\,\tau S}. \qquad \square$$

Note. In (iv) and (v), the set T (which contains the set of places of L ramified in L^{ab}/K) may be inadequate. Indeed, consider the following example (with $K = \mathbb{Q}$): $L = \mathbb{Q}(\mu_4, \sqrt[4]{18})$, $L' = \mathbb{Q}(\mu_4)$, $T = \{2\}$, for which $L^{\mathrm{ab}'} = L$, $L^{\mathrm{ab}} = L'\mathbb{Q}(\sqrt{18}) = \mathbb{Q}(\mu_8)$, $L'^{\mathrm{ab}} = L'$; the ideal $\mathfrak{a}' = (3)$ is prime to T but its Artin Symbol in L/L' does not exist; the Artin Symbol of $\mathfrak{a} = (3)$ exists in L^{ab}/K but not its transfer. However, this is not annoying since any element (of the above abelian Galois groups) is the Artin Symbol of a suitable ideal.

4.5.1 Corollary. *We have:*

$$\mathcal{C}\!\ell^{\mathrm{res}}_{\mathfrak{m}}(\mathrm{N}_{L/K}(I_{L,T})) = \mathcal{C}\!\ell^{\mathrm{res}}_{\mathfrak{m}}(\mathrm{N}_{L^{\mathrm{ab}}/K}(I_{L^{\mathrm{ab}},T})). \qquad \square$$

4.5.2 Corollary. *We have the exact sequences:*

$$1 \longrightarrow \mathcal{C}\!\ell^{\mathrm{res}}_{\mathfrak{m}}(\mathrm{N}_{L/K}(I_{L,T})) \longrightarrow \mathcal{C}\ell^{\mathrm{res}}_{\mathfrak{m}} \xrightarrow{\alpha^{\mathrm{res}}_{L/K}} \mathrm{Gal}(L^{\mathrm{ab}}/K) \longrightarrow 1,$$

$$1 \longrightarrow \mathcal{C}\!\ell^{\mathrm{ord}}_{\mathfrak{m}}(\mathrm{N}_{L/K}(I_{L,T})) \longrightarrow \mathcal{C}\ell^{\mathrm{ord}}_{\mathfrak{m}} \xrightarrow{\alpha^{\mathrm{ord}}_{L/K}} \mathrm{Gal}(L^{\mathrm{ab\,nc}}/K) \longrightarrow 1,$$

where $L^{\mathrm{ab\,nc}}/K$ is the maximal Pl^{r}_∞-split (i.e., noncomplexified) abelian subextension of L/K. $\square$

4.5.3 Example. In the particular case where the extension L^{ab}/K has conductor $\mathfrak{f} = 1$ (i.e., L^{ab}/K is unramified but may be complexified), we obtain, by taking $T = \emptyset$, the exact sequences:

[22] Note that $(P_{T\setminus\{v\}, \frac{\mathfrak{m}}{\mathfrak{m}_v}, \mathrm{pos}} \cdot \langle\, \mathfrak{p}_v \,\rangle) \cap I_T = \{(x)\, \mathfrak{p}_v^{-\mathrm{v}(x)},\ \ x \in K^\times_{T\setminus\{v\}, \frac{\mathfrak{m}}{\mathfrak{m}_v}, \mathrm{pos}}\}$.

$$1 \longrightarrow \mathcal{C}\ell^{\mathrm{res}}(\mathrm{N}_{L/K}(I_L)) \longrightarrow \mathcal{C}\ell^{\mathrm{res}} \xrightarrow{\alpha^{\mathrm{res}}_{L/K}} \mathrm{Gal}(L^{\mathrm{ab}}/K) \longrightarrow 1,$$

$$1 \longrightarrow \mathcal{C}\ell^{\mathrm{ord}}(\mathrm{N}_{L/K}(I_L)) \longrightarrow \mathcal{C}\ell^{\mathrm{ord}} \xrightarrow{\alpha^{\mathrm{ord}}_{L/K}} \mathrm{Gal}(L^{\mathrm{ab\,nc}}/K) \longrightarrow 1,$$

which give a description of $\mathrm{Gal}(L^{\mathrm{ab}}/K)$ (resp. of $\mathrm{Gal}(L^{\mathrm{ab\,nc}}/K)$) in terms of usual ideal classes; this occurs only if the base field is not principal in the restricted sense (resp. in the ordinary sense). □

We return to the general setting of Theorem 4.5. For this, let $\mathfrak{m}$ with support contained in T be a multiple of the conductor $\mathfrak{f}$ of L/K, and let A_T be the Artin group of L/K in I_T which we can take to be equal to $T_{T,\mathfrak{m}} = P_{T,\mathfrak{m},\mathrm{pos}}\mathrm{N}_{L/K}(I_{L,T})$. For each $v \in T$ we set $\mathfrak{m}_v := \mathfrak{p}_v^{\mathrm{v}(\mathfrak{m})}$.

4.5.4 Corollary (decomposition law of places in L^{ab}/K). *We have:*

(i) (ramification groups). From the Artin isomorphism $I_T/A_T \simeq G^{\mathrm{ab}}$, we obtain, for each place $v \in T$ and all $i \geq 1$, the isomorphisms:[23]

$$P_{T,\frac{\mathfrak{m}}{\mathfrak{m}_v}\mathfrak{p}_v^i,\mathrm{pos}}A_T/A_T \simeq D_v^i(L^{\mathrm{ab}}/K) \ ;$$

in particular, $v \in T$ is unramified in L^{ab}/K if and only if we have:

$$P_{T,\frac{\mathfrak{m}}{\mathfrak{m}_v},\mathrm{pos}} \subseteq A_T. \ ^{24}$$

Similarly, v is tamely ramified in L^{ab}/K if and only if:

$$P_{T,\frac{\mathfrak{m}}{\mathfrak{m}_v}\mathfrak{p}_v,\mathrm{pos}} \subseteq A_T.$$

(ii) (decomposition groups). For $v \in T$, let $\mathfrak{a}_v$ be prime to T and such that $\mathfrak{a}_v = \mathfrak{p}_v(u_{\frac{\mathfrak{m}}{\mathfrak{m}_v}})$, $u_{\frac{\mathfrak{m}}{\mathfrak{m}_v}} \in K^{\times}_{T\backslash\{v\},\frac{\mathfrak{m}}{\mathfrak{m}_v},\mathrm{pos}}$ (see I.5.1.2); we then have the isomorphism:

$$\langle\, \mathfrak{a}_v \,\rangle P_{T,\frac{\mathfrak{m}}{\mathfrak{m}_v},\mathrm{pos}}A_T/A_T \simeq D_v(L^{\mathrm{ab}}/K) \ ;$$

if $v \in Pl_0\backslash T$ (v is thus unramified), we have:

$$\langle\, \mathfrak{p}_v \,\rangle A_T/A_T \simeq D_v(L^{\mathrm{ab}}/K),$$

and the residue degree $f_v(L^{\mathrm{ab}}/K)$ of v in L^{ab}/K is equal to the order of the class of $\mathfrak{p}_v$ in I_T/A_T.

(iii) If $v \in Pl_\infty^{\mathrm{r}}$, then v is noncomplexified in L^{ab}/K if and only if:

$$P_{T,\mathfrak{m},\mathrm{pos}}\langle\, v \,\rangle := P_{T,\mathfrak{m},Pl_\infty^{\mathrm{r}}\backslash\{v\}} \subseteq A_T. \qquad \square$$

[23] See 1.1.1 for some notations about higher ramification.

[24] Relate this with the characterization of the conductor given in 4.4.1.

4.5.5 Remark. For the noncomplexification of a real place v, a necessary and sufficient condition is that for an arbitrary element $u_{\mathfrak{m},v} \in K^{\times}_{T,\mathfrak{m},Pl^{\mathrm{r}}_{\infty}\backslash\{v\}}$ such that $i_v(u_{\mathfrak{m},v}) < 0$ then $(u_{\mathfrak{m},v}) \in A_T$.

The case of an infinite place $v \in Pl^{\mathrm{r}}_{\infty}$ can be treated directly if L^{ab} is known (but it is not anymore a class field theoretic proof): we have $f_v = 1$ (resp. 2) if the extension $i_{w_0}(L^{\mathrm{ab}})$ is real (resp. complex) for an arbitrary $w_0|v$ in L^{ab} (we have $i_v(K) \subset \mathbb{R}$ since v is real). □

4.6 Density Theorem (1926). The surjectivity of $\alpha_{L/K}$ can be shown without using analytic arguments, and one can even prove a little more [e, Ko3, Ch. 2, § 4.4, Th. 2.70]; however, in practice it is better to consider it through the density theorem which asserts that every class $\mathfrak{a}P_{R,\mathfrak{f},\mathrm{pos}}$, $\mathfrak{a}$ prime to $\mathfrak{f}$, contains an infinity of prime ideals, with density:

$$\frac{1}{|\mathcal{C}\ell^{\mathrm{res}}_{\mathfrak{f}}|}.$$

Thus, for any $\sigma \in G^{\mathrm{ab}}$, we can have the equality:

$$\left(\frac{L^{\mathrm{ab}}/K}{\mathfrak{p}}\right) = \sigma,$$

for an infinite number of prime ideals $\mathfrak{p}$ of K, unramified in L^{ab}/K, with density equal to:

$$\frac{1}{[L^{\mathrm{ab}} : K]}.$$

Note. One can find in [d, Lang1, Ch. VIII, § 4] the general Galois statement (the Čebotarev theorem) which is in fact deduced, after an argument of Deuring (1934), from the above abelian density theorem; see also [c, Nar1, Ch. 7] and [e, Ko3, Ch. 1, § 6.7]. More precisely, this theorem was conjectured by Frobenius (1896), proved by Čebotarev (1926), with a simplified proof by Schreier (1927); these proofs (using cyclotomic fields) originate, as we have mentioned in 4.4.2, the fundamental proof by Artin of his reciprocity law. The Čebotarev theorem is the following. Let L/K be Galois with Galois group G, and let $t \in G$; then the set of unramified primes $\mathfrak{p}$ of K such that $t = \left(\frac{L/K}{\mathfrak{P}}\right)$, for a $\mathfrak{P}|\mathfrak{p}$ in L, has a density equal to $\frac{1}{[L:K]}\,|\{sts^{-1},\ s \in G\}|$ (note that $s\left(\frac{L/K}{\mathfrak{P}}\right)s^{-1} = \left(\frac{L/K}{s\mathfrak{P}}\right)$ as usual).

In terms of generalized class groups, the existence theorem takes the following form (we do not state once more the four usual properties which characterize this correspondence; see 3.5, (i) to (iv)).

4.7 Theorem (global existence). *Let $\mathfrak{m}$ be a modulus of K built from $T \subset Pl_0$. Then there exists a bijective Galois correspondence between the set of* *subgroups $\mathcal{C}_{\mathfrak{m}}$ of $\mathcal{C}\ell_{\mathfrak{m}}^{\mathrm{res}}$ (or of subgroups $N_{\mathfrak{m}}$ of I_T containing $P_{T,\mathfrak{m},\mathrm{pos}}$) and the set of abelian extensions M of K, of conductor $\mathfrak{f}$ dividing $\mathfrak{m}$. The Artin map yields the equivalent two exact sequences:*

$$1 \longrightarrow \mathcal{C}_{\mathfrak{m}} \longrightarrow \mathcal{C}\ell_{\mathfrak{m}}^{\mathrm{res}} \xrightarrow{\alpha_{M/K}^{\mathrm{res}}} \mathrm{Gal}(M/K) \longrightarrow 1,$$

$$1 \longrightarrow N_{\mathfrak{m}} \longrightarrow I_T \xrightarrow{\alpha_{M/K}^{\mathrm{res}}} \mathrm{Gal}(M/K) \longrightarrow 1,$$

with $\mathcal{C}_{\mathfrak{m}} := c\ell_{\mathfrak{m}}^{\mathrm{res}}(\mathrm{N}_{M/K}(I_{M,T}))$ and $N_{\mathfrak{m}} := P_{T,\mathfrak{m},\mathrm{pos}}\,\mathrm{N}_{M/K}(I_{M,T})$. □

The group $\mathcal{C}_{\mathfrak{m}}$ (resp. $N_{\mathfrak{m}}$) is called the class group (resp. the congruence group) corresponding to M/K (but it is also a norm group in terms of ideal classes).

4.7.1 Remarks. (i) The Artin group of the decomposition subfield of v in M (with Artin group A_T) is given by:

$$\langle\, \mathfrak{a}_v \,\rangle P_{T,\frac{\mathfrak{m}}{\mathfrak{m}_v},\mathrm{pos}} A_T, \quad \langle\, \mathfrak{p}_v \,\rangle A_T, \quad \langle\, (u_{\mathfrak{m},v}) \,\rangle A_T,$$

depending on the situation (by 4.5.4, (ii), (iii), and 4.5.5); that of the inertia subfield is given by:

$$P_{T,\frac{\mathfrak{m}}{\mathfrak{m}_v},\mathrm{pos}} A_T .$$

(ii) The Artin group of the maximal v-tamely ramified subextension is:

$$P_{T,\frac{\mathfrak{m}}{\mathfrak{m}_v}\mathfrak{p}_v,\mathrm{pos}} A_T .$$

(iii) If we want S-decomposition, we replace $N_{\mathfrak{m}} := P_{T,\mathfrak{m},\mathrm{pos}}\,\mathrm{N}_{M/K}(I_{M,T})$ by:

$$P_{T,\mathfrak{m},\mathrm{pos}} \langle\, S \,\rangle \mathrm{N}_{M/K}(I_{M,T}) := P_{T,\mathfrak{m},\Delta_\infty} \,.\, \langle\, S_0 \,\rangle \mathrm{N}_{M/K}(I_{M,T}),$$

where $\Delta_\infty := Pl_\infty^{\mathrm{r}} \backslash S_\infty$. □

4.7.2 Corollary (norm lifting theorem). *Let L/K be a finite extension of number fields, let M/K be an abelian extension, and let $\mathfrak{m}$ with support contained in T be a modulus of K multiple of the conductor of M/K. Then any modulus $\mathfrak{m}'$ of L, with support contained in the set of places of L above those of T and such that:*

$$\mathrm{N}_{L/K}(P_{L,T,\mathfrak{m}',\mathrm{pos}}) \subseteq P_{T,\mathfrak{m},\mathrm{pos}},$$

is a multiple of the conductor of LM/L. If $\mathcal{C}$ is the subgroup of $\mathcal{C}\ell_{\mathfrak{m}}^{\mathrm{res}}$ corresponding to M, then the subgroup $\mathcal{C}'$ of $\mathcal{C}\ell_{L,\mathfrak{m}'}^{\mathrm{res}}$ corresponding to LM over L is given by:

$$\mathcal{C}' = \{c\ell_{L,\mathfrak{m}'}^{\mathrm{res}}(\mathfrak{a}'),\ \mathfrak{a}' \in I_{L,T},\ c\ell_{\mathfrak{m}}^{\mathrm{res}}(\mathrm{N}_{L/K}(\mathfrak{a}')) \in \mathcal{C}\} =: \mathrm{N}_{L/K}^{-1}(\mathcal{C}).$$

Proof. We check that the given condition is equivalent to:

$$\mathrm{N}_{L/K}(L^\times U^{\mathrm{res}}_{L,\mathfrak{m}'}) \subseteq K^\times U^{\mathrm{res}}_{\mathfrak{m}}.$$

It follows by 3.3, (iv), that the image of $\rho_{LM/L}(U^{\mathrm{res}}_{L,\mathfrak{m}'})$ in $\mathrm{Gal}(M/K)$ is:

$$\rho_{M/K}(\mathrm{N}_{L/K}(U^{\mathrm{res}}_{L,\mathfrak{m}'})) \subseteq \rho_{M/K}(K^\times U^{\mathrm{res}}_{\mathfrak{m}}) = 1 ;$$

so that $U^{\mathrm{res}}_{L,\mathfrak{m}'} \subseteq \mathrm{Ker}(\rho_{LM/L})$, which indeed shows that $\mathfrak{m}'$ is a multiple of $\mathfrak{f}_{LM/L}$. The rest is then only a translation of global norm lifting Theorem 3.5.3 in terms of class groups. □

We will return in 5.7 to the action of the norm in this context of generalized class groups.

It is clear that the fields corresponding to $\mathcal{C}_{\mathfrak{m}} = 1$ (i.e., $N_{\mathfrak{m}} = P_{T,\mathfrak{m},\mathrm{pos}}$ in terms of ideal groups, $N_{\mathfrak{m}} = K^\times U^{\mathrm{res}}_{\mathfrak{m}}$ in terms of idèle groups) play a crucial role in the correspondence of class field theory; hence we are going to look in more detail at these ray class fields.

§5 Ray Class Fields — Hilbert Class Fields

Let K be a number field and let $\mathfrak{m}$ be a modulus of K built on $T \subset Pl_0$. The unique abelian extension of K corresponding to $K^\times U^{\mathrm{res}}_{\mathfrak{m}}$ in the idelic version, in other words to $P_{T,\mathfrak{m},\mathrm{pos}}$ in the ideal group version, is called the restricted (or narrow) ray class field modulo $\mathfrak{m}$, and is denoted:

$$K(\mathfrak{m}) =: K(\mathfrak{m})^{\mathrm{res}} ;$$

thus, for this field we have $\mathrm{Gal}(K(\mathfrak{m})^{\mathrm{res}}/K) \simeq \mathcal{C}\ell^{\mathrm{res}}_{\mathfrak{m}}$ and:

$$\mathrm{N}_{K(\mathfrak{m})^{\mathrm{res}}/K}(J_{K(\mathfrak{m})^{\mathrm{res}}}) \subset K^\times U^{\mathrm{res}}_{\mathfrak{m}} \quad \text{and} \quad \mathrm{N}_{K(\mathfrak{m})^{\mathrm{res}}/K}(I_{K(\mathfrak{m})^{\mathrm{res}},T}) \subset P_{T,\mathfrak{m},\mathrm{pos}}.$$

For $\mathfrak{m} = 1$, we obtain $K(1)^{\mathrm{res}}$, denoted H^{res}, and called the restricted Hilbert class field. Hilbert had very early conjectured the existence of the absolute (or wide) class field H^{ord} (for us the maximal Pl^{r}_∞-split subextension of H^{res}, called the ordinary class field), and in this context, in which most of the proofs are due to Furtwängler, had predicted the main principles of class field theory.

The extension H^{res}/K (resp. H^{ord}/K) is thus the maximal unramified (resp. unramified and noncomplexified) abelian extension of K, and we have:

$$\mathrm{Gal}(H^{\mathrm{res}}/K) \simeq \mathcal{C}\ell^{\mathrm{res}}, \quad \mathrm{Gal}(H^{\mathrm{ord}}/K) \simeq \mathcal{C}\ell^{\mathrm{ord}}.$$

a) Elementary Properties — Decomposition Law

We start by giving a number of elementary remarks which we divide in five statements 5.1.1 to 5.1.5.

5.1 PROPERTIES OF RAY CLASS FIELDS. In the sequel we fix a modulus $\mathfrak{m}$ of K, with support T.

5.1.1 CONDUCTOR OF A RAY CLASS FIELD. The existence of a (norm or Artin) conductor for any abelian extension of K implies that the conductor $\mathfrak{f}$ of $K(\mathfrak{m})^{\mathrm{res}}$ divides $\mathfrak{m}$ (and is possibly not equal to it); we thus have $U_{\mathfrak{f}}^{\mathrm{res}} \subseteq K^\times U_{\mathfrak{m}}^{\mathrm{res}}$, hence $K^\times U_{\mathfrak{f}}^{\mathrm{res}} = K^\times U_{\mathfrak{m}}^{\mathrm{res}}$, which means (by uniqueness in the correspondence of class field theory) that $K(\mathfrak{m})^{\mathrm{res}} = K(\mathfrak{f})^{\mathrm{res}}$, and is also equivalent to the condition $P_{T,\mathfrak{m},\mathrm{pos}} = P_{T,\mathfrak{f},\mathrm{pos}}$. In fact, it is simpler to say that $\mathfrak{m}$ is the conductor of $K(\mathfrak{m})^{\mathrm{res}}$ if and only if, for all $v \in T$, we have $\mathcal{C}\ell_{\frac{\mathfrak{m}}{\mathfrak{p}_v}}^{\mathrm{res}} \neq \mathcal{C}\ell_{\mathfrak{m}}^{\mathrm{res}}$, which yields:

$$(E_{\frac{\mathfrak{m}}{\mathfrak{p}_v}}^{\mathrm{res}} : E_{\mathfrak{m}}^{\mathrm{res}}) < \varphi(\mathfrak{m})\varphi\Big(\frac{\mathfrak{m}}{\mathfrak{p}_v}\Big)^{-1} \text{ for all } v \in T,$$

using formula I.4.5.1 (recall that $\varphi(\mathfrak{m})\varphi(\frac{\mathfrak{m}}{\mathfrak{p}_v})^{-1} = q_v$ or $q_v - 1$ depending on whether $\mathrm{v}(\mathfrak{m}) > 1$ or $\mathrm{v}(\mathfrak{m}) = 1$).

5.1.1.1 Example. For $K = \mathbb{Q}(\sqrt{3})$ and $\mathfrak{m} = \mathfrak{l}_{11}$ (a prime ideal above 11), we find that $[K(\mathfrak{m})^{\mathrm{res}} : K(1)^{\mathrm{res}}] = 1$: this is immediate from the fact that $E^{\mathrm{res}} = \langle\, \varepsilon \,\rangle$ with $\varepsilon := 2 + \sqrt{3}$, that $E_{\mathfrak{l}_{11}}^{\mathrm{res}} = \langle\, \varepsilon^{10} \,\rangle$, and $\varphi(\mathfrak{l}_{11}) = 10$ (see I.4.5.6, (i)). Here, we have $\mathfrak{f} = 1$, in other words $K(\mathfrak{l}_{11})^{\mathrm{res}}$ is equal to the restricted Hilbert class field which is of degree 2 over K. □

5.1.1.2 Exercise. Assume that K is such that E^{res} is finite (so that K is equal to $\mathbb{Q}$ or to an imaginary quadratic field). Characterize the moduli $\mathfrak{m}$ which are not conductors of any abelian extension of K.

Answer. We first note that $\mathfrak{m}$ is a conductor if and only if $K(\mathfrak{m})^{\mathrm{res}}$ has conductor $\mathfrak{m}$; hence the following general case (valid without any assumption on K):

(0) If $\mathfrak{p}_2 | 2$ has residue degree equal to 1 in $K/\mathbb{Q}$ and if $\mathfrak{n}$ is any modulus not divisible by $\mathfrak{p}_2$, then $\mathfrak{m} := \mathfrak{p}_2\,\mathfrak{n}$ is not a conductor.

Indeed, we have $[K(\mathfrak{m})^{\mathrm{res}} : K(\mathfrak{n})^{\mathrm{res}}] = 1$.

The criterion giving the nonconductors $\mathfrak{m}$ can be written: there exists $\mathfrak{p}_v | \mathfrak{m}$ such that:

(1) $\mathrm{v}(\mathfrak{m}) = 1$ and $q_v - 1 \leq u_v$,

or:

(2) $\mathrm{v}(\mathfrak{m}) > 1$ and $q_v \leq u_v$,

with $u_v := (E^{\mathrm{res}}_{\frac{\mathfrak{m}}{\mathfrak{p}_v}} : E^{\mathrm{res}}_{\mathfrak{m}})$.

For $u_v = 1$, the only possible solution corresponds to (1) and is relative to case (0); this is the case for the field $\mathbb{Q}$ for which the nonconductors are the $2n\mathbb{Z}$ with n odd. Thus, we only need to consider the case $u_v > 1$.

Assume now that K is an imaginary quadratic field different from $\mathbb{Q}(\mu_4)$ and $\mathbb{Q}(\mu_3)$. Since $u_v = 2$, this is equivalent to $E_{\frac{\mathfrak{m}}{\mathfrak{p}_v}} = \langle -1 \rangle, \ \ E_{\mathfrak{m}} = 1$, or to $\frac{\mathfrak{m}}{\mathfrak{p}_v} | 2, \ \ \mathfrak{m} \nmid 2$. Case (1) yields the following moduli, in addition to those given in case (0):

$$\mathfrak{p}_3,\ \mathfrak{p}'_3,\ 2\mathfrak{p}_3,\ 2\mathfrak{p}'_3,\ \text{ if 3 is split and 2 is not split,}$$
$$\mathfrak{p}_3,\ 2\mathfrak{p}_3,\ \text{ if 3 is ramified and 2 is not split.}$$

Similarly, case (2) yields the additional moduli:

$$\mathfrak{p}_2^2,\ \mathfrak{p}_2'^2,\ \text{if 2 is split,}$$
$$\mathfrak{p}_2^3,\ \text{if 2 is ramified.}$$

For $K = \mathbb{Q}(\mu_4)$ or $\mathbb{Q}(\mu_3)$, we proceed in the same way and we obtain the following conductors (in addition to those coming from case (0) for $\mathbb{Q}(\mu_4)$):

$$\mathfrak{p}_2^2,\ \mathfrak{p}_2^3,\ \mathfrak{p}_5,\ \mathfrak{p}'_5,\ \text{ for } K = \mathbb{Q}(\mu_4),$$
$$(2),\ \mathfrak{p}_3,\ \mathfrak{p}_3^2,\ 2\mathfrak{p}_3,\ \mathfrak{p}_7,\ \mathfrak{p}'_7,\ \text{ for } K = \mathbb{Q}(\mu_3). \qquad \square$$

See also in [j, Coh2, Ch. 3, § 5.2] an original algorithmic expression for conductors, discriminants and signatures of abelian extensions of a number field K.

5.1.2 Artin Conductor of an Abelian Field. More generally, by uniqueness in the correspondence of class field theory and by definition of the norm conductor for an abelian extension M of K, the smallest modulus $\mathfrak{n}$ such that:

$$M \subseteq K(\mathfrak{n})^{\mathrm{res}}$$

is again the conductor of M/K, which gives a third definition of the conductor widely used in the case of abelian extensions of $\mathbb{Q}$ (see 5.5, 5.5.1), and which can be expressed as follows. Let $\mathfrak{m}$ be a modulus with support T such that $M \subseteq K(\mathfrak{m})^{\mathrm{res}}$ (which in terms of Artin groups is equivalent to $P_{T,\mathfrak{m},\mathrm{pos}} \subseteq A_T := A_{M/K,T}$); then $\mathfrak{m}$ is the conductor of M if and only if for each $v \in T$, A_T does not contain $P_{T,\frac{\mathfrak{m}}{\mathfrak{p}_v},\mathrm{pos}}$.

5.1.3 S-Decomposition. For any finite set S of noncomplex places of K which is disjoint from T, the maximal S-split subextension $K(\mathfrak{m})^S$ of $K(\mathfrak{m})^{\mathrm{res}}$ corresponds to $K^\times \langle S \rangle U^{\mathrm{res}}_{\mathfrak{m}} = K^\times U^S_{\mathfrak{m}}$ (in the idelic version), to $P_{T,\mathfrak{m},\mathrm{pos}}\langle S \rangle := P_{T,\mathfrak{m},Pl^{\mathrm{r}}_\infty \backslash S_\infty} \cdot \langle S_0 \rangle$ (in the ideal group version), and hence we have:

$$\mathrm{Gal}(K(\mathfrak{m})^S/K) \simeq \mathcal{C}\!\ell_{\mathfrak{m}}^S \quad \text{and} \quad \mathrm{Gal}(K(\mathfrak{m})^{\mathrm{res}}/K(\mathfrak{m})^S) \simeq \langle\, c\ell_{\mathfrak{m}}^{\mathrm{res}}(S)\,\rangle,$$

in the same sense as in I.4.4.1.

When $S = Pl_\infty^{\mathrm{r}}$, we obtain the field $K(\mathfrak{m})^{Pl_\infty^{\mathrm{r}}} =: K(\mathfrak{m})^{\mathrm{ord}}$ which is the ray class field modulo $\mathfrak{m}$ in the ordinary sense, in other words the maximal noncomplexified subextension of $K(\mathfrak{m})^{\mathrm{res}}$; it corresponds respectively to $K^\times U_{\mathfrak{m}}^{\mathrm{ord}}$ or to $P_{T,\mathfrak{m}}$, and we have:

$$\mathrm{Gal}(K(\mathfrak{m})^{\mathrm{ord}}/K) \simeq \mathcal{C}\!\ell_{\mathfrak{m}}^{\mathrm{ord}}\ ;$$

in certain contexts, we can also denote it by $K(\mathfrak{m})^{\mathrm{nc}}$.

As in 5.1.1, the conductor $\mathfrak{f}$ of $K(\mathfrak{m})^S$ is a divisor of $\mathfrak{m}$ which can be characterized in an analogous manner; we simply replace $E^{\mathrm{res}}_{\frac{\mathfrak{m}}{\mathfrak{p}_v}}$ and $E^{\mathrm{res}}_{\mathfrak{m}}$ by $E^S_{\frac{\mathfrak{m}}{\mathfrak{p}_v}}$ and $E^S_{\mathfrak{m}}$.

When $\mathfrak{m} = 1$, we denote by H^S the field $K(1)^S$; it is the maximal S-split subextension of the restricted Hilbert class field H^{res}. We will call it the S-split Hilbert class field.

5.1.4 Norm Groups. We have the following general diagram in which, besides each field M, we have indicated the ideal group corresponding to it by class field theory, then the idèle group, and for which the Artin (or reciprocity) map induces the isomorphism $\mathrm{Gal}(M/K) \simeq I_T/N$ (or J/N):

$P_{T,\mathrm{pos}}$ $K^\times U^{\mathrm{res}}$	H^{res}	——	$H^{\mathrm{res}}K(\mathfrak{m})^S$	——	$K(\mathfrak{m})^{\mathrm{res}}$	$P_{T,\mathfrak{m},\mathrm{pos}}$ $K^\times U^{\mathrm{res}}_{\mathfrak{m}}$
	|		|			
$P_{T,\mathrm{pos}}\langle\, S\,\rangle$ $K^\times U^S$	H^S	——	$K(\mathfrak{m})^S$		$P_{T,\mathfrak{m},\mathrm{pos}}\langle\, S\,\rangle$ $K^\times U^S_{\mathfrak{m}}$	
	|					
I_T J	K					

Recall that $P_{T,\mathfrak{m},\mathrm{pos}}\langle\, S\,\rangle := P_{T,\mathfrak{m},Pl_\infty^{\mathrm{r}}\backslash S_\infty} \cdot \langle\, S_0\,\rangle$. Recall also the four exact sequences induced by the reciprocity or Artin map, in the particular case of ray class fields:

$$1 \longrightarrow K^\times U_{\mathfrak{m}}^{\mathrm{res}} \longrightarrow J \xrightarrow{\ \rho^{\mathrm{res}}\ } \mathrm{Gal}(K(\mathfrak{m})^{\mathrm{res}}/K) \longrightarrow 1,$$

$$1 \longrightarrow K^\times U_{\mathfrak{m}}^{S} \longrightarrow J \xrightarrow{\ \rho^{S}\ } \mathrm{Gal}(K(\mathfrak{m})^{S}/K) \longrightarrow 1,$$

$$1 \longrightarrow P_{T,\mathfrak{m},\mathrm{pos}} \longrightarrow I_T \xrightarrow{\ \alpha^{\mathrm{res}}\ } \mathrm{Gal}(K(\mathfrak{m})^{\mathrm{res}}/K) \longrightarrow 1,$$

$$1 \longrightarrow P_{T,\mathfrak{m},\mathrm{pos}}\langle\, S\,\rangle \longrightarrow I_T \xrightarrow{\ \alpha^{S}\ } \mathrm{Gal}(K(\mathfrak{m})^{S}/K) \longrightarrow 1.$$

5.1.5 Intersection and Compositum of Ray Class Fields. As already remarked, for $\mathfrak{m}_1$, $\mathfrak{m}_2$ with supports contained in T, we have:

$$P_{T,\mathfrak{m}_1,\mathrm{pos}}\,P_{T,\mathfrak{m}_2,\mathrm{pos}} = P_{T,\mathrm{g.c.d.}(\mathfrak{m}_1,\mathfrak{m}_2),\mathrm{pos}},$$

or in (clearer) idelic terms:

$$K^\times U_{\mathfrak{m}_1}^{\mathrm{res}}\,K^\times U_{\mathfrak{m}_2}^{\mathrm{res}} = K^\times U_{\mathrm{g.c.d.}(\mathfrak{m}_1,\mathfrak{m}_2)}^{\mathrm{res}},$$

showing, by the usual Galois correspondence, that we always have:

$$K(\mathfrak{m}_1)^{\mathrm{res}} \cap K(\mathfrak{m}_2)^{\mathrm{res}} = K(\mathrm{g.c.d.}(\mathfrak{m}_1,\mathfrak{m}_2))^{\mathrm{res}},$$

which is still true with S-splitting. On the contrary, the trivial inclusion:

$$K(\mathfrak{m}_1)^{\mathrm{res}}\,K(\mathfrak{m}_2)^{\mathrm{res}} \subseteq K(\mathrm{l.c.m.}(\mathfrak{m}_1,\mathfrak{m}_2))^{\mathrm{res}}$$

may not be an equality, as is shown by the following example.

5.1.5.1 Example. Let $K = \mathbb{Q}(\sqrt{2})$, $\mathfrak{m}_1 = (4)$, $\mathfrak{m}_2 = (3)$. Then we have $\mathcal{C}\ell^{\mathrm{res}} = \mathcal{C}\ell^{\mathrm{ord}} = 1$, $E =: E^{\mathrm{res}} = \langle\, \varepsilon\,\rangle$ with $\varepsilon = 3 + 2\sqrt{2}$, and in particular $\varepsilon^2 = 1 + 16 + 12\sqrt{2}$, which implies:

$$E_{\mathfrak{m}_1} = E^2\ ,\ E_{\mathfrak{m}_2} = E^4\ ,\ E_{\mathfrak{m}_1\mathfrak{m}_2} = E^4,$$

and yields (see I.4.5.6, (i)):

$$[K(\mathfrak{m}_1)^{\mathrm{res}} : K] = 4,\ [K(\mathfrak{m}_2)^{\mathrm{res}} : K] = 2,\ [K(\mathfrak{m}_1\mathfrak{m}_2)^{\mathrm{res}} : K] = 16,$$

thus showing that $[K(\mathfrak{m}_1\mathfrak{m}_2)^{\mathrm{res}} : K(\mathfrak{m}_1)^{\mathrm{res}}K(\mathfrak{m}_2)^{\mathrm{res}}] = 2$. □

5.1.5.2 Exercise. Check that for arbitrary moduli $\mathfrak{m}_1$ and $\mathfrak{m}_2$, the general formula is (denoting to simplify notations by $\wedge$ and $\vee$ the g.c.d. and l.c.m. operators):

$$[K(\mathfrak{m}_1\vee\mathfrak{m}_2)^{\mathrm{res}} : K(\mathfrak{m}_1)^{\mathrm{res}}K(\mathfrak{m}_2)^{\mathrm{res}}] = \frac{(E^{\mathrm{res}}_{\mathfrak{m}_1\wedge\mathfrak{m}_2} : E^{\mathrm{res}}_{\mathfrak{m}_1})}{(E^{\mathrm{res}}_{\mathfrak{m}_2} : E^{\mathrm{res}}_{\mathfrak{m}_1\vee\mathfrak{m}_2})} = \frac{(E^{\mathrm{res}}_{\mathfrak{m}_1\wedge\mathfrak{m}_2} : E^{\mathrm{res}}_{\mathfrak{m}_2})}{(E^{\mathrm{res}}_{\mathfrak{m}_1} : E^{\mathrm{res}}_{\mathfrak{m}_1\vee\mathfrak{m}_2})}.$$

It is clear that there exists an identical formula in terms of S-split ray class fields and S-units. □

It is useful to relate the above fact 5.1.5.2 with Proposition 4.1.1. In particular, we see that if $\mathfrak{m}_1$ and $\mathfrak{m}_2$ are the conductors of $K(\mathfrak{m}_1)^{\mathrm{res}}$ and $K(\mathfrak{m}_2)^{\mathrm{res}}$, then $\mathfrak{m}_1 \vee \mathfrak{m}_2$ is the conductor of their compositum.

5.2 Decomposition Law of Places in a Ray Class Field. The decomposition law of places in $K(\mathfrak{m})^{\mathrm{res}}/K$ is especially simple and typical of class field theory since it relates this information to questions about ideal classes (see 4.5.4, and compare also with the idelic formulation in 3.3.5); recall that here T is the support of $\mathfrak{m}$:

• If v is a finite place not belonging to T (hence unramified), then its residue degree in $K(\mathfrak{m})^{\rm res}/K$ is equal to the order of the class of $\mathfrak{p}_v$ in $\mathcal{C}\ell_{\mathfrak{m}}^{\rm res} = I_T/P_{T,\mathfrak{m},\rm pos}$; it is totally split if and only if $\mathfrak{p}_v \in P_{T,\mathfrak{m},\rm pos}$.

• If $v \in T$ is unramified, this means that $K(\mathfrak{m})^{\rm res} = K(\frac{\mathfrak{m}}{\mathfrak{m}_v})^{\rm res}$, where $\mathfrak{m}_v := \mathfrak{p}_v^{\mathrm{v}(\mathfrak{m})}$, and the preceding statement is still valid if we perform the computations in $\mathcal{C}\ell^{\rm res}_{\frac{\mathfrak{m}}{\mathfrak{m}_v}} = I_{T\backslash\{v\}}/P_{T\backslash\{v\},\frac{\mathfrak{m}}{\mathfrak{m}_v},\rm pos} \simeq \mathcal{C}\ell_{\mathfrak{m}}^{\rm res}$, the place v being totally split if and only if $\mathfrak{p}_v \in P_{T\backslash\{v\},\frac{\mathfrak{m}}{\mathfrak{m}_v},\rm pos}$.

In particular (case $\mathfrak{m} = 1$) the residue degree of a finite place v in $H^{\rm res}/K$ is equal to the order of the restricted class (i.e., in $\mathcal{C}\ell^{\rm res}$) of $\mathfrak{p}_v$; its residue degree in $H^{\rm ord}/K$ is equal to the order of the ordinary class (i.e., in $\mathcal{C}\ell^{\rm ord}$) of $\mathfrak{p}_v$. Hence, the prime ideals which are totally split in $H^{\rm res}/K$ (resp. $H^{\rm ord}/K$) are those which are principal in the restricted (resp. ordinary) sense (see below the example concerning $\mathbb{Q}(\sqrt{-23}\,)$).

• If v is a real place at infinity, then v is totally split in $K(\mathfrak{m})^{\rm res}/K$ if and only if $P_{T,\mathfrak{m},Pl_\infty^{\rm r}\backslash\{v\}} \subseteq P_{T,\mathfrak{m},\rm pos}$. Thus this occurs if and only if there exists $\varepsilon_{\mathfrak{m}} \in E_{\mathfrak{m}}^{\rm ord}$ such that:

$$\begin{aligned} &i_u(\varepsilon_{\mathfrak{m}}) > 0, \ \text{for each real infinite place } u \neq v,\\ &i_v(\varepsilon_{\mathfrak{m}}) < 0 \end{aligned}$$

(see I.4.5.8, (i) for $S = \emptyset$ and $\delta_\infty = \{v\}$).

• If v is a ramified finite place (i.e., dividing the conductor), we simply perform the computations in the inertia field of v which is given explicitly in Exercise 5.2.2; since this field is a ray class field, this reduces the computation of the residue degree of v to the preceding situation.

5.2.1 Example. Let $K = \mathbb{Q}(\sqrt{-23}\,)$ and $H = K(\theta_0)$, where $\mathrm{Irr}(\theta_0, K) = X^3 - X - 1$. The K-conjugates of θ_0 are:

$$\begin{aligned} \theta_0, \quad \theta_1 &:= \frac{1}{\sqrt{-23}}\Big(3\,\theta_0^2 - \frac{9+\sqrt{-23}}{2}\,\theta_0 - 2\Big),\\ \theta_2 &:= \frac{1}{\sqrt{-23}}\Big(-3\,\theta_0^2 + \frac{9-\sqrt{-23}}{2}\,\theta_0 + 2\Big). \end{aligned}$$

We know from I.6.3.3 that H is the Hilbert class field of K and, by 3.6.1, that $\mathrm{Gal}(H/\mathbb{Q})$ is the dihedral group of order 6. We will illustrate the fact that the Frobenius $\left(\frac{H/K}{\mathfrak{p}}\right)$ of a prime ideal $\mathfrak{p}$ depends only on its class in the class group of K (of order 3). For this, we select the generator σ of $\mathrm{Gal}(H/K)$ such that $\sigma(\theta_0) = \theta_1$.

It is easily checked that, for $\mathfrak{p} \neq (\sqrt{-23})$, $\left(\frac{H/K}{\mathfrak{p}}\right)$ is characterized by the congruence:

$$\left(\frac{H/K}{\mathfrak{p}}\right)\theta_0 \equiv \theta_0^{\mathrm{N}\mathfrak{p}} \bmod \mathfrak{p},$$

so that $\left(\frac{H/K}{\mathfrak{p}}\right) = 1,\ \sigma,\ \sigma^2$, according as $\theta_0^{\mathrm{N}\mathfrak{p}} \equiv \theta_0,\ \theta_1,\ \theta_2 \bmod \mathfrak{p}$. If $\mathfrak{p}$ is inert in $K/\mathbb{Q}$, it is trivial that $\left(\frac{H/K}{\mathfrak{p}}\right) = 1$ since $H/\mathbb{Q}$ is not cyclic, and we always have $\theta_0^{\mathrm{N}\mathfrak{p}} \equiv \theta_0 \bmod \mathfrak{p}$; but such an ideal is principal for trivial reasons.

Suppose now that $\mathfrak{p}$ is split in $K/\mathbb{Q}$. If $\mathfrak{p}'$ is the conjugate of $\mathfrak{p}$, the Galois operation 3.6.1 gives $\left(\frac{H/K}{\mathfrak{p}'}\right) = \left(\frac{H/K}{\mathfrak{p}}\right)^{-1}$ (which is in accordance with the principality of $\mathfrak{p}\mathfrak{p}'$).

For $\mathfrak{p}_2 = (2, \frac{1+\sqrt{-23}}{2})$ we find $\theta_0^2 \equiv \theta_1 \bmod \mathfrak{p}_2$ since $\frac{9+\sqrt{-23}}{2} \in \mathfrak{p}_2$, and for $\mathfrak{p}_3 = (3, \frac{1+\sqrt{-23}}{2})$ we find $\theta_0^3 \equiv \theta_2 \bmod \mathfrak{p}_3$ using the congruence $\sqrt{-23} \equiv -1 \bmod \mathfrak{p}_3$. Thus $\left(\frac{H/K}{\mathfrak{p}_2}\right) = \sigma$ and $\left(\frac{H/K}{\mathfrak{p}_3}\right) = \sigma^2$. In other words, the Artin symbol $\left(\frac{H/K}{\mathfrak{p}_2\mathfrak{p}_3}\right)$ is trivial, and indeed, we have $\mathfrak{p}_2\mathfrak{p}_3 = (\frac{1+\sqrt{-23}}{2})$, a principal ideal.

With $\mathfrak{p}_{13} = (13, 4 + \sqrt{-23})$ we find $\theta_0^{13} \equiv \theta_1 \bmod \mathfrak{p}_{13}$, so that $\left(\frac{H/K}{\mathfrak{p}_{13}}\right) = \sigma$. We must verify that $\mathfrak{p}_2\mathfrak{p}'_{13}$ is principal, which is indeed the case since $(\frac{9+\sqrt{-23}}{2}) = \mathfrak{p}_2\mathfrak{p}'_{13}$ (and not $\mathfrak{p}_2\mathfrak{p}_{13}$ since $\sqrt{-23} \equiv -4 \bmod \mathfrak{p}_{13}$ or, equivalently, $\sqrt{-23} \equiv 4 \bmod \mathfrak{p}'_{13}$).

For all the split prime ideals $\mathfrak{p}$ with $\mathrm{N}\mathfrak{p} < 59$ we find $\left(\frac{H/K}{\mathfrak{p}}\right) \in \{\sigma, \sigma^2\}$ and we verify that $\mathfrak{p}$ is always in the "good" nontrivial class.

For the prime number 59, we find $\theta_0^{59} \equiv \theta_0 \bmod \mathfrak{p}_{59}$. This means that $\left(\frac{H/K}{\mathfrak{p}_{59}}\right) = \left(\frac{H/K}{\mathfrak{p}'_{59}}\right) = 1$ and that the prime ideals above 59 are principal (we have $\mathrm{N}(6 + \sqrt{-23}) = 59$ which proves the claim). Note that 59 is the least example giving nontrivial principal ideals.

This gives a good idea of a reciprocity law since the splitting of the polynomial $f = X^3 - X - 1$ (into one (f_3), two $(f_1 f'_2)$, or three $(f_1 f'_1 f''_1)$ irreducible factors in $\mathbb{Q}_p[X]$) has been characterized by means of ray classes (i.e., multiplicative congruences). More precisely:

- $(\frac{p}{23}) = -1$ implies $f = f_1 f'_2$ (indeed, this is equivalent to $(\frac{-23}{p}) = -1$ (first reciprocity !), and therefore $\mathfrak{p} = (p)$ is inert in $K/\mathbb{Q}$ and split in H/K);
- $(\frac{p}{23}) = +1$ and $\mathfrak{p}|p$ principal imply $f = f_1 f'_1 f''_1$ ($\mathfrak{p}$ is split in $K/\mathbb{Q}$ and in H/K);
- $(\frac{p}{23}) = +1$ and $\mathfrak{p}|p$ nonprincipal imply $f = f_3$ ($\mathfrak{p}$ is split in $K/\mathbb{Q}$ and inert in H/K).

One verifies that the first case is equivalent to:

$$\mathfrak{p} \in j_{K/\mathbb{Q}}((a_i)\, P_{\mathbb{Q},(23),\mathrm{pos}}),\quad a_i \in \{5,\ 7,\ 10,\ 11,\ 14,\ 15,\ 17,\ 19,\ 20,\ 21,\ 22\},$$

and that the last one is equivalent to:

$$\mathfrak{p} \in \mathfrak{p}_2 P_K \cup \mathfrak{p}_2^2 P_K.$$

The problem has been "linearized" in an obvious way.

In terms of quadratic forms, we check that the norm form $x^2 + xy + 6y^2$ represents $p \neq 23$ if and only if $X^3 - X - 1$ has three roots in $\mathbb{Q}_p$ (indeed, this is equivalent to $\mathfrak{p}$ split and principal).

The power of class field theory comes from the fact that it is impossible to deduce the above rules from elementary properties of number fields and/or polynomials. See [Wy] for other examples and comments. □

5.2.2 Exercise (study of ramification in a ray class field). Let K be a number field together with sets of places T and S.

(i) Let $\mathfrak{m}$ be a modulus of K with support T, and let $v \in T$. Show that if we set $\mathfrak{m}_v := \mathfrak{p}_v^{\mathrm{v}(\mathfrak{m})}$, then $K(\frac{\mathfrak{m}}{\mathfrak{m}_v})^S$ is the inertia field of v in the extension $K(\mathfrak{m})^S/K$.

Deduce a formula for the ramification index of v in $K(\mathfrak{m})^S/K$.

Generalize by giving a description of $\mathrm{Gal}(K(\mathfrak{m})^S/K(\mathfrak{n})^{S\cup\delta_\infty})$, where $\delta_\infty \subseteq Pl_\infty^{\mathrm{r}} \backslash S_\infty$, and where $\mathfrak{n} = \prod_{v\in t} \mathfrak{m}_v$ for $t \subseteq T$.

Compute also the residue degree of v in $K(\mathfrak{m})^S/K$.

(ii) Show that the maximal T-tamely ramified abelian extension (i.e., T-ramified and such that for every place $v \in T$, the ramification index of v in this extension is prime to the residue characteristic of v) is equal to $K(\mathfrak{m}_{\mathrm{ta}})^{\mathrm{res}}$ for $\mathfrak{m}_{\mathrm{ta}} = \prod_{v\in T} \mathfrak{p}_v$.

Answer. (i) We start with the case $S = \emptyset$, and give several approaches.

By the conductor theorem, v is unramified in $K(\frac{\mathfrak{m}}{\mathfrak{m}_v})^{\mathrm{res}}$, hence the inertia field M of v in $K(\mathfrak{m})^{\mathrm{res}}/K$ contains $K(\frac{\mathfrak{m}}{\mathfrak{m}_v})^{\mathrm{res}}$; since M/K is unramified at v, its conductor $\mathfrak{f}$ is not divisible by $\mathfrak{p}_v$, and since $M \subseteq K(\mathfrak{m})^{\mathrm{res}}$, we have $\mathfrak{f} \,|\, \mathfrak{m}$ (see 5.1.2) hence $\mathfrak{f} \,|\, \frac{\mathfrak{m}}{\mathfrak{m}_v}$, and we have $M \subseteq K(\mathfrak{f})^{\mathrm{res}} \subseteq K(\frac{\mathfrak{m}}{\mathfrak{m}_v})^{\mathrm{res}}$.

We can base a proof on the characterization of $I_v(K(\mathfrak{m})^{\mathrm{res}}/K)$ given in 4.5.4, (i), so that in this case the inertia field corresponds to the group $P_{T,\frac{\mathfrak{m}}{\mathfrak{m}_v},\mathrm{pos}}$.

The inertia group of v in $K(\mathfrak{m})^S/K$ is $\mathrm{Gal}(K(\mathfrak{m})^S/K(\frac{\mathfrak{m}}{\mathfrak{m}_v})^S)$; indeed, use 1.2.1 and the fact that:

$$\mathrm{Gal}(K(\mathfrak{m})^{\mathrm{res}}/K(\mathfrak{m})^S) \ \text{ and } \ \mathrm{Gal}(K(\tfrac{\mathfrak{m}}{\mathfrak{m}_v})^{\mathrm{res}}/K(\tfrac{\mathfrak{m}}{\mathfrak{m}_v})^S)$$

are generated by the decomposition groups of the $v \in S$ in the corresponding extensions.

In idelic terms, the inertia field of v in $K(\mathfrak{m})^S/K$ corresponds to $K^\times U_{\mathfrak{m}}^S U_v$ by 3.3, (iii), or 3.5.1, (ii); but clearly we have $U_{\mathfrak{m}}^S U_v = U_{\frac{\mathfrak{m}}{\mathfrak{m}_v}}^S$.

Formula I.4.5.1, for $\mathfrak{n} = \frac{\mathfrak{m}}{\mathfrak{m}_v}$ (in which case $\varphi(\mathfrak{m}) = \varphi(\mathfrak{n})\varphi(\mathfrak{m}_v)$ since $\mathfrak{n}$ and $\mathfrak{m}_v$ are coprime) and $\delta_\infty = \emptyset$, immediately yields:

$$e_v(K(\mathfrak{m})^S/K) = \frac{\varphi(\mathfrak{m}_v)}{(E_{\frac{\mathfrak{m}}{\mathfrak{m}_v}}^S : E_{\mathfrak{m}}^S)},$$

for any $v \in T$, which, for $S = \emptyset$ yields:

$$e_v(K(\mathfrak{m})^{\mathrm{res}}/K) = \frac{\varphi(\mathfrak{m}_v)}{(E^{\mathrm{res}}_{\frac{\mathfrak{m}}{\mathfrak{m}_v}} : E^{\mathrm{res}}_{\mathfrak{m}})}.$$

The same arguments show that for $t \subseteq T$, $\mathfrak{n} = \prod_{v \in t} \mathfrak{m}_v$, and $\delta_\infty \subseteq Pl^{\mathrm{r}}_\infty \backslash S_\infty$, the ray class field $K(\mathfrak{n})^{S \cup \delta_\infty}$ is the subfield of $K(\mathfrak{m})^S$ fixed under the subgroup generated by the inertia groups of the places of $T \backslash t$ and the decomposition groups of the places of δ_∞.

If $v \in Pl^{\mathrm{r}}_\infty$, by I.4.5.1 for $\mathfrak{n} = \mathfrak{m}$ and $\delta_\infty = \{v\}$, we also obtain:

$$f_v(K(\mathfrak{m})^S/K) = \frac{2}{(E^{S \cup \{v\}}_{\mathfrak{m}} : E^S_{\mathfrak{m}})} = \frac{2}{|\mathrm{sgn}_v(E^{S \cup \{v\}}_{\mathfrak{m}})|}.$$

To obtain a formula for the residue degree of a finite place v not belonging to $T \cup S$, we use directly 3.3.5 by computing the image of $K_v^\times$ in $J/K^\times U^S_{\mathfrak{m}}$, so that we obtain:

$$f_v(K(\mathfrak{m})^S/K) = (K_v^\times : i_v(E^{S \cup \{v\}}_{\mathfrak{m}})U_v),$$

but it is still possible to perform a direct computation using I.4.5.1, from which we recover (see 5.2):

$$f_v(K(\mathfrak{m})^S/K) = \frac{|\mathcal{C}\ell^S_{\mathfrak{m}}|}{|\mathcal{C}\ell^{S \cup \{v\}}_{\mathfrak{m}}|} = |\langle\, c\ell^S_{\mathfrak{m}}(\mathfrak{p}_v)\,\rangle|.$$

The idelic formulation is more convenient if the $S \cup \{v\}$-units are known, the other one relies on a computation of generalized ideal classes (here the class of $\mathfrak{p}_v$); as always, the correspondence is justified by I.5.1.

See also III.1.1.6, (ii) for a slightly more general context.

(ii) The formula for e_v immediately shows that $K(\mathfrak{m}_{\mathrm{ta}})^{\mathrm{res}}$ is T-tamely ramified. Let $L \supseteq K(\mathfrak{m}_{\mathrm{ta}})^{\mathrm{res}}$ be the maximal T-tamely ramified abelian extension of K. Let M be a finite extension of $K(\mathfrak{m}_{\mathrm{ta}})^{\mathrm{res}}$ in L and let $\mathfrak{m} \in \langle\, T\,\rangle_{\mathbb{N}}$ be a multiple of its conductor, with support T ($\mathfrak{m}_{\mathrm{ta}}$ is always the tame part of $\mathfrak{m}$); if N corresponds to M, we have $K^\times U^{\mathrm{res}}_{\mathfrak{m}} \subseteq N$. Since by 3.5.1, (ii) the idèle group corresponding to M is also equal to

$$\prod_v U_v^1 \,.\, N,$$

we have $\prod_v U_v^1 \,.\, K^\times U^{\mathrm{res}}_{\mathfrak{m}} \subseteq \prod_v U_v^1 \,.\, N = N$, giving $K^\times U^{\mathrm{res}}_{\mathfrak{m}_{\mathrm{ta}}} \subseteq N$, and showing that we have $M \subseteq K(\mathfrak{m}_{\mathrm{ta}})^{\mathrm{res}}$, which does not depend on the choice of $\mathfrak{m}$. This shows that $L = \bigcup_M M$ is finite and equal to the ray class field $K(\mathfrak{m}_{\mathrm{ta}})^{\mathrm{res}}$ (the finiteness of L/K comes from 1.3.3.1 and of the finiteness of H^{res}/K, which also implies that of ray class fields). In the same way, $K(\mathfrak{m}_{\mathrm{ta}})^S$ is the maximal T-tamely ramified S-split abelian extension of K. □

b) Rank Formulas — The Reflection Theorem

When we take limits on $\mathfrak{m}$, we must use slightly different notations. Let K be a number field together with sets of places T and S, and let $\langle T \rangle_{\mathbb{N}}$ be the monoid generated by the $\mathfrak{p}_v$ for $v \in T$. By reference to the notion of Hilbert class field when we use sets T and S which are not necessarily empty, we put the following.

5.3 Notations. (i) From 5.1.2, we set:

$$H_T^S := \bigcup_{\mathfrak{m} \in \langle T \rangle_{\mathbb{N}}} K(\mathfrak{m})^S,$$

which is the maximal T-ramified S-split abelian extension of K. We also use the notation $H_T^{S_0\,\mathrm{res}}$ (resp. $H_T^{S_0\,\mathrm{ord}}$) when $S_\infty = \emptyset$ (resp. $S_\infty = Pl_\infty^{\mathrm{r}}$).[25]

(ii) From 5.2.2, (ii), we define:

$$H_{\mathrm{ta}}^S := \bigcup_{\mathfrak{m}_{\mathrm{ta}}} K(\mathfrak{m}_{\mathrm{ta}})^S,$$

which is the maximal tamely ramified S-split abelian extension of K (the tame moduli $\mathfrak{m}_{\mathrm{ta}}$ have an arbitrary support, but are squarefree). □

For a fixed finite T, the groups:

$$\mathrm{Gal}(H_T^S/K) \simeq \varprojlim_{\mathfrak{m} \in \langle T \rangle_{\mathbb{N}}} \mathcal{C}\ell_{\mathfrak{m}}^S =: \mathcal{C}\ell_T^S,$$

will be studied in great detail in Chapter III. Meanwhile, we can prove a number of properties on the p-ranks of these groups which had been mentioned at the end of Section 4 of Chapter I.

5.4 Rank Formulas. We start, in the following exercise, with the simplest situation (i.e., without any Galois structure) which is an essential prelude to the reflection theorem.

5.4.1 Exercise (Šafarevič's formula (1964), reflection formula (1998)). The notations are those of I.4.5, I.4.6. For finite and disjoint sets T, $S = S_0 \cup S_\infty$, and for $\mathfrak{m} \in \langle T \rangle_{\mathbb{N}}$, $\Delta_\infty := Pl_\infty^{\mathrm{r}} \backslash S_\infty$, we set:

$$Y_{T,\mathfrak{m}}^S := \{\alpha \in K_T^{\times p} K_{T,\mathfrak{m},\Delta_\infty}^\times,\ (\alpha) = \mathfrak{a}^p \mathfrak{a}_{S_0},\ \mathfrak{a} \in I_T,\ \mathfrak{a}_{S_0} \in \langle S_0 \rangle\},$$

$$V_T^S := \{\alpha \in K_T^{\times p} K_{T,\Delta_\infty}^\times,\ (\alpha) = \mathfrak{a}^p \mathfrak{a}_{S_0},$$
$$\mathfrak{a} \in I_T,\ \mathfrak{a}_{S_0} \in \langle S_0 \rangle,\ \ i_v(\alpha) \in K_v^{\times p}\ \ \forall v \in T\},$$

[25] In these definitions, T is not assumed finite. When $T = \emptyset$, we recover the S-split Hilbert class field H^S.

$\delta_v := 1$ or 0 according as K_v contains μ_p or not, $\delta := 1$ or 0 according as K contains μ_p or not.

(i) Prove the formula:

$$\mathrm{rk}_p(Y_T^{S_0\,\mathrm{ord}}/K_T^{\times p}) = \mathrm{rk}_p(\mathcal{C}\ell^{S_0\,\mathrm{ord}}) + \delta + |S_0| + r_1 + r_2 - 1.$$

(ii) Show that for any sufficiently large modulus $\mathfrak{m} \in \langle T \rangle_{\mathbb{N}}$, we have $Y_{T,\mathfrak{m}}^S = V_T^S$, and deduce Šafarevič's rank formula:

$$\mathrm{rk}_p(\mathcal{C}\ell_T^S) = \mathrm{rk}_p(V_T^S/K_T^{\times p}) + \sum_{v\in T_p}[K_v : \mathbb{Q}_p] \\ + \sum_{v\in T}\delta_v - \delta - |S_0| + 1 - r_1 - r_2 + \delta_{2,p}|\Delta_\infty|,$$

where $\delta_{2,p}$ is the Kronecker symbol equal to 1 if $p = 2$ and to 0 otherwise.

(iii) We now assume that $\mu_p \subset K$ and $T_p \cup S_p = Pl_p$. Show that V_T^S is the radical of the maximal elementary S_0-ramified $T \cup \Delta_\infty$-split abelian p-extension, and deduce the formula:

$$\mathrm{rk}_p(\mathcal{C}\ell_T^{S_0\cup S_\infty}) - \mathrm{rk}_p(\mathcal{C}\ell_{S_0}^{T\cup\Delta_\infty}) = |T| - |S_0| + \sum_{v\in T_p}[K_v : \mathbb{Q}_p] - r_1 - r_2 + \delta_{2,p}|\Delta_\infty|.$$

Find the corresponding formula when one removes the assumption $T_p \cup S_p = Pl_p$.

(iv) (asked in [j, Coh2, Ch. 3, § 6, Exer. 16]). Let K be a number field such that $(\mathcal{C}\ell_{(4)}^{\mathrm{ord}})_2 = 1$; show that K is totally real.

Answer. (i) If $\alpha \in Y_T^{S_0\,\mathrm{ord}}$ ($\mathfrak{m} = 1$, $\Delta_\infty = \emptyset$), we have $\alpha \in K_T^\times$ and $(\alpha) =: \mathfrak{a}^p\mathfrak{a}_{S_0}$; if we send α to the class of $\mathfrak{a}$ in $\mathcal{C}\ell^{S_0\,\mathrm{ord}}$, we obtain the exact sequence:

$$1 \longrightarrow E^{S_0\,\mathrm{ord}}/(E^{S_0\,\mathrm{ord}})^p \longrightarrow Y_T^{S_0\,\mathrm{ord}}/K_T^{\times p} \longrightarrow {}_p\mathcal{C}\ell^{S_0\,\mathrm{ord}} \longrightarrow 1,$$

where ${}_p\mathcal{C}\ell^{S_0\,\mathrm{ord}}$ is the subgroup of $\mathcal{C}\ell^{S_0\,\mathrm{ord}}$ formed by classes killed by p. By the Dirichlet Theorem I.3.7.1, the rank formula follows.

(ii) Let $\alpha \in K_T^\times$; it is clear, by using the chinese remainder theorem, that if $i_v(\alpha) \in K_v^{\times p}$ for each $v \in T$, we have $\alpha \in K_T^{\times p}K_{T,\mathfrak{m}}^\times$ for all $\mathfrak{m}$ with support T; to have equivalence, it is enough to choose $\mathfrak{m}$ such that $i_T(K_{T,\mathfrak{m}}^\times) \subset \bigoplus_{v\in T}(U_v)^p$ and we then have $Y_{T,\mathfrak{m}}^S = V_T^S$.

Note. If $\mu_p \subset K$, by I.6.3.4, (iii), we can choose $\mathfrak{m} = \prod_{v\in T\setminus T_p}\mathfrak{p}_v \prod_{v\in T_p}\mathfrak{p}_v^{pe_v+1}$.

Let $H_T^S[p]$ be the maximal elementary p-subextension of H_T^S (i.e., fixed under $(\mathcal{C}\ell_T^S)^p$); by I.4.5.1, $H_T^S[p]/K$ is finite. Assume that $\mathfrak{m} \in \langle T \rangle_{\mathbb{N}}$ is such that $Y_{T,\mathfrak{m}}^S = V_T^S$ and is a multiple of the conductor of $H_T^S[p]$, so that $\mathrm{rk}_p(\mathcal{C}\ell_T^S) = \mathrm{rk}_p(\mathcal{C}\ell_\mathfrak{m}^S)$. By I.4.5, (ii) applied to $\mathfrak{n} = 1$ and $\delta_\infty = \Delta_\infty$, we obtain:

$$\begin{aligned}\mathrm{rk}_p(\mathcal{C}\ell_T^S) &= \mathrm{rk}_p(\mathcal{C}\ell^{S_0\,\mathrm{ord}}) + \sum_{v\in T}\mathrm{rk}_p(U_v) + \delta_{2,p}|\Delta_\infty| \\ &\qquad -\mathrm{rk}_p(Y_T^{S_0\,\mathrm{ord}}/K_T^{\times p}) + \mathrm{rk}_p(V_T^S/K_T^{\times p}) \\ &= \mathrm{rk}_p(V_T^S/K_T^{\times p}) + \sum_{v\in T}\mathrm{rk}_p(U_v) - |S_0| - r_1 - r_2 + 1 - \delta + \delta_{2,p}|\Delta_\infty|\end{aligned}$$

(using (i)). Since $\mathrm{rk}_p(U_v) = \delta_v$ (resp. $\delta_v + [K_v : \mathbb{Q}_p]$) if $v \nmid p$ (resp. $v|p$) by I.3.1.1, we obtain Šafarevič's formula with decomposition.

(iii) We have $\alpha \in V_T^S$ if and only if $i_v(\alpha) \in K_v^{\times p}$ for each $v \in T \cup \Delta_\infty$ and if $(\alpha) = \mathfrak{a}^p\mathfrak{a}_{S_0}$; this gives the $T\cup\Delta_\infty$-splitting and the $Pl_p \cup S_0$-ramification (see I.6.3); the equality $Pl_p = T_p \cup S_p$ implies the S_0-ramification. The converse is trivial after noting that if $\alpha K^{\times p}$ is an element of the radical of $H_{S_0}^{T\cup\Delta_\infty}[p]$, we may assume that α is prime to T. The rank formula of (ii) then gives the result.

More generally, set $\Delta_p := Pl_p \backslash (T_p \cup S_p)$ and consider:

$$\mathfrak{m}^* := \prod_{v\in S_0\backslash S_p}\mathfrak{p}_v \prod_{v\in S_p}\mathfrak{p}_v^{pe_v+1} \prod_{v\in\Delta_p}\mathfrak{p}_v^{pe_v},$$

where, for $v|p$, e_v denotes the ramification index of v in $K/\mathbb{Q}(\mu_p)$. Let us check that $K\left(\sqrt[p]{V_T^S}\right)$ is the maximal elementary p-subextension of $K(\mathfrak{m}^*)^{T\cup\Delta_\infty}$. If $\alpha \in V_T^S$, the v-conductor computations made in 1.6.3 show that if v is tame (i.e., if $v \in S_0 \setminus S_p$) the v-conductor of $K(\sqrt[p]{\alpha})/K$ is $\mathfrak{f}_v = (1)$ or $\mathfrak{p}_v$, and that otherwise (i.e., if $v \in Pl_p\backslash T_p = S_p \cup \Delta_p$) then $\mathfrak{f}_v = \mathfrak{p}_v^{pe_v+1-r}$, where $r = 0$ is equivalent to $\mathrm{v}(\alpha) \not\equiv 0 \bmod (p)$; we thus obtain the inclusion:

$$K\left(\sqrt[p]{V_T^S}\right) \subseteq K(\mathfrak{m}^*)^{T\cup\Delta_\infty}.$$

If $K(\sqrt[p]{\alpha}) \subseteq K(\mathfrak{m}^*)^{T\cup\Delta_\infty}$, analogous considerations show that $\alpha \in V_T^S$. We thus obtain the formula that we have already mentioned:

$$\begin{aligned}&\mathrm{rk}_p(\mathcal{C}\ell_T^{S_0\cup S_\infty}) - \mathrm{rk}_p(\mathcal{C}\ell_{\mathfrak{m}^*}^{T\cup\Delta_\infty}) = \\ &\qquad |T| - |S_0| + \sum_{v\in T_p}[K_v : \mathbb{Q}_p] - r_1 - r_2 + \delta_{2,p}|\Delta_\infty|.\end{aligned}$$

(iv) For $p = 2$, $T = S = \emptyset$, we have $\mathfrak{m}^* = (4)$ and:

$$\mathrm{rk}_2(\mathcal{C}\ell^{\mathrm{res}}) - \mathrm{rk}_2(\mathcal{C}\ell_{(4)}^{\mathrm{ord}}) = -r_2\ ;$$

under the assumption of the question, we thus obtain the stronger result:

$$r_2 = 0 \ \text{ and } \ \mathrm{rk}_2(\mathcal{C}\ell^{\mathrm{res}}) = 0. \qquad \square$$

The formulas of (iii) are only a particular case of the reflection theorem whose statement we are going to give below. However, they already show the symmetry which, in the Kummer case, roughly speaking exchanges ramification and decomposition, except that for $p = 2$ and the places at infinity, we obtain for instance (assuming that $T_2 \cup S_2 = Pl_2$):

$$\mathrm{rk}_2(\mathcal{C}\ell_T^{S_0\,\mathrm{res}}) - \mathrm{rk}_2(\mathcal{C}\ell_{S_0}^{T\,\mathrm{ord}}) = |T| - |S_0| + \sum_{v \in T_2} [K_v : \mathbb{Q}_2] - r_2.$$

5.4.2 Reflection Principle. The most general statement assumes the following definitions and facts, borrowed from the language of group representations (use [Se4] and in particular Section 12 for rationality questions), and which we only recall briefly:

• Let g be a finite group of order prime to p. We denote by $\mathfrak{X}_p(g)$ the set of $\mathbb{F}_p$-irreducible characters of g, and for $\chi \in \mathfrak{X}_p(g)$ we set:

$$e_\chi := \frac{\psi(1)}{|g|} \sum_{s \in g} \chi(s^{-1})\, s,$$

where ψ is an absolutely irreducible character such that χ is equal to the sum of the distinct $\mathbb{F}_p$-conjugates of ψ: an $\mathbb{F}_p$-conjugate of ψ is of the form ψ^{p^i}, where $\psi^{p^i}(s) := \psi(s^{p^i})$ for all $s \in g$. We thus obtain a fundamental system of central orthogonal idempotents of the algebra $\mathbb{F}_p[g]$, thanks to the assumption that $p \nmid |g|$.

We denote by V_χ the $\mathbb{F}_p$-irreducible representation with character χ. If g is commutative, $\psi(1) = 1$, V_χ is an $\mathbb{F}_p$-vector space of dimension equal to the order of p modulo the order of ψ.

• For any $\mathbb{Z}[g]$-module M of finite type and any $\chi \in \mathfrak{X}_p(g)$ we set:

$$M_\chi := (M \otimes \mathbb{F}_p)^{e_\chi},$$

and we call χ-rank of M the integer $r := \mathrm{rk}_\chi(M)$ such that:

$$M_\chi \simeq r\, V_\chi := \bigoplus_{i=1}^{r} V_\chi.$$

Therefore, we have $\mathrm{rk}_p(M_\chi) = \chi(1)\mathrm{rk}_\chi(M)$.

• Assume that g, of order prime to p, is an automorphism group of K (K containing μ_p), and let $k := K^g$. If T and S are sets of places of K stable under g, we denote by T_k and S_k the sets of places of k below those of T and S. Finally, for any place u of k we denote by abuse of notation by d_u the decomposition group, in K/k, of a place v of K above u (thus d_u is only defined up to conjugation).

• Let ω be the Teichmüller character, i.e., the character defined by the action of g on μ_p (if $s \in g$, $\omega(s)$ is the unique element $a \in \mathbb{F}_p^\times$ such that $s(\zeta) = \zeta^a$ for all $\zeta \in \mu_p$). If $\chi \in \mathfrak{X}_p(g)$ we set:

$$\chi^* := \omega\chi^{-1}\ ;$$

we still have $\chi^* \in \mathfrak{X}_p(g)$ and this defines the fundamental involution attached to the reflection principle (the mirror involution).

- We then define for all $\chi \in \mathfrak{X}_p(g)$ (with $\psi|\chi$):

$$\begin{aligned}\rho_\chi(T,S) &:= \psi(1)r_2(k) + \sum_{u\in Pl^{\mathrm{r}}_{k,\infty}} \rho_{u,\chi} + \sum_{u\in T_k} \rho_{u,\chi} + \delta_{\omega,\chi} - \delta_{1,\chi}\\ &\quad - \sum_{u\in S_{0,k}} \rho_{u,\chi^*} - \psi(1) \sum_{u\in S_{p,k}\cup\Delta_{p,k}} [k_u : \mathbb{Q}_p] - \delta_{2,p}\psi(1)|S_{\infty,k}|\\ &= \sum_{u\in T_k} \rho_{u,\chi} + \psi(1)\sum_{u\in T_{p,k}} [k_u : \mathbb{Q}_p] - \sum_{u\in S_{0,k}} \rho_{u,\chi^*} + \delta_{\omega,\chi} - \delta_{1,\chi}\\ &\quad - \psi(1)r_2(k) - \sum_{u\in Pl^{\mathrm{r}}_{k,\infty}} \rho_{u,\chi^*} + \delta_{2,p}\psi(1)|\Delta_{\infty,k}|,\end{aligned}$$

where the $\delta_{a,b}$ denote Kronecker symbols and where:

$$\rho_{u,\chi} := \frac{1}{|d_u|}\sum_{t\in d_u}\psi(t),\ \ \rho_{u,\chi^*} := \frac{1}{|d_u|}\sum_{t\in d_u}\omega\psi^{-1}(t),$$

$$\Delta_{p,k} := Pl_{k,p}\backslash(T_{p,k}\cup S_{p,k}),\ \ \Delta_{\infty,k} := Pl^{\mathrm{rnc}}_{k,\infty}\backslash S_{\infty,k},$$

where $Pl^{\mathrm{rnc}}_{k,\infty}$ is the set of real places of k noncomplexified in K. Be careful to distinguish between $Pl^{\mathrm{r}}_{k,\infty}$ and $Pl^{\mathrm{r}}_{\infty,k}$; in particular:

$$|S_{\infty,k}| + |\Delta_{\infty,k}| = r_1(k) - r_1^{\mathrm{c}}(k),$$

where $r_1^{\mathrm{c}}(k)$ is the number of real places of k which are complexified in K.

5.4.3 Remark. It is now important to comment on the relationship between Kummer theory and class field theory which will give the reflection theorem with characters. We use the notations of (Ch. I; § 6), where the symbol * also denotes the dual of a group. Let L/K be a p-elementary Kummer extension of radical W and Galois group A. Then we have the "Spiegelungsrelation", which comes directly from I.6.2, in view of the g-module action on a dual, and which can be stated as follows for any $\chi \in \mathfrak{X}_p(g)$:

$$W_{\chi^*} \simeq (A^*)_{\chi^*} \simeq (A_\chi)^*$$

(canonical isomorphisms of g-modules), and yields the relation:

$$\mathrm{rk}_{\chi^*}(W) = \mathrm{rk}_\chi(A).$$

We suppose that K is given together with sets of places T and S; we do not assume that $Pl_p = T_p \cup S_p$. This leads to the context of Exercise 5.4.1, (iii), for which we put $\mathfrak{m}^* := \prod_{v\in S_0\backslash S_p} \mathfrak{p}_v \prod_{v\in S_p} \mathfrak{p}_v^{pe_v+1} \prod_{v\in\Delta_p} \mathfrak{p}_v^{pe_v}$, with $\Delta_p := Pl_p\backslash(T_p\cup S_p)$. The reflection theorem then consists in the application of the above to the extension $L := K(\mathfrak{m}^*)^{T\cup\Delta_\infty}[p]$ for which the χ^*-component W_{χ^*} of the radical $W = V_T^S$ will be computed using the χ^*-component of the class group $\mathcal{C}\!\ell_T^S$ (see below), the χ-component of A_χ then being nothing else than the χ-component of $\mathcal{C}\ell_{\mathfrak{m}^*}^{T\cup\Delta_\infty}$ under the isomorphism of class field theory. □

5.4.4 Proposition. *Let K be any number field together with sets of places T and S, and p a prime. We have the exact sequences of $\mathbb{F}_p$-vector spaces:*

$$1 \longrightarrow E^{S_0\,\mathrm{ord}}/(E^{S_0\,\mathrm{ord}})^p \longrightarrow Y_T^{S_0\,\mathrm{ord}}/K_T^{\times p} \longrightarrow {}_p\mathcal{C}\ell^{S_0\,\mathrm{ord}} \longrightarrow 1,$$

$$1 \longrightarrow Y_T^{S_0\,\mathrm{ord}}/V_T^S \longrightarrow \bigoplus_{v\in T} U_v/(U_v)^p \bigoplus_{v\in\Delta_\infty} (\{\pm1\})_p \longrightarrow X \longrightarrow 1,$$

where X is the kernel of the surjective map:

$$\mathcal{C}\ell_T^S/(\mathcal{C}\ell_T^S)^p \longrightarrow \mathcal{C}\ell^{S_0\,\mathrm{ord}}/(\mathcal{C}\ell^{S_0\,\mathrm{ord}})^p.$$

Proof. The first exact sequence is given in 5.4.1, (i). For the second one, see the proof of I.4.5, (ii) with $\mathfrak{n} = 1$, $\delta_\infty = \Delta_\infty$, and with $\mathfrak{m} = \prod_{v\in T} \mathfrak{p}_v^{m_v}$ sufficiently large in order to have $U_v^{m_v} \subseteq (U_v)^p$ for all $v \in T$, $\mathcal{C}\ell_{\mathfrak{m}}^S/(\mathcal{C}\ell_{\mathfrak{m}}^S)^p = \mathcal{C}\ell_T^S/(\mathcal{C}\ell_T^S)^p$, and $Y_{T,\mathfrak{m}}^S = V_T^S$ (see 5.4.1, (ii)). □

Suppose now that K is given together with a group of automorphisms g of order prime to p. We do not assume that $\mu_p \subset K$. Then, using 5.4.2, the above exact sequences give immediately a generalization of Šafarevič's formula, with characters:

$$\mathrm{rk}_{\chi^*}(\mathcal{C}\ell_T^S) - \mathrm{rk}_{\chi^*}(V_T^S) = \mathrm{rk}_{\chi^*}\Big(\bigoplus_{v\in T} U_v \bigoplus_{v\in\Delta_\infty} \{\pm1\}\Big) - \mathrm{rk}_{\chi^*}(E^{S_0\,\mathrm{ord}}).$$

The right hand side is a straightforward computation using representation theory in the context of 2.3 and the S-unit Dirichlet–Herbrand Theorem I.3.7 (see in 5.4.7 the value of this expression which is a little more complicated than $\rho_\chi(T,S)$ given above corresponding to the case $\mu_p \subset K$). If $\mu_p \subset K$, we can use the Kummer interpretation of Remark 5.4.3, giving the reflection theorem for which we recall the main notations.

Notations. For a number field K containing μ_p, given together with sets of places T and $S = S_0 \cup S_\infty$, we put:

$$T_p := T \cap Pl_p,\ S_p := S \cap Pl_p,\ \Delta_p := Pl_p \backslash (T_p \cup S_p),\ \Delta_\infty := Pl_\infty^{\mathrm{r}} \backslash S_\infty,$$

$$T^* := T \cup \Delta_\infty,\ S^* := S_0 \cup \Delta_p,\ \mathfrak{m}^* := \prod_{v\in S_0\backslash S_p} \mathfrak{p}_v \prod_{v\in S_p} \mathfrak{p}_v^{pe_v+1} \prod_{v\in\Delta_p} \mathfrak{p}_v^{pe_v},$$

where, for $v|p$, e_v denotes the ramification index of v in $K/\mathbb{Q}(\mu_p)$. We use also the notations given in Paragraph 5.4.2. □

5.4.5 Theorem (of T-S-reflection [Gr10, Ch. I, Th. 5.18] (1998)). *Let K be a number field containing the group μ_p of pth roots of unity, and g an* *automorphism group of K of order prime to p. We assume that T and S are g-invariant sets. Then for all $\chi \in \mathfrak{X}_p(g)$, we have:*

$$\mathrm{rk}_{\chi^*}(\mathcal{C}\ell_T^S) - \mathrm{rk}_\chi(\mathcal{C}\ell_{\mathfrak{m}^*}^{T^*}) = \rho_\chi(T, S),$$

and if, in addition, $\Delta_p = \emptyset$, then $\mathcal{C}\ell_{\mathfrak{m}^*}^{T^*} = \mathcal{C}\ell_{S^*}^{T^*}$ and we obtain:

$$\mathrm{rk}_{\chi^*}(\mathcal{C}\ell_T^{S_0 \cup S_\infty}) - \mathrm{rk}_\chi(\mathcal{C}\ell_{S_0}^{T \cup \Delta_\infty}) = \rho_\chi(T, S). \qquad \square$$

When $\Delta_p = \emptyset$, reflection is perfect in that the operation sending (T, S) to (S^*, T^*) is an involution. When $\Delta_p \neq \emptyset$, we only have the inequality $\mathrm{rk}_\chi(\mathcal{C}\ell_{\mathfrak{m}^*}^{T^*}) \le \mathrm{rk}_\chi(\mathcal{C}\ell_{S^*}^{T^*})$ since $\mathfrak{m}^*$ is not necessarily equal to the conductor of $H_{S^*}^{T^*}[p]$ (the maximal p-elementary subextension of $H_{S^*}^{T^*}$). A simple sufficient condition for equality is that $\rho_{u,\chi^*} = 0$ for all $u \in \Delta_{p,k}$.

Similarly, using the inequality:

$$\mathrm{rk}_\chi(\mathcal{C}\ell_{\mathfrak{m}^*}^{T^*}) \ge \mathrm{rk}_\chi(\mathcal{C}\ell_{S_0}^{T^*}),$$

which easily yields, using the involution $(T, S, \chi) \mapsto (S_0, T \cup \Delta_\infty, \chi^*)$ for the upper bound:

$$\rho_\chi(T, S) \le \mathrm{rk}_{\chi^*}(\mathcal{C}\ell_T^S) - \mathrm{rk}_\chi(\mathcal{C}\ell_{S_0}^{T \cup \Delta_\infty}) \le -\rho_{\chi^*}(S_0, T \cup \Delta_\infty),$$

we obtain classical inequalities which we indicate in the case $T = S_0 = \emptyset$.

5.4.6 Corollary (classical "Spiegelungssätze": $k = \mathbb{Q}$). *Let K be a Galois extension of $\mathbb{Q}$, containing μ_p, of degree not divisible by p, and let $\chi \in \mathfrak{X}_p(\mathrm{Gal}(K/\mathbb{Q}))$. For $\chi \neq 1$, ω, we have the following inequalities:*

(i) Case $p \neq 2$ (Leopoldt's "Spiegelungssatz" [Le2] (1958)):

$$\frac{\psi(c) - \psi(1)}{2} \le \mathrm{rk}_{\chi^*}(\mathcal{C}\ell) - \mathrm{rk}_\chi(\mathcal{C}\ell) \le \frac{\psi(c) + \psi(1)}{2},$$

where c is the restriction to K of complex conjugation and $\psi|\chi$; if in addition $K/\mathbb{Q}$ is abelian and χ is even, we have:

$$0 \le \mathrm{rk}_{\chi^*}(\mathcal{C}\ell) - \mathrm{rk}_\chi(\mathcal{C}\ell) \le 1.$$

(ii) Case $p = 2$ (Armitage–Fröhlich, Taylor, Oriat):

$$0 \le \mathrm{rk}_{\chi^{-1}}(\mathcal{C}\ell^{\mathrm{res}}) - \mathrm{rk}_{\chi^{-1}}(\mathcal{C}\ell^{\mathrm{ord}}) + \mathrm{rk}_\chi(\mathcal{C}\ell^{\mathrm{res}}) - \mathrm{rk}_\chi(\mathcal{C}\ell^{\mathrm{ord}}) \le \psi(1)\ ;$$

if $K/\mathbb{Q}$ is abelian, then when $\chi \neq \chi^{-1}$ we have the following two possibilities:

$$\mathrm{rk}_{\chi^{-1}}(\mathcal{C}\ell^{\mathrm{res}}) = \mathrm{rk}_{\chi^{-1}}(\mathcal{C}\ell^{\mathrm{ord}}) \quad \textit{and} \quad \mathrm{rk}_\chi(\mathcal{C}\ell^{\mathrm{res}}) = \mathrm{rk}_\chi(\mathcal{C}\ell^{\mathrm{ord}}) + 1,$$
$$\mathrm{rk}_{\chi^{-1}}(\mathcal{C}\ell^{\mathrm{res}}) = \mathrm{rk}_{\chi^{-1}}(\mathcal{C}\ell^{\mathrm{ord}}) + 1 \quad \textit{and} \quad \mathrm{rk}_\chi(\mathcal{C}\ell^{\mathrm{res}}) = \mathrm{rk}_\chi(\mathcal{C}\ell^{\mathrm{ord}}),$$

and when $\chi = \chi^{-1}$, we have the equality:

$$\mathrm{rk}_\chi(\mathcal{C}\ell^{\mathrm{res}}) = \mathrm{rk}_\chi(\mathcal{C}\ell^{\mathrm{ord}}). \qquad \square$$

5.4.6.1 Remark. If $\mathcal{C}$ is (for example) a generalized p-class group of K, the χ-component $\mathcal{C}^{e_\chi}$ depends only on the faithful character χ' corresponding to χ and on the subfield K' of K fixed under the kernel of χ; in other words, we have the following relation:

$$\mathcal{C}^{e_\chi} \simeq (\mathrm{N}_{K/K'}\mathcal{C})^{e_{\chi'}} = \mathcal{C}'^{e_{\chi'}},$$

in which the analogous generalized p-class group $\mathcal{C}'$ of K' enters only via its χ'-component. All this is valid only in the semi-simple case $p \nmid |g|$. For the proof, use the relation $(\mathrm{N} \circ j)(\mathcal{C}') = \mathcal{C}'^{[K:K']} = \mathcal{C}'$, yielding the surjectivity of $\mathrm{N} := \mathrm{N}_{K/K'}$ and the injectivity of $j := j_{K/K'}$.

For instance, this applies to $\mathcal{C} = (c\ell^{\mathrm{res}}_{\mathfrak{m}}(S))_p$, in the p-Sylow of $\mathcal{C}\ell^S_{\mathfrak{m}} := \mathcal{C}\ell^{\mathrm{res}}_{\mathfrak{m}}/c\ell^{\mathrm{res}}_{\mathfrak{m}}(S)$, whose χ-component may be simplified according to the decomposition of the $v \in S$ in K'/k (e.g., $\chi \neq 1$ and v nonsplit in K'/k). $\square$

5.4.6.2 Example 1 (case of the Scholz theorem). Let $K = \mathbb{Q}(\sqrt{d}, \sqrt{-3})$, $d > 0$, $d \notin \mathbb{Q}^{\times 2}$, $p = 3$, and $\mathcal{C}_K := (\mathcal{C}\ell_K)_3$. If χ is the quadratic character whose kernel fixes $\mathbb{Q}(\sqrt{d})$, then the kernel of χ^* fixes $\mathbb{Q}(\sqrt{-3d})$; thus $\mathcal{C}_K^{e_\chi}$ (for instance) is isomorphic to $\mathcal{C}^{e_{\chi'}}_{\mathbb{Q}(\sqrt{d})} \simeq \mathcal{C}_{\mathbb{Q}(\sqrt{d})}$ since $\mathbb{Q}(\sqrt{d})$ has only two characters (χ' and 1) for which $\mathcal{C}_{\mathbb{Q}(\sqrt{d})} = \mathcal{C}^{e_{\chi'}}_{\mathbb{Q}(\sqrt{d})} \oplus \mathcal{C}^{e_1}_{\mathbb{Q}(\sqrt{d})}$ with (in a similar way) $\mathcal{C}^{e_1}_{\mathbb{Q}(\sqrt{d})} \simeq \mathcal{C}^{e_{1'}}_{\mathbb{Q}} = 1$! Of course, $\mathcal{C}_K^{e_{\chi^*}} \simeq \mathcal{C}_{\mathbb{Q}(\sqrt{-3d})}$. $\square$

In general, the $\mathcal{C}_K^{e_\chi}$ are only particular components of the p-class groups of the subfields of K.

5.4.6.3 Example 2. Let $K = \mathbb{Q}(\mu_p)$, $p \neq 2$, and let $g = \mathrm{Gal}(K/\mathbb{Q})$. We easily obtain from 5.4.5 (with $S = Pl_p$, $T = \emptyset$, $\chi = 1$):

$$\mathrm{rk}_\omega(\mathcal{C}\ell^{Pl_p}_{\mathbb{Q}(\mu_p)}) = 0$$

(since the unit character 1 leads to invariants of $\mathbb{Q}$, we have $\mathrm{rk}_1(\mathcal{C}\ell_{Pl_p}) = \mathrm{rk}_p(\mathcal{C}\ell_{\mathbb{Q},Pl_p}) = 1$, and we check that $\rho_1(\emptyset, Pl_p) = 1$); but $c\ell(Pl_p) = 1$ since $Pl_p = \{(1-\zeta)\}$. Hence:

$$\mathrm{rk}_\omega(\mathcal{C}\ell_{\mathbb{Q}(\mu_p)}) = 0$$

for all prime numbers p. $\square$

For additional concrete examples, see [Gr10, Ch. II].

To be complete on this representation-theoretic aspect, we also give the generalization with characters of Šafarevič's formula proved in 5.4.1 which does not need the assumption $\mu_p \subset K$.

5.4.7 Proposition (Šafarevič's formula with characters). *Let p be a prime number. Let g be an automorphism group of K of order prime to p, with fixed field k. Let T and $S = S_0 \cup S_\infty$ be two disjoint finite g-invariant sets of finite and noncomplex places of K. Then for any $\chi \in \mathfrak{X}_p(g)$ we have:*

$$\begin{aligned}\mathrm{rk}_\chi(\mathcal{C}\ell_T^S) = \mathrm{rk}_\chi(V_T^S/K_T^{\times p}) + \psi(1) \sum_{u \in T_{p,k}} [k_u : \mathbb{Q}_p] + \sum_{u \in T_k} \delta_u \rho_{u,\omega_u \chi^{-1}} \\ - \delta_{\omega,\chi}\delta - \sum_{u \in Pl_{k,\infty}^{\mathrm{r}} \cup S_{0,k}} \rho_{u,\chi} + \delta_{1,\chi} - \psi(1) r_2(k) + \delta_{2,p}\psi(1)|\Delta_{\infty,k}|,\end{aligned}$$

where:

$$\begin{aligned}V_T^S := \{\alpha \in K_T^{\times p} K_{T,\Delta_\infty}^{\times}, \ \ (\alpha) = \mathfrak{a}^p \mathfrak{a}_{S_0}, \\ \mathfrak{a} \in I_T, \ \mathfrak{a}_{S_0} \in \langle\, S_0 \,\rangle, \ \ i_v(\alpha) \in K_v^{\times p} \ \ \forall v \in T\},\end{aligned}$$

where the $\delta_{a,b}$ denote Kronecker symbols, where ω_u is the local Teichmüller character (possibly trivial) given by the action of $d_u \simeq \mathrm{Gal}(K_v/k_u)$ on $\mu_p(K_v)$ (for v above u), $\delta_u := 1$ or 0 according as K_v contains μ_p or not, and $\delta := 1$ or 0 according as K contains μ_p or not. □

5.4.8 Exercise. Let K be a number field containing μ_n for some integer $n \geq 2$; assume that K is given together with sets of places T and S such that $T \cup S_0$ contains all the places above the prime divisors of n. Consider:

$$\begin{aligned}V_T^S := \{\alpha \in K_T^{\times n} K_{T,\Delta_\infty}^{\times}, \ \ (\alpha) = \mathfrak{a}^n \mathfrak{a}_{S_0}, \\ \mathfrak{a} \in I_T, \ \mathfrak{a}_{S_0} \in \langle\, S_0 \,\rangle, \ \ i_v(\alpha) \in K_v^{\times n} \ \ \forall v \in T\}.\end{aligned}$$

Show that the norm group of $L := K\Big(\sqrt[n]{V_T^S}\Big)$ is:

$$N = P_{S_0,\mathfrak{n},S_\infty} \,.\, \langle\, T \,\rangle \,.\, I_{S_0}^n =: P_{S_0,\mathfrak{n},\mathrm{pos}} \langle\, T \cup \Delta_\infty \,\rangle \,.\, I_{S_0}^n,$$

for any sufficiently large modulus $\mathfrak{n} \in \langle\, S_0 \,\rangle_{\mathbb{N}}$, where $\Delta_\infty := Pl_\infty^{\mathrm{r}} \setminus S_\infty$.

Answer. Using arguments analogous to those of 5.4.1, (iii), one can show that L is the maximal S_0-ramified $T \cup \Delta_\infty$-split extension of K, with exponent dividing n. It is a finite extension. Since $N = P_{S_0,\mathfrak{n},S_\infty} \,.\, \langle\, T \,\rangle \,.\, \mathrm{N}_{L/K}(I_{L,S_0})$, for any $\mathfrak{n}$ multiple of the conductor of L/K it is clear that:

$$N_0 := P_{S_0,\mathfrak{n},S_\infty} \,.\, \langle\, T \,\rangle \,.\, I_{S_0}^n \subseteq N \ ;$$

since N_0 corresponds to an S_0-ramified $T \cup \Delta_\infty$-split abelian extension with exponent dividing n, the maximality of L gives the result. □

This Kummer situation is the starting point for the proof of the existence theorem of global class field theory; for this, one shows that if K is an arbitrary number field and M an abelian extension of exponent n of K then, for $K' := K(\mu_n)$, we have (for suitable T and S and with self-explanatory notations):

$$M\,K' \subseteq K'\Big(\sqrt[n]{V_{T'}^{S'}}\Big)\ ;$$

we then descend to the extension M/K thanks to the type of reasoning used in 3.6, (ii). Even though we have assumed the truth of the existence theorem, this aspect is still interesting for us since it can be used algorithmically to find M concretely starting from the class field data (conductor, Artin group); this is one of the objectives of [j, Coh2, Ch. 5] to which we refer.

The reader will have noted that we are in a reflection situation and that if we want to come back to the usual situation, we must start from the radical defined by $V_{S_0}^{T\cup\Delta_\infty}$.

c) Class Field Theory Over $\mathbb{Q}$

We come back to ray class fields by looking at the case where the base field is $\mathbb{Q}$.

5.5 Ray Class Fields on the Field of Rational Numbers. If $K = \mathbb{Q}$, any modulus is of the form $m\mathbb{Z}$ for $m \geq 1$, and the ray class field $\mathbb{Q}(m)^{\mathrm{res}}$ is simply the cyclotomic field $\mathbb{Q}(\mu_m)$ of mth roots of unity (see 5.5.1). Note that (except for $m = 1,\ 2$, where $\mathbb{Q}(m)^{\mathrm{res}} = \mathbb{Q}$), the place at infinity of $\mathbb{Q}$ is complexified in $\mathbb{Q}(m)^{\mathrm{res}}/\mathbb{Q}$; the maximal real subfield of $\mathbb{Q}(m)^{\mathrm{res}}$ is the field $\mathbb{Q}(m)^{\{\infty\}} =: \mathbb{Q}(m)^{\mathrm{ord}} =: \mathbb{Q}(m)^{\mathrm{nc}}$.

The case of ray class fields over $\mathbb{Q}$ gives the simplest example in which $m\mathbb{Z}$ is not always equal to the conductor of $\mathbb{Q}(m)^{\mathrm{res}}$: indeed, $m\mathbb{Z}$ (or simply m) is the conductor of $\mathbb{Q}(m)^{\mathrm{res}}$ if and only if m is odd or divisible by 4 (this follows from 5.1.1.2).

Using classical properties of cyclotomic fields [c, Wa, Ch. 2], we can now prove that $\mathbb{Q}(m)^{\mathrm{res}} = \mathbb{Q}(\mu_m)$.

5.5.1 Proposition. *For any rational integer $m \geq 1$, we have:*

$$\mathbb{Q}(m)^{\mathrm{res}} = \mathbb{Q}(\mu_m).$$

Proof. Indeed, we already have:

$$\mathrm{Gal}(\mathbb{Q}(m)^{\mathrm{res}}/\mathbb{Q}) \simeq \mathrm{Gal}(\mathbb{Q}(\mu_m)/\mathbb{Q}) \simeq (\mathbb{Z}/m\mathbb{Z})^\times,$$

so an inclusion will be sufficient. We will check that $\mathbb{Q}(\mu_m) \subseteq \mathbb{Q}(m)^{\mathrm{res}}$; for this we may assume that m is a conductor. By 4.1.1, we are reduced to compute the conductor of $\mathbb{Q}(\mu_{\ell^n})$ where ℓ^n is the ℓ-part of m ($n \geq 1$, $n \geq 2$ if $\ell = 2$). Then, since ℓ is the only ramified place, we see that this conductor is that of $\mathbb{Q}_\ell(\mu_{\ell^n})$. Part (i) of Exercise 3.4.3 computes the norm group, and hence the conductor, equal to ℓ^n. □

5.5.2 Remark. At the level of generalized class groups and of the Artin map, we do not obtain directly the classical isomorphism:

$$\mathrm{Gal}(\mathbb{Q}(\mu_m)/\mathbb{Q}) \simeq (\mathbb{Z}/m\mathbb{Z})^\times .$$

Indeed, if we denote by T the support of $m\mathbb{Z}$ we obtain:

$$\begin{aligned}\mathrm{Gal}(\mathbb{Q}(m)^{\mathrm{res}}/\mathbb{Q}) &\simeq I_T/P_{T,m\mathbb{Z},\mathrm{pos}}\\ &= \{a\mathbb{Z},\ a \in \mathbb{Q}_T^\times\}/\{u\mathbb{Z},\ u \in \mathbb{Q}_T^\times,\ u \equiv 1 \bmod m\mathbb{Z},\ u>0\}\ ;\end{aligned}$$

now consider the map (which is well defined):

$$\begin{aligned} I_T &\longrightarrow (\mathbb{Z}/m\mathbb{Z})^\times\ ;\\ a\mathbb{Z} &\longmapsto\ \ |a| + m\mathbb{Z}\end{aligned}$$

its kernel is the set of the $a\mathbb{Z}$, $a \in \mathbb{Q}_T^\times$ such that $|a| \equiv 1 \bmod m\mathbb{Z}$, hence is equal to $P_{T,m\mathbb{Z},\mathrm{pos}}$; since surjectivity is trivial, we obtain the expected isomorphism (which is specific to the base field $\mathbb{Q}$).

We recover the Artin map:

$$I_T/P_{T,m\mathbb{Z},\mathrm{pos}} \longrightarrow \mathrm{Gal}(\mathbb{Q}(m)^{\mathrm{res}}/\mathbb{Q}),$$

which sends $\ell\mathbb{Z} =: (\ell)$, with ℓ positive prime number not dividing m, to the Frobenius:

$$\left(\frac{\mathbb{Q}(m)^{\mathrm{res}}/\mathbb{Q}}{(\ell)}\right)$$

(which acts via $\zeta \longrightarrow \zeta^\ell$ for any $\zeta \in \mu_m$), by composing the above isomorphism:

$$I_T/P_{T,m\mathbb{Z},\mathrm{pos}} \longrightarrow (\mathbb{Z}/m\mathbb{Z})^\times,$$

with the one sending $a + m\mathbb{Z} \in (\mathbb{Z}/m\mathbb{Z})^\times$ for $a \in \mathbb{Q}_T^\times$ to the Artin symbol:

$$\left(\frac{\mathbb{Q}(m)^{\mathrm{res}}/\mathbb{Q}}{(b)}\right),$$

where b is a *positive representative* of $a + m\mathbb{Z}$ (and not $|a|$!). More generally, we would like to insist on the fact that a Frobenius $\left(\frac{L^{\mathrm{ab}}/K}{v}\right)$, or $\left(\frac{L^{\mathrm{ab}}/K}{\mathfrak{p}_v}\right)$, involves $q_v := |F_v|$, in other words a positive generator of $\mathrm{N}\mathfrak{p}_v$ (and similarly, by multiplicativity for the Artin symbol); for instance, for $m = 7$, $\left(\frac{\mathbb{Q}(7)^{\mathrm{res}}/\mathbb{Q}}{(-2)}\right)$ would be the Frobenius of 2, of order 3, while, choosing 5 as representative of the class of -2 modulo (7), $\left(\frac{\mathbb{Q}(7)^{\mathrm{res}}/\mathbb{Q}}{(5)}\right)$ is the Frobenius of 5, of order 6, which is indeed the automorphism $\zeta \rightarrow \zeta^{-2}$. Going from $a \in \mathbb{Q}_T^\times$ to $b > 0$ is not necessary if we set:

$$\left(\frac{\mathbb{Q}(m)^{\mathrm{res}}/\mathbb{Q}}{a}\right) := \left(\frac{\mathbb{Q}(m)^{\mathrm{res}}/\mathbb{Q}}{(a)}\right)$$

for any $a > 0$ prime to m, and:

$$\left(\frac{\mathbb{Q}(m)^{\mathrm{res}}/\mathbb{Q}}{-1}\right) := c \text{ (complex conjugation)},$$

but this is a trick only valid for the base field $\mathbb{Q}$. □

5.6 Class Field Theory Correspondence. Thus, in the case of the base field $\mathbb{Q}$, the class field theory correspondence is the well-known Kronecker–Weber theorem, of which a direct proof is not too difficult ([c, Wa, Ch. 14], [Neum1]). In other words, in the cyclotomic field $\mathbb{Q}(\mu_m)$, there is a bijective Galois correspondence between the set of subfields and the set of subgroups of $(\mathbb{Z}/m\mathbb{Z})^\times$.

5.6.1 Artin Group of Abelian Extensions of $\mathbb{Q}$. If L is an abelian extension of $\mathbb{Q}$ with conductor $f\mathbb{Z}$, for which we know $H := \mathrm{Gal}(\mathbb{Q}(f)^{\mathrm{res}}/L)$ as a subgroup of $(\mathbb{Z}/f\mathbb{Z})^\times$, then (denoting by R the support of f) the Artin group of $L/\mathbb{Q}$ is equal to:

$$A_{L/\mathbb{Q}} = \{a\mathbb{Z},\ a \in \mathbb{Q}_R^\times,\ a > 0,\ a + f\mathbb{Z} \in H\}.$$

It indeed contains $P_{R,f\mathbb{Z},\mathrm{pos}}$.

5.6.2 Decomposition Law of Prime Numbers in $\mathbb{Q}(m)^{\mathrm{res}}/\mathbb{Q}$. By 5.2, we need not consider prime divisors ℓ of m since the inertia field is $\mathbb{Q}(n)^{\mathrm{res}}$, where n is the largest divisor of m prime to ℓ.

The residue degree of ℓ is then the order of the class $\ell + m\mathbb{Z}$ in $(\mathbb{Z}/m\mathbb{Z})^\times$ and the decomposition field is the field fixed under $\langle \ell + m\mathbb{Z} \rangle$. Thus, ℓ is totally split in $\mathbb{Q}(m)^{\mathrm{res}}/\mathbb{Q}$ if and only if $\ell \equiv 1 \bmod m\mathbb{Z}$.

If the conductor of an abelian extension L is $f\mathbb{Z}$, the residue degree of $\ell \nmid f$ in $L/\mathbb{Q}$ is then the order of $\ell + f\mathbb{Z}$ modulo H.

5.6.3 Abelian Closure of $\mathbb{Q}$. The Galois group of $\overline{\mathbb{Q}}^{\mathrm{ab}} = \mathbb{Q}(\mu)$ (the field generated by all the roots of unity) is:

$$\overline{G}^{\mathrm{ab}} \simeq \varprojlim_{m \geq 1} (\mathbb{Z}/m\mathbb{Z})^\times \simeq \widehat{\mathbb{Z}}^\times = \prod_{p\,\mathrm{prime}} \mathbb{Z}_p^\times,$$

whose structure is well known (we will find again this result in III.4.1.11 in an idelic way). The inertia groups correspond to each $\mathbb{Z}_p^\times$.

Since $\mathrm{Gal}(\mathbb{Q}(m)^{\mathrm{res}}/\mathbb{Q})$ is the direct sum of the inertia groups of the ramified primes, it is clear that it may be convenient to use duality to express the class field theory correspondence and the law of decomposition of the places. This leads to the notion of Dirichlet characters. We leave this to the reader (see [c, Wa, Ch. 3]).

d) Congruence Groups

Before coming back to generalized class groups, we explain in this short subsection a classical formalism which is necessary when we cannot take quotients by a suitable ray group, situation which we avoid since we assume the Artin reciprocity law (or any equivalent statement) from the start.

The notion of congruence groups, used instead of idèle class groups or generalized class groups, introduced without any knowledge of the existence of a norm or Artin conductor, is the following. Consider the groups $N_{\mathfrak{m}}$ (the congruence groups) which are the subgroups of I_T containing $P_{T,\mathfrak{m},\mathrm{pos}}$, where $\mathfrak{m}$ is a modulus of K with support $T \subset Pl_0$; we then define an equivalence relation by:

$$N_{\mathfrak{m}_1} \sim N_{\mathfrak{m}_2},$$

for $\mathfrak{m}_1$, $\mathfrak{m}_2$ with supports T_1, T_2, if and only if:

$$N_{\mathfrak{m}_2} \cap I_{T_1} = N_{\mathfrak{m}_1} \cap I_{T_2}.$$

In this context, class field theory consists in proving that there exists a bijective Galois correspondence between finite abelian extensions of K and equivalence classes of congruence groups; the conductor (of the extension corresponding to the class) is then the g.c.d. of the $\mathfrak{m}$ belonging to the class. From the point of view that we have adopted here, this fact is quite clear for the following reason: if $\mathfrak{f}$ is the conductor of the abelian extension L/K and if $\mathfrak{m}$, built on T, is a multiple of $\mathfrak{f}$, the congruence group relative to $\mathfrak{m}$ is the Takagi group:

$$N_{\mathfrak{m}} := P_{T,\mathfrak{m},\mathrm{pos}} \mathrm{N}_{L/K}(I_{L,T}).$$

The equivalence relation is a simple translation of the invariance of the quotients $I_T/N_{\mathfrak{m}}$ when $\mathfrak{m}$ ranges over all the multiples of $\mathfrak{f}$ (see 4.3.2, 4.4.1).

However, this point of view can be convenient to define a priori subgroups of I_T (containing $P_{T,\mathfrak{m},\mathrm{pos}}$), for instance by asking that certain ideals should be norms; we must then algorithmically compute the conductor of this congruence group and find the structure of the corresponding quotient group. This is the point of view used in [j, Coh2, Ch. 3 and 4].

e) Norm Action on Generalized Class Groups

Let L/K be a finite extension of number fields. It is useful to give the action of the arithmetic norm for L/K on generalized class groups, using the fact that it corresponds to restriction of automorphisms under the Artin map. By perversity, we are going to start by looking at class groups in idelic terms (in fact this is technically simpler).

Notation and assumption. Let $\mathfrak{m} \in \langle T \rangle_{\mathbb{N}}$, and let $\mathfrak{m}'$ be a modulus of L built from the set of places of L above those of T, such that:

$$\mathrm{N}_{L/K}(U^{\mathrm{res}}_{L,\mathfrak{m}'}) \subseteq K^\times U^{\mathrm{res}}_{K,\mathfrak{m}},$$

where $\mathrm{N}_{L/K}$ is the norm map from J_L to J_K. □

We then have:

5.7 Proposition. *For any finite set S of places of K, disjoint from T, we have the following commutative diagram, where S' denotes the set of places of L above those of S:*

$$\begin{array}{ccc} \mathcal{C}\ell^{S'}_{L,\mathfrak{m}'} & \xrightarrow{\ \rho_L\ } & \mathrm{Gal}(L(\mathfrak{m}')^{S'}/L) \\ \downarrow{\scriptstyle \mathrm{N}_{L/K}} & & \downarrow{\scriptstyle restriction} \\ \mathcal{C}\ell^{S}_{K,\mathfrak{m}} & \xrightarrow{\ \rho_K\ } & \mathrm{Gal}(K(\mathfrak{m})^{S}/K) \end{array}$$

Proof. Indeed, in the idelic formulation (see I.5.1):

$$\mathcal{C}\ell^{\mathrm{res}}_{L,\mathfrak{m}'} \simeq J_L/L^\times U^{\mathrm{res}}_{L,\mathfrak{m}'},\ \ \mathcal{C}\ell^{\mathrm{res}}_{K,\mathfrak{m}} \simeq J_K/K^\times U^{\mathrm{res}}_{K,\mathfrak{m}}\ ;$$

by assumption we have the inclusion $\mathrm{N}_{L/K}(L^\times U^{\mathrm{res}}_{L,\mathfrak{m}'}) \subseteq K^\times U^{\mathrm{res}}_{K,\mathfrak{m}}$ or, equivalently, $\mathrm{N}_{L/K}(P_{L,T,\mathfrak{m}',\mathrm{pos}}) \subseteq P_{K,T,\mathfrak{m},\mathrm{pos}}$, showing that $\mathrm{N}_{L/K}$ is defined as a map from $\mathcal{C}\ell^{\mathrm{res}}_{L,\mathfrak{m}'}$ to $\mathcal{C}\ell^{\mathrm{res}}_{K,\mathfrak{m}}$ and is also given by:

$$\mathrm{N}_{L/K}(c\ell^{\mathrm{res}}_{L,\mathfrak{m}'}(\mathfrak{a}')) := c\ell^{\mathrm{res}}_{K,\mathfrak{m}}(\mathrm{N}_{L/K}(\mathfrak{a}')),$$

for any ideal $\mathfrak{a}'$ of L prime to T. Furthermore, applying norm lifting Theorem 4.7.2 to the extension $M := K(\mathfrak{m})^{\mathrm{res}}$ and to the norm group $N := K^\times U^{\mathrm{res}}_{K,\mathfrak{m}}$, the idèle group corresponding to $LK(\mathfrak{m})^{\mathrm{res}}$ on L, is equal to $\mathrm{N}^{-1}_{L/K}(K^\times U^{\mathrm{res}}_{K,\mathfrak{m}})$ which contains $L^\times U^{\mathrm{res}}_{L,\mathfrak{m}'}$, once again by assumption; thus, we have $LK(\mathfrak{m})^{\mathrm{res}} \subseteq L(\mathfrak{m}')^{\mathrm{res}}$ and hence the restriction:

$$\mathrm{Gal}(L(\mathfrak{m}')^{\mathrm{res}}/L) \longrightarrow \mathrm{Gal}(K(\mathfrak{m})^{\mathrm{res}}/K)$$

makes sense.

Finally, since $\mathrm{N}_{L/K}\Big(\bigoplus_{w\in S'} L^\times_w\Big) \subseteq \bigoplus_{v\in S} K^\times_v$ (i.e., $\mathrm{N}_{L/K}(\langle S'\rangle) \subseteq \langle S\rangle$) and using the decomposition properties 1.2.5, all the above statements are still valid in terms of S and S'-splitting.

The above diagram is thus well defined, and its commutativity again comes from applying 4.5, (iv) to the extension $L(\mathfrak{m}')^{S'}/K$. □

5.7.1 Corollary. *We have the diagram:*

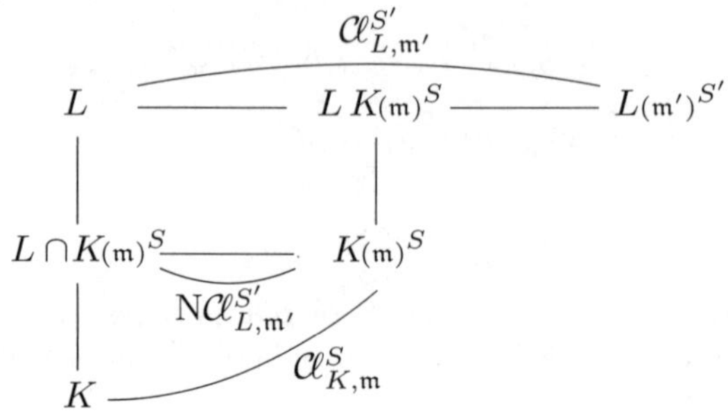

where $\mathrm{Gal}(K(\mathfrak{m})^S/L \cap K(\mathfrak{m})^S) \simeq \mathrm{N}_{L/K}(\mathcal{C}\ell^{S'}_{L,\mathfrak{m}'})$. *In particular the norm map is surjective if and only if* L *and* $K(\mathfrak{m})^S$ *are linearly disjoint over* K*; when this is the case,* $|\mathcal{C}\ell^{S}_{K,\mathfrak{m}}|$ *divides* $|\mathcal{C}\ell^{S'}_{L,\mathfrak{m}'}|$. □

5.7.2 Remarks. (i) We also have $\mathrm{Gal}(L(\mathfrak{m}')^{S'}/L\,K(\mathfrak{m})^S) \simeq {}_{\mathrm{N}}\mathcal{C}\ell^{S'}_{L,\mathfrak{m}'}$ (the kernel of the arithmetic norm $\mathrm{N}_{L/K}$), equal to:

$$\{c\ell^{S'}_{L,\mathfrak{m}'}(\mathfrak{a}'),\ c\ell^{S}_{K,\mathfrak{m}}(\mathrm{N}_{L/K}(\mathfrak{a}')) = 1\}.$$

(ii) Going to the limit but keeping T and S fixed (see 5.3), we obtain a similar diagram, in which $\mathrm{N}_{L/K} : J_L \longrightarrow J_K$ yields:

$$\mathrm{N}_{L/K} : \mathcal{C}\ell^{S'}_{L,T'} := \varprojlim_{\mathfrak{m}\in\langle T\rangle_{\mathbb{N}}} J_L/L^\times U^{\mathrm{res}}_{L,(\mathfrak{m})} \longrightarrow \mathcal{C}\ell^{S}_{K,T} := \varprojlim_{\mathfrak{m}\in\langle T\rangle_{\mathbb{N}}} J_K/K^\times U^{\mathrm{res}}_{K,\mathfrak{m}},$$

where, to simplify, we have chosen $\mathfrak{m}' := (\mathfrak{m})$, the modulus obtained by extending $\mathfrak{m}$ to L, which is always suitable. □

5.7.3 Remarks. (i) Without any difficulty we can even go completely to the limit (for instance with $S = \emptyset$) so as to obtain the corresponding abelian closures $\overline{K}^{\mathrm{ab}}$ and $\overline{L}^{\mathrm{ab}}$, the norm, still coming from $\mathrm{N}_{L/K} : J_L \longrightarrow J_K$, giving:

$$\mathrm{N}_{L/K} : C_L/D_L \longrightarrow C_K/D_K.$$

In this case $\mathrm{N}_{L/K}(C_L/D_L) = \mathrm{N}_{L/K}(C_L)/D_K \simeq \mathrm{Gal}(\overline{K}^{\mathrm{ab}}/L^{\mathrm{ab}})$ and the norm is surjective if and only if $L^{\mathrm{ab}} = K$. But this is nothing else than class field theory in L/K.

(ii) The usual particular cases ($T = \emptyset$, $S = \emptyset$, and $S = Pl^{\mathrm{r}}_\infty$) can be represented by an analogous diagram of finite extensions, such as:

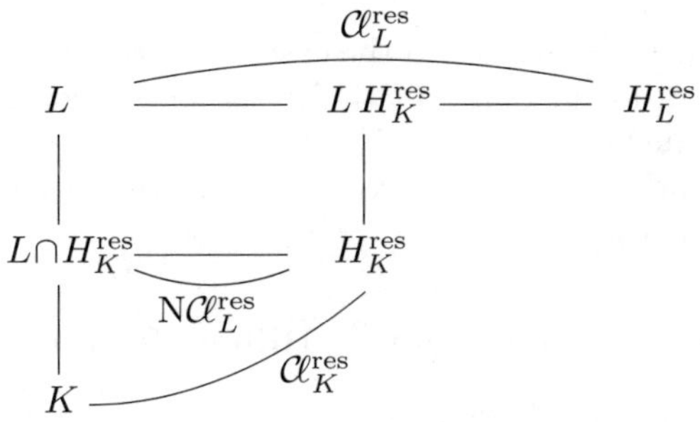

Here $L \cap H_K^{\mathrm{res}}/K$ is the maximal abelian subextension of L/K which is unramified (at the finite places), and the norm map is surjective if and only if this extension reduces to K. For the ordinary sense, we replace everywhere "res" by "ord"; $L \cap H_K^{\mathrm{ord}}/K$ is then the maximal abelian subextension which is unramified and noncomplexified, whose Galois group measures the surjectivity defect of the norm on the ordinary class group of L. $\square$

5.7.4 Example. Assume that L is a quadratic extension of K and that there exists a finite ramified place in L/K or a real place of K which is complexified in L; consider the ordinary sense ($T = \emptyset$, $S = Pl_\infty^{\mathrm{r}}$):

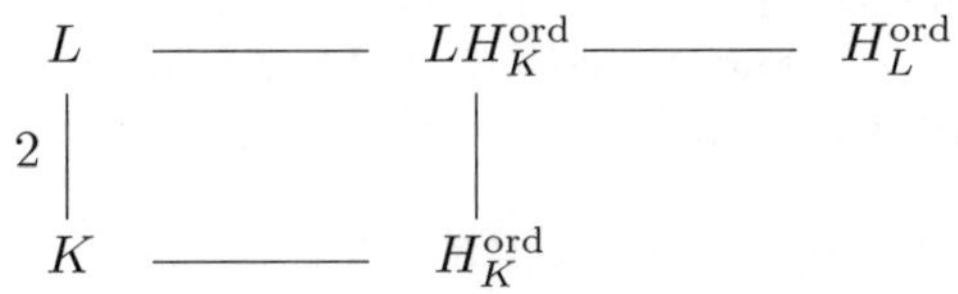

Since L/K is ramified or complexified in at least one place, we have $L \cap H_K^{\mathrm{ord}} = K$; hence the ordinary class number of K divides the ordinary class number of L. Thus, we have the exact sequence:

$$1 \longrightarrow {}_{\mathrm{N}}\mathcal{C}\ell_L^{\mathrm{ord}} \longrightarrow \mathcal{C}\ell_L^{\mathrm{ord}} \xrightarrow{\mathrm{N}_{L/K}} \mathcal{C}\ell_K^{\mathrm{ord}} \longrightarrow 1$$

which is a simple translation of the classical equality (with notations which are themselves classical):

$$h_L = h_L^* \, h_K,$$

in which h_L^* is called the relative class number. $\square$

5.7.5 Remark. Note that this result is often stated for a field with complex conjugation (or a CM field), in other words a totally complex field L which is a quadratic extension of a totally real field K; in this context, h_L^* and h_K are often denoted h_L^- and h_L^+ (relative class number and real class number), but this notation is ambiguous since we do not necessarily have $\mathcal{C}\ell_L = \mathcal{C}\ell_L^+ \oplus \mathcal{C}\ell_L^-$, in the usual Galois meaning for which $\mathcal{C}\ell_L^\pm := \{c\ell(\mathfrak{a}) \in \mathcal{C}\ell_L,\ c\ell(\mathfrak{a}^c) = c\ell(\mathfrak{a})^{\pm 1}\}$, c denoting complex conjugation (the obstruction coming from the 2-Sylow subgroups).

This is particularly interesting in the case of cyclotomic fields $L = \mathbb{Q}(\mu_m)$, since in this case, although $\mathbb{Q}(\mu_m)/\mathbb{Q}(\mu_m)^{\mathrm{nc}}$ is unramified as soon as m (assumed to be a conductor) is not a prime power, we always have the relation "$h = h^- \, h^+$" (this nonramification property is proved in III.1.4.2 but can be checked in an elementary way). □

f) The Principal Ideal Theorem — Hilbert Towers

5.8 The Principal Ideal Theorem in the Tame Case. Let K be a number field together with sets of places T and S. For the tame modulus $\mathfrak{m} = \prod_{v \in T} \mathfrak{p}_v$, we consider the S-split ray class field $K' := K(\mathfrak{m})^S$, and we denote by T' and S' the sets of places of K' above those of T and S. Let $K'' := K'(\mathfrak{m}')^{S'}$ for $\mathfrak{m}' = \prod_{v' \in T'} \mathfrak{p}_{v'}$ be the analogous ray class field over K'. By 3.6, it is a Galois extension of K.

5.8.1 Lemma 1. *The maximal abelian subextension K''^{ab} of K'' in K''/K is equal to K'.*

Proof. Set $L := K''^{\mathrm{ab}} \supseteq K'$. The extension K''/K is T-tamely ramified (use 5.2.2, (ii) for the extensions K'/K and K''/K', and multiplicativity of ramification indices); it follows that L/K is also T-tamely ramified and, once again using 5.2.2, (ii), we have $L \subseteq K'$, proving equality. □

If $G' := \mathrm{Gal}(K'/K)$ and $G'' := \mathrm{Gal}(K''/K)$, we have:

$$G' = G''^{\mathrm{ab}} \text{ and } \mathrm{Gal}(K''/K') = [G'', G''].$$

We now state a purely algebraic but nontrivial classical result [26] which was the prelude to further development of the subject which, after Suzuki ([Su], [i, Miy0, Suzuki]), has been renewed by Gruenberg–Weiss [GW] whose main result we give a little later.

5.8.2 Lemma 2. *Let G be a finite group whose commutator subgroup $[G\,,\,G]$ is commutative. Then the transfer map from G to $[G\,,\,G]$ is trivial.* □

It then immediately follows, from property 4.5, (v) of the Artin map, that for a number field K given together with sets of places T and S, we have:

5.8.3 Theorem (principal ideal). *For $\mathfrak{m} = \prod_{v \in T} \mathfrak{p}_v$, consider the S-split ray class field $K' := K(\mathfrak{m})^S$, and denote by T' and S' the sets of places of K' above those of T and S. Then for the ideal extension map $j_{K'/K} : I_{K,T} \longrightarrow I_{K',T'}$,*

[26] Artin–Furtwängler (1930), Magnus (1934), Iyanaga (1930, 1934); see [d, AT, Ch. 13, § 4] and [c, Neu1, Ch. VI, 7.6].

we have $j_{K'/K}(I_{K,T}) \subset P_{K',T',\mathfrak{m}',\mathrm{pos}}\langle\, S'\,\rangle$. *In other words, the natural map* $j_{K'/K} : \mathcal{C}\ell^S_{K,\mathfrak{m}} \longrightarrow \mathcal{C}\ell^{S'}_{K',\mathfrak{m}'}$ *is zero.* □

When $T = \emptyset$ and $S = Pl^{\mathrm{r}}_\infty$ (for instance), the field K' is the ordinary Hilbert class field H_K^{ord}, and we obtain the famous result (which was conjectured by Hilbert from the very beginning of the theory), called the principal ideal theorem, which states that the extension to H_K^{ord} of an ideal of K is principal.

This theorem does not say precisely how this principalization takes place; historically (see for instance Olga Taussky's account in [i, Tau]), the Hilbert Theorem 94 asserted that in any unramified cyclic extension M of K, the capitulation kernel (i.e., the kernel of the transfer map for M/K or that of extension of classes from K to M) is of order a multiple of $[M : K]$. Many partial results were then given (for instance those of Tannaka–Terada, Furuya, Thiébaud), and we refer to [Miy3] and [i, Miy0, Suzuki] for a detailed account of the main results on these problems, which seem to have reached their optimal formulation with the result of [GW] which we simply state in a less general situation.

5.8.4 Definition (Gruenberg–Weiss). Let G be a finite abelian group. We say that a finite abelian group X is a transfer kernel for G if there exists an exact sequence of the form:

$$1 \longrightarrow A \longrightarrow H \longrightarrow G \longrightarrow 1,$$

with A a finite abelian group, such that:

$$X \simeq \mathrm{Ker}(\mathrm{Ver} : H/[H,H] \longrightarrow A),$$

where as usual Ver denotes the transfer map (see 1.4.1). □

5.8.5 Theorem (Gruenberg–Weiss [GW] (2000)). *Let X be a finite abelian group of exponent dividing $|G|$. Then X is a transfer kernel for G if and only if $|G|$ divides $|X|$.* □

We apply this to the following data: L/K is a subextension of H_K^{ord}/K, $H = \mathrm{Gal}(H_L^{\mathrm{ord}}/K)$, $A = \mathrm{Gal}(H_L^{\mathrm{ord}}/L)$, $G = \mathrm{Gal}(L/K)$, so the exact sequence:

$$1 \longrightarrow X \longrightarrow \mathcal{C}\ell_K^{\mathrm{ord}} \simeq H/[H,H] \xrightarrow{\mathrm{Ver}} \mathcal{C}\ell_L^{\mathrm{ord}} \simeq A,$$

implies that $|G|$ divides $|X|$ (i.e., in more colorful terms, the capitulation kernel in L/K is of order a multiple of $|G| = [L : K]$). The condition that the exponent of X is a divisor of $|G|$ is here trivially satisfied since $\mathrm{N}_{L/K} \circ j_{L/K} = [L : K]$ in $\mathcal{C}\ell_K^{\mathrm{ord}}$. For $L = H_K^{\mathrm{ord}}$ we find again the principal ideal theorem.

More generally:

5.8.6 Corollary. *Let L/K be an abelian T-tamely ramified S-split extension, then for $\mathfrak{m} = \prod_{v\in T} \mathfrak{p}_v$ and $\mathfrak{m}' = \prod_{v'\in T'} \mathfrak{p}_{v'}$ in L, the capitulation kernel:*

$$\mathrm{Ker}(\mathcal{C}\ell_{K,\mathfrak{m}}^S \longrightarrow \mathcal{C}\ell_{L,\mathfrak{m}'}^{S'})$$

has order a multiple of $[L : K]$.

Proof. Replace H_K^{ord} and H_L^{ord} by $K(\mathfrak{m})^S$ and $L(\mathfrak{m}')^{S'}$ respectively. □

5.9 Hilbert Towers. Of course the ideals of H_K^{ord} are not necessarily principal and one may ask if, by iteration, the tower of number fields:

$$K^{(0)} := K \subseteq K^{(1)} \subseteq \cdots \subseteq K^{(\infty)} := \bigcup_{i\geq 0} K^{(i)},$$

inductively defined by $K^{(i+1)} := H_{K^{(i)}}^{\mathrm{ord}}$, is finite or not (finiteness being equivalent to the existence of $n_0 \geq 0$ such that the class group of $K^{(n_0)}$ is trivial). The field $K^{(\infty)}$ is called the Hilbert class fields tower (in the ordinary sense). In a similar way, for any prime number p, we define the p-Hilbert class fields tower $K^{(\infty)}(p) := \bigcup_{i\geq 0} K^{(i)}(p)$, with $K^{(i+1)}(p) := H_{K^{(i)}(p)}^{\mathrm{ord}}(p)$, and ask for the same question. The notation $K^{(\infty)}(p)$ is legitimate since the maximal pro-p-subextension of $K^{(\infty)}$ is solvable and thus coincide with the p-tower. This problem, which is just as famous, was solved in the negative in 1964, thanks to a group-theoretical result of Šafarevič. [27]

5.9.1 Theorem (Golod–Šafarevič–Gaschütz–Vinberg). *Let G be a pro-p-group of finite rank (i.e., with a finite number of generators); let $d(G)$ and $r(G)$ be respectively the minimal number of generators and of relations defining the group G.* [28] *If G is a finite group, then:*

$$r(G) > \tfrac{1}{4}\,(d(G))^2.$$ □

It is then sufficient to exhibit an example for which $d(G)$ is sufficiently large compared with $r(G)$, for $G := \mathrm{Gal}(K^{(\infty)}(p)/K)$, which class field theory easily gives (see Exercise 5.9.5).

This result thus showed the complexity of the unramified Galois closure of a number field, and showed that the historical utopia of finding a finite extension of K whose principality would reduce computations in K to ordinary element arithmetic was doomed (see in 5.9.3 the proof that the existence of such an extension is equivalent to the finiteness of the class fields tower).

[27] See in [d, CF, Ch. IX], [g, Se3, Ch. I, Ann. 3; NSW, Ch. III, § 9], the Golod–Šafarevič theorem, which was later improved in a number of ways, such as the Gaschütz–Vinberg theorem and some results of Koch.

[28] [e, Ko3, Ch. 3, §§ 1.16, 2.7].

Many infinite Hilbert class fields towers have been constructed (for instance by Matsumura [Mat], Martinet [Mar1], Schmithals [Schm], Schoof [Scho], Maire [Mai1]). Recently, tamely ramified class fields towers (or p-towers) have also been studied [Mai1], both for number fields and for function fields, and have renewed the study of the Martinet constants on number field discriminants, for which upper bounds are obtained from infinite towers (see in the introduction of [HM1] a good historical review of the subject). It is in the same paper [HM1] that Hajir–Maire give improvements on these constants, in this wider context, and for which many interesting questions can be asked (see [HM2, HM3], [HM4] for additional results). We will give in 5.9.4, (iii) a detailed example after [HM2] of such a computation, showing also some links with genera theory.

Note (Martinet's constants). Let K be a number field of signature (r_1, r_2). The infinity type of K is the rational number $\frac{r_1}{[K:\mathbb{Q}]}$ and its root discriminant is $\mathrm{rd}_K := (\mathrm{d}_{K/\mathbb{Q}})^{1/[K:\mathbb{Q}]}$, where $\mathrm{d}_{K/\mathbb{Q}}$ is the absolute value of the discriminant of $K/\mathbb{Q}$. For fixed $t \in \mathbb{Q} \cap [0, 1]$ and integers $n \geq 1$ such that number fields of degree n and infinity type t exist (i.e., such that $tn \in \mathbb{N}$ and $tn \equiv n \bmod (2)$), we let (after [Mar1], [HM1, § 1.1]):

$$\alpha_n(t) := \min_K \{\mathrm{rd}_K,\ [K:\mathbb{Q}] = n,\ \tfrac{r_1}{[K:\mathbb{Q}]} = t\},$$
$$\alpha(t) := \liminf_n\ \alpha_n(t).$$

Let T be a finite set of finite places of K and let $K =: K_0 \subseteq K_1 \subseteq \cdots \subseteq K_\infty := \bigcup_i K_i$ be a tower of T-tamely ramified, noncomplexified extensions of K (the noncomplexification insures that the infinity type is constant in the tower). Then, if K_∞/K is infinite, we easily obtain $\alpha(t) \leq \mathrm{rd}_K \cdot \prod_{v \in T} (\mathrm{N}\mathfrak{p}_v)^{1/[K:\mathbb{Q}]}$.

In the nontame case for p-towers, the moduli $\mathfrak{m}$ (with fixed support T such that $T_p \neq \emptyset$) can take an infinite number of values, and the only canonical tower is then that of the $H_{T}^{S}(p)$ (see 5.3) which are, in general, infinite extensions of the base field; in addition, we will see in Section 2 of Chapter IV that, in this context, the transfer map is on the contrary injective (under the assumption that $Pl_p \subseteq T$, $S_0 = \emptyset$, for the ordinary sense, and assuming the Leopoldt conjecture for p). It is however possible to ask that ramification is bounded by observing that because of the reciprocity law we have a correspondence between the natural filtration of the local unit groups with that of higher ramification groups (in upper numbering); this is the study which has been started in [HM3]. Hence this is quite a different context (even though the problem of principalization under extensions can be asked in complete generality), and which leads to difficult questions related to the theory of pro-p-groups (for instance the conjecture of Fontaine–Mazur stated in [g, NSW, Ch. X, § 8]) which we will not describe (see [Haj] for an introduction to these problems in the particular case of p-Hilbert towers).

5.9.2 Remarks. (i) The Hilbert class field is a particular solution to the principalization problem of the ideal group of a field K; we will not expand on this, but it is clear that the classes of K can principalize in many other abelian extensions of K, and we now have quite a precise understanding of the ideal extension map for the extension $\overline{K}^{\mathrm{ab}}/K$ ([Gr9, 0.1, 0.4], [Kur]). In addition, if we do not request any Galois conditions, it is easy to principalize $\mathcal{C}\ell_K^{\mathrm{ord}}$ in a brutal way; setting $\mathcal{C}\ell_K^{\mathrm{ord}} = \langle\, c\ell^{\mathrm{ord}}(\mathfrak{a}_i)\,\rangle_{1\le i\le r}$, we simply consider:

$$L := K(\sqrt[n_1]{\alpha_1}, \dots, \sqrt[n_r]{\alpha_r}),$$

where $\mathfrak{a}_i^{n_i} = (\alpha_i)$, $\alpha_i \in K^\times$, $1 \le i \le r$, but this is far from class field theory considerations.

(ii) Finally, concerning the construction of unramified extensions (for instance), we have only mentioned the abelian case (Hilbert class fields) or briefly the soluble case (p-Hilbert class fields towers); however, it is important to note that a *principal* number field can have an *infinite* unramified extension (Galois or non-Galois) (see various examples in [Mai3]; we reproduce such an example in Exercise 5.9.7). □

5.9.3 Proposition. *Let K be a number field and let L be an arbitrary finite extension of K such that $\mathcal{C}\ell_L^{\mathrm{ord}} = 1$. Then L contains the (finite) ordinary Hilbert class fields tower of K.*

Proof. Consider the extension $H_K^{\mathrm{ord}}L$ of L which is abelian, unramified and noncomplexified, hence equal to L.

By induction, seen as an extension of $K^{(i)}$, L contains $K^{(i+1)}$, hence we have:

$$K^{(\infty)} =: K^{(n_0)} \subseteq L.$$

The extension $K^{(n_0)}$ is the minimal solution to the problem. □

Note that if there are several floors in the tower, then $K^{(\infty)}$ is not contained in L^{ab}.

There is an analogous result with sets T and S for the corresponding tower, relative to the existence of L such that $\mathcal{C}\ell_{L,T'}^{S'} = 1$.

5.9.4 Examples (from [Mai1] and [HM2, § 3.2] (1999/2000)). (i) The field $\mathbb{Q}(\sqrt{53 \times 131})$ has an infinite restricted Hilbert class fields tower but a finite ordinary Hilbert class fields tower (see Exercise 5.9.6).

(ii) The number field (totally complex of degree 10):

$$\mathbb{Q}(\xi, \sqrt{-36\xi^4 + 125\xi^3 - 221\xi^2 + 182\xi - 80}),$$

where ξ is a root of the polynomial:

$$X^5 - 2X^4 + 3X^3 - 3X^2 - X + 1,$$

has an infinite 2-class fields tower whose root discriminant is equal to $84.37\cdots$ (see Exercise 5.9.8).

(iii) The number field $\mathbb{Q}(\theta)$ (totally complex of degree 12), where θ is a root of the polynomial:

$$X^{12} + 339X^{10} - 19752X^8 - 2188735X^6 + 284236829X^4 \\ + 4401349506X^2 + 15622982921,$$

has an infinite tower of number fields (tamely ramified at a place dividing 3) with root discriminant bounded by $82.2\cdots$.

I thank F. Hajir and C. Maire for the authorization to reproduce the details for this example and to use their source text (the notations being the same as ours). We will see that cohomological computations of the Appendix and genera theory (Chapter IV) are needed for the proof.

"The number field arithmetic which is at the heart of our construction takes place in degree 6 number fields; computer packages such as PARI and KANT make it easy to carry out these calculations. However, we would like to present the examples in such a way that a reader armed with an ordinary calculator can verify all of our claims. To this end, we provide (at the cost of lengthening the presentation slightly) much supplementary data and a method for verifying each step in the reasoning. We also provide some data (such as class number, generators for the unit group) whose validity need not be verified but which would aid the reader who wishes to check our claims independently.

Let $k = \mathbb{Q}(\xi)$ where ξ is a root of $f = x^6 + x^4 - 4x^3 - 7x^2 - x + 1$. The prime factorization of the discriminant of f is $\mathrm{d}_f = -23\cdot 35509$; thus, $\mathrm{d}_f = \mathrm{d}_k$ is also the discriminant of k, and $Z_k = \mathbb{Z}[\xi]$. The roots of f are:

$$\begin{aligned}
\xi_1 &= -0.76166245384468100791784609 7\cdots \\
\xi_2 &= -0.69953796284372129907057255 3\cdots \\
\xi_3 &= +0.29522571317729963668939709 8\cdots \\
\xi_4 &= +1.83015782341636731046020011 5\cdots \\
\xi_5 &= -0.33209155995263232008058928 1\cdots \\
&\qquad + 1.83394227605082629317069415 2\cdots \sqrt{-1} \\
\xi_6 &= -0.33209155995263232008058928 1\cdots \\
&\qquad - 1.83394227605082629317069415 2\cdots \sqrt{-1}.
\end{aligned}$$

Thus, k has signature $(4, 1)$. The restricted class number of k is 1. The unit group of k is generated by $\{\xi,\ 4\xi^5 - 3\xi^4 + 6\xi^3 - 20\xi^2 - 13\xi + 6,\ 6\xi^5 - 4\xi^4 + 9\xi^3 - 30\xi^2 - 21\xi + 8,\ -\xi^5 + \xi^4 - 2\xi^3 + 6\xi^2 + \xi - 1,\ -1\}$.

Generators for some Z_k-ideals of small norm are listed in the table below where $\pi_r = a_5\xi^5 + a_4\xi^4 + a_3\xi^3 + a_2\xi^2 + a_1\xi + a_0$ generates a prime ideal

$\pi_r Z_k$ of norm r, and the coefficients of h_{π_r}, the minimal polynomial of π_r, are listed in descending powers.

π_r	$a_5, a_4, a_3, a_2, a_1, a_0$	h_{π_r}
π_3	$-6, 4, -9, 30, 21, -7$	$1, 0, -5, 2, 5, -5, 3$
π_7	$-9, 6, -13, 44, 31, -12$	$1, 1, -29, 98, 624, -449, -7$
π_{13}	$-7, 5, -11, 36, 23, -9$	$1, 3, -4, -24, -23, 7, 13$
π_{19}	$5, -4, 8, -26, -15, 6$	$1, 11, 50, 120, 151, 89, 19$
π'_{19}	$5, -3, 7, -24, -20, 6$	$1, -3, -10, 13, 29, -8, -19$
π_{23}	$-5, 4, -8, 26, 15, -9$	$1, 7, 20, 30, 16, -20, -23$
π'_{23}	$6, -4, 9, -30, -22, 6$	$1, 6, 11, 0, -30, -46, -23$
π_{29}	$11, -8, 17, -56, -35, 16$	$1, -7, 3, 52, -82, 55, -29$
π_{31}	$7, -5, 11, -36, -22, 7$	$1, 9, 22, 13, -15, -38, -31$

The fact that $19Z_k$ has two prime factors of residue degree 1, can be seen, for instance, from the factorization of f over $\mathbb{F}_{19}$: $f(x) \equiv (x+7)(x-2)(x^4+14x^3+2x^2+11x+4) \bmod 19$. Similarly, f factors over $\mathbb{F}_{23}$ as $f(x) \equiv (x+10)^2(x-5)(x^3+8x^2+19x+4) \bmod 23$. To see that the pairs π_{19}, π'_{19} and π_{23}, π'_{23} generate different prime ideals, one can check that the minimal polynomials of π_{19}/π'_{19} and π_{23}/π'_{23} are not integral.

The element $\eta = -671\xi^5 + 467\xi^4 - 994\xi^3 + 3360\xi^2 + 2314\xi - 961 \in Z_k$ is totally negative. Its minimal polynomial is $g(y) = y^6 + 339y^5 - 19752y^4 - 2188735y^3 + 284236829y^2 + 4401349506y + 15622982921$. The ideal (η) factors into eight prime ideals of Z_k; in fact, one can check that $\eta = \pi_7\pi_{13}\pi_{19}\pi'_{19}\pi_{23}\pi'_{23}\pi_{29}\pi_{31}$. We let $K = k(\sqrt{\eta})$, a totally complex field of degree 12. A defining polynomial for K is $g(y^2)$. We note that η is congruent to a square modulo $4Z_k$; explicitly, $\eta = \beta^2 + 4\gamma$ with $\beta = \xi^5 + \xi^4 + \xi^3 + 1$ and $\gamma = -173\xi^5 + 112\xi^4 - 270\xi^3 + 815\xi^2 + 576\xi - 237$. Thus, the relative discriminant $\mathrm{d}_{K/k}$ is simply (η), and K/k is complexified at the four infinite real places of k, ramified at the eight primes dividing η, and nowhere else. The root discriminant of K is:

$$\mathrm{rd}_K = \mathrm{rd}_k . (\mathrm{N}_{K/k}\mathrm{d}_{K/k})^{1/12} = (23 \cdot 35509)^{1/6}(7 \cdot 13 \cdot 19^2 \cdot 23^2 \cdot 29 \cdot 31)^{1/12} = 68.363 \cdots$$

Consider $t := \{\mathfrak{p}\}$ for the prime $\mathfrak{p} = \pi_3 Z_k$ of k above 3; $\mathfrak{p}$ is inert in K/k. We put $T = \{\mathfrak{p}Z_K\}$."

Now we leave the text of [HM2] and give a direct (but similar) reasoning for the infiniteness of the group $\mathcal{G}_{K,T} = \mathcal{G}^{\mathrm{res}}_{K,T} = \mathcal{G}^{\mathrm{ord}}_{K,T}$ whose abelianization is $(\mathcal{C}\ell_{K,T})_2$. Suppose that $\mathcal{G}_{K,T}$ is finite.

Recall that $d := d(\mathcal{G}_{K,T}) = \mathrm{rk}_2(H^1(\mathcal{G}_{K,T}, \mathbb{Z}/2\mathbb{Z})) = \mathrm{rk}_2(\mathcal{C}\ell_{K,T})$, and that $r := r(\mathcal{G}_{K,T}) = \mathrm{rk}_2(H^2(\mathcal{G}_{K,T}, \mathbb{Z}/2\mathbb{Z}))$. Then, from Corollary 3.8.2 of the Appendix, we obtain:

$$d < 2 + 2\sqrt{6} < 6.9$$

since $r_1 = 0$, $r_2 = 6$, and $-1 \not\equiv 1 \bmod \mathfrak{p}Z_K$.

Now we apply Corollary IV.4.5.1 to the extension K/k for the sets t and $s = Pl^{\mathrm{r}}_{k,\infty}$; since all elements of s are complexified in K, this gives:

$$1 \longrightarrow E^{\mathrm{ord}}_{k,\mathfrak{p}} / E^{\mathrm{ord}}_{k,\mathfrak{p}} \cap \mathrm{N}_{K/k}(J_K) \stackrel{\nu}{\longrightarrow} \Omega^s_t(K/k) \stackrel{\pi}{\longrightarrow} (\mathrm{Gal}(H_{K/k,t}/K\,H^{\mathrm{ord}}_{k,t}))_2 \longrightarrow 1,$$

where:

$$E^{\mathrm{ord}}_{k,\mathfrak{p}} := \{\varepsilon \in E^{\mathrm{ord}}_k,\ \ \varepsilon \equiv 1 \bmod \mathfrak{p}\},$$

$$\Omega^s_t(K/k) \simeq \Big\{(\sigma_u)_{u\notin t} \in \bigoplus_{u\in s} D_v(K/k) \bigoplus_{u\notin t\cup s} I_u(K/k),\ \prod_{u\notin t} \sigma_u = 1\Big\} \simeq (\mathbb{Z}/2\mathbb{Z})^{11}$$

since, in K, four infinite real places of k are complexified and eight finite places of k are ramified. Since $d = \mathrm{rk}_2(\mathrm{Gal}(H_{K,T}/K)) \geq \mathrm{rk}_2(\mathrm{Gal}(H_{K/k,t}/K\,H^{\mathrm{ord}}_{k,t}))$, we have:

$$d \geq \mathrm{rk}_2(\Omega^s_t(K/k)) - \mathrm{rk}_2(E^{\mathrm{ord}}_{k,\mathfrak{p}}) = 11 - 4 = 7$$

because the numerical data above show that $\mathrm{rk}_2(E^{\mathrm{ord}}_{k,\mathfrak{p}}) = 4$, a contradiction.

Therefore, K has an infinite tower of number fields (tamely ramified at a place dividing 3, unramified elsewhere) with a root discriminant bounded by $\mathrm{rd}_K \times 9^{1/12} = 82.1\cdots$, giving $\alpha(0) < 82.2\cdots$.

Note that the classical reasoning with class groups (i.e., $t = \emptyset$) and corresponding genera theory does not succeed in this case. □

5.9.5 Exercise (Golod–Šafarevič's first example). Show that the 2-class fields tower of $K = \mathbb{Q}(\sqrt{-2\times 3\times 5\times 7\times 11\times 13})$ is infinite (hint: use the Example of Corollary IV.4.5.1 in $K/\mathbb{Q}$, and Corollary 3.8.2 of the Appendix). □

5.9.6 Exercise. Consider the fields $k = \mathbb{Q}(\sqrt{-131})$ whose class number is 5, and put $K = \mathbb{Q}(\sqrt{53\times 131})$. Let $M = H_k(\sqrt{-53})$, where H_k is the Hilbert class field of k. Note that $\mathcal{C}\!\ell^{\mathrm{res}}_K \simeq \mathbb{Z}/2\mathbb{Z} \times \mathbb{Z}/2\mathbb{Z}$ and $\mathcal{C}\!\ell^{\mathrm{ord}}_K \simeq \mathbb{Z}/2\mathbb{Z}$.

(i) Check that 53 is totally split in $H_k/\mathbb{Q}$ and that 2 is totally split in H_k/k.

(ii) Deduce that $\mathrm{rk}_2(\mathcal{C}\!\ell_M) \geq 9$ (apply IV.4.5.1 in M/H_k).

(iii) Prove that the 2-class fields tower of M is infinite. Therefore, the class fields tower of K, in the restricted sense, is infinite since M/K is unramified (but not necessarily its 2-class fields tower!).

(iv) Using Corollary 3.8.1 of the Appendix, show that the class fields tower of K in the ordinary sense is finite. Therefore, it is $K(\sqrt{53})$.

For generalizations of such examples, see [Mai1]. □

5.9.7 Exercise (after [Mai3] using PARI). Consider the totally real field F associated to the irreducible polynomial:

$$X^7 - 3X^6 - 13X^5 + 28X^4 + 42X^3 - 47X^2 - 31X + 12,$$

whose discriminant is the prime number $\ell = 17380678572159893$ (the Galois group of the Galois closure of $F/\mathbb{Q}$ is S_7).

Let $q = 1051$, $r = 16747$, $K = \mathbb{Q}(\sqrt{\ell . q . r})$, and $M = FK$.

(i) Check the decomposition of ℓ, q, r in $F/\mathbb{Q}$ (hint: ℓ splits into five places of residue degree 1 and one place of ramification index 2; q splits into six places; r is totally split).

(ii) Deduce that M has an infinite 2-Hilbert class fields tower $\overline{H}_M^{\rm ord}(2)$ and that M/K is unramified. Therefore, the Galois closure of $\overline{H}_M^{\rm ord}(2)$ over K is an infinite unramified Galois extension of K.

(iii) Prove that the Hilbert tower $\overline{H}_K^{\rm ord}$ of K is equal to the genus field $\mathbb{Q}(\sqrt{\ell}, \sqrt{q . r})$ of K (hint: check that the fields $\mathbb{Q}(\sqrt{\ell})$ and $\mathbb{Q}(\sqrt{q . r})$ are principal and that the class number of K is equal to 2). Note that the genus field of K is thus principal and admits an infinite unramified extension.

For other similar constructions, see [Mai3]. □

5.9.8 Exercise (after [HM2] using PARI). Consider the Example 5.9.4, (ii). Let $k = \mathbb{Q}(\xi)$, $K = k(\sqrt{\eta})$, with $\eta = -36\xi^4 + 125\xi^3 - 221\xi^2 + 182\xi - 80$. Show that the discriminant of k is -31391, that its signature is $(3, 1)$, that η is totally negative and such that $(\eta) = \pi_7\pi_7'\pi_{11}\pi_{11}'\pi_{13}\pi_{19}\pi_{19}'\pi_{23}\pi_{29}$, with the principle of notations of the Example 5.9.4, (iii). Deduce that K/k is ramified at nine finite places and complexified at three real places, so that $d \geq 7$. Conclude that the 2-Hilbert tower of K cannot be finite. □

In the totally complex case, the historical example of Martinet [Mar1] (1978) yields a root discriminant less than 92.4 in the following way.

Consider $k = \mathbb{Q}(\mu_{11}, \sqrt{2})^{\rm nc}$, a totally real field of degree 10 in which 23 is totally split, and use $K/k = k(\sqrt{-23})/k$ which is ramified at ten places $v|23$ and complexified at ten real places.

§6 The Hasse Principle — For Norms — For Powers

6.1 An Index Computation. The equality:

$$(J_K : K^\times \mathrm{N}_{L/K}(J_L)) = [L^{\rm ab} : K],$$

which comes from the fundamental properties of the global reciprocity map for a finite extension L/K will allow us to give a nontrivial example of the local-global principle mentioned in the Introduction, the Hasse norm theorem. For this, we will interpret this index as a product of suitable norm indices; this computation, which is useful in practice only when L/K is a *cyclic* extension, involves interesting arithmetic invariants (such as the order of the group of ambiguous classes). For a cohomological approach see [d, Lang1, Ch. IX].

6.1.1 Notations. Let L/K be a cyclic extension with Galois group $G =: \langle \sigma \rangle$, and write $\mathrm{N}_{L/K} =: \mathrm{N}$. Recall that for $S = \emptyset$, $U^S = U$, $E^S = E$, and $\mathcal{C}\!\ell^S = \mathcal{C}\!\ell$ denote respectively the group of unit idèles in the restricted sense

($U^{\rm res}$), the group of units in the restricted sense ($E^{\rm res}$), and the restricted class group ($\mathcal{C}\!\ell^{\rm res}$) of K. In the computations below, to simplify notations we omit these superscripts. □

Finally, it will be necessary to check that each of the indices that we wiil write below is finite (in particular by using the fact that $J_L/L^\times U_L \simeq \mathcal{C}\!\ell_L$ and $J_K/K^\times U_K \simeq \mathcal{C}\!\ell_K$ are finite).

Because of the inclusions $K^\times{\rm N}(U_L) \subseteq K^\times{\rm N}(J_L) \subseteq J_K$, we have the equality:

$$(J_K : K^\times{\rm N}(J_L)) = \frac{(J_K : K^\times{\rm N}(U_L))}{(K^\times{\rm N}(J_L) : K^\times{\rm N}(U_L))}$$

(the finiteness of the numerator comes from that of $J_K/K^\times U_K \simeq \mathcal{C}\!\ell_K$ and from that of $U_K/{\rm N}(U_L)$ by local class field theory 1.4.3). We have the exact sequence:

$$1 \longrightarrow K^\times \cap {\rm N}(J_L)/K^\times \cap {\rm N}(L^\times U_L) \longrightarrow {\rm N}(J_L)/{\rm N}(L^\times U_L) \longrightarrow K^\times{\rm N}(J_L)/K^\times{\rm N}(U_L) \longrightarrow 1,$$

since the kernel is equal to:

$$\begin{aligned} K^\times{\rm N}(U_L) \cap {\rm N}(J_L)/{\rm N}(L^\times U_L) &\simeq (K^\times \cap {\rm N}(J_L)){\rm N}(U_L)/{\rm N}(L^\times U_L) \\ &\simeq K^\times \cap {\rm N}(J_L)/K^\times \cap {\rm N}(L^\times U_L)\ ; \end{aligned}$$

thus, ${\rm N}(J_L)/{\rm N}(L^\times U_L)$ being finite as a quotient of $\mathcal{C}\!\ell_L$, we obtain the formula:

$$(J_K : K^\times{\rm N}(J_L)) = \frac{(J_K : K^\times{\rm N}(U_L))(K^\times \cap {\rm N}(J_L) : K^\times \cap {\rm N}(L^\times U_L))}{({\rm N}(J_L) : {\rm N}(L^\times U_L))}\ ;$$

furthermore, the exact sequence:

$$1 \longrightarrow {}_{\rm N}J_L\, L^\times U_L \longrightarrow J_L \longrightarrow {\rm N}(J_L)/{\rm N}(L^\times U_L) \longrightarrow 1$$

where ${}_{\rm N}J_L = \{\boldsymbol{x} \in J_L,\ {\rm N}\boldsymbol{x} = 1\}$, allows us to interpret the finite index $({\rm N}(J_L) : {\rm N}(L^\times U_L))$ in the form:

$$(J_L : {}_{\rm N}J_L\, L^\times U_L) = \frac{(J_L : J_L^{1-\sigma} L^\times U_L)}{({}_{\rm N}J_L L^\times U_L : J_L^{1-\sigma} L^\times U_L)}.$$

It is here that class groups enter since, in the exact sequence:

$$1 \longrightarrow J_L^{1-\sigma} L^\times U_L/L^\times U_L \longrightarrow J_L/L^\times U_L \longrightarrow J_L/J_L^{1-\sigma} L^\times U_L \longrightarrow 1,$$

$J_L/L^\times U_L \simeq \mathcal{C}\!\ell_L$ and $J_L^{1-\sigma} L^\times U_L/L^\times U_L \simeq (\mathcal{C}\!\ell_L)^{1-\sigma}$; we thus have:

$$1 \longrightarrow (\mathcal{C}\!\ell_L)^{1-\sigma} \longrightarrow \mathcal{C}\!\ell_L \longrightarrow J_L/J_L^{1-\sigma} L^\times U_L \longrightarrow 1$$

and thanks to:

$$1 \longrightarrow (\mathcal{C}\ell_L)^G \longrightarrow \mathcal{C}\ell_L \longrightarrow (\mathcal{C}\ell_L)^{1-\sigma} \longrightarrow 1,$$

this allows us to write $(J_L : J_L^{1-\sigma} L^\times U_L) = |(\mathcal{C}\ell_L)^G|$, which is equal to the number of invariant classes for the cyclic extension L/K (also called the number of ambiguous classes); the (delicate) computation of $|(\mathcal{C}\ell_L)^G|$ yields the following result.

6.1.2 Lemma. *For any cyclic extension L/K with Galois group G, we have:*

$$|(\mathcal{C}\ell_L)^G| = \frac{|\mathcal{C}\ell_K| \prod_{v \in Pl_0} e_v}{[L:K]\ (E_K : E_K \cap \mathrm{N}L^\times)},$$

where e_v is the ramification index of v in L/K.[29] □

Therefore, we have obtained:

$$(J_K : K^\times \mathrm{N}(J_L)) = \frac{(J_K : K^\times \mathrm{N}(U_L))}{|(\mathcal{C}\ell_L)^G|} \times$$
$$(K^\times \cap \mathrm{N}(J_L) : K^\times \cap \mathrm{N}(L^\times U_L))({}_{\mathrm{N}}J_L\, L^\times U_L : J_L^{1-\sigma} L^\times U_L)\ ;$$

however, we can write:

$$\begin{aligned}(J_K : K^\times \mathrm{N}(U_L)) &= (J_K : K^\times U_K)(K^\times U_K : K^\times \mathrm{N}(U_L)) \\ &= |\mathcal{C}\ell_K| \frac{(U_K : \mathrm{N}(U_L))}{(E_K : E_K \cap \mathrm{N}(U_L))},\end{aligned}$$

since we have the exact sequence:

$$1 \longrightarrow E_K/E_K \cap \mathrm{N}(U_L) \longrightarrow U_K/\mathrm{N}(U_L) \longrightarrow K^\times U_K/K^\times \mathrm{N}(U_L) \longrightarrow 1.$$

By local class field theory (see 1.4.3, (ii)), the numerator is:

$$(U_K : \mathrm{N}(U_L)) = \prod_{v \in Pl_0} e_v,$$

and the denominator can be written in the form:

$$(E_K : E_K \cap \mathrm{N}(U_L)) = \frac{(E_K : E_K \cap \mathrm{N}L^\times)}{(E_K \cap \mathrm{N}(U_L) : E_K \cap \mathrm{N}L^\times)},$$

the inclusion $E_K \cap \mathrm{N}L^\times \subseteq E_K \cap \mathrm{N}(U_L)$ being an easy consequence of 2.5.4 (we have here $E_K \cap \mathrm{N}(U_L) = E_K \cap \mathrm{N}(J_L)$). Thus, the index $(J_K : K^\times \mathrm{N}(U_L))$ can be written:

[29] Recall that classes and units are taken in the restricted sense.

$$(J_K : K^\times \mathrm{N}(U_L)) = \frac{|\mathcal{C}\!\ell_K| \prod_{v\in Pl_0} e_v}{(E_K : E_K \cap \mathrm{N}L^\times)} (E_K \cap \mathrm{N}(U_L) : E_K \cap \mathrm{N}L^\times)$$
$$= |(\mathcal{C}\!\ell_L)^G| \; [L:K] \; (E_K \cap \mathrm{N}(U_L) : E_K \cap \mathrm{N}L^\times)$$

using 6.1.2. Coming back to the expression for $(J_K : K^\times \mathrm{N}J_L)$, we obtain:

$$(J_K : K^\times \mathrm{N}J_L) = [L:K] \; (E_K \cap \mathrm{N}U_L : E_K \cap \mathrm{N}L^\times) \times$$
$$(K^\times \cap \mathrm{N}(J_L) : K^\times \cap \mathrm{N}(L^\times U_L))({}_{\mathrm{N}}J_L \, L^\times U_L : J_L^{1-\sigma} L^\times U_L) \, ;$$

this index being equal to $[L:K]$ (by the fundamental equality of global class field theory), we therefore obtain:

(i) $E_K \cap \mathrm{N}(U_L) = E_K \cap \mathrm{N}L^\times$,
(ii) $K^\times \cap \mathrm{N}(J_L) = K^\times \cap \mathrm{N}(L^\times U_L)$,
(iii) ${}_{\mathrm{N}}J_L \, L^\times U_L = J_L^{1-\sigma} L^\times U_L$.

Statement (iii) does not tell us much since ${}_{\mathrm{N}}J_L = J_L^{1-\sigma}$ by the Hilbert Theorem 90 for the idèle group in a cyclic extension; this fact has an analog for an arbitrary Galois extension L/K and can be written $H^1(G, J_L) = 1$ (see 2.4). But one checks that (ii) can be written:

$$\begin{aligned} K^\times \cap \mathrm{N}(J_L) &= K^\times \cap \mathrm{N}(L^\times U_L) \\ &= \mathrm{N}L^\times \, (E_K \cap \mathrm{N}(U_L)) \\ &= \mathrm{N}L^\times \, (E_K \cap \mathrm{N}L^\times) \;\; \text{(by (i))} \\ &= \mathrm{N}L^\times . \end{aligned}$$

Thus we have proved the following result.

6.2 Theorem (Hasse's norm theorem (1930)). *Let L/K be a cyclic extension of number fields. Then a necessary and sufficient condition for an $x \in K^\times$ to be the norm of an element of $L^\times$ is that x be a local norm everywhere for L/K or, equivalently, $i_v(x) =: \mathrm{N}_{L_v/K_v}(y_v)$, $y_v \in L_v^\times$, for all $v \in Pl$. The product formula tells us that this is equivalent to $\left(\frac{x\,,\,L/K}{v}\right) = 1$ for all noncomplex places v except an arbitrarily chosen one.* □

6.2.1 Remark. Since by 4.4.3 we know how to compute the symbols $\left(\frac{x\,,\,L/K}{v}\right)$, in the cyclic case it is numerically possible to know whether or not $x \in \mathrm{N}_{L/K}(L^\times)$; for this, we can omit the computation at an arbitrary place v_0. Recall also that it is sufficient to check this for the (finite) places which are ramified in L/K (except one), as soon as we know that $\mathrm{v}(x) \equiv 0 \bmod (f_v)$ for every place v (in particular for $v \in Pl_\infty^{\mathrm{r}}$), where f_v is the residue degree of v for L/K (see 1.4.3).

Note that Hasse's theorem does not give us the solution $y \in L^\times$ such that $\mathrm{N}_{L/K}(y) = x$, but that if y_0 is one of them, the others will be given by

$y = y_0\, z^{1-\sigma}$, $z \in L^\times$. For an algorithmic point of view, see [Sim] or [j, Coh2, Ch. 7, § 5] where S-unit groups play a fundamental role. □

6.2.2 Exercise. Let L be a totally imaginary number field. It is known (claimed by Hilbert (1902), proved by Siegel (1919)) that -1 is the sum of 1, 2, or 4 squares in L (with evident condition of minimality). Thus, if we suppose that $\sqrt{-1} \notin L$, -1 is the sum of 2 squares if and only if it is a norm in $L(\sqrt{-1})/L$.

Prove that this is the case if and only if for all $w|2$ the local degree $[L_w : \mathbb{Q}_2]$ is even (hint: use the Hasse norm Theorem 6.2 and show that -1 is a local norm at the odd places of L; for $w|2$, use the norm residue symbols $(-1, L_w(\sqrt{-1})/L_w)$, then the local norm lifting Theorem 1.5.4 for $K_v = \mathbb{Q}_2$, $M = \mathbb{Q}_2(\sqrt{-1})$, and finally 1.6.5).

We can omit a place $w_0|2$ and deduce that the corresponding local degree is even! This is not surprising since here the product formula looks like:

$$\sum_{w|2} [L_w : \mathbb{Q}_2] = [L : \mathbb{Q}] = 2r_2(L) \equiv 0 \bmod (2).$$

For instance, there is nothing to do for the field generated by a root of $X^4 + 2X + 2$. □

6.2.3 Remark (Chevalley's ambiguous class formula (1933)). Classically, the above index computations are done in the ordinary sense ($U^{\rm ord}$, $E^{\rm ord}$, and $\mathcal{C}\ell^{\rm ord}$), which lead to a formula involving $|(\mathcal{C}\ell_L^{\rm ord})^G|$, which itself involves the "ramification" of real infinite places, and which for us can be written (still for a cyclic extension L/K):

$$|(\mathcal{C}\ell_L^{\rm ord})^G| = \frac{|\mathcal{C}\ell_K^{\rm ord}| \prod_{v \in Pl_0} e_v \prod_{v \in Pl_\infty^{\rm r}} f_v}{[L : K]\ (E_K^{\rm ord} : E_K^{\rm ord} \cap \mathrm{N}_{L/K}(L^\times))}.$$

This formula occurs for the first time in complete generality in [h, Che1] and relies on work of Herbrand on the unit group, more precisely on the computation of the Herbrand quotient of E_L [d, Lang1, Ch. IX, § 1], which is given by the formula:

$$\frac{(E_K : \mathrm{N}_{L/K}(E_L))}{({}_{\mathrm{N}}E_L : E_L^{1-\sigma})} = \frac{2^{r_1^{\rm c}}}{[L : K]},$$

where $r_1^{\rm c}$ is the number of real places of K complexified in L, and which is the key of Chevalley's formula.

It has been extended to the case of S-decomposition in [Ja2]. □

We also give without proof a more general formula which allows to perform computations of invariant classes in cyclic extensions (see [Gr8]).

6.2.4 Proposition (invariant class formula with unramified modulus). *Let L/K be a cyclic extension of number fields with Galois group G, and let $\mathfrak{m} = \prod_{v \in T} \mathfrak{p}_v^{m_v}$, $m_v \geq 0$, T disjoint from the set of places which are ramified in L/K. Let $\mathfrak{m}'$ be the extension of $\mathfrak{m}$ to L. Let $\mathcal{C}' := I' P_{L,T,\mathfrak{m}',\mathrm{pos}} / P_{L,T,\mathfrak{m}',\mathrm{pos}} \subseteq \mathcal{C}\ell^{\mathrm{res}}_{L,\mathfrak{m}'}$, where I' is an arbitrary sub-G-module of $I_{L,T}$; we then have:*

$$|(\mathcal{C}\ell^{\mathrm{res}}_{L,\mathfrak{m}'}/\mathcal{C}')^G| = \frac{|\mathcal{C}\ell^{\mathrm{res}}_{\mathfrak{m}}| \prod_{v \in Pl_0} e_v}{[L:K]\, |\mathrm{N}_{L/K}(\mathcal{C}')|\, (\Lambda : \Lambda \cap \mathrm{N}_{L/K}(L^\times))},$$

where $\Lambda := \{x \in K^\times_{T,\mathfrak{m},\mathrm{pos}},\ (x) \in \mathrm{N}_{L/K}(I')\}$. □

6.2.5 Remarks. (i) Recall that the action of $\mathrm{N}_{L/K}$ on $\mathcal{C}\ell^{\mathrm{res}}_{L,\mathfrak{m}'}$ is given in 5.7, and implies that $\mathrm{N}_{L/K}(\mathcal{C}')$ makes sense.

(ii) By the Hasse norm Theorem 6.2, the term $(\Lambda : \Lambda \cap \mathrm{N}_{L/K}(L^\times))$ is of a *local nature*; it can be written $(\Lambda : \Lambda \cap \mathrm{N}_{L/K}(J_L))$ and depends only on the norm residue symbols at the ramified places since the condition $(x) \in \mathrm{N}_{L/K}(I')$ defining Λ implies that these norm conditions are already satisfied at the unramified places (by 1.4.3).

This remark is of course valid for $E_K \cap \mathrm{N}_{L/K}(L^\times) = E_K \cap \mathrm{N}_{L/K}(J_L)$ in the various ambiguous class formulas.

(iii) We obtain an expression for $|(\mathcal{C}\ell^{S'}_{L,\mathfrak{m}'})^G|$, where $S' \subset Pl_L$ is stable under G, by choosing for $\mathcal{C}'$ the G-module $\langle\, c\ell^{\mathrm{res}}_{L,\mathfrak{m}'}(S')\, \rangle$ (in the sense explained in I.4.4.1, (ii)). □

The Hasse principle is a key tool in the proof of the Hasse–Minkowski theorem on quadratic forms over number fields (see also the direct proofs of [a, BŠa; D; Se1]). The three-variable case is the object of Exercise 6.4 which can also be found in [d, CF, Exer. 4] and which is given in all the books dealing with the Hasse principle. The four-variable case is solved in 7.3.2.

When L/K is not cyclic, the Hasse principle is in general not true and its defect group, which is essentially given in cohomological terms, is in fact an algebraic invariant, related to the family of decomposition groups of ramified places. To be complete on this, we simply give the following results of Scholz-Tate involving Schur multipliers. [30]

Let L/K be Galois with Galois group G; for any noncomplex place v of K we denote by w a place of L above v, fixed arbitrarily, and we denote by D_w the decomposition group of w in L/K.

6.2.6 Proposition. *We have a canonical isomorphism:*

$$K^\times \cap \mathrm{N}_{L/K}(J_L)/\mathrm{N}_{L/K}(L^\times) \simeq H^{-3}(G,\mathbb{Z}) \Big/ \mathrm{Inf}\Big(\bigoplus_v H^{-3}(D_w,\mathbb{Z})\Big),$$

[30] [d, CF, Ch. VII, § 11.4], [Scholz], [Jeh], as well as the work of Razar [Ra] which is the culmination of several approaches (Garbanati, Gerth, Gurak, ...).

where $H^{-3} := H_2$ and where, for an element $(\alpha_w)_v$ of $\bigoplus_v H^{-3}(D_w, \mathbb{Z})$:

$$\mathrm{Inf}((\alpha_w)_v) := \prod_v \mathrm{Inf}_w(\alpha_w),$$

Inf_w *denoting the inflation map* $H^{-3}(D_w, \mathbb{Z}) \longrightarrow H^{-3}(G, \mathbb{Z})$. □

6.2.7 Remarks. (i) Since the group G is finite, by duality we can express that the dual of $K^\times \cap \mathrm{N}_{L/K}(J_L)/\mathrm{N}_{L/K}(L^\times)$ is isomorphic to:

$$\mathrm{Ker}\Big(\mathrm{Res} : H^2(G, \mathbb{Q}/\mathbb{Z}) \longrightarrow \bigoplus_v H^2(D_w, \mathbb{Q}/\mathbb{Z})\Big),$$

where $\mathrm{Res} := (\mathrm{Res}_w)_v$ is the family of restriction maps. Since $\mathbb{Q}$ is uniquely divisible, its cohomology is trivial and we may replace the $H^2(\bullet, \mathbb{Q}/\mathbb{Z})$ by the $H^3(\bullet, \mathbb{Z})$. When v is unramified in L/K, D_w is cyclic (generated by the Frobenius of w), and $H^3(D_w, \mathbb{Z}) = H^1(D_w, \mathbb{Z}) = 1$, so that only the finite ramified places enter in the definition of Inf and of Res.

(ii) Finally, when L/K is abelian, Razar [Ra] has shown that we may replace $H^{-3}(G, \mathbb{Z})$ and $H^{-3}(D_w, \mathbb{Z})$ by $\bigwedge^2 G$ and $\bigwedge^2 D_w$, respectively, which in this case enables us to perform explicit computations and to immediately construct counterexamples to the Hasse principle (the simplest being $\mathbb{Q}(\sqrt{13}, \sqrt{17})/\mathbb{Q}$ of Scholz [Scholz], given in [d, CF, Ch. VII, § 11.4], for which -1 is a local norm everywhere without being a global norm).

(iii) For example, if L/K is abelian and if there exists a place v of K such that $D_w = G$, then the Hasse principle for the norm is true in L/K.

In the case where $G \simeq \mathbb{Z}/p\mathbb{Z} \times \mathbb{Z}/p\mathbb{Z}$, the Hasse principle for the norm holds in L/K if and only if there exists v such that $D_w = G$. □

We will again meet the default group $K^\times \cap \mathrm{N}_{L/K}(J_L)/\mathrm{N}_{L/K}(L^\times)$ in the section dealing with central class fields ((Ch. IV; (c)) and Exercise IV.4.10). This group is also called the knot group of the extension L/K, notion which was introduced and studied by Scholz, then by Jehne.

6.3 Local-Global Principle for Powers. Starting with Chapter III, we will come back to the fine study of the elementary parts of class field theory so as to obtain the structure of $\mathrm{Gal}(\overline{K}^{\mathrm{ab}}/K)$, and deduce a number of little-known consequences. Meanwhile, in the following Theorem 6.3.3 the reader will find the solution of an important local-global problem (the local-global principle for powers) which only relies on the surjectivity of the Artin map and on simple Kummer theory arguments which can found for the first time in [SchFK] and in [Che5, I], and which should not be considered as a result of class field theory, although it is an essential tool for it.[31] This result will be crucial to explain in detail certain elements of the structure of $\mathrm{Gal}(\overline{K}^{\mathrm{ab}}/K)$

[31] [d, AT, Ch. X, § 1] (see also [e, Ko3, Ch. 2, § 1.12, Th. 2.21]).

(for instance by means of the Schmidt–Chevalley Theorem III.4.3 and the Grunwald–Wang Theorem III.4.16.4); note that it is the starting point for the p-adic class field theory of Jaulent and for the study of the connected component of the unit element of C done by Artin–Tate and Weil. Finally, we mention that it exhibits the famous special case (which is an obstruction at 2 of the corresponding Hasse principle), which shows once more to those who are not yet convinced, that 2 is the most "interesting" prime number.

6.3.1 Notations. Let K be a number field[32], and let p^e for some $e \geq 1$ be a fixed power of a prime number p. We set (in a suitable algebraic closure):

$$\mu_{p^e} =: \langle\, \zeta_e \,\rangle \ \text{ and } \ \mu_{p^k} =: \langle\, \zeta_k \,\rangle, \ \text{ where } \ \zeta_k := \zeta_e^{p^{e-k}},$$

for $0 \leq k \leq e$. Denote by K' the field $K(\mu_{p^e})$, and set $G' := \mathrm{Gal}(K'/K)$ which is isomorphic to a subgroup of $(\mathbb{Z}/p^e\mathbb{Z})^\times$. □

6.3.2 Theorem. *Let $x \in K^\times$ be such that $x =: x'^{p^e}$, $x' \in K'^\times$. Then $x = y^{p^e}$ for an $y \in K^\times$, except in the following exceptional case:*

- $p = 2, \quad e \geq 2,$
- $K \cap \mathbb{Q}(\mu_{2^e}) = \mathbb{Q}(\zeta_n + \zeta_n^{-1})$, *for* $2 \leq n \leq e$,
- $x = (-1)^{2^{e-n}} x_0 \,.\, y^{2^e}$, *with* $x_0 := (2 + \zeta_n + \zeta_n^{-1})^{2^{e-1}}$ *and* $y \in K^\times$.

In this case, $(-1)^{2^{e-n}} x_0 = (1 + \zeta_n)^{2^e}$. □

6.3.3 Theorem (local-global principle for powers). *Let Σ be a finite set of places of K. Let $x \in K^\times$ be such that $i_v(x) \in K_v^{\times p^e}$ for all places $v \notin \Sigma$. Then $x = y^{p^e}$ for an $y \in K^\times$, except in the following Σ-special case:*

- $p = 2, \quad e \geq 3,$
- $K \cap \mathbb{Q}(\mu_{2^e}) = \mathbb{Q}(\zeta_n + \zeta_n^{-1})$, *for* $2 \leq n < e$,
- *for all places* $v \in Pl_2 \backslash (\Sigma \cap Pl_2)$, K_v *contains one of the numbers:*

$$1 + \zeta_n, \ \ \zeta_{n+1} + \zeta_{n+1}^{-1}, \ \ \sqrt{-1}(\zeta_{n+1} + \zeta_{n+1}^{-1}),$$

- $x = x_0 \,.\, y^{2^e}$, *with* $x_0 := (2 + \zeta_n + \zeta_n^{-1})^{2^{e-1}}$ *and* $y \in K^\times$. □

We give detailed proofs by the way of the following exercise in which the reader will find many other properties and examples, as well as the study of an idèle $\boldsymbol{s}$ such that $x_0 = \boldsymbol{s}^{2^e}$ in the special case (i.e., the Σ-special case above for $\Sigma = \emptyset$). The notations are those of 6.3.1.

6.3.4 Exercise. Let $x \in K^\times$ be such that $x =: x'^{p^e}$, for an $x' \in K'^\times$.

(α) (case $p \neq 2$). In this case, $G' =: \langle\, \sigma \,\rangle$ is cyclic.

[32] Results 6.3.2 is valid for any field of characteristic equal to 0; in particular it will be used for the completions of the field K.

(i) Show that $H^1(G', \mu_{p^e}) = 1$ (since μ_{p^e} is finite, its Herbrand quotient is trivial [d, Se2, Ch. VIII, § 4] and we have $|H^1(G', \mu_{p^e})| = |\mu_{p^e}^{G'}/\mathrm{N}_{K'/K}(\mu_{p^e})|$; thus, one can show the equality $\mu_{p^e}^{G'} = \mathrm{N}_{K'/K}(\mu_{p^e})$, but one can also show by a direct computation that ${}_{\mathrm{N}}\mu_{p^e} = \mu_{p^e}^{1-\sigma}$, where ${}_{\mathrm{N}}\mu_{p^e}$ is the kernel of $\mathrm{N}_{K'/K}$ in μ_{p^e}).

(ii) From the equality $x = x'^{p^e}$ and the above results, deduce that $x = y^{p^e}$ for $y \in K^\times$.

(β) (case $p = 2$, $e \geq 2$). In this case, G' is isomorphic to a subgroup of $\langle -\bar{1} \rangle \oplus \langle \bar{5} \rangle$ in $(\mathbb{Z}/2^e\mathbb{Z})^\times$. Set $Q := K \cap \mathbb{Q}(\mu_{2^e})$.

(i) Show that $x \in K^{\times 2^e}$ is still true if K contains $\mathbb{Q}(\mu_4)$ or if for some $n \geq 3$, Q is equal to the subfield Q'_{n-2} of $\mathbb{Q}(\mu_{2^n})$, of relative degree equal to 2, different from $Q_{n-2} := \mathbb{Q}(\zeta_n + \zeta_n^{-1})$ (see Fig. 6.1) (thus there will only remain the case where $Q = Q_{n-2}$ for some $n \geq 2$).

(ii) Show that in all other cases, we have $x \in \langle -1 \rangle K^{\times 2^{e-1}}$.

From now on, we assume that $Q = Q_{n-2}$ for $e \geq n \geq 2$; in particular $K_1 := K(\sqrt{-1}\,)$ is a quadratic extension of K containing μ_{2^n} and not $\mu_{2^{n+1}}$.

(iii) Show that, if the set of counterexamples (to $x \in K^{\times 2^e}$) is not empty, it is of the form $x_0 K^{\times 2^e}$ for an arbitrary solution x_0 (by abuse of language we will say that the counterexample is unique).

(iv) Show that, for all $n \geq 2$:

$$x_n := (1 + \zeta_n)^{2^n} \in Q_{n-2}^\times \cap (\mathbb{Q}(\mu_{2^n}))^{\times 2^n}, \text{ where } Q_{n-2} := \mathbb{Q}(\zeta_n + \zeta_n^{-1}),$$

and that it is also an element of $-Q_{n-2}^{\times 2^{n-1}}$.

(v) Conclude by giving a characterization of the cases where K contains a counterexample, and give its value. This will prove Theorem 6.3.2.

(γ) (Hasse principle for powers). In this question, p is once again an arbitrary prime number and e an integer which is ≥ 1. Let Σ be a finite set of noncomplex places of K, and let $x \in K^\times$ be such that $i_v(x) \in K_v^{\times p^e}$ for all places v not belonging to Σ. Consider the Kummer extension $K'(\sqrt[p^e]{x})/K'$.

(i) Check that there exists a place v'_0 of K', unramified in $K'(\sqrt[p^e]{x})/K'$, which is not above a place of Σ, whose Frobenius in $K'(\sqrt[p^e]{x})/K'$ is a generator (density Theorem 4.6).

(ii) Deduce that $x \in K'^{\times p^e}$, and then that $x \in K^{\times p^e}$, except perhaps in the case $p = 2$, for a particular case which one asks to characterize: it is the Σ-special case, where nonetheless we have $x \in K^{\times 2^{e-1}}$.

(iii) Let ${}_{p^e}J$ (resp. ${}_{p^e}C$) be the set of idèles (resp. of idèle classes) of order a divisor of p^e, and let $\mathcal{cl}$ be the canonical map $J \longrightarrow C$. Deduce from the above that ${}_{p^e}C = \mathcal{cl}({}_{p^e}J)$ except in the Σ-special case for $\Sigma = \emptyset$ (then simply called the special case), in which case ${}_{2^e}C = \langle \mathcal{cl}(\boldsymbol{s}) \rangle . \mathcal{cl}({}_{2^e}J)$, with $\mathcal{cl}(\boldsymbol{s}^2) \in \mathcal{cl}({}_{2^{e-1}}J)$, $\mathcal{cl}(\boldsymbol{s}) \notin \mathcal{cl}({}_{2^e}J)$ for a suitable idèle $\boldsymbol{s}$ (see the data in the proof of (ii) above). In other words, ${}_{2^e}C/\mathcal{cl}({}_{2^e}J)$ has order 2 in the special case.

Answer. For statements (α) and (β), we may assume that $\mu_{p^e} \not\subset K$; note also that $\mu_{p^e}^{G'} = \mu_p(K)$ is of the form μ_{p^k} for $0 \leq k < e$, with $k \geq 1$ for $p = 2$.

(α) (i) If $k = 0$, the result is trivial. If $k \geq 1$, $K \cap \mathbb{Q}(\mu_{p^e}) = \mathbb{Q}(\mu_{p^k})$ (indeed, between $\mathbb{Q}(\mu_{p^k})$ and $\mathbb{Q}(\mu_{p^e})$ there exist only the fields $\mathbb{Q}(\mu_{p^{k+i}})$ for $0 \leq i \leq e - k$ since $k \geq 1$), G' has order p^{e-k}, and we have:

$$\mathrm{Irr}(\zeta_e, K) = X^{p^{e-k}} - \zeta_k,$$

hence $\mu_{p^k} \subseteq \mathrm{N}_{K'/K}(\mu_{p^e})$.

(ii) If $x = x'^{p^e}$, we have $1 = (x'^{1-\sigma})^{p^e}$, hence $x'^{1-\sigma} =: \zeta' \in {}_{\mathrm{N}}\mu_{p^e}$; since $H^1(G', \mu_{p^e}) = 1$ (i.e., ${}_{\mathrm{N}}\mu_{p^e} = \mu_{p^e}^{1-\sigma}$), there exists $\xi \in \mu_{p^e}$ such that $\zeta' = \xi^{1-\sigma}$, and there exists $y \in K^\times$ such that $x' = \xi y$. We thus have $x = y^{p^e}$.

We can also say that the exact sequence:

$$1 \longrightarrow \mu_{p^e} \longrightarrow K'^\times \xrightarrow{\ p^e\ } K'^{\times p^e} \longrightarrow 1,$$

yields, since $H^1(G', \mu_{p^e}) = 1$, the surjective map:

$$K'^{\times G} = K^\times \xrightarrow{\ p^e\ } K'^{\times p^e G} = K^\times \cap K'^{\times p^e},$$

so that $K^{\times p^e} = K^\times \cap K'^{\times p^e}$.

(β) (i) If K contains $\mathbb{Q}(\mu_4)$, we have $Q = \mathbb{Q}(\mu_{2^k})$ for $k \geq 2$, G' is cyclic of order 2^{e-k}, and we still have $\mathrm{Irr}(\zeta_e, K) = X^{2^{e-k}} - \zeta_k$ and the same result (here we find that $-\zeta_k \in \mathrm{N}_{K'/K}(\mu_{2^e})$, but $\zeta_k = (-\zeta_k)^{1+2^{k-1}}$). The following field diagram gives the structure of $\mathbb{Q}(\mu_{2^\infty})/\mathbb{Q}$:

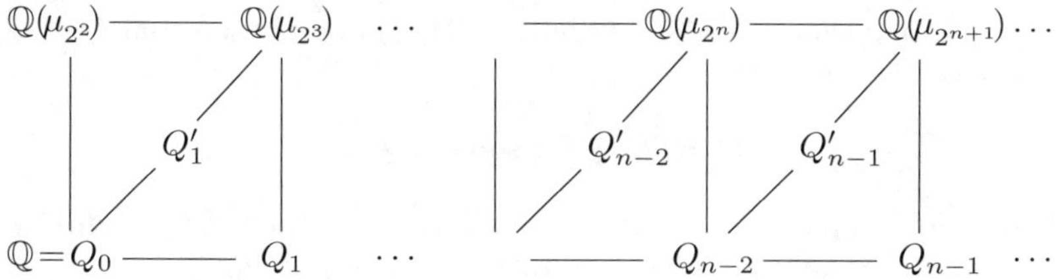

Fig. 6.1

If for some $n \geq 3$, $Q = Q'_{n-2}$ (which is the subfield of $\mathbb{Q}(\mu_{2^e})$ fixed under $\langle -5^{2^{n-3}} \rangle$), then necessarily $n \leq e$, the group G' is cyclic of order 2^{e-n+1} and we have:

$$\mathrm{Irr}(\zeta_e, K) = X^{2^{e-n+1}} - \omega X^{2^{e-n}} - 1,$$

where $\omega := \zeta_n - \zeta_n^{-1}$, so that $\mathrm{N}_{K'/K}(\mu_{2^e}) = \langle -1 \rangle = \mu_2(K)$. In this case, we still have $x = y^{2^e}$.

(ii) If we consider the equality $x = x'^{2^e}$ in $K_1 := K(\sqrt{-1}) \subseteq K'$, fact (i) shows that there exists $y_1 \in K_1^\times$ such that $x = y_1^{2^e}$; hence, $x^2 = \mathrm{N}_{K_1/K}(x) = (\mathrm{N}_{K_1/K}(y_1))^{2^e}$, proving the existence of $y := \mathrm{N}_{K_1/K}(y_1) \in K^\times$ such that $x = \pm y^{2^{e-1}}$.

(iii) From the following exact sequence of $\langle\, \tau \,\rangle$-modules:

$$1 \longrightarrow \mu_{2^n} \longrightarrow K_1^\times \xrightarrow{\;2^e\;} K_1^{\times 2^e} \longrightarrow 1,$$

by taking invariants under $\langle\, \tau \,\rangle$, we obtain:

$$1 \longrightarrow \mu_2 \longrightarrow K^\times \xrightarrow{\;2^e\;} K^\times \cap K_1^{\times 2^e} \longrightarrow H^1(\langle\, \tau \,\rangle, \mu_{2^n}) \longrightarrow 1$$

since $H^1(\langle\, \tau \,\rangle, K_1^\times) = 1$ (Theorem 90). Thus:

$$K^\times \cap K_1^{\times 2^e} / K^{\times 2^e} \simeq {}_{\mathrm{N}}\mu_{2^n} / \mu_{2^n}^{1-\tau} \simeq \mu_2 / \mathrm{N}_{K_1/K}(\mu_{2^n}),$$

and so "the" counterexample happens if and only if $\mathrm{N}_{K_1/K}(\zeta_n) = 1$, hence if and only if $\tau(\zeta_n) = \zeta_n^{-1}$, which means that $\mathbb{Q}(\zeta_n + \zeta_n^{-1}) \subseteq K$, i.e., $Q = \mathbb{Q}(\zeta_n + \zeta_n^{-1})$, which is indeed the case.

(iv) If $n \geq 2$, we have $x_n := (1 + \zeta_n)^{2^n} \in Q_{n-2}^\times$ since $(1 + \zeta_n^{-1})^{2^n} = (1 + \zeta_n)^{2^n}$; furthermore, we have:

$$(1 + \zeta_n)^2 = 1 + \zeta_n^2 + 2\zeta_n = \zeta_n(\zeta_n^{-1} + \zeta_n + 2),$$

which shows that:

$$x_n = -y_n^{2^{n-1}}, \text{ with } y_n := 2 + \zeta_n + \zeta_n^{-1} \in Q_{n-2}^\times.$$

Note that x_n is not even a square in $Q_{n-2}^\times$

(v) Assume that K does not contain $\mathbb{Q}(\mu_4)$ and is such that $Q \neq Q'_{n-2}$ for all $n \geq 3$. We thus have:

$$Q = Q_{n-2} \text{ for some } n \geq 2.$$

Recall that $K_1^\times \cap \mu_{2^e} = \mu_{2^n}$ since $K_1 \cap \mathbb{Q}(\mu_{2^e})$ contains $\mathbb{Q}(\mu_4)$ and Q_{n-2}, hence $\mathbb{Q}(\mu_{2^n})$, the only possible quadratic extension of Q_{n-2}.

If $n = e$, by "uniqueness" the counterexample in K is equal to:

$$x := x_e = -y_e^{2^{e-1}} \text{ (equal to } (1 + \zeta_e)^{2^e} \text{ in } K')$$

(indeed, if we had $x = y^{2^e}$ with $y \in K^\times$, then -1 would be a square in K, which is not the case).

If $n < e$, we have at our disposal the element $x_n \in Q_{n-2}^\times$ (see (iv)) such that:

$$x_n = -y_n^{2^{n-1}}, \; y_n \in Q_{n-2}^\times,$$

so that we can consider:

$$x := x_n^{2^{e-n}} = y_n^{2^{e-1}} \text{ (equal to } (1+\zeta_n)^{2^e} \text{ in } K') ;$$

x is the desired counterexample in K: indeed, if $y_n^{2^{e-1}} = y^{2^e}$, $y \in K^\times$, then $y_n = \xi y^2$, $\xi \in \mu_{2^{e-1}}$; but $\xi \in K^\times$, hence $\xi = \pm 1$ so we easily obtain:

$$y = \pm(\zeta_{n+1} + \zeta_{n+1}^{-1}), \text{ if } \xi = 1,$$
$$y = \pm\sqrt{-1}(\zeta_{n+1} + \zeta_{n+1}^{-1}), \text{ if } \xi = -1.$$

But the field $Q_{n-2}(y) \subseteq K$ is respectively equal to Q_{n-1} and to Q'_{n-1}, which means (since here $n < e$) that $K \cap \mathbb{Q}(\mu_{2^e})$ is not equal to Q_{n-2}, a contradiction.

To summarize, the counterexample of Theorem 6.3.2 takes place if and only if:

$$K \cap \mathbb{Q}(\mu_{2^e}) = \mathbb{Q}(\zeta_n + \zeta_n^{-1}), \;\; 2 \le n \le e,$$

and it is given by:

$$x := (-1)^{2^{e-n}} (2 + \zeta_n + \zeta_n^{-1})^{2^{e-1}}$$

(equal to $(1+\zeta_n)^{2^e}$ in K').

6.3.4.1 Examples. (i) If K does not contain $\mathbb{Q}(\mu_4)$ and if we take $e = 2$, we always obtain $Q = \mathbb{Q} = Q_0$ (i.e., $n = e = 2$), and:

$$x = x_2 = -4 = (1 + \sqrt{-1}\,)^4,$$

which is a 4th power in $K(\sqrt{-1}\,)$ but is not a square in K.

(ii) For $e = 3$, if K contains $Q_1 = \mathbb{Q}(\sqrt{2}\,)$ but does not contain $\mathbb{Q}(\sqrt{-1}\,)$, we obtain $Q = Q_1$ (i.e., $n = e = 3$), and:

$$x = x_3 = -(2 + \sqrt{2}\,)^4 = (1 + \zeta_3)^8,$$

with $\zeta_3^2 = \sqrt{-1}$, $\zeta_3 + \zeta_3^{-1} = \sqrt{2}$.

(iii) Finally note that, when K does not contain $\mathbb{Q}(\sqrt{2}\,)$ and $e = 3$, $Q = Q_0$, hence $n = 2$, we have:

$$x = x_2^2 = 16 = (1 + \sqrt{-1}\,)^8 = (\sqrt{2}\,)^8 = (\sqrt{-2}\,)^8,$$

which is an 8th power in $\mathbb{Q}(\mu_8)$ and only a 4th power in K. □

Note. The case $n = e \ge 2$ is the only case where there exists a counterexample with K'/K cyclic (of degree 2); we will see that the special case assumes the noncyclicity, hence $n < e$ with $n \ge 2$.

(γ) (i) Since $K'(\sqrt[p^e]{x}\,)/K'$ is cyclic, there exists an infinite number of such places by the density theorem or simply the surjectivity of the Artin map (but the finiteness assumption on Σ is essential).

(ii) Let v be the place of K below v'_0. The extension $K'(\sqrt[p^e]{x})/K'$ is split at v'_0 (since $i_v(x) \in K_v^{\times p^e}$, a fortiori $i'_{v'_0}(x) \in K'^{\times p^e}_{v'_0}$), hence the Frobenius of v'_0 is equal to 1, and $K'(\sqrt[p^e]{x}) = K'$, hence $x \in K'^{\times p^e}$. We thus have $x \in K^{\times p^e}$, except if $p = 2$, $e \geq 2$, $Q = Q_{n-2}$ for some n such that $2 \leq n \leq e$, and if (up to an element of $K^{\times 2^e}$) $x = (-1)^{2^{e-n}} y_n^{2^{e-1}}$. But since $e \geq 2$, in the case $x = -y_e^{2^{e-1}}$ we would have $-1 \in K_v^{\times 2}$ for all places $v \notin \Sigma$ (since $y_e^{2^{e-1}} \in K^{\times 2}$); this is impossible since $\sqrt{-1} \notin K$ (choose $v_1 \notin \Sigma$ such that the Frobenius of v_1 for $K(\sqrt{-1})/K$ is of order 2, or simply note that this is the reasoning used in (i) above for $e = 1$ with $x = -1$). We are thus in the case where (see (β), (v)):

$$Q = Q_{n-2}, \quad 2 \leq n < e,$$

which corresponds to the counterexample $x = x_n^{2^{e-n}} = y_n^{2^{e-1}}$. It follows that the extension K'/K contains the biquadratic subextension $K\mathbb{Q}(\mu_{2^{n+1}})/K$ since $e \geq n+1$. In K' we have:

$$x = (1 + \zeta_n)^{2^e} = (\zeta_{n+1} + \zeta_{n+1}^{-1})^{2^e} = (\sqrt{-1}\,(\zeta_{n+1} + \zeta_{n+1}^{-1}))^{2^e},$$

so that $1 + \zeta_n$, $\zeta_{n+1} + \zeta_{n+1}^{-1}$, $\sqrt{-1}\,(\zeta_{n+1} + \zeta_{n+1}^{-1})$ are generators over K of $K\mathbb{Q}(\mu_{2^n})$, KQ_{n-1}, KQ'_{n-1}, respectively.

But if v is any place of K not dividing 2, it is split in at least one of the three extensions $K\mathbb{Q}(\mu_{2^n})$, KQ_{n-1}, and KQ'_{n-1} (since v is unramified in $K\mathbb{Q}(\mu_{2^{n+1}})/K$ and $\mathrm{Gal}(K\mathbb{Q}(\mu_{2^{n+1}})/K) \simeq (\mathbb{Z}/2\mathbb{Z})^2$). It follows that for such places $i_v(x) \in K_v^{\times 2^e}$ (for instance if $v|\infty$, the splitting takes place in KQ_{n-1}/K). It follows that our initial assumption ($i_v(x) \in K_v^{\times p^e}$ for all $v \notin \Sigma$) is satisfied for x, except perhaps for the even places not belonging to Σ.

Finally, if v divides 2, $i_v(x) \in K_v^{\times 2^e}$ if and only if K_v contains one of the three quadratic extensions of Q_{n-2} in $\mathbb{Q}(\mu_{2^{n+1}})$, in other words if v splits at least partially in $K\mathbb{Q}(\mu_{2^{n+1}})/K$ (apply the results of (α) and (β) to K_v by discussing over $K_v \cap \mathbb{Q}(\mu_{2^e})$).

This defines the Σ-special case of Theorem 6.3.3, which is especially tricky:

- $p = 2, \quad e \geq 3$,
- $K \cap \mathbb{Q}(\mu_{2^e}) = \mathbb{Q}(\zeta_n + \zeta_n^{-1})$, for some n such that $2 \leq n < e$,
- for all places $v \in Pl_2 \backslash \Sigma_2$, K_v contains one of the numbers: [33]

$$1 + \zeta_n, \quad \zeta_{n+1} + \zeta_{n+1}^{-1}, \quad \sqrt{-1}\,(\zeta_{n+1} + \zeta_{n+1}^{-1}),$$

the defect to the Hasse principle relative to Σ being due to:

$$x \in x_0 K^{\times 2^e}, \quad \text{with} \quad x_0 = (2 + \zeta_n + \zeta_n^{-1})^{2^{e-1}},$$

[33] where the first one may be replaced by $\sqrt{-1}$ since $Q_{n-2}(\sqrt{-1}) = \mathbb{Q}(\mu_{2^n})$.

where $\zeta_n = \zeta_{n+1}^2$ is a generator of μ_{2^n}.

The minimal example is for $K = \mathbb{Q}$, $e = 3$, $n = 2$, $\Sigma = Pl_2$, which is the example $x = 16$ given in 6.3.4.1, (iii).

6.3.4.2 Notation. We introduce the set Pl_2^{ns} of the ("nonsplit") places $v|2$ such that $\text{Gal}(K_v(\mu_{2^{n+1}})/K_v)$ is isomorphic to $\mathbb{Z}/2\mathbb{Z}\times\mathbb{Z}/2\mathbb{Z}$ (or, equivalently, $K_v \cap \mathbb{Q}(\mu_{2^e}) = Q_{n-2}$). □

We can say that the third condition of the Σ-special case is equivalent to the condition:

- $Pl_2^{\text{ns}} \subseteq \Sigma_2$,

according to the following diagram:

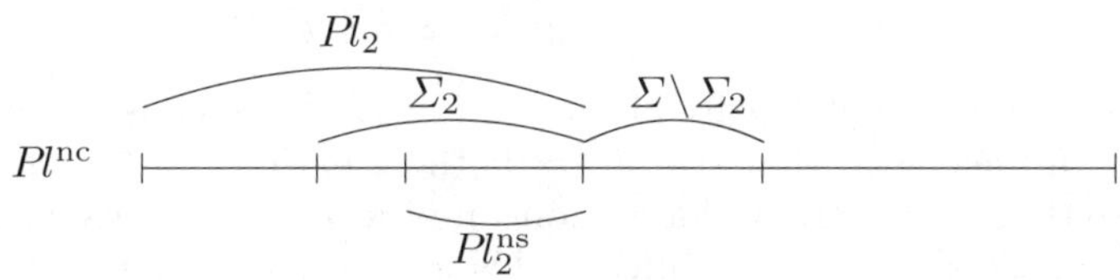

This diagram means that we have (unfortunately, when $Pl_2^{\text{ns}} \neq \emptyset$) discarded the places $v|2$ which would have told us (for a local reason) that x is not a 2^eth power. In other words, to choose $\Sigma_2 \neq \emptyset$ when $Pl_2^{\text{ns}} \neq \emptyset$ is in practice artificial, and the true special case corresponds to the fields K for which:

$$K \cap \mathbb{Q}(\mu_{2^e}) = \mathbb{Q}(\zeta_n + \zeta_n^{-1}), \quad \text{for some } n \text{ such that } 2 \leq n < e,$$

and whose places $v|2$ are all partially split in $K\mathbb{Q}(\mu_{2^{n+1}})/K$.

6.3.4.3 Remark (special case: the idèle $\boldsymbol{s}$). The Hasse principle for powers consists precisely in choosing $\Sigma = \emptyset$, in which case the last condition (corresponding to the existence of the special case, hence to the fact that the principle is false) can be written:

- $Pl_2^{\text{ns}} = \emptyset$.

In this case $i(2 + \zeta_n + \zeta_n^{-1})^{2^n}$ is of the form $\boldsymbol{s}^{2^{n+1}}$, where $\boldsymbol{s}$ is an idèle whose components s_v are, at each place, one of the numbers (considered in K_v):

$$1 + \zeta_n, \quad \zeta_{n+1} + \zeta_{n+1}^{-1}, \quad \sqrt{-1}\,(\zeta_{n+1} + \zeta_{n+1}^{-1}).$$

For any $e > n$, we also have $i(x_0) := i(2 + \zeta_n + \zeta_n^{-1})^{2^{e-1}} = \boldsymbol{s}^{2^e}$. The idèle $\boldsymbol{s}$ is not unique and is clearly defined only up to some element of ${}_{2^{n+1}}J$ (this comes from the fact that if v is totally split in $K\mathbb{Q}(\mu_{2^{n+1}})/K$, we can choose the component s_v among three possibilities; this total splitting is equivalent to $\zeta_{n+1} \in K_v$).

One should keep in mind that for all $e > n$, the idèle group $\langle\, \boldsymbol{s}\, \rangle\,.\,{}_{2^e}J$ does not depend on the choice of $\boldsymbol{s}$. □

(iii) Note (still in the special case with $Pl_2^{\rm ns} = \emptyset$) that $\boldsymbol{s}^2$ is of the form $\boldsymbol{\zeta}\, i(2 + \zeta_n + \zeta_n^{-1})$, where $\boldsymbol{\zeta} \in {}_{2^n}J$, $\boldsymbol{\zeta} \notin {}_{2^{n-1}}J$: indeed, we have $\boldsymbol{\zeta} =: (\zeta_v)_v$ with $\zeta_v = \zeta_n$, or $\zeta_v = 1$, or $\zeta_v = -1$ since:

$$(1 + \zeta_n)^2 = \zeta_n y_n, \quad (\zeta_{n+1} + \zeta_{n+1}^{-1})^2 = y_n, \quad (\sqrt{-1}\,(\zeta_{n+1} + \zeta_{n+1}^{-1}))^2 = -y_n,$$

with $y_n := 2 + \zeta_n + \zeta_n^{-1}$; but by the Čebotarev theorem, we have equidistribution of all possibilities. In other words:

$$\mathcal{cl}(\boldsymbol{s}) \in {}_{2^{n+1}}C, \quad \mathcal{cl}(\boldsymbol{s}) \notin {}_{2^n}C.$$

Finally, for all $e > n$:

$$\mathcal{cl}(\boldsymbol{s}^2) \in \mathcal{cl}({}_{2^{e-1}}J), \quad \mathcal{cl}(\boldsymbol{s}) \notin \mathcal{cl}({}_{2^e}J)$$

(indeed, otherwise we would easily obtain $y_n \in \pm K^{\times 2}$, which is absurd).

Let $\boldsymbol{x}$ be an idèle such that $\mathcal{cl}(\boldsymbol{x})^{p^e} = 1$; there exists $x \in K^\times$ such that $\boldsymbol{x}^{p^e} = i(x)$, so that $i_v(x) \in K_v^{\times p^e}$ for all noncomplex places v. Thus, in general we have $\boldsymbol{x}^{p^e} =: i(y)^{p^e}$, $y \in K^\times$, which yields $\boldsymbol{x} =: \boldsymbol{\zeta}\; i(y)$, where $\boldsymbol{\zeta} \in {}_{p^e}J$ is an idèle of the form $(\zeta_v)_v$, with $\zeta_v \in {}_{p^e}\mu(K_v)$; the result follows in this case.

The special case gives the additional solutions $\boldsymbol{x}^{2^e} = \boldsymbol{s}^{2^e}\,.\,i(y^{2^e})$, which proves the final result (independently of the choice of $\boldsymbol{s}$). This finishes the question (γ) and the exercise. □

For use in Exercise 6.4 on quadratic forms below, we note that we have shown that $a \in K^\times$ is a square in $K^\times$ if and only if it is a square locally at almost every place of K since troubles can start only for 8th powers.

6.3.5 Remarks. (i) The existence of a counterexample to the equality:

$$K^\times \cap K(\mu_{p^e})^{\times\, p^e} = K^{\times\, p^e}$$

(Theorem 6.3.2) should not be mistaken with a Σ-special case (Theorem 6.3.3), which is a counterexample to the Hasse principle for powers and which is relative to the choice of Σ; the former case is characterized by the following conditions:

- $p = 2, \quad e \geq 2,$
- $K \cap \mathbb{Q}(\mu_{2^e}) = \mathbb{Q}(\zeta_n + \zeta_n^{-1})$, for $2 \leq n \leq e$,

and concerns $x := (-(2 + \zeta_n + \zeta_n^{-1})^{2^{n-1}})^{2^{e-n}}$, which yields the minimal example $x = -4$ (for $K = \mathbb{Q}$ and $e = 2$) which is never a special case. To avoid any misunderstanding, this case has been called instead the *exceptional case*.

(ii) Note also that $\pm(2 + \zeta_n + \zeta_n^{-1})^{2^{n-1}}$ is a 2^nth power in each of the three quadratic subextensions of $K\mathbb{Q}(\mu_{2^{n+1}})/K$ (the minus sign occurs only for $K\mathbb{Q}(\mu_{2^n})/K$). □

Finally, we give the following numerical example of a true special case.

6.3.6 Example. Let $K = \mathbb{Q}(\sqrt{7})$, $p = 2$, and $e = 3$. Then $K \cap \mathbb{Q}(\mu_8) = \mathbb{Q}$ (i.e., $n = 2$), so that we have the exceptional case with $x = 16$ ($16 = (1+\sqrt{-1})^8 = (\sqrt{2}\,)^8 = (\sqrt{-2}\,)^8$ in $\mathbb{Q}(\mu_8)$). From the above, we know that $i_v(x) \in K_v^{\times 8}$ for any place not dividing 2, and for the unique place v_0 of K above 2, we have $K_{v_0} = \mathbb{Q}_2(\sqrt{7}) = \mathbb{Q}_2(\sqrt{-1})$ (i.e., $Pl_2^{\rm ns} = \emptyset$), which implies that $i_{v_0}(x) \in K_{v_0}^{\times 8}$ (we could also have chosen $K = \mathbb{Q}(\sqrt{\pm 14})$). We thus have a special case (i.e., $\Sigma = \emptyset$); therefore it is an absolute counterexample to the Hasse principle for powers. □

We will speak of the *special case* only in this type of situation.

6.3.7 Exercise. Consider the special case of Example 6.3.6 above. Let ℓ be the residue characteristic of $v \in Pl_0$. Show that the idèle $\boldsymbol{s} =: (s_v)_v$ (see 6.3.4.3) can be chosen as follows. If $\ell = 2$, $s_v = 1 + \sqrt{-1}$; if $\ell = 7$, $s_v = \sqrt{2}$; if $\ell \equiv 1 \bmod (4)$, $s_v = 1 + \sqrt{-1}$, and if $\ell \equiv -1 \bmod (4)$, $s_v = \sqrt{(-1)^{\frac{\ell+1}{4}} 2}$.

Check that $1+\sqrt{-1}$, $\sqrt{2}$, $\sqrt{-2}$ are all possible for s_v if and only if $|F_v^\times| \equiv 1 \bmod (8)$ (i.e., $\ell \equiv 1 \bmod (8)$, or $\ell \not\equiv 1 \bmod (8)$ and $\ell \equiv \pm 5,\ \pm 11,\ \pm 13 \bmod (28)$). □

6.3.8 Exercise (another criterion for pth powers). Let S be a finite set of noncomplex places of K such that $(\langle\, c\ell_{K'}^{\rm res}(S')\,\rangle)_p = (\mathcal{C}\ell_{K'}^{\rm res})_p$ (in the sense of I.4.4.1, (ii)), where S' is the set of places of $K' := K(\mu_p)$ above those of S. Let $x \in K^\times$ satisfying the following conditions:

- $(x) = \mathfrak{a}^p$, for an ideal $\mathfrak{a}$ of K,
- $i_v(x) \in K_v^{\times p}$ for all $v \in S \cup Pl_p$.

Show that $x \in K^{\times p}$.

Answer. The assumption on S' in K' is equivalent to $H_{K'(p)}^{S'} = K'$. But the assumption $(x) = \mathfrak{a}^p$ implies that $K'(\sqrt[p]{x}\,)/K'$ is unramified outside p, and the assumption $i_v(x) \in K_v^{\times p}$ for all $v \in S \cup Pl_p$ implies that it is S'-split and unramified; thus $K'(\sqrt[p]{x}\,) \subseteq H_{K'}^{S'}$, hence $x \in K^{\times p}$ by the above and 6.3.2 for $e = 1$.

Here, the number of local conditions is finite, but the required conditions assume the knowledge of the p-class group of K'.

We can also replace the conditions $i_v(x) \in K_v^{\times p}$ for $v|p$ by the Kummer nonramification conditions for places above p (use I.6.3, (ii) in K'). □

6.3.9 Remark. Let K be a number field and p a prime. Suppose we need to prove that some $\alpha \in K^\times$ is not a pth power in K. Then it is sufficient to find a place v such that α is not congruent to a pth power modulo $\mathfrak{p}_v$. The local-global principle for powers implies that such a place always exists but it does not give a bound for the number of tests. □

6.4 Exercise (quadratic forms in three variables). For $a,\ b \in K^\times$, consider the quadratic form:

$$(q) \quad X^2 - a\,Y^2 - b\,Z^2.$$

We will say that it represents 0 in K if there exist $x,\ y,\ z \in K$ (not all zero) such that:

$$x^2 - ay^2 - bz^2 = 0.$$

Set $L = K(\sqrt{a})$.

(i) Check that (q) represents 0 in K if and only if $b \in \mathrm{N}_{L/K}(L^\times)$.

(ii) Characterize the finite places of K which are ramified in L/K (see I.6.3).

(iii) Check that, for unramified places v (finite or not), the local norm condition $i_v(b) \in \mathrm{N}_{L_v/K_v}(L_v^\times)$ (where $L_v = L_w = K_v(\sqrt{a})$, for any $w|v$) is equivalent to $\mathrm{v}(b) \equiv 0 \bmod (f_v)$. For a concrete use of this, note that it is necessary to check the local conditions only when $\mathrm{v}(b)$ is odd, and that then we must have $f_v = 1$, which is the case if and only if $i_v(a) \in K_v^{\times 2}$.

To apply the Hasse principle, we can now assume that we are in the case where the bad places are the finite ramified places (where one may be omitted). It is then necessary to compute the symbols $(i_v(b), L_v/K_v)$ which can be identified with the quadratic Hilbert symbols $\left(\frac{a,\,b}{v}\right)$ over K; the case of odd places corresponds to regular Hilbert symbols and is given by a formula, so there essentially remains the case of even places. These symbols are also the Hasse symbols $\left(\frac{b\,,\,L/K}{v}\right)$ that we know how to compute in terms of Frobenius' thanks to the global approach explained in 4.4.3 which thus reduces to the techniques of question (iii) since here the conductor $\mathfrak{f}_{L/K}$ is known by 1.6.3.

(iv) Let $K = \mathbb{Q}(\sqrt{2})$, $a = 2 + 3\sqrt{2}$, and $b = -15(1 + \sqrt{2})$; does the form (q) represent 0 in K ?

(v) Specialize all the above to $K = \mathbb{Q}$, taking into account the important simplifications that we have in this case, and show that, for $v \neq 2$, we have $\left(\frac{a,\,b}{v}\right) = \left(\frac{u}{v}\right)$ (quadratic residue symbol in $F_v^\times$), where $u = (-1)^{\mathrm{v}(a)\mathrm{v}(b)}\, a^{\mathrm{v}(b)}\, b^{-\mathrm{v}(a)}$ (a particular case of the general formula given in 1.6.8 and proved in 7.1.5). Although, because of the product formula it is not really necessary to give a formula for the Hilbert symbol at $v = (2)$, show that:

$$\left(\frac{a,\,b}{v}\right) = \left(-1\right)^{\frac{a'-1}{2}\,\frac{b'-1}{2}} \left(\frac{2}{u'}\right),$$

where $a' = 2^{-\mathrm{v}(a)}\, a$, $b' = 2^{-\mathrm{v}(b)}\, b$, $u' = a^{\mathrm{v}(b)}\, b^{-\mathrm{v}(a)} = a'^{\mathrm{v}(b)}\, b'^{-\mathrm{v}(a)}$, and where:

$$\left(\frac{2}{u'}\right) := \left(-1\right)^{\frac{u'^2-1}{8}},$$

for a 2-adic unit u'.

Answer. The case where a is a square in $K^\times$ being solved trivially by a direct study, we will implicitly assume that a is not a square; in this case, for any solution, z is nonzero.

(i) We write:

$$b = \frac{x^2 - ay^2}{z^2} = \mathrm{N}_{L/K}\Big(\frac{x + y\sqrt{a}}{z}\Big).$$

(ii) An odd place v is ramified if and only if $\mathrm{v}(a) \equiv 1 \bmod (2)$, an even place is ramified if and only if $\frac{a}{t^2} \equiv 1 \bmod (4)$ is not soluble for $t \in K^\times$.

(iii) This is Corollary 1.4.3: if $i_v(b) \in \mathrm{N}_{L_v/K_v}(L_v^\times) = \mathrm{N}_{L_v/K_v}(\pi_w^{\mathbb{Z}} U_w) \subset \pi_v^{f_v \mathbb{Z}} U_v$, we indeed have $\mathrm{v}(b) \equiv 0 \bmod (f_v)$ (including the case $\pi_v = -1$ and $f_v = 2$ corresponding to a complexified real place v), and the converse comes from the fact that, in the unramified case, $U_v \subseteq \mathrm{N}_{L_w/K_v}(U_w)$.

(iv) We have $(a) = (\sqrt{2})(3 + \sqrt{2}) = \mathfrak{p}_2\mathfrak{p}_7$ (the place 2 is ramified in $K/\mathbb{Q}$ and the place 7 is split). It follows that the places of K which are ramified in L/K are the places $\mathfrak{p}_2$ and $\mathfrak{p}_7$.

Since $b = -15(1 + \sqrt{2})$ has even valuation at every place except perhaps at places above ∞, 3, and 5, we must see whether or not $\mathrm{v}(b) \equiv 0 \bmod (f_v)$ is true for these places; we will simply check that if $\mathrm{v}(b) \equiv 1 \bmod (2)$, then $f_v = 1$ (splitting):

- the place ∞ splits in $K/\mathbb{Q}$ into two places v_1, v_2, and we have $i_{v_1}(b) < 0$ (i.e., $\mathrm{v}_1(b) = 1$), $i_{v_2}(b) > 0$ (i.e., $\mathrm{v}_2(b) = 0$). But we have $i_{v_1}(a) > 0$ (v_1 is split in L/K, hence $f_{v_1} = 1$);
- the place 3 is inert in $K/\mathbb{Q}$, hence $F_v = \mathbb{F}_9$, and since $a \equiv -1 \bmod (3)$, the residual image of a is a square (i.e., $i_v(a) \in K_v^{\times 2}$, hence $f_v = 1$);
- the place 5 is also inert, $F_v = \mathbb{F}_{25}$, and we find that $a^3 \equiv 1 \bmod (5)$, hence that a is also a square at 5.

The remaining local norm conditions are for the even place $\mathfrak{p}_2$ (which we can omit), and for $v = \mathfrak{p}_7$. We can compute the Hasse symbol:

$$\Big(\frac{b\,,\,L/K}{\mathfrak{p}_7}\Big) = \Big(\frac{-15(1+\sqrt{2})\,,\,L/K}{\mathfrak{p}_7}\Big).$$

In fact, it defines a regular Hilbert symbol of order 2 which we will learn how to compute in 7.1.5; with this result, we would obtain:

$$\Big(\frac{a,\,b}{\mathfrak{p}_7}\Big) \equiv \Big((-1)^{\mathrm{v}(a)\mathrm{v}(b)}\, a^{\mathrm{v}(b)}\, b^{-\mathrm{v}(a)}\Big)^{\frac{q_v-1}{2}} \equiv b^{-3} \equiv 1 \bmod \mathfrak{p}_7.$$

Let us nonetheless directly compute this Hasse symbol knowing that the conductor of L/K is (because of 1.6.3 which yields $r = 0$ in the computation of the 2-part of the conductor):

$$\mathfrak{f} = \mathfrak{p}_2^{2\times 2+1}\mathfrak{p}_7 = (4)\mathfrak{p}_2\mathfrak{p}_7.$$

A $\mathfrak{p}_7$-associate b' of b must satisfy (see 4.4.3, (γ)):

$$\begin{aligned} b' &\equiv b \bmod \mathfrak{p}_7 \text{ (since here } \mathrm{v}(b)=0), \\ b' &\equiv 1 \bmod (4)\mathfrak{p}_2, \\ i_{v_2}(b') &> 0, \end{aligned}$$

which for example yields $b' = 17 + 12\sqrt{2}$. Here b' happens to be a unit (this was not done on purpose!); the ideal $\mathfrak{b}$ (in the general formula $(b') =: \mathfrak{p}_7^{\mathrm{v}(b)}\mathfrak{b}$) whose Artin symbol we must compute is thus the unit ideal, hence b is indeed a local norm at $\mathfrak{p}_7$.

The product formula tells us that b is also a local norm at $\mathfrak{p}_2$. To double-check this, we want to compute the symbol:

$$\left(\frac{b,\, L/K}{\mathfrak{p}_2}\right) = \left(\frac{-15(1+\sqrt{2}),\, L/K}{\mathfrak{p}_2}\right),$$

which does not have any simple formula. We proceed as above, and we check that $9 + 5\sqrt{2}$ is a $\mathfrak{p}_2$-associate which yields $\mathfrak{b} = \mathfrak{p}_{31}$ split in L/K since the residual image of a is a square.

Thus, the given quadratic form represents 0 in K.

(v) The norm conditions on b at the unramified odd places $v = (\ell)$ (i.e., such that $\mathrm{v}(a) \equiv 0 \bmod (2)$) are therefore (by (iii)) $\left(\frac{a\ell^{-\mathrm{v}(a)}}{v}\right)^{\mathrm{v}(b)} = 1$, including for $v = \infty$; we thus indeed obtain $\left(\frac{u}{v}\right) = 1$ in this case.

There remain the odd places $v = (\ell)$ such that $\mathrm{v}(a) \equiv 1 \bmod (2)$ (i.e., the ramified finite odd places) for which we must check that b is in the norm group of $\mathbb{Q}_\ell(\sqrt{a})/\mathbb{Q}_\ell$. Set $a' := a\ell^{-\mathrm{v}(a)}$ and $b' := b\ell^{-\mathrm{v}(b)}$. We use Exercise 1.6.5 which shows that b is a local norm at ℓ if and only if $b \in N_2$ (resp. N_3) when $\left(\frac{-a'}{\ell}\right) = 1$ (resp. -1) (we see whether $\mathbb{Q}_\ell(\sqrt{a}) = \mathbb{Q}_\ell(\sqrt{-\ell})$ or $\mathbb{Q}_\ell(\sqrt{-\ell\zeta})$). But $b \in N_2$ (resp. N_3) is equivalent to $\left(\frac{b'}{\ell}\right) = 1$ (resp. $(-1)^{\mathrm{v}(b)}$). It follows that b is a local norm at ℓ if and only if:

$$\left(\frac{-a'}{\ell}\right)^{\mathrm{v}(b)} = \left(\frac{b'}{\ell}\right),$$

hence $\left(\frac{u}{\ell}\right) = 1$ with the given formula for u since $\mathrm{v}(a)$ is odd.

The case of the place $v = 2$ can be obtained in analogous way by noting that the Hilbert symbol and the explicit formulas that we must prove are $\mathbb{F}_2$-bilinear and symmetrical in a and b, and thus we are reduced to the computation of the six symbols:

$$\left(\frac{2\,,\,2}{v}\right),\ \left(\frac{2\,,\,q}{v}\right),\ \left(\frac{2\,,\,-1}{v}\right),\ \left(\frac{q\,,\,q}{v}\right),\ \left(\frac{q\,,\,-1}{v}\right),\ \left(\frac{-1\,,\,-1}{v}\right),$$

where q is an odd prime. The computation of these symbols is immediate from the results of 1.6.5. In fact, the general properties of symbols (see 7.1.1 below) show that it is sufficient to compute:

$$\left(\frac{2\,,\,q}{v}\right),\ \left(\frac{q\,,\,-1}{v}\right),\ \left(\frac{-1\,,\,-1}{v}\right).$$ □

6.4.1 Remark. If the quadratic form $X^2 - a\,Y^2 - b\,Z^2$ does not represent 0 in K, it does not represent 0 in a nonzero even number of completions of K (consider for instance $X^2 + Y^2 + Z^2$ in $\mathbb{Q}$). □

§7 Symbols Over Number Fields — Hilbert and Regular Kernels

The notion of symbol, which can be set in the very general context of Milnor's K-theory, where the letter $\mathrm{K} \neq K$ does not denote a field but a functor ([Mil], [Sil]), is directly inspired from the Hilbert symbols that we have already encountered in 1.6.7, 1.6.8, 6.4; hence we will start by giving more completely their properties, obtain from this the general definition of symbols over a field, and ask whether or not we know all the symbols over a number field.

Thanks to this, we will see that class field theory is only an (essential) prelude to a larger theory which involves many invariants which we have not met up to now; as already said, the only unified point of view on these questions is of a cohomological nature ([Schn], [Ta2]), and we refer to the enormous bibliography devoted to higher K-theory.

7.1 Definitions (local Hilbert symbol). Let K be a number field. Then for any place v of K we define the local Hilbert symbol at v:

$$(\,\bullet\,,\,\bullet\,)_v \ : \ K_v^\times \times K_v^\times \longrightarrow \mu(K_v),$$

by:

$$(x\,,\,y)_v := \frac{\left(y\,,\,K_v(\sqrt[m_v]{x}\,)/K_v\right)\ \sqrt[m_v]{x}}{\sqrt[m_v]{x}}$$

for all $x,\ y \in K_v^\times$, where $m_v := |\mu(K_v)|$ and where $(\,\bullet\,,\,K_v(\sqrt[m_v]{x}\,)/K_v)$ is the norm residue symbol for the cyclic extension $K_v(\sqrt[m_v]{x}\,)/K_v$ (see 1.4). □

Note, once and for all, that if v is a complex place at infinity, then $(\,\bullet\,,\,\bullet\,)_v = 1$.

7.1.1 Proposition. *The local Hilbert symbol* $(\,\bullet\,,\,\bullet\,)_v$ *has the following properties:*

(i) it is $\mathbb{Z}$-bilinear, nondegenerate as a bilinear map on $K_v^\times/K_v^{\times m_v} \times K_v^\times/K_v^{\times m_v}$, and continuous as a map on $K_v^\times \times K_v^\times$;

(ii) it satisfies:

$$(x\,,\,1-x)_v = 1 \text{ for all } x \in K_v^\times\backslash\{1\},$$
$$(x\,,\,-x)_v = 1 \text{ for all } x \in K_v^\times,$$
$$(x\,,\,y)_v = (y\,,\,x)_v^{-1} \text{ for all } x,\ y \in K_v^\times \text{ (antisymmetry) ;}$$

(iii) we have $(x\,,\,y)_v = 1$ if and only if y is a norm in $K_v(\sqrt[m_v]{x})/K_v$ (or x is a norm in $K_v(\sqrt[m_v]{y})/K_v$);

(iv) in an extension L/K, for any $w|v$ in L we have, with evident notations:

$$(x^{\frac{m_w}{m_v}}\,,\,y')_w = (x\,,\,\mathrm{N}_{L_w/K_v}(y'))_v$$

for all $x \in K_v^\times$ and $y' \in L_w^\times$;

(v) for any isomorphism τ of K, we have $(\tau x\,,\,\tau y)_{\tau v} = \tau(x\,,\,y)_v$ for all $x,\ y \in K_v^\times$;

(vi) if v is unramified in $K_v(\sqrt[m_v]{x})/K_v$, we have:

$$(x\,,\,y)_v = \left(\frac{(K_v(\sqrt[m_v]{x})/K_v)\ \sqrt[m_v]{x}}{\sqrt[m_v]{x}}\right)^{\mathrm{v}(y)}$$

for all $y \in K_v^\times$, where $(K_v(\sqrt[m_v]{x})/K_v)$ is the Frobenius of $K_v(\sqrt[m_v]{x})/K_v$.

Note. In (v), τv is the place of τK for which $|\tau a|_{\tau v} = |a|_v$ for all $a \in K$; afterwards, by abuse of notation, τ also denotes the isomorphism $\tau : K_v \longrightarrow (\tau K)_{\tau v}$ coming from $K \subset \bigoplus_{v'|\ell} K_{v'} \longrightarrow \tau K \subset \bigoplus_{v'|\ell} (\tau K)_{\tau v'}$, by density (the generalization of the situation of 2.3.1), the embeddings $\bigoplus_{v'|\ell} i_{v'}$ on K and $\bigoplus_{v'|\ell} i_{\tau v'}$ on τK being understood; then it is also the extension by continuity of $i_{\tau v} \circ \tau \circ i_v^{-1}$ on $i_v(K)$. Thanks to this, the expressions τx, τy, and $\tau(x\,,\,y)_v$ make sense.

Proof of the proposition. (i) We have $(x\,,\,yz)_v = (x\,,\,y)_v(x\,,\,z)_v$ because of the multiplicativity of the norm residue symbol and of the isomorphism of Kummer duality (see I.6.1).

We have $(xy\,,\,z)_v = \dfrac{\tau(\sqrt[m_v]{xy})}{\sqrt[m_v]{xy}}$, where $\tau := (z\,,\,K_v(\sqrt[m_v]{xy})/K_v)$; but τ is the restriction to $K_v(\sqrt[m_v]{xy})$ of $\sigma := (z\,,\,K_v(\sqrt[m_v]{x},\ \sqrt[m_v]{y})/K_v)$ by 1.4, (ii), and we have:

$$\frac{\sigma(\sqrt[m_v]{xy})}{\sqrt[m_v]{xy}} = \frac{\sigma(\sqrt[m_v]{x})}{\sqrt[m_v]{x}} \times \frac{\sigma(\sqrt[m_v]{y})}{\sqrt[m_v]{y}},$$

thus giving the result since the restrictions of σ to $K_v(\sqrt[m_v]{x})$ and to $K_v(\sqrt[m_v]{y})$ are the corresponding norm residue symbols of z.

Assume that $(x\,,\,y)_v = 1$ for all $y \in K_v^\times$; the surjectivity of the norm residue symbol implies that $K_v(\sqrt[m_v]{x}) = K_v$, so that $x \in K_v^{\times m_v}$.

If $(x\,,\,y)_v = 1$ for all $x \in K_v^\times$, we have $(y,\,K_v(\sqrt[m_v]{K_v^\times})/K_v) = 1$ hence y is a norm in $K_v(\sqrt[m_v]{K_v^\times})$, hence $y \in K_v^{\times m_v}$ (see 1.6.6).

Continuity comes from the fact that, if u and u' are sufficiently close to 1 in U_v, then u and u' are m_vth powers in $K_v^\times$, and we have trivially $(xu\,,\,yu')_v = (x\,,\,y)_v$.

To prove some of the above properties, we may use the antisymmetry that we will prove in full generality in 7.2.1.

(ii) Let us show that $1-x$ is the norm of an element of $M := K_v(\sqrt[m_v]{x}\,)$. If $d|m_v$ is the degree of M/K_v, Kummer theory shows that there exists $t \in K_v^\times$ such that $x = t^{\frac{m_v}{d}}$, and we have $M = K_v(\sqrt[d]{t}\,)$. Since for all $\xi \in \mu(K_v)$ we have:

$$\mathrm{N}_{M/K_v}(1-\xi\sqrt[d]{t}\,) = 1-\xi^d t,$$

then denoting by ζ_v a generator of $\mu(K_v)$, it follows that:

$$\mathrm{N}_{M/K_v}\Big(\prod_{i=1}^{\frac{m_v}{d}}(1-\zeta_v^i\sqrt[d]{t}\,)\Big) = \prod_{i=1}^{\frac{m_v}{d}}(1-\zeta_v^{di}t) = 1-t^{\frac{m_v}{d}} = 1-x.$$

The relation $(x\,,\,-x)_v = 1$ as well as antisymmetry then follow from this (see 7.2.1).

Facts (iii), (iv), (v), and (vi) follow trivially from the corresponding properties 1.4 of the norm residue symbol. □

7.1.2 Remark. If m is a divisor of m_v, the symbol $(\,\bullet\,,\,\bullet\,)_v^{(m)}$ defined by:

$$(x\,,\,y) \in K_v^\times \times K_v^\times \longmapsto \frac{(y\,,\,K_v(\sqrt[m]{x}\,)/K_v)\ \sqrt[m]{x}}{\sqrt[m]{x}}$$

for all $x,\ y \in K_v^\times$, is equal to $(\,\bullet\,,\,\bullet\,)_v^{\frac{m_v}{m}}$ since $\sqrt[m_v]{x^{\frac{m_v}{m}}} = \sqrt[m]{x}$. By abuse of language, it is called the local Hilbert symbol of order m. In common usage, we write simply $(x\,,\,y)_v$ instead of $(x\,,\,y)_v^{(m)}$, the context being in general sufficient to give the order of the symbols under study (for example $m=2$ for the usual quadratic Hilbert symbol). The Hilbert symbol defined in 7.1 has maximal order with an evident meaning. □

Before going any further, it is necessary to introduce the regular Hilbert symbol (which has the advantage of being explicit), and for this we give some notations and definitions.

7.1.3 Notations. (i) Let $v \in Pl_0$ and let ℓ be the residue characteristic of v. We know (see I.3.1.1) that we have the decomposition:

$$\mu(K_v) = \mu_{q_v-1} \oplus \mu_\ell(K_v),$$

where $q_v := |F_v|$ is a power of ℓ and where $\mu_\ell(K_v) = \mathrm{tor}_{\mathbb{Z}}(U_v^1)$ is also of order a power of ℓ, which we will write here in the more descriptive form:

$$\mu(K_v) =: \mu(K_v)^{\mathrm{reg}} \oplus \mu(K_v)^1.$$

(ii) For $v \in Pl_\infty^{\mathrm{r}}$, we have $\mu(K_v) = \mu_2$ and we set, in accordance with the fact that $U_v^1 = \mathbb{R}^{\times +}$:

$$\mu(K_v)^{\mathrm{reg}} := \mu_2, \ \ \mu(K_v)^1 := 1.$$

(iii) As is easily checked, $\mu(K_v)^1 = 1$ for almost all places of K (indeed, $\mu(K_v)^1 \neq 1$ implies that K_v contains μ_ℓ, hence $\mathbb{Q}_\ell(\mu_\ell)$, which implies that $\ell - 1 \leq [K : \mathbb{Q}]$). The places v for which $\mu(K_v)^1 \neq 1$ will be called the irregular places of K. □

7.1.4 Definitions (regular Hilbert symbol). We define the regular or tame Hilbert symbol at a noncomplex place v as the Hilbert symbol of order $|\mu(K_v)^{\text{reg}}|$, in other words as the symbol $(\bullet\,,\,\bullet)_v^{\text{reg}} := (\bullet\,,\,\bullet)_v^{m_v^1}$, where $m_v^1 := |\mu(K_v)^1|$ (equal to $\frac{m_v}{q_v-1}$ in the finite case). □

7.1.5 Proposition (regular Hilbert symbol formula). *For any $x,\ y \in K_v^\times$, $(x\,,\,y)_v^{\text{reg}}$ is the component on $\mu(K_v)^{\text{reg}}$ of:*

$$(-1)^{\text{v}(x)\text{v}(y)} x^{\text{v}(y)} y^{-\text{v}(x)}.$$

Note. For computations and when v is finite, it is equivalent to take the residual image of the above expression since $\mu(K_v)^{\text{reg}} = \mu_{q_v-1} \simeq F_v^\times$ (canonically), and for a real infinite place v, it is the sign of this same expression, equal to $(-1)^{\text{v}(x)\text{v}(y)}$.

Proof of the proposition. The case of an infinite place v being trivial directly, we assume that v is a finite place. We have:

$$(x\,,\,y)_v^{m_v^1} = (x^{m_v^1}\,,\,y)_v = \frac{(y\,,\,K_v(\sqrt[q_v-1]{x}\,)/K_v)\ \sqrt[q_v-1]{x}}{\sqrt[q_v-1]{x}}.$$

To identify this Hilbert symbol, it is sufficient to compute (using bilinearity and antisymmetry):

$$(u\,,\,u')_v^{m_v^1},\quad (u\,,\,\pi)_v^{m_v^1},\quad (\pi\,,\,\pi)_v^{m_v^1},$$

for $u,\ u' \in U_v$ and for a uniformizer π of K_v:

- by 7.1.1, (vi), we have $(u\,,\,u')_v^{m_v^1} = 1$ since $K_v(\sqrt[q_v-1]{u}\,)/K_v$ is unramified;
- in the same way, $(u\,,\,\pi)_v^{m_v^1} = \frac{\sigma(\sqrt[q_v-1]{u})}{\sqrt[q_v-1]{u}}$, where σ is the Frobenius of

$K_v(\sqrt[q_v-1]{u})/K_v$; it follows that we have:

$$\frac{\sigma(\sqrt[q_v-1]{u})}{\sqrt[q_v-1]{u}} \equiv \frac{(\sqrt[q_v-1]{u})^{q_v}}{\sqrt[q_v-1]{u}} \equiv u \bmod (\pi),$$

which shows that $(u\,,\,\pi)_v^{m_v^1}$ is the component of u on μ_{q_v-1};

- by 7.1.1, (ii), we have $(\pi\,,\,-\pi)_v = 1$, hence $(\pi\,,\,\pi)_v^{m_v^1} = (-1\,,\,\pi)_v^{m_v^1}$, so

that we are reduced to the preceding situation with $u = -1$.

By "gluing the pieces together", we obtain the desired formula. □

Note. It is also possible to use symbols having an order dividing $q_v - 1$ (v finite); this is for instance the case for quadratic Hilbert symbols which are given, for any odd place v, by the residual image of $((-1)^{\mathrm{v}(a)\mathrm{v}(b)}\, a^{\mathrm{v}(b)}\, b^{-\mathrm{v}(a)})^{\frac{q_v-1}{2}}$. By abuse of language, we will also say that they are regular symbols.

Note that the basic symbols $(\bullet\,,\,\bullet)_v^{\mathrm{reg}}$ and $(\bullet\,,\,\bullet)_v$ coincide for almost every place (the regular places).

We will see in 7.5 how to compute in practice the irregular Hilbert symbols, which are finite in number; the numerical computations done in 6.4 for some quadratic symbols are illustrations of this.

The above study motivates the general definition of a symbol on a field k, with values in an abelian group A.

7.2 Definition (symbols on a field). Let k be a field, and let A be an abelian group. A symbol (on k, with values in A) is a $\mathbb{Z}$-bilinear map:

$$(\bullet\,,\,\bullet) \;:\; k^\times \times k^\times \longrightarrow A,$$

such that $(x\,,\,y) = 1$ for all $x,\ y \in k^\times \backslash \{1\}$ such that $x + y = 1$. □

If we consider the quotient group:

$$\mathrm{K}_2(k) := k^\times \otimes_{\mathbb{Z}} k^\times / \langle\, x \otimes y\,;\ x,\ y \in k^\times \backslash \{1\},\ x + y = 1\,\rangle$$

(second Milnor's K-group of the field k), it is immediate to check that this object satisfies the following universal property. For any symbol:

$$(\bullet\,,\,\bullet) \;:\; k^\times \times k^\times \longrightarrow A,$$

there exists a unique group homomorphism $h : \mathrm{K}_2(k) \longrightarrow A$, such that the following diagram commutes:

$$\begin{array}{ccc} k^\times \times k^\times & \xrightarrow{(\bullet\,,\,\bullet)} & A \\ {\scriptstyle \{\bullet\,,\,\bullet\}}\downarrow & \nearrow{\scriptstyle h} & \\ \mathrm{K}_2(k) & & \end{array}$$

the vertical arrow being the canonical map sending the pair $(x,\ y)$ to the image of $x \otimes y$; this map is not necessarily surjective, but its image generates $\mathrm{K}_2(k)$.

We have denoted $\{x\,,\,y\}$ the canonical image of $x \otimes y$ in $\mathrm{K}_2(k)$; by construction, it is clear that $\{\bullet\,,\,\bullet\}$ is itself a symbol with values in $\mathrm{K}_2(k)$.

7.2.1 Exercise. Show that any symbol $(\bullet\,,\,\bullet) \;:\; k^\times \times k^\times \longrightarrow A$ satisfies $(x\,,\,-x) = 1$ for all $x \in k^\times$. Deduce that any symbol is antisymmetric.

Answer. If $x = 1$, we have $(1\,,\,-1) = 1$ by linearity on the first component. Thus we may assume that $x \neq 1$. We then have the equalities:

$$1 = \Big(\frac{1}{x}\,,\,1 - \frac{1}{x}\Big) = \Big(x\,,\,1 - \frac{1}{x}\Big)^{-1} = \Big(x\,,\,\frac{1-x}{-x}\Big)^{-1}$$
$$= \Big(x\,,\,1 - x\Big)^{-1}\Big(x\,,\,\frac{1}{-x}\Big)^{-1} = (x\,,\,-x).$$

We then apply this property to the product xy, and we obtain:

$$1 = (xy\,,\,-xy) = (x\,,\,-xy)(y\,,\,-xy)$$
$$= (x\,,\,-x)(x\,,\,y)(y\,,\,x)(y\,,\,-y) = (x\,,\,y)(y\,,\,x),$$

so the antisymmetry follows. □

7.2.2 Examples. (i) If $k = K_v$ (the completion of the number field K at a finite place v), the local Hilbert symbol $(\bullet,\,\bullet)_v$ defines the group homomorphism:

$$h_v : \mathrm{K}_2(K_v) \longrightarrow \mu(K_v),$$

which is in fact an isomorphism (a result of Moore [e, Ko3, Ch. 2, § 6.6]).

(ii) If $k = K$, we can consider the symbol obtained by globalizing the family of local Hilbert symbols in the following way:

$$\begin{aligned}(\bullet,\,\bullet)\ :\ K^\times \times K^\times &\longrightarrow \bigoplus_{v\in Pl^{\mathrm{nc}}} \mu(K_v),\\ (x\,,\,y) &\longmapsto ((i_v(x)\,,\,i_v(y))_v)_{v\in Pl^{\mathrm{nc}}}\end{aligned}$$

where $Pl^{\mathrm{nc}} := Pl \setminus Pl_\infty^{\mathrm{c}}$ is the set of noncomplex places of K. This symbol indeed takes its values in the direct sum since for any place v such that $\mathrm{v}(x) = \mathrm{v}(y) = 0$, $i_v(y)$ is a unit and the extension $K_v(\sqrt[m_v]{i_v(x)}\,)/K_v$ is unramified except perhaps if the residue characteristic ℓ of v divides m_v, which happens only for a finite number of places (the irregular places, i.e., those for which K_v contains μ_ℓ; see 7.1.3, (iii)). For convenience, this symbol will be called the global Hilbert symbol. It defines the homomorphism:

$$h : \mathrm{K}_2(K) \longrightarrow \bigoplus_{v\in Pl^{\mathrm{nc}}} \mu(K_v).$$

(iii) We can also define the global regular Hilbert symbol, from the local regular Hilbert symbols (see 7.1.4, 7.1.5):

$$(\bullet,\,\bullet)^{\mathrm{reg}}\ :\ K^\times \times K^\times \longrightarrow \bigoplus_{v\in Pl^{\mathrm{nc}}} \mu(K_v)^{\mathrm{reg}}$$

which sends any pair $(x,\ y)$ to $((i_v(x)\,,\,i_v(y))_v^{\mathrm{reg}})_{v\in Pl^{\mathrm{nc}}}$. Recall that for $v \in Pl_0$:

$$(\bullet,\,\bullet)_v^{\mathrm{reg}} := (\bullet,\,\bullet)_v^{m_v^1},\ \ \text{where } m_v^1 := |\mu(K_v)^1| = \frac{m_v}{q_v - 1}\cdot$$

The resulting homomorphism $\mathrm{K}_2(K) \longrightarrow \bigoplus_{v \in Pl^{\mathrm{nc}}} \mu(K_v)^{\mathrm{reg}}$ is denoted h^{reg}. □

7.3 Theorem. *For all $x,\ y \in K^\times$, we have in $\mu(K)$ the product formula:*

$$\prod_{v \in Pl^{\mathrm{nc}}} i_v^{-1}\Big((i_v(x)\,,\, i_v(y))_v^{\frac{m_v}{m}} \Big) = 1,$$

where $m := |\mu(K)|$, $m_v := |\mu(K_v)|$.

Proof. We have $(i_v(x)\,,\, i_v(y))_v^{\frac{m_v}{m}} = (i_v(x^{\frac{m_v}{m}})\,,\, i_v(y))_v$, and the norm residue symbol:

$$\Big(i_v(y)\,,\, K_v\Big(\sqrt[m_v]{i_v(x^{\frac{m_v}{m}})}\Big)\Big/K_v\Big) = \Big(i_v(y)\,,\, K_v\Big(\sqrt[m]{i_v(x)}\Big)\Big/K_v\Big)$$

being the norm residue symbol in L_v/K_v for $L := K(\sqrt[m]{x})$, which is abelian over K, in terms of Hasse symbols 3.1.2 we obtain:

$$i_v^{-1}\Big((i_v(x)\,,\, i_v(y))_v^{\frac{m_v}{m}} \Big) = \Big(\frac{y\,,\, K(\sqrt[m]{x})/K}{v}\Big) \sqrt[m]{x} \Big/ \sqrt[m]{x} \in \mu(K)\ ;$$

hence, by the isomorphism of Kummer duality I.6.1, the product formula (in $G := \mathrm{Gal}(K(\sqrt[m]{x})/K)$):

$$\prod_v \Big(\frac{y\,,\, K(\sqrt[m]{x})/K}{v}\Big) = 1$$

is transformed into the analogous formula on the $i_v^{-1}\Big((i_v(x)\,,\, i_v(y))_v^{\frac{m_v}{m}} \Big)$ (in $\mu(K)$). □

7.3.1 Definitions (Hilbert symbols of K — product formula). If we set:

$$\Big(\frac{x\,,\, y}{v}\Big) := i_v^{-1}\Big((i_v(x)\,,\, i_v(y))_v^{\frac{m_v}{m}} \Big) = \Big(\frac{y\,,\, K(\sqrt[m]{x})/K}{v}\Big) \sqrt[m]{x} \Big/ \sqrt[m]{x} \in \mu(K)$$

for all $x,\ y \in K^\times$, then this defines $\Big(\frac{\bullet\,,\,\bullet}{v}\Big)$, which we call the v-Hilbert symbol of K (defined on $K^\times \times K^\times$); its order m is maximal. We then have the simpler expression:

$$\prod_{v \in Pl^{\mathrm{nc}}} \Big(\frac{x\,,\, y}{v}\Big) = 1, \text{ for all } x,\ y \in K^\times,$$

which is called the product formula for Hilbert symbols on K. □

Note. Do not confuse the symbol $(\bullet,\ \bullet)_v$ defined on K_v with values in $\mu(K_v)$, with the symbol $\Big(\frac{\bullet\,,\,\bullet}{v}\Big)$ defined on K with values in $\mu(K)$. We have:

$$i_v \circ \left(\frac{\bullet\,,\,\bullet}{v}\right) = \left(\bullet\,,\,\bullet\right)_v^{\frac{m_v}{m}} \circ i_v \quad \text{on } K^\times \times K^\times.$$

7.3.2 Exercise (prescribed Hilbert symbols). Let K be a number field and let $m := |\mu(K)|$.

(i) For $i = 1, \ldots, r$, let a_i be fixed elements of $K^\times$, and let $(\zeta_{i,v})_v$ with $\zeta_{i,v} \in \mu(K)$ be r families with finite support in Pl^{nc}; assume that for each $v \in Pl^{\mathrm{nc}}$ there exists $x(v) \in K^\times$ such that:

$$\left(\frac{a_i\,,\,x(v)}{v}\right) = \zeta_{i,v}, \quad i = 1, \ldots, r, \text{[34]}$$

and assume that:

$$\prod_v \zeta_{i,v} = 1, \quad i = 1, \ldots, r.$$

Show that there exists $x \in K^\times$ such that for all $v \in Pl^{\mathrm{nc}}$:

$$\left(\frac{a_i\,,\,x}{v}\right) = \zeta_{i,v}, \quad i = 1, \ldots, r.$$

(ii) Deduce the Hasse–Minkowski theorem for quadratic forms in four variables over K (hint: let $a\,Y^2 + b\,Z^2 - (c\,T^2 + d\,U^2)$, a, b, c, $d \in K^\times$, be such a quadratic form; check that there exists $x \in K^\times$ for which $x\,X^2 - a\,Y^2 - b\,Z^2$ and $x\,X^2 - c\,T^2 - d\,U^2$ represent 0 in K).

Answer. (i) Let $A := \langle\, a_1, \ldots, a_r \,\rangle$, $L := K(\sqrt[m]{A})$, and $G := \mathrm{Gal}(L/K)$. Denote by $s_v \in G$ the Hasse symbol $\left(\frac{x(v)\,,\,L/K}{v}\right)$; we thus have $s_v \in D_v := D_v(L/K)$. We have:

$$s_v(\sqrt[m]{a_i}) = \left(\frac{x(v)\,,\,K(\sqrt[m]{a_i})/K}{v}\right) \sqrt[m]{a_i} = \zeta_{i,v}\,\sqrt[m]{a_i}$$

for all v and $i = 1, \ldots, r$, so that $(s_v)_v$ has finite support. Set $s := \prod_v s_v$; we easily obtain $s(\sqrt[m]{a_i}) = \left(\prod_v \zeta_{i,v}\right)\sqrt[m]{a_i} = \sqrt[m]{a_i}$, $i = 1, \ldots, r$, hence $s = 1$. We can thus apply Theorem 3.4.4 on the converse of the product formula, and so there exists $x \in K^\times$ such that:

$$\left(\frac{x\,,\,L/K}{v}\right) = s_v$$

for all $v \in Pl^{\mathrm{nc}}$. By construction we immediately have:

$$\left(\frac{a_i\,,\,x}{v}\right) = s_v(\sqrt[m]{a_i})/\sqrt[m]{a_i} = \zeta_{i,v}, \quad i = 1, \ldots, r$$

[34] A necessary condition is that the order of $\zeta_{i,v}$ must be a divisor of that of the decomposition group of v in $K(\sqrt[m]{a_i})/K$ since this decomposition group is an image under the Hasse symbol and that $K(\sqrt[m]{a_i})/K$ is cyclic; it is sufficient only for $r = 1$: for $K = \mathbb{Q}(\sqrt{-1})$, $v|2$, $a_1 = 2$, $\zeta_{1,v} = -1$, $\zeta_{2,v} = 1$, $x(v)$ does not exist.

for all $v \in Pl^{\mathrm{nc}}$.

(ii) Consider the quadratic form:

$$(q) \quad aY^2 + bZ^2 - (cT^2 + dU^2), \quad a,\ b,\ c,\ d \in K^\times,$$

and assume that it represents 0 in all the K_v. For all v there exist y_v, z_v, t_v, $u_v \in K_v$ (not all zero) such that:

$$x_v := ay_v^2 + bz_v^2 = ct_v^2 + du_v^2,$$

in K_v (we have omitted the embeddings i_v).

If $x_v = 0$, it is easy to see that the forms aY^2+bZ^2 and cT^2+dU^2 represent any element of K_v, so we can find a solution y'_v, z'_v, t'_v, u'_v which yields $x_v = 1$, which we will now assume (for instance, since y_v and t_v cannot be equal to zero, we take $y'_v := (a^{-1}+1)/2$, $t'_v := (c^{-1}+1)/2$, $z'_v := z_v(a^{-1}-1)/2y_v$, and $u'_v := u_v(c^{-1}-1)/2t_v$).

Since $ax_v = (ay_v)^2 + abz_v^2$, we have (see 6.4):

$$1 = (ax_v,\ -ab)_v = (a,\ -a)_v(a,\ b)_v(x_v,\ -ab)_v,$$

so that we obtain:

$$(-ab,\ x_v)_v = (a,\ b)_v =: \left(\frac{a,\ b}{v}\right)$$

for all v, and similarly:

$$(-cd,\ x_v)_v = (c,\ d)_v =: \left(\frac{c,\ d}{v}\right).$$

For all v set:

$$\zeta_{1,v} := (-ab,\ x_v)_v, \quad \zeta_{2,v} := (-cd,\ x_v)_v,$$

which satisfies the first assumption of (i) for the $\zeta_{i,v}$, $i = 1,\ 2$ (the fact that $x_v \in K_v^\times$ does not matter since there also exists $x(v) \in K^\times$ by approximation at v). Since for all v:

$$(-ab,\ x_v)_v = \left(\frac{a,\ b}{v}\right), \quad (-cd,\ x_v)_v = \left(\frac{c,\ d}{v}\right),$$

by the product formula we have:

$$\prod_v \zeta_{1,v} = \prod_v \left(\frac{a,\ b}{v}\right) = 1, \quad \prod_v \zeta_{2,v} = \prod_v \left(\frac{c,\ d}{v}\right) = 1.$$

It follows that there exists $x \in K^\times$ such that:

$$\left(\frac{-ab,\ x}{v}\right) = \left(\frac{a,\ b}{v}\right) = \left(\frac{-ab,\ a}{v}\right)^{-1}, \quad \left(\frac{-cd,\ x}{v}\right) = \left(\frac{c,\ d}{v}\right) = \left(\frac{-cd,\ c}{v}\right)^{-1}$$

for all v, which can also be written:

$$\left(\frac{-ab,\, ax}{v}\right) = 1, \quad \left(\frac{-cd,\, cx}{v}\right) = 1$$

for all v. The forms $xX^2 - aY^2 - bZ^2$ and $xX^2 - cT^2 - dU^2$ thus represent 0 in all the K_v and hence in K. The result follows by equality of the two expressions for x which one obtains from this.

Beware that we are not allowed to exclude a place in the statement; for example, for $K = \mathbb{Q}$:

$$Y^2 + Z^2 + 3T^2 + 5U^2$$

represents 0 in all the completions of $\mathbb{Q}$ except in $\mathbb{R}$. Indeed, for $\ell \neq 2$ the given form represents 0 in $\mathbb{Q}_\ell$ (in each case, we put a suitable variable T or U equal to zero, and write that -5 or -3 is a norm in $\mathbb{Q}_\ell(\sqrt{-1})/\mathbb{Q}_\ell$ since it is then a unit in an unramified local extension). For $\ell = 2$, we check that -3 is a norm in $\mathbb{Q}_2(\sqrt{-1})/\mathbb{Q}_2$. □

7.4 Power Residue Symbol, nth Power Reciprocity Law. Let K be a number field, m the order of $\mu(K)$ and n a divisor of m. We have just defined the v-Hilbert symbol of K:

$$\begin{aligned} \left(\frac{\bullet\,,\,\bullet}{v}\right) \ :\ K^\times \times K^\times &\longrightarrow \mu(K) \\ (x\,,\,y) &\longmapsto i_v^{-1}\Big(\big(i_v(x)\,,\,i_v(y)\big)_v^{\frac{m_v}{m}}\Big) \end{aligned}$$

which is essentially the same as the local Hilbert symbol of order m restricted to $i_v(K)$, and which is also given by the action of the Hasse symbol $\left(\frac{y\,,\,K(\sqrt[m]{x})/K}{v}\right)$ on $\sqrt[m]{x}$. We will be using the Hilbert symbol of order n:

$$\left(\frac{\bullet\,,\,\bullet}{v}\right)_n := \left(\frac{\bullet\,,\,\bullet}{v}\right)^{\frac{m}{n}}.$$

7.4.1 Definition. For $x \in K^\times$, we denote by R_x the set of finite places of K which are ramified in $K(\sqrt[n]{x})/K$ (R_x is contained in the set of places v such that $\mathrm{v}(x) \not\equiv 0 \bmod (n)$, or dividing n). For $v \notin R_x$ we define via the Frobenius symbol:

$$\left(\frac{x}{v}\right)_n := \left(\frac{K(\sqrt[n]{x})/K}{v}\right)\sqrt[n]{x} \,\big/ \sqrt[n]{x}. \qquad \square$$

This defines an nth power residue symbol (including when $v \in Pl_\infty^{\mathrm{r}}$) since, if $v \notin R_x$, we have:

$$\left(\frac{x}{v}\right)_n = 1 \text{ if and only if } i_v(x) \in K_v^{\times n}.$$

7.4.2 Remarks. (i) If $v \notin R_x$ is a finite place, we also set $\left(\frac{x}{v}\right)_n =: \left(\frac{x}{\mathfrak{p}_v}\right)_n$, so that we can define by multiplicativity:

$$\left(\frac{x}{\mathfrak{b}}\right)_n := \prod_{v \in Pl_0} \left(\frac{x}{v}\right)_n^{\mathrm{v}(\mathfrak{b})} \quad \text{for all } \mathfrak{b} \in I_{R_x}.$$

Hence, in terms of Artin symbols, we have (since $\mu_n \subset K$):

$$\left(\frac{x}{\mathfrak{b}}\right)_n = \left(\frac{K(\sqrt[n]{x})/K}{\mathfrak{b}}\right) \sqrt[n]{x} \,/\, \sqrt[n]{x}.$$

(ii) We can also define an idelic version $\left(\dfrac{x}{(y_v)_v}\right)_n$, in a completely clear and analogous way, so that, for all $y \in K^\times$ prime to R_x (identifying $i(y)$ with y) we get:

$$\left(\frac{x}{y}\right)_n := \prod_{v \in Pl^{\mathrm{nc}}} \left(\frac{x}{v}\right)_n^{\mathrm{v}(y)},$$

which thus involves the real infinite places. □

Note that in the literature, $\left(\dfrac{x}{y}\right)_n$ usually means $\left(\dfrac{x}{(y)}\right)_n$.

7.4.3 Proposition. *We have the following functorial properties of the nth power residue symbol, where x, y, are elements of $K^\times$:*

(i) $\left(\dfrac{x}{\mathfrak{b}}\right)_n \left(\dfrac{y}{\mathfrak{b}}\right)_n = \left(\dfrac{x\,y}{\mathfrak{b}}\right)_n$, *if* $\mathfrak{b} \in I_{R_x} \cap I_{R_y}$;

(ii) $\left(\dfrac{x}{\mathfrak{b}}\right)_n \left(\dfrac{x}{\mathfrak{c}}\right)_n = \left(\dfrac{x}{\mathfrak{b}\,\mathfrak{c}}\right)_n$, *if* $\mathfrak{b},\ \mathfrak{c} \in I_{R_x}$;

(iii) for any isomorphism τ of K we have $\left(\dfrac{\tau(x)}{\tau(\mathfrak{b})}\right)_n = \tau\left(\dfrac{x}{\mathfrak{b}}\right)_n$, *if* $\mathfrak{b} \in I_{R_x}$;

(iv) for any divisor d of n we have $\left(\dfrac{x}{\mathfrak{b}}\right)_n^d = \left(\dfrac{x}{\mathfrak{b}}\right)_{\frac{n}{d}}$, *if* $\mathfrak{b} \in I_{R_x}$;

(v) let L be a finite extension of K; we have $\left(\dfrac{x}{\mathfrak{b}'}\right)_{L,n} = \left(\dfrac{x}{\mathrm{N}_{L/K}(\mathfrak{b}')}\right)_n$, *if* $\mathfrak{b}' \in I_{L,R_x}$;

(vi) for any prime ideal $\mathfrak{p}$, prime to x and n, we have $\left(\dfrac{x}{\mathfrak{p}}\right)_n \equiv x^{\frac{\mathrm{N}\mathfrak{p}-1}{n}} \bmod \mathfrak{p}$;

(vii) let $N_x \subset I_{R_x}$ be the norm group corresponding to $K(\sqrt[n]{x})/K$; then, for $\mathfrak{b},\ \mathfrak{c} \in I_{R_x}$, we have $\left(\dfrac{x}{\mathfrak{b}}\right)_n = \left(\dfrac{x}{\mathfrak{c}}\right)_n$ *if and only if* $\mathfrak{b}\,\mathfrak{c}^{-1} \in N_x$;

(viii) for $v \notin R_x$, we have $\left(\dfrac{x\,,\,y}{v}\right)_n = \left(\dfrac{x}{\mathfrak{p}_v}\right)_n^{\mathrm{v}(y)}$ *for all* $y \in K^\times$.

Proof. Use properties 4.5 of the Artin map and/or properties 7.1.1 of the local Hilbert symbol, noting that $i_v\!\left(\left(\dfrac{x}{v}\right)_n\right) = (x,\,\pi_v)_v$ if $v \notin R_x$. □

Then we can state:

7.4.4 Theorem. *The nth power reciprocity law (n dividing $|\mu(K)|$) is given by the relation:*

$$\left(\frac{y}{x}\right)_n \left(\frac{x}{y}\right)_n^{-1} = \prod_{v|n} \left(\frac{x\,,\,y}{v}\right)_n$$

for all $x,\ y \in K^\times$, x and y coprime (i.e., for any place v, we have $\mathrm{v}(x) = 0$ or $\mathrm{v}(y) = 0$, including the case $v|\infty$) and prime to n.

Proof. We compute the left hand side by using the definition of the symbols $\left(\frac{\bullet}{\bullet}\right)_n$ and by noting that the products can be restricted to the places not dividing n (because of the relations $\mathrm{v}(x) = \mathrm{v}(y) = 0$ for $v|n$):

$$\left(\frac{y}{x}\right)_n \left(\frac{x}{y}\right)_n^{-1} = \prod_{v\nmid n} \left(\frac{y}{v}\right)_n^{\mathrm{v}(x)} \prod_{v\nmid n} \left(\frac{x}{v}\right)_n^{-\mathrm{v}(y)};$$

then, treating the cases $\mathrm{v}(x) \neq 0$ and $\mathrm{v}(y) \neq 0$ for $v \nmid n$ separately and coming back to Hilbert symbols because of 7.4.3, (viii), this yields:

$$\left(\frac{y}{x}\right)_n \left(\frac{x}{y}\right)_n^{-1} = \prod_{\substack{v\nmid n \\ \mathrm{v}(x)\neq 0}} \left(\frac{y\,,\,x}{v}\right)_n \prod_{\substack{v\nmid n \\ \mathrm{v}(y)\neq 0}} \left(\frac{x\,,\,y}{v}\right)_n^{-1} = \prod_{\substack{v\nmid n \\ \mathrm{v}(xy)\neq 0}} \left(\frac{y\,,\,x}{v}\right)_n = \prod_{v\nmid n} \left(\frac{y\,,\,x}{v}\right)_n$$

since $\left(\frac{y\,,\,x}{v}\right)_n = 1$ if $\mathrm{v}(xy) = 0$, $v \nmid n$; the theorem follows by using the product formula. □

The nth power reciprocity law for the number field K is thus explicit as soon as the right hand side is computed. Because of 7.5 below and the continuity of the Hilbert symbols, using suitable representatives $x,\ y \in K^\times$ of the finite groups $U_v/(U_v)^n$, $v|n$, we only need a *finite number* of numerical computations (once for all). We thus obtain in particular the quadratic reciprocity law of Jacobi ($K = \mathbb{Q}$, $n = 2$; the signed form that we have obtained being a slight generalization) by checking that:

$$\left(\frac{y}{x}\right)\left(\frac{x}{y}\right) = \left(\frac{x\,,\,y}{2}\right)_2 = \left(-1\right)^{\frac{x-1}{2}\frac{y-1}{2}}$$

for all rational $x,\ y$, coprime, odd, not both negative (see 6.4, (v)).

7.4.5 Remark. The formula is false when x and y are both negative; for example we have:

$$\left(\frac{-3}{-5}\right)\left(\frac{-5}{-3}\right) = \left(\frac{-3}{\infty}\right)\left(\frac{-3}{5}\right)\left(\frac{-5}{\infty}\right)\left(\frac{-5}{3}\right) = \left(\frac{15}{\infty}\right)\left(\frac{-3}{5}\right)\left(\frac{-5}{3}\right) = -1,$$

although $\frac{x-1}{2}\frac{y-1}{2} = 6$ which would give the value $+1$; in the general case, we must multiply the right hand side of the formula by:

$$(-1)^{\frac{\mathrm{sgn}(x)-1}{2}\frac{\mathrm{sgn}(y)-1}{2}} = (-1)^{\mathrm{v}(x)\mathrm{v}(y)},$$

where v is the valuation corresponding to $v = \infty$. □

7.4.6 Proposition. *If $z \in K^\times$ is such that $\mathrm{v}(z) = 0$ for any place v not dividing n, and if $x \in K^\times$ is prime to n, we have the supplementary formula:*

$$\left(\frac{z}{x}\right)_n = \prod_{v|n} \left(\frac{x\,,\,z}{v}\right)_n.$$

Proof. We have:

$$\left(\frac{z}{x}\right)_n = \prod_{v} \left(\frac{z}{v}\right)_n^{\mathrm{v}(x)} = \prod_{v\nmid n} \left(\frac{z}{v}\right)_n^{\mathrm{v}(x)} = \prod_{v\nmid n} \left(\frac{z\,,\,x}{v}\right)_n = \prod_{v|n} \left(\frac{x\,,\,z}{v}\right)_n. \quad \square$$

This can be applied to the quadratic case to prove that:

$$\left(\frac{2}{x}\right)_2 = \left(\frac{x\,,\,2}{2}\right)_2 = (-1)^{\frac{x^2-1}{8}},$$

for any odd rational x.

For additional material on these reciprocity laws aspects, see [a, Ko1], [f, Lem], [Kub2], [KubO], [Wy].

7.5 Computation of a Hilbert Symbol by Global Means. As already mentioned in 1.6.8, assume that we want to compute a local Hilbert symbol $(x\,,\,y)$ of order n, with $x,\ y \in k^\times$, where k is a finite extension of $\mathbb{Q}_\ell$ containing the group μ_n of nth roots of unity. The method consists in looking for a number field K containing μ_n, and such that $K_v = k$ for some place $v|\ell$ of K.

In general it is an irregular symbol (i.e., $\ell|n$), and by localizing the problem, we are reduced to the case where $n = \ell^h$ for $h \geq 1$. Then, if $(K,\ v|\ell)$ is a solution, we consider here K as a subfield of k and, by density, we are reduced to the case where $x,\ y \in K^\times$ (this only involves the properties of $k^\times$ deduced from the knowledge of $k^\times/k^{\times\ell^h}$); we then have $(x\,,\,y) = \left(\frac{x\,,\,y}{v}\right)$.

We can even construct K containing μ_{ℓ^h} and having a *unique* ℓ-adic place v, in which case the Hilbert symbol $\left(\frac{x\,,\,y}{v}\right)$ can be computed, because of the product formula, by using only regular symbols: if $k =: \mathbb{Q}_\ell(\mu_{\ell^h})(\alpha)$, $\alpha \in \overline{\mathbb{Q}}_\ell$, of degree d over $\mathbb{Q}_\ell(\mu_{\ell^h})$, we may assume that $\alpha \in \overline{\mathbb{Q}}$ and that it has degree d over $\mathbb{Q}(\mu_{\ell^h})$ (Krasner's lemma [b, Rob, Ch. 3, § 1.5]), hence $K := \mathbb{Q}(\mu_{\ell^h})(\alpha)$ is a suitable field since $v|\ell$ is split neither in $K/\mathbb{Q}(\mu_{\ell^h})$ (by our choice of α) nor in $\mathbb{Q}(\mu_{\ell^h})/\mathbb{Q}$ (totally ramified). However, if this construction is numerically too delicate, it is still very much possible to compute the Hasse symbol $\left(\frac{y\,,\,L/K}{v}\right)$, with $L = K(\sqrt[\ell^h]{x})$ for K containing μ_{ℓ^h}, if necessary by brutally adjoining these roots of unity, by the usual method explained in 4.4.3; in this case, we do not need to assume that $v|\ell$ is unique (and in general we cannot). We then obtain an Artin symbol of the form $\left(\frac{L/K}{\mathfrak{b}}\right)^{-1}$, for some ideal $\mathfrak{b}$ of K prime to a modulus $\mathfrak{m}$, multiple of the conductor of L/K and divisible by all the prime ideals dividing (x) and (ℓ), from which we obtain:

$$(x\,,\,y) = \left(\frac{L/K}{\mathfrak{b}}\right)^{-1} \sqrt[\ell^h]{x} \Big/ \sqrt[\ell^h]{x} = \left(\frac{x}{\mathfrak{b}}\right)_{\ell^h}^{-1},$$

in L/K. If $\mathfrak{p}$ is a prime ideal of K dividing $\mathfrak{b}$, by definition of a Frobenius, and since x and ℓ^h are prime to $\mathfrak{p}$, we know that:

$$\left(\frac{x}{\mathfrak{p}}\right)_{\ell^h} \equiv x^{\frac{q-1}{\ell^h}} \bmod \mathfrak{p},$$

where $q := \mathrm{N}\mathfrak{p}$, which identifies $\left(\frac{x}{\mathfrak{p}}\right)_{\ell^h}$ in μ_{ℓ^h} (global). We thus obtain $\left(\frac{x}{\mathfrak{b}}\right)_{\ell^h}$ by multiplicativity.

If we are *a priori* in a given (global) field K containing μ_{ℓ^h}, and that we want to compute $\left(\frac{x\,,\,y}{v}\right)$, $x, y \in K^\times$, where $v|\ell$ is not unique, we can either use the Hasse symbol, or change the global field K (note that x and y must be reinterpreted in the new field by means of the common completion!). The reader can practice on the field $K := \mathbb{Q}\left(\sqrt{-3+\sqrt{2}},\, \sqrt{-3-\sqrt{2}},\, \mu_8\right)$.

Let us give an example in the nonsplit case so as to apply both points of view.

7.5.1 Example. Take $k = \mathbb{Q}_2(i)$, where $i = \sqrt{-1}$ and let us compute the symbol of order 4:

$$(6\,,\,3+i).$$

We will choose $K = \mathbb{Q}(i)$, $L = K(\sqrt[4]{6})$ and begin by computing the Hasse symbol $\left(\frac{3+i\,,\,L/K}{v}\right)$, where v is the place of K above 2.

Since $6 = -3i(1+i)^2$, $K(\sqrt{6}) = K(\sqrt{-3i})$, and by Kummer theory ($-3i \equiv i \bmod (4)$, which is not congruent to a square modulo (4)) we see that v is ramified in $K(\sqrt{6})/K$, hence totally ramified in L/K. We then check that:

$$\pi = \frac{1+\sqrt{-3i}}{\sqrt[4]{6}} - 1$$

is a uniformizer of L_v. Using the higher ramification groups we find (by computing the valuations of $\pi^{\sigma-1}$ and π^{σ^2-1} for a generator σ of $\mathrm{Gal}(L/K)$) that the conductor of L/K is equal to (24) (see 1.6.2). We then look for $y' \in K$ such that:

$$\frac{y'}{3+i} \equiv 1 \bmod (8)$$

$$y' \equiv 1 \bmod (3),$$

and we obtain for instance $y' = -5 + 9i = (1 + i)(2 + 7i)$ which yields $\mathfrak{b} = (2 + 7i)$ (a prime ideal above 53). Thus, we have $\left(\frac{6}{\mathfrak{b}}\right)_4 \equiv 6^{13} \equiv -1$ modulo $\mathfrak{b}$; hence:

$$(6\,,\,3 + i) = (-1)^{-1} = -1.$$

Beware that if $\mathfrak{b}$ is not a prime ideal, we must come back to the $\left(\frac{x}{\mathfrak{p}}\right)_4$ for $\mathfrak{p}|\mathfrak{b}$, and conclude by multiplicativity.

The product formula is here reduced to (using prime ideals instead of places):

$$\Big(\frac{3+i,\, L/K}{\mathfrak{p}_2}\Big)\Big(\frac{3+i,\, L/K}{\mathfrak{p}_5}\Big)\Big(\frac{3+i,\, L/K}{(3)}\Big) = 1,$$

where $\mathfrak{p}_2 = (1 + i)$, $\mathfrak{p}_5 = (2 - i)$, which can be written:

$$\Big(\frac{6,\, 3+i}{\mathfrak{p}_2}\Big)\Big(\frac{6,\, 3+i}{\mathfrak{p}_5}\Big)\Big(\frac{6,\, 3+i}{(3)}\Big) = 1,$$

in terms of Hilbert symbols of order 4. But the computation of regular symbols yields:

$$\Big(\frac{6,\, 3+i}{\mathfrak{p}_5}\Big) \equiv 6^1 \equiv 1 \bmod \mathfrak{p}_5,$$
$$\Big(\frac{6,\, 3+i}{(3)}\Big) \equiv ((3+i)^{-1})^2 \equiv -1 \bmod (3),$$

giving once again the result in a nicer way.

However the first method is useful when x $(= 6)$ is fixed and y varies: setting $y =: (1+i)^m z$, $m \in \mathbb{Z}$, z prime to $\mathfrak{p}_2$, and using the chinese remainder theorem, we obtain the general solution:

$$\mathfrak{b} =: (z'), \quad z' := 9z - 8(i-1)^m,$$

where z' is defined modulo (24). The solution is then given by the Artin symbol:

$$\Big(\frac{L/K}{(z')}\Big)^{-1}.$$

Thus, let $A \subset I_T$ for $T = \{\mathfrak{p}_2, (3)\}$ be the Artin group of $K(\sqrt[4]{6})$; we check that $I_T = \langle\, \mathfrak{p}_{13}\,\rangle\, A$, for $\mathfrak{p}_{13} := (3+2i)$, whose Frobenius sends $\sqrt[4]{6}$ to $-i\,\sqrt[4]{6}$, so that $(6,\, y) = 1,\ -1,\ i,\ -i$, according to whether:

$$(z') \in A,\ (3+2i)^2\, A,\ (3+2i)\, A,\ (3+2i)^3\, A = (3-2i)\, A.$$

If $y = 9 + 32i$, we find $z' \equiv 1 \bmod (24)$, hence $(6,\, y) = 1$, although method using the product formula needs the computation of four regular symbols.

The generalized class group $I_T/P_{T,(24)}$ has order 64, so that $A/P_{T,(24)}$ has order 16 and can easily be computed.

Finally, since any class modulo $P_{T,(24)}$ contains an infinity of prime ideals, we can look for $\mathfrak{b} =: \mathfrak{q}$ prime, and then $(6,\, y)$ is given by the residual image of $\left(6^{\frac{\mathrm{N}\mathfrak{q}-1}{4}}\right)^{-1}$ mod $\mathfrak{q}$. For example, if $y = 3 + 2i$ then $z' = 19 + 18i$ which is composite, but $z' - 24(1+i) = -5 - 6i$ yields $\mathfrak{p}_{61}$, and $(6^{15})^{-1} \equiv 11 \bmod \mathfrak{p}_{61}$ implies that $(6,\, 3+2i) = -i$. □

Note that the case $n = 2$ is particularly simple since K is easier to find and that all the computations lead to quadratic residue symbols.

Let us come back to the general theory of symbols for a number field K. The homomorphism:

$$h : \mathrm{K}_2(K) \longrightarrow \bigoplus_{v \in Pl^{\mathrm{nc}}} \mu(K_v)$$

which comes from the global Hilbert symbol (see 7.2.2, (ii)) is therefore the most precise possible for identifying $\mathrm{K}_2(K)$ as a function of known symbols. The two main questions which can be asked about h are what are its kernel and image. One is deep (Garland's theorem [Ga] for the finiteness of the kernel, which we will assume), the other is a nontrivial application of the techniques of class field theory that we have developed. This is Moore's theorem on the characterization of the image, which we are going to prove by following a paper of Jaulent written in [Ja1] from the paper of Chase–Waterhouse [ChaW]; see also [Mil, Th. 16.1]. With the above notations and definitions 7.1.3, 7.2.2, we can then state:

7.6 Theorem (fundamental diagram of the K_2 (1971/1972)). *We have, for any number field K, the following commutative diagram:*

$$\begin{array}{ccccccccc}
1 \longrightarrow & \mathrm{WK}_2(K) & \longrightarrow & \mathrm{K}_2(K) & \xrightarrow{\ h\ } & \bigoplus_{v \in Pl^{\mathrm{nc}}} \mu(K_v) & \xrightarrow{\ \pi\ } & \mu(K) & \longrightarrow 1 \\
& \big\downarrow \cap & & \big\| & & \big\downarrow \oplus m_v^1 & & \big\downarrow m & \\
1 \longrightarrow & \mathrm{R}_2^{\mathrm{ord}}(K) & \longrightarrow & \mathrm{K}_2(K) & \xrightarrow{\ h^{\mathrm{reg}}\ } & \bigoplus_{v \in Pl^{\mathrm{nc}}} \mu(K_v)^{\mathrm{reg}} & \longrightarrow & 1 &
\end{array}$$

where $\pi((\xi_v)_v) := \prod_v i_v^{-1}(\xi_v^{\frac{m_v}{m}})$ for all $(\xi_v)_v \in \bigoplus_{v \in Pl^{\mathrm{nc}}} \mu(K_v)$, $m := |\mu(K)|$, $m_v := |\mu(K_v)|$, $m_v^1 := |\mu(K_v)^1| = \frac{m_v}{q_v - 1}$ for v finite, and where $\mathrm{WK}_2(K)$ (resp. $\mathrm{R}_2^{\mathrm{ord}}(K)$) denotes the kernel of the global Hilbert symbol h (resp. of the global regular Hilbert symbol h^{reg}).

Proof. The crucial point is to show that the set of families:

$$(\xi_v)_v \in \bigoplus_{v\in Pl^{\mathrm{nc}}} \mu(K_v),$$

such that $\prod_v i_v^{-1}(\xi_v^{\frac{m_v}{m}}) = 1$, is contained in the image of h. By localizing the problem, we can fix a prime number p and reduce to families in $\bigoplus_{v\in Pl^{\mathrm{nc}}} \mu_p(K_v)$ satisfying the product formula. Denote by Σ the set formed by the places of K dividing p, the infinite real places, and the places v such that $m_v^1 \neq 1$ (Σ is finite).

Thus, let $(\xi_v)_v \in \bigoplus_{v\in Pl^{\mathrm{nc}}} \mu_p(K_v)$ such that $\prod_v i_v^{-1}(\xi_v^{\frac{m_v}{m}}) = 1$. The first step consists in reducing to a situation where the (bad) places of Σ do not occur. For each $v \in \Sigma$, there exist $x_v,\ y_v \in K_v^\times$ such that $(x_v\,,\,y_v)_v = \xi_v$ (we choose x_v such that $K_v(\sqrt[m_v]{x_v})/K_v$ has degree m_v, and we use the surjectivity of the norm residue symbol to find y_v). By approximation on Σ, we can find $x,\ y \in K^\times$ such that:

$$(i_v(x)\,,\,i_v(y))_v = (x_v\,,\,y_v)_v = \xi_v \text{ for all } v \in \Sigma.$$

Since $h(\{x\,,\,y\})$ belongs to a direct sum, replacing if necessary x (for example) by a suitable power (prime to p), we may assume that $h(\{x\,,\,y\}) \in \bigoplus_{v\in Pl^{\mathrm{nc}}} \mu_p(K_v)$ and that, on Σ, we still have $(i_v(x)\,,\,i_v(y))_v = \xi_v$. If we consider:

$$(\xi'_v)_v := \frac{(\xi_v)_v}{h(\{x\,,\,y\})},$$

it is an element of $\bigoplus_{v\in Pl^{\mathrm{nc}}} \mu_p(K_v)$ satisfying the product formula, and it is such that $\xi'_v = 1$ for all $v \in \Sigma$. Hence, for the support Σ' of $(\xi'_v)_v$ the Hilbert symbols are regular, and we are going to obtain the equality $h(\{x'\,,\,y'\}) = (\xi'_v)_v$ for suitable x' and y' in $K^\times$.

For $e := \mathrm{v}_p(m) \geq 0$, consider now the cyclotomic field $K_{e+1} := K(\zeta_{e+1})$, where as usual ζ_{e+1} is a primitive p^{e+1}th root of unity; it is a nontrivial cyclic extension of K (of degree p if $e \geq 1$, of degree dividing $p-1$ if $e = 0$). Let:

$$\mathfrak{a}' := \prod_{v\in\Sigma'} \mathfrak{p}_v,$$

which is an ideal of K prime to p (since $\xi'_v \neq 1$ implies that $v \notin \Sigma$), hence prime to the ramified places of K_{e+1}/K. By the Čebotarev Theorem 4.6, there exists a prime ideal $\mathfrak{q}$ of K, prime to $\Sigma \cup \Sigma'$, such that the Artin symbol $\left(\frac{K_{e+1}/K}{\mathfrak{a}'\mathfrak{q}}\right)$ generates $\mathrm{Gal}(K_{e+1}/K)$. Since on Σ' we have regular symbols, we have:

$$(\xi'_v\,,\,\pi_v)_v = \xi'_v \text{ for all } v \in \Sigma',$$

the result being independent of the choice of a uniformizer π_v of K_v (see 7.1.5). By approximation on $\Sigma' \cup \{\mathfrak{q}\}$, we can find $x' \in K^\times$ such that:

$$x' \equiv 1 \bmod \mathfrak{q},$$

$$i_v(x')\,\xi_v'^{-1} \in U_v^1 \text{ for all } v \in \Sigma'.$$

We then have (again by 7.1.5):

$$(i_v(x')\,,\,\pi_v)_v = (\xi_v'\,,\,\pi_v)_v = \xi_v' \text{ for all } v \in \Sigma'.$$

Let T be the union of $\Sigma_0 := \Sigma \cap Pl_0$ with the set of finite places dividing x' (T is prime to $\mathfrak{q}$ and to Σ'). Consider the modulus $\mathfrak{n} = \prod_{v \in T} \mathfrak{p}_v^n$, for a sufficiently large integer n (in particular, we can assume that $\mathfrak{n}$ is a multiple of the conductor of K_{e+1}/K). By the Čebotarev theorem, there exists a prime ideal $\mathfrak{l}$ prime to $T \cup \Sigma'$ and such that $cl_{\mathfrak{n}}^{\text{res}}(\mathfrak{l}) = cl_{\mathfrak{n}}^{\text{res}}(\mathfrak{a}'\mathfrak{q})$, in $\mathcal{C}\ell_{\mathfrak{n}}^{\text{res}}$, so that we can write:

$$\mathfrak{a}'\mathfrak{q} = \mathfrak{l}\,(y'), \quad y' \in K^{\times}_{T,\mathfrak{n},\text{pos}}.$$

Now consider $h(\{x'\,,\,y'\}) = ((i_v(x')\,,\,i_v(y'))_v)_v$:

- if v is an infinite place, we have $(i_v(x')\,,\,i_v(y'))_v = 1$ since $i_v(y') > 0$;
- if $v \in \Sigma'$, because of the congruences imposed on x', we have:

$$(i_v(x')\,,\,i_v(y'))_v = (\xi_v'\,,\,i_v(y'))_v,$$

but $i_v(y')$ is a uniformizer of K_v since $\mathrm{v}(y') = \mathrm{v}(\mathfrak{a}') = 1$, and by what we have seen above:

$$(i_v(x')\,,\,i_v(y'))_v = \xi_v'\;;$$

- if v corresponds to $\mathfrak{q}$, we obtain $(i_v(x')\,,\,i_v(y'))_v = 1$ since we have chosen $x' \equiv 1 \bmod \mathfrak{q}$;
- if $v \in T$ (the symbol is then not necessarily regular), we have $(i_v(x')\,,\,i_v(y'))_v = 1$ since $i_v(y') \in U_v^n$ (a local norm for n sufficiently large);
- if v is none of the above and if v does not correspond to $\mathfrak{l}$, we have $(i_v(x')\,,\,i_v(y'))_v = 1$ since $i_v(x')$ and $i_v(y')$ are local units by definition of T and y', and the symbol is regular;
- finally, assume that v is the place corresponding to $\mathfrak{l}$; we have:

$$\left(\frac{K_{e+1}/K}{\mathfrak{a}'\mathfrak{q}}\right) = \left(\frac{K_{e+1}/K}{\mathfrak{l}}\right)\left(\frac{K_{e+1}/K}{(y')}\right) = \left(\frac{K_{e+1}/K}{\mathfrak{l}}\right)$$

since by assumption $y' \equiv 1 \bmod \mathfrak{n}$, and $\mathfrak{n}$ is a multiple of the conductor of K_{e+1}/K. Hence, $\left(\frac{K_{e+1}/K}{\mathfrak{l}}\right) = \left(\frac{K_{e+1}/K}{\mathfrak{a}'\mathfrak{q}}\right)$ is by assumption a generator of $\mathrm{Gal}(K_{e+1}/K)$, which shows that $\mathfrak{l}$ is not split in K_{e+1}/K, hence that K_v does not contain ζ_{e+1}, or, equivalently that $\mathrm{v}_p(\frac{m_v}{m}) = 0$. Since $\frac{m_v}{m}$ is a p-adic unit, we deduce from this and the product formula that $(i_v(x')\,,\,i_v(y'))_v = 1$.

We have thus proved the first exact sequence of the diagram.

The surjectivity of h^{reg} is equivalent to the fact that any element of $\bigoplus_{v \in Pl^{\text{nc}}} \mu(K_v)^{\text{reg}}$ (which can be written $((\zeta_v^{\text{reg}})^{m_v^1})_v$ since m_v^1 is prime to the order of $\zeta_v^{\rm reg}$) is the image under the surjection $\oplus m_v^1$ of an element $(\zeta_v)_v =: (\zeta_v^{\rm reg} \,.\, \zeta_v^1)_v$ of $\bigoplus_{v \in Pl^{\rm nc}} \mu(K_v)$ satisfying the product formula. Since the component $(\zeta_v^1)_v$ belongs to the kernel of $\oplus m_v^1$, it is sufficient to check that it is possible to find it so that $(\zeta_v)_v \in {\rm Ker}(\pi)$, and this is equivalent to be able to solve, for any $\zeta \in \mu(K)$:

$$\prod_v i_v^{-1}(\zeta_v^1)^{\frac{m_v}{m}} = \zeta, \quad \text{with } (\zeta_v^1)_v \in \bigoplus_{v \in Pl^{\rm nc}} \mu(K_v)^1.$$

Let us localize at a prime divisor p of m (the case $p \nmid m$ being trivial), take $\zeta \in \mu_p(K)$, and consider $v_0 | p$; the inclusion $i_{v_0}(\mu_p(K)) \subseteq \mu(K_{v_0})^1$ shows that:

$$\left(\frac{m_{v_0}}{m}\right)_p = \frac{m_{v_0}^1}{m_p},$$

and implies the existence of $\zeta_{v_0}^1$ such that $(\zeta_{v_0}^1)^{\frac{m_{v_0}}{m}} = i_{v_0}(\zeta)$; we then set $\zeta_v^1 = 1$ for all $v \neq v_0$.

This proves the two exact sequences of the diagram. □

7.6.1 Corollary. *We have the exact sequence:*

$$1 \longrightarrow {\rm R}_2^{\rm ord}(K)/{\rm WK}_2(K) \xrightarrow{\ \alpha\ } \bigoplus_v \mu(K_v)^1 \xrightarrow{\ \beta\ } \mu(K) \longrightarrow 1,$$

in which α is obtained from the restriction of h to ${\rm R}_2^{\rm ord}(K)$ and β is the restriction of π to $\bigoplus_v \mu(K_v)^1$. Thus $({\rm R}_2^{\rm ord}(K) : {\rm WK}_2(K)) = \frac{1}{m} \prod_v m_v^1$. □

This result indicates that everything can be reduced to the fundamental invariant ${\rm WK}_2(K)$, even though ${\rm R}_2^{\rm ord}$ can be more easily interpreted arithmetically (see below).

We note that ${\rm WK}_2(K) = {\rm R}_2^{\rm ord}(K)$ if and only if for all prime number p:

$$\mu_p(K) \simeq \bigoplus_{v|p} \mu_p(K_v)\ ;$$

for applying this, it is sufficient to check the primes p for which there exists an irregular place $v|p$. We will often encounter this condition (see for example III.4.2.5).

7.6.2 Definitions (Hilbert and regular kernels). The kernel of h (denoted ${\rm H}_2(K)$ or ${\rm WK}_2(K)$) is called the Hilbert or wild kernel (in ${\rm K}_2(K)$)[35], and the kernel of $h^{\rm reg}$ (denoted ${\rm R}_2^{\rm ord}(K)$), is called the regular or tame kernel. □

Recall that ${\rm R}_2^{\rm ord}(K)$ is also equal to ${\rm K}_2^{\rm ord}(Z_K)$ and that this interpretation of ${\rm R}_2(K)$ as the ${\rm K}_2$ of the ring of integers Z_K of K is due to Quillen (1973)

[35] The notation ${\rm WK}_2$ is to be preferred, instead of ${\rm H}_2$, to avoid confusion with homology groups, but we will continue to speak of the Hilbert kernel.

and must be understood with the language of the general K-theory of rings (see an arithmetic study of the regular and Hilbert kernels in [Keu], as a prelude to numerous developments on this subject).

7.6.3 Remarks. (i) The notation $\mathrm{R}_2^{\mathrm{ord}}(K) = \mathrm{K}_2^{\mathrm{ord}}(Z_K)$ represents the modification (introduced in 1986 in [Gr6]) coming from the consideration of the real places at infinity in the definition of h^{reg}; in other words, the classical kernel $\mathrm{R}_2(K) = \mathrm{K}_2(Z_K)$ must be understood as $\mathrm{R}_2^{\mathrm{res}}(K) = \mathrm{K}_2^{\mathrm{res}}(Z_K)$, which fortunately is compatible with the general system of notations that we have adopted here. The difference between these two definitions is given precisely by the following trivial exact sequence:

$$1 \longrightarrow \mathrm{K}_2^{\mathrm{ord}}(Z_K) \longrightarrow \mathrm{K}_2(Z_K) \longrightarrow (\mathbb{Z}/2\mathbb{Z})^{r_1} \longrightarrow 1,$$

but the existence of $\mathrm{K}_2^{\mathrm{ord}}(Z_K)$ (and also probably the existence of more general groups of the form $\mathrm{K}_2^S(Z_{K,T})$ for our usual sets of places T and S, $Z_{K,T}$ being the ring of T-integers of K) is essential.

(ii) The equality $\pi \circ h = 1$ follows of course from the product formula 7.3 for Hilbert symbols, but the exactness that is obtained (Moore's theorem) says that this product formula is the unique relation between Hilbert symbols. This property is called "uniqueness of reciprocity laws".

(iii) The existence of $\mathrm{WK}_2(K)$ (a finite group which is in general nontrivial) means that there can exist symbols on K which do not come from Hilbert symbols (and called exotic symbols because of this); but, although class field theory gives quite good information on these kernels (see below), up to now it has not been possible to exhibit (numerically) a single exotic symbol!

It is easy to find fields for which $\mathrm{WK}_2(K) \neq 1$; on the contrary, it is more difficult to characterize the cases where, for a given p, the p-Sylow subgroup of this kernel is trivial. See for example the results of Kolster–Movahhedi [KM1] dealing (after a few particular cases of Thomas) with the case of biquadratic fields for $p = 2$, the case of quadratic fields having been treated before by Browkin–Schinzel, then revisited by Jaulent–Soriano. Many other papers are concerned with the case $p = 2$ (Conner–Hurrelbrink, Candiotti–Kramer, Berger, Hettling, ...). In [KM2] is given a characterization of the p-extensions of $\mathbb{Q}$ such that the p-Sylow subgroup of the Hilbert kernel is trivial ($p \neq 2$). In [Sor] is given an approach of the structure of the Hilbert kernel in a Kummer situation. □

Thus many new questions related to the K-theory of number fields can be asked, which are not the subject of this book. However, we will mention some of the most classical results, since as mentioned in the introduction to this section, they involve invariants which are directly linked with class field theory (in particular through the reflection theorem).

7.7 Links Between Class Field Theory and $K_2(K)$. The relationships which exist between these K-theory kernels and class field theory are the following.

7.7.1 Logarithmic Class Group. Concerning $WK_2(K)$, under the fundamental assumption $\mu_{2p} \subset K$, the results of Jaulent can be summarized by the relation:

$$WK_2(K)/WK_2(K)^p \simeq \mu_p \otimes \widetilde{\mathcal{C}\ell}_K,$$

where the p-group $\widetilde{\mathcal{C}\ell}_K$ (of logarithmic classes) is an invariant of class field theory, related to Gross's conjecture which we will state in III.4.13, which can be defined from the usual arithmetic of the number field K.[36] This represents the best practical approach to the Hilbert kernel since we have at our disposal the corresponding formalism, which is completely parallel with the (better-known) one for class groups "$\mathcal{C}\ell$" or (a little less known) for the torsion groups "$\mathcal{T}$". It is therefore possible to perform numerical computations [DS]. See (Ch. III, § 7) for a direct approach of the definition of the logarithmic class group and the proof of the above property.

7.7.2 Tate's Results. For $R_2^{\mathrm{ord}}(K)$, still when $\mu_p \subset K$, we have a Kummer interpretation coming from the results of Tate published in 1976 in [Ta2], which is given by the exact sequence:

$$1 \longrightarrow \mu_p \otimes N_2(K) \longrightarrow \mu_p \otimes W_{K,Pl_p,\mathrm{pos}} \xrightarrow{\;f\;} {}_pR_2^{\mathrm{ord}}(K) \longrightarrow 1,$$

where $W_{K,Pl_p,\mathrm{pos}} := \mathrm{Rad}(H^{\mathrm{ord}}_{Pl_p}[p]/K)$ is the radical of the maximal abelian p-ramified noncomplexified elementary p-extension of K, f being defined by:

$$f(\zeta \otimes x) := \{\zeta\,,\,x\}$$

for all $\zeta \in \mu_p$, $x \in W_{K,Pl_p,\mathrm{pos}}$, and where:

$$N_2(K) := \{x \in K^\times,\ \{\zeta_1\,,\,x\} = 1\}/K^{\times p}$$

(Tate's kernel, where ζ_1 is a generator of μ_p) is such that:

$$\mu_p \otimes N_2(K) \simeq (\mu_p \otimes \mu_p) \oplus \mu_p^{r_2}.$$

Note. When the number field K is given together with an automorphism group g, the fact that in these statements we write $\mu_p \otimes X$ (instead of ${}_pX$ or X/X^p) allows us to have canonical isomorphisms of g-modules.

Tate's exact sequence already yields:

7.7.2.1 Proposition. *When K contains μ_p we have:*

---

36 [Ja4; Ja5; Ja6], [JaSor], [Sor], [JaMi], [JaMai].

$$\mathrm{rk}_p(\mathrm{R}_2^{\mathrm{ord}}(K)) = \mathrm{rk}_p(\mathcal{C}\ell_{Pl_p}^{\mathrm{ord}}) - (r_2 + 1). \qquad \square$$

Note. Recall that $H_{Pl_p}^{\mathrm{ord}}[p]$ has a conductor which divides $\mathfrak{m} = \prod_{v|p} \mathfrak{p}_v^{pe_v+1}$, where e_v is the ramification index of v in $K/\mathbb{Q}(\mu_p)$. Therefore, $\mathrm{rk}_p(\mathcal{C}\ell_{Pl_p}^{\mathrm{ord}}) = \mathrm{rk}_p(\mathcal{C}\ell_{\mathfrak{m}}^{\mathrm{ord}})$.

Assuming the Leopoldt conjecture for p, this rank is also equal to the p-rank of the torsion group of $\mathrm{Gal}(H_{Pl_p}^{\mathrm{ord}}(p)/K)$ (see III.4.2.2).

The reflection theorem I.4.6, (ii), applied to $\mathcal{C}\ell_{Pl_p}^{\mathrm{ord}}$, implies:

7.7.2.2 Corollary. *In the Kummer case we have:*

$$\mathrm{rk}_p(\mathrm{R}_2^{\mathrm{ord}}(K)) = \mathrm{rk}_p(\mathcal{C}\ell^{Pl_p\,\mathrm{res}}) + |Pl_p| - 1. \qquad \square$$

7.7.3 TATE'S RESULTS IN THE NON-KUMMER CASE. In the general case, we must introduce $K' := K(\mu_p)$, use formulas with characters and the reflection principle. More precisely, starting from Tate's exact sequence, the reflection theorem allows us to prove the following general formula.

7.7.3.1 Theorem. *For any number field K, we have:*

$$\mathrm{rk}_p(\mathrm{R}_2^{\mathrm{ord}}(K)) = \mathrm{rk}_{\omega^{-1}}(\mathcal{C}\ell_{K'}^{Pl'_p\,\mathrm{res}}) + |\{v|p,\ d_v = 1\}| - \delta,$$

where ω is the Teichmüller character, d_v is the decomposition group of v in K'/K, and $\delta = 1$ or 0 according as $\mu_p \subset K$ or not. $\square$

7.8 p-REGULAR FIELDS. We have introduced in 1989, in [GrJ], the following definition.

7.8.1 Definition. Number fields for which $(\mathrm{R}_2^{\mathrm{ord}}(K))_p = 1$ are called p-regular. $\square$

These fields have a much simpler arithmetic since deep invariants vanish, and for these fields we can even compute higher K-groups (as in [RØ] and a few others). We will see in III.4.1.10, III.4.2.6, (i), and especially in (Ch. IV; § 3, (b)), the similar notion of p-rational fields and what class field theory says about them. The above general formula shows that the p-regularity depends on the ω^{-1}-component of the Pl'_p-class group (in the restricted sense) of $K' := K(\mu_p)$. We will see that the ω-component is concerned with the p-rationality and that the two notions coincide if and only if $\omega^2 = 1$. Then, If K contains the maximal real subfield of $\mathbb{Q}(\mu_p)$, p-regularity and p-rationality will be equivalent notions (this condition is always satisfied for $p = 2$ and $p = 3$).

7.8.1.1 Example. The number field $K \supset \mu_p$ is p-regular (or p-rational) if and only if p does not split in $K/\mathbb{Q}$ and $(\mathcal{C}\ell^{\mathrm{res}})_p$ is generated by means of the single prime ideal of K above p (use 7.7.2.2). □

We will now prove that $\mathbb{Q}$ is p-regular for all p. Thus for $K = \mathbb{Q}$, we will have:

$$\mathrm{WK}_2(\mathbb{Q}) = \mathrm{R}_2^{\mathrm{ord}}(\mathbb{Q}) = 1, \quad \mathrm{R}_2^{\mathrm{res}}(\mathbb{Q}) \simeq \mathbb{Z}/2\mathbb{Z}.$$

The proof given below is a direct one and does not use the class field theory results we have seen up to now; but it is also possible to check that the rank formula above gives the result by analytical means.

7.8.1.2 Theorem (Gauss's first proof of the quadratic reciprocity law, revisited by Tate). *The global regular Hilbert symbol induces the isomorphism:*

$$\mathrm{K}_2(\mathbb{Q}) \simeq \mu(\mathbb{R}) \bigoplus_{\substack{\ell \text{ prime} \\ \ell \neq 2}} \mu(\mathbb{Q}_\ell) \simeq \{\pm 1\} \bigoplus_{\substack{\ell \text{ prime} \\ \ell \neq 2}} \mathbb{F}_\ell^\times .$$

Proof. [37] The letters ℓ, p, q denote prime numbers, a, b, c denote nonzero elements of $\mathbb{Z}$, and $\{a, b\}$ is the image of (a, b) in $\mathrm{K}_2(\mathbb{Q})$. We use the regular Hilbert's symbol $h^{\mathrm{reg}} : \mathrm{K}_2(\mathbb{Q}) \longrightarrow \{\pm 1\} \bigoplus_\ell \mu(\mathbb{Q}_\ell)^{\mathrm{reg}}$, where we replace $\mu(\mathbb{Q}_\ell)^{\mathrm{reg}}$ by $\mathbb{F}_\ell^\times$ (note that $\mu(\mathbb{Q}_2)^{\mathrm{reg}} = \mathbb{F}_2^\times = 1$); the first factor corresponds to the place ∞. Put:

$$\begin{aligned} t_\infty &= \langle \{-1, -1\} \rangle, \\ t_0 &= \langle \{a, b\}, \ \ a, b \in [1, \infty[\rangle, \\ t_\ell &= \langle \{a, b\}, \ \ a, b \in [1, \ell] \rangle, \ \ \ell \text{ prime}. \end{aligned}$$

Note that $\mathrm{K}_2(\mathbb{Q})$ is generated by the $\{u, v\}$, $u, \ v \in \mathbb{Z} \setminus \{0\}$.

Lemma 1. *We have* $\mathrm{K}_2(\mathbb{Q}) = t_\infty \oplus t_0$.

Proof. The bilinearity gives $\{a, b\} = \{|a|, |b|\} . s$, where s is an element of $\langle \{-1, -1\}; \ \{c, -1\}, \ c > 0 \rangle$. But $\{c, -c\} = 1$, which yields $\{c, -1\} = \{c, c\} \in t_0$. The sum is direct since we have $\left(\frac{-1,\,-1}{\infty}\right) = -1$ and $\left(\frac{|a|,\,|b|}{\infty}\right) = 1$, proving the result. □

Since $h^{\mathrm{reg}}(t_\infty) = \{\pm 1\}$ and $h^{\mathrm{reg}}(t_0) \subseteq \bigoplus_\ell \mathbb{F}_\ell^\times$, it is equivalent to prove that the restriction of h^{reg} to t_0 yields $t_0 \simeq \bigoplus_\ell \mathbb{F}_\ell^\times$.

---

[37] Inspired by a conference of Tate (Grenoble 1968): "Sur la première démonstration par Gauss de la loi de réciprocité". See also [Ta1, § 3, (17)], [Mil, Th. 11.6], [f, Lem, Th. 2.30].

Lemma 2. *Let p be fixed. Then the restriction of h^{reg} to t_p yields an isomorphism of t_p onto $\bigoplus_{\ell\leq p}\mathbb{F}_\ell^\times$.*

Proof. We note that $t_2 \simeq \mathbb{F}_2^\times$ (indeed, $\{2,\,2\} = \{2,\,-2\}\{2,\,1-2\} = 1$). For $p \neq 2$, let q be the greatest prime number such that $q < p$; we suppose, by induction, that $t_q \simeq \bigoplus_{\ell\leq q}\mathbb{F}_\ell^\times$. Thus it is sufficient to prove that $t_p/t_q \simeq \mathbb{F}_p^\times$ under h^{reg}.

Consider the following two maps:

$$\varphi : t_p/t_q \longrightarrow \mathbb{F}_p^\times, \quad \theta : \mathbb{F}_p^\times \longrightarrow t_p/t_q,$$

for which:

$$\varphi(\{a,\,b\}) := \Big(\frac{a,\,b}{p}\Big)^{\text{reg}} \in \mathbb{F}_p^\times, \quad \theta(\overline{c}) = \{c,\,p\} \bmod t_q,$$

where c is the representative of $\overline{c} \in \mathbb{F}_p^\times$ in $[1,\,p[$. Since:

$$\Big(\frac{a,\,b}{p}\Big)^{\text{reg}} = (-1)^{\mathrm{v}_p(a)\mathrm{v}_p(b)}\; a^{\mathrm{v}_p(b)}\, b^{-\mathrm{v}_p(a)} \bmod p,$$

φ is trivial on t_q.

We will prove that $\varphi\circ\theta$ and $\theta\circ\varphi$ are the corresponding identity maps (it is not a priori evident that θ is a group homomorphism).

We have $\varphi\circ\theta(\overline{c}) = \varphi(\{c,\,p\}) = (\frac{c,\,p}{p})^{\text{reg}} = \overline{c}$ since $\mathrm{v}_p(c) = 0$. Then $\theta\circ\varphi(\{a,\,b\}) = \theta((\frac{a,\,b}{p})^{\text{reg}}) = \theta(\overline{c})$, where $\overline{c} = (-\overline{1})^{\mathrm{v}_p(a)\mathrm{v}_p(b)}\;\overline{a}^{\mathrm{v}_p(b)}\,\overline{b}^{-\mathrm{v}_p(a)}$. The case where $a,\,b \in [1,\,p[$ is immediate since $\{a,\,b\} \in t_q$. If $a < p$ and $b = p$, then $c = a$, thus $\theta(\overline{c}) \equiv \{a,\,p\} \bmod t_q$. If $a = p$ and $b < p$, then $\overline{c} = \overline{b}^{-1}$, and we have to verify that $\{p,\,b\}\{c,\,p\}^{-1} = \{p,\,bc\} \in t_q$. Put $bc = 1 + dp$; the case $b = 1$ being trivial, suppose $b > 1$, thus $0 < d < p$. Then we have $1 = \{-dp,\,1+dp\} = \{-dp,\,bc\} = \{-d,\,b\}\{-d,\,c\}\{p,\,bc\}$ giving the result since $\{-1,\,u\} = \{u,\,u\}$. If $a = b = p$, then $\overline{c} = -\overline{1}$, $c = p-1$, but we have $\{p-1,\,p\} = \{-1,\,p\} = \{p,\,p\}$. This finishes the proof of the lemma. □

Taking the direct limit (i.e., the union $t_0 = \bigcup_\ell t_\ell$), the theorem follows. □

We now consider the Hilbert symbol $(\frac{\bullet,\,\bullet}{2})$ for the place 2; it takes its values in $\{\pm1\}$. Since this Hilbert symbol is of order 2, we can write $(\frac{\bullet,\,\bullet}{2}) =: g\circ\{\bullet,\,\bullet\}$, where $g : \{\pm1\}\bigoplus_{\ell\neq2}\mathbb{F}_\ell^\times/\mathbb{F}_\ell^{\times2} \longrightarrow \{\pm1\}$, is such that:

$$g((u_v)_{v\neq2}) = g_\infty(u_\infty)\prod_{\ell\neq2} g_\ell(u_\ell),$$

with $g_\infty := g|_{\{\pm 1\}}$, $g_\ell := g|_{\mathbb{F}_\ell^\times/\mathbb{F}_\ell^{\times 2}}$. We then have $g_v(\bullet) = (\bullet)^{\delta_v}$, $\delta_v = 0$ or 1 depending only on v. Let $a, b \in \mathbb{Z}\backslash\{0\}$. Taking $u_\infty = (\frac{a,\,b}{\infty})$ and, $u_\ell = (\frac{a,\,b}{\ell})$ mod $\mathbb{F}_\ell^{\times 2}$ for all $\ell \neq 2$, which implies that $(\frac{a,\,b}{\ell})$ mod $\mathbb{F}_\ell^{\times 2}$ is the quadratic Hilbert symbol $(\frac{a,\,b}{\ell})_2$, we get:

$$\Big(\frac{a,\,b}{2}\Big) = \Big(\frac{a,\,b}{\infty}\Big)^{\delta_\infty} \prod_{\ell \neq 2} \Big(\frac{a,\,b}{\ell}\Big)_2^{\delta_\ell} \quad \text{for all } a,\, b \in \mathbb{Z}\backslash\{0\}.$$

The uniqueness of the product formula readily gives $\delta_v = 1$ for all $v \neq 2$, but of course, the point of view of Gauss was to *prove* the product formula, giving easily the reciprocity law as we know it. For this, the computation of $(\frac{-1,\,-1}{2})$ gives $\delta_\infty = 1$; if $\ell \equiv 3 \bmod (4)$, the computation of $(\frac{\ell,\,-1}{2})$ gives $\delta_\ell = 1$; if $\ell \equiv 5 \bmod (8)$, the computation of $(\frac{\ell,\,2}{2})$ gives $\delta_\ell = 1$; for $\ell \equiv 1 \bmod (8)$, the computation is not easy since $(\frac{\ell,\,b}{2}) = 1$ for any b. In this case, let p be the least prime $\ell \equiv 1 \bmod (8)$ such that $\delta_\ell = 0$. Gauss proved the existence of a prime $q < p$ such that $(\frac{p}{q}) = -1$ (see [f, Lem, Th. 2.30]), yielding a contradiction to the computation of $(\frac{p,\,q}{2}) = \prod_{\ell \neq 2} (\frac{p,\,q}{\ell})_2^{\delta_\ell} = (\frac{p,\,q}{q})_2\,(\frac{p,\,q}{p})_2^0 = (\frac{p}{q})$ since $(\frac{p,\,q}{2}) = 1$.

Note. The isomorphism of Theorem 7.8.1.2 is not canonical, the image of h being $\{(\zeta_v)_v \in \bigoplus_{v \in Pl_\mathbb{Q}} \mu(\mathbb{Q}_v),\ \prod_v \zeta_v^{\frac{m_v}{2}} = 1\}$. Since $\frac{m_\infty}{2} = \frac{m_2}{2} = \frac{m_3}{2} = 1$, we can express ζ_∞, ζ_2, or ζ_3 by means of the other ζ_v. Here, we "eliminate" the wild place.

7.8.2 Remark. Recall that the main theorem of Mazur–Wiles–Kolyvagin on abelian extensions K of $\mathbb{Q}$ yields, in the real case, the following analytical expression for $|\mathrm{R}_2^{\mathrm{ord}}(K)|$ which had been conjectured by Birch–Tate (on this subject, see [Grt]):

$$|\mathrm{R}_2^{\mathrm{ord}}(K)| = \frac{w_2}{2^{[K\,:\,\mathbb{Q}]}}\,|\zeta_K(-1)| =: w_2\,|\zeta_K^{\mathrm{ord}}(-1)|,$$

where ζ_K is the Dedekind zeta function of K, which must be interpreted as the restricted zeta function ζ_K^{res} (see III.2.6.5), and w_2 is the largest integer n such that $\mathrm{Gal}(K(\mu_n)/K)$ is killed by 2.

For $K = \mathbb{Q}$, we have $w_2 = 24$, $\zeta_\mathbb{Q}^{\mathrm{ord}}(-1) = \frac{1}{2}\zeta_\mathbb{Q}^{\mathrm{res}}(-1) = -\frac{1}{24}$, thus giving $\mathrm{R}_2^{\mathrm{ord}}(\mathbb{Q}) = 1$. □

III. Abelian Extensions with Restricted Ramification — Abelian Closure

This chapter deals with the correspondence of class field theory both for finite and infinite extensions; this second aspect, obtained by limiting processes, will enable us to understand the structure of the maximal abelian extension of a number field K (Section 4 of the present chapter). Indeed, since any finite abelian extension of K is contained in a ray class field $K(\mathfrak{m})^{\mathrm{res}}$, we have $\overline{K}^{\mathrm{ab}} = \bigcup_{\mathfrak{m}} K(\mathfrak{m})^{\mathrm{res}}$, where $\mathfrak{m}$ ranges in the set of moduli of K.

If for a fixed $T \subset Pl_0$ we restrict to moduli $\mathfrak{m}$ with support equal to T, we obtain by the conductor theorem the maximal abelian T-ramified extension H_T^{res} of K already introduced in II.5.3. If we restrict to tame moduli, i.e., of the form $\mathfrak{m} = \prod_v \mathfrak{p}_v$ (squarefree finite product), we obtain the maximal tamely ramified abelian extension $H_{\mathrm{ta}}^{\mathrm{res}}$ of K (see II.5.2.2, (ii)).

The corresponding Galois groups are therefore inverse limits of the generalized class groups $C\ell_{\mathfrak{m}}^{\mathrm{res}}$; the study of these inverse limits (and also of their p-Sylow subgroups) is the crucial point of this chapter. The idelic point of view, in the sense of II.3.8, will therefore temporarily be put aside; nevertheless, we will recover some results directly given by II.3.8.

As usual the above can also be seen with S-decomposition, which gives for instance the extension H_T^S, the maximal T-ramified S-split abelian extension of K; for $T = \emptyset$, we obtain the S-split Hilbert class field H^S of K.

§1 Generalities on H_T^S/H^S and its Subextensions

a) Description of $\mathrm{Gal}(K(\mathfrak{m})^S/H^S)$

The Hilbert class field $H^{\mathrm{res}} := K(1)^{\mathrm{res}}$ of the field K (corresponding to $T = S = \emptyset$) is a pivotal element in the description of $\mathrm{Gal}(\overline{K}^{\mathrm{ab}}/K)$ since $\mathrm{Gal}(H^{\mathrm{res}}/K)$ is isomorphic to the class group $C\ell^{\mathrm{res}}$ which is a measure of the nonprincipality in the restricted (or narrow) sense of the ideals of K, while for an arbitrary modulus $\mathfrak{m}$, the group $\mathrm{Gal}(K(\mathfrak{m})^{\mathrm{res}}/H^{\mathrm{res}})$ involves the other aspect of the arithmetic of the field K, that which comes from the unit groups, as the following result of class field theory shows.

Let K together with sets of places T and S, and let $\mathfrak{m} = \prod_{v\in T} \mathfrak{p}_v^{m_v}$, $m_v \geq 0$, $\Delta_\infty := Pl_\infty^{\mathrm{r}} \backslash S_\infty$.

1.1 Theorem. *Let $t \subseteq T$, let $\mathfrak{n} = \prod_{v \in t} \mathfrak{p}_v^{m_v}$ be the corresponding pure divisor of $\mathfrak{m}$, and let $\delta_\infty \subseteq \Delta_\infty$. We have the following exact sequence:*

$$1 \longrightarrow E_{\mathfrak{n}}^{S \cup \delta_\infty} / E_{\mathfrak{m}}^{S} \xrightarrow{i_{T \backslash t, \delta_\infty}} \bigoplus_{v \in T \backslash t} U_v / U_v^{m_v} \bigoplus_{v \in \delta_\infty} \{\pm 1\} \xrightarrow{\rho^S}$$
$$\mathrm{Gal}(K(\mathfrak{m})^S / K(\mathfrak{n})^{S \cup \delta_\infty}) \longrightarrow 1,$$

where ρ^S is the global reciprocity map.

1.1.1 Corollary. *We have the exact sequence:*

$$1 \longrightarrow E^{\mathrm{res}} / E_{\mathfrak{m}}^{\mathrm{res}} \xrightarrow{i_T} \bigoplus_{v \in T} U_v / U_v^{m_v} \xrightarrow{\rho^{\mathrm{res}}}$$
$$\mathrm{Gal}(K(\mathfrak{m})^{\mathrm{res}} / H^{\mathrm{res}}) = \langle\, I_v(K(\mathfrak{m})^{\mathrm{res}} / K)\,\rangle_{v \in T} \longrightarrow 1. \quad \square$$

1.1.2 Corollary. *(i) For all $v \in T$ we have the exact sequence, where we set $\mathfrak{m}_v := \mathfrak{p}_v^{m_v}$:*

$$1 \longrightarrow E_{\frac{\mathfrak{m}}{\mathfrak{m}_v}}^{\mathrm{res}} / E_{\mathfrak{m}}^{\mathrm{res}} \xrightarrow{i_v} U_v / U_v^{m_v} \xrightarrow{\rho^{\mathrm{res}}}$$
$$\mathrm{Gal}(K(\mathfrak{m})^{\mathrm{res}} / K(\tfrac{\mathfrak{m}}{\mathfrak{m}_v})^{\mathrm{res}}) = I_v(K(\mathfrak{m})^{\mathrm{res}} / K) \longrightarrow 1,$$

(ii) for $\delta_\infty \subset Pl_\infty^{\mathrm{r}}$ we have the exact sequence:

$$1 \longrightarrow E_{\mathfrak{m}}^{\delta_\infty} / E_{\mathfrak{m}}^{\mathrm{res}} \xrightarrow{i_{\delta_\infty}} \bigoplus_{v \in \delta_\infty} \{\pm 1\} \xrightarrow{\rho^{\mathrm{res}}}$$
$$\mathrm{Gal}(K(\mathfrak{m})^{\mathrm{res}} / K(\mathfrak{m})^{\delta_\infty}) = \langle\, D_v(K(\mathfrak{m})^{\mathrm{res}} / K)\,\rangle_{v \in \delta_\infty} \longrightarrow 1. \quad \square$$

Proof of the theorem. Although the statement is of idelic nature (see 1.1.4, (ii)), the point of view of ideal classes is also natural since we are dealing here with principal ideals. By the class field theory correspondence given by the Artin map II.5.1.4, we have:

$$P_{T,\mathfrak{n},\mathrm{pos}}\langle\, S \cup \delta_\infty \,\rangle / P_{T,\mathfrak{m},\mathrm{pos}}\langle\, S \,\rangle \simeq \mathrm{Gal}(K(\mathfrak{m})^S / K(\mathfrak{n})^{S \cup \delta_\infty}),$$

and I.4.5, (i) (with $n_v = m_v$ for $v \in t$, $n_v = 0$ for $v \in T \backslash t$) proves the result because of the following homomorphism described in I.4.5.2:

$$\bigoplus_{v \in T \backslash t} U_v / U_v^{m_v} \bigoplus_{v \in \delta_\infty} \{\pm 1\} \xrightarrow{h_{\mathfrak{m}}^S} P_{T,\mathfrak{n},\mathrm{pos}}\langle\, S \cup \delta_\infty \,\rangle / P_{T,\mathfrak{m},\mathrm{pos}}\langle\, S \,\rangle.$$

The resulting map is the composition of the map $h_{\mathfrak{m}}^S$ and of the Artin map (it is therefore the global reciprocity map composed with the inversion automorphism ι of the Galois group). $\square$

From this point of view, we should write $\iota \circ \rho^S$ in the exact sequence above; we will not do this since the idelic version 1.1.4 gives ρ^S.

1.1.3 Remark. Note that we have, by II.5.2.2, (i):

$$\mathrm{Gal}(K(\mathfrak{m})^S/K(\mathfrak{n})^{S\cup\delta_\infty}) = \langle\, I_v(K(\mathfrak{m})^S/K)\,\rangle_{v\in T\backslash t} \cdot \langle\, D_v(K(\mathfrak{m})^S/K)\,\rangle_{v\in\delta_\infty},$$

the group generated by the inertia groups of the places $v \in T\backslash t$ and the decomposition groups of the places $v \in \delta_\infty$, in $K(\mathfrak{m})^S/K$.

This interpretation is valid only if $\mathfrak{n}$ is a pure divisor of $\mathfrak{m}$, in other words if $\mathfrak{n}$ and $\frac{\mathfrak{m}}{\mathfrak{n}}$ are coprime (see more general situations in 1.1.6 and 1.1.8). □

1.1.4 Exercise. Give an idelic proof of the above theorem.

Answer. (i) For a first idelic proof, we begin with the analogous isomorphism:

$$K^\times U_{\mathfrak{n}}^{S\cup\delta_\infty}/K^\times U_{\mathfrak{m}}^S \simeq \mathrm{Gal}(K(\mathfrak{m})^S/K(\mathfrak{n})^{S\cup\delta_\infty})$$

(use II.5.1.4 and the property II.3.5, (iv) of the class field theory correspondence), and we use I.5.1.3; we now obtain the reciprocity map ρ^S.

(ii) If we use the Remark 1.1.3 and the property of the reciprocity map given in II.3.3, (iii), it is sufficient, from the exact sequence:

$$1 \longrightarrow K^\times U_{\mathfrak{m}}^S \longrightarrow J \xrightarrow{\ \rho^S\ } \mathrm{Gal}(K(\mathfrak{m})^S/K) \longrightarrow 1,$$

to restrict ρ^S to $\displaystyle\bigoplus_{v\in T\backslash t} U_v \bigoplus_{v\in\delta_\infty} \mathbb{R}^\times$ and to compute the kernel:

$$\Big(\bigoplus_{v\in T\backslash t} U_v \bigoplus_{v\in\delta_\infty} \mathbb{R}^\times\Big) \cap (K^\times U_{\mathfrak{m}}^S).$$

□

1.1.5 Comparison of the Ordinary and Restricted Senses. The exact sequence of 1.1.2, (ii) shows for instance (with $\delta_\infty = Pl_\infty^{\mathrm{r}}$) that $K(\mathfrak{m})^{\mathrm{res}}$ is a noncomplexified extension of K if and only if $(E_{\mathfrak{m}}^{\mathrm{ord}} : E_{\mathfrak{m}}^{\mathrm{res}}) = 2^{r_1}$ (i.e., the group $\mathrm{sgn}_\infty(E_{\mathfrak{m}}^{\mathrm{ord}})$ is equal to $\mathrm{sgn}_\infty(K^\times) \simeq \{\pm 1\}^{r_1}$); in the same way (with $\delta_\infty = \{v\}$, $v \in Pl_\infty^{\mathrm{r}}$), we see that v is complexified in $K(\mathfrak{m})^{\mathrm{res}}/K$ if and only if $E_{\mathfrak{m}}^{\{v\}} = E_{\mathfrak{m}}^{\mathrm{res}}$ (i.e., there do not exist any unit congruent to 1 modulo $\mathfrak{m}$, positive on $Pl_\infty^{\mathrm{r}}\backslash\{v\}$, and negative at v).

In particular (with $T = \emptyset$, $\delta_\infty = Pl_\infty^{\mathrm{r}}$), we have:

$$1 \longrightarrow E^{\mathrm{ord}}/E^{\mathrm{res}} \longrightarrow \bigoplus_{v\in Pl_\infty^{\mathrm{r}}} \{\pm 1\} \xrightarrow{\ \rho^{\mathrm{res}}\ } \mathrm{Gal}(H^{\mathrm{res}}/H^{\mathrm{ord}}) = \langle\, D_v(H^{\mathrm{res}}/K)\,\rangle_{v\in Pl_\infty^{\mathrm{r}}} \longrightarrow 1,$$

which classically says that the two notions of Hilbert class fields (or of class groups) coincide if and only if the unit group in the ordinary sense can represent all possible signatures. In the case of a real quadratic field, this condition is equivalent to the fact that the fundamental unit has norm -1.

On the contrary, assuming that K is not totally complex, we have $-1 \notin E^{\mathrm{res}}$, hence it is never true in this case that $\mathrm{Gal}(H^{\mathrm{res}}/H^{\mathrm{ord}}) \simeq (\mathbb{Z}/2\mathbb{Z})^{r_1}$ but at best $\mathrm{Gal}(H^{\mathrm{res}}/H^{\mathrm{ord}}) \simeq (\mathbb{Z}/2\mathbb{Z})^{r_1-1}$ (when $E^{\mathrm{ord}} = \langle -1 \rangle E^{\mathrm{res}}$). For a real quadratic field, this group is at most of order 2.

If K is totally real ($r_2 = 0$), then $E^{\mathrm{ord}}/E^{\mathrm{res}} \simeq (\mathbb{Z}/2\mathbb{Z})^{r_1}$ is equivalent to $E^{\mathrm{res}} = (E^{\mathrm{ord}})^2$. This is not true if $r_2 > 0$: for instance, for $K = \mathbb{Q}(\sqrt[3]{2})$ we have $E^{\mathrm{ord}} = \langle -1, \varepsilon \rangle$ with $\varepsilon = 1 - \sqrt[3]{2}$, but $E^{\mathrm{res}} = \langle -\varepsilon \rangle$ (consider $\mathfrak{p}_{17} = (1 + 2\sqrt[3]{2})$; we have $-\varepsilon \equiv 7 \bmod \mathfrak{p}_{17}$, but 7 is not a square modulo $\mathfrak{p}_{17}$; thus $-\varepsilon$ is not a square).

1.1.6 Exercise (generalization of 1.1 — relative inertia and decomposition groups). Let K be a number field together with sets of places T and S. Let $\mathfrak{m} = \prod_{v \in T} \mathfrak{p}_v^{m_v}$, $m_v \geq 0$, and let $\mathfrak{n}$ be an arbitrary divisor of $\mathfrak{m}$; set $\mathfrak{n} = \prod_{v \in T} \mathfrak{p}_v^{n_v}$ with $n_v \leq m_v$ for all $v \in T$. Finally, let $\delta_\infty \subseteq Pl_\infty^{\mathrm{r}} \backslash S_\infty$.

(i) (α) Show that there exists an exact sequence:

$$1 \longrightarrow E_{\mathfrak{n}}^{S \cup \delta_\infty}/E_{\mathfrak{m}}^{S} \xrightarrow{i_{T,\delta_\infty}} \bigoplus_{v \in T} U_v^{n_v}/U_v^{m_v} \quad \bigoplus_{v \in \delta_\infty} \{\pm 1\} \xrightarrow{\rho^S} \mathrm{Gal}(K(\mathfrak{m})^S/K(\mathfrak{n})^{S \cup \delta_\infty}) \longrightarrow 1.$$

(β) Check that $\mathrm{Gal}(K(\mathfrak{m})^S/K(\mathfrak{n})^{S \cup \delta_\infty})$ is generated by the higher ramification groups of index n_v (in upper numbering) of the $v \in T$, and by the decomposition groups of the $v \in \delta_\infty$, in $K(\mathfrak{m})^S/K$ (use the method of 1.1.4, (ii) based on II.3.3, (iii)).

(γ) Check that we have the shifting property expressed by the following exact sequence:

$$1 \longrightarrow E^{S \cup \delta_\infty}/E_{\frac{\mathfrak{m}}{\mathfrak{n}}}^{S} \xrightarrow{i_{T,\delta_\infty}} \bigoplus_{v \in T} U_v/U_v^{m_v - n_v} \quad \bigoplus_{v \in \delta_\infty} \{\pm 1\} \xrightarrow{\rho^S} \mathrm{Gal}(K(\tfrac{\mathfrak{m}}{\mathfrak{n}})^S/H^{S \cup \delta_\infty}) \longrightarrow 1\ ;$$

but the corresponding maps ρ^S are relative to different extensions.

(ii) (α) Let v be a fixed place of K not belonging to S, and set $\mathfrak{m}_v := \mathfrak{p}_v^{m_v}$ and $\mathfrak{n}_v := \mathfrak{p}_v^{n_v}$ (with $m_v = n_v = 0$ if $v \notin T$). Using II.3.3, (iii), or II.3.3.5 with $N := K^\times U_{\mathfrak{m}}^S$, show that under the reciprocity isomorphism we have:

$$D_v(K(\mathfrak{m})^S/K) \simeq K_v^\times/i_v(E_{\frac{\mathfrak{m}}{\mathfrak{m}_v}}^{S \cup \{v\}})U_v^{m_v},$$
$$I_v(K(\mathfrak{m})^S/K) \simeq U_v/i_v(E_{\frac{\mathfrak{m}}{\mathfrak{m}_v}}^{S})U_v^{m_v}\ ;$$

for this, check that:

$$(K^\times U_{\mathfrak{m}}^S) \cap K_v^\times = i_v(E_{\frac{\mathfrak{m}}{\mathfrak{m}_v}}^{S \cup \{v\}})U_v^{m_v}, \quad (K^\times U_{\mathfrak{m}}^S) \cap U_v = i_v(E_{\frac{\mathfrak{m}}{\mathfrak{m}_v}}^{S})U_v^{m_v}.$$

(β) Using II.3.5.4 deduce that if $v \notin S \cup \delta_\infty$, the image of the relative decomposition group $D_v(K(\mathfrak{m})^S/K(\mathfrak{n})^{S \cup \delta_\infty})$ under the above isomorphism is equal to:

$$i_v(E^{S\cup\{v\}\cup\delta_\infty}_{\frac{\mathfrak{n}}{\mathfrak{n}_v}})U_v^{n_v}/i_v(E^{S\cup\{v\}}_{\frac{\mathfrak{m}}{\mathfrak{m}_v}})U_v^{m_v},$$

and that of the relative inertia group $I_v(K(\mathfrak{m})^S/K(\mathfrak{n})^{S\cup\delta_\infty})$ is equal to:

$$i_v(E^{S\cup\delta_\infty}_{\frac{\mathfrak{n}}{\mathfrak{n}_v}})U_v^{n_v}/i_v(E^{S}_{\frac{\mathfrak{m}}{\mathfrak{m}_v}})U_v^{m_v}.$$

(γ) Deduce from the above that, when $v\in Pl_0$ and $\mathfrak{m}_v=\mathfrak{n}_v$ (i.e., $m_v=n_v$, which is always the case if $v\notin T$), we have still for $v\notin S\cup\delta_\infty$:

$$D_v(K(\mathfrak{m})^S/K(\mathfrak{n})^{S\cup\delta_\infty})\simeq E^{S\cup\{v\}\cup\delta_\infty}_{\frac{\mathfrak{n}}{\mathfrak{n}_v}}/E^{S\cup\{v\}}_{\frac{\mathfrak{m}}{\mathfrak{m}_v}}E^{S\cup\delta_\infty}_{\mathfrak{n}},$$
$$I_v(K(\mathfrak{m})^S/K(\mathfrak{n})^{S\cup\delta_\infty})\simeq E^{S\cup\delta_\infty}_{\frac{\mathfrak{n}}{\mathfrak{n}_v}}/E^{S}_{\frac{\mathfrak{m}}{\mathfrak{m}_v}}E^{S\cup\delta_\infty}_{\mathfrak{n}}.$$

(δ) Interpret the case where $v\in Pl_\infty^{\mathrm{r}}\backslash(S_\infty\cup\delta_\infty)$ (i.e., $\mathfrak{m}_v=\mathfrak{n}_v=1$, $K_v^\times=\mathbb{R}^\times, U_v=\mathbb{R}^{\times+}$) by checking that:

$$D_v(K(\mathfrak{m})^S/K(\mathfrak{n})^{S\cup\delta_\infty})\simeq \mathrm{sgn}_v(E_{\mathfrak{n}}^{S\cup\{v\}\cup\delta_\infty})/\mathrm{sgn}_v(E_{\mathfrak{m}}^{S\cup\{v\}}).$$ □

1.1.7 Example. Let $K=\mathbb{Q}(\sqrt{7})$. The prime numbers 5 and 17 are inert in $K/\mathbb{Q}$ and 2 and 7 are ramified. Consider the following moduli:

$$\mathfrak{m}=(5\times 17),\ \mathfrak{n}=(17),$$

and choose for v the place corresponding to (17) in K. The fundamental unit of K is $\varepsilon:=8+3\sqrt{7}$ (totally positive) and $|\mathcal{C}\ell^{\mathrm{res}}|=2$. The formula of I.4.5.1 yields:

$$[K(\mathfrak{m})^{\mathrm{res}}:K]=2\,\frac{(5^2-1)(17^2-1)}{(E^{\mathrm{res}}:E^{\mathrm{res}}_{\mathfrak{m}})}=\frac{2^9\times 3^3}{6}=2^8\times 3^2,$$
$$[K(\mathfrak{n})^{\mathrm{res}}:K]=2\,\frac{(17^2-1)}{(E^{\mathrm{res}}:E^{\mathrm{res}}_{\mathfrak{n}})}=\frac{2^6\times 3^2}{3}=2^6\times 3,$$

since:

$$(\varepsilon\ -1)=(\sqrt{7}\,(3+\sqrt{7}))=\mathfrak{p}_2\,\mathfrak{p}_7,$$
$$(\varepsilon^2-1)=(2\times 3\times\sqrt{7}\,(8+3\sqrt{7}))=(2)\,(3)\,\mathfrak{p}_7,$$
$$(\varepsilon^3-1)=(17\times\sqrt{7}\,(45+17\sqrt{7}))=(2)\,\mathfrak{p}_7\,(17),$$
$$(\varepsilon^4-1)=(2^5\times 3\times\sqrt{7}\,(127+48\sqrt{7}))=(2)^5\,(3)\,\mathfrak{p}_7,$$
$$(\varepsilon^6-1)=(2\times 3^2\times 5\times 17\times\sqrt{7}\,(2024+765\sqrt{7}))=(2)\,(3)^2\,(5)\,\mathfrak{p}_7\,(17).$$

Since $E^{\mathrm{res}}_{\frac{\mathfrak{m}}{\mathfrak{m}_v}}=E^{\mathrm{res}}_{(5)}=\langle\,\varepsilon^6\,\rangle$, $E^{\mathrm{res}}_{\frac{\mathfrak{n}}{\mathfrak{n}_v}}=E^{\mathrm{res}}=\langle\,\varepsilon\,\rangle$, the image of ε^6 in $F_v^\times$ is trivial, hence $I_v(K(\mathfrak{m})^{\mathrm{res}}/K)$ is cyclic of order $17^2-1=2^5\times 3^2$; the image of ε in $F_v^\times$ is of order 3, hence $I_v(K(\mathfrak{n})^{\mathrm{res}}/K)$ is cyclic of order $2^5\times 3$.

It follows that the ramification index of v in $K(\mathfrak{m})^{\mathrm{res}}/K(\mathfrak{n})^{\mathrm{res}}$, is equal to 3, which can be checked directly by noting that we have:

$$E^{\mathrm{res}}_{\frac{\mathfrak{n}}{\mathfrak{n}_v}} = \langle\, \varepsilon \,\rangle, \;\; E^{\mathrm{res}}_{\frac{\mathfrak{m}}{\mathfrak{m}_v}} E^{\mathrm{res}}_{\mathfrak{n}} = \langle\, \varepsilon^6 \,\rangle\langle\, \varepsilon^3 \,\rangle = \langle\, \varepsilon^3 \,\rangle,$$

so that $I_v(K(\mathfrak{m})^{\mathrm{res}}/K(\mathfrak{n})^{\mathrm{res}}) \simeq \mathbb{Z}/3\mathbb{Z}$. We can also use the ramification index computations done in II.5.2.2, (i).

We thus have ramification of v in $K(17)^{\mathrm{res}}$ (index $2^5 \times 3$) and again ramification of v in $K(5{\times}17)^{\mathrm{res}}/K(17)^{\mathrm{res}}$ (index 3), the 3-part of its ramification in $K(5{\times}17)^{\mathrm{res}}/K$ then being the maximum possible value. This property will be explained later by the Schmidt–Chevalley Theorem 4.3 on the behavior of units. □

1.1.8 Remark. We can of course add finite decomposition in the exact sequence 1.1.6, (i), (α). If $\delta := \delta_0 \cup \delta_\infty$ is disjoint from S, and δ_0 is disjoint from the support of $\mathfrak{n}$, we have the following exact sequence:

$$1 \longrightarrow E^{S\cup\delta}_{\mathfrak{n}}/E^S_{\mathfrak{m}} \longrightarrow \bigoplus_{v\in T, v\notin\delta_0} U^{n_v}_v/U^{m_v}_v \;\bigoplus_{v\in\delta_0} K^\times_v/U^{m_v}_v \;\bigoplus_{v\in\delta_\infty} \{\pm 1\} \xrightarrow{\rho^S}$$
$$\mathrm{Gal}(K(\mathfrak{m})^S/K(\mathfrak{n})^{S\cup\delta}) \longrightarrow 1,$$

with $m_v = 0$ if $v \in \delta_0$ and $v \notin T$, and where the image under the reciprocity map is the subgroup of $\mathrm{Gal}(K(\mathfrak{m})^S/K)$ generated by the decomposition groups of the $v \in \delta$ and by the higher ramification groups (with upper indices n_v) of the $v \in T$, $v \notin \delta_0$, in $K(\mathfrak{m})^S/K$. However it involves the infinite groups $K^\times_v/U^{m_v}_v$ (for $v \in \delta_0$) hence also the infinite group $E^{S\cup\delta}_{\mathfrak{n}}/E^S_{\mathfrak{m}}$. □

The exact sequence of 1.1, which comes directly from the global reciprocity map, will be fundamental for the practical study of $\mathrm{Gal}(K(\mathfrak{m})^S/H^S)$, and afterwards of $\mathrm{Gal}(H^S_T/H^S)$, which we will do after a limiting process for which the localization of the problem is needed.

b) The Case of p-Extensions

1.2 Localization of Class Field Theory. The study of $K(\mathfrak{m})^S/K$ and of H^S_T/K can be reduced to that of their maximal p-subextensions denoted $K(\mathfrak{m})^S(p)/K$ and $H^S_T(p)/K$, for every prime number p. At the level of the correspondence of class field theory, to obtain the Galois groups of these p-extensions (or pro-p-extensions) it is sufficient to localize at p all the exact sequences that we have obtained, and since $\mathbb{Z}_p$ is a flat $\mathbb{Z}$-module, a convenient way is to replace each abelian group A by $A \otimes_{\mathbb{Z}} \mathbb{Z}_p$. When A is *finite*, we thus obtain its p-Sylow subgroup. We will see below that in the case of infinite groups we must be careful, and also note that in that case the result of their tensoring by $\mathbb{Z}_p$ has generally nothing to do with a p-Sylow subgroup, a notion which makes sense for $\mathbb{Z}$-torsion abelian groups and for profinite groups which for us will be commutative.[1]

[1] In the first case, we prefer to speak of the p-primary subgroup since this subgroup is simply the subset of elements of p-power order. For a profinite group, see [g, Se3, Ch. I].

In this context, it may be useful to briefly recall the rules which govern the use of tensor products. The flatness of $\mathbb{Z}_p$ means that for any exact sequence of abelian groups:

$$1 \longrightarrow B \longrightarrow A \xrightarrow{\ f\ } C,$$

we have the exact sequence:

$$1 \longrightarrow B \otimes \mathbb{Z}_p \longrightarrow A \otimes \mathbb{Z}_p \xrightarrow{f\otimes 1} C \otimes \mathbb{Z}_p,$$

the tensor products being taken over $\mathbb{Z}$ and the map $f \otimes 1$ being defined by:

$$(f \otimes 1)(\alpha \otimes a) := f(\alpha) \otimes a,$$

on the generating system of the $\alpha \otimes a$, $\alpha \in A$, $a \in \mathbb{Z}_p$. Since tensor products trivially preserve surjections, we will only write short exact sequences:

$$1 \longrightarrow B \longrightarrow A \xrightarrow{\ f\ } C \longrightarrow 1,$$

which thus yield:

$$1 \longrightarrow B \otimes \mathbb{Z}_p \longrightarrow A \otimes \mathbb{Z}_p \xrightarrow{f\otimes 1} C \otimes \mathbb{Z}_p \longrightarrow 1.$$

1.2.1 Example 1. The exact sequence:

$$1 \longrightarrow E^{\mathrm{ord}} \longrightarrow K^\times \xrightarrow{(\,\bullet\,)} P \longrightarrow 1,$$

yields:

$$1 \longrightarrow E^{\mathrm{ord}} \otimes \mathbb{Z}_p \longrightarrow K^\times \otimes \mathbb{Z}_p \xrightarrow{(\,\bullet\,)\otimes 1} P \otimes \mathbb{Z}_p \longrightarrow 1,$$

which means that the "ideal" generated by $\prod_i \alpha_i \otimes a_i$ (finite product, $\alpha_i \in K^\times$, $a_i \in \mathbb{Z}_p$) is equal to $\prod_i (\alpha_i) \otimes a_i$, and that this ideal is the unit ideal if and only if $\prod_i \alpha_i \otimes a_i$ is a "unit", i.e., $\prod_i \alpha_i \otimes a_i = \prod_j \varepsilon_j \otimes b_j$ ($\varepsilon_j \in E^{\mathrm{ord}}$, $b_j \in \mathbb{Z}_p$). □

Flatness allows us to prove (almost) everything which seems natural like:

1.2.2 Example 2. If B_1 and B_2 are subgroups of A, we have:

$$\begin{aligned}(B_1/B_1 \cap B_2) \otimes \mathbb{Z}_p &\simeq B_1 \otimes \mathbb{Z}_p/(B_1 \otimes \mathbb{Z}_p) \cap (B_2 \otimes \mathbb{Z}_p)\\ &\simeq B_1 \otimes \mathbb{Z}_p/(B_1 \cap B_2) \otimes \mathbb{Z}_p,\end{aligned}$$

since we have the exact sequence:

$$1 \longrightarrow B_2 \longrightarrow B_1B_2 \longrightarrow B_1/B_1 \cap B_2 \longrightarrow 1,$$

and the trivial equality $(B_1 \,.\, B_2) \otimes \mathbb{Z}_p = B_1 \otimes \mathbb{Z}_p \,.\, B_2 \otimes \mathbb{Z}_p$, which proves the first isomorphism, the second coming from:

$$1 \longrightarrow B_1 \cap B_2 \longrightarrow B_1 \longrightarrow B_1/B_1 \cap B_2 \longrightarrow 1.$$

We thus obtain that $(B_1 \cap B_2) \otimes \mathbb{Z}_p = (B_1 \otimes \mathbb{Z}_p) \cap (B_2 \otimes \mathbb{Z}_p)$. □

We will see in 2.3.1 that this is not true anymore for infinite intersections nor, more generally, for limiting processes.

1.2.3 Remark. What has just been said about tensoring with $\mathbb{Z}_p$ (over $\mathbb{Z}$) is also true about tensoring (over a field Q) of a Q-vector space by a field extension of Q; on the other hand, $\mathbb{F}_p$ is not flat for tensoring over $\mathbb{Z}$. □

Let us now check the following essential property mentioned at the beginning: if A is finite (or only torsion), we have $A \otimes \mathbb{Z}_p \simeq (A)_p$, where $(A)_p$ can be seen either as the p-Sylow (or p-primary) subgroup of A or, more importantly for Galois theory, as the quotient A/A' (which is a p-torsion group), where:

$$A' := \{\alpha \in A,\ \alpha \text{ of order prime to } p\}.$$

The exact sequence:

$$1 \longrightarrow A' \longrightarrow A \longrightarrow A/A' \longrightarrow 1$$

yields $A \otimes \mathbb{Z}_p \simeq (A/A') \otimes \mathbb{Z}_p$ since $A' \otimes \mathbb{Z}_p = 1$: indeed, if $\alpha' \in A$ has order n prime to p, we have $n \in \mathbb{Z}_p^\times$ hence we can write, for all $a \in \mathbb{Z}_p$:

$$\alpha' \otimes a = \alpha' \otimes (an^{-1}n) = \alpha'^n \otimes (an^{-1}) = 1 \otimes (an^{-1}) = 1.$$

In particular, we will have $(A/A') \otimes \mathbb{Z}_p \simeq A/A'$ if we show that for any p-torsion abelian group A, we have $A \otimes \mathbb{Z}_p \simeq A$. To prove this isomorphism, we use the universal property of the tensor product which implies the existence of a $\mathbb{Z}$-linear map g giving the following commutative diagram:

$$\begin{array}{ccc} A \times \mathbb{Z}_p & \xrightarrow{\ b\ } & A \\ \big\downarrow & \nearrow_{g} & \\ A \otimes \mathbb{Z}_p & & \end{array}$$

where $b(\alpha, a) := \alpha^a$ (which makes sense since if α has order p^e, we have $\alpha^a := \alpha^{a \bmod (p^e)} \in A$); we then check that the map g, which is such that $g(\alpha \otimes a) := \alpha^a$, is equal to the inverse of the canonical map:

$$f : A \longrightarrow A \otimes \mathbb{Z}_p.$$

1.2.4 Remarks. (i) If A is not a torsion $\mathbb{Z}$-module, $A \otimes \mathbb{Z}_p$ is the p-completion of A. We give it the topology for which a fundamental system of neighbourhoods of the unit element is formed by the $A \otimes p^n\mathbb{Z}_p$, for $n \geq 0$; in particular this topology may be different from that of the subgroups of finite index. We also have the profinite p-completion of A:

$$\widehat{A} := \varprojlim_{U} A/U,$$

for the subgroups U of A having finite index equal to a power of p; by definition it is a profinite group (i.e., a pro-p-group).

(ii) We may ask under which condition the p-completion of A plays the role of a profinite p-completion. Let us check that $A \otimes \mathbb{Z}_p$ is homeomorphic to a profinite group if and only if it is a $\mathbb{Z}_p$-module of finite type. One direction being clear, the converse easily follows from the fact that $A \otimes \mathbb{Z}_p$ is then compact, hence $A \otimes \mathbb{Z}_p / A \otimes p\mathbb{Z}_p$ is finite since $A \otimes p\mathbb{Z}_p$ is an open set. Write:

$$A \otimes \mathbb{Z}_p =: \langle\, x_1, \ldots , x_r \,\rangle_{\mathbb{Z}_p} \,.\, A \otimes p\mathbb{Z}_p,$$

$x_i \in A \otimes \mathbb{Z}_p$; thus $A \otimes \mathbb{Z}_p = \langle\, x_1, \ldots , x_r \,\rangle_{\mathbb{Z}_p} \,.\, A \otimes p^n\mathbb{Z}_p$ for all n, which easily yields (using I.5.4.1):

$$A \otimes \mathbb{Z}_p = \langle\, x_1, \ldots , x_r \,\rangle_{\mathbb{Z}_p}.$$

When the above condition is satisfied, we have:

$$A \otimes \mathbb{Z}_p \simeq \varprojlim_{n\geq 0} \ A \otimes \mathbb{Z}_p/A \otimes p^n\mathbb{Z}_p \simeq \varprojlim_{n\geq 0} \ A/A^{p^n},$$

which is equal to the profinite p-completion of A. In particular the condition is true when A is a $\mathbb{Z}$-module of finite type since we then have $A \simeq \mathrm{tor}_{\mathbb{Z}}(A) \times \mathbb{Z}^r$, with $\mathrm{tor}_{\mathbb{Z}}(A)$ finite, $r \in \mathbb{N}$, and since tensor products preserve direct sums, we obtain $A \otimes \mathbb{Z}_p \simeq (\mathrm{tor}_{\mathbb{Z}}(A))_p \times \mathbb{Z}_p^r$.

(iii) Finally, if A is a nontorsion commutative profinite group, the p-Sylow subgroup of A is not anymore given by $A \otimes \mathbb{Z}_p$, so that in this case we need to use the notation $(A)_p$. For example $\mathbb{Z}_p \otimes \mathbb{Z}_p$ is not equal to $\mathbb{Z}_p$; for $q \neq p$, $\mathbb{Z}_q \otimes \mathbb{Z}_p$ is not trivial, and a fortiori $\widehat{\mathbb{Z}} \otimes \mathbb{Z}_p$ is not equal to the p-Sylow subgroup $(\mathbb{Z}_p)$ of $\widehat{\mathbb{Z}}$ (the results are uninteresting $\mathbb{Z}_p$-modules of infinite rank), but however we have:

$$\varprojlim_{m\geq 1} \ ((\mathbb{Z}/m\mathbb{Z}) \otimes \mathbb{Z}_p) \simeq \varprojlim_{n\geq 0} \ (\mathbb{Z}/p^n\mathbb{Z}) = \mathbb{Z}_p.$$

(iv) We can illustrate the above notions with the following groups, constructed from multiplicative groups of the residue fields of a number field:

$$A = \bigoplus_v F_v^\times \ \text{ and } \ A = \prod_v F_v^\times,$$

where $\bigoplus_v F_v^\times$ is torsion but not profinite, while $\prod_v F_v^\times$ is profinite but not torsion. □

Let us come back to the study of $K_{(\mathfrak{m})}{}^S(p)$. It is necessary to say that the localization process is not only for convenience; often it is technically necessary because tame and wild ramifications play fundamentally distinct roles. Recall the corresponding definitions.

1.3 Tame and Wild Ramifications: Definitions and Notations. (i) Let L/K be a finite Galois extension. We say that a place $v \in Pl_0$ is tamely ramified in L/K if the residue characteristic of v does not divide the ramification index of v in L/K; otherwise it is said to be wildly ramified. By definition, the infinite places are always tame (even in the case of complexification, which implies that $[L : K]$ is even), as are more generally the unramified finite places. In the case where L/K is a p-extension, the wild places can only be places above p in K (review properties II.1.3).

(ii) When a prime number p occurs because of a localization, the set Pl_p of places of K above p plays a special role. In addition, for certain objects of the form $\mathcal{O}^S_T$, to simplify notations when $T = Pl_p$, we will write $\mathcal{O}^S_p$.[2]

Note. In certain contexts it is understood that the fact that L/K is tamely ramified at v implies that it is ramified (i.e., $e_v(L/K) > 1$), but tamely.

Recall some basic facts about the structure of the p-groups $(U_v)_p$, and $(U_v/U_v^{m_v})_p$, $v \in Pl_0$, $m_v \geq 1$, relative to a fixed prime number p (since U_v is a profinite group, the notion of a p-Sylow subgroup makes sense for U_v as for its closed subgroups and quotient groups by a closed subgroup).

1.3.1 Proposition (structure of the local unit groups). *(i) For $v \nmid p$, we have $(U_v/U_v^1)_p \simeq (F_v^\times)_p$, where $F_v^\times$ is the multiplicative group of the residue field of K at v, and for all $i \geq 1$, $(U_v^1/U_v^i)_p = 1$. We thus have:*

$$(U_v)_p \simeq (U_v/U_v^{m_v})_p \simeq (F_v^\times)_p \text{ for all } m_v \geq 1.$$

(ii) For $v|p$, we have $(U_v/U_v^1)_p = 1$, while for all $i \geq 1$, U_v^1/U_v^i is a finite p-group such that $U_v^j/U_v^{j+1} \simeq F_v$ for all $j \geq 1$. Thus, in this case:

$$(U_v/U_v^{m_v})_p \simeq U_v^1/U_v^{m_v} \text{ for all } m_v \geq 1, \text{ and } (U_v)_p = U_v^1.$$

(iii) For $v \in Pl_\infty^{\mathrm{r}}$, we have by definition $(U_v)_p = (\mathbb{R}^{\times +})_p = 1$ for all p, but $(F_v^\times)_p = 1$ for $p \neq 2$, $(F_v^\times)_2 \simeq \{\pm 1\}$. □

1.3.2 Proposition. *Let p be a fixed prime number and let $\mathfrak{m} = \prod_{v \in T} \mathfrak{p}_v^{m_v}$, $m_v \geq 0$. Then the maximal p-subextension of $K_{(\mathfrak{m})}{}^S/K$ coincides with that of $K_{(\mathfrak{m}_0)}{}^S/K$, where $\mathfrak{m}_0|\mathfrak{m}$ is given by:*

[2] There is no possible confusion with the notation $(\mathcal{O}^S)_p$ which denotes a p-Sylow subgroup, nor with $\mathcal{O}^S_{(p)}$ relative to the modulus $\mathfrak{m} = (p)$.

$$\mathfrak{m}_0 := \prod_{v\in T_p,\, m_v\geq 2} \mathfrak{p}_v^{m_v} \prod_{\substack{v\in T_{\rm ta},\, m_v\geq 1\\ \mathrm{N}\mathfrak{p}_v\equiv 1 \bmod (p)}} \mathfrak{p}_v,$$

where $T_p := T \cap Pl_p$ *and* $T_{\rm ta} := T \backslash T_p$.

Proof. This result follows from the formula of I.4.5.5, (i) which gives:

$$[K(\mathfrak{m})^S : K(\mathfrak{m}_0)^S] \times (E_{\mathfrak{m}_0}^S : E_{\mathfrak{m}}^S) = \frac{\varphi(\mathfrak{m})}{\varphi(\mathfrak{m}_0)}\ ;$$

then $[K(\mathfrak{m})^S : K(\mathfrak{m}_0)^S]$ is a divisor of $\frac{\varphi(\mathfrak{m})}{\varphi(\mathfrak{m}_0)}$, and, using the properties of the generalized Euler φ-function, we check that this integer is prime to p. □

To study $K(\mathfrak{m})^{\rm res}(p)/K$ and its subextensions we may assume that T and $\mathfrak{m}$ are such that:

$$\mathfrak{m} =: \mathfrak{m}_p \mathfrak{m}_{\rm ta}$$

(decomposition of $\mathfrak{m}$ into its wild and tame parts), where:

$$\mathfrak{m}_p := \prod_{v\in T_p} \mathfrak{p}_v^{m_v}, \ \ \text{with } m_v \geq 2 \text{ for all } v \in T_p,$$

$$\mathfrak{m}_{\rm ta} := \prod_{v\in T_{\rm ta}} \mathfrak{p}_v, \ \ \text{with } \mathrm{N}\mathfrak{p}_v \equiv 1 \bmod (p) \text{ for all } v \in T_{\rm ta}.$$

1.3.3 Remarks. (i) We can check that $K(\mathfrak{m}_{\rm ta})^{\rm res}(p)$ is the maximal tamely ramified p-extension of K in $H_T^{\rm res}$ (see II.5.2.2, (ii)).

(ii) In any case, the conductors of $K(\mathfrak{m}_{\rm ta})^{\rm res}(p)$ and of $K(\mathfrak{m})^{\rm res}(p)$ can be strict divisors of $\mathfrak{m}_{\rm ta}$ and $\mathfrak{m}$ (see II.5.1.1); note that the support of $\mathfrak{m}_{\rm ta}$ is here equal to $T_{\rm ta}$ and not to T since the moduli $\prod_{v\in T} \mathfrak{p}_v$ and $\prod_{v\in T_{\rm ta}} \mathfrak{p}_v$ yield the same p-subextension.

(iii) Choosing moduli $\mathfrak{m}$ of the above form is useful in practice, but we will not always assume that this is the case in our general reasonings. For instance, in Proposition 1.4 below, we may or may not assume that $\mathfrak{m}$ is of this form; when it is not the case, we easily recover the above discussion since certain quantities become trivial ($U_v^1/U_v^{m_v} = 1$ for $m_v = 1$, and $(F_v^\times)_p = 1$ for $\mathrm{N}\mathfrak{p}_v \not\equiv 1 \bmod (p)$). □

We can rewrite the exact sequence of 1.1 in the following form.

1.4 Proposition. *Let* $t \subseteq T$, *let* $\mathfrak{n} = \prod_{v\in t} \mathfrak{p}_v^{m_v}$ *be the corresponding pure divisor of* $\mathfrak{m}$, *and let* $\delta_\infty \subseteq \Delta_\infty$. *We then have the exact sequence:*

$$1 \longrightarrow (E_{\mathfrak{n}}^{S\cup\delta_\infty}/E_{\mathfrak{m}}^S)_p \xrightarrow{\ i_{T\backslash t,\delta_\infty}\ }$$

$$\bigoplus_{v\in T_p\backslash t_p} U_v^1/U_v^{m_v} \bigoplus_{v\in T_{\rm ta}\backslash t_{\rm ta}} (F_v^\times)_p \bigoplus_{v\in\delta_\infty} (\{\pm 1\})_p \xrightarrow{\ \rho^S\ } \mathrm{Gal}(K(\mathfrak{m})^S(p)/K(\mathfrak{n})^{S\cup\delta_\infty}(p))$$

$$= \langle\, I_v(K(\mathfrak{m})^S(p)/K)\,\rangle_{v\in T\backslash t} \langle\, D_v(K(\mathfrak{m})^S(p)/K)\,\rangle_{v\in\delta_\infty} \longrightarrow 1, \ \square$$

1.4.1 Corollary. *We have the exact sequences:*

$$1 \longrightarrow (E^{\rm res}/E^{\rm res}_{\mathfrak m})_p \xrightarrow{\ i_T\ } \bigoplus_{v\in T_p} U_v^1/U_v^{m_v} \bigoplus_{v\in T_{\rm ta}} (F_v^\times)_p \xrightarrow{\ \rho^{\rm res}\ }$$
$$\mathrm{Gal}(K(\mathfrak m)^{\rm res}(p)/H^{\rm res}(p)) = \langle\, I_v(K(\mathfrak m)^{\rm res}(p)/K)\,\rangle_{v\in T} \longrightarrow 1,$$
$$1 \longrightarrow (E^{\rm ord}/E^{\rm ord}_{\mathfrak m})_p \xrightarrow{\ i_T\ } \bigoplus_{v\in T_p} U_v^1/U_v^{m_v} \bigoplus_{v\in T_{\rm ta}} (F_v^\times)_p \xrightarrow{\ \rho^{\rm ord}\ }$$
$$\mathrm{Gal}(K(\mathfrak m)^{\rm ord}(p)/H^{\rm ord}(p)) = \langle\, I_v(K(\mathfrak m)^{\rm ord}(p)/K)\,\rangle_{v\in T} \longrightarrow 1. \square$$

In certain cases, however, such a localization is useless.

1.4.2 Example (use of the set S). In 1.1 for $t = T\backslash\{v\}$, $\delta_\infty = \emptyset$, we choose successively $S = \emptyset$ and $Pl_\infty^{\rm r}$, which yields the exact sequences:

$$1 \longrightarrow E^{\rm res}_{\frac{\mathfrak m}{\mathfrak m_v}}/E^{\rm res}_{\mathfrak m} \xrightarrow{\ i_v\ } U_v/U_v^{m_v} \xrightarrow{\ \rho^{\rm res}\ } I_v(K(\mathfrak m)^{\rm res}/K) \longrightarrow 1,$$
$$1 \longrightarrow E^{\rm ord}_{\frac{\mathfrak m}{\mathfrak m_v}}/E^{\rm ord}_{\mathfrak m} \xrightarrow{\ i_v\ } U_v/U_v^{m_v} \xrightarrow{\ \rho^{\rm ord}\ } I_v(K(\mathfrak m)^{\rm ord}/K) \longrightarrow 1\ ;$$

we deduce that $K(\mathfrak m)^{\rm res}/K(\mathfrak m)^{\rm ord}$ is unramified at $v \in T$ if and only if the groups generated by the signatures of $E^{\rm ord}_{\frac{\mathfrak m}{\mathfrak m_v}}$ and of $E^{\rm ord}_{\mathfrak m}$ coincide. Note that, by I.4.5.7, the order of $I_v(K(\mathfrak m)^{\rm res}/K(\mathfrak m)^{\rm ord})$ is a divisor of:

$$[K(\mathfrak m)^{\rm res} : K(\mathfrak m)^{\rm ord}] = \frac{2^{r_1}}{(E^{\rm ord}_{\mathfrak m} : E^{\rm res}_{\mathfrak m})},$$

which is a priori any divisor of 2^{r_1}. This result can also be obtained from Exercise 1.1.6, (ii), (γ) for $\mathfrak n = \mathfrak m$, $S = \emptyset$, $\delta_\infty = Pl_\infty^{\rm r}$, by expressing the fact that the group $I_v(K(\mathfrak m)^{\rm res}/K(\mathfrak m)^{\rm ord})$ is trivial.

For example, if $K = \mathbb Q$ and $\mathfrak m = m\mathbb Z$, we recover a classical result, which is easy to prove directly using properties of cyclotomic fields but which is not a generalizable reasoning of class field theory: we have $E^{\rm ord} = \langle -1 \rangle$, $E^{\rm res} = 1$, hence (assuming m odd and different from 1, or divisible by 4), $E^{\rm ord}_{\mathfrak m} = 1$ since $-1 \equiv 1 \bmod m\mathbb Z$ implies $m = 1$ or 2, $E^{\rm ord}_{\frac{\mathfrak m}{\mathfrak m_v}} = \langle -1 \rangle$ if and only if $m = m_v$ (i.e., m is the power of an arbitrary prime number). Thus, if m is a nontrivial power of the prime number ℓ ($m \equiv 0 \bmod (4)$ if $\ell = 2$), then ℓ is ramified in $\mathbb Q(\mu_m)/\mathbb Q(\mu_m)^{\rm nc}$ since it is totally ramified in $\mathbb Q(\mu_m)/\mathbb Q$, and elsewhere the extension $\mathbb Q(\mu_m)/\mathbb Q(\mu_m)^{\rm nc}$ is unramified for the finite places (it is of course of degree 2 in every case). $\square$

1.4.3 Proposition (deployment criterion). *Keeping the notations of Theorem 1.1, where $\mathfrak n$ is a pure divisor of $\mathfrak m$, and setting $\mathfrak m_v := \mathfrak p_v^{m_v}$, we have:*

$$\mathrm{Gal}(K(\mathfrak m)^S/K(\mathfrak n)^{S\cup\delta_\infty}) = \bigoplus_{v\in T\backslash t} I_v(K(\mathfrak m)^S/K) \bigoplus_{v\in\delta_\infty} D_v(K(\mathfrak m)^S/K)$$

if and only if:

$$(E_{\mathfrak{n}}^{S\cup\delta_\infty} : E_{\mathfrak{m}}^S) = \prod_{v\in T\backslash t} (E^S_{\frac{\mathfrak{m}}{\mathfrak{m}_v}} : E^S_{\mathfrak{m}}) \,.\, \prod_{v\in\delta_\infty} (E_{\mathfrak{m}}^{S\cup\{v\}} : E_{\mathfrak{m}}^S).$$

Proof. Indeed, direct sum takes place if and only if (with simplified notations):

$$\prod_{v\in T\backslash t} |I_v| \,.\, \prod_{v\in\delta_\infty} |D_v| = [K(\mathfrak{m})^S : K(\mathfrak{n})^{S\cup\delta_\infty}] \;;$$

but, by 1.1.2, we have:

$$|I_v| = \frac{(U_v : U_v^{\mathfrak{m}_v})}{(E^S_{\frac{\mathfrak{m}}{\mathfrak{m}_v}} : E^S_{\mathfrak{m}})} \ \text{ for } v \in T\backslash t, \ \ |D_v| = \frac{2}{(E_{\mathfrak{m}}^{S\cup\{v\}} : E_{\mathfrak{m}}^S)} \ \text{ for } v \in \delta_\infty,$$

and the result easily follows by using I.4.5.1.

Example. For $K = \mathbb{Q}$, $S = \delta_\infty = \emptyset$, we have $E^{\text{res}} = 1$, in which case the deployment condition is true for any $\mathfrak{m} = m\mathbb{Z}$, with $\mathfrak{n} = 1$. We thus obtain in terms of the cyclotomic fields $\mathbb{Q}(\mu_m) = \mathbb{Q}(m)^{\text{res}}$ the well-known fact (review II.5.5) that:

$$\mathrm{Gal}(\mathbb{Q}(m)^{\text{res}}/\mathbb{Q}) = \bigoplus_{v|m} I_v(\mathbb{Q}(m)^{\text{res}}/\mathbb{Q}),$$

the inertia groups I_v being the subgroups:

$$\begin{aligned}\mathrm{Gal}(\mathbb{Q}(m)^{\text{res}}/\mathbb{Q}(\tfrac{m}{m_v})^{\text{res}}) &\simeq \{\overline{a} \in (\mathbb{Z}/m\mathbb{Z})^\times,\ a \equiv 1 \bmod (m/m_v)\}\\ &\simeq (\mathbb{Z}/m_v\mathbb{Z})^\times.\end{aligned}$$

□

c) The Structure of Gal(H_T^S/H^S) — p-Adic Ranks

In all these computations, which only use the congruential properties of the S-units of K, we see that a limiting process is essential to have a clearer view of the structure of the Galois groups in the spirit of Chapter II, Section 3, (b). We also see that the wild part plays a crucial role which will be governed by the conjecture of Leopoldt–Jaulent–Roy on the p-adic behavior of these units.

In addition to the topological Lemma I.5.5 which will often be used, we will need the following definitions.

1.5 Definition of the $\widehat{\mathbb{Z}}$ or $\mathbb{Z}_p$-Module Structure. (i) We consider a profinite abelian group G (for instance the Galois group of an abelian extension) as a $\widehat{\mathbb{Z}}$-module. Indeed, since such a group is of the form $\varprojlim_H G/H$, for the set of closed subgroups of finite index H of G, the map:

$$\begin{array}{ccc}\widehat{\mathbb{Z}} \times G & \longrightarrow & G \\ (a\,,\sigma) & \longmapsto & \sigma^a := ((\sigma H)^{a \bmod (G:H)})_H\end{array}$$

defines a continuous scalar map which extends by continuity the $\mathbb{Z}$-module structure.[3]

(ii) If G is an abelian pro-p-group, we will only consider G as a $\mathbb{Z}_p$-module (in the same way): indeed, if we identify $\widehat{\mathbb{Z}}$ with $\prod_\ell \mathbb{Z}_\ell$, we see that an element a of $\mathbb{Z}_\ell$ (seen as an element of $\widehat{\mathbb{Z}}$), for $\ell \neq p$, acts on G by $\sigma^a = 1$ for all $\sigma \in G$. In other words, if G is a profinite abelian group written as the direct product $\prod_p (G)_p$ of its p-Sylow subgroups, the group $\widehat{\mathbb{Z}} = \prod_\ell \mathbb{Z}_\ell$ acts in the following way on G: if $a =: (a_\ell)_\ell \in \widehat{\mathbb{Z}}$ and $\sigma =: (\sigma_p)_p \in G$, then $\sigma^a = (\sigma_p^{a_p})_p$.

(iii) We will say that an abelian pro-p-extension is $\mathbb{Z}_p$-free if its Galois group, considered as a $\mathbb{Z}_p$-module, has this property. A $\mathbb{Z}_p$-extension is a Galois extension whose Galois group is isomorphic to $\mathbb{Z}_p$. If $G = \mathrm{Gal}(L/K)$ is a $\mathbb{Z}_p$-module of finite type, we will use the structure theorem for $\mathbb{Z}_p$-modules of finite type. In particular, if $\mathrm{tor}(G)$ is the torsion $\mathbb{Z}_p$-module of G (which is a finite group), the field fixed under $\mathrm{tor}(G)$ is a direct compositum over K of $\mathbb{Z}_p$-extensions of K. Conversely, the compositum of all the $\mathbb{Z}_p$-extensions of K is equal to $L^{\mathrm{tor}(G)}$ (indeed, since a $\mathbb{Z}_p$-extension is free, it must be fixed under $\mathrm{tor}(G)$).

1.5.1 Exercise (the connected component D_0). Let us come back to the context of Exercise I.4.2.8, (iii) dealing with reduced idèles. Show that:

$$\mathrm{adh}_0(E^{\mathrm{ord}}) := \bigcap_{\mathfrak{m}} (E^{\mathrm{ord}} U^{\mathrm{res}}_{0,\mathfrak{m}})$$

(the closure of $i_0(E^{\mathrm{ord}})$ in U_0^{ord}) is the $\widehat{\mathbb{Z}}$-module generated by $i_0(E^{\mathrm{ord}})$.

Deduce another expression of the connected component D_0 of the unit element of $U_0^{\mathrm{ord}}/E^{\mathrm{ord}}$.

Answer. The group U_0^{ord} is profinite; hence the same holds for its closed subgroups, which are thus sub-$\widehat{\mathbb{Z}}$-modules. It is therefore sufficient to show that, in a topological $\widehat{\mathbb{Z}}$-module A (neighbourhoods: $A^{n\widehat{\mathbb{Z}}}$, $n \geq 1$), the closure of any subgroup B of finite type (as a $\mathbb{Z}$-module) is equal to the $\widehat{\mathbb{Z}}$-module generated by B. Indeed, the inclusion $\langle B \rangle_{\widehat{\mathbb{Z}}} \subseteq \mathrm{adh}(B)$ uses the density of $\mathbb{Z}$ in $\widehat{\mathbb{Z}}$, and the inclusion $\mathrm{adh}(B) \subseteq \langle B \rangle_{\widehat{\mathbb{Z}}}$ comes from the fact that $\langle B \rangle_{\widehat{\mathbb{Z}}}$ is closed (since it is of finite type as a $\widehat{\mathbb{Z}}$-module, it is the image under a continuous map of the compact set $\widehat{\mathbb{Z}}^r$ for a suitable r). The result follows, with here $A = U_0^{\mathrm{ord}}$ and $B = i_0(E^{\mathrm{ord}})$.

---

[3] It makes sense to write $a \bmod (G:H)$ since $\widehat{\mathbb{Z}} = \varprojlim_{m \geq 1} \mathbb{Z}/m\mathbb{Z}$: if $a =: (a_m)_m$, then a modulo $(G:H)$ is given by $a_{\lambda(G:H)} \bmod (G:H)$ for any $\lambda \geq 1$.

Since the connected component D_0 is closed in $U_0^{\text{ord}}/E^{\text{ord}}$, it contains the closure of the unit element which is equal to $\text{adh}_0(E^{\text{ord}})/E^{\text{ord}}$; the quotient $U_0^{\text{ord}}/\text{adh}_0(E^{\text{ord}})$ being totally discontinuous since it is profinite we indeed have (see II.3.7):

$$D_0 = \text{adh}_0(E^{\text{ord}})/E^{\text{ord}} = \langle\, i_0(E^{\text{ord}})\,\rangle_{\widehat{\mathbb{Z}}}/E^{\text{ord}},$$

whose structure will be given in 4.4.6. □

Let us consider the exact sequence of 1.1 for $t = \delta_\infty = \emptyset$ (where we fix $T = T_p \cup T_{\text{ta}}$, but not $\mathfrak{m}$), which we can write in the following slightly different form where $i_T := (i_v)_{v\in T}$:

$$1 \longrightarrow i_T(E^S)\,.\bigoplus_{v\in T} U_v^{m_v}\Big/\bigoplus_{v\in T} U_v^{m_v} \longrightarrow \bigoplus_{v\in T} U_v\Big/\bigoplus_{v\in T} U_v^{m_v} \xrightarrow{\ \rho^S\ } \text{Gal}(K(\mathfrak{m})^S/H^S) \longrightarrow 1,$$

which yields the canonical isomorphism:

$$\text{Gal}(K(\mathfrak{m})^S/H^S) \simeq \bigoplus_{v\in T} U_v\Big/ i_T(E^S)\,.\bigoplus_{v\in T} U_v^{m_v}.$$

When $\mathfrak{m} = \prod_{v\in T} \mathfrak{p}_v^{m_v}$ ranges in $\langle\, T\,\rangle_{\mathbb{N}}$, the $\bigoplus_{v\in T} U_v^{m_v}$ form a fundamental system of open neighbourhoods of the unit element in the compact group $\bigoplus_{v\in T} U_v$, and the groups $i_T(E^S)\,.\bigoplus_{v\in T} U_v^{m_v}$ are open subgroups of $\bigoplus_{v\in T} U_v$; by going to the limit, it follows (by I.5.5) that:

$$\begin{aligned}\text{Gal}(H_T^S/H^S) &\simeq \varprojlim_{\mathfrak{m}} \Big(\bigoplus_{v\in T} U_v\Big/ i_T(E^S)\,.\bigoplus_{v\in T} U_v^{m_v}\Big)\\ &\simeq \bigoplus_{v\in T} U_v\Big/\bigcap_{\mathfrak{m}}\Big(i_T(E^S)\,.\bigoplus_{v\in T} U_v^{m_v}\Big) = \bigoplus_{v\in T} U_v\Big/\text{adh}_T(E^S),\end{aligned}$$

where H_T^S is the maximal T-ramified S-split abelian extension of K, H^S is the S-split Hilbert class field of K, and where $\text{adh}_T(E^S)$ is the closure in $\bigoplus_{v\in T} U_v$ of the diagonal embedding $i_T(E^S)$. As we have already said, the entire difficulty is concentrated in the nature of this closure; we will come back to this in Section 2.

1.5.2 Remark and Definition (T-principal S-units). To obtain the p-Sylow subgroup of:

$$\bigoplus_{v\in T} U_v\Big/\text{adh}_T(E^S),$$

it is convenient to replace E^S by its subgroup E'^S, of finite index prime to p in E^S, defined as being the inverse image:

$$E'^S := \Big\{\varepsilon \in E^S,\ i_T(\varepsilon) \in \bigoplus_{v\in T_p} U_v^1 \bigoplus_{v\in T_{\rm ta}} (F_v^\times)_p\Big\}\ ;$$

the group E'^S is thus the group of S-units congruent to 1 modulo $\prod_{v\in T_p} \mathfrak{p}_v$ and whose image in $\bigoplus_{v\in T_{\rm ta}} F_v^\times$ have order a power of p. It is called the group of T-principal S-units of K. □

Note. The group E'^S depends explicitly on T; we will see however a little later that the nice object is not E'^S (which is numerically useful), but $E^S \otimes \mathbb{Z}_p$ whose image under a natural extension of i_T is $\mathrm{adh}_T(E'^S)$.

1.5.3 Notations. Denote by $H_T^S(p)$ (resp. $H^S(p)$) the maximal T-ramified S-split abelian pro-p-extension of K (resp. the S-split p-Hilbert class field of K). Set $T_p := T \cap Pl_p$ and $T_{\rm ta} := T\backslash T_p$. □

We have obtained in complete generality:

1.6 Theorem ($\mathbb{Z}_p$-structure of $\mathrm{Gal}(H_T^S(p)/H^S(p))$). *We have:*

$$\mathrm{Gal}(H_T^S(p)/H^S(p)) \simeq \Big(\bigoplus_{v\in T_p} U_v^1 \bigoplus_{v\in T_{\rm ta}} (F_v^\times)_p\Big)\Big/\mathrm{adh}_T(E'^S),$$

where $\mathrm{adh}_T(E'^S)$ *is the topological closure of* $i_T(E'^S)$ *in* $\bigoplus_{v\in T_p} U_v^1 \bigoplus_{v\in T_{\rm ta}} (F_v^\times)_p$. *This Galois group is a* $\mathbb{Z}_p$*-module of finite type.* □

Using the conventions of notation given in 1.3, (ii) when $T = Pl_p$, we obtain for the maximal p-ramified noncomplexified abelian pro-p-extension $H_p^{\rm ord}(p)$ of K:

1.6.1 Corollary (usual p-ramification: $T = Pl_p$, $S = Pl_\infty^{\rm r}$). *We have:*

$$\mathrm{Gal}(H_p^{\rm ord}(p)/H^{\rm ord}(p)) \simeq \bigoplus_{v|p} U_v^1\Big/\mathrm{adh}_p(E'^{\rm ord}),$$

which can also be written in terms of the following so-called ordinary p-ramified class field exact sequence:

$$1 \longrightarrow \mathrm{adh}_p(E'^{\rm ord}) \longrightarrow \bigoplus_{v|p} U_v^1 \xrightarrow{\ \rho^{\rm ord}\ } \mathrm{Gal}(H_p^{\rm ord}(p)/K) \longrightarrow (\mathcal{C}\!\ell^{\rm ord})_p \longrightarrow 1. \quad \square$$

Since $\bigoplus_{v\in T_{\rm ta}} (F_v^\times)_p$ is finite and since the $\mathbb{Z}_p$-rank of $\mathrm{adh}_T(F)$ is constant for any subgroup F of finite index of E^S, the $\mathbb{Z}_p$-rank r_T^S of $\mathrm{adh}_T(E^S)$ does not depend on the sets $T_{\rm ta}$ and S_∞, and the following definition makes sense (see also 2.2, (ii)).

1.6.2 Definition (T_p-adic rank). The T_p-adic rank of E^S is by definition:

$$\mathrm{r}_{T_p}^{S_0} := \mathrm{rk}_{\mathbb{Z}_p}(\mathrm{adh}_{T_p}(E'^{S_0\,\mathrm{ord}})).$$

For $T_p = Pl_p$:

$$\mathrm{r}_p^{S_0} := \mathrm{rk}_{\mathbb{Z}_p}(\mathrm{adh}_p(E'^{S_0\,\mathrm{ord}}))$$

is called the p-adic rank of E^S. □

Since $\mathrm{Gal}(H_T^S(p)/H^S(p))$ has finite index in $\mathrm{Gal}(H_T^S(p)/K)$, the latter is a $\mathbb{Z}_p$-module of finite type and with the same $\mathbb{Z}_p$-rank. Since for each $v \in T_p$ we have $\mathrm{rk}_{\mathbb{Z}_p}(U_v^1) = [K_v : \mathbb{Q}_p]$, we obtain from 1.6:

1.6.3 Corollary. *The $\mathbb{Z}_p$-rank $\widetilde{\mathrm{r}}_T^S$ of the $\mathbb{Z}_p$-module $\mathrm{Gal}(H_T^S(p)/K)$ (i.e., the maximal number of independent T-ramified S-split $\mathbb{Z}_p$-extensions of K) depends only on T_p and S_0, and is equal to:*

$$\widetilde{\mathrm{r}}_{T_p}^{S_0} = \sum_{v\in T_p} [K_v : \mathbb{Q}_p] - \mathrm{r}_{T_p}^{S_0}.$$

For $T_p = Pl_p$, we have:

$$\widetilde{\mathrm{r}}_p^{S_0} = [K : \mathbb{Q}] - \mathrm{r}_p^{S_0}.$$ □

In Sections 2 and 3 we will see what is the conjectural value of this p-adic rank; recall simply that for $S_0 = \emptyset$, Leopoldt has conjectured in 1962 in [Le3] the following.

1.6.4 Conjecture (the Leopoldt conjecture). *The p-adic rank of the unit group of K coincides with its $\mathbb{Z}$-rank $r_1 + r_2 - 1$; in other words, $\widetilde{\mathrm{r}}_p = r_2 + 1$ (i.e., the maximal number of independent $\mathbb{Z}_p$-extensions of K is equal to $r_2 + 1$).* □

In fact, the conjecture was stated only for real abelian number fields, through the nonvanishing of the p-adic regulator; its translation in terms of p-adic rank or of $\mathbb{Q}_p$-independence of the p-adic logarithms of the fundamental units gives the general statement (we will explain this in 3.1.8, in the more general context of the p-adic Schanuel conjecture).

1.6.5 Exercise. Using the Notations 1.5.3 and that of Theorem 1.6 for $p = 2$, we introduce a set $\delta_\infty \subseteq \Delta_\infty := Pl_\infty^{\mathrm{r}} \backslash S_\infty$. From the exact sequence of 1.1 or 1.4, prove the following isomorphism:

$$\mathrm{Gal}(H_T^S(2)/H^{S\cup\delta_\infty}(2)) \simeq \Big(\bigoplus_{v\in T_2} U_v^1 \bigoplus_{v\in T_{\mathrm{ta}}\cup\delta_\infty} (F_v^\times)_2\Big)\Big/\mathrm{adh}_{T\cup\delta_\infty}(E'^{S\cup\delta_\infty}).$$ □

In the next exercise, we show how to express the closures of unit groups in a more intrinsic way than in 1.5.2; for this, we need that the embeddings i_v take their values in the p-Sylow subgroups of the groups U_v.

1.6.6 Exercise (definition of the embeddings $\bar{i}_v$). Let K be a number field.

(i) We fix a prime number p. Show that if v is a finite place, there exists on $K^\times_{\{v\}} \otimes \mathbb{Z}_p$ a canonical $\mathbb{Z}_p$-module homomorphism $\bar{i}_v$, with values in the $\mathbb{Z}_p$-module $(U_v)_p$ (equal to U_v^1 if $v|p$, and to $(\mu_{q_v-1})_p \simeq (F_v^\times)_p$ otherwise), and defined on $K^\times_{\{v\}} \otimes 1$ by means of the composition of i_v with the projection pr_v on the factor $(U_v)_p$ of U_v, by:

$$\bar{i}_v(x \otimes 1) := \mathrm{pr}_v \circ i_v(x)$$

for all $x \in K^\times$, x prime to v. Show that this embedding can also be obtained by setting:

$$\bar{i}_v(x \otimes 1) := (i_v(x^\lambda))^{\lambda^{-1}},$$

for any $\lambda \in \mathbb{Z}$ prime to p such that $i_v(x^\lambda) \in (U_v)_p$ (for $v \nmid p$, by abuse of notation i_v denotes in this case the composition of the usual embedding with reduction modulo $\mathfrak{p}_v$ under the identification of μ_{q_v-1} with $F_v^\times$).

If v is a real place at infinity and $p = 2$, show that $\mathrm{pr}_v \circ i_v : K^\times \longrightarrow \{\pm 1\}$ (which we identify with the function sgn_v) can be extended to $K^\times \otimes \mathbb{Z}_2$.

(ii) By referring to 1.5, show that there exists on $E^{\mathrm{ord}} \otimes \widehat{\mathbb{Z}}$ a canonical embedding $\bar{i}_0$ with values in the $\widehat{\mathbb{Z}}$-module $U_0^{\mathrm{ord}} := \prod_{v \in Pl_0} U_v \prod_{v \in Pl_\infty^{\mathrm{r}}} \{\pm 1\}$; give it explicitly.

(iii) Let T, $S = S_0 \cup S_\infty$, $\delta_\infty \subseteq Pl_\infty^{\mathrm{r}} \backslash S_\infty$ be as usual. Check that we have:

$$\bar{i}_{T,\delta_\infty}(E^S \otimes \mathbb{Z}_p) = \mathrm{adh}_{T \cup \delta_\infty}(E'^S)$$

in terms of T-principal S-units, where $\bar{i}_{T,\delta_\infty} := (\bar{i}_v)_{v \in T \cup \delta_\infty}$.

Similarly, show that in U_0^{ord} we have:

$$\bar{i}_0(E^{\mathrm{ord}} \otimes \widehat{\mathbb{Z}}) = \mathrm{adh}_0(E^{\mathrm{ord}}).$$

Answer. Using the universal property of tensor product for the bilinear map $(x, \alpha) \in K^\times_{\{v\}} \times \mathbb{Z}_p \longrightarrow (\mathrm{pr}_v \circ i_v(x))^\alpha$, the *existence* of $\bar{i}_v$ in (i) is immediate. For example, for $K = \mathbb{Q}$, $v = 7$ and $x = 3$, we have:

- for $p = 3$, $\lambda = 2$ is suitable and $\bar{i}_7(3) = (\bar{3}^2)^{\frac{1}{2}} = \bar{2}^{\frac{1}{2}} = \bar{2}^2 = \bar{4}$ in $(F_7^\times)_3$ (or, lifting in $(\mu_{q_v-1})_3 = (\mu_6)_3$, the unique cube root $\zeta \equiv 4 \bmod (7)$);
- for $p = 7$, $\bar{i}_7(3) = (3^6)^{\frac{1}{6}} = (1 + 7 \times 104)^{\frac{1}{6}}$, which is given by the usual binomial formula $(1 + z)^k$, $z \in (7)$, $k \in \mathbb{Z}_7^\times$.

Note that the $\bar{i}_v$ are not injective maps.

As for (ii), since $E^{\text{ord}} \otimes \widehat{\mathbb{Z}} \simeq \prod_p E^{\text{ord}} \otimes \mathbb{Z}_p$ (canonically)[4], it is sufficient to work on the p-Sylow subgroups to show that this embedding is equal to $\prod_p (\bar{i}_v^{(p)})_v$, each $(\bar{i}_v^{(p)})_v$ being the family of the embeddings above, with values in $\prod_{v \nmid p} (F_v^\times)_p \prod_{v|p} U_v^1$.

In fact, the *existence* of $\bar{i}_0$ is evident as in (i) (universal property) and is the extension by $\widehat{\mathbb{Z}}$-linearity of the map which sends $\varepsilon \otimes 1$ to $i_0(\varepsilon)$, but it is useful to give its p-components $\bar{i}_0^{(p)} := (\bar{i}_v^{(p)})_v$ precisely (see the reason in the remark of 4.4.8). In particular, $\bar{i}_0(\varepsilon) = 1$ for $\varepsilon \in E^{\text{ord}} \otimes \widehat{\mathbb{Z}}$ means that each component $\varepsilon^{(p)}$ of ε is such that, for every place v, $\bar{i}_v^{(p)}(\varepsilon^{(p)}) = 1$ in $(F_v^\times)_p$ if $v \nmid p$ (resp. U_v^1 if $v|p$) or, globally, that the residual image of ε in $\prod_{v \nmid \infty} F_v^\times \prod_{v|\infty} \{\pm 1\}$ is trivial, as well as its canonical image in $\prod_{v \nmid \infty} U_v^1$. It is not too difficult to deduce the *injectivity* of $\bar{i}_0$ by applying the local-global principle for powers (this will be done in Theorem 4.4 and used in Theorem 4.4.6 to give the structure of the connected component D_0).

Fact (iii) is analogous to what we have done in 1.5.1: here $\bar{i}_{T,\delta_\infty}(E^S \otimes \mathbb{Z}_p)$ is the $\mathbb{Z}_p$-module generated by $i_{T,\delta_\infty}(E'^S)$, and $\bar{i}_0(E^{\text{ord}} \otimes \mathbb{Z}_p)$ is the $\widehat{\mathbb{Z}}$-module generated by $i_0(E^{\text{ord}})$ (immediate); but these unit groups are of finite type. □

Globalizing Theorem 1.6, we obtain:

1.6.7 Corollary. *We have the exact sequence:*

$$1 \longrightarrow \text{adh}_{T \cup \delta_\infty}(E^{S \cup \delta_\infty}) \longrightarrow \bigoplus_{v \in T} U_v \bigoplus_{v \in \delta_\infty} \{\pm 1\} \longrightarrow \mathcal{C}\ell_T^S \longrightarrow \mathcal{C}\ell^{S \cup \delta_\infty} \longrightarrow 1,$$

which gives as particular cases:

$$1 \longrightarrow \text{adh}_T(E^{\text{res}}) \longrightarrow \bigoplus_{v \in T} U_v \longrightarrow \mathcal{C}\ell_T^{\text{res}} \longrightarrow \mathcal{C}\ell^{\text{res}} \longrightarrow 1,$$

$$1 \longrightarrow \text{adh}_T(E^{\text{ord}}) \longrightarrow \bigoplus_{v \in T} U_v \longrightarrow \mathcal{C}\ell_T^{\text{ord}} \longrightarrow \mathcal{C}\ell^{\text{ord}} \longrightarrow 1,$$

$$1 \longrightarrow \text{adh}_{T \cup Pl_\infty^r}(E^{\text{ord}}) \longrightarrow \bigoplus_{v \in T} U_v \bigoplus_{v \in Pl_\infty^r} \{\pm 1\} \longrightarrow \mathcal{C}\ell_T^{\text{res}} \longrightarrow \mathcal{C}\ell^{\text{ord}} \longrightarrow 1. \quad \square$$

This being said, it is advantageous to study the $\mathcal{C}\ell_T^S$ locally because of the deep p-adic aspects which appear independently for each p; this is the object of the next section in which we will modify the system of notation, starting from the expression:

$$\mathcal{A}_T^S := \text{Gal}(H_T^S(p)/K) \simeq (\mathcal{C}\ell_T^S)_p,$$

since other objects are going to appear which would not be amenable to the notations that we have used up to now.

---

[4] The group $E^{\text{ord}} \otimes \widehat{\mathbb{Z}} \simeq \text{tor}(E^{\text{ord}}) \times \widehat{\mathbb{Z}}^{r_1+r_2-1}$ being a commutative profinite group, it is equal to the direct product of its p-Sylow subgroups which are the $E^{\text{ord}} \otimes \mathbb{Z}_p$.

§2 Computation of $\mathcal{A}_T^S := \mathrm{Gal}(H_T^S(p)/K)$ and $\mathcal{T}_T^S := \mathrm{tor}_{\mathbb{Z}_p}(\mathcal{A}_T^S)$

Section 1 showed that the study of the Galois groups of the extensions of H^{res} which are abelian over K (for example $K(\mathfrak{m})^{\mathrm{res}}/H^{\mathrm{res}}$, $H_T^{\mathrm{res}}/H^{\mathrm{res}}$, $\overline{K}^{\mathrm{ab}}/H^{\mathrm{res}}$, with or without decomposition), can be reduced to congruential properties of units and can be expressed elementarily in algebraic terms (including in limiting processes for infinite extensions); this comes from the fact that when we are above H^{res}, class group questions are not involved. This is of course not true for the study of these same extensions over K itself as well as the corresponding limiting processes, and for this we need some additional tools (logarithms for instance). A large part of the present Chapter III will deal with this.

Let K together with sets of places T and S. Recall that for p prime, $H_T^S(p)$ is the maximal T-ramified S-split abelian pro-p-extension of K.

Set $T_p := T \cap Pl_p$ and $T_{\mathrm{ta}} := T \backslash T_p$. The aim of Section 2 is also the study of the influence of wild ramification, in other words we implicitly assume that $T_p \neq \emptyset$ (if $T_p = Pl_p$, we are dealing with classical p-ramification; if $T_p \neq Pl_p$, we will speak of incomplete p-ramification).[5] We will have other occasions to come back to the study of tamely ramified p-extensions, all the more so that, when $T_p = \emptyset$, $H_T^S(p)$ is simply the p-ray class field $K(\mathfrak{m})^S(p)$, where $\mathfrak{m} = \prod_{v \in T} \mathfrak{p}_v$; this defines the maximal T-tamely ramified S-split abelian p-extension, of finite degree over K, which does not need the p-adic techniques that we are going to introduce.

a) $\mathbb{Z}_p$-Free-Extensions — Logarithms

We give a number of notations with the conventions of 1.3, (ii) when $T = Pl_p$; we use the above definitions and 1.5.

2.1 Notations. (i) For a fixed prime number p we set:

$$\mathcal{A}_T^S := \mathrm{Gal}(H_T^S(p)/K), \quad \mathcal{B}_T^S := \mathrm{Gal}(H_T^S(p)/H^S(p)).$$

(ii) We denote by:

$$\widetilde{K}_p$$

the compositum of all the $\mathbb{Z}_p$-extensions of K; since by II.1.3.3 only the places above p can ramify in $\widetilde{K}_p/K$ and since the infinite places are split, we

[5] The expression "p-ramification" is improper here if $T_{\mathrm{ta}} \neq \emptyset$ since it should mean "unramified outside p"; thus one should say "complete or incomplete ramification at p", but this abuse of language will be frequently made since it is T_p which plays the central role.

have $\widetilde{K}_p \subseteq H_p^{\mathrm{ord}}(p)$ (the maximal p-ramified noncomplexified abelian pro-p-extension of K).

(iii) We denote by:

$$\widetilde{K}_T^S := \widetilde{K}_p \cap H_T^S,$$

the maximal T-ramified S-split subextension of $\widetilde{K}_p/K$, and by:

$$\widetilde{K}_T^{S\,\mathrm{fr}}$$

the compositum of all the T-ramified S-split $\mathbb{Z}_p$-extensions of K (i.e., the maximal $\mathbb{Z}_p$-free subextension of $\widetilde{K}_T^S$). The fields $\widetilde{K}_T^S$ and $\widetilde{K}_T^{S\,\mathrm{fr}}$ depend only on T_p and S_0 and will be denoted $\widetilde{K}_{T_p}^{S_0}$ and $\widetilde{K}_{T_p}^{S_0\,\mathrm{fr}}$.

(iv) We set:

$$\mathcal{Z}_{T_p}^{S_0} := \mathrm{Gal}(\widetilde{K}_{T_p}^{S_0\,\mathrm{fr}}/K), \quad \mathcal{T}_T^S := \mathrm{Gal}(H_T^S(p)/\widetilde{K}_{T_p}^{S_0\,\mathrm{fr}}). \qquad \square$$

Note. The notations are chosen so that for instance $\widetilde{K}_{T_p}^{S_0}$ (resp. $\widetilde{K}_{T_p}^{\mathrm{fr}\,S_0}$) always denotes the maximal S_0-split subextension of the corresponding extension $\widetilde{K}_{T_p}$ (resp. $\widetilde{K}_{T_p}^{\mathrm{fr}}$); similarly for $\widetilde{K}_{T_p}^{\mathrm{fr}}$ and for $\widetilde{K}_{T_p}^{S_0\,\mathrm{fr}}$ which denote the corresponding maximal free subextensions. Thus $\widetilde{K}_{T_p}^{\mathrm{fr}}$ denotes the compositum of the T_p-ramified $\mathbb{Z}_p$-extensions of K (we recover the compositum of all the $\mathbb{Z}_p$-extensions of K by taking $T_p = Pl_p$, case in which $\widetilde{K}_p = \widetilde{K}_p^{\mathrm{fr}}$). Finally, for $T_p = \emptyset$, $\widetilde{K}^{S_0}$ denotes $\widetilde{K}_p \cap H^{S_0\,\mathrm{ord}}$ (the maximal S_0-split subextension of the Hilbert class field of K, contained in the compositum of the $\mathbb{Z}_p$-extensions of K), $\widetilde{K}^{S_0\,\mathrm{fr}}$ being always the base field K.

2.1.1 Remark. The group $\mathcal{A}_T^S$ is a $\mathbb{Z}_p$-module of finite type; this justifies the above notations, it shows that $\mathcal{T}_T^S$ is its torsion $\mathbb{Z}_p$-module (thus finite), and that we can write $\mathcal{A}_T^S =: \mathcal{T}_T^S \oplus \Gamma$, where Γ (nonunique), isomorphic to $\mathcal{Z}_{T_p}^{S_0}$, is a free $\mathbb{Z}_p$-module of rank $\widetilde{\mathrm{r}}_{T_p}^{S_0}$ computed in 1.6.3. $\square$

These definitions can be summarized by the diagram:

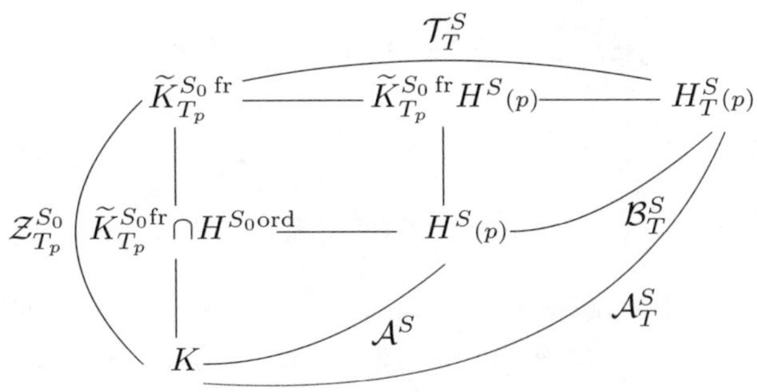

Fig. 2.1

in which $\mathcal{T}_T^S$ and $\mathcal{A}^S \simeq (\mathcal{C}\ell^S)_p$ are finite p-groups.

2.1.2 Notation. Going to the limit, the Artin maps:

$$\mathcal{C}\!\ell_{\mathfrak{m}}^{S} \longrightarrow \mathrm{Gal}(K(\mathfrak{m})^{S}/K), \quad \mathfrak{m} \in \langle T \rangle_{\mathbb{N}},$$

yield the homeomorphism of profinite groups:

$$\alpha_T^S \; : \; \varprojlim_{\mathfrak{m} \in \langle T \rangle_{\mathbb{N}}} (\mathcal{C}\!\ell_{\mathfrak{m}}^{S})_p \longrightarrow \mathcal{A}_T^S,$$

which will now simply be denoted Art since, whatever the way in which we write the ideal class group object on the left hand side, Art factors the map which sends an unramified prime ideal to its Frobenius in the extension under consideration (see II.4.3.1). □

2.2 Logarithms: Definitions and Properties. (i) We consider the $\mathbb{Q}_p$-vector space $\bigoplus_{v \in T_p} K_v$, and we denote by $\log_{T_p}$ the partial p-adic logarithm (with support T_p):

$$\begin{aligned} K^{\times} &\longrightarrow \bigoplus_{v \in T_p} K_v \\ x &\longmapsto (\log_v(x))_{v \in T_p} \end{aligned}$$

where $\log_v(x) := \log(i_v(x))$ for all $v \in T_p$, and where $\log \; : \; \mathbb{C}_p^{\times} \longrightarrow \mathbb{C}_p$ is the Iwasawa extension of the usual p-adic logarithm defined on the group of principal units $u =: 1 + x$ of $\mathbb{C}_p$ by the convergent serie:

$$\log(1+x) := \sum_{i \geq 1} (-1)^{i+1} \frac{x^i}{i},$$

and by $\log(p) := 0$ [c, Wa, Ch. 5, § 1]. We still denote by log the map $\bigoplus_{v \in T_p} K_v^{\times} \longrightarrow \bigoplus_{v \in T_p} K_v$ defined by $\log((x_v)_v) := (\log(x_v))_v$. We then have $\log_{T_p} = \log \circ i_{T_p}$ on $K^{\times}$, yielding another definition of $\log_{T_p}$.

(ii) We then define the quotient $\mathbb{Q}_p$-vector (logarithmic) space:

$$\mathcal{L}_{T_p}^{S_0} := \Big(\bigoplus_{v \in T_p} K_v \Big) \Big/ \mathbb{Q}_p \log_{T_p}(E^{S_0}).$$

By the Definition 1.6.2 of the T_p-adic rank $\mathrm{r}_{T_p}^{S_0}$ of E^{S_0}, we have:

$$\dim_{\mathbb{Q}_p}(\mathcal{L}_{T_p}^{S_0}) = \widetilde{\mathrm{r}}_{T_p}^{S_0}$$

(the $\mathbb{Z}_p$-rank of $\mathcal{Z}_{T_p}^{S_0}$) because $\log(\mathrm{adh}_{T_p}(E'^{S_0})) = \mathbb{Z}_p \log_{T_p}(E'^{S_0})$ has the same $\mathbb{Z}_p$-rank as $\mathrm{adh}_{T_p}(E'^{S_0})$ since, by [c, Wa, Prop. 5.6], the kernel of log on $\bigoplus_{v \in T_p} U_v^1$ is the group $\bigoplus_{v \in T_p} \mu_p(K_v)$ which is finite.

(iii) Finally, we denote by $\mathrm{Log}_{T_p}^{S_0}$ the map from I_T to $\mathcal{L}_{T_p}^{S_0}$ sending $\mathfrak{a} \in I_T$ to $\mathrm{Log}_{T_p}^{S_0}(\mathfrak{a})$ defined as follows.

If m is such that $\mathfrak{a}^m = (\alpha)$ with $\alpha \in K_T^\times$, we set:

$$\mathrm{Log}_{T_p}^{S_0}(\mathfrak{a}) := \tfrac{1}{m}\log_{T_p}(\alpha) \bmod \mathbb{Q}_p \log_{T_p}(E^{S_0}).$$

It is clear that if we also have $\mathfrak{a}^{m'} = (\alpha')$ with $\alpha' \in K_T^\times$, then $(\alpha)^{m'} = (\alpha')^m$, hence $\alpha^{m'} = \alpha'^m \varepsilon$ with $\varepsilon \in E^{\mathrm{ord}}$, so that:

$$\tfrac{1}{m}\log_{T_p}(\alpha) = \tfrac{1}{m'}\log_{T_p}(\alpha') + \tfrac{1}{m\,m'}\log_{T_p}(\varepsilon),$$

showing that $\mathrm{Log}_{T_p}^{S_0}$ is defined in $\mathcal{L}_{T_p}^{S_0}$ since $\mathbb{Q}_p\log_{T_p}(E^{\mathrm{ord}}) \subseteq \mathbb{Q}_p\log_{T_p}(E^{S_0})$.

Note. Although the map $\mathrm{Log}_{T_p}^{S_0}$ is defined on I_{T_p}, it is here restricted to I_T. We will later need to consider the function $\mathrm{Log}_{T_p}^{S_0}$ on $I_T \otimes \mathbb{Z}_p$, which defines a map from $I_T \otimes \mathbb{Z}_p$ to $\mathcal{L}_{T_p}^{S_0}$ having as image the $\mathbb{Z}_p$-module generated by $\mathrm{Log}_{T_p}^{S_0}(I_T)$.

2.2.1 Proposition. *If g is an automorphism group of K and if T_p is stable under g, the map $\log_{T_p}$ is a g-module homomorphism, for the canonical g-module structure of $\bigoplus_{v\in T_p} K_v$ (see (Ch. II; § 2, (c)).*

Proof. We first use the following facts: $\log_{T_p} = \log \circ i_{T_p}$ where i_{T_p} is a g-module homomorphism; for any $u \in \bigoplus_{v\in T_p} U_v^1$, $\log(u)$ is approximated by a polynomial in u with coefficients in $\mathbb{Q}$, so that log on $\bigoplus_{v\in T_p} U_v^1$ is a g-module homomorphism. Then, for $x \in \bigoplus_{v\in T_p} K_v^\times$, we write $\log(x) = \frac{1}{n}\log(x^n)$ with n such that $x^n = (p^{h_v})_v \,.\, u$, $u \in \bigoplus_{v\in T_p} U_v^1$ (i.e., $\log(x^n) = \log(u)$). □

b) $\mathcal{A}_T^S$ as an Infinitesimal Ray Class Group

Consider now $\mathcal{C}\ell_{\mathfrak{m}}^S := I_T/P_{T,\mathfrak{m},\mathrm{pos}}\langle S\rangle$, $\mathfrak{m} \in \langle T\rangle_{\mathbb{N}}$, where we recall that $P_{T,\mathfrak{m},\mathrm{pos}}\langle S\rangle := P_{T,\mathfrak{m},\Delta_\infty} \,.\, \langle S_0\rangle$, with $\Delta_\infty := Pl_\infty^{\mathrm{r}} \backslash S_\infty$. We have by 2.1.2:

$$\mathcal{A}_T^S = \mathrm{Gal}(H_T^S(p)/K) \simeq \varprojlim_{\mathfrak{m}\in\langle T\rangle_{\mathbb{N}}} (\mathcal{C}\ell_{\mathfrak{m}}^S)_p \;;$$

by 1.3.2 we can take the inverse limit over the moduli $\mathfrak{m}(n) := \mathfrak{m}_{p(n)}\mathfrak{m}_{\mathrm{ta}}$, $n \geq 0$, where:

$$\mathfrak{m}_{p(n)} = \prod_{v\in T_p} \mathfrak{p}_v^n, \quad \mathfrak{m}_{\mathrm{ta}} = \prod_{v\in T_{\mathrm{ta}}} \mathfrak{p}_v,$$

with $T_p := T \cap Pl_p$ and $T_{\mathrm{ta}} := T\backslash T_p$. We may also assume that the $v \in T_{\mathrm{ta}}$ satisfy $\mathrm{N}\mathfrak{p}_v \equiv 1 \bmod (p)$, otherwise they are uninteresting. We thus have:

$$\mathcal{A}_T^S \simeq \varprojlim_{n\geq 0} (\mathcal{C}\ell_{\mathfrak{m}(n)}^S)_p.$$

The following lemma gives a convenient way to study inverse limits of the above form that are involved in class field theory over K. It introduces the p-completion of a group, such as $\mathcal{I}_T := I_T \otimes \mathbb{Z}_p$ together with the topology of the $\mathcal{I}_T^{p^n}$, which will be our starting point. However $\mathcal{I}_T \simeq \bigoplus_{v \in Pl_0 \backslash T} \mathbb{Z}_p$ is Hausdorff but noncompact, and the situation has some similarities with the idelic description which consists in finding the kernel in J_0 of the reciprocity map (here it will be that of the Artin map in $\mathcal{I}_T$).

2.3 Fundamental lemma. *Let p be a prime number. Let X be a free abelian group and let $(Y_n)_{n\geq 0}$ be a family of subgroups of finite index of X such that:*

(i) $Y_{n+1} \subseteq Y_n$ for all $n \geq 0$,

(ii) $\dim_{\mathbb{F}_p}(X/X^pY_n) = r$ (a nonnegative constant), for all sufficiently large n.[6]

We set $\mathcal{X} := X \otimes \mathbb{Z}_p$ (with the topology of the $\mathcal{X}^{p^i}$, $i \in \mathbb{N}$), $\mathcal{Y}_n := Y_n \otimes \mathbb{Z}_p$, and $\mathcal{Y}_\infty := \bigcap_n \mathcal{Y}_n$. Then there exist $x_1, \ldots, x_r \in X$ such that:

$$\mathcal{X} = \langle\, x_1, \ldots, x_r \,\rangle \otimes \mathbb{Z}_p \,.\, \mathcal{Y}_\infty$$

and we have the homeomorphism:

$$\varprojlim_{n\geq 0} \, (X/Y_n)_p \simeq \mathcal{X}/\mathcal{Y}_\infty.$$

Proof. Let n_0 be such that the p-rank of X/Y_n is equal to r for all $n \geq n_0$ (see (ii)); we set:

$$(X/Y_{n_0})_p =: \langle\, x_1, \ldots, x_r \,\rangle Y_{n_0}/Y_{n_0}, \quad x_1, \ldots, x_r \in X,$$

and:

$$\mathcal{C} := \langle\, x_1, \ldots, x_r \,\rangle \otimes \mathbb{Z}_p.$$

We thus have $\mathcal{X} = \mathcal{C}\mathcal{Y}_{n_0}$. By (ii), the canonical maps $\mathcal{X}/\mathcal{X}^p\mathcal{Y}_n \longrightarrow \mathcal{X}/\mathcal{X}^p\mathcal{Y}_{n_0}$ are isomorphisms for all $n \geq n_0$, and, from the equality $\mathcal{X}^p\mathcal{Y}_n = \mathcal{X}^p\mathcal{Y}_{n_0}$ for all $n \geq n_0$, we deduce that $\mathcal{X} = \mathcal{C}\mathcal{Y}_{n_0} = \mathcal{C}\mathcal{X}^p\mathcal{Y}_{n_0} = \mathcal{C}\mathcal{X}^p\mathcal{Y}_n$ for all $n \geq n_0$, which by iteration (with fixed $n \geq n_0$) yields $\mathcal{X} = \mathcal{C}\mathcal{X}^{p^h}\mathcal{Y}_n$ for all $h \geq 0$. Thus $\mathcal{X} = \mathcal{C}\mathcal{Y}_n$ as soon as p^h is equal to the exponent of $\mathcal{X}/\mathcal{Y}_n$.

By I.5.4 applied to $A := \mathcal{X}$ (for the topology of the $\mathcal{X}^{p^i}$, $i \in \mathbb{N}$, $\mathcal{X}$ is Hausdorff), to the $A_n := \mathcal{Y}_n$ (which are open), and to the compact set $C := \mathcal{C}$, we have $\mathcal{X} = \mathcal{C}\mathcal{Y}_\infty$, and by I.5.5 for $B := \mathcal{Y}_\infty$, we obtain the homeomorphism:

$$\mathcal{X}/\mathcal{Y}_\infty \simeq \varprojlim_{n\geq 0} \, (\mathcal{X}/\mathcal{Y}_n) \simeq \varprojlim_{n\geq 0} \, (X/Y_n)_p. \qquad \square$$

[6] i.e., the p-rank of X/Y_n is constant for n large enough; since this p-rank is nondecreasing as a function of n, it is sufficient to ask that it is bounded.

2.3.1 Remark. Note that $\left(\bigcap_n Y_n\right) \otimes \mathbb{Z}_p$ and $\bigcap_n \left(Y_n \otimes \mathbb{Z}_p\right) = \bigcap_n \mathcal{Y}_n$ are different; for example, if X is the group of rational numbers prime to p and if:

$$Y_n = \{a \in X,\ a \equiv 1 \bmod (p^n)\},$$

we have $\bigcap_n Y_n = 1$ but $(1+2p) \otimes \log(1-2p) \cdot \frac{1}{1-2p} \otimes \log(1+2p)$ is a nontrivial element of $\bigcap_n \mathcal{Y}_n$ (it will be later called a p-infinitesimal of $\mathbb{Q}$). □

2.3.2 Notations. For the following statements, we use a system of notation analogous to that of the above lemma, in other words for number or ideal groups we denote by the corresponding calligraphic letter their p-completion:

$$\mathcal{E} := E \otimes \mathbb{Z}_p,\ \mathcal{K}^\times := K^\times \otimes \mathbb{Z}_p,$$
$$\mathcal{I} := I \otimes \mathbb{Z}_p,\ \mathcal{P} := P \otimes \mathbb{Z}_p,\ \mathcal{S}_0 := \langle\, S_0 \,\rangle \otimes \mathbb{Z}_p,$$

the upper and lower indices being unchanged. □

Note. To understand clearly the relationships between these p-completions, keep in mind the flatness of $\mathbb{Z}_p$.

2.3.3 Exercise. Let X be an abelian group and let $\mathcal{X} := X \otimes \mathbb{Z}_p$ be given the topology of the $\mathcal{X}^{p^n}$, $n \in \mathbb{N}$.

(i) Give an example for which $\mathcal{X}$ is not Hausdorff.

(ii) Show that $\mathcal{E}^{\mathrm{ord}}$, $\mathcal{I}$, and $\mathcal{K}^\times$ are Hausdorff.

Answer. The topological group $\mathcal{X}$ is Hausdorff if and only if $\bigcap_{n\in\mathbb{N}} \mathcal{X}^{p^n} = 1$.

(i) Consider $X = \mathbb{Z}_p/\mathbb{Z}$, which is p-divisible since for all $n \in \mathbb{N}$, any $\alpha \in \mathbb{Z}_p$ can be written as $a_n + p^n\alpha_n$, $a_n \in \mathbb{Z}$, and $\alpha_n \in \mathbb{Z}_p$. We then have:

$$X \otimes \mathbb{Z}_p = (p^n X) \otimes \mathbb{Z}_p = p^n (X \otimes \mathbb{Z}_p)$$

which is also p-divisible, so that $\bigcap_{n\in\mathbb{N}} \mathcal{X}^{p^n} = \mathcal{X}$; but $\mathcal{X} \neq 1$ and is even a $\mathbb{Z}_p$-module of infinite rank.

(ii) The Hausdorff property is clear as soon as $\mathcal{X}$ is a $\mathbb{Z}_p$-module of finite type or a free $\mathbb{Z}_p$-module (case of $\mathcal{E}^{\mathrm{ord}}$ and of $\mathcal{I}$ respectively). Let $x \in \bigcap_{n\in\mathbb{N}} \mathcal{K}^{\times p^n}$. We have $(x) \in \bigcap_{n\in\mathbb{N}} \mathcal{I}^{p^n}$, so that $x \in \mathcal{E}^{\mathrm{ord}}$; but $\mathcal{E}^{\mathrm{ord}} \cap \mathcal{K}^{\times p^n} = (\mathcal{E}^{\mathrm{ord}})^{p^n}$, giving the result (we can also say directly that $x \in \mathcal{E}^S$ for a suitable finite S). □

If $\mathcal{X}$ is Hausdorff, so are its subgroups, and also its quotients by closed subgroups.

2.3.4 Remarks. (i) These p-completions $\mathcal{X}$ will be given the topology of the $\mathcal{X}^{p^n}$, except if specified otherwise (as in 4.18).

(ii) The group $\mathcal{X}$ is homeomorphic to a profinite group if and only if it is compact, hence if and only if it is a $\mathbb{Z}_p$-module of finite type (see 1.2.4, (ii)).

(iii) The group $X \otimes 1$ is dense in $\mathcal{X}$. □

2.3.5 Notations. Let K be a number field together with sets of places T and S, and let $T_p := T \cap Pl_p$, $T_{\text{ta}} := T \backslash T_p$, $\Delta_\infty := Pl_\infty^{\text{r}} \backslash S_\infty$. For every $n \geq 0$, put $\mathfrak{m}_{(n)} := \prod_{v \in T_p} \mathfrak{p}_v^n \prod_{v \in T_{\text{ta}}} \mathfrak{p}_v$ and $\mathcal{P}_{T,\mathfrak{m}(n),\Delta_\infty} := P_{T,\mathfrak{m}(n),\Delta_\infty} \otimes \mathbb{Z}_p$. □

The group $\mathcal{I}_T / \mathcal{P}_{T,\mathfrak{m}(n),\Delta_\infty} \simeq \mathcal{C}\ell_{\mathfrak{m}(n)}^{S_\infty} \otimes \mathbb{Z}_p$ being finite, $\mathcal{P}_{T,\mathfrak{m}(n),\Delta_\infty}$ is an open subgroup of $\mathcal{I}_T$. Hence, we have the following result, where we recall that $\mathcal{A}_T^S := \text{Gal}(H_T^S(p)/K)$ and $\mathcal{B}_T^S := \text{Gal}(H_T^S(p)/H^S(p))$.

2.4 Theorem (description of $\mathcal{A}_T^S$). *Consider* $\mathcal{P}_{T,\infty,\Delta_\infty} := \bigcap_n \mathcal{P}_{T,\mathfrak{m}(n),\Delta_\infty}$. *We then have the homeomorphism:*

$$\mathcal{A}_T^S \simeq \mathcal{I}_T / \mathcal{P}_{T,\infty,\text{pos}}\langle S \rangle,$$

with $\mathcal{P}_{T,\infty,\text{pos}}\langle S \rangle := \mathcal{P}_{T,\infty,\Delta_\infty} \,.\, \mathcal{S}_0$. *Under this homeomorphism we have:*

$$\mathcal{B}_T^S \simeq \mathcal{P}_{T,\Delta_\infty} \,.\, \mathcal{S}_0 / \mathcal{P}_{T,\infty,\Delta_\infty} \,.\, \mathcal{S}_0 \simeq \mathcal{P}_{T,\Delta_\infty} / \mathcal{P}_{T,\infty,\Delta_\infty} \,.\, (\mathcal{E}^S),$$

where $(\mathcal{E}^S) := \{(\varepsilon),\ \varepsilon \in \mathcal{E}^S\}$.

Proof. Indeed, for $X = I_T$, $Y_n = P_{T,\mathfrak{m}(n),\text{pos}}\langle S \rangle := P_{T,\mathfrak{m}(n),\Delta_\infty} \,.\, \langle S_0 \rangle$, the hypotheses of Lemma 2.3 are satisfied, the p-rank of the groups $\mathcal{C}\ell_{\mathfrak{m}(n)}^S$ being bounded because of I.4.5.4 which yields the upper bound:

$$\text{rk}_p(\mathcal{C}\ell_{\mathfrak{m}(n)}^S) \leq \text{rk}_p(\mathcal{C}\ell^{\text{res}}) + \sum_{v \in T_p} [K_v : \mathbb{Q}_p] + |T|.$$

We again use I.5.4 with the compact set $\mathcal{S}_0$, and this yields:

$$\bigcap_n (\mathcal{P}_{T,\mathfrak{m}(n),\Delta_\infty} \,.\, \mathcal{S}_0) = \mathcal{P}_{T,\infty,\Delta_\infty} \,.\, \mathcal{S}_0.$$

We finally obtain:

$$\mathcal{A}_T^S \simeq \varprojlim_{n \geq 0} (\mathcal{C}\ell_{\mathfrak{m}(n)}^S)_p \simeq \mathcal{I}_T / \mathcal{P}_{T,\infty,\Delta_\infty} \,.\, \mathcal{S}_0.$$

Since the subgroup of I_T corresponding to H^S is $P_{T,\Delta_\infty} \,.\, \langle S_0 \rangle$, it immediately follows that, under the above isomorphism, we have:

$$\mathcal{B}_T^S \simeq \mathcal{P}_{T,\Delta_\infty} \,.\, \mathcal{S}_0 / \mathcal{P}_{T,\infty,\Delta_\infty} \,.\, \mathcal{S}_0 \simeq \mathcal{P}_{T,\Delta_\infty} / \mathcal{P}_{T,\infty,\Delta_\infty} \,.\, (\mathcal{E}^S)$$

since $(P_{T,\Delta_\infty} \otimes \mathbb{Z}_p) \cap (\langle S_0 \rangle \otimes \mathbb{Z}_p) = (P_{T,\Delta_\infty} \cap \langle S_0 \rangle) \otimes \mathbb{Z}_p$, and $P_{T,\Delta_\infty} \cap \langle S_0 \rangle$ is the group of ideals generated by an S-unit. □

The expression of $\mathcal{B}_T^S$ given by 1.6 is simpler than the one obtained here which allows us to link the two points of view; on the other hand, as we have already said, the class group is an obstruction to an elementary description of $\mathcal{A}_T^S$.

The following result gives a convenient characterization of $\mathcal{P}_{T,\infty,\Delta_\infty}$.

2.4.1 Proposition. *We have:*

$$\mathcal{P}_{T,\infty,\Delta_\infty} = \{(x),\ \ x \in \mathcal{K}_{T,\Delta_\infty}^\times,\ \ \bar{i}_T(x) = 1\},$$

where $\bar{i}_T$ is the embedding $\mathcal{K}_T^\times \longrightarrow \bigoplus_{v\in T_p} U_v^1 \bigoplus_{v\in T_{\mathrm{ta}}} (F_v^\times)_p$ (see 1.6.6, (i)).

Proof. If $(x) \in \mathcal{P}_{T,\infty,\Delta_\infty}$, we may assume that $x \in \mathcal{K}_{T,\mathfrak{m}(1),\Delta_\infty}^\times$; we then have $x \in \bigcap_n (\mathcal{K}_{T,\mathfrak{m}(n),\Delta_\infty}^\times . \mathcal{E}_{\mathfrak{m}(1)}^{S_\infty})$ and by I.5.4 applied to the group $\mathcal{K}_T^\times$, to the open sets $\mathcal{K}_{T,\mathfrak{m}(n),\Delta_\infty}^\times$, and to the compact set $\mathcal{E}_{\mathfrak{m}(1)}^{S_\infty}$, we have $x = x_\infty\, \varepsilon$ with $x_\infty \in \bigcap_n \mathcal{K}_{T,\mathfrak{m}(n),\Delta_\infty}^\times$ (thus $\bar{i}_T(x_\infty) = 1$) and $\varepsilon \in \mathcal{E}_{\mathfrak{m}(1)}^{S_\infty}$. An inclusion follows since $(x) = (x\,\varepsilon^{-1}) = (x_\infty)$.[7] The other inclusion comes from the exact sequence of I.4.3.2 for $\mathfrak{n} = 1$, $\delta_\infty = \emptyset$:

$$1 \longrightarrow K_{T,\mathfrak{m}(n),\Delta_\infty}^\times \longrightarrow K_{T,\Delta_\infty}^\times \xrightarrow{\ i_T\ } \bigoplus_{v\in T_p} U_v/U_v^n \bigoplus_{v\in T_{\mathrm{ta}}} F_v^\times \longrightarrow 1,$$

which yields, after tensoring by $\mathbb{Z}_p$:

$$1 \longrightarrow \mathcal{K}_{T,\mathfrak{m}(n),\Delta_\infty}^\times \longrightarrow \mathcal{K}_{T,\Delta_\infty}^\times \xrightarrow{\ \bar{i}_T\ } \bigoplus_{v\in T_p} U_v^1/U_v^n \bigoplus_{v\in T_{\mathrm{ta}}} (F_v^\times)_p \longrightarrow 1,$$

so that we easily obtain from I.5.5, the exact sequence:

$$1 \longrightarrow \bigcap_n \mathcal{K}_{T,\mathfrak{m}(n),\Delta_\infty}^\times \longrightarrow \mathcal{K}_{T,\Delta_\infty}^\times \xrightarrow{\ \bar{i}_T\ } \bigoplus_{v\in T_p} U_v^1 \bigoplus_{v\in T_{\mathrm{ta}}} (F_v^\times)_p \longrightarrow 1,$$

showing that the kernel of $\bar{i}_T$ in $\mathcal{K}_{T,\Delta_\infty}^\times$ is equal to:

$$\mathcal{K}_{T,\infty,\Delta_\infty}^\times := \bigcap_n \mathcal{K}_{T,\mathfrak{m}(n),\Delta_\infty}^\times.$$

The proposition follows. □

2.4.2 Remarks. (i) The group $\mathcal{K}_{T,\infty,\Delta_\infty}^\times$ has been called by Jaulent the group of T-infinitesimals Δ_∞-positive of K; these groups will occur again several times (in the Chapter IV and the Appendix for instance).

(ii) In terms of infinitesimals, we have the exact sequence:

[7] Two such generators x_∞ differ by an element of $\mathcal{E}_T^{S_\infty} := \{\eta \in \mathcal{E}^{S_\infty},\ \ \bar{i}_T(\eta) = 1\}$.

$$1 \longrightarrow \mathcal{E}^S \,.\, \mathcal{K}^\times_{T,\infty,\Delta_\infty} \longrightarrow \mathcal{K}^\times_{T,\Delta_\infty} \xrightarrow{\bar{i}_T} \bigoplus_{v\in T_p} U_v^1 \bigoplus_{v\in T_{\rm ta}} (F_v^\times)_p \Big/ \mathrm{adh}_T(E'^S) \longrightarrow 1,$$

which also gives an interpretation of $\mathcal{B}_T^S$ (see 1.6 or 2.4).

(iii) If we put $\mathcal{R}_{T,\infty,\mathrm{pos}} := \Big(\bigcap_n P_{T,\mathfrak{m}(n),\mathrm{pos}}\Big) \otimes \mathbb{Z}_p$, we have $\mathcal{R}_{T,\infty,\mathrm{pos}} \subseteq \mathcal{P}_{T,\infty,\mathrm{pos}} \subseteq \mathcal{I}_T$, and $\mathcal{P}_{T,\infty,\mathrm{pos}}/\mathcal{R}_{T,\infty,\mathrm{pos}}$ is the connected component of 1 in $\mathcal{I}_T/\mathcal{R}_{T,\infty,\mathrm{pos}}$. Thus, we have a similar formalism to that of the idèle class group $J/K^\times$ when we look at $(J/K^\times)\Big/\Big(\bigcap_{\mathfrak{m}}(K^\times U^{\mathrm{res}}_{\mathfrak{m}})\Big/K^\times\Big)$ instead of $J\Big/\bigcap_{\mathfrak{m}}(K^\times U^{\mathrm{res}}_{\mathfrak{m}})$. We will not go any further in this direction since we have introduced these objects only for computational aspects. The reader may check that, for $K = \mathbb{Q}(\sqrt{2})$, $p = 2$, $T = \{(\sqrt{2})\}$, the ideal $\Big(\frac{11+6\sqrt{2}}{7}\Big)$ is in $\bigcap_n P_{T,(\sqrt{2})^n,\mathrm{pos}}$ and that $\mathcal{R}_{T,\infty,\mathrm{pos}}$ is different from $\mathcal{P}_{T,\infty,\mathrm{pos}}$. □

We now come to the study of the natural extension of the function $\mathrm{Log}^{S_0}_{T_p}$ to $\mathcal{I}_T$ (see 2.2), and we look at its values on $\mathcal{P}_{T,\infty,\Delta_\infty} \,.\, \mathcal{S}_0$. We already know that $\mathrm{Log}^{S_0}_{T_p}(\mathcal{S}_0) = 0$. If $\mathfrak{a} \in \mathcal{P}_{T,\infty,\Delta_\infty}$, the above proposition implies immediately $\mathrm{Log}^{S_0}_{T_p}(\mathfrak{a}) = 0$. Thus $\mathrm{Log}^{S_0}_{T_p}$ factors, and we still denote by $\mathrm{Log}^{S_0}_{T_p}$ the map:

$$\mathcal{I}_T/\mathcal{P}_{T,\infty,\Delta_\infty} \,.\, \mathcal{S}_0 \longrightarrow \mathcal{L}^{S_0}_{T_p}.$$

The map:

$$\mathcal{A}^S_T \longrightarrow \mathcal{L}^{S_0}_{T_p},$$

which is obtained from it, is thus $\mathrm{Log}^{S_0}_{T_p} \circ \mathrm{Art}^{-1}$ which we still denote $\mathrm{Log}^{S_0}_{T_p}$ *by abuse of notation.*

Let us find the kernel of the above map (for $T_p \neq \emptyset$). Assume that $\mathrm{Log}^{S_0}_{T_p}(\mathfrak{a}) = 0$ in $\mathcal{L}^{S_0}_{T_p}$ for some $\mathfrak{a} \in \mathcal{I}_T$. There exists an integer $h \neq 0$ such that:

$$\mathfrak{a}^h = (\alpha), \quad \alpha \in \mathcal{K}^\times_{T,\mathfrak{m}(n_1),\Delta_\infty},$$

where n_1 is chosen such that the $U_v^{n_1}$ for $v \in T_p$ have no torsion, so we obtain, by the natural extension of $\log_{T_p}$ to $\mathcal{K}^\times_{T,\mathfrak{m}(n_1),\mathrm{pos}}$ (which is $\log \circ \bar{i}_{T_p}$):

$$\log_{T_p}(\alpha) \in \mathbb{Q}_p\log_{T_p}(E^{S_0}) = \mathbb{Q}_p\log_{T_p}(E^S_{\mathfrak{m}(n_1)})\ ;$$

thus, there exist $c \in \mathbb{Z}_p$ and $\varepsilon \in \mathcal{E}^S_{\mathfrak{m}(n_1)}$ such that $c\log_{T_p}(\alpha) = \log_{T_p}(\varepsilon)$. This equality yields (in $\mathcal{K}^\times_{T,\mathfrak{m}(n_1),\Delta_\infty}$):

$$\log_{T_p}(\alpha^c \,.\, \varepsilon^{-1}) = 0.$$

However, we have the following general result.

2.4.3 Lemma 1. *The kernel of* $\log_{T_p}$ *on* $\mathcal{K}_{T_p}^\times$ *is equal to the set of* x *such that* $\bar{i}_{T_p}(x) \in \bigoplus_{v\in T_p} \mu_p(K_v)$.

Proof. We have $\log_{T_p} = \log \circ \bar{i}_{T_p}$, which proves the result since the kernel of log on $\bigoplus_{v\in T_p} U_v^1$ is equal to $\bigoplus_{v\in T_p} \mu_p(K_v)$ by [c, Wa, Prop. 5.6]. □

Since here $\alpha^c\,.\,\varepsilon^{-1} \in \mathcal{K}_{T,\mathfrak{m}(n_1),\Delta_\infty}^\times$, we have $\bar{i}_T(\alpha^c\,.\,\varepsilon^{-1}) = 1$, and 2.4.1 yields:

$$\mathfrak{a}^{hc} \in \mathcal{P}_{T,\infty,\Delta_\infty}\,.\,\mathcal{S}_0,$$

which can be written $\mathrm{Art}(\mathfrak{a}^{hc}) = 1$ in $\mathcal{A}_T^S$, in other words $\mathrm{Art}(\mathfrak{a}) \in \mathcal{T}_T^S$, the converse being clear.

2.4.4 Lemma 2. *We have the equalities:*

$$\mathbb{Z}_p\mathrm{Log}_{T_p}^{S_0}(I_T) = \mathbb{Z}_p\mathrm{Log}_{T_p}^{S_0}(I_{T_p}) \ \text{ and } \ \mathbb{Z}_p\mathrm{Log}_{T_p}^{S_0}(P_{T,\Delta_\infty}) = \mathbb{Z}_p\mathrm{Log}_{T_p}^{S_0}(P_{T_p}).$$

Proof. This comes from the fact that (as Galois groups):

$$\mathbb{Z}_p\mathrm{Log}_{T_p}^{S_0}(I_T) = \mathcal{Z}_{T_p}^{S_0},$$

independently of the tame places of T, and that:

$$\mathbb{Z}_p\mathrm{Log}_{T_p}^{S_0}(P_{T,\Delta_\infty}) \text{ fixes } \widetilde{K}_{T_p}^{S_0\,\mathrm{fr}} \cap H^S = \widetilde{K}_{T_p}^{S_0\,\mathrm{fr}} \cap H^{S_0\,\mathrm{ord}}.$$

This can be seen directly by noting that, because of I.5.1.2, $I_{T_p} = I_T P_{T_p,\mathfrak{m}_p(n)}$ for all n, so that:

$$\mathrm{Log}_{T_p}^{S_0}(\mathcal{I}_{T_p}) = \mathrm{Log}_{T_p}^{S_0}(\mathcal{I}_T) + \mathrm{Log}_{T_p}^{S_0}(\mathcal{P}_{T_p,\mathfrak{m}_p(n)}) \ \text{ for all } n,$$

which proves the first equality because of I.5.4 since $\mathrm{Log}_{T_p}^{S_0}(\mathcal{I}_T)$ is compact. The chinese remainder theorem shows that $K_{T_p}^\times = K_{T,\Delta_\infty}^\times K_{T_p,\mathfrak{m}_p(n)}^\times$ for all n, so that:

$$\mathcal{P}_{T_p} = \mathcal{P}_{T,\Delta_\infty}\mathcal{P}_{T_p,\mathfrak{m}_p(n)} \text{ for all } n,$$

which in analogous way proves the second equality. □

We have thus obtained the following result giving a simple numerical invariant which canonically describes $\mathrm{Gal}(\widetilde{K}_{T_p}^{S_0\,\mathrm{fr}}/K)$, $\widetilde{K}_{T_p}^{S_0\,\mathrm{fr}}$ being the compositum of the T_p-ramified S_0-split $\mathbb{Z}_p$-extensions of K, and which characterizes the torsion $\mathbb{Z}_p$-module $\mathcal{T}_T^S$ of $\mathcal{A}_T^S$. Recall that $H_T^S(p)$ (resp. $H^S(p)$) denotes the maximal T-ramified S-split abelian pro-p-extension of K (resp. the S-split p-Hilbert class field of K) (see Fig. 2.1).

2.5 Theorem ($\mathbb{Z}_p$-structure of $\mathcal{A}_T^S$ [Gr1; Gr2] (1982/1983)). *(i) The function* $\mathrm{Log}_{T_p}^{S_0}$ *(which is composed with* Art^{-1}*) induces on* $\mathcal{A}_T^S := \mathrm{Gal}(H_T^S(p)/K)$ *the exact sequence:*

$$1 \longrightarrow \mathcal{T}_T^S \longrightarrow \mathcal{A}_T^S \xrightarrow{\mathrm{Log}_{T_p}^{S_0}} \mathbb{Z}_p\mathrm{Log}_{T_p}^{S_0}(I_{T_p}) \longrightarrow 0,$$

in which $\mathbb{Z}_p\mathrm{Log}_{T_p}^{S_0}(I_{T_p})$ *is homeomorphic to* $\mathcal{Z}_{T_p}^{S_0} := \mathrm{Gal}(\widetilde{K}_{T_p}^{S_0\,\mathrm{fr}}/K)$.

(ii) By restriction to $\mathcal{B}_T^S := \mathrm{Gal}(H_T^S(p)/H^S(p))$, *we have the exact sequence:*

$$1 \longrightarrow \mathrm{tor}_{\mathbb{Z}_p}\Big(\bigoplus_{v\in T_p} U_v^1 \bigoplus_{v\in T_{\mathrm{ta}}} (F_v^\times)_p \Big/ \mathrm{adh}_T(E'^S)\Big) \longrightarrow \mathcal{B}_T^S \xrightarrow{\mathrm{Log}_{T_p}^{S_0}} \mathbb{Z}_p\mathrm{Log}_{T_p}^{S_0}(P_{T_p}) \longrightarrow 0,$$

in which $\mathbb{Z}_p\mathrm{Log}_{T_p}^{S_0}(P_{T_p}) \simeq \mathrm{Gal}(\widetilde{K}_{T_p}^{S_0\,\mathrm{fr}}/\widetilde{K}_{T_p}^{S_0\,\mathrm{fr}} \cap H^{S_0\,\mathrm{ord}})$, *and where* E'^S *is the group of* T*-principal* S*-units (see 1.5.2).*

(iii) If v *is a finite place of* K *not belonging to* T_p, *the image in* $\mathbb{Z}_p\mathrm{Log}_{T_p}^{S_0}(I_{T_p})$ *of the Frobenius* $\left(\dfrac{\widetilde{K}_{T_p}^{S_0\,\mathrm{fr}}/K}{v}\right)$ *is* $\mathrm{Log}_{T_p}^{S_0}(\mathfrak{p}_v)$. □

2.5.1 Corollary (usual p-ramification: $T = Pl_p$, $S = Pl_\infty^{\mathrm{r}}$). *We have (with the usual conventions of notation when* $T = Pl_p$*) the exact sequence:*

$$1 \longrightarrow \mathcal{T}_p^{\mathrm{ord}} \longrightarrow \mathcal{A}_p^{\mathrm{ord}} \xrightarrow{\mathrm{Log}_p} \mathbb{Z}_p\mathrm{Log}_p(I_p) \longrightarrow 0,$$

in which $\mathcal{Z}_p := \mathrm{Gal}(\widetilde{K}_p/K) \simeq \mathbb{Z}_p\mathrm{Log}_p(I_p)$.

If $v \nmid p$, *the image in* $\mathbb{Z}_p\mathrm{Log}_p(I_p)$ *of* $\left(\dfrac{\widetilde{K}_p/K}{v}\right)$ *is* $\mathrm{Log}_p(\mathfrak{p}_v)$. □

c) Computation of $\mathcal{T}_T^S$

The above theorem gives an explicit way for computing these torsion groups.

Note. In the following statements, $\mathbb{Z}_p\mathrm{Log}_{T_p}^{S_0}(P_{T_p})$, being the image of $\mathbb{Z}_p\mathrm{log}_{T_p}(K_{T_p}^\times)$ in $\mathcal{L}_{T_p}^{S_0}$, is equal to $\bigoplus_{v\in T_p} \log(U_v^1) \bmod \mathbb{Q}_p\mathrm{log}_{T_p}(E^{S_0})$.

2.6 Theorem. *We have the following formulas (see Fig. 2.1).*

(i) (part of the S_0*-split Hilbert class field contained in the free quotient):*

$$[\widetilde{K}_{T_p}^{S_0\,\mathrm{fr}} \cap H^{S_0\,\mathrm{ord}} : K] = (\mathbb{Z}_p\mathrm{Log}_{T_p}^{S_0}(I_{T_p}) : \mathbb{Z}_p\mathrm{Log}_{T_p}^{S_0}(P_{T_p}))\ ;$$

(ii) (order of the torsion):

$$|\mathcal{T}_T^S| = \Big|\mathrm{tor}_{\mathbb{Z}_p}\Big(\bigoplus_{v\in T_p} U_v^1 \bigoplus_{v\in T_{\mathrm{ta}}} (F_v^\times)_p \Big/ \mathrm{adh}_T(E'^S)\Big)\Big| \times \frac{|(\mathcal{C}\ell^S)_p|}{(\mathbb{Z}_p\mathrm{Log}_{T_p}^{S_0}(I_{T_p}) : \mathbb{Z}_p\mathrm{Log}_{T_p}^{S_0}(P_{T_p}))}. \quad \square$$

Note that in (ii) the expression for $|\mathcal{T}_T^S|$ really depends on the tame places of T and on the infinite places of S.

2.6.1 Corollary (usual p-ramification: $T = Pl_p$, $S_0 = \emptyset$). *We have:*

(i) $[\widetilde{K}_p \cap H^{\mathrm{ord}} : K] = (\mathbb{Z}_p\mathrm{Log}_p(I_p) : \mathbb{Z}_p\mathrm{Log}_p(P_p))$*;*

(ii$_1$) (restricted sense):

$$|\mathcal{T}_p^{\mathrm{res}}| = \Big|\mathrm{tor}_{\mathbb{Z}_p}\Big(\bigoplus_{v|p} U_v^1 \Big/ \mathrm{adh}_p(E'^{\mathrm{res}})\Big)\Big| \times \frac{|(\mathcal{C}\ell^{\mathrm{res}})_p|}{(\mathbb{Z}_p\mathrm{Log}_p(I_p) : \mathbb{Z}_p\mathrm{Log}_p(P_p))},$$

(ii$_2$) (ordinary sense):

$$|\mathcal{T}_p^{\mathrm{ord}}| = \Big|\mathrm{tor}_{\mathbb{Z}_p}\Big(\bigoplus_{v|p} U_v^1 \Big/ \mathrm{adh}_p(E'^{\mathrm{ord}})\Big)\Big| \times \frac{|(\mathcal{C}\ell^{\mathrm{ord}})_p|}{(\mathbb{Z}_p\mathrm{Log}_p(I_p) : \mathbb{Z}_p\mathrm{Log}_p(P_p))}. \quad \square$$

For a study of the p-rank of the groups $\mathcal{T}_T^S$, we refer to 4.2. For the relations between the various $\mathcal{T}_T^{S_\infty}$, when T contains Pl_p, see the important Theorem 4.1.5. Indeed, we first need to prove some important results on the structure of $\mathcal{A}_T^S$, such as deployment Theorem 4.1 which will simplify the first factor in formula (ii) of Theorem 2.6 above, when $Pl_p \subseteq T$.

In the basic case of usual p-ramification in the ordinary sense, we obtain the diagram:

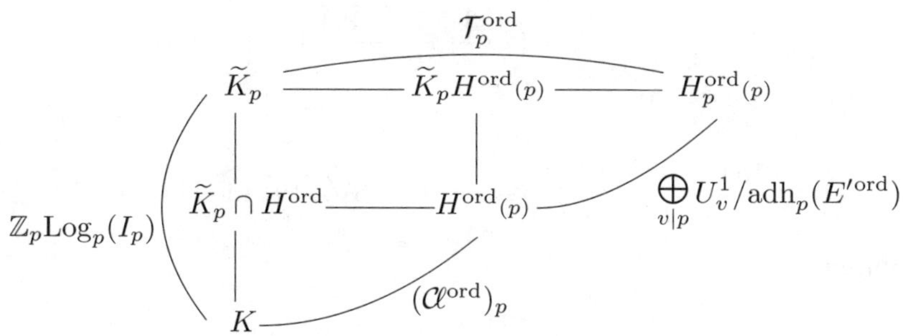

Fig. 2.2

(see Fig. 2.1) in which:

$$\mathrm{Gal}(H_p^{\mathrm{ord}}(p)/\widetilde{K}_p H^{\mathrm{ord}}(p)) \simeq \mathrm{tor}_{\mathbb{Z}_p}\Big(\bigoplus_{v|p} U_v^1 \big/ \mathrm{adh}_p(E'^{\mathrm{ord}})\Big),$$

$$\mathrm{Gal}(\widetilde{K}_p/\widetilde{K}_p \cap H^{\mathrm{ord}}(p)) \simeq \mathbb{Z}_p\mathrm{Log}_p(P_p) = \bigoplus_{v|p} \log(U_v^1) \bmod \mathbb{Q}_p\mathrm{log}_p(E).$$

This diagram shows clearly how the three fundamental arithmetic invariants occur: the class group, the unit group, and the torsion group of the p-ramification.

2.6.2 Remark (practical computations). Since there exist ideals $\mathfrak{a}_1, \ldots, \mathfrak{a}_r$ prime to T_p, such that:

$$\mathcal{C}\ell^{S_0\,\mathrm{ord}} = \langle\, c\ell^{S_0\,\mathrm{ord}}(\mathfrak{a}_1), \ldots, c\ell^{S_0\,\mathrm{ord}}(\mathfrak{a}_r)\,\rangle,$$

we obtain that the index $(\mathbb{Z}_p\mathrm{Log}_{T_p}^{S_0}(I_{T_p}) : \mathbb{Z}_p\mathrm{Log}_{T_p}^{S_0}(P_{T_p}))$ is equal to:

$$\begin{aligned}(\mathrm{Log}_{T_p}^{S_0}\langle\, \mathfrak{a}_1, \ldots, \mathfrak{a}_r\,\rangle + \mathbb{Z}_p\mathrm{Log}_{T_p}^{S_0}(P_{T_p}) : \mathbb{Z}_p\mathrm{Log}_{T_p}^{S_0}(P_{T_p})) &= \\ \Big(\mathrm{Log}_{T_p}^{S_0}\langle\, \mathfrak{a}_1, \ldots, \mathfrak{a}_r\,\rangle + \bigoplus_{v\in T_p} \log(U_v^1) : \bigoplus_{v\in T_p} \log(U_v^1)\Big)&,\end{aligned}$$

where each term $\bigoplus_{v\in T_p} \log(U_v^1)$ of the above index is considered modulo $\mathbb{Q}_p\mathrm{log}_{T_p}(E^{S_0})$. □

These expressions are easily computable, as soon as we know the class and unit group of K (in the sense given by the software packages PARI, KANT,...).

2.6.3 Examples. (i) $K = \mathbb{Q}(\sqrt{-191}\,)$, $p = 13$ (split in $K/\mathbb{Q}$). We have $\mathcal{C}\ell = \langle\, c\ell(\mathfrak{l})\,\rangle$ of order 13, where $\mathfrak{l}$ is a prime ideal above 2. We have:

$$\mathfrak{l}^{13} = \Big(\frac{153 + 7\sqrt{-191}}{2}\Big),$$

and since:

$$\Big(\frac{153 + 7\sqrt{-191}}{2}\Big)^{12} \equiv 1 + 4 \times 13\sqrt{-191} \bmod (13^2),$$

we obtain:

$$\mathrm{Log}_{13}(\mathfrak{l}) = \frac{1}{13}\,\mathrm{Log}_{13}\Big(\frac{153 + 7\sqrt{-191}}{2}\Big) = (u, \log(2) - u),$$

where u is a 13-adic unit. Since $\mathrm{Log}_{13}(P_{13}) = 13\mathbb{Z}_{13} \oplus 13\mathbb{Z}_{13}$, we obtain, since $I_{13} = \langle\, \mathfrak{l}\rangle P_{13}$ (in other words $\mathbb{Z}_{13}\mathrm{Log}_{13}(I_{13}) = \mathbb{Z}_{13}\mathrm{Log}_{13}(\mathfrak{l}) + \mathbb{Z}_{13}\mathrm{Log}_{13}(P_{13})$):

$$(\mathbb{Z}_{13}\mathrm{Log}_{13}(I_{13}) : \mathbb{Z}_{13}\mathrm{Log}_{13}(P_{13})) = 13,$$

which yields (since $\bigoplus_{v|13} U_v^1 = \mathbb{Z}_{13}^\times \oplus \mathbb{Z}_{13}^\times$ has no 13-torsion):

$$\mathcal{T}_{13} = 1 \ \text{ and } \ H \subset \widetilde{K}_{13}.$$

The Hilbert class field of K is contained in the compositum of the $\mathbb{Z}_{13}$-extensions of K.

(ii) $K = \mathbb{Q}(\sqrt{-383}\,)$, $p = 17$ (split in $K/\mathbb{Q}$). We have $\mathcal{C}\ell = \langle\, c\ell(\mathfrak{l})\,\rangle$ of order 17, where $\mathfrak{l}$ is a prime ideal above 2. We have:

$$\mathfrak{l}^{17} = \Big(\frac{711 + 7\sqrt{-383}}{2}\Big),$$

and we obtain here:

$$(\mathbb{Z}_{17}\mathrm{Log}_{17}(I_{17}) : \mathbb{Z}_{17}\mathrm{Log}_{17}(P_{17})) = 1,$$

because $\mathrm{Log}_{17}(\mathfrak{l}) \in 17(\mathbb{Z}_{17} \oplus \mathbb{Z}_{17})$ since:

$$\Big(\frac{711 + 7\sqrt{-383}}{2}\Big)^{16} \equiv 1 \bmod (17^2)\ ;$$

so that:

$$|\mathcal{T}_{17}| = 17 \ \text{ and } \ H \cap \widetilde{K}_{17} = K.$$

Here, H is linearly disjoint from the compositum of the $\mathbb{Z}_{17}$-extensions of K. □

For another detailed numerical example of such computations, see 5.2.2.

Contrary to the general case where logarithms are necessary, the groups $\mathcal{T}_p^{\mathrm{ord}}$ are more predictable in the totally real case; this comes from the fact that, assuming the Leopoldt conjecture for p, $\mathbb{Q}_p\mathrm{log}_p(E)$ is a *hyperplane* of $\bigoplus_{v|p} K_v$, and hence that Log_p is "almost trivial". More precisely, we have $\dim_{\mathbb{Q}_p}(\mathbb{Q}_p\mathrm{log}_p(E)) = [K : \mathbb{Q}] - 1$, which implies that $\widetilde{K}_p$ is the cyclotomic $\mathbb{Z}_p$-extension $K\mathbb{Q}^{\mathrm{cycl}}{}^{(p)}$ of K (to avoid any confusion with an extension of $\mathbb{Q}_p$, the cyclotomic $\mathbb{Z}_p$-extension $\widetilde{\mathbb{Q}}_p$ of $\mathbb{Q}$ is denoted $\mathbb{Q}^{\mathrm{cycl}}{}^{(p)}$).

2.6.4 Theorem (totally real case). *Let K be a totally real number field satisfying the Leopoldt conjecture for p. Then we have:*

$$[\widetilde{K}_p \cap H^{\mathrm{ord}} : K] = \frac{(q\mathbb{Z}_p : \log(\mathrm{N}_{K/\mathbb{Q}}(U_K)))}{[K \cap \mathbb{Q}^{\mathrm{cycl}}{}^{(p)} : \mathbb{Q}]} = \frac{(\check{e}_p^{\mathrm{ab}}(K/\mathbb{Q}))_p}{t\,[K \cap \mathbb{Q}^{\mathrm{cycl}}{}^{(p)} : \mathbb{Q}]},$$

with $t = 1$ except if $p = 2$ and $-1 \notin \mathrm{N}_{K/\mathbb{Q}}\Big(\bigoplus_{v|2} U_v\Big)$ in which case $t = 2$, and $q = p$ except if $p = 2$, in which case $q = 4$.

Note. In the general case, $\check{e}_p^{\rm ab}(K/\mathbb{Q})$ is the ramification index of the local extension $\check{K}_p^{\rm ab}/\mathbb{Q}_p$ (see II.2.6.3). For $p=2$, -1 is a norm of local units in $K/\mathbb{Q}$ if and only if -1 is a norm in $\check{K}_2^{\rm ab}/\mathbb{Q}_2$. If $K/\mathbb{Q}$ is Galois, $\check{K}_p^{\rm ab}/\mathbb{Q}_p$ is equal to $K_{v_0}^{\rm ab}/\mathbb{Q}_p$ for an arbitrary $v_0|p$.

Proof of the theorem. Let Q_{n_0} (resp. Q_n) be the field $K \cap \mathbb{Q}^{\rm cycl}(p)$ (resp. $H^{\rm ord} \cap \mathbb{Q}^{\rm cycl}(p)$), of degree p^{n_0} (resp. p^n) over $\mathbb{Q}$, and let $L := Q_n K = H^{\rm ord} \cap \widetilde{K}_p$. For any field M we *temporarily* set $U_M := \bigoplus_{w \in Pl_{M,p}} U_w$.

Consider $\mathrm{N}_{L/\mathbb{Q}}(U_L)$; since L/K is unramified, we have:

$$\mathrm{N}_{L/K}(U_L) = U_K \ \text{ and } \ \mathrm{N}_{L/\mathbb{Q}}(U_L) = \mathrm{N}_{K/\mathbb{Q}}(U_K),$$

whose index in $U_{\mathbb{Q}}$ is equal to $\check{e}_p^{\rm ab}(K/\mathbb{Q})$. We also have:

$$\mathrm{N}_{L/\mathbb{Q}}(U_L) = \mathrm{N}_{Q_n/\mathbb{Q}}(\mathrm{N}_{L/Q_n}(U_L)) \subseteq V_n := \mathrm{N}_{Q_n/\mathbb{Q}}(U_{Q_n}) \ ;$$

by local class field theory, we have (see II.3.4.3):

$$V_n = \mu_{p-1} \oplus \langle\, (1+p)^{p^n} \,\rangle_{\mathbb{Z}_p} \quad \text{if } p \neq 2,$$
$$V_n = \langle\, -1 \,\rangle \oplus \langle\, 5^{2^n} \,\rangle_{\mathbb{Z}_2} \quad \text{if } p = 2,$$

which is of index p^n in $U_{\mathbb{Q}}$.

Let us show that $\mathrm{N}_{L/\mathbb{Q}}(U_L) \not\subseteq V_{n+1}$. Assume the contrary, and let σ be a nontrivial element of $\mathrm{Gal}(Q_{n+1}L/L)$ which is cyclic of order p. Since $Q_{n+1}L/L$ is ramified in at least one place above p, there exists $u \in U_L$ such that $\sigma = \rho(u)$, where ρ is the global reciprocity map in $Q_{n+1}L/L$ (see II.3.3, (iii)); restricting to Q_{n+1} we obtain σ' with order p in $\mathrm{Gal}(Q_{n+1}/\mathbb{Q})$, and by II.3.3, (iv) and (ii), we have $\sigma' = \rho'(\mathrm{N}_{L/\mathbb{Q}}(u))$ for the reciprocity map ρ' in $Q_{n+1}/\mathbb{Q}$. Since $\mathrm{N}_{L/\mathbb{Q}}(u) \in V_{n+1}$, we obtain $\sigma' = 1$, a contradiction.

We therefore have:

$$\mathrm{N}_{K/\mathbb{Q}}(U_K) \subseteq V_n, \ \ \mathrm{N}_{K/\mathbb{Q}}(U_K) \not\subseteq V_{n+1};$$

since $V_n/V_{n+1} \simeq \mathbb{Z}/p\mathbb{Z}$ and since $V_{n+1}\mathrm{N}_{K/\mathbb{Q}}(U_K)/V_{n+1}$ is a nontrivial subgroup, we have:

$$V_{n+1}\mathrm{N}_{K/\mathbb{Q}}(U_K) = V_n,$$

which leads to the following discussion.

(i) $p \neq 2$. Then $\mathrm{N}_{K/\mathbb{Q}}(U_K)$ has index prime to p in V_n, hence:

$$(\check{e}_p^{\rm ab}(K/\mathbb{Q}))_p = p^n = [Q_{n_0} : \mathbb{Q}][L : K],$$

giving the theorem in this case.

(ii) $p = 2$. The group $\mathrm{N}_{K/\mathbb{Q}}(U_K)$ is one of the three groups:

$$\langle\, -1 \,\rangle \oplus \langle\, 5^{2^n} \,\rangle_{\mathbb{Z}_2}, \ \ \langle\, 5^{2^n} \,\rangle_{\mathbb{Z}_2}, \ \ \langle\, -5^{2^n} \,\rangle_{\mathbb{Z}_2} \ ;$$

and the result follows in terms of the torsion of $\mathrm{N}_{K/\mathbb{Q}}(U_K)$.

Note that we have:

$$\frac{(\check{e}_p^{\mathrm{ab}}(K/\mathbb{Q}))_p}{t} = \frac{(U_\mathbb{Q} : \mathrm{N}_{K/\mathbb{Q}}(U_K))}{t} = (\log(U_\mathbb{Q}) : \log(\mathrm{N}_{K/\mathbb{Q}}(U_K))),$$

in every case, giving the formula:

$$[\mathbb{Q}^{\mathrm{cycl}}(p) \cap H^{\mathrm{ord}} : \mathbb{Q}] = (q\mathbb{Z}_p : \log(\mathrm{N}_{K/\mathbb{Q}}(U_K))). \qquad \square$$

2.6.5 Remarks. (i) We deduce, assuming the assumptions and notations of 2.6.4, the value of $|\mathcal{T}_p^{\mathrm{ord}}|$ which, after a few transformations from 2.6.1, (ii$_2$), can be written in the form:

$$|\mathcal{T}_p^{\mathrm{ord}}| \overset{p}{=} \frac{p\,[K \cap \mathbb{Q}^{\mathrm{cycl}}(p) : \mathbb{Q}]\,|\mathcal{C}\ell^{\mathrm{ord}}|}{\prod_{v|p} \mathrm{N}\mathfrak{p}_v} \times \frac{\mathrm{Reg}}{\sqrt{\mathrm{Disc}}},$$

where $\overset{p}{=}$ means equality up to a p-adic unit factor, where Reg and Disc are the p-adic regulator and the discriminant of K ([Coa, App. 1], [Gr1, Th. IV 1]). For a consistent definition of $\mathrm{Reg}/\sqrt{\mathrm{Disc}}$, see [AF] and [Col, 5.2 à 5.4]; in [Col] one can also find the proof of the residue formula for the p-adic zeta function of K which we can write in the form:

$$\lim_{s\to 1}\,(s-1)\,\frac{1}{2^{[K\,:\,\mathbb{Q}]}}\,\zeta_{K,p}(s) = $$

$$|\mathcal{C}\ell^{\mathrm{ord}}| \times \frac{\prod_{v|p}(\mathrm{N}\mathfrak{p}_v - 1)}{2\prod_{v|p}\mathrm{N}\mathfrak{p}_v} \times \frac{\mathrm{Reg}}{\sqrt{\mathrm{Disc}}} \overset{p}{=} \frac{|\mathcal{T}_p^{\mathrm{ord}}|}{q\,[K \cap \mathbb{Q}^{\mathrm{cycl}}(p) : \mathbb{Q}]}.$$

(ii) The fact that we consider:

$$\zeta_{K,p}^{\mathrm{ord}} := \frac{1}{2^{[K\,:\,\mathbb{Q}]}}\,\zeta_{K,p},$$

comes from the imprimitivity at the infinite places of the usual p-adic zeta function which should be written $\zeta_{K,p}^{\mathrm{res}}$.[8] This is consistent with the relation:

$$|\mathcal{T}_2^{\mathrm{res}}| = 2^{[K\,:\,\mathbb{Q}]}\,|\mathcal{T}_2^{\mathrm{ord}}|,$$

deduced from Theorem 4.1.5 which we will state later from deployment Theorem 4.1, and more generally when one introduces a tame part T_{ta} which

[8] Each factor "2" in the above relation must be considered as the Euler factor $1 - \frac{1}{\mathrm{N}\mathfrak{p}_v}$, with "$\mathrm{N}\mathfrak{p}_v = -1$" for the infinite real places v; then the "integral" $1-(-1)^s$ makes sense, at least for $s \in \mathbb{Z}_2$, in the context of 2-adic measures [Gr7, § 3].

yields $|\mathcal{T}_T^{\mathrm{ord}}|$ for $T := Pl_p \cup T_{\mathrm{ta}}$, except that we must remove the Euler factors at $v \in T_{\mathrm{ta}}$ from the zeta function and consider:

$$\zeta_{K,T}^{\mathrm{ord}}(s) := \zeta_{K,p}^{\mathrm{ord}}(s) \times \prod_{v \in T_{\mathrm{ta}}} \Big(1 - \frac{1}{(\mathrm{N}\mathfrak{p}_v)^s}\Big),$$

whose residue is (up to a p-adic unit factor) equal to:

$$\frac{|\mathcal{T}_T^{\mathrm{ord}}|}{q\,[K \cap \mathbb{Q}^{\mathrm{cycl}}(p) : \mathbb{Q}]}$$

since $1 - \dfrac{1}{\mathrm{N}\mathfrak{p}_v} \overset{p}{=} \mathrm{N}\mathfrak{p}_v - 1 = |F_v^\times|$ for all $v \in T_{\mathrm{ta}}$.

(iii) Finally, the relationship between the analytic and class field theoretic aspects of these formulas is mainly given by the p-adic measures used in 1978 in [Se6], where the group $\mathcal{T}_p^{\mathrm{ord}}$ plays a crucial role, still in the context of the Leopoldt conjecture and residue formulas. □

2.6.6 Exercise. Let K be an imaginary quadratic field. Prove that the 2-Hilbert class field of K is contained in the compositum of its $\mathbb{Z}_2$-extensions if and only if K is one of the following fields:

- $\mathbb{Q}(\sqrt{-1}\,)$, $\mathbb{Q}(\sqrt{-2}\,)$;
- $\mathbb{Q}(\sqrt{-\ell}\,)$, ℓ prime, $\ell \equiv 3,\ 5,\ 7 \bmod 8$;
- $\mathbb{Q}(\sqrt{-2\ell}\,)$, ℓ prime, $\ell \equiv 3,\ 5 \bmod 8$;
- $\mathbb{Q}(\sqrt{-\ell q}\,)$, ℓ, q primes, $\ell \equiv -q \equiv 3 \bmod 8$.

Answer. Use formulas 2.6.1 or see [Gr3, Theorem 2.3]. □

d) Class Field Theory Correspondence in $H_T^{\mathrm{res}}(p)/K$

We recall that $H_T^{\mathrm{res}}(p)$ is the maximal abelian T-ramified pro-p-extension of K and $\widetilde{K}_{T_p}^{\mathrm{fr}}$ the compositum of the T_p-ramified $\mathbb{Z}_p$-extensions of K. We first remark that Theorem 2.5 can be stated in the following slightly more general form.

2.7 Proposition. *Let M/K be a finite subextension of $H_T^{\mathrm{res}}(p)/K$ and let $A \subseteq I_T$ be its Artin group. We have the isomorphism:*

$$\mathrm{Gal}(H_T^{\mathrm{res}}(p)/M) \simeq A \otimes \mathbb{Z}_p/\mathcal{P}_{T,\infty,\mathrm{pos}},$$

and the exact sequence:

$$1 \longrightarrow \mathrm{Gal}(H_T^{\mathrm{res}}(p)/M\widetilde{K}_{T_p}^{\mathrm{fr}}) \longrightarrow \mathrm{Gal}(H_T^{\mathrm{res}}(p)/M) \xrightarrow{\mathrm{Log}_{T_p}} \mathbb{Z}_p\mathrm{Log}_{T_p}(A) \longrightarrow 0,$$

in which $\mathbb{Z}_p\mathrm{Log}_{T_p}(A) \simeq \mathrm{Gal}(\widetilde{K}_{T_p}^{\mathrm{fr}}/M \cap \widetilde{K}_{T_p}^{\mathrm{fr}})$, *and where the kernel is the torsion* $\mathbb{Z}_p$*-module of* $\mathrm{Gal}(H_T^{\mathrm{res}}(p)/M)$. □

In the general case of abelian pro-p-extensions with restricted ramification, and mainly when $T_p \neq \emptyset$, we can state the correspondence of class field theory in the following way.

2.8 Theorem. *There is a bijective Galois correspondence between the subextensions* M/K *of* $H_T^{\mathrm{res}}(p)/K$ *and the closed subgroups* $\mathcal{N}$ *of* $\mathcal{I}_T$ *containing* $\mathcal{P}_{T,\infty,\mathrm{pos}}$. *For a given* M, *we have* $\mathcal{N} = \bigcap_{\substack{M' \subseteq M \\ M'/K \text{ finite}}} A_{M'} \otimes \mathbb{Z}_p$, *where* $A_{M'}$ *is the Artin group of* M' *in* I_T. □

The finite extensions correspond to closed subgroups of finite index of the form $A \otimes \mathbb{Z}_p$ as in 2.7 above (for instance $A = P_{T,\mathfrak{m},\mathrm{pos}}$, $\mathfrak{m} \in \langle T \rangle_{\mathbb{N}}$, for ray class fields).

2.8.1 Example. Take $T = Pl_p$. Then it is immediate to check that the subgroup $\mathcal{N}$ of $\mathcal{I}_T$, corresponding to the cyclotomic p-extension $K\mathbb{Q}(\mu_{p^\infty})(p)$ of K, is the subgroup $\{\mathfrak{a} \in \mathcal{I}_p,\ \mathrm{N}_{K/\mathbb{Q}}(\mathfrak{a}) \in \mathcal{P}_{\mathbb{Q},p,\infty,\mathrm{pos}}\}$, in other words $\mathcal{N} = \{\mathfrak{a} \in \mathcal{I}_p,\ \bar{i}_p(\mathrm{N}\mathfrak{a}) = 1 \text{ in } \mathbb{Z}_p^\times\}$, where N now denotes the absolute norm taking values in $\mathcal{Q}_{\mathrm{pos}}^\times = \mathbb{Q}_{\mathrm{pos}}^\times \otimes \mathbb{Z}_p$. □

We leave the reader to establish the functorial properties of the above correspondence (restriction, decomposition and inertia groups, norm lifting, transfer, etc...), in the spirit of II.4.5, where ray groups mod $\mathfrak{m}$ or $\frac{\mathfrak{m}}{\mathfrak{m}_v}$ become infinitesimal ray groups for ∞_T or $\infty_{T\backslash\{v\}}$ in an evident meaning. For instance, the following result will be needed for the study of the logarithmic class group in Section 7.

2.8.2 Exercise. Let $v \in T_p$. Show that the inertia field (resp. the decomposition field) of v in $H_T^{\mathrm{ord}}(p)/K$ corresponds to the norm group $\mathcal{P}_{T,\infty_t}$ (resp. $(\mathcal{P}_{t,\infty_t} \cdot \langle\, \mathfrak{p}_v \,\rangle \otimes \mathbb{Z}_p) \cap \mathcal{I}_T$), where ∞_t is connected with $t := T\backslash\{v\}$. □

Then, the function Log_{T_p} allows us to characterize the $\mathbb{Z}_p$-torsion and the $\mathbb{Z}_p$-free part of any relative extension, and hence of any subextension (see 2.7). In the above example 2.8.1 for $p = 2$, using $K\mathbb{Q}^{\mathrm{cycl}}(2)/K$ instead of $K\mathbb{Q}(\mu_{2^\infty})/K$, we obtain:

$$\mathrm{Gal}(\widetilde{K}_2/K\mathbb{Q}^{\mathrm{cycl}}(2)) \simeq \{\mathrm{Log}_2(\mathfrak{a}) \in \mathrm{Log}_2(\mathcal{I}_2),\ \ \mathrm{Log}_{\mathbb{Q},2}(\mathrm{N}_{K/\mathbb{Q}}(\mathfrak{a})) = 0\}.$$

We always have $\mathrm{N}_{K/\mathbb{Q}}(\mathfrak{a}) = (\mathrm{N}\mathfrak{a})$ but possibly with $\bar{i}_2(-\mathrm{N}\mathfrak{a}) = 1$, which means that the Artin symbol of $\mathfrak{a}$ in $H_2^{\mathrm{res}}(2)/K$ fixes $K\mathbb{Q}^{\mathrm{cycl}}(2)$ but not $K\mathbb{Q}(\mu_{2^\infty})$.

Finally, if we want to have S-decomposition, we simply add $\langle S \rangle$, saying as usual that:

$$\mathcal{P}_{T,\infty,\mathrm{pos}}\langle S \rangle := \mathcal{P}_{T,\infty,\Delta_\infty} \cdot \mathcal{S}_0,$$

and we use the function $\mathrm{Log}_{T_p}^{S_0}$ to study the corresponding $\mathbb{Z}_p$-torsion.

We will study S-decomposition questions in the compositum $\widetilde{K}_p$ of the $\mathbb{Z}_p$-extensions of K. We will split the problem into two parts; the first one (which will be Section 3 below) will look at the tame places, the case of decomposition of the wild places in $\widetilde{K}_p/K$, much more delicate, will be delayed until Section 5, in which an explicit computational method will be given.

Section 4 will be devoted to the study of the structure of $\mathrm{Gal}(\overline{K}^{\mathrm{ab}}(p)/K)$ (obtained by going to the limit on the above groups $\mathcal{A}_T^{\mathrm{res}}$), and this study will use all the deep arguments of class field theory that we have seen up to now. Questions connected with the decomposition of wild places in $\overline{K}^{\mathrm{ab}}(p)/K$ will also be examined in Section 4 from a theoretical angle, and will lead to new p-adic questions (such as Conjecture 4.12).

§3 Compositum of the S-Split $\mathbb{Z}_p$-Extensions — The p-Adic Conjecture

Assumption. Let $S = S_0 \cup S_\infty$ be a finite set of noncomplex places of K, where S_0 is assumed to be prime to p.

a) p-Adic Ranks: The Leopoldt–Jaulent–Roy Conjecture

The value of the $\mathbb{Z}_p$-rank $\widetilde{\mathrm{r}}_p^{S_0}$ of $\mathcal{Z}_p^{S_0}$ (i.e., the maximal number of independent S_0-split $\mathbb{Z}_p$-extensions of K) is given by the formula:

$$\widetilde{\mathrm{r}}_p^{S_0} = [K : \mathbb{Q}] - \mathrm{r}_p^{S_0},$$

where $\mathrm{r}_p^{S_0} := \mathrm{rk}_{\mathbb{Z}_p}(\mathrm{adh}_p(E'^{S_0}))$ is the p-adic rank of E^{S_0} (see 1.6.2, 1.6.3). The value of this p-adic rank is based on a generalization of the Leopoldt Conjecture 1.6.4; this conjecture, first stated by Jaulent then generalized by Roy ([Ja2; Ja3], [Roy1; Roy2]), is Galois in nature, and is related to a p-adic transcendence conjecture, consequence of the p-adic Schanuel conjecture (see Remark 3.1.8). Nonetheless, recall that the Leopoldt conjecture has been proved in the case of abelian extensions of $\mathbb{Q}$ (Ax–Brumer), thanks to the deep results of Baker on the independence of logarithms of algebraic numbers, and in a few other cases thanks to the transcendence results of Waldschmidt which contain, among others, those of Baker; in any case, one obtains lower bounds for the p-adic rank which, for certain Galois groups (i.e., those whose irreducible characters have "small" degrees), lead to an equality with the $\mathbb{Z}$-rank $r_1 + r_2 - 1$.[9]

[9] [Wal1; Wal2], [EKW], [Roy1; Roy2], [Ja3], [Lau].

In 4.12, we will give for the unit group (i.e., $S_0 = \emptyset$) a statement which is more precise than the Leopoldt conjecture.

We should also cite a large number of sufficient conditions of algebraic-arithmetic nature (i.e., without any transcendence) for the validity of the Leopoldt conjecture (we can give many such results thanks to the results of class field theory); we will not do so because we believe that this aspect is a dead-end.

3.1 THE p-ADIC RANK AND THE REPRESENTATION THEORY. To formulate and study these conjectures, we are led to use the language of representations of a finite group Γ, here the Galois group of a Galois extension $N/\mathbb{Q}$ (we refer to [Se4]).

3.1.1 Notations (characters). We denote by Ψ_N the set of absolutely irreducible characters ψ of Γ, and by V_ψ the corresponding absolutely irreducible representation.

The unit character will be denoted indifferently ψ_0 or 1. □

For example, the regular representation $\mathbb{C}_p[\Gamma]$ is isomorphic to the representation $\bigoplus_{\psi \in \Psi_N} \psi(1) V_\psi$, and $\bigoplus_{w|p} N_w \simeq N \otimes_{\mathbb{Q}} \mathbb{Q}_p$ (see Ch. I, § 2 and Ch. II, § 2) is isomorphic to $\mathbb{Q}_p[\Gamma]$, the regular representation over $\mathbb{Q}_p$ (this is also the normal basis theorem in $N/\mathbb{Q}$).

Recall the definition of the p-adic rank in a slightly different context and introduce the notion of monogeneous group.

3.1.2 Definitions. (i) For any subgroup F of finite type of $K_p^\times := K_{Pl_p}^\times$, we set:

$$\mathrm{r}_p(F) := \dim_{\mathbb{Q}_p}(\mathbb{Q}_p \log_{K,p}(F)),$$

in $\bigoplus_{v|p} K_v$; by 2.2, (i), (ii), this number is equal to $\mathrm{rk}_{\mathbb{Z}_p}(\mathrm{adh}_p(F'))$, the p-adic rank of F, where $F' := \Big\{x \in F,\ i_p(x) \in \bigoplus_{v|p} U_v^1\Big\}$ (of finite index in F).[10]

(ii) We say that a subgroup F of $K^\times$ is monogeneous if $F \otimes_{\mathbb{Z}} \mathbb{Q}$ is contained in a monogeneous $\mathbb{Q}[\mathrm{Gal}(\overline{\mathbb{Q}}/\mathbb{Q})]$-module (in other words, of the form $\langle \eta \rangle_{\mathrm{Gal}(\overline{\mathbb{Q}}/\mathbb{Q})} \otimes_{\mathbb{Z}} \mathbb{Q}$, for some algebraic number η). □

3.1.3 Conjecture (Jaulent (1984/1985)). *Let p be a fixed prime number and let K be a number field. Let F be a subgroup of finite type of $K_p^\times$. Then $\mathrm{r}_p(F) = \dim_{\mathbb{Q}}(F \otimes_{\mathbb{Z}} \mathbb{Q})$ (i.e., the p-adic rank of F is equal to its $\mathbb{Z}$-rank) as soon as F is monogeneous.* □

[10] Recall that $\mathrm{adh}_p(F')$ is also equal to $\bar{\imath}_p(F \otimes \mathbb{Z}_p)$ (see 1.6.6).

It is easy to check that if N is a Galois extension of $\mathbb{Q}$ containing K, then F is monogeneous if and only if there exists $\eta_0 \in N$ such that $F \otimes_{\mathbb{Z}} \mathbb{Q} \subseteq \langle\, \eta_0 \,\rangle_\Gamma \otimes_{\mathbb{Z}} \mathbb{Q}$, with $\Gamma := \mathrm{Gal}(N/\mathbb{Q})$: consider η in a Galois extension L of $\mathbb{Q}$ containing N, and replace η by its norm in L/N. In practice we can thus choose for N the Galois closure of $K/\mathbb{Q}$.

It is clear that F is monogeneous if and only if the Γ-module $\langle\, F \,\rangle_\Gamma$ generated by F is monogeneous (it is then not necessarily a subgroup of $K^\times$), so that we may indifferently state the conjecture for the field N, either in terms of subgroups, or in terms of sub-Γ-modules of $N_p^\times$.

In the most general situation, Roy has given a formulation (which contains statement 3.1.3) which theoretically allows the determination of the p-adic rank of F when F is not assumed to be monogeneous, as follows.

3.1.3′ Conjecture (Roy [Roy1] (1991)). *Let K be a number field, N its Galois closure over $\mathbb{Q}$, and $\Gamma := \mathrm{Gal}(N/\mathbb{Q})$. If F is a subgroup of finite type of $K_p^\times$, then $\mathrm{r}_p(F)$ is the largest integer r such that there exists a $\mathbb{Q}[\Gamma]$-module homomorphism $\varphi : \langle\, F \,\rangle_\Gamma \otimes_{\mathbb{Z}} \mathbb{Q} \longrightarrow \mathbb{Q}[\Gamma]$ such that $\dim_{\mathbb{Q}}(\varphi(F \otimes_{\mathbb{Z}} \mathbb{Q})) = \mathrm{r}$.* □

Note. To simplify, we will call the above statements the p-adic conjecture. Note that, as for the Leopoldt conjecture, if the p-adic conjecture is true for a field, it is also true for its subfields.

3.1.4 Proposition. *When F is a monogeneous sub-Γ-module of $N_p^\times$, the conjecture implies that the representations $F \otimes_{\mathbb{Z}} \mathbb{Q}$ (of Γ over $\mathbb{Q}$) and $\mathbb{Q}_p \log_{N,p}(F)$ (of Γ over $\mathbb{Q}_p$) have the same character.*

Proof. We have (because of the equality of the $\mathbb{Q}_p$-dimensions) that $\log_{N,p}$ induces the canonical isomorphism of Γ-modules:

$$F \otimes_{\mathbb{Z}} \mathbb{Q}_p \simeq \mathbb{Q}_p \log_{N,p}(F),$$

whence the result since the representations $F \otimes_{\mathbb{Z}} \mathbb{Q}$ and $F \otimes_{\mathbb{Z}} \mathbb{Q}_p$ have the same character. Thus, they are subrepresentations of the corresponding regular representations $\mathbb{Q}[\Gamma]$ and $\mathbb{Q}_p[\Gamma]$. □

If F is a nonmonogeneous sub-Γ-module of $N_p^\times$, the representation $F \otimes_{\mathbb{Z}} \mathbb{Q}_p$ is not contained in the regular representation $\mathbb{Q}_p[\Gamma]$, and it is impossible to have the above isomorphism since we always have:

$$\mathbb{Q}_p \log_{N,p}(F) \subseteq \bigoplus_{w|p} N_w \simeq \mathbb{Q}_p[\Gamma]\ ;$$

the p-adic rank of F is then strictly less than its $\mathbb{Z}$-rank. Thus, in terms of Γ-modules in N, the conjecture gives a necessary and sufficient condition.

Note that if F is not a Γ-module in N, the p-adic rank of F can be equal to its $\mathbb{Z}$-rank, even when F is not monogeneous (see Exercise 3.1.9).

If K is an arbitrary number field, we will say that it satisfies the p-adic conjecture if this conjecture is true for any monogeneous subgroup F of $K_p^\times$.

3.1.5 Remarks. (i) The first typical example of a monogeneous Γ-module is given by $\langle \eta \rangle_\Gamma$, where (η) is a nontrivial principal power of a prime ideal $\mathfrak{l}$ of N, totally split in $N/\mathbb{Q}$; we then have $\langle \eta \rangle_\Gamma \otimes_\mathbb{Z} \mathbb{Q} \simeq \mathbb{Q}[\Gamma]$, and $F = \langle \eta \rangle_\Gamma$ satisfies the hypotheses of the conjecture. In this case, the p-adic rank is (conjecturally) equal to $[N : \mathbb{Q}]$. The same is true when $\mathfrak{l}$ is not totally split, but we then obtain a subrepresentation of $\mathbb{Q}[\Gamma]$ (the permutation representation of Γ modulo the decomposition group D of $\mathfrak{l}$ in $N/\mathbb{Q}$) and the p-adic rank becomes the index of D in Γ.

(ii) The case of E_N^{ord} is the second typical case, for which the p-adic conjecture is the Leopoldt conjecture itself (the monogeneity of E_N^{ord} follows from the Dirichlet–Herbrand Theorem I.3.7.2). □

Thus, the unit group E^{ord} of K is monogeneous with p-adic rank conjecturally equal to $r_1 + r_2 - 1$.

The case of S-unit groups E^S of K (for $S_0 \neq \emptyset$) mixes both aspects, and it is necessary to explain how the conjecture can be applied. Note first the following result which explains how to proceed in general for the case of a Γ-module (and which is an immediate consequence of the statements 3.1.3, 3.1.3′).

3.1.6 Proposition. *Let F be a sub-Γ-module of finite type of $N_p^\times$. Then, assuming the p-adic conjecture, $\mathrm{r}_p(F)$ is equal to the $\mathbb{Z}$-rank of any maximal monogeneous submodule of F. This p-adic rank is thus equal to the $\mathbb{Q}$-dimension of the largest representation common to $F \otimes_\mathbb{Z} \mathbb{Q}$ and to the regular representation of Γ. It follows that we have:*

$$\mathbb{Q}_p \mathrm{log}_{N,p}(F) = \mathbb{Q}_p \mathrm{log}_{N,p}(F_0)$$

for any such maximal monogeneous submodule F_0 of F. □

3.1.7 Remarks. Let again K be a number field, N the Galois closure of K over $\mathbb{Q}$, $\Gamma := \mathrm{Gal}(N/\mathbb{Q})$, and $H := \mathrm{Gal}(N/K)$. Let $S = S_0 \cup S_\infty$ be a finite set of noncomplex places of K, and let $S' := S'_0 \cup S'_\infty$ be the set of (noncomplex) places of N above those of S.

In the case of S-units of K, $E^S \otimes \mathbb{Q}$ only depends on S_0.

We have two interesting particular cases in which the value of the p-adic rank of E^{S_0} is given by a formula:

(i) the case where S_0 is the set of prime ideals of K above a set s_0 of prime numbers prime to p since then $E_N^{S'_0}$ is a Γ-module, and representation theory allows us to conclude thanks to a Galois descent (see 3.3.4);

(ii) the case where E^{S_0} is directly monogeneous (see 3.5), which happens when S_0 is sufficiently small (see 3.4). □

We can give now some evidence for the above p-adic conjectures.

3.1.8 Remark (the p-adic Schanuel conjecture (see [Wal2, § 2], [Wal3])). Let N be a finite normal extension of $\mathbb{Q}$ with Galois group Γ, and let $F = \langle\, \eta \,\rangle_\Gamma$ be a monogeneous sub-Γ-module of $N_p^\times$; we suppose that $F \otimes_{\mathbb{Z}} \mathbb{Q} \simeq \mathbb{Q}[\Gamma]$ (if it is not the case, we know that $F \otimes_{\mathbb{Z}} \mathbb{Q}$ is a direct summand in such a maximal monogeneous $\mathbb{Q}[\Gamma]$-module). Then the $\sigma(\eta)$, for $\sigma \in \Gamma$, are $\mathbb{Z}$-multiplicatively independent (in other words, the $\mathbb{Z}$-rank of F is equal to $[N : \mathbb{Q}]$), which implies that the $\log(\sigma(\eta))$, $\sigma \in \Gamma$, are $\mathbb{Q}$-linearly independent in $\mathbb{C}_p$. But the Schanuel p-adic conjecture precisely implies that these $\log(\sigma(\eta))$ are *algebraically* independent. We can thus apply this to the following.

Let $\mathrm{R}(X) := \mathrm{Det}(X_{\tau\sigma})_{\tau,\sigma\in\Gamma}$, where the X_σ are indeterminates. It is well known that $\mathrm{R}(X)$ is an explicit nonzero polynomial of $\mathbb{Z}[(X_\sigma)_\sigma]$. The above implies (conjecturally):

$$\mathrm{Reg}(\eta) := \mathrm{Det}(\log(\tau\sigma(\eta)))_{\tau,\sigma\in\Gamma} \neq 0.$$

Then the vectors $(\log(\tau\sigma(\eta)))_{\tau\in\Gamma}$, for $\sigma \in \Gamma$, are $\mathbb{Q}_p$-linearly independent. If $\sum_\sigma a_\sigma \log_{N,p}(\sigma(\eta)) = 0$ in $\bigoplus_{w|p} N_w$, for some $a_\sigma \in \mathbb{Q}_p$, then by 2.2.1, for all $\tau \in \Gamma$ we have $\sum_\sigma a_\sigma \log_{N,p}(\tau\sigma(\eta)) = 0$, which yields $\sum_\sigma a_\sigma \log(\tau\sigma(\eta)) = 0$ (by projection on N_{w_0}, viewing N as a subfield of N_{w_0}); thus $a_\sigma = 0$ for all $\sigma \in \Gamma$, the vectors $\log_{N,p}(\sigma(\eta))$ are $\mathbb{Q}_p$-linearly independent, therefore the p-adic rank of F has the maximum value $[N : \mathbb{Q}]$.

If, for instance, we consider the group E_N^{ord} of units, we can find $\eta \in N_p^\times$ as above such that $E_N^{\mathrm{ord}} \otimes_{\mathbb{Z}} \mathbb{Q}$ is a direct summand in $\langle\, \eta \,\rangle_\Gamma \otimes_{\mathbb{Z}} \mathbb{Q}$. It is then clear that the p-adic rank of E_N^{ord} is exactly equal to its $\mathbb{Z}$-rank.

From this point of view, the Leopoldt conjecture appears as a particular (but important) case of p-adic independence of logarithms of algebraic numbers. □

3.1.9 Exercise. Consider $K = N = \mathbb{Q}(\sqrt{-1})$, and assume that $p = 2$. Set $\sqrt{-1} =: i$ and:

$$F := \langle\, \varepsilon,\ \eta \,\rangle_{\mathbb{Z}},$$

where $\varepsilon = 1 - 2(1+i)$, $\eta = 1 + 2(1-i)$.

(i) Check that:

$$\langle\, F \,\rangle_\Gamma \otimes_{\mathbb{Z}} \mathbb{Q} = E^{S_0} \otimes_{\mathbb{Z}} \mathbb{Q},$$

where S_0 is the set of places of K above 5 and 13. We thus have $|S_0| = 4$, so that $\langle\, F \,\rangle_\Gamma$ cannot be monogeneous; in fact, we have $\langle\, F \,\rangle_\Gamma \otimes \mathbb{Q} \simeq 2V_{\psi_0} \oplus 2V_{\psi_1}$ with evident notations using the two irreducible representations of $G = \mathrm{Gal}(\mathbb{Q}(i)/\mathbb{Q})$.

(ii) Check that, in $\mathbb{Q}_2(i)$, $\log_2(\varepsilon)$ and $\log_2(\eta)$ are $\mathbb{Q}_2$-linearly independent (we thus have $\mathrm{r}_2(F) = 2 = \dim_{\mathbb{Q}}(F \otimes \mathbb{Q})$).

(iii) Find two elements which generate the kernel of:

$$E^{S_0} \otimes_{\mathbb{Z}} \mathbb{Q}_2 \xrightarrow{\log_2} \mathbb{Q}_2(i).$$

Answer. (i) F is generated by a generator of $\mathfrak{p}_5$ and by a generator of $\mathfrak{p}_{13}$; since 5 and 13 are split in $\mathbb{Q}(i)$ and that E is torsion, point (i) is clear.

(ii) It is enough to compute in $\mathbb{Q}_2(i)$ a sufficiently good approximation to $\log_2(\varepsilon)$ and to $\log_2(\eta)$; we easily find that:

$$\tfrac{1}{2}\log_2(\varepsilon) \equiv -1 - 3i \bmod (8),$$
$$\tfrac{1}{2}\log_2(\eta) \equiv 1 + i \bmod (8).$$

If these two quantities are $\mathbb{Q}_2$-dependent, they are $\mathbb{Z}_2$-dependent hence there exist $a,\ b \in \mathbb{Z}_2$, not both divisible by 2, such that:

$$a(-1-3i) + b(1+i) \equiv 0 \bmod (8),$$

which yields $a \equiv b \equiv 0 \bmod (2)$, a contradiction.

(iii) We have $E^{S_0} \otimes \mathbb{Q}_2 =: \langle \varepsilon,\ \varepsilon',\ \eta,\ \eta' \rangle \otimes \mathbb{Q}_2$, where $\varepsilon',\ \eta'$ are the conjugates of ε and of η. For the unit representation, we have the generator:

$$(\varepsilon\varepsilon') \otimes (\log_2(13)) \,.\, (\eta\eta') \otimes (-\log_2(5)) \ ;$$

this means that the logarithms of $\varepsilon\varepsilon' = 5$ and of $\eta\eta' = 13$ are linearly dependent over $\mathbb{Q}_2$. For the nonunit representation, we are going to find a dependence relation by noting that $\log_2(\frac{\varepsilon}{\varepsilon'})$ and $\log_2(\frac{\eta}{\eta'})$ are, in $\mathbb{Q}_2(i)$, elements of zero trace (since $\frac{\varepsilon}{\varepsilon'}$ and $\frac{\eta}{\eta'}$ are of norm 1). Set:

$$\log_2\Big(\frac{\varepsilon}{\varepsilon'}\Big) =: a\,i,\ \ \log_2\Big(\frac{\eta}{\eta'}\Big) =: b\,i,\ \ a,\ b \in \mathbb{Q}_2 \ ;$$

the second generator is thus:

$$\Big(\frac{\varepsilon}{\varepsilon'}\Big) \otimes b \,.\, \Big(\frac{\eta}{\eta'}\Big) \otimes (-a).$$

We can check by a direct computation that we have $a \equiv 4 \bmod (16)$ and $b \equiv 4 \bmod (16)$. □

We come back to general considerations in view of the study of the p-adic rank of E^{S_0}.

When S_0', the set of places of N above those of S_0, is not stable under Γ, $E_N^{S_0'}$ is not a Γ-module, and it can be useful to introduce S_0'', the set of conjugates of elements of S_0', which is equal to the set of places of N above the elements of the set s_0 of residue characteristics of the elements of S_0, in which case $E_N^{S_0''} \otimes_{\mathbb{Z}} \mathbb{Q}$ is the smallest Γ-module containing $E_N^{S_0'} \otimes_{\mathbb{Z}} \mathbb{Q}$. However, $E_N^{S_0'}$ is always an H-module.

3.2 EXTENSION OF SCALARS. To study $E^{S_0} \otimes_{\mathbb{Z}} \mathbb{Q}$ and $\mathbb{Q}_p \log_{K,p}(E^{S_0})$, as respective subspaces of the $\mathbb{Q}$ and $\mathbb{Q}_p$-representations $E_N^{S_0''} \otimes_{\mathbb{Z}} \mathbb{Q}$ and $\bigoplus_{w|p} N_w$ of Γ, we can use an extension of scalars since we are only interested in the dimensions; moreover, in such a way we may use the absolutely irreducible representations of Γ, which is more convenient. Thus, we will study:

$$E^{S_0} \otimes_{\mathbb{Z}} \mathbb{C}_p \subseteq E_N^{S_0''} \otimes_{\mathbb{Z}} \mathbb{C}_p,$$
$$\mathbb{Q}_p \log_{K,p}(E^{S_0}) \otimes_{\mathbb{Q}_p} \mathbb{C}_p \subseteq \Big(\bigoplus_{w|p} N_w\Big) \otimes_{\mathbb{Q}_p} \mathbb{C}_p,$$

the function $\log_{K,p} =: \log_p$ being considered as having values in $\bigoplus_{w|p} N_w$ through the injective diagonal embedding $\bigoplus_{v|p} K_v \longrightarrow \bigoplus_{v|p} \bigoplus_{w|v} N_w$.

We thus have, for the $\mathbb{Z}_p$-rank $\widetilde{\mathrm{r}}_p^{S_0}$ of $\mathcal{Z}_p^{S_0} := \mathrm{Gal}(\widetilde{K}_p^{S_0\,\mathrm{fr}}/K)$ (see 2.2, (ii)):

$$\widetilde{\mathrm{r}}_p^{S_0} = \dim_{\mathbb{C}_p}(\mathcal{L}_p^{S_0} \otimes_{\mathbb{Q}_p} \mathbb{C}_p)$$

since $\widetilde{\mathrm{r}}_p^{S_0} = [K : \mathbb{Q}] - \mathrm{r}_p^{S_0}$, where now $\mathrm{r}_p^{S_0}$ also means $\mathrm{r}_p(E^{S_0})$ (see 3.1.2, (i)).

For any $\mathbb{Q}_p[H]$-module M we have $M^H = e_H M$, where $e_H := \frac{1}{|H|} \sum_{t \in H} t$. We easily check that:

$$\mathcal{L}_p^{S_0} \simeq (\,\mathcal{L}_{N,p}^{S_0'}\,)^H$$

since:

$$e_H(\mathbb{Q}_p \log_{N,p}(E_N^{S_0'})) = \mathbb{Q}_p \log_{N,p}(\mathrm{N}_{N/K}(E_N^{S_0'})) \simeq \mathbb{Q}_p \log_p(E^{S_0}).$$

Note that this isomorphism is in terms of H-modules and, even when $\mathcal{L}_{N,p}^{S_0'}$ is a Γ-module, $(\mathcal{L}_{N,p}^{S_0'})^H$ is not necessarily one. Of course, by flatness we have:

$$\mathcal{L}_p^{S_0} \otimes \mathbb{C}_p \simeq (\mathcal{L}_{N,p}^{S_0'} \otimes \mathbb{C}_p)^H.$$

b) The Galois Case

Assume here that S_0 is the set of places of K above a finite set s_0 of prime numbers different from p (case (i) of 3.1.7).

Recall first some useful results on permutation representations corresponding to the Galois group Γ of a Galois extension $N/\mathbb{Q}$. We use Notations 3.1.1. For any place u of $\mathbb{Q}$, let D_u be the decomposition group of a place w_0 of N above u in $N/\mathbb{Q}$.

3.3 Proposition. *(i) The permutation representation of Γ modulo D_u has a character equal to $\mathrm{Ind}_{D_u}^{\Gamma}(1_{D_u})$ and is isomorphic to:*

$$\bigoplus_{\psi\in\Psi_N} \rho_{u,\psi} V_\psi, \quad \text{where} \quad \rho_{u,\psi} := \frac{1}{|D_u|}\sum_{t\in D_u}\psi(t).$$

If N^ψ is the field fixed under the kernel of ψ,[11] *and if D_u^ψ is a decomposition group of u in $N^\psi/\mathbb{Q}$, we also have $\rho_{u,\psi} = \frac{1}{|D_u^\psi|}\sum_{t\in D_u^\psi}\psi'(t)$, where ψ' is the faithful character corresponding to ψ.*

(ii) If s_0 is a finite set of prime numbers and S'_0 is the set of places of N above s_0, then the character of $E_N^{S'_0}\otimes_{\mathbb{Q}_p}\mathbb{C}_p$ is $\sum_{u\in s_0\cup\{\infty\}} \mathrm{Ind}_{D_u}^{\Gamma}(1_{D_u}) - 1_\Gamma$, and we have:

$$E_N^{S'_0}\otimes_{\mathbb{Z}}\mathbb{C}_p \simeq \bigoplus_{\psi\in\Psi_N} n_\psi V_\psi, \quad \text{where} \quad n_\psi := \sum_{u\in s_0\cup\{\infty\}} \rho_{u,\psi} - \delta_{1,\psi},$$

with $\delta_{1,\psi} := 1$ or 0 according as $\psi = 1$ or not (S-unit Dirichlet–Herbrand Theorem I.3.7). □

3.3.1 Notation. Let V and V' be two $\mathbb{C}_p$-representations of Γ, with respective characters φ, φ'. We denote by $\varphi\wedge\varphi'$ the character of the maximal subrepresentation common to V and V'; in other words, if $\varphi = \sum_i n_i\,\psi_i$, $\varphi' = \sum_i n'_i\,\psi_i$, n_i, $n'_i \geq 0$, ψ_i distinct irreducibles, we have:

$$\varphi\wedge\varphi' = \sum_i (n_i\wedge n'_i)\psi_i,$$

where by abuse of notation $n\wedge n'$ denotes $\min(n,n')$ for n, $n'\in\mathbb{Z}$. □

3.3.2 Lemma 1. *Assuming the p-adic conjecture in N, and with the notations of 3.3 and 3.3.1, we have:*

$$\mathbb{Q}_p\mathrm{log}_{N,p}(E_N^{S'_0})\otimes_{\mathbb{Q}_p}\mathbb{C}_p \simeq \bigoplus_{\psi\in\Psi_N}(\psi(1)\wedge n_\psi)V_\psi,$$

$$\mathcal{L}_{N,p}^{S'_0}\otimes_{\mathbb{Q}_p}\mathbb{C}_p \simeq \bigoplus_{\psi\in\Psi_N}(\psi(1)-\psi(1)\wedge n_\psi)V_\psi.$$

Proof. We use 3.1.6 since $E_N^{S'_0}$ is a Γ-module: any maximal monogeneous submodule of $E_N^{S'_0}\otimes_{\mathbb{Z}}\mathbb{C}_p$ is isomorphic to $\bigoplus_{\psi\in\Psi_N}(\psi(1)\wedge n_\psi)V_\psi$; furthermore, since $\Big(\bigoplus_{w|p} N_w\Big)\otimes_{\mathbb{Q}_p}\mathbb{C}_p \simeq \mathbb{C}_p[\Gamma]$, we obtain the second isomorphism of representations. □

To come back to K, we consider the sub-vector space fixed under H:

$$\Big(\bigoplus_{\psi\in\Psi_N}(\psi(1)-\psi(1)\wedge n_\psi)V_\psi\Big)^H;$$

[11] In other words by the set of $s\in\Gamma$ such that $\psi(s)=\psi(1)$ [Se4, § 6.5, Ex. 2].

since H acts on every irreducible representation V_ψ, we have $V_\psi^H \subseteq V_\psi$, hence:

$$\Big(\bigoplus_{\psi\in\Psi_N}(\psi(1)-\psi(1)\wedge n_\psi)V_\psi\Big)^H = \bigoplus_{\psi\in\Psi_N}(\psi(1)-\psi(1)\wedge n_\psi)V_\psi^H.$$

3.3.3 Lemma 2. *The $\mathbb{C}_p$-dimension of V_ψ^H is equal to $\rho_\psi^H := \frac{1}{|H|}\sum_{t\in H}\psi(t)$.*

Proof. Consider V_ψ as a representation of H; its character is equal to $\mathrm{Res}_H(\psi)$. Since V_ψ^H is the largest sub-H-module of V_ψ on which H acts trivially, the character of V_ψ^H is equal to ρ_ψ^H times the unit character 1_H, where ρ_ψ^H is the $\mathbb{C}_p$-dimension that we need. By computing the corresponding scalar product, we thus have $\rho_\psi^H = \langle\,\mathrm{Res}_H(\psi), 1_H\,\rangle_H = \frac{1}{|H|}\sum_{t\in H}\psi(t)$. □

We therefore obtain a first general result on the value of $\widetilde{\mathrm{r}}_p^{S_0}$.

3.3.4 Theorem (Galois expression for $\widetilde{\mathrm{r}}_p^{S_0}$). *Let S_0 be the set of places of K above a finite set of prime numbers different from p. Let N be the Galois closure of K over $\mathbb{Q}$ and $H := \mathrm{Gal}(N/K)$. Then if N satisfies the p-adic conjecture 3.1.3, the $\mathbb{Z}_p$-rank of the Galois group of the compositum of the S_0-split $\mathbb{Z}_p$-extensions of K is:*

$$\widetilde{\mathrm{r}}_p^{S_0} = \sum_{\psi\in\Psi_N}(\psi(1)-\psi(1)\wedge n_\psi)\rho_\psi^H,$$

where n_ψ, ρ_ψ^H are given in 3.3 and 3.3.3. If $K/\mathbb{Q}$ is Galois, we obtain:

$$\widetilde{\mathrm{r}}_p^{S_0} = \sum_{\psi\in\Psi_K}(\psi(1)-\psi(1)\wedge n_\psi)\psi(1) = \sum_{\psi\in\Psi_K}\psi(1)\max(\psi(1)-n_\psi, 0). \quad \square$$

3.3.5 Remark (totally real fields). If K is a totally real number field, its Galois closure N is real, the place at infinity of $\mathbb{Q}$ is thus totally split, $\rho_{\infty,\psi} = \psi(1)$ for all $\psi \in \Psi_N$, and we have (still if S_0 is the set of places of K above a set s_0 of prime numbers different from p):

$$n_\psi \geq \psi(1) \ \text{ for all } \psi \neq 1,\ n_1 = |s_0|\ ;$$

assuming the p-adic conjecture in N, we thus have:

$$\widetilde{\mathrm{r}}_p^{S_0} = \max(1-|s_0|,\ 0).$$

In the totally real case, the above Galois context is thus useless and we can even refine this result by only assuming the validity of the Leopoldt conjecture for p in K. Indeed, this already proves that the $\mathbb{Q}_p$-dimension of $\mathbb{Q}_p\log_p(E)$ is equal to $[K:\mathbb{Q}]-1$. Furthermore, if $q \in s_0$, $\log_p(q) \notin \mathbb{Q}_p\log_p(E)$ (this last subspace is the kernel of the trace map: $\bigoplus_{v|p} K_v \longrightarrow \mathbb{Q}_p$); hence

$\mathbb{Q}_p\log_p(E^{S_0}) \simeq \bigoplus_{v|p} K_v$ and $\widetilde{\mathrm{r}}_p^{S_0} = 0$ if $s_0 \neq \emptyset$. We recover the fact that in the unique $\mathbb{Z}_p$-extension of K (the cyclotomic $\mathbb{Z}_p$-extension), no finite place can be totally split. □

3.3.6 Corollary. *Let $K/\mathbb{Q}$ be a complex Galois extension such that the decomposition group of ∞ is a normal subgroup of* $\mathrm{Gal}(K/\mathbb{Q})$. *Introducing the set Ψ_K^- of odd characters of K, we have:*

$$\widetilde{\mathrm{r}}_p^{S_0} = \max(1 - |s_0|, 0) + \sum_{\psi \in \Psi_K^-} \psi(1)\max\Big(\psi(1) - \sum_{u \in s_0} \rho_{u,\psi}, 0\Big) \ ;$$

if $K/\mathbb{Q}$ is abelian, we have:

$$\widetilde{\mathrm{r}}_p^{S_0} = \max(1 - |s_0|, 0) + \sum_{\psi \in \Psi_K^-} \max\Big(1 - \sum_{u \in s_0} d_{u,\psi}, 0\Big),$$

with $d_{u,\psi} = 1$ or 0 according as u is totally split in K^ψ or not. □

3.3.7 Example. Let us take for $K/\mathbb{Q}$ a noncyclic cubic extension. We have $\Gamma := \mathrm{Gal}(N/\mathbb{Q}) \simeq D_6$, semidirect product of a cyclic group $\langle \sigma \rangle$ of order 3 and of a cyclic group $H =: \langle \tau \rangle$ of order 2 (which fixes K). The characters of D_6 are:

- ψ_0, such that: $\psi_0(s) = 1$ for all $s \in \Gamma$,
- ψ_1, such that: $\psi_1(1) = \psi_1(\sigma) = \psi_1(\sigma^2) = 1$,
$\psi_1(\tau) = \psi_1(\tau\sigma) = \psi_1(\tau\sigma^2) = -1$,
- ψ_2, such that: $\psi_2(1) = 2$,
$\psi_2(\sigma) = \psi_2(\sigma^2) = -1$,
$\psi_2(\tau) = \psi_2(\tau\sigma) = \psi_2(\tau\sigma^2) = 0$.

We thus obtain for the dimensions of the V_ψ^H (see 3.3.3):

$$\rho_{\psi_0}^H = 1, \ \rho_{\psi_1}^H = 0, \ \rho_{\psi_2}^H = 1.$$

Let D_u, $u \in Pl_\mathbb{Q}$, be a decomposition subgroup in $N/\mathbb{Q}$ (fixed up to conjugation), and let $\rho_{u,\psi} = \frac{1}{|D_u|} \sum_{t \in D_u} \psi(t)$ for all ψ (see 3.3, (i)):

(i) $D_u = 1$ yields: $\rho_{u,\psi_0} = \rho_{u,\psi_1} = 1, \ \rho_{u,\psi_2} = 2$,

(ii) $D_u = H$ (or a conjugate) yields: $\rho_{u,\psi_0} = 1, \ \rho_{u,\psi_1} = 0, \ \rho_{u,\psi_2} = 1$,

(iii) $D_u = \langle \sigma \rangle$ yields: $\rho_{u,\psi_0} = 1, \ \rho_{u,\psi_1} = 1, \ \rho_{u,\psi_2} = 0$,

(iv) $D_u = \Gamma$ yields: $\rho_{u,\psi_0} = 1, \ \rho_{u,\psi_1} = \rho_{u,\psi_2} = 0$;

the formula for $\widetilde{\mathrm{r}}_p^{S_0}$ becomes here (in the context of 3.3.4):

$$\widetilde{\mathrm{r}}_p^{S_0} = 1 - 1 \wedge n_{\psi_0} + 2 - 2 \wedge n_{\psi_2} = 3 - 1 \wedge n_{\psi_0} - 2 \wedge n_{\psi_2}.$$

If we assume that $D_\infty = H$ (K real and not totally real; thus $0 \le \widetilde{\mathrm{r}}_p^{S_0} \le 2$), it follows that, by 3.3, (ii):

$$n_{\psi_0} = 1 + \sum_{u \in s_0} \rho_{u,\psi_0} - 1 = |s_0|,$$
$$n_{\psi_2} = 1 + \sum_{u \in s_0} \rho_{u,\psi_2} = 1 + 2 \times |\{u \in s_0,\ D_u = 1\}| + |\{u \in s_0,\ D_u = H\}|,$$

where the relation $D_u = H$ must be understood as: a conjugate of D_u is equal to H; hence:

$$\begin{aligned}\widetilde{\mathrm{r}}_p^{S_0} &= 3 - 1 \wedge |s_0| - 2 \wedge (1 + 2 \times |\{u \in s_0,\ D_u = 1\}| + |\{u \in s_0,\ D_u = H\}|) \\ &= 2 - 1 \wedge |s_0| - 1 \wedge |\{u \in s_0,\ D_u \subseteq H\}|.\end{aligned}$$

If $s_0 \neq \emptyset$, the term $1 \wedge |s_0|$ is equal to 1 and is relative to the cyclotomic $\mathbb{Z}_p$-extension which thus disappears; the case $D_u \subseteq H$ (which means that there exists a place of N above u which is totally split in K) allows us to make $\widetilde{K}_p^{S_0}/K$ finite ($\widetilde{\mathrm{r}}_p^{S_0} = 0$). □

c) The Monogeneous Case

Let us now study the value of $\widetilde{\mathrm{r}}_p^{S_0}$ in the case where we do not assume anymore that S_0 is the set of places of K above a set of prime numbers, but we assume that E^{S_0} is monogeneous (case (ii) of 3.1.7). We first have the following reduction in which the prime number p does not enter.

3.4 Lemma. *If E^{S_0} is monogeneous with $S_0 \neq \emptyset$, there exists a prime number q such that $S_0 \subseteq Pl_q$, and we have $|S_0| \le r_2 + 1$.*

Proof. Assume that S_0 contains $\mathfrak{q}|q$ and $\mathfrak{l}|\ell$, for $q \neq \ell$, and let $\varepsilon_{\mathfrak{q}}$, $\varepsilon_{\mathfrak{l}}$ be nontrivial $\{\mathfrak{q}\}$ and $\{\mathfrak{l}\}$-units (i.e., not units) of K; since there exists $\eta \in N^\times$ such that $E^{S_0} \otimes_{\mathbb{Z}} \mathbb{Q} \subseteq \langle \eta \rangle_\Gamma \otimes_{\mathbb{Z}} \mathbb{Q}$, it follows that for the extensions σ_i to N of the $[K : \mathbb{Q}]$ $\mathbb{Q}$-isomorphisms of K, we have $\sigma_i(\varepsilon_{\mathfrak{q}})$, $\sigma_i(\varepsilon_{\mathfrak{l}}) \in \langle \eta \rangle_\Gamma \otimes_{\mathbb{Z}} \mathbb{Q}$, and in particular $\langle \eta \rangle_\Gamma \otimes_{\mathbb{Z}} \mathbb{Q}$ contains $\mathrm{N}_{K/\mathbb{Q}}(\varepsilon_{\mathfrak{q}}) = \pm q^a$, $\mathrm{N}_{K/\mathbb{Q}}(\varepsilon_{\mathfrak{l}}) = \pm \ell^b$, with a, $b \neq 0$, a contradiction since the representation $\langle \eta \rangle_\Gamma \otimes_{\mathbb{Z}} \mathbb{Q}$ would then contain twice the unit representation. We then have:

$$E^{S_0} \otimes_{\mathbb{Z}} \mathbb{Q} \subseteq (\langle \eta \rangle_\Gamma \otimes_{\mathbb{Z}} \mathbb{Q})^H.$$

Since $\langle \eta \rangle_\Gamma \otimes_{\mathbb{Z}} \mathbb{Q}$ is isomorphic to a submodule of $\mathbb{Q}[\Gamma]$, $(\langle \eta \rangle_\Gamma \otimes_{\mathbb{Z}} \mathbb{Q})^H$ is isomorphic to a subspace of $\mathbb{Q}[\Gamma]^H = e_H \mathbb{Q}[\Gamma]$ (where $e_H := \frac{1}{|H|} \sum_{t \in H} t$) which has dimension $(\Gamma : H) = [K : \mathbb{Q}]$ (immediate). Thus, by the Dirichlet Theorem I.3.7.1:

$$\mathrm{rk}_{\mathbb{Z}}(E^{S_0}) = r_1 + r_2 - 1 + |S_0| \le r_1 + 2r_2,$$

which yields $|S_0| \le r_2 + 1$. □

3.4.1 Example. Assume that $K/\mathbb{Q}$ is Galois and real (i.e., $r_2 = 0$), then the case $S_0 \neq \emptyset$ implies that $S_0 = \{\mathfrak{q}\}$, where $\mathfrak{q}$ is a prime ideal of K above a prime number q. Since $\langle\, E^{\{\mathfrak{q}\}} \otimes_{\mathbb{Z}} \mathbb{Q}\,\rangle_\Gamma = E^{Pl_q} \otimes_{\mathbb{Z}} \mathbb{Q}$ is also monogeneous, we must have $|Pl_q| = 1$, in other words q not split in K, which (apart in the cyclic case) is usually impossible (for example for $K = \mathbb{Q}(\sqrt{5}, \sqrt{29})$).

The condition that E^{S_0} is monogeneous can thus essentially only be applied to the case $S_0 = \emptyset$ (in the totally real case), or to the case of nontotally real fields, which have more possibilities (for example $K = \mathbb{Q}(\sqrt{5}, \sqrt{-11})$ and q not split in $\mathbb{Q}(\sqrt{5})$). However, in the imaginary Galois case, the monogeneity of E^{S_0} (S_0 nonempty and contained in Pl_q) implies that of E^{Pl_q}, and the condition $|Pl_q| = r_2 + 1$ can only take place if $r_2 + 1$ divides $2r_2$, which is equivalent to $r_2 = 1$ (K is an imaginary quadratic field where q splits). □

Since $\widetilde{\mathrm{r}}_p^{S_0} = \dim_{\mathbb{Q}_p}\Big(\bigoplus_{v|p} K_v \big/ \mathbb{Q}_p \mathrm{log}_p(E^{S_0})\Big)$ and that E^{S_0} is monogeneous, assuming the p-adic conjecture, we have the equality:

$$\begin{aligned}\widetilde{\mathrm{r}}_p^{S_0} &= [K:\mathbb{Q}] - \dim_{\mathbb{Q}}(E^{S_0} \otimes_{\mathbb{Z}} \mathbb{Q}) \\ &= [K:\mathbb{Q}] + 1 - |S_0 \cup Pl_\infty| = r_2 + 1 - |S_0|,\end{aligned}$$

by the Dirichlet theorem. We have thus proved:

3.5 Theorem. *Let S_0 be the empty set or a subset of Pl_q, where q is a prime number different from p; when S_0 is not empty, we assume that E^{S_0} is monogeneous. If K satisfies the p-adic conjecture, the p-rank of the Galois group of the compositum of the S_0-split $\mathbb{Z}_p$-extensions of K satisfies:*

$$\widetilde{\mathrm{r}}_p^{S_0} = r_2 + 1 - |S_0|.$$

□

3.5.1 Remark (sufficient condition for monogeneity). Note that E^{S_0} is monogeneous if and only if the Γ-module $\langle\, E^{S_0}\,\rangle_\Gamma$ generated by E^{S_0} is monogeneous; but the character of $\langle\, E^{S_0}\,\rangle_\Gamma \otimes_{\mathbb{Z}} \mathbb{C}_p = \langle\, E^{S_0} \otimes_{\mathbb{Z}} \mathbb{C}_p\,\rangle_\Gamma$ can be computed in practice for any given numerical example (under the necessary conditions that $S_0 \subseteq Pl_q$ and $|S_0| \leq r_2+1$), hence we have a convenient way of checking whether E^{S_0} is monogeneous or not.

In the general case, we can find a sufficient condition as follows, where we recall that S'_0 is the set of places of N above those of S_0. Let $S''_0 := Pl_{N,q}$ be the set of places of N above q (we thus have $S'_0 \subseteq S''_0$). We know that $E_N^{S''_0} \otimes_{\mathbb{Z}} \mathbb{C}_p = \bigoplus_\psi n_\psi V_\psi$, where $n_\psi = \sum_{u \in \{q,\infty\}} \rho_{u,\psi} - \delta_{1,\psi}$ (see 3.3); but:

$$\begin{aligned}&\langle\, E^{S_0} \otimes_{\mathbb{Z}} \mathbb{C}_p\,\rangle_\Gamma \subseteq \langle\, (E_N^{S''_0} \otimes_{\mathbb{Z}} \mathbb{C}_p)^H\,\rangle_\Gamma = \\ &\qquad \Big\langle \bigoplus_\psi n_\psi V_\psi^H \Big\rangle_\Gamma = \bigoplus_\psi n_\psi \langle\, V_\psi^H\,\rangle_\Gamma = \bigoplus_\psi n_\psi\, \delta_\psi^H\, V_\psi,\end{aligned}$$

where $\delta_\psi^H = 0$ or 1 according as $\rho_\psi^H = 0$ or not. Thus, a sufficient condition for the monogeneity of E^{S_0} is:

$$\sum_{u \in \{q,\infty\}} \rho_{u,\psi} - \delta_{1,\psi} \leq \psi(1) \text{ for all } \psi \in \Psi_N \text{ such that } \rho_\psi^H \neq 0. \qquad \square$$

3.5.2 Example. If K is a totally real non-Galois cubic field, we can use 3.3.7, here with $D_\infty = 1$. Let us choose q such that $D_q = \langle\, \sigma \,\rangle$ (there is a unique place of K above q which is thus the only element of S_0); we have:

$$E_N^{S_0''} \otimes_{\mathbb{Z}} \mathbb{C}_p = V_{\psi_0} \oplus 2V_{\psi_1} \oplus 2V_{\psi_2}$$

(meaning that $E_N^{S_0''}$ is not monogeneous), but:

$$E^{S_0} \otimes_{\mathbb{Z}} \mathbb{C}_p \subseteq (E_N^{S_0''} \otimes_{\mathbb{Z}} \mathbb{C}_p)^H = V_{\psi_0}^H \oplus 2V_{\psi_2}^H$$

is therefore monogeneous and such that:

$$\langle\, E^{S_0} \otimes_{\mathbb{Z}} \mathbb{C}_p \,\rangle_\Gamma \subseteq V_{\psi_0} \oplus 2V_{\psi_2}.$$

Since E^{S_0} has $\mathbb{Z}$-rank 3, the above inclusions are in fact equalities. $\square$

Thus, the monogeneity of $E_N^{S_0''}$ (or equivalently, that of $E_N^{S_0'}$) is a sufficient condition for that of E^{S_0}, weaker than that given above.

3.5.3 Example. We take once again the situation of Example 3.3.7 for K not totally real (i.e., $D_\infty = H$). Since $\rho_{\infty,\psi_0} = 1$, $\rho_{\infty,\psi_1} = 0$, $\rho_{\infty,\psi_2} = 1$, the sufficient condition for monogeneity of E^{S_0} (in the case where $s_0 = \{q\}$) is here equivalent to $\rho_{q,\psi_2} \leq 1$, which yields $D_q \neq 1$, in other words q not totally split in N. For example, in $K = \mathbb{Q}(\sqrt[3]{2})$ we can choose $q = 7$ (split in $\mathbb{Q}(j)$, inert in $N/\mathbb{Q}(j)$), or $q = 2$ (inert in $\mathbb{Q}(j)$, ramified in $N/\mathbb{Q}(j)$), or $q = 5$ (inert in $\mathbb{Q}(j)$, split in $N/\mathbb{Q}(j)$), but not $q = 43$. $\square$

The sufficient condition of 3.5.1 is not necessary (in fact it does not involve the cardinality of S_0).

3.5.4 Exercise. In the case $D_q = 1$ and $S_0 \neq \emptyset$ in 3.5.3, show that E^{S_0} is monogeneous if and only if $|S_0| = 1$; for this, show that the representation $\langle\, E^{S_0} \otimes_{\mathbb{Z}} \mathbb{C}_p \,\rangle_\Gamma$ is given by $V_{\psi_0} \oplus 3V_{\psi_2}$ if $|S_0| > 1$ and by $V_{\psi_0} \oplus 2V_{\psi_2}$ if $|S_0| = 1$.

Answer. By 3.3, the representation defined by $\langle\, E \otimes_{\mathbb{Z}} \mathbb{C}_p \,\rangle_\Gamma$ is equal to:

$$\langle\, (E_N \otimes_{\mathbb{Z}} \mathbb{C}_p)^H \,\rangle_\Gamma = \langle\, (V_{\psi_2})^H \,\rangle_\Gamma = V_{\psi_2},$$

and the representation $\langle\, Pl_q \,\rangle_\Gamma \otimes_{\mathbb{Z}} \mathbb{C}_p$ is equal to:

$$\langle ((\langle Pl_{N,\mathbb{Q}} \rangle \otimes_{\mathbb{Z}} \mathbb{C}_p)^H \rangle_\Gamma = \langle V_{\psi_0}^H \oplus 2V_{\psi_2}^H \rangle_\Gamma = V_{\psi_0} \oplus 2V_{\psi_2},$$

since $D_q = 1$; the representation:

$$\langle E^{S_0} \otimes_{\mathbb{Z}} \mathbb{C}_p \rangle_\Gamma \simeq (\langle E \otimes_{\mathbb{Z}} \mathbb{C}_p \rangle_\Gamma) \bigoplus (\langle S_0 \rangle_\Gamma \otimes_{\mathbb{Z}} \mathbb{C}_p)$$

is therefore contained in $V_{\psi_0} \oplus 3V_{\psi_2}$.

If $S_0 = \{\mathfrak{q}\}$, we check that $\langle S_0 \rangle_\Gamma = \langle \mathfrak{q},\ \mathfrak{q}^\sigma,\ \mathfrak{q}^{\sigma^2} \rangle$ has rank 3 and that $\langle S_0 \rangle_\Gamma \otimes_{\mathbb{Z}} \mathbb{C}_p = V_{\psi_0} \oplus V_{\psi_2}$; hence $\langle E^{S_0} \otimes_{\mathbb{Z}} \mathbb{C}_p \rangle_\Gamma = V_{\psi_0} \oplus 2V_{\psi_2}$, showing that E^{S_0} is monogeneous.

If $S_0 = \{\mathfrak{q},\ \mathfrak{q}'\}$, this means that $(\mathfrak{q}) = \mathfrak{Q}\,\mathfrak{Q}^\tau = \mathfrak{Q}^{1+\tau}$ and $(\mathfrak{q}') = \mathfrak{Q}^\sigma\,\mathfrak{Q}^{\tau\sigma} = \mathfrak{Q}^{\sigma+\tau\sigma}$ (for some ideal $\mathfrak{Q}$ of N above q); we thus obtain:

$$\langle S_0 \rangle_\Gamma = \langle \mathfrak{Q}^{1+\tau},\ \mathfrak{Q}^{\sigma+\tau\sigma},\ \mathfrak{Q}^{\sigma+\tau\sigma^2}, \mathfrak{Q}^{\sigma^2+\tau},\ \mathfrak{Q}^{\sigma^2+\tau\sigma},\ \mathfrak{Q}^{1+\tau\sigma^2} \rangle,$$

which has rank 5 and yields the representation $V_{\psi_0} \oplus 2V_{\psi_2}$, hence:

$$\langle E^{S_0} \otimes_{\mathbb{Z}} \mathbb{C}_p \rangle_\Gamma = V_{\psi_0} \oplus 3V_{\psi_2},$$

showing that E^{S_0} is not monogeneous. The case where $|S_0| = 3$ is clear since the condition $|S_0| \leq r_2 + 1 = 2$ is not anymore satisfied. □

The following study is quite important since it gives statements which are equivalent to the p-adic conjecture and are of constant use. In this direction there is an abundant literature, and we can cite for instance in addition to the results obtained by the transcendence arguments mentioned at the beginning of Section 3, (a): [c, Wa, Ch. 5, § 5], [San], [a, BŠa], and [g, NSW, Ch. X, § 3] for a cohomological interpretation of the Leopoldt conjecture.

3.6 The p-Adic Conjecture for Monogeneous S-Unit Groups. Let K be a number field, let S be a finite set of noncomplex places of K prime to p, and assume that E^S is *monogeneous* (see 3.4), which is the case in particular when $S_0 = \emptyset$.[12] Thus, the reader who prefers to restrict to the case of the Leopoldt conjecture must take $S_0 = \emptyset$ everywhere.

We are going to prove a number of equivalent formulations to the relation:

$$\mathrm{r}_p^{S_0} := \dim_{\mathbb{Q}_p}(\mathbb{Q}_p \log_p(E^S)) = \dim_{\mathbb{Q}}(E^S \otimes \mathbb{Q}) = r_1 + r_2 - 1 + |S_0|$$

(the p-adic conjecture).

3.6.1 Notations. We set $E_n^S := \{\varepsilon \in E^S,\ \varepsilon \equiv 1 \bmod (p^{\delta_{2,p}+n})\},\ \ n \geq 1$, where $\delta_{2,p}$ is the Kronecker symbol, and we put $r := r_1 + r_2 - 1 + |S_0|$. It is easily checked that E_1^S has no torsion and is such that $\mathrm{v}(\log_v(\varepsilon)) = \mathrm{v}(\varepsilon - 1)$ for all $\varepsilon \in E_1^S$ and $v|p$. We denote by $\{\varepsilon_1, \ldots, \varepsilon_r\}$ a $\mathbb{Z}$-basis of E_1^S. □

[12] Here we use the letter S so as to have completely general notations, but the values of the ranks which are involved only depend on S_0.

3.6.2 Theorem. *When E^S is monogeneous, the p-adic conjecture is equivalent to each of the following statements:*

(i) $\dim_{\mathbb{Q}_p}(\mathbb{Q}_p \log_p(E_1^S)) = r$;

(ii) $\{\log_p(\varepsilon_1), \ldots, \log_p(\varepsilon_r)\}$ *is a* $\mathbb{Z}_p$*-basis of* $\mathbb{Z}_p \log_p(E_1^S)$;

(iii) there exists $n_0 \geq 1$ *such that* $E_{n_0}^S \subseteq (E_1^S)^p$;

(iv) there exists $n_0 \geq 1$ *such that* $E_n^S \subseteq (E_{n_0}^S)^{p^{n-n_0}}$ *for all* $n \geq n_0$;

(v) the canonical map $\mathcal{E}_1^S := E_1^S \otimes \mathbb{Z}_p \longrightarrow \mathbb{Z}_p \log_p(E_1^S)$ *is injective;*[13]

(vi) the canonical map $\bar{i}_p := (\bar{i}_v)_{v|p} : \mathcal{E}^S := E^S \otimes \mathbb{Z}_p \longrightarrow \bigoplus_{v|p} U_v^1$ *is injective (see 1.6.6).*

Proof. Since E_1^S is of finite index in E^S, the p-adic conjecture implies the equality (i).

(i) $\Rightarrow$ (ii): clear.

(ii) $\Rightarrow$ (iii): by contradiction, assume that for all $n \geq 1$, $E_n^S \not\subseteq (E_1^S)^p$; it follows that there exists a sequence $\varepsilon(n) \in E_n^S$ such that $\varepsilon(n) \notin (E_1^S)^p$. Set:

$$\varepsilon(n) =: \prod_{i=1}^{r} \varepsilon_i^{a_i^n}, \ a_i^n \in \mathbb{Z} \ ;$$

we thus have $\log_p(\varepsilon(n)) = \sum_{i=1}^{r} a_i^n \log_p(\varepsilon_i)$ for all n, and there exists a limit point $(a_i)_i \in \mathbb{Z}_p^r$ of $(a_i^n)_i$ such that $0 = \sum_{i=1}^{r} a_i \log_p(\varepsilon_i)$. Hence $a_i = 0$ for all i, and for a suitable n all the a_i^n are divisible by p so $\varepsilon(n) \in (E_1^S)^p$, a contradiction.

(iii) $\Rightarrow$ (iv): let $n \geq n_0$ and let $\varepsilon \in E_n^S$; since $E_n^S \subseteq E_{n_0}^S$, we have $\varepsilon \in (E_1^S)^p$ and we denote by k the largest integer such that $\varepsilon \in (E_k^S)^p$ (we have $k \geq 1$). Set $\varepsilon =: \eta^p$, with $\eta = 1+\alpha p^{\delta_{2,p}+k}$, $\alpha \notin (p)$; we know that $\eta^p = 1+\alpha' p^{\delta_{2,p}+k+1}$, $\alpha' \notin (p)$, hence $k \geq n-1$ and $\eta \in E_{n-1}^S$. The result follows by induction (limited to n_0).

(iv) $\Rightarrow$ (v): we have $\mathcal{E}_1^S = \bigoplus_{i=1}^{r}(\langle \varepsilon_i \rangle \otimes \mathbb{Z}_p)$. Let $\varepsilon =: \prod_{i=1}^{r} \varepsilon_i \otimes a_i$, $a_i \in \mathbb{Z}_p$, such that $\sum_{i=1}^{r} a_i \log_p(\varepsilon_i) = 0$. For all $n \geq n_0$ we can find $a_i' \in \mathbb{Z}$, close to a_i (at least modulo (p^n)), such that:

$$\sum_{i=1}^{r} a_i' \log_p(\varepsilon_i) \equiv 0 \bmod (p^n),$$

showing that:

$$\varepsilon' := \prod_{i=1}^{r} \varepsilon_i^{a_i'} \in E_n^S \subseteq (E_{n_0}^S)^{p^{n-n_0}} \ ;$$

[13] It is surjective and, since $E_1^S \simeq \mathbb{Z}^r$, we have $\mathbb{Z}_p \log_p(E_1^S) \simeq \mathbb{Z}_p^r$.

the a_i' are therefore divisible by p^{n-n_0}, and the a_i, arbitrarily close to 0, are equal to zero. Thus $\varepsilon = 1$.

(v) $\Rightarrow$ (vi): if $\varepsilon \in \mathcal{E}^S$ is such that $\bar{i}_p(\varepsilon) = 1$, then $\log_p(\varepsilon^m) = 0$, the integer $m > 0$ being chosen such that $\varepsilon^m \in \mathcal{E}_1^S$; by assumption it follows that $\varepsilon^m = 1$, hence:

$$\varepsilon \in \mathrm{tor}_{\mathbb{Z}_p}(\mathcal{E}^S) = \mathrm{tor}_{\mathbb{Z}}(E^S) \otimes \mathbb{Z}_p \subseteq \mu_p(K) \otimes 1.$$

The result follows since on $\mu_p(K) \otimes 1$ the map $\bar{i}_p$ is injective.

Finally (vi) implies the p-adic conjecture since the $\mathbb{Z}_p$-rank of $\mathbb{Z}_p\log_p(E^S)$ is then equal to r (on $\bigoplus_{v|p} U_v^1$ the kernel of log is finite and $\mathbb{Z}_p\log_p(E^S) = \log(\bar{i}_p(\mathcal{E}^S))$ with $\bar{i}_p(\mathcal{E}^S)$ of rank r). We can also use the fact that $\bar{i}_p(\mathcal{E}^S) = \mathrm{adh}_p(E'^S)$ for the group of p-principal S-units. □

Recall in the same way that $E'^{S_0\,\mathrm{ord}}$ is the group of p-principal ordinary S_0-units (note that $E'^{S_0\,\mathrm{ord}}$ is not always equal to $E_1^{S_0\,\mathrm{ord}}$), and that $\mu_p(K)$ is the group of roots of unity of order a power of p contained in K.

3.6.3 Corollary 1. *Assuming the p-adic conjecture, we have:*

$$\mathrm{tor}_{\mathbb{Z}_p}(\bar{i}_p(\mathcal{E}^{S_0\,\mathrm{ord}})) = \mathrm{tor}_{\mathbb{Z}_p}(\mathrm{adh}_p(E'^{S_0\,\mathrm{ord}})) = i_p(\mu_p(K)),$$

in other words $\Big\{\varepsilon \in \mathcal{E}^{S_0\,\mathrm{ord}},\ \bar{i}_p(\varepsilon) \in \bigoplus_{v|p} \mu_p(K_v)\Big\} = \mu_p(K) \otimes 1.$

Proof. The injectivity of $\bar{i}_p$ immediately implies the result since we are reduced to computing the p-torsion of $\mathcal{E}^{S_0\,\mathrm{ord}}$ which is here isomorphic to $\mathrm{tor}_{\mathbb{Z}}(E^{S_0\,\mathrm{ord}}) \otimes \mathbb{Z}_p \simeq \mu_p(K)$. □

3.6.4 Corollary 2. *Assuming the p-adic conjecture, we have* $\mathcal{S}_0 \cap \mathcal{P}_{p,\infty} = 1$, *where* $\mathcal{S}_0 = \langle\, S_0 \,\rangle \otimes \mathbb{Z}_p$ *(see 2.4).*

Proof. Let $\mathfrak{a} \in \mathcal{S}_0 \cap \mathcal{P}_{p,\infty}$. We can write $\mathfrak{a} = (\eta)$, $\eta \in \mathcal{E}^{S_0\,\mathrm{ord}}$ with $\bar{i}_p(\eta) = 1$ by 2.4.1; hence $\eta = 1$ by (vi). □

3.6.5 Corollary 3. *If* $|S_0| = 1$ *then, assuming the Leopoldt conjecture for* p, *we have* $\mathrm{r}_p(E^{S_0\,\mathrm{ord}}) = r_1 + r_2$.

Proof. We have $E^{S_0\,\mathrm{ord}} =: \langle\, \pi \,\rangle E^{\mathrm{ord}}$, where π is a suitable S_0-unit; we thus have:

$$\mathbb{Q}_p\log_p(E^{S_0\,\mathrm{ord}}) = \mathbb{Q}_p\log_p(\pi) + \mathbb{Q}_p\log_p(E^{\mathrm{ord}}),$$

where $\mathbb{Q}_p\log_p(E^{\mathrm{ord}})$ has dimension equal to $r_1 + r_2 - 1$ by assumption; but $\log_p(\pi) \notin \mathbb{Q}_p\log_p(E^{\mathrm{ord}})$ (this subspace is contained in the kernel of the trace, contrary to $\log_p(\pi)$). □

In this corollary, it is not necessary to assume that $E^{S_0\mathrm{ord}}$ is monogeneous. If $|S_0| > 1$, we only have $\mathrm{r}_p(E^{S_0\,\mathrm{ord}}) \geq r_1 + r_2$.

The following result depends on the important forthcoming Theorem 4.4, but it seems preferable to state it now.

3.6.6 Corollary 4 (local-global characterization of the Leopoldt conjecture [Ja2; Ja7]). *The Leopoldt conjecture for p is equivalent to:*

$$\bar{i}(\mathcal{E}^{\mathrm{ord}}) \cap \prod_{v \in Pl^{\mathrm{nc}}} \mu_p(K_v) = i(\mu_p(K)),$$

where, by abuse of notation in this statement, $\bar{i}$ takes its values in $\prod_v (U_v)_p$ where $(U_v)_p = \mu_p(K_v) \simeq (F_v^\times)_p$ for the tame places.

Proof. In one direction (i.e., assuming the Leopoldt conjecture) use Corollary 1 to the p-component, $\bar{i}_p(\varepsilon)$, of $\bar{i}(\varepsilon) \in \prod_v \mu_p(K_v)$, then, for the other, suppose $\bar{i}_p(\varepsilon) = 1$ which implies $\bar{i}(\varepsilon) \in \bar{i}(\mu_p(K) \otimes 1)$, use the injectivity of $\bar{i}_{\mathrm{ta},\infty}$ on $\mathcal{E}^{\mathrm{ord}}$ (see 4.4) and the injectivity of $\bar{i}_p$ on $\mu_p(K) \otimes 1$. □

Finally, in [Fl] the following conjecture, weaker than the Leopoldt conjecture, is stated.

3.7 Conjecture (Fleckinger (1986)). *Let K be a number field containing a primitive pth root of unity ζ. Then, for $\varepsilon \in \mathcal{E}^{\mathrm{ord}}$:*

$$\bar{i}_p(\varepsilon)^p = 1 \textit{ if and only if } \bar{i}_p(\varepsilon) \in i_p(\langle\, \zeta \,\rangle). \qquad \square$$

It is proved in [Fl, Th. 1] that the two following conditions are equivalent:

(i) The Leopoldt conjecture for p is true *for all* number fields K.

(ii) *For all* number fields K, ${}_p(\bar{i}_p(\mathcal{E}^{\mathrm{ord}})) = i_p({}_p\mu(K))$.

The assumption (ii) has the following analytical statement for a fixed number field K containing ζ.

(ii′) There exists $c_K > 0$ such that for all $\varepsilon \in E^{\mathrm{ord}}$ we have:

$$\prod_{(\zeta_v)_v} \Big(\sum_{v|p} |i_v(\varepsilon) - \zeta_v|_{K_v} \Big) \geq c_K,$$

where $(\zeta_v)_v$ varies in $\bigoplus_{v|p} {}_p\mu(K_v) \setminus i_p(\langle\, \zeta \,\rangle)$.

§4 Structure Theorems for the Abelian Closure of K

We fix an algebraic closure $\overline{K}$ of K and consider its maximal abelian subextension $\overline{K}^{\mathrm{ab}}$ in which the correspondence of global class field theory takes place.

a) Deployment of $\mathrm{Gal}(\overline{K}^{\mathrm{ab}}(p)/H_p^{\mathrm{ord}}(p))$

To begin, let us prove a fundamental deployment theorem which is quite nontrivial since it is based on the general p-adic conjecture, and which will allow us to obtain essential information on the structure of $(\overline{G}^{\mathrm{ab}})_p := \mathrm{Gal}(\overline{K}^{\mathrm{ab}}(p)/K)$ for any prime number p.

Notations and assumptions. Let T be a finite set of finite places of K containing Pl_p and $T_{\mathrm{ta}} := T \backslash Pl_p$. Let $S_0 \subset Pl_0$ be prime to T; when S_0 is not empty, we assume that $E^{S_0\,\mathrm{ord}}$ is monogeneous (see 3.4). As usual we denote by $H_T^{\mathrm{res}}(p)$ (resp. $H_p^{S_0\,\mathrm{ord}}(p)$) the maximal T-ramified (resp. p-ramified S_0-split noncomplexified) abelian pro-p-extension of K. □

4.1 Theorem (deployment above $H_p^{\mathrm{ord}}(p)$). *Assuming the p-adic conjecture, we have:*

$$\mathrm{Gal}(H_T^{\mathrm{res}}(p)/H_p^{S_0\,\mathrm{ord}}(p)) = \bigoplus_{v\in T_{\mathrm{ta}}} I_v \bigoplus_{v\in Pl_\infty^{\mathrm{r}}} D_v \bigoplus_{v\in S_0} D_v$$
$$\simeq \bigoplus_{v\in T_{\mathrm{ta}}\cup Pl_\infty^{\mathrm{r}}} (F_v^\times)_p \bigoplus \mathbb{Z}_p^{|S_0|},$$

where I_v is the inertia group of $v \in T_{\mathrm{ta}}$ and D_v is the decomposition group of $v \in Pl_\infty^{\mathrm{r}} \cup S_0$ in $H_T^{\mathrm{res}}(p)/K$.

In particular if $S_0 = \emptyset$, assuming the Leopoldt conjecture for p, we have:

$$\mathrm{Gal}(H_T^{\mathrm{res}}(p)/H_p^{\mathrm{ord}}(p)) = \bigoplus_{v\in T_{\mathrm{ta}}} I_v \bigoplus_{v\in Pl_\infty^{\mathrm{r}}} D_v \simeq \bigoplus_{v\in T_{\mathrm{ta}}\cup Pl_\infty^{\mathrm{r}}} (F_v^\times)_p.$$

Proof. By the correspondence of infinite class field theory 2.8, we have immediately:

$$\mathrm{Gal}(H_T^{\mathrm{res}}(p)/H_p^{S_0\,\mathrm{ord}}(p)) \simeq \mathcal{P}_{T,\infty_p} \,.\, \mathcal{S}_0/\mathcal{P}_{T,\infty,\mathrm{pos}},$$

where we have set $\mathcal{P}_{T,\infty_p} := \bigcap_n \mathcal{P}_{T,\mathfrak{m}_p(n)} = \mathcal{P}_{p,\infty} \cap \mathcal{I}_T$. We have:

$$\mathcal{P}_{T,\infty_p} \,.\, \mathcal{S}_0/\mathcal{P}_{T,\infty,\mathrm{pos}} = (\mathcal{P}_{T,\infty_p}/\mathcal{P}_{T,\infty,\mathrm{pos}}) \bigoplus (\mathcal{S}_0 \,.\, \mathcal{P}_{T,\infty,\mathrm{pos}}/\mathcal{P}_{T,\infty,\mathrm{pos}})$$

since, by our assumption of p-adic independence of S_0-units, we have (see 3.6.4):

$$\mathcal{S}_0 \cap \mathcal{P}_{T,\infty_p} = 1.$$

We thus obtain the direct summand corresponding to the subgroup generated by the decomposition groups of the places of S_0:

$$\mathrm{Gal}(H_T^{\mathrm{res}}(p)/H_T^{S_0\mathrm{res}}(p)) \simeq \mathcal{S}_0/\mathcal{S}_0 \cap \mathcal{P}_{T,\infty,\mathrm{pos}} \simeq \mathcal{S}_0 \simeq \mathbb{Z}_p^{|S_0|}$$

since, a fortiori, $\mathcal{S}_0 \cap \mathcal{P}_{T,\infty,\mathrm{pos}} = 1$. The other direct summand:

$$\mathrm{Gal}(H_T^{\mathrm{res}}(p)/H_p^{\mathrm{ord}}(p)) \simeq \mathcal{P}_{T,\infty_p}/\mathcal{P}_{T,\infty,\mathrm{pos}},$$

can be computed thanks to 2.4.1, but it is interesting to understand the behavior under the limiting process, from the following property of ray class fields (afterwards, see 4.1.6).

4.1.1 Lemma. *Let* $\mathfrak{m} = \mathfrak{m}_p \mathfrak{m}_{\mathrm{ta}}$, *with* $\mathfrak{m}_p \in \langle\, Pl_p \,\rangle_{\mathbb{N}}$ *and* $\mathfrak{m}_{\mathrm{ta}} = \prod_{v \in T_{\mathrm{ta}}} \mathfrak{p}_v$. *Then, for sufficiently large* $\mathfrak{m}_p$, $H_T^{\mathrm{res}}(p)$ *is the direct compositum of* $K(\mathfrak{m})^{\mathrm{res}}(p)$ *with* $H_p^{\mathrm{ord}}(p)$ *over* $K(\mathfrak{m}_p)^{\mathrm{ord}}(p)$.

Proof. We have $H_p^{\mathrm{ord}}(p) = \bigcup_{\mathfrak{n} \in \langle Pl_p \rangle_{\mathbb{N}}} K(\mathfrak{n})^{\mathrm{ord}}(p)$ so that, for sufficiently large $\mathfrak{n}$, we can write because of II.5.1.5:

$$K(\mathfrak{m})^{\mathrm{res}}(p) \cap H_p^{\mathrm{ord}}(p) = K(\mathfrak{m})^{\mathrm{res}}(p) \cap K(\mathfrak{n})^{\mathrm{ord}}(p) = K(\mathfrak{m}_p)^{\mathrm{ord}}(p).$$

But, by 1.4 (for $t = Pl_p$, $\delta_\infty = Pl_\infty^{\mathrm{r}}$), $[K(\mathfrak{m})^{\mathrm{res}}(p) : K(\mathfrak{m}_p)^{\mathrm{ord}}(p)]$ is nondecreasing and bounded from above by $\prod_{v \in T_{\mathrm{ta}} \cup Pl_\infty^{\mathrm{r}}} |(F_v^\times)_p|$. The lemma follows since $H_T^{\mathrm{res}}(p) = \bigcup_{\mathfrak{m}} K(\mathfrak{m})^{\mathrm{res}}(p)$ (see 1.3.2). □

We thus have:

$$\mathrm{Gal}(H_T^{\mathrm{res}}(p)/H_p^{\mathrm{ord}}(p)) \simeq \varprojlim_{n \geq 0} \mathrm{Gal}(K(\mathfrak{m}(n))^{\mathrm{res}}(p)/K(\mathfrak{m}_p(n))^{\mathrm{ord}}(p)),$$

taking here $\mathfrak{m}(n) := \mathfrak{m}_p(n)\mathfrak{m}_{\mathrm{ta}}$ with $\mathfrak{m}_p(n) = \prod_{v|p} \mathfrak{p}_v^n$ and $\mathfrak{m}_{\mathrm{ta}} = \prod_{v \in T_{\mathrm{ta}}} \mathfrak{p}_v$. The exact sequence of 1.4 can then be written:

$$1 \longrightarrow (E_{\mathfrak{m}_p(n)}^{\mathrm{ord}}/E_{\mathfrak{m}(n)}^{\mathrm{res}})_p \longrightarrow \bigoplus_{v \in T_{\mathrm{ta}} \cup Pl_\infty^{\mathrm{r}}} (F_v^\times)_p \xrightarrow{\ \rho\ }$$
$$\mathrm{Gal}(K(\mathfrak{m}(n))^{\mathrm{res}}(p)/K(\mathfrak{m}_p(n))^{\mathrm{ord}}(p)) \longrightarrow 1;$$

therefore, by 3.6.2, (iv), it follows that $E_{\mathfrak{m}_p(n)}^{\mathrm{ord}} \subseteq E^{p^{h(n)}}$, $h(n) \longrightarrow \infty$ with n, so that $(E_{\mathfrak{m}_p(n)}^{\mathrm{ord}}/E_{\mathfrak{m}(n)}^{\mathrm{res}})_p = 1$ for all sufficiently large n (i.e., such that $p^{h(n)}$ kills $\bigoplus_{v \in T_{\mathrm{ta}} \cup Pl_\infty^{\mathrm{r}}} (F_v^\times)_p$). We could also have performed the computations in idelic terms from:

$$(K^\times U_{\mathfrak{m}_p(n)}^{\mathrm{ord}}/K^\times U_{\mathfrak{m}(n)}^{\mathrm{res}})_p \simeq (U_{\mathfrak{m}_p(n)}^{\mathrm{ord}}/E_{\mathfrak{m}_p(n)}^{\mathrm{ord}} U_{\mathfrak{m}(n)}^{\mathrm{res}})_p \simeq (U_{\mathfrak{m}_p(n)}^{\mathrm{ord}}/U_{\mathfrak{m}(n)}^{\mathrm{res}})_p$$

for all sufficiently large n. We thus have:

$$\mathrm{Gal}(H_T^{\mathrm{res}}(p)/H_p^{\mathrm{ord}}(p)) \simeq \bigoplus_{v \in T_{\mathrm{ta}} \cup Pl_\infty^{\mathrm{r}}} (F_v^\times)_p\ ;$$

but we have $\mathrm{Gal}(H_T^{\mathrm{res}}(p)/H_p^{\mathrm{ord}}(p)) = \langle\, I_v \,\rangle_{v \in T_{\mathrm{ta}}} \langle\, D_v \,\rangle_{v \in Pl_\infty^{\mathrm{r}}}$, hence the order of this group is sufficient to insure deployment: indeed, each I_v for $v \in T_{\mathrm{ta}}$ (resp.

D_v for $v \in Pl_\infty^{\mathrm{r}}$) is a quotient of $(F_v^\times)_p$ (resp. of $(\{\pm 1\})_p$), therefore we have $I_v \simeq (F_v^\times)_p$ (resp. $D_v \simeq (\{\pm 1\})_p$) for each of the above places, proving the theorem. □

4.1.2 Remark. This approach to the deployment theorem shows that it can be realized at a finite level (which can be made explicit in practice), more precisely that for a given T_{ta}:

$$\mathrm{Gal}(K(\mathfrak{m})^{\mathrm{res}}(p)/K(\mathfrak{m}_p)^{\mathrm{ord}}(p)) \simeq \bigoplus_{v \in T_{\mathrm{ta}} \cup Pl_\infty^{\mathrm{r}}} (F_v^\times)_p$$

for all sufficiently large $\mathfrak{m}_p$. □

4.1.3 Exercise. Take $K = \mathbb{Q}(\sqrt{2})$, $p = 2$, $T_{\mathrm{ta}} = \{\mathfrak{p}_7, \mathfrak{p}_7'\}$, $S_0 = \emptyset$. Find the smallest modulus $(\sqrt{2})^n$ for which there is deployment on $T_{\mathrm{ta}} \cup Pl_\infty^{\mathrm{r}}$ for any larger one.

Answer. We have $E^{\mathrm{ord}} = \langle -1, \varepsilon \rangle$, with $\varepsilon = 1 + \sqrt{2}$, and:

$$\bigoplus_{v \in T_{\mathrm{ta}} \cup Pl_\infty^{\mathrm{r}}} (F_v^\times)_2 \simeq (\mathbb{Z}/2\mathbb{Z})^4.$$

It is thus sufficient to start with a value of n such that:

$$E^{\mathrm{ord}}_{(\sqrt{2})^n} \subseteq (E^{\mathrm{ord}})^2 = \langle \varepsilon^2 \rangle,$$

which yields an *upper bound* for the solution.

For $n = 4$ we find $E^{\mathrm{ord}}_{(\sqrt{2})^4} = \langle \varepsilon^4 \rangle$, which is suitable; on the other hand, $n = 3$ yields $E^{\mathrm{ord}}_{(\sqrt{2})^3} = \langle -\varepsilon^2 \rangle$ which is not suitable since $-\varepsilon^2$ has even order in $E^{\mathrm{ord}}_{(\sqrt{2})^3}/E^{\mathrm{res}}_{7(\sqrt{2})^3}$ (or, more simply: the diagonal residual image of -1 has order 2).

Note that $n = 4$ is suitable for any T_{ta} such that $\bigoplus_{v \in T_{\mathrm{ta}}} (F_v^\times)_2$ has exponent 2. □

Deployment Theorem 4.1 has the following consequence (with Notations 2.1).

4.1.4 Corollary. *The field $H_T^{\mathrm{res}}(p)$ is the direct compositum of $H_T^{S_0\mathrm{res}}$ with $\widetilde{K}_p$ over $\widetilde{K}_p^{S_0}$; in other words, we have:*

$$\begin{aligned} \mathrm{Gal}(H_T^{S_0\mathrm{res}}(p)/\widetilde{K}_p^{S_0}) &\simeq \mathrm{Gal}(H_T^{\mathrm{res}}(p)/\widetilde{K}_p) = \mathcal{T}_T^{\mathrm{res}}, \\ \mathrm{Gal}(\widetilde{K}_p/\widetilde{K}_p^{S_0}) &\simeq \mathrm{Gal}(H_T^{\mathrm{res}}(p)/H_T^{S_0\mathrm{res}}(p)) \simeq \mathbb{Z}_p^{|S_0|}. \end{aligned}$$

□

Note that $\widetilde{K}_p^{S_0}$ is generally not $\mathbb{Z}_p$-free.

Furthermore, the deployment theorem shows that $\mathcal{T}_T^{\text{res}} = \text{Gal}(H_T^{\text{res}}(p)/\widetilde{K}_p)$ is a finite p-group whose structure (or at least the order) is accessible and can essentially be reduced to that of the group $\mathcal{T}_p^{\text{ord}}$, so that we obtain strengthenings of the results of 2.6 and its corollaries.

4.1.5 Theorem. *Let K be a number field satisfying the Leopoldt conjecture for p. Let T be a finite set of finite places of K containing Pl_p, and let $S_\infty \subseteq Pl_\infty^{\text{r}}$. Put $T_{\text{ta}} := T\backslash Pl_p$, $\Delta_\infty := Pl_\infty^{\text{r}}\backslash S_\infty$. We then have the exact sequence:*

$$1 \longrightarrow \bigoplus_{v\in T_{\text{ta}}\cup\Delta_\infty} (F_v^\times)_p \longrightarrow \mathcal{T}_T^{S_\infty} \longrightarrow \mathcal{T}_p^{\text{ord}} \longrightarrow 1,$$

giving the following particular cases:

$$1 \longrightarrow \bigoplus_{v\in T_{\text{ta}}} (F_v^\times)_p \longrightarrow \mathcal{T}_T^{\text{ord}} \longrightarrow \mathcal{T}_p^{\text{ord}} \longrightarrow 1,$$

$$1 \longrightarrow \bigoplus_{v\in T_{\text{ta}}\cup Pl_\infty^{\text{r}}} (F_v^\times)_p \longrightarrow \mathcal{T}_T^{\text{res}} \longrightarrow \mathcal{T}_p^{\text{ord}} \longrightarrow 1,$$

$$1 \longrightarrow \{\pm 1\}^{r_1} \longrightarrow \mathcal{T}_2^{\text{res}} \longrightarrow \mathcal{T}_2^{\text{ord}} \longrightarrow 1. \qquad \square$$

4.1.6 Exercise (direct proof of the deployment theorem) (after [Gr7, § 2]). We start from 1.6, 1.6.5, which yield the exact sequence (still with T containing Pl_p but assuming $S_0 = \emptyset$):

$$1 \longrightarrow \text{Gal}(H_T^{\text{res}}(p)/H_p^{\text{ord}}(p)) \longrightarrow$$
$$\Big(\bigoplus_{v|p} U_v^1 \bigoplus_{v\in T_{\text{ta}}\cup Pl_\infty^{\text{r}}} (F_v^\times)_p\Big)\Big/\bar{\imath}_{T,\infty}(\mathcal{E}^{\text{ord}}) \longrightarrow \bigoplus_{v|p} U_v^1\Big/\bar{\imath}_p(\mathcal{E}^{\text{ord}}) \longrightarrow 1,$$

where $\mathcal{E}^{\text{ord}} := E^{\text{ord}} \otimes \mathbb{Z}_p$.

(i) Deduce that we have:

$$\text{Gal}(H_T^{\text{res}}(p)/H_p^{\text{ord}}(p)) \simeq \Big(\bar{\imath}_p(\mathcal{E}^{\text{ord}}) \bigoplus_{v\in T_{\text{ta}}\cup Pl_\infty^{\text{r}}} (F_v^\times)_p\Big)\Big/\bar{\imath}_{T,\infty}(\mathcal{E}^{\text{ord}})$$
$$\simeq \bigoplus_{v\in T_{\text{ta}}\cup Pl_\infty^{\text{r}}} (F_v^\times)_p\Big/\bar{\imath}_{T,\infty}(\mathcal{E}^{\text{ord}}) \cap \Big(\bigoplus_{v\in T_{\text{ta}}\cup Pl_\infty^{\text{r}}} (F_v^\times)_p\Big).$$

(ii) Consider the following composition of maps:

$$\mathcal{E}^{\text{ord}} \xrightarrow{\ \bar{\imath}_{T,\infty}\ } \bar{\imath}_{T,\infty}(\mathcal{E}^{\text{ord}}) \xrightarrow{\ \text{pr}_p\ } \bigoplus_{v|p} U_v^1,$$

where pr_p is the projection on the summand $\bigoplus_{v|p} U_v^1$; check that it is equal to the restriction of $\bar{\imath}_p$.

(iii) Prove the injectivity of pr_p on $\bar{\imath}_{T,\infty}(\mathcal{E}^{\text{ord}})$, and conclude. Hint: the p-adic conjecture is equivalent to the injectivity of $\bar{\imath}_p$ (see 3.6.2, (vi));

but this injectivity implies that of pr_p, which means that $\bar{i}_{T,\infty}(\mathcal{E}^{\mathrm{ord}}) \cap \Big(\bigoplus_{v\in T_{\mathrm{ta}}\cup Pl_\infty^{\mathrm{r}}} (F_v^\times)_p\Big) = 1$. □

Let $\overline{K}^{\mathrm{ab}}$ be the abelian closure of K. Deployment Theorem 4.1 already gives some information on the structure of $\mathrm{Gal}(\overline{K}^{\mathrm{ab}}(p)/K)$ (assuming the Leopoldt conjecture for p) since $\overline{K}^{\mathrm{ab}}(p)$ can be considered as the union of the fields $H_T^{\mathrm{res}}(p)$, when T containing Pl_p varies. We thus obtain:

$$\begin{aligned}\mathrm{Gal}(\overline{K}^{\mathrm{ab}}(p)/H_p^{\mathrm{ord}}(p)) &\simeq \varprojlim_T \mathrm{Gal}(H_T^{\mathrm{res}}(p)/H_p^{\mathrm{ord}}(p))\\ &\simeq \varprojlim_{T_{\mathrm{ta}}} \Big(\bigoplus_{v\in T_{\mathrm{ta}}\cup Pl_\infty^{\mathrm{r}}} (F_v^\times)_p\Big) = \prod_{v\nmid p} (F_v^\times)_p,\end{aligned}$$

giving:

4.1.7 Corollary. *Assuming the Leopoldt conjecture for p, we have the exact sequences:*

$$1 \longrightarrow \prod_{v\nmid p}(F_v^\times)_p \longrightarrow \mathrm{Gal}(\overline{K}^{\mathrm{ab}}(p)/K) \longrightarrow \mathcal{T}_p^{\mathrm{ord}} \times \mathbb{Z}_p^{r_2+1} \longrightarrow 1,$$

$$1 \longrightarrow \prod_{v\nmid p}(F_v^\times)_p \longrightarrow \mathrm{Gal}(\overline{K}^{\mathrm{ab}}(p)/H^{\mathrm{ord}}(p)) \longrightarrow$$

$$\mathrm{tor}_{\mathbb{Z}_p}\Big(\bigoplus_{v\mid p} U_v^1 \big/ \bar{i}_p(\mathcal{E}^{\mathrm{ord}})\Big) \times \mathbb{Z}_p^{r_2+1} \longrightarrow 1,$$

in which $\prod_{v\nmid p}(F_v^\times)_p \simeq \mathrm{Gal}(\overline{K}^{\mathrm{ab}}(p)/H_p^{\mathrm{ord}}(p))$. □

The group $\mathcal{T}_p^{\mathrm{ord}} := \mathrm{Gal}(H_p^{\mathrm{ord}}(p)/\widetilde{K}_p)$ appears as an important arithmetical p-invariant of class field theory over K since it involves both ideal classes and unit aspects; we have given its order in 2.6.1, (ii$_2$), and we will study a little later the numerical problems posed by the part of $\mathcal{T}_p^{\mathrm{ord}}$ which comes from the units, and which suggest that we should go a little further than the p-adic conjecture. Such a group is also dual to certain ideal class and K-theory groups, as is shown by general reflection theorem.

In the Appendix at the end of the book, we give a detailed proof that $\mathcal{T}_p^{\mathrm{ord}}$ is the arithmetical interpretation of a natural cohomological invariant. Books on Galois cohomology only keep the cohomological formalism, but here we have wanted to show, on the one hand the basically p-adic character of this invariant, and, on the other hand, the fact that it is arithmetically quite computable as we have seen in all of Section 2.

See Subsection (b) below which gives more details on this and gives some additional results concerning the torsion groups $\mathcal{T}_T^S$.

Globalizing the results of 4.1.7, we obtain the following result which will be made more precise in diagrams 4.2 and 4.2′.

4.1.8 Theorem. *Let K be a number field of signature (r_1, r_2) and let $\overline{K}^{\rm ab}$ be the abelian closure of K. Then, assuming the Leopoldt conjecture in K for all p, we have the exact sequence:*

$$1 \longrightarrow \prod_{v \in Pl_0} F_v^\times \times \{\pm 1\}^{r_1} \longrightarrow \mathrm{Gal}(\overline{K}^{\rm ab}/K) \longrightarrow \prod_p \mathcal{T}_p^{\rm ord} \times \widehat{\mathbb{Z}}^{r_2+1} \longrightarrow 1,$$

where, for each p, $\mathcal{T}_p^{\rm ord} := \mathrm{Gal}(H_p^{\rm ord}(p)/\widetilde{K}_p)$ is the p-torsion group corresponding to abelian p-ramification over K. [14] □

The term $\prod_p \mathcal{T}_p^{\rm ord}$ defines a global object obtained indirectly; thus, for example we have the problem of its cardinality. We will try to see in 4.14 what can be done about this.

The following lemma gives a little more precise information on the structure of abelian pro-p-groups, seen as $\mathbb{Z}_p$-modules not necessarily of finite type, but the result is not canonical.

4.1.9 Lemma. *Let L/K be an abelian pro-p-extension and let M/K be a $\mathbb{Z}_p$-free subextension of rank $r \geq 1$ (i.e., $\mathrm{Gal}(M/K) \simeq \mathbb{Z}_p^r$). Then L is the direct compositum over K of M with a (nonunique) extension F of K.*

Proof. Let $(\sigma_1, \ldots, \sigma_r)$ be a $\mathbb{Z}_p$-basis of $\mathrm{Gal}(M/K)$. Let $(\sigma'_1, \ldots, \sigma'_r) \in \mathrm{Gal}(L/K)^r$ extending $(\sigma_1, \ldots, \sigma_r)$; by construction $(\sigma'_1, \ldots, \sigma'_r)$ projects on $(\sigma_1, \ldots, \sigma_r)$, which implies that the $\mathbb{Z}_p$-module A generated by $\sigma'_1, \ldots, \sigma'_r$ is free with $\mathbb{Z}_p$-rank r (write a relation between the σ'_i and project it). The field F fixed under A is a solution. □

4.1.10 Example. Let K be a number field satisfying the Leopoldt conjecture for p. If $\mathcal{T}_p^{\rm ord} = 1$ (i.e., if K is p-rational in the sense of [MoNg], [GrJ]; see Ch. IV, § 3, (b)), by the above lemma we have the noncanonical isomorphism:

$$\mathrm{Gal}(\overline{K}^{\rm ab}(p)/K) \simeq \prod_{v \nmid p} (F_v^\times)_p \times \mathbb{Z}_p{}^{r_2+1}.$$

□

4.1.11 Case of the Base Field $\mathbb{Q}$. For $K = \mathbb{Q}$, we have $\mathcal{T}_{\mathbb{Q},p}^{\rm ord} = 1$ for all p: indeed, $\mathcal{C}\ell_{\mathbb{Q}}^{\rm ord} = 1$ and $\mathrm{tor}_{\mathbb{Z}_p}(U_p/\langle -1 \rangle)_p = \mathrm{tor}_{\mathbb{Z}_p}(\mathbb{Z}_p^\times/\langle -1 \rangle)_p = 1$ for all p. Thus:

$$\mathrm{Gal}(\overline{\mathbb{Q}}^{\rm ab}(p)/\mathbb{Q}) \simeq \prod_{\ell \neq p} (\mathbb{F}_\ell^\times)_p \times (\{\pm 1\})_p \times \mathbb{Z}_p.$$

---

[14] i.e., $H_p^{\rm ord}(p)$ is the maximal p-ramified noncomplexified abelian pro-p-extension of K and $\widetilde{K}_p$ the compositum of the $\mathbb{Z}_p$-extensions of K.

For $p \neq 2$, the subgroup corresponding to $\prod_{\ell \neq p} (\mathbb{F}_\ell^\times)_p$ fixes the cyclotomic $\mathbb{Z}_p$-extension $\mathbb{Q}^{\mathrm{cycl}}(p)$ of $\mathbb{Q}$, and we can take for F the compositum $H_{\mathrm{ta}}^{\mathrm{res}}(p)$ of the fields $\mathbb{Q}(\mu_\ell)(p)$, $\ell \neq p$ (the maximal tamely ramified subextension of $\overline{\mathbb{Q}}^{\mathrm{ab}}(p)$).

For $p = 2$, we have the following diagram, where $F := H_{\mathrm{ta}}^{\mathrm{res}}(2)(\mu_4)$:

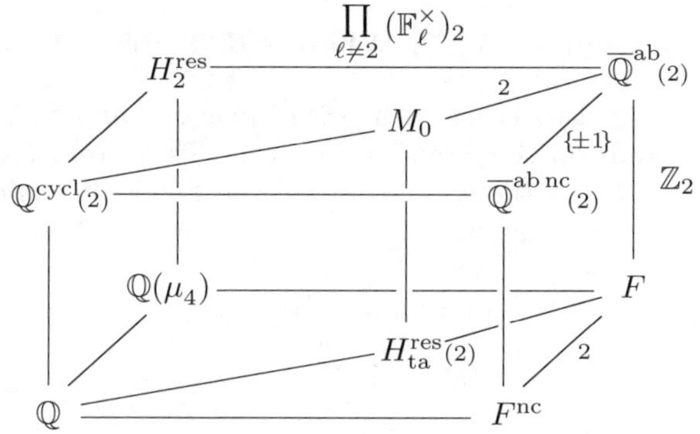

Fig. 4.1

In this diagram, the subgroup corresponding to $\prod_{\ell \neq 2} (\mathbb{F}_\ell^\times)_2$ fixes $H_2^{\mathrm{res}} = \mathbb{Q}(\mu_{2^\infty})$ (the maximal 2-ramified 2-extension of $\mathbb{Q}$), but the subgroup corresponding to $\{\pm 1\}$ (relative to the place at infinity) does not fix the compositum M_0 of $\mathbb{Q}^{\mathrm{cycl}}(2) = H_2^{\mathrm{ord}}$ with the compositum $H_{\mathrm{ta}}^{\mathrm{res}}(2)$ of the fields $\mathbb{Q}(\mu_\ell)(2)$, $\ell \neq 2$ (the maximal tamely ramified subextension of $\overline{\mathbb{Q}}^{\mathrm{ab}}(2)$, which is complexified), but the maximal real subfield of $\overline{\mathbb{Q}}^{\mathrm{ab}}(2)$ (compare with II.5.6.3).

4.1.11.1 Remark. On the simplest possible example, we see that class field theory does not yield in this approach the canonical decomposition that we could expect, in other words $\overline{\mathbb{Q}}^{\mathrm{ab}}(2)$ as a direct compositum of H_2^{res} (totally wildly ramified) and of $H_{\mathrm{ta}}^{\mathrm{res}}(2)$ (maximal tamely ramified). The reason for this is that there are two independent canonical elements of order 2: complex conjugation, which fixes $\overline{\mathbb{Q}}^{\mathrm{ab\,nc}}(2)$, and the generator of $\mathrm{Gal}(\overline{\mathbb{Q}}^{\mathrm{ab}}(2)/M_0)$ which corresponds to the diagonal embedding of -1 in $\prod_{\ell \neq 2} (\mathbb{F}_\ell^\times)_2 \times \{\pm 1\}$.

Note that $(U_0^{\mathrm{ord}})_2 \simeq \prod_{\ell \neq 2} (\mathbb{F}_\ell^\times)_2 \times \{\pm 1\} \times \mathbb{Z}/2\mathbb{Z} \times \mathbb{Z}_2$, where the additional $\mathbb{Z}/2\mathbb{Z}$ is the torsion group of $U_2 = \mathbb{Z}_2^\times$, so that one should not mistake the diagonal embedding of -1 in $(U_0^{\mathrm{ord}})_2$ (which gives the identity under the reciprocity map ρ) with the one above (see 4.4.5.1 for the general case of this phenomenon). □

4.1.11.2 Remark. For $K = \mathbb{Q}$, the connected component D_0 is trivial; this can already be seen because of II.3.8.2, and yields:

$$\mathrm{Gal}(\overline{\mathbb{Q}}^{\mathrm{ab}}/\mathbb{Q}) \simeq U_0^{\mathrm{ord}}/\langle -1 \rangle \simeq U_0^{\mathrm{res}},$$

from which we recover the above observations. □

b) Triviality Criterion for $\mathcal{T}_T^{\mathrm{ord}}$: When is $\mathcal{G}_T^{\mathrm{ord}}$ Pro-p-Free ?

Let K be a number field together with sets of places T and S. We will first give the general formula for the p-rank of the group $\mathcal{T}_T^S := \mathrm{tor}_{\mathbb{Z}_p}(\mathcal{A}_T^S)$ without the condition $Pl_p \subseteq T$ nor the assumption of the Leopoldt conjecture, and without any Kummer hypothesis.

4.2 Theorem (p-rank of torsion groups). *We have:*

$$\begin{aligned}\mathrm{rk}_p(\mathcal{T}_T^S) &= \mathrm{rk}_p(V_T^S/K_T^{\times p}) + \sum_{v\in T} \delta_v - \delta \\ &\quad - (r_1 + r_2 - 1 + |S_0| - \mathrm{r}_{T_p}^{S_0}) + \delta_{2,p}|\Delta_\infty|,\end{aligned}$$

where:

$$\begin{aligned}V_T^S &:= \{\alpha \in K_T^{\times p} K_{T,\Delta_\infty}^{\times},\ (\alpha) = \mathfrak{a}^p \mathfrak{a}_{S_0}, \\ &\qquad \mathfrak{a} \in I_T,\ \mathfrak{a}_{S_0} \in \langle\, S_0 \,\rangle,\ \ i_v(\alpha) \in K_v^{\times p}\ \ \forall v \in T\}, \\ \mathrm{r}_{T_p}^{S_0} &:= \mathrm{rk}_{\mathbb{Z}_p}(\mathrm{adh}_{T_p}(E'^{S_0})) = \dim_{\mathbb{Q}_p}(\mathbb{Q}_p \mathrm{log}_{T_p}(E^{S_0}))\end{aligned}$$

(the T_p-adic rank of E^{S_0}), $\delta_v := 1$ or 0 according as K_v contains μ_p or not, $\delta := 1$ or 0 according as K contains μ_p or not, where $\delta_{2,p}$ is the Kronecker symbol, and $\Delta_\infty := Pl_\infty^{\mathrm{r}} \backslash S_\infty$.

Proof. We can always write $\mathcal{A}_T^S \simeq \mathcal{T}_T^S \times \mathbb{Z}_p^{\widetilde{\mathrm{r}}_{T_p}^{S_0}}$, where $\widetilde{\mathrm{r}}_{T_p}^{S_0}$, given in 1.6.3 as being equal to $\widetilde{\mathrm{r}}_{T_p}^{S_0} = \sum_{v\in T_p} [K_v : \mathbb{Q}_p] - \mathrm{r}_{T_p}^{S_0}$, is by 2.5, (i) the $\mathbb{Z}_p$-rank of $\mathbb{Z}_p \mathrm{Log}_{T_p}^{S_0}(I_{T_p})$; see also 2.2, (ii).

By definition, we have $\mathcal{A}_T^S \simeq (\mathcal{C}\ell_T^S)_p$ whose p-rank is given by Šafarevič's formula I.4.6. The result easily follows. □

Note that, by the Dirichlet Theorem I.3.7.1:

$$r_1 + r_2 - 1 + |S_0| - \mathrm{r}_{T_p}^{S_0} = \dim_{\mathbb{Q}_p}(E^{S_0} \otimes \mathbb{Q}_p) - \dim_{\mathbb{Q}_p}(\mathbb{Q}_p \mathrm{log}_{T_p}(E^{S_0}))$$

is nonnegative and measures a p-adic rank defect in a general situation (we will come back to it in Section 5).

4.2.1 Remarks. (i) For S and T_p fixed, and for $T_{\rm ta}$ sufficiently large, we have $V_T^S = K_T^{\times p}$. Indeed, the group:

$$Y_T^S/K_T^{\times p} := \{\alpha \in K_T^{\times p} K_{T,\Delta_\infty}^\times,\ (\alpha) = \mathfrak{a}^p \mathfrak{a}_{S_0}\}/K_T^{\times p},$$

which is canonically isomorphic to:

$$Y^S/K^{\times p} := \{\alpha \in K^{\times p} K_{\Delta_\infty}^\times,\ (\alpha) = \mathfrak{a}^p \mathfrak{a}_{S_0}\}/K^{\times p},$$

is independent of T (use I.5.1.2); it is then easy to find a finite set t of tame places for which the only $\alpha \in Y^S$ such that $i_v(\alpha) \in K_v^{\times p}$ for all $v \in t$ are the elements of $K^{\times p}$. Therefore, for $T_{\rm ta}$ sufficently large:

$$\mathrm{rk}_p(\mathcal{T}_T^S) = \sum_{v\in T} \delta_v - \delta - (r_1 + r_2 - 1 + |S_0| - \mathrm{r}_{T_p}^{S_0}) + \delta_{2,p}|\Delta_\infty|.$$

(ii) Recall that, under the assumption $\mu_p \subset K$ and $T_p \cup S_p = Pl_p$, we have by II.5.4.1, (iii):

$$\mathrm{rk}_p(V_T^S/K_T^{\times p}) = \mathrm{rk}_p(\mathcal{C}\ell_{S_0}^{T\cup\Delta_\infty}). \qquad \square$$

More generally, by using the reflection Theorem II.5.4.5 for $\mathcal{A}_T^S \simeq (\mathcal{C}\ell_T^S)_p$, we obtain the following result which involves the class group of $K' := K(\mu_p)$.

4.2.2 Proposition. *If $T_p \cup S_p = Pl_p$, we have, with the notations of 4.2:*

$$\mathrm{rk}_p(\mathcal{T}_T^S) = \mathrm{rk}_\omega(\mathcal{C}\ell_{K',S_0'}^{T'\cup\Delta'_\infty}) + \sum_{v\in T} \delta_v - \delta - (r_1 + r_2 - 1 + |S_0| - \mathrm{r}_{T_p}^{S_0}) + \delta_{2,p}|\Delta_\infty|,$$

where ω is the Teichmüller character ($\omega = 1$ if and only if $\mu_p \subset K$). In the case $\mu_p \subset K$, $T = Pl_p$, $S = Pl_\infty^{\rm r}$, and assuming the Leopoldt conjecture for p, we obtain (see II.7.7.2.2):

$$\mathrm{rk}_p(\mathcal{T}_p^{\rm ord}) = \mathrm{rk}_p(\mathrm{R}_2^{\rm ord}(K)) = \mathrm{rk}_p(\mathcal{C}\ell^{Pl_p\,\rm res}) + |Pl_p| - 1. \qquad \square$$

4.2.3 Corollary. *If T contains Pl_p and if K satisfies the Leopoldt conjecture for p, we have:*

$$\mathrm{rk}_p(\mathcal{T}_T^{\rm ord}) = \mathrm{rk}_p(V_T^{\rm ord}/K_T^{\times p}) + \sum_{v\in T} \delta_v - \delta,$$

$$\mathrm{rk}_p(\mathcal{T}_T^{\rm res}) = \mathrm{rk}_p(V_T^{\rm res}/K_T^{\times p}) + \sum_{v\in T} \delta_v - \delta + \delta_{2,p} r_1. \qquad \square$$

This last result is essentially that of Haberland [g, Hab], quoted in [e, Ko3, Ch. 3, Th. 3.74] since, using a particular case of Theorem 2.2 of the Appendix:

$$\mathrm{rk}_p(\mathcal{T}_T^{\rm res}) = \mathrm{rk}_p(H^2(\mathcal{G}_T^{\rm res}, \mathbb{F}_p)),$$

where $\mathcal{G}_T^{\rm res}$ is the Galois group of the maximal T-ramified pro-p-extension of K whose abelianization is $\mathcal{A}_T^{\rm res}$ (see also [Ng1]). But we come back to the case of $\mathcal{G}_T^{\rm ord}$, which corresponds instead to the maximal T-ramified noncomplexified pro-p-extension, for reasons which will become clear shortly. Still in the case $Pl_p \subseteq T$ and assuming the Leopoldt conjecture, we thus know the minimal number of generators and the minimal number of relations (in the corresponding free pro-p-group) [15] of the pro-p-group $\mathcal{G}_T^{\rm ord}$:

$$\mathrm{rk}_p(H^1(\mathcal{G}_T^{\rm ord}, \mathbb{F}_p)) = \mathrm{rk}_p(\mathcal{C}\ell_T^{\rm ord}) = \mathrm{rk}_p(V_T^{\rm ord}/K_T^{\times p}) + \sum_{v\in T} \delta_v - \delta + 1 + r_2,$$
$$\mathrm{rk}_p(H^2(\mathcal{G}_T^{\rm ord}, \mathbb{F}_p)) = \mathrm{rk}_p(\mathcal{T}_T^{\rm ord}) = \mathrm{rk}_p(V_T^{\rm ord}/K_T^{\times p}) + \sum_{v\in T} \delta_v - \delta.$$

Under these assumptions, the group $\mathcal{G}_T^{\rm ord}$ is pro-p-free if and only if $\mathcal{T}_T^{\rm ord} = 1$, and using the deployment theorem in the form given in 4.1.5 and formula 2.6.1, (ii$_2$), this is equivalent to:

$$\sum_{v\in T_{\rm ta}} \delta_v = 0, \quad H^{\rm ord}(p) \subset \widetilde{K}_p, \quad \text{and} \quad \mathrm{tor}_{\mathbb{Z}_p}\Big(\bigoplus_{v|p} U_v^1 \Big/ \mathrm{adh}_p(E'^{\rm ord})\Big) = 1.$$

It is then pro-p-free on $r_2 + 1$ generators. The first condition can be written in the form $T = Pl_p$, if we accept not to consider the tame places v for which $\mathrm{N}\mathfrak{p}_v \not\equiv 1 \bmod (p)$ (i.e., $\delta_v = 0$), places which are useless, and the last condition can be made more precise thanks to the following general result which has some interest in itself (and which is independent of any assumption).

4.2.4 Lemma. *We have the exact sequence:*

$$1 \longrightarrow \mathrm{tor}_{\mathbb{Z}_p}\Big(\bigoplus_{v|p} U_v^1\Big) \Big/ \mathrm{tor}_{\mathbb{Z}_p}(\mathrm{adh}_p(E'^{\rm ord})) \longrightarrow$$
$$\mathrm{tor}_{\mathbb{Z}_p}\Big(\bigoplus_{v|p} U_v^1 \Big/ \mathrm{adh}_p(E'^{\rm ord})\Big) \xrightarrow{\ \log\ } \mathrm{tor}_{\mathbb{Z}_p}\Big(\log\Big(\bigoplus_{v|p} U_v^1\Big) \Big/ \mathbb{Z}_p \log_p(E'^{\rm ord})\Big) \longrightarrow 0,$$

where $\mathbb{Z}_p\log_p(E'^{\rm ord}) = \log(\mathrm{adh}_p(E'^{\rm ord}))$. [16]

Proof. The surjectivity comes from the fact that if $u \in \bigoplus_{v|p} U_v^1$ is such that $p^n \log(u) = \log(\varepsilon)$, $\varepsilon \in \mathrm{adh}_p(E'^{\rm ord})$, we have $u^{p^n} =: \zeta\varepsilon$ for $\zeta \in \mathrm{tor}_{\mathbb{Z}_p}\Big(\bigoplus_{v|p} U_v^1\Big)$, hence there exists $m \geq n$ such that $u^{p^m} \in \mathrm{adh}_p(E'^{\rm ord})$; u is a preimage.

If u is such that $\log(u) \in \log(\mathrm{adh}_p(E'^{\rm ord}))$, $u =: \zeta\varepsilon$ as above, giving the kernel. □

[15] See [e, Ko3, Ch. 3, §§ 1.12, 1.13] for the definitions of these classical notions.

[16] $\log_p := (\log \circ i_v)_{v|p}$ where log is the Iwasawa logarithm on $\mathbb{C}_p^\times$; $\log\Big(\bigoplus_{v|p} U_v^1\Big)$ means $\bigoplus_{v|p} \log(U_v^1)$ (see 2.2).

Recall that by 3.6.3 we have, assuming the Leopoldt conjecture for p:

$$\mathrm{tor}_{\mathbb{Z}_p}(\mathrm{adh}_p(E'^{\mathrm{ord}})) = i_p(\mu_p(K)).$$

It is not difficult to generalize this exact sequence to the case of the the study of:

$$\mathrm{tor}_{\mathbb{Z}_p}\Big(\bigoplus_{v\in T_p} U_v^1 \bigoplus_{v\in T_{\mathrm{ta}}} (F_v^\times)_p \Big/ \mathrm{adh}_T(E'^S)\Big),$$

via the map log on $\bigoplus_{v\in T_p} U_v^1$.

Assumptions. Let T be a finite set of finite places of K, T containing Pl_p and not containing tame places v such that $\mathrm{N}\mathfrak{p}_v \not\equiv 1 \bmod (p)$. We assume that K satisfies the Leopoldt conjecture for p.

We have obtained the following.

4.2.5 Theorem. *The Galois group $\mathcal{G}_T^{\mathrm{ord}}$ of the maximal T-ramified noncomplexified pro-p-extension of K is pro-p-free if and only if the following four conditions are satisfied:*

- $T_{\mathrm{ta}} = \emptyset$ *(i.e., $T = Pl_p$),*
- $\bigoplus_{v|p} \mu_p(K_v) = i_p(\mu_p(K))$,
- $\mathbb{Z}_p\log_p(E'^{\mathrm{ord}})$ *is a direct summand in* $\bigoplus_{v|p}\log(U_v^1)$,
- *the p-Hilbert class field (in the ordinary sense) is contained in the compositum of the $\mathbb{Z}_p$-extensions of K.*

In that case, it is free on $r_2 + 1$ generators. □

Note. We can replace the second condition by the equivalent condition:
- $(\mathrm{R}_2^{\mathrm{ord}}(K))_p = (\mathrm{WK}_2(K))_p$ (see II.7.6.1),

and the last two conditions (under the second one!) by:
- $V_p^{\mathrm{ord}} := \{\alpha \in K_p^\times,\ (\alpha) = \mathfrak{a}^p,\ i_v(\alpha) \in K_v^{\times p}\ \ \forall v|p\} = K_p^{\times p}$.

4.2.6 Remarks. (i) When K (satisfying the Leopoldt conjecture for p) is such that $\mathcal{T}_p^{\mathrm{ord}} = 1$ (which is equivalent to the last three conditions above), we will say that K is p-rational (we will come back to this notion in Chapter IV, Section 3, (b)).

(ii) Theorem 4.1.5 shows that $\mathcal{T}_2^{\mathrm{res}}$ always has a subgroup isomorphic to $(\mathbb{Z}/2\mathbb{Z})^{r_1}$, so that for a field K which is not totally complex, the 2-group $\mathcal{G}_2^{\mathrm{res}}$ is never pro-2-free. □

It is clear that the groups $\mathcal{T}_T^S$ and V_T^S are closely related, but, even if V_T^S seems to be more accessible, it is really $\mathcal{T}_T^S$ which has the nice properties.

4.2.7 Exercise (generalization of the relationship between $\mathcal{T}_T^S$ and V_T^S). Let T and S be the usual sets of places and let p be a prime number (*we do not assume any hypothesis*). Set:

$$\begin{aligned}\mathcal{V}_T^S &:= \{\alpha \in K_T^{\times p} K_{T,\Delta_\infty}^\times,\ (\alpha) = \mathfrak{a}^p\,\mathfrak{a}_{S_0},\\ &\qquad \mathfrak{a} \in I_T,\ \mathfrak{a}_{S_0} \in \langle\, S_0 \,\rangle,\ i_v(\alpha) \in K_v^{\times p}\ \ \forall v \in T\} \otimes \mathbb{Z}_p,\\ \mathcal{K}_T^\times &:= K_T^\times \otimes \mathbb{Z}_p,\ \ \mathcal{K}_{T,\Delta_\infty}^\times := K_{T,\Delta_\infty}^\times \otimes \mathbb{Z}_p,\\ \mathcal{E}_T^S &:= \{\varepsilon \in \mathcal{E}^S := E^S \otimes \mathbb{Z}_p,\ \ \bar{i}_T(\varepsilon) = 1\}\end{aligned}$$

(see 1.6.6). It is easily checked that $\mathcal{V}_T^S/\mathcal{K}_T^{\times p}$ is also given by:

$$\{\alpha \in \mathcal{K}_T^{\times p}\mathcal{K}_{T,\Delta_\infty}^\times,\ \ (\alpha) = \mathfrak{a}^p\,\mathfrak{a}_{S_0},\ \ \mathfrak{a} \in \mathcal{I}_T,\ \mathfrak{a}_{S_0} \in \mathcal{S}_0,\ \ \bar{i}_T(\alpha) = 1\}\mathcal{K}_T^{\times p}/\mathcal{K}_T^{\times p}\ ;$$

thus, the notation $\mathcal{V}_T^S$ may be understood in terms of infinitesimals.

Show (using p-adic techniques of class field theory) that we have the exact sequences:

$$1 \longrightarrow {}_p(\bar{i}_T(\mathcal{E}^S)) \longrightarrow \bigoplus_{v\in T} {}_p\mu(K_v) \longrightarrow {}_p\mathcal{T}_T^S \longrightarrow \mathcal{V}_T^S/\mathcal{E}_T^S\,.\,\mathcal{K}_T^{\times p} \longrightarrow 1,$$

$$1 \longrightarrow \mathcal{E}_T^S/\mathcal{E}_T^S \cap (\mathcal{E}^{S_0\,\mathrm{ord}})^p \longrightarrow V_T^S/K_T^{\times p} \longrightarrow \mathcal{V}_T^S/\mathcal{E}_T^S\,.\,\mathcal{K}_T^{\times p} \longrightarrow 1.$$

Answer. Let $\tau \in {}_p\mathcal{T}_T^S$; we have $\tau =: \mathrm{Art}(\mathfrak{a})$, $\mathfrak{a} \in \mathcal{I}_T$, with $\mathfrak{a}^p =: (u_\infty)\,\mathfrak{a}_{S_0}$, $u_\infty \in \mathcal{K}_{T,\Delta_\infty}^\times$ such that $\bar{i}_T(u_\infty) = 1$ (see 2.4, 2.4.1), $\mathfrak{a}_{S_0} \in \mathcal{S}_0$. If $\tau =: \mathrm{Art}(\mathfrak{b})$ with similarly $\mathfrak{b}^p =: (v_\infty)\mathfrak{b}_{S_0}$, we have $\mathfrak{b} =: \mathfrak{a}\,(w_\infty)\,\mathfrak{c}_{S_0}$, hence $(v_\infty)\,\mathfrak{b}_{S_0} = (u_\infty)\,\mathfrak{a}_{S_0}\,(w_\infty^p)\,\mathfrak{c}_{S_0}^p$ which implies that:

$$v_\infty =: u_\infty\, w_\infty^p\, \varepsilon_\infty,\ \ \varepsilon_\infty \in \mathcal{E}_T^S.$$

We can send τ to the image of u_∞ in $\mathcal{V}_T^S/\mathcal{E}_T^S\,.\,\mathcal{K}_T^{\times p}$.

(i) Surjectivity. Let $u \in V_T^S$; we have $(u) =: \mathfrak{a}^p\mathfrak{a}_{S_0}$ and, by approximation, $u =: x(n)^p u_n$, $x(n) \in K_T^\times$, $u_n \in K_{T,\mathfrak{m}(n),\Delta_\infty}^\times$ (this being possible for all n). We have:

$$(u_n) = (u\,x(n)^{-p}) = (\mathfrak{a}\,(x(n))^{-1})^p\,\mathfrak{a}_{S_0} = \mathfrak{a}(n)^p\,\mathfrak{a}_{S_0}.$$

We set $\tau_n := \mathrm{Art}(\mathfrak{a}(n)) \in \mathcal{A}_T^S$; if τ is a limit point of $(\tau_n)_n$, we then have $\tau^p = 1$ (i.e., $\tau \in {}_p\mathcal{T}_T^S$) and we check that τ is a preimage.

(ii) Kernel. If $\tau =: \mathrm{Art}(\mathfrak{a})$ with $\mathfrak{a}^p =: (u_\infty)\,\mathfrak{a}_{S_0}$ and $u_\infty =: \varepsilon_\infty\, x^p$, $\varepsilon_\infty \in \mathcal{E}_T^S$, $x \in \mathcal{K}_T^\times$, we have $(\mathfrak{a}\,(x)^{-1})^p = (\varepsilon_\infty)\,\mathfrak{a}_{S_0} =: \mathfrak{b}_{S_0}^p$, hence $\mathfrak{a} = (x)\,\mathfrak{b}_{S_0}$, so that we obtain $\tau = \mathrm{Art}(x)$. But $x^p = u_\infty\,\varepsilon_\infty^{-1}$ and $\bar{i}_T(x) \in \bigoplus_{v\in T_p} {}_p\mu(K_v)$. The kernel is thus equal to the image under Art of $\bigoplus_{v\in T_p} {}_p\mu(K_v)$ in $\mathcal{B}_T^S$, and it is more convenient to use the fact that it is the image in $\bigoplus_{v\in T_p} U_v^1 \bigoplus_{v\in T_{\mathrm{ta}}} (F_v^\times)_p\Big/\bar{i}_T(\mathcal{E}^S)$; the result follows. □

4.2.8 Remark. When $S = Pl_\infty^{\mathrm{r}}$, $Pl_p \subseteq T$, and assuming the Leopoldt conjecture for p, we recover the rank formula given in 4.2.3 since we have:

$$1 \longrightarrow {}_p\mu(K) \longrightarrow \bigoplus_{v\in T} {}_p\mu(K_v) \longrightarrow {}_p\mathcal{T}_T^{\mathrm{ord}} \longrightarrow V_T^{\mathrm{ord}}/K_T^{\times p} \longrightarrow 1.$$

When we write the dual exact sequence, the two central terms can be interpreted in terms of cohomology groups as above, and we obtain:

$$1 \longrightarrow (V_T^{\mathrm{ord}}/K_T^{\times p})^* \longrightarrow H^2(\mathcal{G}_T^{\mathrm{ord}}, \mathbb{F}_p) \longrightarrow \bigoplus_{v\in T} H^2(\mathcal{G}_v, \mathbb{F}_p) \longrightarrow ({}_p\mu(K))^* \longrightarrow 1,$$

where $\mathcal{G}_v$ is the Galois group of the maximal pro-p-extension of K_v.The exact sequence of Haberland–Koch [e, Ko3, Ch. 3, § 2.6, Th. 3.74] corresponds in fact to the case $S = \emptyset$ (the restricted sense). As already mentioned, the reflection phenomena imply that, if the restricted sense is canonical for class groups, the ordinary sense thus becomes canonical for the groups $\mathcal{A}_T$, $\mathcal{T}_T$, and $\mathcal{G}_T$, which corresponds to the fact that we do not complexify (i.e., ramify in the old terminology) the real places of K. □

To conclude, we can say that we have described these cohomology groups in terms of relatively simple arithmetical invariants (which are effectively computable). But, as the reasoning made in the Appendix (Section 3) shows, it is difficult to interpret $H^2(\mathcal{G}_T^S, \mathbb{F}_p)$ in this way without the usual assumptions ($Pl_p \subseteq T$, $S \subseteq Pl_\infty^{\mathrm{r}}$ and the Leopoldt conjecture). However, in the general case for T and S, but still assuming the Leopoldt conjecture, in the Appendix we will prove a number of new results.

We now come back to the general theory and study the inertia and decomposition groups of a place of K in the extension $\overline{K}^{\mathrm{ab}}(p)/K$. We will see that there are other deployment theorems for finite sets of places (including wild places) which do not assume the validity of the Leopoldt conjecture (we will call them weak deployment theorems).

c) The Schmidt–Chevalley Theorem — Inertia Groups in $\overline{K}^{\mathrm{ab}}(p)/K$

We begin by an important result ((i) of the theorem given below) which uses the local-global principle for powers, which was outlined in 1930 by Schmidt [SchFK] then given again in 1951 by Chevalley [Che5, I]. It has been used by Weil [h, We2] as well as by Artin–Tate [d, AT, Ch. 10] in the study of the properties of the connected component of the unit element of C. Recently it has been considered in [KubO] for some insights in foundation of class field theory. An illustration is (modestly) given by Example 1.1.7. It can be seen as a tame analog of the Leopoldt conjecture (in the spirit of Lemma 4.1.1 and Exercise 4.1.3 which are based on the characterization 3.6.2, (iv) of the Leopoldt conjecture).

4.3 Theorem (Schmidt–Chevalley). *Let K be a number field together with sets of places T_1 and S, and let δ_∞ be a subset of $Pl_\infty^{\mathrm{r}} \backslash S_\infty$. Let p be a fixed prime number.*

(i) For all $n \geq 1$ we can find a set t of tame places, disjoint from $T_1 \cup S_0$, and such that:

$$E_{\mathfrak{n}}^{S \cup \delta_\infty} \subseteq (E^{S_0 \,\mathrm{ord}})^{p^n},$$

for $\mathfrak{n} = \prod_{v \in t} \mathfrak{p}_v$.

(ii) If $\mathfrak{m}_1 = \prod_{v \in T_{1,p}} \mathfrak{p}_v^{m_{1,v}} \prod_{v \in T_{1,\mathrm{ta}}} \mathfrak{p}_v$ is a modulus with support T_1, we can find a set t of tame places, disjoint from $T_1 \cup S_0$, and such that:

$$\begin{aligned}\mathrm{Gal}(K(\mathfrak{m})^S(p)/K(\mathfrak{n})^{S \cup \delta_\infty}(p)) &= \bigoplus_{v \in T_1} I_v(K(\mathfrak{m})^S(p)/K) \bigoplus_{v \in \delta_\infty} D_v(K(\mathfrak{m})^S(p)/K) \\ &\simeq \bigoplus_{v \in T_{1,p}} U_v^1/U_v^{m_{1,v}} \bigoplus_{v \in T_{1,\mathrm{ta}} \cup \delta_\infty} (F_v^\times)_p,\end{aligned}$$

for $\mathfrak{m} := \mathfrak{m}_1 \mathfrak{n}$, $\mathfrak{n} = \prod_{v \in t} \mathfrak{p}_v$.

Note. The result (i) is a fortiori true if we add other places to t, wild or not, and if we take for $\mathfrak{n}$ an arbitrary modulus with support t.

Proof of the theorem. (i) Let $e \geq n+1$ and $K' := K(\mu_{p^e})$. Consider the extension:

$$K'(\sqrt[p^e]{E^{S \cup \delta_\infty}})/K',$$

and choose for t a set of tame finite places, prime to $T_1 \cup S_0$, and such that the Frobenius' of the places of t' (the set of places of K' above those of t) generate $\mathrm{Gal}(K'(\sqrt[p^e]{E^{S \cup \delta_\infty}})/K')$. The subextension $K'(\sqrt[p^e]{E_{\mathfrak{n}}^{S \cup \delta_\infty}})/K'$ is t'-split, which implies, by our choice of the Frobenius', its triviality and the fact that $E_{\mathfrak{n}}^{S \cup \delta_\infty} \subset K'^{p^e}$, hence, by II.6.3.2:

$$E_{\mathfrak{n}}^{S \cup \delta_\infty} \subseteq \langle -1 \rangle (E^{S_0 \,\mathrm{ord}})^{p^{e-1}} \subseteq \langle -1 \rangle (E^{S_0 \,\mathrm{ord}})^{p^n}.$$

We check that, in the case $p = 2$ and for the exceptional case, we can manage to have $E_{\mathfrak{n}}^{S \cup \delta_\infty} \subseteq (E^{S_0 \,\mathrm{ord}})^{2^n}$ (for this it is sufficient that t contains a place in which -1 is not a local square), proving (i).

(ii) Afterwards, by Proposition 1.4, to have the given deployment it is sufficient to choose n so that p^n kills the group:

$$\bigoplus_{v \in T_{1,p}} U_v^1/U_v^{m_{1,v}} \bigoplus_{v \in T_{1,\mathrm{ta}} \cup \delta_\infty} (F_v^\times)_p,$$

since the image of $E_{\mathfrak{n}}^{S \cup \delta_\infty}$ then becomes trivial; in other words, we have $(E_{\mathfrak{n}}^{S \cup \delta_\infty}/E_{\mathfrak{m}}^S)_p = 1$. □

The proof of (i) shows that $|t|$ can be bounded by the p-rank of $E^{S \cup \delta_\infty}$.

Globalizing, we obtain for all integers m the inclusion $E_{\mathfrak{n}}^{S\cup\delta_\infty} \subseteq (E^{S_0\,\mathrm{ord}})^m$ (the modulus $\mathfrak{n}$ (depending on m) is then equal to the l.c.m. of the moduli corresponding to each localization).

Note. One may wonder whether there could exist a link between the p-adic Leopoldt conjecture (or the p-adic conjecture in monogenic groups) and the Schmidt–Chevalley theorem. It could be that, for diophantine reasons (in the spirit of the "$a\,b\,c$-conjecture"), if a unit ε is such that $\varepsilon - 1 \in (p^n)$, for a large n, then the ideal $(\varepsilon - 1)$ is divisible by "many" tame prime ideals. This statement is actually strengthened by the fact that, if the Leopoldt conjecture is not satisfied, for any n there exist *infinitely many* such ε for which the image in E/E^p is *constant* regarding ε and n (see the proof of (ii) $\Rightarrow$ (iii) in 3.6.2), and ε not a pth power would become incompatible with the Schmidt–Chevalley theorem ...

Fact 4.3, (ii) above is a deployment theorem at the finite level. By going to the limit, we will obtain deployment results of the inertia groups which will be called weak deployment theorems, meaning that they can hold only for *finite* sets of places (these sets however being arbitrary, without any assumption of p-adic independence, and not dealing only with tame places).

In Subsection (h), we will prove by idelic means a similar result for *decomposition groups* (see Theorem 4.16.7 which of course gives again the case of inertia groups); this result, which is the source of nontrivial practical applications, is one of the deepest global properties of $\overline{K}^{\mathrm{ab}}/K$.

Let $H_{\mathrm{ta}}^{\mathrm{res}}(p)$ be the maximal tamely ramified pro-p-subextension of $\overline{K}^{\mathrm{ab}}$. We will use here the notation H_T for T infinite ($T = Pl_0 \backslash \Sigma$ or $Pl_{\mathrm{ta}} \backslash \Sigma_{\mathrm{ta}}$); this still defines the maximal T-ramified abelian extension (see II.5.3), and tautologically, we even have $\overline{K}^{\mathrm{ab}} = H_{Pl_0}^{\mathrm{res}}$.

4.3.1 Corollary (weak deployment theorem). *Let Σ be a finite set of finite places of K, and let $\delta_\infty \subseteq Pl_\infty^{\mathrm{r}}$. We have:*

$$\begin{aligned}\mathrm{Gal}(\overline{K}^{\mathrm{ab}}(p)/H_{Pl_0\backslash\Sigma}^{\delta_\infty}(p)) &= \bigoplus_{v\in\Sigma} I_v(\overline{K}^{\mathrm{ab}}(p)/K) \bigoplus_{v\in\delta_\infty} D_v(\overline{K}^{\mathrm{ab}}(p)/K) \\ &\simeq \bigoplus_{v\in\Sigma_p} U_v^1 \bigoplus_{v\in\Sigma_{\mathrm{ta}}\cup\delta_\infty} (F_v^\times)_p \ ;\end{aligned}$$

$$\begin{aligned}\mathrm{Gal}(H_{\mathrm{ta}}^{\mathrm{res}}(p)/H_{Pl_{\mathrm{ta}}\backslash\Sigma_{\mathrm{ta}}}^{\delta_\infty}(p)) &= \bigoplus_{v\in\Sigma_{\mathrm{ta}}} I_v(H_{\mathrm{ta}}^{\mathrm{res}}(p)/K) \bigoplus_{v\in\delta_\infty} D_v(H_{\mathrm{ta}}^{\mathrm{res}}(p)/K) \\ &\simeq \bigoplus_{v\in\Sigma_{\mathrm{ta}}\cup\delta_\infty} (F_v^\times)_p .\end{aligned}$$

4.3.2 Corollary (inertia groups). *For any place $v|p$ we have:*

$$I_v(\overline{K}^{\mathrm{ab}}(p)/K) \simeq U_v^1 .$$

For any tame finite place v we have:

$$I_v(\overline{K}^{\rm ab}(p)/K) \simeq I_v(H_{\rm ta}^{\rm res}(p)/K) \simeq (F_v^\times)_p,$$

and for any real place at infinity v:

$$D_v(\overline{K}^{\rm ab}(p)/K) \simeq D_v(H_{\rm ta}^{\rm res}(p)/K) \simeq (\{\pm 1\})_p.$$

Proof of the corollaries. For convenience, we return to the notations of the Schmidt–Chevalley Theorem 4.3 for $T_1 := \Sigma$. Consider:

$$\mathrm{Gal}(\overline{K}^{\rm ab}(p)/K) = \varprojlim_{\mathfrak{r}} \mathrm{Gal}(K(\mathfrak{r})^{\rm res}(p)/K)$$

($\mathfrak{r}$ ranges in the set of moduli of K whose support contains T_1; we then denote by $\mathfrak{r}_1$ the part of $\mathfrak{r}$ with support T_1). By II.1.2.3.1, we can thus write:

$$\begin{aligned}\mathrm{Gal}(\overline{K}^{\rm ab}(p)/H^{\delta_\infty}_{Pl_0\setminus T_1}(p)) &= \langle\, I_v(\overline{K}^{\rm ab}(p)/K)\,\rangle_{v\in T_1} \cdot \langle\, D_v(\overline{K}^{\rm ab}(p)/K)\,\rangle_{v\in\delta_\infty}\\ &\simeq \varprojlim_{\mathfrak{r}} \left(\langle\, I_v(K(\mathfrak{r})^{\rm res}(p)/K)\,\rangle_{v\in T_1} \cdot \langle\, D_v(K(\mathfrak{r})^{\rm res}(p)/K)\,\rangle_{v\in\delta_\infty}\right).\end{aligned}$$

For each $\mathfrak{r}$ we can choose integers $m_{1,v}$ for $v \in T_{1,p}$, sufficiently large so as to make the modulus $\mathfrak{m}_1$ a multiple of $\mathfrak{r}_1$; we then obtain a modulus $\mathfrak{n}$, according to 4.3, (ii), for which we assume that $\mathfrak{m} = \mathfrak{m}_1\mathfrak{n}$ is a multiple of $\mathfrak{r}$ (if necessary by enlarging $\mathfrak{n}$, which goes in the right direction for the conclusion). The moduli $\mathfrak{m}$ thus obtained form a cofinal subset, hence we have:

$$\begin{aligned}&\langle\, I_v(\overline{K}^{\rm ab}(p)/K)\,\rangle_{v\in T_1} \cdot \langle\, D_v(\overline{K}^{\rm ab}(p)/K)\,\rangle_{v\in\delta_\infty} \simeq\\ &\varprojlim_{(m_{1,v})_v} \Big(\bigoplus_{v\in T_{1,p}} U_v^1/U_v^{m_{1,v}} \bigoplus_{v\in T_{1,\rm ta}\cup\delta_\infty} (F_v^\times)_p\Big) = \bigoplus_{v\in T_{1,p}} U_v^1 \bigoplus_{v\in T_{1,\rm ta}\cup\delta_\infty} (F_v^\times)_p.\end{aligned}$$

The deployment of the groups I_v and D_v follows easily, proving the case of $\overline{K}^{\rm ab}(p)/K$ in each corollary.

In case of $H_{\rm ta}^{\rm res}(p)/K$, we have $\mathrm{Gal}(H_{\rm ta}^{\rm res}(p)/K) = \varprojlim_{\mathfrak{r}} \mathrm{Gal}(K(\mathfrak{r})^{\rm res}(p)/K)$, where $\mathfrak{r}$ ranges in the set of tame (i.e., squarefree) moduli; the rest of the reasoning is then similar in the extension $H_{\rm ta}^{\rm res}(p)/H^{\delta_\infty}_{Pl_{\rm ta}\setminus T_{1,\rm ta}}(p)$, and is in fact simpler since here $\mathfrak{m}_1 = \mathfrak{r}_1$. In the tame case, deployment already takes place in $K(\mathfrak{m}_1\mathfrak{n})^{\rm res}/K(\mathfrak{n})^{\delta_\infty}$, for a modulus $\mathfrak{n}$ corresponding to $\mathfrak{m}_1 = \prod_{v\in T_{1,\rm ta}} \mathfrak{p}_v$ (relate this with 1.1.7). □

For $\Sigma = \{v\}$, the above Corollary 4.3.1 says that the inertia field of v in $\overline{K}^{\rm ab}(p)/K$ is $H^{\rm res}_{Pl_0\setminus\{v\}}(p)$. If v is tame, the inertia field in $H_{\rm ta}^{\rm res}(p)/K$ is $H^{\rm res}_{Pl_{\rm ta}\setminus\{v\}}(p)$.

Analogous remark for an infinite place and its decomposition group.

4.3.3 Remarks. (i) The proof of the Schmidt–Chevalley Theorem 4.3 does not need the Leopoldt conjecture, but on the other hand it is not possible to deduce from 4.3.1 that we have $\mathrm{Gal}(\overline{K}^{\mathrm{ab}}(p)/H_p^{\mathrm{ord}}(p)) \simeq \prod_{v \nmid p}(F_v^\times)_p$.

(ii) Symmetrically, the weak deployment theorem for the tame places in $H_{\mathrm{ta}}^{\mathrm{res}}(p)/K$ does not follow by projection from Theorem 4.1 on deployment above $H_p^{\mathrm{ord}}(p)$. In any case, as we will see in Subsection (d) and in Chapter V, the group $\mathrm{Gal}(H_{\mathrm{ta}}^{\mathrm{res}}(p)/H^{\mathrm{ord}}(p))$ is generally not the direct product of the groups I_v for $v \in Pl_{\mathrm{ta}}$ and D_v for $v \in Pl_\infty^{\mathrm{r}}$.

(iii) The weak deployment Theorem 4.3.1 should further the study of p-adic L-functions of totally real fields K, in the direction indicated in [Gr7]. □

4.3.4 Exercise. Let $\mathfrak{m}_1 = \prod_{v \in T_1} \mathfrak{p}_v^{m_{1,v}}$ with $m_{1,v} \geq 0$ be a fixed modulus. Show that for all sufficiently large $\mathfrak{n}$ prime to $\mathfrak{m}_1$, we have:

$$[K(\mathfrak{m}_1\mathfrak{n})^{\mathrm{res}} : K(\mathfrak{m}_1)^{\mathrm{res}} K(\mathfrak{n})^{\mathrm{res}}] = (E^{\mathrm{res}} : E_{\mathfrak{m}_1}^{\mathrm{res}}).$$

Answer. In the globalization of the Schmidt–Chevalley Theorem 4.3, it is enough to have (for $S = \delta_\infty = \emptyset$):

$$E_{\mathfrak{n}}^{\mathrm{res}} \subseteq (E^{\mathrm{ord}})^m,$$

where m is an integer which kills the finite group $\bigoplus_{v \in T_1} U_v/U_v^{m_{1,v}}$, in which case $E_{\mathfrak{n}}^{\mathrm{res}} = E_{\mathfrak{m}_1\mathfrak{n}}^{\mathrm{res}}$, and the formula of Exercise II.5.1.5.2 proves the result by taking $\mathfrak{m}_2 := \mathfrak{n}$. □

This shows that the degree defect under study may be arbitrarily large, except if K is equal to $\mathbb{Q}$ or to an imaginary quadratic field where this index is equal to 1 for $\mathbb{Q}$ and is bounded by $|\mu(K)|$ for an imaginary quadratic field. We are even going to see below that this phenomenon due to units goes to the limit.

d) Galois Diagram for $\overline{K}^{\mathrm{ab}}(p)/K$ — Structure of the Connected Component D_0 — The Fundamental Equality: $\overline{K_v}^{\mathrm{ab}} = (\overline{K}^{\mathrm{ab}})_v$

Notations. (i) To obtain a general Galois diagram, we introduce the maximal tamely ramified pro-p-subextension $H_{\mathrm{ta}}^{\mathrm{res}}(p)$ of $\overline{K}^{\mathrm{ab}}(p)$ and consider the compositum:

$$M_0 := H_p^{\mathrm{ord}}(p) H_{\mathrm{ta}}^{\mathrm{res}}(p)$$

(direct compositum over the ordinary p-Hilbert class field $H^{\mathrm{ord}}(p)$).

(ii) Generally we set, for any modulus $\mathfrak{m}$ and $S \subseteq Pl_\infty^{\mathrm{r}}$:

$$\mathcal{E}_{\mathfrak{m}}^S := E_{\mathfrak{m}}^S \otimes \mathbb{Z}_p.$$

□

By 1.6, 1.6.5 applied in the tame case (i.e., $T_p = \emptyset$), and for $S = \delta_\infty = Pl_\infty^{\mathrm{r}}$, we have:

$$\begin{aligned}\mathrm{Gal}(M_0/H_p^{\mathrm{ord}}(p)) &\simeq \mathrm{Gal}(H_{\mathrm{ta}}^{\mathrm{res}}(p)/H^{\mathrm{ord}}(p)) \\ &\simeq \varprojlim_{T_{\mathrm{ta}}} \Big(\bigoplus_{v\in T_{\mathrm{ta}}\cup Pl_\infty^{\mathrm{r}}} (F_v^\times)_p \Big/ \bar{\imath}_{T_{\mathrm{ta}},\infty}(\mathcal{E}^{\mathrm{ord}})\Big),\end{aligned}$$

where $\bar{\imath}_{T_{\mathrm{ta}},\infty} := (\bar{\imath}_v)_{v\in T_{\mathrm{ta}}\cup Pl_\infty^{\mathrm{r}}}$ (see 1.6.6), the inverse limit being over the finite T_{ta} contained in the set Pl_{ta} of tame places ordered by inclusion. We have:

$$\bigoplus_{v\in T_{\mathrm{ta}}\cup Pl_\infty^{\mathrm{r}}} (F_v^\times)_p \Big/ \bar{\imath}_{T_{\mathrm{ta}},\infty}(\mathcal{E}^{\mathrm{ord}}) \simeq \prod_{v\nmid p} (F_v^\times)_p \Big/ \prod_{v\in Pl_{\mathrm{ta}}\backslash T_{\mathrm{ta}}} (F_v^\times)_p \,.\, \bar{\imath}_{\mathrm{ta},\infty}(\mathcal{E}^{\mathrm{ord}}),$$

with $\bar{\imath}_{\mathrm{ta},\infty} := (\bar{\imath}_v)_{v\nmid p}$, and since $\bar{\imath}_{\mathrm{ta},\infty}(\mathcal{E}^{\mathrm{ord}})$ is compact we have, by I.5.4, $\bigcap_{T_{\mathrm{ta}}\subset Pl_{\mathrm{ta}}} \Big(\prod_{v\in Pl_{\mathrm{ta}}\backslash T_{\mathrm{ta}}} (F_v^\times)_p \,.\, \bar{\imath}_{\mathrm{ta},\infty}(\mathcal{E}^{\mathrm{ord}})\Big) = \bar{\imath}_{\mathrm{ta},\infty}(\mathcal{E}^{\mathrm{ord}})$. By I.5.5 we obtain the following result:

$$\mathrm{Gal}(M_0/H_p^{\mathrm{ord}}(p)) \simeq \prod_{v\nmid p} (F_v^\times)_p \Big/ \bar{\imath}_{\mathrm{ta},\infty}(\mathcal{E}^{\mathrm{ord}}).$$

Assuming the Leopoldt conjecture for p, we know by 4.1.7 that:

$$\mathrm{Gal}(\overline{K}^{\mathrm{ab}}(p)/H_p^{\mathrm{ord}}(p)) \simeq \prod_{v\nmid p} (F_v^\times)_p,$$

so that we have, under this isomorphism:

$$\mathrm{Gal}(\overline{K}^{\mathrm{ab}}(p)/M_0) \simeq \bar{\imath}_{\mathrm{ta},\infty}(\mathcal{E}^{\mathrm{ord}}) \ ;$$

but everything which is before this interpretation is valid without assuming the Leopoldt conjecture.

Afterwards, the crucial fact is given by the following result (valid without any assumption).

4.4 Theorem. *The map* $\bar{\imath}_{\mathrm{ta},\infty} : \mathcal{E}^{\mathrm{ord}} \longrightarrow \prod_{v\nmid p} (F_v^\times)_p$ *is injective and we then have the canonical homeomorphisms:*

$$\mathrm{Gal}(\overline{K}^{\mathrm{ab}}(p)/H_p^{\mathrm{ord}}(p)H_{\mathrm{ta}}^{\mathrm{res}}(p)) \simeq \mathcal{E}^{\mathrm{ord}} := E^{\mathrm{ord}} \otimes \mathbb{Z}_p \simeq \mu_p(K) \oplus \mathbb{Z}_p^{r_1+r_2-1}.$$

Proof. Let $\varepsilon \in \mathcal{E}^{\mathrm{ord}}$ be such that $\bar{\imath}_{\mathrm{ta},\infty}(\varepsilon) = 1$, and let $e \geq 1$. By definition of $\bar{\imath}_{\mathrm{ta},\infty}$, the residual image of ε in $(F_v^\times)_p$ is trivial for any tame place v. If we write:

$$\varepsilon =: \prod_i \varepsilon_i \otimes a_i =: \prod_i \varepsilon_i \otimes a'_i \,.\, \prod_i \varepsilon_i^{p^e} \otimes b_i =: \varepsilon' \otimes 1 \,.\, \eta^{p^e},$$

with $a_i =: a'_i + p^e b_i$, $a'_i \in \mathbb{Z}$, $b_i \in \mathbb{Z}_p$, we have $\varepsilon' \in E^{\mathrm{ord}}$, $\eta \in \mathcal{E}^{\mathrm{ord}}$, and we easily obtain (Hensel's lemma) $i_v(\varepsilon') \in K_v^{\times p^e}$ for any tame place v, hence $\varepsilon' \in (E^{\mathrm{ord}})^{p^{e-1}}$ (by II.6.3.3 applied to $\Sigma = Pl_p$), so that $\varepsilon \in (\mathcal{E}^{\mathrm{ord}})^{p^{e-1}}$. Thus $\varepsilon = 1$ since e is arbitrary and $\mathcal{E}^{\mathrm{ord}}$ is Hausdorff. □

Note. It is clear that for any finite set Σ of noncomplex places, $\bar{i}_{Pl^{\mathrm{nc}} \setminus \Sigma}$ is also an injective map on $\mathcal{E}^{\mathrm{ord}}$.

We have therefore proved a result of [Ja7, § 2.7] which interprets the p-completion of the unit group as a subgroup of $(\overline{G}^{\mathrm{ab}})_p$.

4.4.1 FUNDAMENTAL DIAGRAM FOR $\overline{K}^{\mathrm{ab}}(p)/K$. This study can be summarized by the following diagram (assuming the Leopoldt conjecture for p):

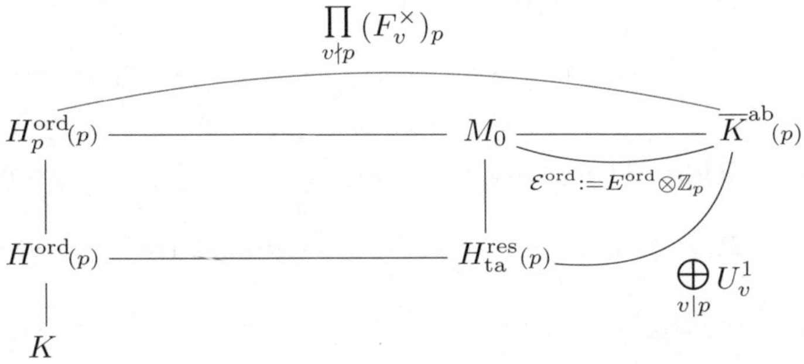

Fig. 4.2

In this diagram, by 4.3.1 applied to $\Sigma = Pl_p$, $\delta_\infty = \emptyset$:

$$\mathrm{Gal}(\overline{K}^{\mathrm{ab}}(p)/H_{\mathrm{ta}}^{\mathrm{res}}(p)) \simeq \bigoplus_{v|p} U_v^1$$

is the subgroup generated by the inertia groups of the $v|p$, which are isomorphic to the U_v^1 (they are deployed). In particular the $\mathbb{Z}_p$-rank of the group $\mathrm{Gal}(\overline{K}^{\mathrm{ab}}(p)/H_{\mathrm{ta}}^{\mathrm{res}}(p))$ is equal to $[K : \mathbb{Q}]$, and its torsion group is isomorphic to $\bigoplus_{v|p} \mu_p(K_v)$.

4.4.2 Exercise. (i) Give a proof of this result on $\mathrm{Gal}(\overline{K}^{\mathrm{ab}}(p)/H_{\mathrm{ta}}^{\mathrm{res}}(p))$, by a limiting process using 1.4, by computing:

$$\varprojlim_{\mathfrak{m}} \Big(\bigoplus_{v|p} U_v^1 \Big/ \bar{i}_p(\mathcal{E}_{\mathfrak{m}_{\mathrm{ta}}}^{\mathrm{res}}) \, . \, \bigoplus_{v|p} U_v^{m_v} \Big).$$

(ii) Give also another proof of the classical isomorphism 1.6.1:

$$\mathrm{Gal}(M_0/H_{\mathrm{ta}}^{\mathrm{res}}(p)) \simeq \bigoplus_{v|p} U_v^1 \Big/ \bar{i}_p(\mathcal{E}^{\mathrm{ord}}),$$

by computing the inverse limit of the:

$$\mathrm{Gal}(K(\mathfrak{m}_p)^{\mathrm{ord}}(p) K(\mathfrak{m}_{\mathrm{ta}})^{\mathrm{res}}(p)/K(\mathfrak{m}_{\mathrm{ta}})^{\mathrm{res}}(p)),$$

for independent $\mathfrak{m}_p$ and $\mathfrak{m}_{\mathrm{ta}}$ (use II.5.1.4). □

The fundamental diagram implies:

4.4.3 Corollary. *We have the exact sequences of global class field theory (assuming the Leopoldt conjecture for p):*

$$1 \longrightarrow \mathcal{E}^{\mathrm{ord}} \longrightarrow \prod_{v \nmid p} (F_v^{\times})_p \longrightarrow \mathrm{Gal}(H_{\mathrm{ta}}^{\mathrm{res}}(p)/K) \longrightarrow (\mathcal{C}\ell^{\mathrm{ord}})_p \longrightarrow 1,$$

$$1 \longrightarrow \mathcal{E}^{\mathrm{ord}} \longrightarrow \bigoplus_{v|p} U_v^1 \longrightarrow \mathrm{Gal}(H_p^{\mathrm{ord}}(p)/K) \longrightarrow (\mathcal{C}\ell^{\mathrm{ord}})_p \longrightarrow 1. \qquad \square$$

4.4.4 Proposition. *The infinite extension $\overline{K}^{\mathrm{ab}}(p)/H_{\mathrm{ta}}^{\mathrm{res}}(p)$ is p-ramified.*

Proof. If $v \in Pl_{\mathrm{ta}}$, we have in terms of inertia groups, the exact sequence:

$$1 \longrightarrow I_v(\overline{K}^{\mathrm{ab}}(p)/H_{\mathrm{ta}}^{\mathrm{res}}(p)) \longrightarrow I_v(\overline{K}^{\mathrm{ab}}(p)/K) \longrightarrow I_v(H_{\mathrm{ta}}^{\mathrm{res}}(p)/K) \longrightarrow 1\ ;$$

but, by 4.3.2, we have $I_v(\overline{K}^{\mathrm{ab}}(p)/K) \simeq I_v(H_{\mathrm{ta}}^{\mathrm{res}}(p)/K) \simeq (F_v^{\times})_p$, hence $\overline{K}^{\mathrm{ab}}(p)/H_{\mathrm{ta}}^{\mathrm{res}}(p)$ is indeed p-ramified. □

Doing the same with the decomposition groups of the real places at infinity, we also see that $\overline{K}^{\mathrm{ab}}(2)/H_{\mathrm{ta}}^{\mathrm{res}}(2)$ is noncomplexified, but it is clear directly that $H_{\mathrm{ta}}^{\mathrm{res}}(2)$ is totally complex.

Here once again, the results do not depend on the Leopoldt conjecture.

For the computation of $D_v(\overline{K}^{\mathrm{ab}}(p)/K)$ in the general case, see 4.5 or 4.12.5.

4.4.5 Complements. We make some observations about the relative extension $\overline{K}^{\mathrm{ab}}(p)/H_p^{\mathrm{ord}}(p)$.

4.4.5.1 Two Isomorphisms. The isomorphisms (assuming the Leopoldt conjecture for the first one):

$$\mathrm{Gal}(\overline{K}^{\mathrm{ab}}(p)/H_p^{\mathrm{ord}}(p)) \simeq \prod_{v \nmid p} (F_v^{\times})_p, \quad \mathrm{Gal}(\overline{K}^{\mathrm{ab}}(p)/H_{\mathrm{ta}}^{\mathrm{res}}(p)) \simeq \bigoplus_{v|p} U_v^1,$$

are not compatible; in particular their restrictions to $\mathrm{Gal}(\overline{K}^{\mathrm{ab}}{}_{(p)}/M_0)$ have images equal respectively to $\bar{i}_{\mathrm{ta},\infty}(\mathcal{E}^{\mathrm{ord}})$ and to $\bar{i}_p(\mathcal{E}^{\mathrm{ord}})$ (analyze the above computations). Concerning this, one should keep in mind the underlying reciprocity map; when for instance we consider $\bar{i}_{\mathrm{ta},\infty}(\varepsilon)$ as an element of $\mathrm{Gal}(\overline{K}^{\mathrm{ab}}{}_{(p)}/M_0)$, we consider the reduced idèle, which is here an element of $(U_0^{\mathrm{ord}})_p \simeq \prod_{v \nmid p} (F_v^\times)_p \prod_{v|p} U_v^1$:

$$\boldsymbol{u} := (\bar{i}_{\mathrm{ta},\infty}(\varepsilon); 1, \ldots, 1)$$

such that $u_v = \bar{i}_v(\varepsilon)$ for all $v \nmid p$, $u_v = 1$ for $v|p$, of which we take the image under the reciprocity map ρ (see II.3.7). In the same way, $\bar{i}_p(\varepsilon)$ corresponds to the idèle:

$$\boldsymbol{u}' := (\ldots, 1, \ldots; \bar{i}_p(\varepsilon)),$$

with support Pl_p, of which we take the image under ρ. But since $\bar{i}_0^{(p)}(\varepsilon)$ gives the identity in $(\overline{G}^{\mathrm{ab}})_p$, the image of $\boldsymbol{u} = (\bar{i}_{\mathrm{ta},\infty}(\varepsilon); 1, \ldots, 1)$ under ρ is equal to that of $\boldsymbol{u}\, \bar{i}_0^{(p)}(\varepsilon)^{-1} = (\ldots, 1, \ldots; \bar{i}_p(\varepsilon)^{-1})$, hence to that of $\boldsymbol{u}'^{-1}$. In other words, on $\mathrm{Gal}(\overline{K}^{\mathrm{ab}}{}_{(p)}/M_0)$ the isomorphisms:

$$\mathcal{E}^{\mathrm{ord}} \xrightarrow{\rho \circ \bar{i}_{\mathrm{ta},\infty}} \mathrm{Gal}(\overline{K}^{\mathrm{ab}}{}_{(p)}/M_0) \ \text{ and } \ \mathcal{E}^{\mathrm{ord}} \xrightarrow{\rho \circ \bar{i}_p} \mathrm{Gal}(\overline{K}^{\mathrm{ab}}{}_{(p)}/M_0)$$

differ by the inversion automorphism on $(\overline{G}^{\mathrm{ab}})_p$.

The isomorphisms:

$$\mathrm{Gal}(M_0/H_p^{\mathrm{ord}}{}_{(p)}) \simeq \mathrm{Gal}(H_{\mathrm{ta}}^{\mathrm{res}}{}_{(p)}/H^{\mathrm{ord}}{}_{(p)}) \simeq \prod_{v \nmid p} (F_v^\times)_p \Big/ \bar{i}_{\mathrm{ta},\infty}(\mathcal{E}^{\mathrm{ord}}),$$

show that the extension $H_{\mathrm{ta}}^{\mathrm{res}}/K$ has quite a complicated structure; even if $\mathcal{C}\ell^{\mathrm{ord}} = 1$ and $E^{\mathrm{ord}} = \langle -1 \rangle$ (case of $\mathbb{Q}$), there is not deployment of the inertia groups of the tame places and of the decomposition groups of the real places (see 4.1.11, (ii)).

4.4.5.2 Leopoldt's Kernel. The isomorphism $\mathrm{Gal}(\overline{K}^{\mathrm{ab}}{}_{(p)}/H_p^{\mathrm{ord}}{}_{(p)}) \simeq \prod_{v \nmid p} (F_v^\times)_p$ is in fact equivalent to the Leopoldt conjecture for p. Indeed, let $\mathcal{E}_{Pl_p}^{\mathrm{ord}} =: \mathcal{E}_p^{\mathrm{ord}}$ be the defect in the Leopoldt conjecture, or Leopoldt's kernel, in other words the kernel of the map:

$$\bar{i}_p : \mathcal{E}^{\mathrm{ord}} \longrightarrow \bigoplus_{v|p} U_v^1 \ ;$$

it is easily checked (as in 4.1) that we have the isomorphism:

$$\mathrm{Gal}(\overline{K}^{\mathrm{ab}}{}_{(p)}/H_p^{\mathrm{ord}}{}_{(p)}) \simeq \prod_{v \nmid p} (F_v^\times)_p \Big/ \bar{i}_{\mathrm{ta},\infty}(\mathcal{E}_p^{\mathrm{ord}}) \ ;$$

the injectivity of $\bar{i}_{\text{ta},\infty}$ (Theorem 4.4) implies our claim. This shows how Figure 4.2 can be modified if we do not assume the Leopoldt conjecture.

4.4.5.3 TAME RELATIVE INERTIA FIELDS. Assuming the Leopoldt conjecture for p, the group $\mathrm{Gal}(\overline{K}^{\text{ab}}(p)/H_p^{\text{ord}}(p))$ is isomorphic to $\prod_{v\nmid p}(F_v^\times)_p$, hence for any tame finite place v the field fixed under $\prod_{v'\neq v}(F_{v'}^\times)_p$ is a cyclic extension of degree $|(F_v^\times)_p| = (q_v-1)_p$ of $H_p^{\text{ord}}(p)$, unramified outside v and p, noncomplexified, and totally ramified at v.

Similarly, if $p=2$ and if v is a real infinite place, the field fixed under $\prod_{v'\neq v}(F_{v'}^\times)_2$ is a quadratic extension of $H_2^{\text{ord}}(2)$, which is 2-ramified, $(Pl_\infty^{\text{r}}\backslash\{v\})$-split and complexified at v. The field $H_2^{\text{res}}(2)$ is of course the direct compositum over $H_2^{\text{ord}}(2)$ of these quadratic extensions, and $\mathrm{Gal}(H_2^{\text{res}}(2)/H_2^{\text{ord}}(2)) \simeq (\mathbb{Z}/2\mathbb{Z})^{r_1}$. Finally, we have $\overline{K}^{\text{ab nc}} = H_{Pl_0}^{\text{ord}}$ and $\mathrm{Gal}(\overline{K}^{\text{ab}}/\overline{K}^{\text{ab nc}}) \simeq (\mathbb{Z}/2\mathbb{Z})^{r_1}$.

Note that the places above p may ramify in $\overline{K}^{\text{ab}}(p)/H_p^{\text{ord}}(p)$ (see 4.9) and in particular in the finite extensions of $H_p^{\text{ord}}(p)$.

4.4.5.4 FUNDAMENTAL DIAGRAM FOR $\overline{K}^{\text{ab}}/K$. The unlocalized analog of Diagram 4.2 can be obtained (assuming the Leopoldt conjecture for all p) by replacing $\otimes\mathbb{Z}_p$ by $\otimes\widehat{\mathbb{Z}}$ and by introducing the extension H_*^{ord}, which is the compositum of the extensions $H_p^{\text{ord}}(p)$, whose Galois group is isomorphic to $\prod_p \mathcal{T}_p^{\text{ord}} \times \widehat{\mathbb{Z}}^{r_2+1}$ (see 4.1.8):

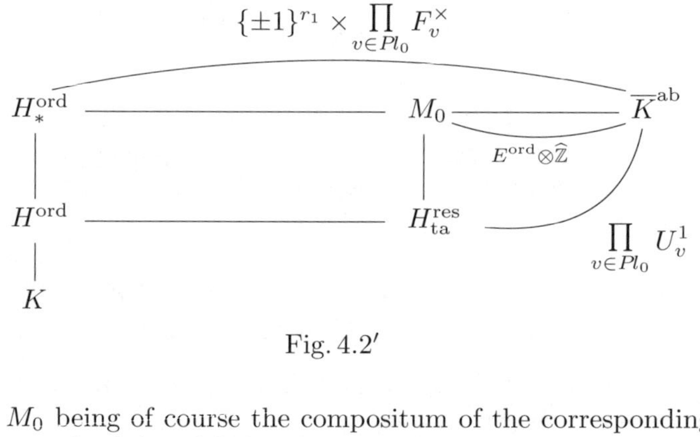

Fig. 4.2′

the field M_0 being of course the compositum of the corresponding localizations. The result of this globalization is not as trivial as one could think, since although this is the case for most of the extensions that are involved, the one giving H_*^{ord} introduces as Galois group the invariant $\prod_p \mathcal{T}_p^{\text{ord}}$ which we have already mentioned, and for which there does not seem to be any direct global definition: indeed, even if it were possible idelically, this would not shed any light on the problem since the groups $\mathcal{T}_p^{\text{ord}}$ are clearly of a p-adic nature, and the numerical study which has been done on them confirms this. We will come back in 4.14 to problems concerning this fundamental invariant.

We know that $\mathrm{adh}_0(E^{\mathrm{ord}}) = \langle\, i_0(E^{\mathrm{ord}})\,\rangle_{\widehat{\mathbb{Z}}} = \overline{i}_0(E^{\mathrm{ord}} \otimes \widehat{\mathbb{Z}})$ in U_0^{ord} considered as $\widehat{\mathbb{Z}}$-module, and that the connected component D_0 of the unit element of $U_0^{\mathrm{ord}}/E^{\mathrm{ord}}$ is equal to $\mathrm{adh}_0(E^{\mathrm{ord}})/E^{\mathrm{ord}}$ (see 1.5.1, 1.6.6, (iii)). Now we can state:

4.4.6 Theorem (structure of the connected component D_0). *We have, for any number field K, in the context of reduced idèles:*

$$\langle\, i_0(E^{\mathrm{ord}})\,\rangle_{\widehat{\mathbb{Z}}} \simeq \mu(K) \times \widehat{\mathbb{Z}}^{r_1+r_2-1}.$$

Therefore, the connected component D_0 in C_0, which is that of the unit element of $U_0^{\mathrm{ord}}/E^{\mathrm{ord}}$, is isomorphic to:

$$(\widehat{\mathbb{Z}}/\mathbb{Z})^{r_1+r_2-1}.$$

Proof. We can prove the injectivity of $\overline{i}_0$ on $E^{\mathrm{ord}} \otimes \widehat{\mathbb{Z}}$ by using Theorem 4.4 for each p, or by showing, in a similar way thanks to the local-global principle for powers, that any $\varepsilon = \prod_i \varepsilon_i \otimes a_i \in E^{\mathrm{ord}} \otimes \widehat{\mathbb{Z}}$ such that $\overline{i}_0(\varepsilon) = 1$ is for all n an nth power in $E^{\mathrm{ord}} \otimes \widehat{\mathbb{Z}}$ (i.e., in the proof of 4.4 replace $\mathbb{Z}_p$ by $\widehat{\mathbb{Z}}$, p^e by $2n$, and consider that $\overline{i}_0(\varepsilon) = 1$ implies that all the residual images of ε are trivial; then use in analogous way the places not dividing $2n$).

We can also work directly in $\langle\, i_0(E^{\mathrm{ord}})\,\rangle_{\widehat{\mathbb{Z}}}$ and check the $\widehat{\mathbb{Z}}$-independence of a fundamental system of units. In any case, we are led to writing that $\mathbb{Z}$ is dense in $\mathbb{Z}_p$ or $\widehat{\mathbb{Z}}$ and, most importantly, to use II.6.3.3.

We thus have $\overline{i}_0(E^{\mathrm{ord}} \otimes \widehat{\mathbb{Z}}) \simeq E^{\mathrm{ord}} \otimes \widehat{\mathbb{Z}} \simeq \mu(K) \times \widehat{\mathbb{Z}}^{r_1+r_2-1}$, so that:

$$D_0 \simeq \mu(K) \times \widehat{\mathbb{Z}}^{r_1+r_2-1}/\mu(K) \times \mathbb{Z}^{r_1+r_2-1} \simeq (\widehat{\mathbb{Z}}/\mathbb{Z})^{r_1+r_2-1}. \qquad \square$$

4.4.7 Remarks (usual connected component D). (i) The computation of D would be similar, but the use of reduced idèles has allowed us to suppress the archimedean factors since we have (after [d, AT, Ch. 9, § 1]):

$$D \simeq ((\widehat{\mathbb{Z}} \oplus \mathbb{R})/\mathbb{Z})^{r_1+r_2-1} \times (\mathbb{R}/\mathbb{Z})^{r_2} \times \mathbb{R},$$

while D_0 is exclusively ultrametric in nature. These two connected components are related by the exact sequence:

$$0 \longrightarrow (\mathbb{R}/\mathbb{Z})^{r_2} \times \mathbb{R}^{r_1+r_2} \longrightarrow D \longrightarrow D_0 \longrightarrow 0.$$

(ii) D_0 is a divisible group since $\widehat{\mathbb{Z}}/\mathbb{Z}$ has this property (simply write that $\mathbb{Z}$ is dense in $\widehat{\mathbb{Z}}$). To compare, we will prove directly the divisibility property of D in 4.15.1 (of course without using the above structure theorem for D which also trivially implies this property).

(iii) D is equal to the product of $\mathbb{R}$ by a compact set, while D_0 is only precompact [17] ($\widehat{\mathbb{Z}}/\mathbb{Z}$ is not Hausdorff).

(iv) D_0 is trivial if and only if E^{ord} is finite (compare with I.4.2.8, (iii)). □

4.4.8 Remark. The fact that the canonical map:

$$\bar{i}_0 \, : \, E^{\mathrm{ord}} \otimes \widehat{\mathbb{Z}} \longrightarrow U_0^{\mathrm{ord}} := \prod_{v \in Pl_0} U_v \prod_{v \in Pl_\infty^{\mathrm{r}}} \{\pm 1\}$$

is injective must not be mistaken with the Leopoldt conjecture on the injectivity (for all p) of:

$$\bar{i}_p \, : \, E^{\mathrm{ord}} \otimes \mathbb{Z}_p \longrightarrow \prod_{v|p} U_v^1$$

(see 3.6.2, (vi)) since in the first case all places occur, and if we want to compare the two situations, we note that the injectivity of $\bar{i}_0$ is equivalent to that of:

$$\bar{i}_0^{(p)} \, : \, E^{\mathrm{ord}} \otimes \mathbb{Z}_p \longrightarrow \prod_{v \nmid p} (F_v^\times)_p \prod_{v|p} U_v^1$$

for all p, where the factor $\prod_{v|p} U_v^1$ *may be omitted* (as in Theorem 4.4) since it is in fact the set of tame places which plays the main role; we can even suppress an arbitrary finite set of places. In other words, in these questions we only use the information given by the *residue fields*.

This being said, the Leopoldt conjecture for all p also implies this result, and we cannot avoid referring to 4.3, Note). □

The above results of Subsection (d) have the following important consequence, which is a particular case of general properties [e, Ko3, Ch. 3, § 3.5]. We give a first proof which, logically, mainly uses the Schmidt–Chevalley Theorem 4.3 and its corollaries; see in 4.12.5 a more direct proof and in 4.16.7 the best idelic form which uses arguments which are common with those which are necessary in the proof of the Grunwald–Wang theorem.

4.5 Theorem. *Let K be a number field. Then for any place v of K, the local extension $(\overline{K}^{\mathrm{ab}})_v$ corresponding to the abelian closure $\overline{K}^{\mathrm{ab}}$ of K is equal to the abelian closure $\overline{K}_v^{\mathrm{ab}}$ of K_v (i.e., $\overline{K}_v^{\mathrm{ab}} = \overline{K}^{\mathrm{ab}} K_v$ if $\overline{K}^{\mathrm{ab}}$ is seen in $\mathbb{C}_\ell$).*

Proof. The case of $v \in Pl_\infty$ being trivial, assume that $v \in Pl_0$. It is sufficient to show that for any prime p we have equality of the corresponding pro-p-subextensions. To simplify, set:

[17] i.e., it satisfies the open cover axiom but it is not necessarily Hausdorff.

$$M := \overline{K}_v^{\mathrm{ab}}(p), \quad M' := (\overline{K}^{\mathrm{ab}}(p))_v = (\overline{K}^{\mathrm{ab}})_{v(p)} \subseteq M.$$

The $\mathbb{Z}_p$-modules $\overline{G}_v^{\mathrm{ab}} := \mathrm{Gal}(M/K_v)$ and $\overline{D}_v^{\mathrm{ab}} := \mathrm{Gal}(M'/K_v)$ are $\mathbb{Z}_p$-modules of finite type (by II.1.8.2) and we have the following lemmas.

4.5.1 Lemma 1. *Let k be a subfield of M' containing K_v; then $M' = M$ if and only if* $\mathrm{Gal}(M/k)$ *and* $\mathrm{Gal}(M'/k)$ *are isomorphic $\mathbb{Z}_p$-modules.*

Proof. Indeed, to say that the $\mathbb{Z}_p$-ranks are equal means that M' contains the maximal $\mathbb{Z}_p$-free subextension M_0 of M over k, in which case $\mathrm{Gal}(M/M_0)$ and $\mathrm{Gal}(M'/M_0)$ are the respective torsion modules of $\mathrm{Gal}(M/k)$ and of $\mathrm{Gal}(M'/k)$, proving the result. □

4.5.2 Lemma 2. *We have $M'^{\mathrm{nr}} = M^{\mathrm{nr}}$ which is a $\mathbb{Z}_p$-extension of K_v.*

Proof. Since $\mathrm{Gal}(M^{\mathrm{nr}}/K_v) \simeq \mathbb{Z}_p$ (see II.1.8.2), it is sufficient to show that M'^{nr}/K_v is not finite. If $q_v := |F_v|$, the abelian extension $K(\mu_{q_v^{p^e}-1})$ of K is unramified at v and we have $K_v(\mu_{q_v^{p^e}-1}) \subset M'^{\mathrm{nr}}$ by definition; since by II.1.1.5, $K_v(\mu_{q_v^{p^e}-1})$ is the unramified extension of degree p^e of K_v, the lemma follows. □

Although this is class field theory, it uses only the elementary (polynomial) theory of cyclotomic fields, as opposed to the following lemma.

4.5.3 Lemma 3. *The inertia groups of M/K_v and of M'/K_v are isomorphic.*

Proof. In the tame case these inertia groups are isomorphic to $(F_v^\times)_p$ (see respectively II.1.8.2, (i), and 4.3.2). In the wild case these inertia groups are isomorphic to U_v^1 (see respectively II.1.8.2, (ii), and 4.3.2). □

The theorem follows by using the above for $k = M^{\mathrm{nr}}$. □

4.5.4 Proposition. *Let $v|p$. Then its residue degree in the p-ramified extension $\overline{K}^{\mathrm{ab}}(p)/H_{\mathrm{ta}}^{\mathrm{res}}(p)$ is equal to 1.*

Proof. It is sufficient to show that the decomposition group of v in $H_{\mathrm{ta}}^{\mathrm{res}}(p)/K$ is not finite (it is then isomorphic to $\mathbb{Z}_p$). This can be done by considering the p-subextensions of the tame extensions $K(\mu_{q_v^{p^e}-1})/K$, whose residue degree is equal to p^e. In fact, 4.5 implies that $(H_{\mathrm{ta}}^{\mathrm{res}}(p))_v = \overline{K}_v^{\mathrm{nr}}(p)$ for all $v|p$.

We can also study the behavior of the decomposition groups of the ray class fields as follows. Let $\mathfrak{m} = \prod_{v' \in T} \mathfrak{p}_{v'}$ be any modulus prime to p; by 1.1.6, (ii), (α) for $\mathfrak{n} = 1$ and $S = \delta_\infty = \emptyset$, we have:

$$D_v(K(\mathfrak{m})^{\rm res}(p)/K) \simeq (K_v^\times/i_v(E_{\mathfrak{m}}^{\{v\}{\rm res}})U_v)_p.$$

By the Schmidt–Chevalley Theorem 4.3, (i), for all $n \geq 1$ we can find $\mathfrak{m}$ such that $E_{\mathfrak{m}}^{\{v\}{\rm res}} \subseteq (E^{\{v\}{\rm ord}})^{p^n}$. Let π_v be a uniformizer of K_v; it is then clear that the order of the image of π_v in $(K_v^\times/i_v(E_{\mathfrak{m}}^{\{v\}{\rm res}})U_v)_p$ can be made arbitrarily large. The result follows, since $K(\mathfrak{m})^{\rm res}(p) \subset H_{\rm ta}^{\rm res}(p)$. □

4.5.5 Proposition. *In $H_{\rm ta}^{\rm res}(p)/K$, the decomposition group of any tame place v is isomorphic to $\mathbb{Z}_p \times (F_v^\times)_p$. Thus v is totally split in $\overline{K}^{\rm ab}(p)/H_{\rm ta}^{\rm res}(p)$.*

Proof. The computation performed in 4.5.4 with ray class fields is valid for any finite tame place v (taking the moduli $\mathfrak{m}$ prime to p and v). Thus $D_v(H_{\rm ta}^{\rm res}(p)/K) \simeq \mathbb{Z}_p \times (F_v^\times)_p$ for such a place. □

This result will be also a consequence of the weak deployment theorem of decomposition groups 4.16.7.

Note. Beware that if $v \nmid p$, then the extension $K(\mu_{q_v^{p^e}-1})(p)/K$ is not necessarily tame (e.g., $\mathbb{Q}(\mu_{3^2-1})/\mathbb{Q}$ for $K = \mathbb{Q}$, $p = 2$, $v = 3$).

e) Decomposition Law of Wild Places in $\overline{K}^{\rm ab}(p)/H_p^{\rm ord}(p)$

This subsection is somewhat connected with difficult questions, such as the study of the free pro-p-extensions of maximal rank of a number field[18], whether or not they are considered cohomologically; it is the question of the law of decomposition of the places above p in $\overline{K}^{\rm ab}(p)/H_p^{\rm ord}(p)$ (a large part of the ramification at p in $\overline{K}^{\rm ab}(p)/K$ taking place in $H_p^{\rm ord}(p)/K$). This study will be based on p-adic considerations which are more precise than those which constitute the p-adic conjecture 3.6, and in 4.12 we will formulate the appropriate conjecture; it is still a p-adic transcendence problem.

For T finite containing Pl_p, consider the extension $H_T^{\rm res}(p)/H_p^{\rm ord}(p)$. For $v|p$ we would like to identify $I_v(H_T^{\rm res}(p)/H_p^{\rm ord}(p))$ and $D_v(H_T^{\rm res}(p)/H_p^{\rm ord}(p))$, then use a limiting process. We assume the Leopoldt conjecture for p.

The starting point is Lemma 4.1.1 since we still want to show what happens at a finite level. Fix a place v above p.

4.6 The Decomposition Groups of Wild Places in $H_T^{\rm res}(p)/H_p^{\rm ord}(p)$. Let $\mathfrak{m} = \prod_{v'|p} \mathfrak{p}_{v'}^{m_{v'}} \prod_{v' \in T_{\rm ta}} \mathfrak{p}_{v'} =: \mathfrak{m}_p\,\mathfrak{m}_{\rm ta}$. It is easy to show from II.1.2.3 that:

$$I_v(H_T^{\rm res}(p)/H_p^{\rm ord}(p)) \simeq I_v(K(\mathfrak{m})^{\rm res}(p)/K(\mathfrak{m}_p)^{\rm ord}(p)),$$

[18] Works of Yamagishi [Y], Hubbard [Hu], and Lannuzel [La], Lannuzel–Nguyen Quang Do [LaNg], for instance.

as soon as $\mathfrak{m}_p$ is sufficiently large, which yields, using Exercise 1.1.6, (ii), (γ) for $\mathfrak{n} = \mathfrak{m}_p$, $S = \emptyset$, and $\delta_\infty = Pl_\infty^{\mathrm{r}}$:

$$I_v(H_T^{\mathrm{res}}(p)/H_p^{\mathrm{ord}}(p)) \simeq \Big(E^{\mathrm{ord}}_{\frac{\mathfrak{m}_p}{\mathfrak{m}_v}}/E^{\mathrm{res}}_{\frac{\mathfrak{m}_p}{\mathfrak{m}_v}\mathfrak{m}_{\mathrm{ta}}}E^{\mathrm{ord}}_{\mathfrak{m}_p}\Big)_p \simeq \Big(E^{\mathrm{ord}}_{\frac{\mathfrak{m}_p}{\mathfrak{m}_v}}/E^{\mathrm{res}}_{\frac{\mathfrak{m}_p}{\mathfrak{m}_v}\mathfrak{m}_{\mathrm{ta}}}\Big)_p$$

for all $\mathfrak{m}_p$ sufficiently large since (assuming the Leopoldt conjecture) we have:

$$(E^{\mathrm{ord}}_{\mathfrak{m}_p} : E^{\mathrm{res}}_{\mathfrak{m}_p\mathfrak{m}_{\mathrm{ta}}})_p = 1.$$

In the same way, we obtain:

$$D_v(H_T^{\mathrm{res}}(p)/H_p^{\mathrm{ord}}(p)) \simeq \Big(E^{\{v\}\mathrm{ord}}_{\frac{\mathfrak{m}_p}{\mathfrak{m}_v}}/E^{\{v\}\mathrm{res}}_{\frac{\mathfrak{m}_p}{\mathfrak{m}_v}\mathfrak{m}_{\mathrm{ta}}}\Big)_p$$

for all $\mathfrak{m}_p$ sufficiently large.

4.6.1 Notation. In $\mathcal{E}^{\mathrm{ord}} := E^{\mathrm{ord}} \otimes \mathbb{Z}_p$ we set:

$$\mathcal{E}^{\mathrm{ord}}_{p\backslash v} := \mathcal{E}^{\mathrm{ord}}_{Pl_p\backslash\{v\}} := \{\varepsilon \in \mathcal{E}^{\mathrm{ord}},\ \ \bar{i}_{v'}(\varepsilon) = 1\ \ \forall v'|p,\ v' \neq v\}\ ;$$

it is also the kernel of the embedding:

$$\bar{i}_{p\backslash v} : \mathcal{E}^{\mathrm{ord}} \longrightarrow \bigoplus_{v'|p, v'\neq v} U^1_{v'}. \qquad \square$$

4.6.2 Remark. Assuming the Leopoldt conjecture for p, $\bar{i}_p$ is injective, hence the $\mathbb{Z}_p$-module $\mathcal{E}^{\mathrm{ord}}_{p\backslash v}$ is isomorphic to $\bar{i}_p(\mathcal{E}^{\mathrm{ord}}) \cap U^1_v$.

If $|Pl_p| = 1$, we obtain $\mathcal{E}^{\mathrm{ord}}_{p\backslash v} = \mathcal{E}^{\mathrm{ord}}$. If $|Pl_p| > 1$, the $\mathbb{Z}_p$-module $\mathcal{E}^{\mathrm{ord}}_{p\backslash v}$ has no torsion since $i_p(\mu_p(K)) \cap U^1_v = 1$. $\square$

In analogous way we define $\mathcal{E}^{\mathrm{res}}_{T\backslash v}$ as being, in $\mathcal{E}^{\mathrm{res}} := E^{\mathrm{res}} \otimes \mathbb{Z}_p$:

$$\mathcal{E}^{\mathrm{res}}_{T\backslash v} := \{\varepsilon \in \mathcal{E}^{\mathrm{res}},\ \bar{i}_{v'}(\varepsilon) = 1\ \ \forall v' \in T,\ v' \neq v\}\ ;$$

note that $\mathcal{E}^{\mathrm{res}}_{T\backslash v}$ is also $\mathcal{E}^{\mathrm{res}}_{T_{\mathrm{ta}}} \cap \mathcal{E}^{\mathrm{ord}}_{p\backslash v}$, where $\mathcal{E}^{\mathrm{res}}_{T_{\mathrm{ta}}}$ is equal to $\mathcal{E}^{\mathrm{res}}_{\mathfrak{m}_{\mathrm{ta}}}$.

Using the formula (with evident notations):

$$I_v(H_T^{\mathrm{res}}(p)/H_p^{\mathrm{ord}}(p)) \simeq \mathcal{E}^{\mathrm{ord}}_{\frac{\mathfrak{m}_p}{\mathfrak{m}_v}}/\mathcal{E}^{\mathrm{res}}_{\frac{\mathfrak{m}_p}{\mathfrak{m}_v}\mathfrak{m}_{\mathrm{ta}}} \simeq \mathcal{E}^{\mathrm{res}}_{T_{\mathrm{ta}}}\mathcal{E}^{\mathrm{ord}}_{\frac{\mathfrak{m}_p}{\mathfrak{m}_v}}/\mathcal{E}^{\mathrm{res}}_{T_{\mathrm{ta}}},$$

which is independent of $\mathfrak{m}_p$ sufficiently large, we are led to the computation of $\bigcap_{\mathfrak{m}_p} \mathcal{E}^{\mathrm{res}}_{T_{\mathrm{ta}}}\mathcal{E}^{\mathrm{ord}}_{\frac{\mathfrak{m}_p}{\mathfrak{m}_v}}$, which is $\mathcal{E}^{\mathrm{res}}_{T_{\mathrm{ta}}}\mathcal{E}^{\mathrm{ord}}_{p\backslash v}$ by I.5.4 applied to the open sets $\mathcal{E}^{\mathrm{ord}}_{\frac{\mathfrak{m}_p}{\mathfrak{m}_v}}$ and to the compact subset $\mathcal{E}^{\mathrm{res}}_{T_{\mathrm{ta}}}$ of $\mathcal{E}^{\mathrm{ord}}$. Thus, we have obtained:

4.6.3 Proposition. *(i) For any $v|p$ we have:*

$$I_v(H_T^{\mathrm{res}}(p)/H_p^{\mathrm{ord}}(p)) \simeq \mathcal{E}^{\mathrm{ord}}_{p\backslash v}/\mathcal{E}^{\mathrm{res}}_{T\backslash v},$$

in other words, we have the exact sequence:

$$1 \longrightarrow \mathcal{E}^{\mathrm{ord}}_{p\backslash v}/\mathcal{E}^{\mathrm{res}}_{T\backslash v} \xrightarrow{\ \bar{i}_{T_{\mathrm{ta}},\infty}\ } \bigoplus_{v' \in T_{\mathrm{ta}} \cup Pl^{\mathrm{r}}_{\infty}} (F^{\times}_{v'})_p \xrightarrow{\ \rho\ } \mathrm{Gal}(M_T(v)/H^{\mathrm{ord}}_p(p)) \longrightarrow 1,$$

where $M_T(v)$ is the inertia field of v and ρ is the global reciprocity map.

(ii) In a similar way we have for $v|p$:

$$D_v(H^{\mathrm{res}}_T(p)/H^{\mathrm{ord}}_p(p)) \simeq \mathcal{E}^{\{v\}\mathrm{ord}}_{p\backslash v}/\mathcal{E}^{\{v\}\mathrm{res}}_{T\backslash v}. \qquad \square$$

4.7 The Decomposition Groups of Wild Places in $\overline{K}^{\mathrm{ab}}(p)/H^{\mathrm{ord}}_p(p)$. Going to the inverse limit on sets T containing Pl_p, using the fact that, when T_{ta} grows, by the Schmidt–Chevalley Theorem 4.3, (i), $E^{\mathrm{res}}_{\mathfrak{m}_{\mathrm{ta}}} \subseteq (E^{\mathrm{ord}})^{p^n}$ where n tends to infinity, we obtain for $I_v(\overline{K}^{\mathrm{ab}}(p)/H^{\mathrm{ord}}_p(p))$, $v|p$:

$$\mathcal{E}^{\mathrm{ord}}_{p\backslash v}\Big/\bigcap_T \mathcal{E}^{\mathrm{res}}_{T\backslash v} = \mathcal{E}^{\mathrm{ord}}_{p\backslash v}\Big/\bigcap_{T_{\mathrm{ta}}}\left(\mathcal{E}^{\mathrm{res}}_{T_{\mathrm{ta}}} \cap \mathcal{E}^{\mathrm{ord}}_{p\backslash v}\right) = \mathcal{E}^{\mathrm{ord}}_{p\backslash v}\Big/\left(\bigcap_{T_{\mathrm{ta}}} \mathcal{E}^{\mathrm{res}}_{T_{\mathrm{ta}}}\right) \cap \mathcal{E}^{\mathrm{ord}}_{p\backslash v} = \mathcal{E}^{\mathrm{ord}}_{p\backslash v},$$

and the exact sequence:

$$1 \longrightarrow \mathcal{E}^{\mathrm{ord}}_{p\backslash v} \xrightarrow{\ \bar{i}_{\mathrm{ta},\infty}\ } \prod_{v' \nmid p} (F^{\times}_{v'})_p \xrightarrow{\ \rho\ } \mathrm{Gal}(M(v)/H^{\mathrm{ord}}_p(p)) \longrightarrow 1,$$

where now the inertia field $M(v)$ of v in $\overline{K}^{\mathrm{ab}}(p)/H^{\mathrm{ord}}_p(p)$ contains:

$$M_0 := H^{\mathrm{ord}}_p(p) H^{\mathrm{res}}_{\mathrm{ta}}(p)$$

since the nonramification of v in $H^{\mathrm{res}}_{\mathrm{ta}}(p)/K$ is carried over to $M_0/H^{\mathrm{ord}}_p(p)$ (see Fig. 4.2).

Assuming the Leopoldt conjecture for p, we have:

4.7.1 Proposition. *The inertia groups $I_v(\overline{K}^{\mathrm{ab}}(p)/H^{\mathrm{ord}}_p(p))$ for $v|p$ correspond to the $\mathcal{E}^{\mathrm{ord}}_{p\backslash v}$ under the isomorphisms:*

$$\mathrm{Gal}(\overline{K}^{\mathrm{ab}}(p)/M_0) \simeq \bar{i}_{\mathrm{ta},\infty}(\mathcal{E}^{\mathrm{ord}}) \simeq \bar{i}_p(\mathcal{E}^{\mathrm{ord}}) \simeq \mathcal{E}^{\mathrm{ord}}.$$

They are deployed and the group corresponding to $\bigoplus_{v|p} \mathcal{E}^{\mathrm{ord}}_{p\backslash v}$ fixes a field M_1, containing M_0, which is therefore the maximal unramified extension of M_0 in $\overline{K}^{\mathrm{ab}}(p)$.

Proof. Indeed, we have seen in 4.4.4 that $\overline{K}^{\mathrm{ab}}(p)/M_0$ is p-ramified. $\square$

This could have been predicted since the U^1_v correspond to the inertia groups in $\overline{K}^{\mathrm{ab}}(p)/H^{\mathrm{res}}_{\mathrm{ta}}(p)$ by 4.3.2 and that $\bar{i}_p(\mathcal{E}^{\mathrm{ord}}) \cap U^1_v \simeq \mathcal{E}^{\mathrm{ord}}_{p\backslash v}$.

We thus have obtained the following diagram:

$$H_p^{\text{ord}}(p) \;\text{———}\; M_0 := H_p^{\text{ord}}(p)\, H_{\text{ta}}^{\text{res}}(p) \xrightarrow{\;unramified\;} M_1 \xrightarrow{\;\bigoplus_{v|p} \mathcal{E}_{p\backslash v}^{\text{ord}}\;} \overline{K}^{\text{ab}}(p),$$

in which:

$$\text{Gal}(\overline{K}^{\text{ab}}(p)/M_0) \simeq \mathcal{E}^{\text{ord}} \;\text{ and }\; \text{Gal}(\overline{K}^{\text{ab}}(p)/H_p^{\text{ord}}(p)) \simeq \prod_{v' \nmid p} (F_{v'}^{\times})_p.$$

Finally, by going to the inverse limit, and again by the fact that for all n we have $E_{\mathfrak{m}_{\text{ta}}}^{\{v\}\text{res}} \subseteq (E^{\{v\}\text{ord}})^{p^n}$ for T_{ta} sufficiently large, we obtain:

4.7.2 Proposition. *For any $v|p$ we have:*

$$D_v(\overline{K}^{\text{ab}}(p)/H_p^{\text{ord}}(p)) \simeq \mathcal{E}_{p\backslash v}^{\{v\}\text{ord}}. \qquad \square$$

4.7.3 Remark. Using the valuation map $\text{v} : \mathcal{E}_{p\backslash v}^{\{v\}\text{ord}} \longrightarrow \mathbb{Z}_p$, we check that there exists a v-unit $\eta_v \in \mathcal{E}_{p\backslash v}^{\{v\}\text{ord}}$ (possibly equal to 1) such that:

$$\mathcal{E}_{p\backslash v}^{\{v\}\text{ord}} = \langle\, \eta_v \,\rangle_{\mathbb{Z}_p} \bigoplus \mathcal{E}_{p\backslash v}^{\text{ord}}.$$

If $\eta_v \neq 1$, we have a Frobenius of infinite order and the residue degree of v in $H_p^{\text{ord}}(p)/K$ is finite (and equal to the p-part of $\text{v}(\eta_v)$), otherwise this Frobenius is equal to 1 in $\overline{K}^{\text{ab}}(p)/H_p^{\text{ord}}(p)$, and of infinite order in $H_p^{\text{ord}}(p)/K$ and thus in $\widetilde{K}_p/K$. The two cases can happen: let K be an imaginary quadratic field; if p is not split in $K/\mathbb{Q}$, $\mathcal{E}_{p\backslash v}^{\{v\}\text{ord}} = \mathcal{E}^{\{v\}\text{ord}}$ (the case $\eta_v \neq 1$); if p is split, $\mathcal{E}_{p\backslash v}^{\{v\}\text{ord}} = 1$. $\square$

4.8 The Decomposition Group of Tame Finite Places. The study of $D_v(\overline{K}^{\text{ab}}(p)/H_p^{\text{ord}}(p))$ for a tame finite place v is immediate. Indeed, by 4.5, we have:

$$D_v(\overline{K}^{\text{ab}}(p)/K) \simeq (F_v^{\times})_p \times \mathbb{Z}_p,$$

and since $D_v(H_p^{\text{ord}}(p)/K) \simeq \mathbb{Z}_p$ (the Frobenius of v in the cyclotomic $\mathbb{Z}_p$-extension of K is nontrivial), the decomposition group of v in $\overline{K}^{\text{ab}}(p)/H_p^{\text{ord}}(p)$ is reduced to the inertia group (isomorphic to $(F_v^{\times})_p$). In other words:

4.8.1 Proposition. *Any finite place v not dividing p is totally split in the extension $H_{Pl_0\backslash\{v\}}^{\text{res}}(p)/H_p^{\text{ord}}(p)$.* $\square$

4.8.2 Remark. For the reasons given above, the place v is also totally split in the extension $H_p^{\text{ord}}(p)/K\mathbb{Q}^{\text{cycl}}(p)$; the decomposition group of v in $\widetilde{K}_p/K$ is then known because of 2.5, (iii). $\square$

To summarize the results 4.6 to 4.8, we have obtained the following, where $\mathcal{E}^{\mathrm{ord}} := E^{\mathrm{ord}} \otimes \mathbb{Z}_p$, $\mathcal{E}^{\{v\}\mathrm{ord}} := E^{\{v\}\mathrm{ord}} \otimes \mathbb{Z}_p$ for any $v|p$.

4.9 Theorem. *Let K be a number field satisfying the Leopoldt conjecture for p. Let $H_p^{\mathrm{ord}}(p)$ be the maximal p-ramified noncomplexified abelian pro-p-extension of K. For any place v of K dividing p, we have the following canonical isomorphisms:*

$$I_v(\overline{K}^{\mathrm{ab}}(p)/H_p^{\mathrm{ord}}(p)) \simeq \mathcal{E}_{p\backslash v}^{\mathrm{ord}} := \{\varepsilon \in \mathcal{E}^{\mathrm{ord}},\ \bar{i}_{p\backslash v}(\varepsilon) = 1\},$$
$$D_v(\overline{K}^{\mathrm{ab}}(p)/H_p^{\mathrm{ord}}(p)) \simeq \mathcal{E}_{p\backslash v}^{\{v\}\mathrm{ord}} := \{\varepsilon \in \mathcal{E}^{\{v\}\mathrm{ord}},\ \bar{i}_{p\backslash v}(\varepsilon) = 1\},$$

where $\bar{i}_{p\backslash v} := (\bar{i}_{v'})_{v'|p, v'\neq v}$ takes its values in $\displaystyle\bigoplus_{v'|p, v'\neq v} U_{v'}^1$.

If v does not divide p, we have:

$$D_v(\overline{K}^{\mathrm{ab}}(p)/H_p^{\mathrm{ord}}(p)) = I_v(\overline{K}^{\mathrm{ab}}(p)/H_p^{\mathrm{ord}}(p)) \simeq (F_v^\times)_p. \qquad \square$$

The group $\mathrm{Gal}(M_1/M_0) \simeq \mathcal{E}^{\mathrm{ord}} \Big/ \displaystyle\bigoplus_{v|p} \mathcal{E}_{p\backslash v}^{\mathrm{ord}}$ depends in quite a complicated way on the splitting of ∞ and of p in the Galois closure N of K over $\mathbb{Q}$. It is evidently trivial if $|Pl_p| = 1$. Before trying to obtain a general statement, we first give some immediate particular cases.

4.9.1 Examples. Let v be a fixed place of K above p. Assume that the Leopoldt conjecture is true for p in K.

(i) Let us show that if the local degree at v is equal to 1, then $\mathcal{E}_{p\backslash v}^{\mathrm{ord}} = 1$ except in the particular case $K = \mathbb{Q}$, $p = 2$, where $\mathcal{E}_{p\backslash v}^{\mathrm{ord}} = \mathcal{E}^{\mathrm{ord}} \simeq \{\pm 1\}$. The case $K = \mathbb{Q}$ and $p \neq 2$ being clear, assume that $K \neq \mathbb{Q}$; thus, we have $|Pl_p| > 1$, and the image of $\varepsilon \in \mathcal{E}_{p\backslash v}^{\mathrm{ord}}$ in $\bigoplus_{v'|p} U_{v'}^1$ is $(\bar{i}_v(\varepsilon), 1, \ldots, 1)$. The global norm of ε being equal to ± 1 (-1 being possible only if $p = 2$), the product of the local norms is also equal to ± 1; the local norms outside v being equal to 1 and $K_v/\mathbb{Q}_p$ being trivial, this yields $\bar{i}_v(\varepsilon) = \pm 1$. For $p = 2$, $\bar{i}_v(\varepsilon) = -1$ is not possible by 3.6.3 since $|Pl_p| > 1$; we therefore have $\bar{i}_p(\varepsilon) = 1$, hence $\varepsilon = 1$.

(ii) If K is totally real then we have $\mathrm{rk}_{\mathbb{Z}_p}(\mathcal{E}_{p\backslash v}^{\mathrm{ord}}) = [K_v : \mathbb{Q}_p] - 1$. Indeed, let $F := \langle\, 1 + p\,\rangle \oplus E^{\mathrm{ord}}$; the map:

$$\log_p : F \otimes \mathbb{Q}_p \longrightarrow \bigoplus_{v'|p} K_{v'}$$

is an isomorphism, and therefore the map:

$$\log_{p\backslash v} : F \otimes \mathbb{Q}_p \longrightarrow \bigoplus_{v'|p, v'\neq v} K_{v'}$$

is surjective, with kernel having $\mathbb{Q}_p$-dimension $[K_v : \mathbb{Q}_p]$. Thus, let $\eta \in F \otimes \mathbb{Q}_p$ such that $\log_p(\eta) = (p, 0, \ldots, 0)$ (where $\log(i_v(\eta)) = p$); since this element does not have zero trace (the sum of the local traces is equal to $p\,[K_v : \mathbb{Q}_p]$), $\eta \notin E^{\rm ord} \otimes \mathbb{Q}_p$, showing that the kernel of $\log_{p\backslash v}$ in $\mathcal{E}^{\rm ord}$ has $\mathbb{Z}_p$-rank equal to $[K_v : \mathbb{Q}_p] - 1$. The conclusion follows since $\mathcal{E}^{\rm ord}_{p\backslash v}$ is of finite index in this kernel.

In this case, $\mathrm{Gal}(\overline{K}^{\rm ab}{}_{(p)}/M_1)$ has $\mathbb{Z}_p$-rank equal to $\sum\limits_{v|p}([K_v : \mathbb{Q}_p] - 1) = [K : \mathbb{Q}] - |Pl_p|$ and $\mathrm{Gal}(M_1/M_0)$ has $\mathbb{Z}_p$-rank equal to $|Pl_p| - 1$.

(iii) If p is totally split in $K/\mathbb{Q}$, and except for the case $K = \mathbb{Q}$ and $p = 2$, we have (by (i)) $M_1 = \overline{K}^{\rm ab}{}_{(p)}$ and the extension $\overline{K}^{\rm ab}{}_{(p)}/M_0$ is unramified (its Galois group is isomorphic to $\mathcal{E}^{\rm ord} \simeq \mu_p(K) \times \mathbb{Z}_p^{r_1+r_2-1}$). In the opposite case where $|Pl_p| = 1$, we have $M_1 = M_0$ and $\mathrm{Gal}(\overline{K}^{\rm ab}{}_{(p)}/M_1) \simeq \mathcal{E}^{\rm ord}$; the case $\mathbb{Q}$ for $p = 2$ falls under this second case for which $M_1 = M_0$ (indeed, $\overline{\mathbb{Q}}^{\rm ab}{}_{(2)}/M_0$ is ramified at 2 (see Fig. 4.1 in 4.1.11).

(iv) Finally, we see that $\mathrm{Gal}(\overline{K}^{\rm ab}{}_{(p)}/M_1)$ is a free $\mathbb{Z}_p$-module if and only if $|Pl_p| > 1$ (we have seen that the $\mathcal{E}^{\rm ord}_{p\backslash v}$ have no torsion in this case). □

It is now clear that the delicate case is that of nontotally real fields, which is the object of the next subsection.

f) The Strong p-Adic Conjecture — Other p-Adic Aspects

We are going to formulate a new conjecture about the $\mathcal{E}^{\rm ord}_{p\backslash v}$ which is a strong form of the p-adic conjecture on units; naturally, it would also follow from the conjecture on algebraic independence of logarithms of algebraic numbers (a particular case of the p-adic Schanuel conjecture already mentioned in 3.1.8). To introduce it, we begin by a number of heuristic considerations which will (as in [Em], [Ja3], [Lau], [Roy1; Roy2]) be set in a context of group representations similar to that of Section 3.

4.10 INTRODUCTION TO THE STRONG p-ADIC CONJECTURE. Let K be a number field and let N be the Galois closure of K over $\mathbb{Q}$; we set $\Gamma := \mathrm{Gal}(N/\mathbb{Q})$ and $H := \mathrm{Gal}(N/K)$. We assume that N satisfies the Leopoldt conjecture for p. We are interested in the study of $\mathcal{E}^{\rm ord}_{K,p\backslash v_0} \simeq \overline{i}_p(\mathcal{E}^{\rm ord}_K) \cap U^1_{v_0}$, for $v_0|p$ fixed.

To simplify, we transform the problem into an additive one by considering the following easy result.

4.10.1 Lemma. *In* $\bigoplus\limits_{v|p} K_v$, *for any place* $v_0|p$ *we have:*

$$\mathbb{Q}_p \mathrm{log}_{K,p}(\mathcal{E}^{\rm ord}_{K,p\backslash v_0}) = (\mathbb{Q}_p \mathrm{log}_{K,p}(E^{\rm ord}_K)) \cap K_{v_0}.$$

Assuming the Leopoldt conjecture, the $\mathbb{Z}_p$-rank of $\mathcal{E}^{\mathrm{ord}}_{K,p\backslash v_0}$ is therefore equal to the $\mathbb{Q}_p$-dimension of the space on the right hand side (see 3.6.2, (v)). $\square$

Let $w_0 \in Pl_{N,v_0}$ be fixed and let D_{w_0} be the decomposition group of w_0 in $N/\mathbb{Q}$.

In $\bigoplus_{w|p} N_w$ consider $\mathbb{Q}_p\log_{N,p}(E^{\mathrm{ord}}_N)$, which we will denote $\mathcal{V}_N$.[19] The idea is that, for p-adic transcendence reasons, this subspace is as arbitrary as possible, even though it defines a perfectly precise representation of Γ.

To characterize such a $\mathcal{V}_N$, we cannot simultaneously define a complex conjugation c and a decomposition group D_{w_0} up to conjugation in Γ since the fact that $N^{D_{w_0}}$ is real or not has a great importance; we proceed instead in the following way. We take an embedding of N in $\mathbb{C}$, and c is then the restriction to N of complex conjugation. We fix a decomposition group of p, and call it D_{w_0}; the set of places $w|p$ is in one-to-one correspondence with Γ/D_{w_0}, and for any $s \in G$, the conjugate $sD_{w_0}s^{-1}$ is D_w for the place $w := sw_0$.

Let φ be an isomorphism of Γ-modules which identifies $\bigoplus_{w|p} N_w$ with $\mathbb{Q}_p[\Gamma]$, so that the components N_w have image equal to the subspaces $s_w\mathbb{Q}_p[D_{w_0}]$, where the s_w (characterized by the relation $s_w w_0 = w$) are representatives of the elements of Γ/D_{w_0}. This is always possible because N_{w_0}, seen as a D_{w_0}-representation, induces the Γ-regular representation $\bigoplus_{w|p} N_w = \bigoplus_{w|p} s_w N_{w_0}$, where s_w also denotes by abuse of notation the corresponding isomorphism:

$$s_w : N_{w_0} \longrightarrow N_w$$

coming from $i_w \circ s_w \circ i_{w_0}^{-1}$ (Theorem II.2.3.2). More precisely, if θ_{w_0} defines a D_{w_0}-normal $\mathbb{Q}_p$-basis of N_{w_0}, then $\theta := (\theta_{w_0}, 0, \ldots, 0)$ defines a Γ-normal $\mathbb{Q}_p$-basis of $\bigoplus_{w|p} N_w$ (i.e., we have $\bigoplus_{w|p} N_w = \mathbb{Q}_p[\Gamma]\,.\,\theta$); we define φ by $\mathbb{Q}_p$-linearity and by $\varphi(\sigma\theta) := \sigma$ for all $\sigma \in \Gamma$. The elements of N_w are characterized by the normal basis defined by $\theta_w := s_w\theta_{w_0}$ (with $\mathrm{Gal}(N_w/K_v) = s_w D_{w_0} s_w^{-1}$) and, to verify that $\varphi(N_w) = s_w\mathbb{Q}_p[D_{w_0}]$, it is sufficient to find $\varphi(\tau(0, \ldots, \theta_w, \ldots, 0))$, for $\tau = s_w t s_w^{-1}$, $t \in D_{w_0}$; but we have:

$$\tau(0, \ldots, \theta_w, \ldots, 0) = s_w t s_w^{-1} s_w \theta = s_w t \theta,$$

so that:

$$\varphi(\tau(0, \ldots, \theta_w, \ldots, 0)) = s_w t \in s_w\mathbb{Q}_p[D_{w_0}].$$

Conversely, any identification φ satisfying the above conditions comes from a θ of the above form. It is thus equivalent to give us an isomorphism φ of Γ-modules such that $\varphi(N_{w_0}) = \mathbb{Q}_p[D_{w_0}]$ (taking $s_{w_0} = 1$). Note that the

[19] Recall that $\mathcal{V}_K := \mathbb{Q}_p\log_{K,p}(E^{\mathrm{ord}}_K) = \mathcal{V}_N^H = e_H\mathcal{V}_N$ with $e_H := \frac{1}{|H|}\sum_{t\in H} t$.

present identification is not the one which is generally used for this type of problem (as in [Em] and [Lau]) since theirs assume that we have performed an extension of scalars. The sequel should explain our choice.

Under the map φ, the image of $\mathbb{Q}_p\log_{N,p}(E_N^{\mathrm{ord}})$ is a sub-Γ-module of $\mathbb{Q}_p[\Gamma]$ which we still denote $\mathcal{V}_N$ by abuse of notation.

4.10.2 Remark. As we have seen in Section 3, the Dirichlet–Herbrand Theorem I.3.7.2 on units shows that $\mathcal{V}_N$ defines, after extension of scalars to $\mathbb{C}_p$, and assuming the Leopoldt conjecture for p, a representation $\mathcal{V}_N \otimes \mathbb{C}_p$ which is isomorphic to:

$$\bigoplus_{\psi \in \Psi_N} n_\psi \, V_\psi,$$

where V_ψ is the absolutely irreducible representation with character ψ, and:

$$n_1 := 0, \quad n_\psi := \tfrac{1}{2}\,(\psi(1) + \psi(c)) \ \text{ for } \psi \neq 1,$$

c being the generator of the decomposition group of a place at infinity in $N/\mathbb{Q}$ (see 3.3); here c is complex conjugation since N is embedded in $\mathbb{C}$.

The respective canonical decompositions of $\mathbb{C}_p[\Gamma]$ and of $\mathcal{V}_N \otimes \mathbb{C}_p$ can be written: [20]

$$\mathbb{C}_p[\Gamma] = \bigoplus_{\psi \in \Psi_N} \mathbb{C}_p[\Gamma]e_\psi, \quad \mathcal{V}_N \otimes \mathbb{C}_p = \bigoplus_{\psi \in \Psi_N} \mathcal{V}_\psi, \quad \mathcal{V}_\psi := e_\psi \mathcal{V}_N,$$

where $e_\psi := \frac{\psi(1)}{|\Gamma|} \sum\limits_{t \in \Gamma} \psi(t^{-1})\, t$ is the central idempotent corresponding to ψ, and give for all ψ:

$$\mathcal{V}_\psi \subseteq \mathbb{C}_p[\Gamma]e_\psi \simeq \psi(1) V_\psi,$$

without necessarily having uniqueness for $\mathcal{V}_\psi$ since $\mathcal{V}_\psi = \bigoplus\limits_{i=1}^{n_\psi} \mathcal{V}_{\psi,i}$ with $\mathcal{V}_{\psi,i} \simeq V_\psi$. We do have uniqueness, however, for the ψ such that $n_\psi = 0$ or $\psi(1)$ (i.e., $\psi(c) = \pm\psi(1)$). Let us check that this uniqueness holds for all ψ if and only if $\langle\, c \,\rangle$ is a normal subgroup of Γ: if $\langle\, c \,\rangle$ is normal, c is central, the components $\frac{1+c}{2} V_\psi$ and $\frac{1-c}{2} V_\psi$ of V_ψ are Γ-modules, hence equal to 0 and to V_ψ (or the opposite) because of the irreducibility of V_ψ; hence $\psi(c) = \pm\psi(1)$ for all ψ; conversely, $\psi(c) = s\psi(1)$ with $s = \pm 1$ (depending on ψ) implies that c acts on V_ψ by means of the eigenvalue s, from which it follows that c is central when this is true for all ψ. □

In the general case, this uniqueness defect can be bypassed in a way, which correctly interprets the above transcendence heuristics, and which will enable us to treat every case, including the relatively trivial case above. Let us describe it in several steps in view of Definition 4.11.1.

[20] See [Se4, Ch. I, § 2.6].

We see that the representation $\mathbb{Q}_p \oplus \mathcal{V}_N$, where $\mathbb{Q}_p$ is identified with its diagonal embedding in $\bigoplus_{w|p} N_w$, is a permutation representation of Γ modulo $\langle c \rangle$ and a subrepresentation of $\mathbb{Q}_p[\Gamma]$.

4.10.3 Lemma. *Any permutation representation of Γ modulo $\langle c \rangle$, which is a subrepresentation of $\mathbb{Q}_p[\Gamma]$, is of the form $\mathbb{Q}_p[\Gamma](1+c)\,e$, for a suitable $e \in \mathbb{Q}_p[\Gamma]$ (i.e., such that the dimension of $\mathbb{Q}_p[\Gamma](1+c)\,e$ is equal to $|\Gamma/\langle c \rangle|$).*

Proof. Indeed, such a representation $\mathcal{V}$ has a $\mathbb{Q}_p$-basis $\{e^t,\ t \in \Gamma_1\}$, where Γ_1 is a complete system of representatives of the elements of $\Gamma/\langle c \rangle$, for which $e^t = t\,e^1 = tce^1$ for all $t \in \Gamma_1$; we thus have $e^1 =: (1+c)\,e$, $e \in \mathbb{Q}_p[\Gamma]$, and $\mathcal{V} = \mathbb{Q}_p[\Gamma](1+c)\,e$. □

4.10.4 Remark. If $c = 1$, the dimension of $\mathcal{V}$ is equal to $|\Gamma|$, hence $\mathcal{V} = \mathbb{Q}_p[\Gamma]$ is the regular representation corresponding to the totally real case that we have settled in 4.9.1, (ii) (e is left-invertible), and the interesting case is of course the case $c \neq 1$. □

The arithmetical interpretation of the above, thanks to the Dirichlet–Herbrand theorem, is the following. Fix a Γ-normal $\mathbb{Q}_p$-basis:

$$\theta := (\theta_{w_0}, 0, \dots, 0),$$

of $\bigoplus_{w|p} N_w$ which is *algebraic over* $\mathbb{Q}$ (we will see in Lemma 4.10.6 that this is always possible). Let $\alpha \in N^\times$ be of the form $(1+p)\,\varepsilon$, where $\varepsilon \in E_N^{\mathrm{ord}}$ is a Minkowski unit[21]; it is a generator in $(\langle 1+p \rangle \otimes \mathbb{Q}_p) \oplus (E_N^{\mathrm{ord}} \otimes \mathbb{Q}_p)$ as a monogeneous $\mathbb{Q}_p[\Gamma]$-module, and there exists $e \in \mathbb{Q}_p[\Gamma]$ such that:

$$\log_{N,p}(\alpha) = (1+c)\,e\,.\,\theta,$$

which yields:

$$\mathbb{Q}_p \oplus \mathcal{V}_N = \mathbb{Q}_p[\Gamma](1+c)\,e\,.\,\theta,$$

whose image is equal to $\mathbb{Q}_p[\Gamma](1+c)\,e$ under the above identification φ.

Note that, if θ' is another basis of the same type, e is right multiplied by an invertible element of $\overline{\mathbb{Q}}[\Gamma]$ (use Lemma 4.10.6 below). Therefore it is e which should be as arbitrary as possible. But its coefficients[22] are linear combinations with coefficients in $\mathbb{Q}(\theta_{w_0})$ (by the same Lemma) of logarithms of algebraic numbers, which are conjecturally algebraically independent. Therefore these coefficients are themselves conjecturally algebraically independent, and this is independent of the choice of the basis θ.

[21] i.e., such that ε and its conjugates generate a subgroup of finite index of E_N^{ord} (see I.3.7.3); we may always assume that ε is a real unit.

[22] Of which there are $\frac{1}{2}|\Gamma|$ only if $c \neq 1$, and which are in $\mathbb{Q}_p$.

Before proving the Lemma on the nature of the coefficients of e, let us give two rather naïve numerical examples (real case), but shown only to illustrate the above.

4.10.5 Examples. (i) If $N = \mathbb{Q}(\sqrt{2})$, $p = 2$, $\varepsilon = 1 + \sqrt{2}$, $\alpha = 3(1 + \sqrt{2})$, we can choose $\theta = \frac{1+\sqrt{2}}{2}$ in $\mathbb{Q}_2(\sqrt{2})$ and we obtain (here with $\Gamma =: \{1, \sigma\}$, $c = 1$):

$$\log_2(\alpha) =: (e_0 + e_1\sigma) \,.\, \theta,$$

where:

$$e_0 = \log(3) + \frac{\sqrt{2}}{2}\log(\varepsilon), \quad e_1 = \log(3) - \frac{\sqrt{2}}{2}\log(\varepsilon) \ ;$$

it is easily checked that $\log(\varepsilon) = u\sqrt{2}$, $u \in \mathbb{Z}_2^\times$, which yields:

$$e_0 = \log(3) + u, \quad e_1 = \log(3) - u \in \mathbb{Q}_2.$$

Finally $\log(3)$ and $\log(\varepsilon)$ (hence e_0 and e_1) are conjecturally algebraically independent.

(ii) If $N = \mathbb{Q}(\sqrt{17})$, $p = 2$, $\varepsilon = 4 + \sqrt{17}$, $\alpha = 3(4 + \sqrt{17})$, $\theta = (1, 0)$ in $\mathbb{Q}_2 \oplus \mathbb{Q}_2$, we have:

$$\log_2(\alpha) = (\log(\alpha), \log(\sigma(\alpha))) =: (e_0 + e_1\sigma) \,.\, \theta,$$

where:

$$e_0 = \log(3) + \log(\varepsilon), \quad e_1 = \log(3) - \log(\varepsilon) \ ;$$

but if we fix $\sqrt{17} = 1 + 2^3 u_0$, $u_0 \in \mathbb{Z}_2^\times$, we obtain $\log(\varepsilon) = 4u$, $u \equiv 1 \bmod (4)$, and:

$$e_0 = \log(3) + 4u, \quad e_1 = \log(3) - 4u \in \mathbb{Q}_2.$$

(iii) We take once again the example of $\mathbb{Q}(\sqrt{2})$ and consider the element:

$$e = \log(3) + u + (\log(3) - u)\, \sigma \ ;$$

it is clear that e is invertible in $\mathbb{Q}_2[\Gamma]$ and that it defines a change of normal basis which is not anymore algebraic over $\mathbb{Q}$ (it is $\theta' := \log_2(\alpha)$). We thus see that the algebraic independence of the coefficients of e is intrinsic only if we use normal bases which are algebraic over $\mathbb{Q}$. □

Consider the elements $\theta := (\theta_{w_0}, 0, \ldots, 0)$ defining a Γ-normal $\mathbb{Q}_p$-basis of $\bigoplus_{w|p} N_w$.

4.10.6 Lemma. *We can always choose for θ_{w_0} a number which is algebraic over $\mathbb{Q}$. In the formula:*

$$\log_{N,p}(\alpha) =: (1 + c) \sum_{s \in \Gamma} e_s \, s\theta, \quad e_s \in \mathbb{Q}_p,$$

we may assume that the e_s *are linear combinations with coefficients in* $\mathbb{Q}(\theta_{w_0})$ *of the* $\log(\tau\alpha)$, $\tau \in \Gamma$.

Proof. It is sufficient to choose for θ_{w_0} an element of N which defines a D_{w_0}-normal $N^{D_{w_0}}$-basis for the extension $N/N^{D_{w_0}}$; θ_{w_0} has degree a divisor of $[N : \mathbb{Q}]$ over $\mathbb{Q}$, and has degree $[N_{w_0} : \mathbb{Q}_p]$ over $\mathbb{Q}_p$; thanks to the Krasner lemma, we may assume that its degree over $\mathbb{Q}$ is still equal to the local degree. The $s_w\theta_{w_0} =: \theta_w$ are then also algebraic over $\mathbb{Q}$.

If $c \neq 1$, the quantity $\sum\limits_{s\in\Gamma} e_s s$ is defined modulo $(1-c)\mathbb{Q}_p[\Gamma]$, so the e_s are not unique. Let us write the equality defining $\log_{N,p}(\alpha)$ in the form:

$$\log_{N,p}(\alpha) =: \sum_{\sigma\in\Gamma} a_\sigma\, \sigma\theta, \ \ a_\sigma \in \mathbb{Q}_p,$$

with $a_\sigma = a_{c\sigma}$ for all σ. Since $\log_{N,p}$ is a Γ-module homomorphism, under conjugation by $\tau \in \Gamma$, we obtain $\tau \log_{N,p}(\alpha) = \log_{N,p}(\tau\alpha) = \sum\limits_{\sigma\in\Gamma} a_\sigma\, \tau\sigma\theta$, which we can write in matrix terms:

$$\Big(\tau\sigma\theta\Big)_{\tau,\sigma}\Big(a_\sigma\Big)_\sigma = \Big(\log_{N,p}(\tau\alpha)\Big)_\tau .$$

But the matrix $\Big(\tau\sigma\theta\Big)_{\tau,\sigma}$ is invertible: indeed, if we send $(x_\sigma)_\sigma$, $x_\sigma \in \mathbb{Q}_p$, to the element $(\tau y)_\tau$ which only depends on $y := \sum\limits_{\sigma\in\Gamma} x_\sigma\, \sigma\theta$, it defines a surjective map of $\mathbb{Q}_p^{[N\,:\,\mathbb{Q}]}$ onto a $\mathbb{Q}_p$-vector space isomorphic to $\bigoplus\limits_{w|p} N_w$ since θ defines a normal basis, proving our claim. Set:

$$\Big(\tau\sigma\theta\Big)_{\tau,\sigma}^{-1} =: \Big(\omega_{\sigma,\tau}\Big)_{\sigma,\tau}, \ \ \omega_{\sigma,\tau} \in \bigoplus_{w|p} N_w \ ;$$

the usual computation of the inverse shows that the components $\omega_{\sigma,\tau}^w$ of $\omega_{\sigma,\tau}$ on the N_w are in $\mathbb{Q}(\theta_w)$. Since $a_\sigma \in \mathbb{Q}_p$, we obtain by projection on the summand N_{w_0}:

$$a_\sigma = \sum_{\tau\in\Gamma} \omega_{\sigma,\tau}^{w_0} \log(\tau\alpha),$$

if we identify the embedding i_{w_0} with the identity map. The result follows. □

In the above numerical example, the inverse matrices are respectively equal to:

$$\begin{pmatrix} \frac{2+\sqrt{2}}{4} & \frac{2-\sqrt{2}}{4} \\ \frac{2-\sqrt{2}}{4} & \frac{2+\sqrt{2}}{4} \end{pmatrix} \ \text{and} \ \begin{pmatrix} (1,0) & (0,1) \\ (0,1) & (1,0) \end{pmatrix}.$$

This being said, we are going to see that it is not necessary to know e *numerically*.

4.11 THE FORMALISM OF THE STRONG p-ADIC CONJECTURE. We leave the arithmetical aspects of the subject and we introduce (in the spirit of the transcendence methods) the following setting.

4.11.1 Definition. From the fact that $(1+c)\,e = (1+c)\sum_{s\in\Gamma_{\rm r}} e_s s,\ e_s \in \mathbb{Q}_p$, where $\Gamma_{\rm r}$ is a complete system of representatives of the elements of $\langle\, c\,\rangle\backslash\Gamma$, we set:

$$e' := \sum_{s\in\Gamma_{\rm r}} e_s s,$$

considering now the e_s as *indeterminates*, and we define $\mathcal{V}'_N$ by the identity:

$$\mathbb{Q}'_p \bigoplus \mathcal{V}'_N := \mathbb{Q}'_p[\Gamma](1+c)\,e', \quad \text{where} \quad \mathbb{Q}'_p := \mathbb{Q}_p((e_s)_{s\in\Gamma_{\rm r}}). \qquad \square$$

This *automatically* defines a permutation representation of Γ modulo $\langle\, c\,\rangle$. In particular, $\mathcal{V}'_N$ has dimension $r_1(N)+r_2(N)-1$ (equal to $[N:\mathbb{Q}]-1$ or to $\frac{1}{2}[N:\mathbb{Q}]-1$ according to whether or not $c=1$) and is supposed to give an interpretation of $\mathbb{Q}_p\log_{N,p}(\mathcal{E}_N^{\rm ord})$. In this context, the Leopoldt conjecture is a tautology, but the point is that we will thus be able to add some new properties.

To interpret (in N) the subspace:

$$\mathbb{Q}_p\log_{N,p}(\mathcal{E}^{\rm ord}_{N,p\backslash w_0}) = (\mathbb{Q}_p\log_{N,p}(E_N^{\rm ord}))\cap N_{w_0},$$

corresponding to the subspace $\mathcal{V}'_N \cap \mathbb{Q}'_p[D_{w_0}]$, we solve the linear system coming from the relation:

$$x'\,(1+c)\,e' \in \mathbb{Q}'_p[D_{w_0}],$$

keeping only the solutions which come from $\mathcal{V}'_N$, in other words the solutions $x' =: \sum_{t\in\Gamma_1} x_t t$ such that $\sum_{t\in\Gamma_1} x_t = 0$. This system can be written:

$$\mathrm{E}X = Y,$$

where E is a $(|\Gamma|,\ |\Gamma/\langle\, c\,\rangle|)$ matrix whose coefficients are suitable e_s, where $X := (x_t)_{t\in\Gamma_1}$, $Y := (y_u)_{u\in\Gamma}$, with $y_u = 0$ for all $u \notin D_{w_0}$. Restricting to the rows of E for which $y_u = 0$, we obtain a linear system of the form:

$$\mathrm{E}_0 X = 0,$$

for a submatrix E_0 of E whose rank over $\mathbb{Q}'_p$ is known. The space of solutions X (of zero trace) allows us, by computing the $\mathrm{E}X$, to compute the dimension of $\mathcal{V}'_N \cap \mathbb{Q}'_p[D_{w_0}]$ as a $\mathbb{Q}'_p$-vector space.

Note. This last computation is reduced to that of the rank of an explicit matrix whose coefficients are in $\mathbb{Q}' := \mathbb{Q}((e_s)_{s\in\Gamma_{\rm r}})$ (and not in $\mathbb{Q}'_p$!). This rank is thus canonical because of the algebraic independence of the e_s, and in some sense the above is independent of p, so that the use of the field $\mathbb{Q}'_p$ (instead of $\mathbb{Q}'$) is only a convenient way since the specialization of the e_s in $\mathbb{Q}_p$ gives the true space $\mathbb{Q}_p \oplus \mathcal{V}_N$.

4.11.2 Remark. When $\langle\, c\,\rangle$ is normal, c is central, $1+c$ and e' commute, and we recover the uniqueness of $\mathbb{Q}'_p \oplus \mathcal{V}'_N$ since we have:

$$\mathbb{Q}'_p[\Gamma](1+c)\,e' = \mathbb{Q}'_p[\Gamma]e'(1+c) = \mathbb{Q}'_p[\Gamma](1+c),$$

by equality of the dimensions. Thus $\mathbb{Q}'_p[\Gamma](1+c) \cap \mathbb{Q}'_p[D_{w_0}] = \mathbb{Q}'_p[D_{w_0}](1+c)$ or 0 according to whether or not $c \in D_{w_0}$, and in the case where $\langle\, c\,\rangle$ is normal, we obtain as expected:

$$\dim_{\mathbb{Q}'_p}(\mathcal{V}'_N \cap \mathbb{Q}'_p[D_{w_0}]) = |D_{w_0}/\langle\, c\,\rangle| - 1$$

if $c \in D_{w_0}$ (i.e., for a real decomposition field), and otherwise:

$$\dim_{\mathbb{Q}'_p}(\mathcal{V}'_N \cap \mathbb{Q}'_p[D_{w_0}]) = 0.$$

□

4.11.3 Example. Let us take (see 3.3.7):

$$\Gamma = D_6 =: \{1,\ \sigma,\ \sigma^2,\ \tau,\ \sigma\tau,\ \sigma^2\tau\}\ ;$$

setting $c := \tau$, $\ e' := e_0 + e_1\sigma + e_2\sigma^2$, $\ x' := x_0 + x_1\sigma + x_2\sigma^2$, we find:

$$\mathrm{E} = \begin{pmatrix} e_0 & e_2 & e_1 \\ e_1 & e_0 & e_2 \\ e_2 & e_1 & e_0 \\ e_0 & e_1 & e_2 \\ e_2 & e_0 & e_1 \\ e_1 & e_2 & e_0 \end{pmatrix}\ ;$$

for $D_{w_0} = \{1,\ \sigma\tau\}$, we thus obtain:

$$\begin{pmatrix} e_1 & e_0 & e_2 \\ e_2 & e_1 & e_0 \\ e_0 & e_1 & e_2 \\ e_1 & e_2 & e_0 \end{pmatrix} \begin{pmatrix} x_0 \\ x_1 \\ x_2 \end{pmatrix} = 0,$$

which yields $x' = 0$.

In fact, we check that only $D_{w_0} = \Gamma$ gives $\mathcal{V}'_N \cap \mathbb{Q}'_p[D_{w_0}] \neq 0$. □

4.11.4 Remark. If (in N) we consider $\mathcal{E}^{\mathrm{ord}}_{N,p\backslash\Sigma}$, for an arbitrary subset Σ of $Pl_{N,p}$, we are led to the determination of:

$$\mathbb{Q}_p\mathrm{log}_{N,p}(\mathcal{E}^{\mathrm{ord}}_{N,p\backslash\Sigma}) = \mathbb{Q}_p\mathrm{log}_{N,p}(E^{\mathrm{ord}}_N) \bigcap \bigoplus_{w\in\Sigma} N_w,$$

hence, in $\mathbb{Q}'_p[\Gamma]$, to the determination of $\mathcal{V}'_N \bigcap \Big(\bigoplus_{w\in\Sigma} s_w \mathbb{Q}'_p[D_{w_0}]\Big)$, in other words, to solve $\mathrm{E}X = Y$, where Y is taken such that:

$$y_u = 0 \text{ for all } u \notin \bigcup_{w\in\Sigma} s_w D_{w_0}. \qquad \square$$

4.11.5 Remark (Galois descent). We have thus created a generic $\mathcal{V}'_N$ for the Galois closure N (recall that $\mathcal{V}'_N$ is $\mathbb{Q}'_p[\Gamma](1+c)\,e'$ minus the unit representation). To come back to K (fixed under H), it is enough to consider:

$$\mathcal{V}'_{K,p\backslash v_0} := \mathcal{V}'_N \bigcap \Big(\bigoplus_{h\in H/H\cap D_{w_0}} h\,\mathbb{Q}'_p[D_{w_0}]\Big)^H$$

which represents $\mathbb{Q}_p\mathrm{log}_{K,p}(E_K^{\mathrm{ord}}) \cap K_{v_0}$ in that we have as usual:

$$\mathbb{Q}_p\mathrm{log}_{K,p}(E_K^{\mathrm{ord}}) = (\mathbb{Q}_p\mathrm{log}_{N,p}(E_N^{\mathrm{ord}}))^H,$$

$$K_{v_0} = \Big(\bigoplus_{w|v_0} N_w\Big)^H = \Big(\bigoplus_{h\in H/H\cap D_{w_0}} N_{hw_0}\Big)^H,$$

the diagonal embeddings being understood. But we still have:

$$\Big(\bigoplus_{h\in H/H\cap D_{w_0}} h\mathbb{Q}'_p[D_{w_0}]\Big)^H = e_H\mathbb{Q}'_p[D_{w_0}],$$

where $e_H := \frac{1}{|H|}\sum_{t\in H} t$; hence it is sufficient to note that $e_H\mathbb{Q}'_p[D_{w_0}]$ has a basis of the form $e_H\,\sigma$, $\sigma \in D_{w_0,\mathrm{r}}$, where $D_{w_0,\mathrm{r}}$ is a complete system of representatives of the elements of $(H\cap D_{w_0})\backslash D_{w_0}$. Therefore we have:

$$\mathcal{V}'_{K,p\backslash v_0} = \mathcal{V}'_N \bigcap \Big(\bigoplus_{\sigma\in (H\cap D_{w_0})\backslash D_{w_0}} \mathbb{Q}'_p\, e_H\,\sigma\Big),$$

which can easily be determined as above, using a similar linear system, but here with an element x' in $\mathcal{V}'^H_N = e_H\,\mathbb{Q}'_p[\Gamma](1+c)\,e'$ which may be written:

$$x' = \sum_{t\in H\backslash\Gamma/\langle c\rangle} x_t\, e_H\, t(1+c)\,e'. \qquad \square$$

4.11.6 Definition. For any place $v|p$, denote by $\mathrm{d}_v(K)$ the $\mathbb{Q}'_p$-dimension of $\mathcal{V}'_{K,p\backslash v} := \mathcal{V}'_N \bigcap \Big(\bigoplus_{\sigma\in(H\cap D_w)\backslash D_w} \mathbb{Q}'_p\, e_H\,\sigma\Big)$, where N is the Galois closure of K over $\mathbb{Q}$, $H := \mathrm{Gal}(N/K)$, and w is a fixed place of N above v. $\square$

4.12 Conjecture (strong p-adic). *Let p be a prime number, $\overline{K}^{\mathrm{ab}}(p)$ the maximal abelian pro-p-extension, and $H_p^{\mathrm{ord}}(p)$ the maximal p-ramified noncomplexified abelian pro-p-extension of K. Let v be a place of K above p. Then the inertia group of v in $\overline{K}^{\mathrm{ab}}(p)/H_p^{\mathrm{ord}}(p)$ is canonically isomorphic to:*

$$\mathcal{E}^{\text{ord}}_{p\backslash v} := \{\varepsilon \in E^{\text{ord}}_K \otimes \mathbb{Z}_p,\ \bar{\imath}_{v'}(\varepsilon) = 1\ \ \forall v'|p,\ v' \neq v\}$$

(see 4.9). Its $\mathbb{Z}_p$-rank depends only on $\Gamma :=$ Gal$(N/\mathbb{Q})$ and on the decomposition groups of the places p and ∞ in $N/\mathbb{Q}$, and is equal to $\mathrm{d}_v(K)$ (see 4.11.6). □

4.12.1 Remark. It is clear that this conjecture can be expressed in a wider manner, and in fact the real strong p-adic conjecture "is" the fact that we can replace $e \in \mathbb{Q}_p[\Gamma]$ by:

$$e' := \sum_{s \in \Gamma_{\mathrm{r}}} e_s\, s,$$

considering the e_s as indeterminates, and deducing any result from the algebraic study of:

$$\mathbb{Q}'_p \oplus \mathcal{V}'_N := \mathbb{Q}'_p[\Gamma](1+c)\, e',$$

where $\mathbb{Q}'_p := \mathbb{Q}_p((e_s)_{s \in \Gamma_{\mathrm{r}}})$. As already said, the Leopoldt conjecture is then true by definition and, for $\Sigma \subseteq Pl_{N,p}$, the dimensions:

$$\mathrm{d}_\Sigma(N) := \dim_{\mathbb{Q}_p}(\mathbb{Q}_p \log_{N,p}(\mathcal{E}^{\text{ord}}_{N,p\backslash\Sigma})),$$

are canonical, the matrix E depending finally only on the group Γ and on the element c. The Galois descent relative to the subfield K of N is done by means of the idempotent e_H. □

4.12.2 Exercise. We again consider the example of the dihedral group of order 6 with $D_{w_0} = \{1, \sigma\tau\}$ and $c = \tau$; the prime number p is thus split in N in three places $w_0,\ \sigma w_0,\ \sigma^2 w_0$. Set:

$$\mathbb{Q}'_p[\Gamma](1+\tau)(e_0 + e_1\sigma + e_2\sigma^2) =: \mathbb{Q}'_p \oplus \mathcal{V}(e_0,\ e_1,\ e_2).$$

What does the strong conjecture give for the dimension of:

$$\mathcal{V}(e_0,\ e_1,\ e_2) \cap (\mathbb{Q}'_p[D_{w_0}] \oplus \sigma\mathbb{Q}'_p[D_{w_0}])\,?$$

Does there exist specializations $a_0,\ a_1,\ a_2 \in \mathbb{Q}_p$ of $e_0,\ e_1,\ e_2$ such that $\mathbb{Q}_p \oplus \mathcal{V}(a_0,\ a_1,\ a_2)$ is a permutation representation of Γ modulo $\langle\, \tau\, \rangle$ for which:

$$\mathcal{V}(a_0,\ a_1,\ a_2) \cap (\mathbb{Q}_p[D_{w_0}] \oplus \sigma\mathbb{Q}_p[D_{w_0}])$$

has nonzero dimension ? Interpret the result.

Answer. We have:

$$\mathbb{Q}'_p[D_{w_0}] \oplus \sigma\mathbb{Q}'_p[D_{w_0}] = \mathbb{Q}'_p \oplus \mathbb{Q}'_p\sigma\tau \oplus \mathbb{Q}'_p\sigma \oplus \mathbb{Q}'_p\sigma^2\tau,$$

and the linear system corresponding to the computation of the intersection $\mathcal{V}(e_0,\ e_1,\ e_2) \cap (\mathbb{Q}'_p[D_{w_0}] \oplus \sigma\mathbb{Q}'_p[D_{w_0}])$ is:

$$\begin{pmatrix} 1 & 1 & 1 \\ e_2 & e_1 & e_0 \\ e_0 & e_1 & e_2 \end{pmatrix} \begin{pmatrix} x_0 \\ x_1 \\ x_2 \end{pmatrix} = 0$$

(we have added a row of 1 so as to have only solutions with zero trace). We have in $\mathbb{Q}'_p$:

$$\begin{vmatrix} 1 & 1 & 1 \\ e_2 & e_1 & e_0 \\ e_0 & e_1 & e_2 \end{vmatrix} = (e_0 - e_2)(e_0 + e_2 - 2e_1) \neq 0,$$

showing that this space of solutions is zero.

Let us specialize by setting $e_0 =: a, \quad e_1 =: b, \quad e_2 =: a, \quad a,\ b \in \mathbb{Q}_p$. In these conditions, the rank of the matrix is strictly less than 3, hence the space of solutions is of nonzero dimension. By 3.3, 3.3.7, (ii), the permutation representation of Γ modulo $\langle\, \tau\,\rangle$ is isomorphic to $V_{\psi_0} \oplus V_{\psi_2}$ and we must check that $\mathbb{Q}_p[\Gamma](1+\tau)(a + b\sigma + a\sigma^2) \simeq V_{\psi_0} \oplus V_{\psi_2}$. But we have:

$$\begin{aligned} (1+\tau)(a + b\sigma + a\sigma^2)\, e_{\psi_0} &= 2(2a+b)e_{\psi_0}, \\ (1+\tau)(a + b\sigma + a\sigma^2)\, e_{\psi_1} &= 0, \\ (1+\tau)(a + b\sigma + a\sigma^2)\, e_{\psi_2} &= (b-a)(1+\tau)\sigma e_{\psi_2}; \end{aligned}$$

hence, we must add the conditions $b \neq -2a$ and $b \neq a$ so as to have such a representation, in which case the space of solutions is equal to:

$$\{(\lambda,\, 0,\, -\lambda), \quad \lambda \in \mathbb{Q}_p\},$$

of dimension 1.

Analysis: assuming the strong p-adic conjecture, the rank of the matrix:

$$\begin{pmatrix} 1 & 1 & 1 \\ e_2 & e_1 & e_0 \\ e_0 & e_1 & e_2 \end{pmatrix}$$

is equal to 3, and in a certain sense we have:

$$\mathbb{Q}_p \log_{N,p}(E_N^{\mathrm{ord}}) \cap (N_{w_0} \oplus N_{\sigma w_0}) = 0$$

in every case. If this conjecture was false, we could have cases satisfying the Leopoldt conjecture (i.e., $\dim_{\mathbb{Q}_p}(\mathbb{Q}_p\log_{N,p}(E_N^{\mathrm{ord}})) = \dim_{\mathbb{Q}_p}(\mathcal{V}(a,\, b,\, a)) = 2$), and such that:

$$\mathbb{Q}_p \log_{N,p}(E_N^{\mathrm{ord}}) \cap (N_{w_0} \oplus N_{\sigma w_0}) \simeq \mathbb{Q}_p\ ;$$

however, we would need to see whether one could find a field K, a normal basis θ and a Minkowski unit ε which give the above situation, which in any case remains completely speculative.

This example shows that the above conjecture is strictely stronger than the Leopoldt conjecture. □

4.12.3 Remarks. (i) If K is totally real, all of the above gives again the result 4.9.1, (ii) which is immediate directly. In other words, in the totally real case the strong p-adic conjecture is equivalent to the Leopoldt conjecture.

(ii) If $K/\mathbb{Q}$ is Galois nonreal, and if complex conjugation c is a central element of $\mathrm{Gal}(K/\mathbb{Q})$, the results only depend on the maximal real subfield K^{nc}, and we again find that:

$$\mathrm{rk}_{\mathbb{Z}_p}(\mathcal{E}^{\mathrm{ord}}_{p\backslash v}) = \frac{|D_v|}{2} - 1 \text{ or } 0$$

according as the decomposition field of v is real or complex.

(iii) If complex conjugation c is a central element of $\mathrm{Gal}(N/\mathbb{Q})$, then $c \notin H$ (otherwise N would not be the Galois closure of K) and K is a complex conjugation field with maximal real subfield $K \cap N^{\mathrm{nc}}$. We still have uniqueness of $\mathcal{V}'_K := (\mathcal{V}'_N)^H$; this is for example the case of $K = \mathbb{Q}(\sqrt{-3+\sqrt{2}})$ whose Galois closure is dihedral of degree 8 and contains the biquadratic field $\mathbb{Q}(\sqrt{2}, \sqrt{7})$. We are thus reduced to the initial problem for the totally real field $K \cap N^{\mathrm{nc}}$.

In particular, the strong p-adic conjecture is a theorem when K (real or complex) is abelian. □

Notations. Recall that $\widetilde{K}_p$ is the compositum of the $\mathbb{Z}_p$-extensions of K, that $\mathrm{r}_{p\backslash v}$ (resp. $\widetilde{\mathrm{r}}_{p\backslash v}$) is the $\mathbb{Z}_p$-rank of $\mathbb{Z}_p\mathrm{log}_{p\backslash v}(E^{\mathrm{ord}})$ (resp. the maximal number of independent $\mathbb{Z}_p$-extensions unramified at v) (see 1.6.2, 1.6.3). □

4.12.4 Corollary (application to p-adic rank computations). *(i) The $\mathbb{Z}_p$-rank of the inertia group of any place $v|p$, in $\widetilde{K}_p/K$, is conjecturally equal to $[K_v : \mathbb{Q}_p] - \mathrm{d}_v$.*

The decomposition group of the place $v|p$, in $\widetilde{K}_p/K$, then has rank either $[K_v : \mathbb{Q}_p] - \mathrm{d}_v$ or $[K_v : \mathbb{Q}_p] - \mathrm{d}_v + 1$, according as the Frobenius of v in $\widetilde{K}_{p\backslash v}/K$ has finite or infinite order (see 4.7.3).[23]

(ii) We have $[K_v : \mathbb{Q}_p] - \mathrm{d}_v \geq 1$ since in the cyclotomic $\mathbb{Z}_p$-extension of K, the inertia groups of the places above p are isomorphic to $\mathbb{Z}_p$.

(iii) We have the conjectural relations:

$$\mathrm{d}_v = r_1 + r_2 - 1 - \mathrm{r}_{p\backslash v} = [K_v : \mathbb{Q}_p] + \widetilde{\mathrm{r}}_{p\backslash v} - (r_2 + 1).$$

(iv) The $\mathbb{Z}_p$-rank of the Galois group of the unramified extension M_0/M_1 (see § 4.9) is conjecturally equal to $r_1 + r_2 - 1 - \sum_{v|p} \mathrm{d}_v$. □

[23] This can then be seen, at least numerically, from the logarithmic interpretation which will be given in detail in Section 5 (see 5.2, (ii)), but would also be relevant to the point of view given in the conjecture for a general result.

We have used the fact that the $\mathbb{Z}_p$-rank of the decomposition group of $v|p$ in $\overline{K}^{\rm ab}(p)/K$ (equal to the sum of the analogous $\mathbb{Z}_p$-ranks in $\overline{K}^{\rm ab}(p)/H_p^{\rm ord}(p)$ and $H_p^{\rm ord}(p)/K$) is equal to $[K_v : \mathbb{Q}_p] + 1$, because of II.1.8.2, (ii), and 4.5, which can be found again directly as we have done in 4.5.4, 4.5.5.

4.12.5 Exercise (another computation of $D_v(\overline{K}^{\rm ab}(p)/K)$). Let v be a finite place of K. By 1.1.6, (ii), (α) for $S = \emptyset$, we have:

$$D_v(K(\mathfrak{m})^{\rm res}/K) \simeq K_v^\times / i_v(E^{\{v\}\rm res}_{\frac{\mathfrak{m}}{\mathfrak{m}_v}}) U_v^{m_v},$$

for any modulus $\mathfrak{m} = \prod\limits_{v'} \mathfrak{p}_{v'}^{m_{v'}}$, where $\mathfrak{m}_v := \mathfrak{p}_v^{m_v}$.

(i) Show that $D_v(K(\mathfrak{m})^{\rm res}(p)/K) \simeq \varprojlim\limits_{n \geq 0} K_v^\times / K_v^{\times p^n} i_v(E^{\{v\}\rm res}_{\frac{\mathfrak{m}}{\mathfrak{m}_v}}) U_v^{m_v}$.

(ii) Writing that $D_v(\overline{K}^{\rm ab}(p)/K) \simeq \varprojlim\limits_{\mathfrak{m}} D_v(K(\mathfrak{m})^{\rm res}(p)/K)$ (see II.1.2.3.1), that $\mathfrak{m} =: \mathfrak{m}_p\, \mathfrak{m}_{\rm ta}$, show that $D_v(\overline{K}^{\rm ab}(p)/K) \simeq \varprojlim\limits_{n \geq 0} \varprojlim\limits_{\mathfrak{m}_p} K_v^\times / K_v^{\times p^n} U_v^{m_v}$.

(iii) Deduce that we have $D_v(\overline{K}^{\rm ab}(p)/K) \simeq \varprojlim\limits_{n \geq 0} K_v^\times / K_v^{\times p^n} \simeq \widehat{K_v^\times}$, and use II.1.8.2 to give another proof of Theorem 4.5.

Answer. (i) Since $K(\mathfrak{m})^{\rm res}/K$ is finite, for n sufficiently large $D_v/D_v^{p^n}$ is isomorphic to the decomposition group of v in $K(\mathfrak{m})^{\rm res}(p)/K$, proving (i).

(ii) The Schmidt–Chevalley Theorem 4.3 shows that for all n we have $i_v(E^{\{v\}\rm res}_{\frac{\mathfrak{m}}{\mathfrak{m}_v}}) \subset K_v^{\times p^n}$, for $\mathfrak{m}_{\rm ta}$ sufficiently large.

(iii) We will have, after a certain point, $U_v^{m_v} \subset K_v^{\times p^n}$, proving the result (if $v \nmid p$, $m_v = 1$, $U_v^1 \subset K_v^{\times p^n}$, and the result does not depend on $\mathfrak{m}_p$). □

This second proof does not use the properties of cyclotomic fields (Lemmas 2 and 3 of 4.5) because it uses more completely the results of class field theory and logically contains its arguments, in particular by the fact that we write $\overline{K}^{\rm ab} = \bigcup\limits_{\mathfrak{m}} K(\mathfrak{m})^{\rm res}$ and that the decomposition law of places in ray class fields is known.

We can generalize this result for an arbitrary finite set $\Sigma =: \Sigma_0 \cup \Sigma_\infty$ of noncomplex places of K by showing, using 1.1.8 for $S = \emptyset$, $\delta = \Sigma$, and $\mathfrak{n} = \prod\limits_{v \in T \setminus \Sigma_0} \mathfrak{p}_v^{m_v}$, that:

$$\begin{aligned} \mathrm{Gal}(\overline{K}^{\rm ab}(p)/\overline{K}^{\rm ab}(p)^\Sigma) &:= \langle\, D_v(\overline{K}^{\rm ab}(p)/K)\,\rangle_{v \in \Sigma} \\ &= \bigoplus_{v \in \Sigma} D_v(\overline{K}^{\rm ab}(p)/K) \simeq \bigoplus_{v \in \Sigma} (\widehat{K_v^\times})_p. \end{aligned}$$

But this will be done later in an idelic way (less computational and more precise) in the weak deployment Theorem 4.16.7, in close relation with the Grunwald–Wang theorem.

The structural properties of $\overline{K}^{\mathrm{ab}}{}_{(p)}/K$ studied in Section 4 cannot easily be descended; however, we have:

4.12.6 Proposition. *Let K be a number field satisfying the Leopoldt conjecture for p, and let $\widetilde{K}_p$ be the compositum of the $\mathbb{Z}_p$-extensions of K. There exists a (nonunique) abelian extension F of K containing the maximal tamely ramified abelian pro-p-extension $H_{\mathrm{ta}}^{\mathrm{res}}{}_{(p)}$ of K, such that $\overline{K}^{\mathrm{ab}}{}_{(p)}$ is the direct compositum of F with $\widetilde{K}_p$ over $\widetilde{K}_p \cap H^{\mathrm{ord}}{}_{(p)}$.*

Proof. Let H_0 be such that $H_p^{\mathrm{ord}}{}_{(p)}$ is the direct compositum of $\widetilde{K}_p H^{\mathrm{ord}}{}_{(p)}$ with H_0 over $H^{\mathrm{ord}}{}_{(p)}$, and let $F_0 := H_0 H_{\mathrm{ta}}^{\mathrm{res}}{}_{(p)}$ (which is a direct compositum over $H^{\mathrm{ord}}{}_{(p)}$). Therefore, since $\mathrm{Gal}(M_0/H_{\mathrm{ta}}^{\mathrm{res}}{}_{(p)}) \simeq \mathrm{Gal}(H_p^{\mathrm{ord}}{}_{(p)}/H^{\mathrm{ord}}{}_{(p)})$, we have $\mathrm{Gal}(M_0/F_0) \simeq \mathrm{Gal}(H_p^{\mathrm{ord}}{}_{(p)}/H_0) \simeq \mathbb{Z}_p^{r_2+1}$. Since M_0/F_0 is free, by 4.1.9 there exists F such that $\overline{K}^{\mathrm{ab}}{}_{(p)}$ is the direct compositum of F with M_0 over F_0. □

Note then that we have $\mathrm{Gal}(F/H_0) \simeq \prod_v (F_v^\times)_p$, and that the groups $I_v(F/H_0)$ for $v \in Pl_{\mathrm{ta}}$, and $D_v(F/H_0)$ for $v \in Pl_\infty^{\mathrm{r}}$ if $p = 2$, are deployed. The extension F_0/H_0 is tame. The extension F/F_0 is p-ramified (with Galois group isomorphic to $\mathcal{E}^{\mathrm{ord}}$), as well as $F_0/H_{\mathrm{ta}}^{\mathrm{res}}{}_{(p)}$.

We now mention the following important conjecture which, as for the Leopoldt conjecture, would follow from p-adic transcendence conjectures. This conjecture is equivalent to the finiteness of the logarithmic class group.[24] Assuming some Kummer conditions, it is related to the theory of the Hilbert kernel expressed in terms of logarithmic class group mentioned in II.7.7.1.

Recall that $K\mathbb{Q}^{\mathrm{cycl}}{}_{(p)}$ denotes the cyclotomic $\mathbb{Z}_p$-extension of the number field K and that $H_p^{\mathrm{ord}}{}_{(p)}$ denotes the maximal p-ramified noncomplexified abelian p-extension of K.

4.13 Conjecture (Gross). *The subgroup of* $\mathrm{Gal}(H_p^{\mathrm{ord}}{}_{(p)}/K)$ *generated by the relative decomposition groups* $D_v(H_p^{\mathrm{ord}}{}_{(p)}/K\mathbb{Q}^{\mathrm{cycl}}{}_{(p)})$*, for $v|p$, fixes a finite extension of* $K\mathbb{Q}^{\mathrm{cycl}}{}_{(p)}$. □

Since the decomposition group of a tame place v in $\overline{K}^{\mathrm{ab}}{}_{(p)}/K\mathbb{Q}^{\mathrm{cycl}}{}_{(p)}$ is isomorphic to the inertia group $(F_v^\times)_p$ (see 4.8.1, 4.8.2), this implies that the maximal locally cyclotomic abelian pro-p-extension H_p^{lc} of K (which is nec-

[24] [Ja3, Ja4, Ja5, Ja6, Ja7, JaSor], [TBS, Ch. VI] for the relationship with the p-adic analogs of Stark's conjectures, and [FG] for some links with Iwasawa theory.

essarily contained in $H_p^{\mathrm{ord}}(p)$) is conjecturally of finite degree over $K\mathbb{Q}^{\mathrm{cycl}}(p)$. In the statement (i.e., for the finiteness), we can of course replace $H_p^{\mathrm{ord}}(p)$ by $\widetilde{K}_p$ and study the groups $D_v(\widetilde{K}_p/K\mathbb{Q}^{\mathrm{cycl}}(p))$, $v|p$; then Gross's conjecture may be studied numerically thanks to the methods that we will develop in Sections 5 and 7.

4.13.1 Remark. More precisely, we will prove in Section 7 that the logarithmic class group $\widetilde{\mathcal{C}\ell}_K$ is canonically isomorphic to $\mathrm{Gal}(H_p^{\mathrm{lc}}/K\mathbb{Q}^{\mathrm{cycl}}(p))$ [Ja5; Ja6; Ja7, Example 2.10]. Thus, thanks to our methods, we will have a way to compute this group. □

It would be interesting to analyze Gross's conjecture in the spirit of Conjecture 4.12, since both conjectures are trivially true for K totally real under the sole assumption of the Leopoldt conjecture.

The following paragraph studies the problem of globalizing the p-torsion groups attached to p-ramification. It has a p-adic nature which is very different from all the above.

4.14 Conjectural Aspects for the Groups $\mathcal{T}_p^{\mathrm{ord}}$ — Heuristics. Although there is no doubt that the Leopoldt conjecture (or the general p-adic conjecture) is true, and depends essentially on progress in p-adic transcendence techniques, the p-adic behavior of unit groups is nonetheless not completely clear. Indeed, consider the term $\prod_p \mathcal{T}_p^{\mathrm{ord}}$ of the global exact sequence of 4.1.8; is it finite or infinite?

If we consider the expression given in 2.6.1, (ii$_2$), we see that the (integral) factor:

$$\frac{|(\mathcal{C}\ell^{\mathrm{ord}})_p|}{(\mathbb{Z}_p\mathrm{Log}_p(I_p) : \mathbb{Z}_p\mathrm{Log}_p(P_p))}$$

is trivial for almost all p since $\mathcal{C}\ell^{\mathrm{ord}}$ is a finite global invariant; the problem is therefore about the group:

$$\mathrm{tor}_{\mathbb{Z}_p}\Big(\bigoplus_{v|p} U_v^1 \Big/ \mathrm{adh}_p(E'^{\mathrm{ord}})\Big)$$

which, in a certain sense, measures the defect of p-adic generation of the group of local principal units by the group of p-principal global units. We now use the exact sequence of Lemma 4.2.4. Assuming the Leopoldt conjecture, $\mathrm{tor}_{\mathbb{Z}_p}(\mathrm{adh}_p(E'^{\mathrm{ord}})) \simeq \mu_p(K)$ (see 3.6.3); hence the kernel is perfectly known and, in any case, it is trivial for almost all p. We are thus reduced to the study of the term:

$$\mathrm{tor}_{\mathbb{Z}_p}\Big(\log\Big(\bigoplus_{v|p} U_v^1\Big) \Big/ \mathbb{Z}_p\mathrm{log}_p(E'^{\mathrm{ord}})\Big)$$

which is a finite p-group whose order is, up to an explicit factor, the p-part of the p-adic regulator of K [c, Wa, Ch. 5, §§ 5, 6]. For a precise formulation in the case of real abelian fields, see [AF]; there is no problem either for the general case of a totally real field, and the nontotally real case needs, as for the statement of Conjecture 4.12, an approach in terms of representations starting from the Galois case. A precise definition of the regulator is useful mainly for analytic formulas, which is not our goal here.

Let us take the simple case of a real quadratic field $K = \mathbb{Q}(\sqrt{d})$. Let ε' be a generating unit of E'^{ord}. If there exists $u \in \bigoplus_{v|p} U_v^1$ such that $\log(u)$ is of order p^h in $\log\Big(\bigoplus_{v|p} U_v^1\Big)\Big/\mathbb{Z}_p\log_p(E'^{\mathrm{ord}})$, with h *greater than or equal to 1*, there exists $a \in \mathbb{Z}_p$ such that $p^h\log(u) = a \log_p(\varepsilon')$; therefore $a \in \mathbb{Z}_p^\times$ and, setting $u' := u^{a^{-1}}$, we finally obtain:

$$\log_p(\varepsilon') = p^h\log(u'),$$

which yields for $p > 3$:

$$\varepsilon' = u'^{p^h}$$

(indeed, there is local torsion only for $p = 2$, or for $p = 3$ when $K_v = \mathbb{Q}_3(\sqrt{-3})$, which occurs when $d \equiv -3 \bmod (9)$). If we assume that p is different from 2 and 3 and that p is unramified in $K/\mathbb{Q}$, we thus have:

$$\varepsilon' \equiv 1 \bmod (p^{1+h}),\ \ h > 0.$$

In practice, if p is split in K this means, for the fundamental unit ε, that we have the necessary condition $\varepsilon^{p-1} \equiv 1 \bmod (p^2)$. If p is inert, this means that $\varepsilon^{p^2-1} \equiv 1 \bmod (p^2)$, noting, for numerical computations, that we are reduced to the case where ε is of norm equal to 1 (if necessary by taking for ε the square of the fundamental unit); in this case, the condition can be written $\varepsilon^{p+1} \equiv 1 \bmod (p^2)$. In other words, if we write ε^{p-1} (resp. ε^{p+1}) in the form $u + v\sqrt{d}$ (u and v integers or half-integers), we have $u + v\sqrt{d} \equiv 1 \bmod (p^2)$ if and only if:

$$v \equiv 0 \bmod (p^2)$$

(this implies the condition $u \equiv 1 \bmod (p^2)$ for a unit).

It seems difficult to say whether this may happen for an infinite number of primes p or not. Numerically we obtain, as examples, the following results:

(i) for $K = \mathbb{Q}(\sqrt{2})$, the only solutions for $p < 2 \times 10^8$ are:

$$p = 13,\ \ p = 31,\ \ p = 1546463\ ;$$

(ii) for $K = \mathbb{Q}(\sqrt{26})$, the only solutions for $p < 2 \times 10^8$ are:

$$p = 2683,\ \ p = 3967,\ \ p = 18587.$$

We can formulate some conjectures; the strongest (implausible) being that $\prod_p \mathcal{T}_p^{\rm ord}$ is a global finite invariant (as the class group or the K_2 of the ring of integers), hence that there are only a finite number of solutions p, and they are of reasonable size (this does not exclude very large values; for instance, recall the relative class numbers of cyclotomic fields, which grow exponentially). On the contrary (and this is what we suggest, as most people do), we conjecture that, for any number field whose unit group is of nonzero $\mathbb{Z}$-rank (i.e., K different from $\mathbb{Q}$ and imaginary quadratic fields), there is an infinite number of nontrivial $\mathcal{T}_p^{\rm ord}$, the behavior of the above two quadratic fields going in this direction.

Note. In the cohomological interpretation of the invariants of class field theory, this conjecture would imply that the group $H^2(\mathcal{G}_p^{\rm ord}, \mathbb{Z}_p(0))$ (whose dual is isomorphic to $\mathcal{T}_p^{\rm ord}$ as is shown in the Appendix) plays a special role with respect to the other twists $H^2(\mathcal{G}_p^{\rm ord}, \mathbb{Z}_p(m))$, $m > 1$ ([Ng1], [Schn]), where $\mathcal{G}_p^{\rm ord}$ is the Galois group of the maximal p-ramified noncomplexified pro-p-extension of K, and $\mathbb{Z}_p(m)$ is the group $\mathbb{Z}_p$ together with the Galois action given by the mth power of the cyclotomic character. These groups $H^2(\mathcal{G}_p^{\rm ord}, \mathbb{Z}_p(m))$ are isomorphic to the finite groups $\mathrm{K}_{2m-2}(Z_K) \otimes \mathbb{Z}_p$, of higher K-theory, assuming the Quillen–Lichtembaum conjecture (except for $p = 2$ where there is an explicit kernel and cokernel).

This problem is indeed difficult since it it related to the problem of:

$$\text{“}\,2^{p-1} \equiv 1 \bmod (p^2)\,\text{”},$$

for which the only two solutions known are:

$$p = 1093,\ \ p = 3511.$$

In fact, at the level of a group of S-units it is exactly the same problem. Indeed, choose $K = \mathbb{Q}$, $p \neq 2$, and $S = \{2\}$; if we consider $H_{\mathbb{Q},p^{(p)}}$ (which is here equal to the cyclotomic $\mathbb{Z}_p$-extension $\mathbb{Q}^{{\rm cycl}\,(p)}$ of $\mathbb{Q}$), using 2.6, (ii) for $K = \mathbb{Q}$ and $T = \{p\}$, the torsion $\mathbb{Z}_p$-module of $\mathrm{Gal}(\mathbb{Q}^{{\rm cycl}\,(p)S}/\mathbb{Q})$ has order:

$$|\mathcal{T}_{\mathbb{Q},p}^S| = |\mathrm{tor}_{\mathbb{Z}_p}((1 + p\mathbb{Z}_p)/\mathrm{adh}_p(\langle\, 2\,\rangle'))|$$

since $E_{\mathbb{Q}}^S = \langle\, 2\,\rangle$, where $\langle\, 2\,\rangle'$ (the group of p-principal 2-units in the restricted sense!) will be for example given by $\langle\, 2^{p-1}\,\rangle$. We thus immediately obtain that $\mathcal{T}_{\mathbb{Q},p}^S$ is nontrivial if and only if:

$$2^{p-1} \equiv 1 \bmod (p^2).$$

Note that here $\mathcal{T}_{\mathbb{Q},p}^S = \mathrm{Gal}(\mathbb{Q}^{{\rm cycl}\,(p)S}/\mathbb{Q})$ since 2 cannot be totally split in the cyclotomic $\mathbb{Z}_p$-extension of $\mathbb{Q}$, which means that the Frobenius of 2 in $\mathbb{Q}^{{\rm cycl}\,(p)}/\mathbb{Q}$ is a topological generator (except for the two values given above), at least within the known numerical bounds. This interpretation also suggests that we have an infinite number of solutions: a natural heuristic reasoning, based on the idea that the probability that $\frac{2^{p-1}-1}{p} \equiv 0 \bmod (p)$ is $\frac{1}{p}$, leads to this conclusion since the serie of the $\frac{1}{p}$ is divergent, but only as $\log(\log(p))$. However, even though the same heuristic is valid for ordinary units (this time for a field K with units), the interpretation in terms of class field theory is a little different. Numerical experimentations on the distributions of the Fermat quotients of algebraic numbers have been performed by Hatada and go in the right direction. If we take into account the p-adic conjecture (3.1.3, 3.1.3′) and the above, the fact that we have units does not seem to be essential, and it is reasonable to think that we have an analogous behavior for the groups E^S in the monogeneous case.

As the infinite cardinality of $\prod_p \mathcal{T}_p^{\mathrm{ord}}$ is probably the rule, this says that, except for $\mathbb{Q}$ and the imaginary quadratic fields, the structure of $\mathrm{Gal}(\overline{K}^{\mathrm{ab}}/K)$ is quite peculiar since there exists a sequence of prime numbers (clearly of zero density) which plays a privileged part; note that $\log(\log(2 \times 10^8)) = 2.95\cdots$, which gives the order of magnitude of the number of solutions that we have found for $\mathbb{Q}(\sqrt{2})$ and $\mathbb{Q}(\sqrt{26})$.

Proposition 4.12.6 and the results of Subsection (e) then tell us that the structure of $\overline{K}^{\mathrm{ab}}(p)$ is usually simple. Let K satisfying the Leopoldt conjecture for p. Let $\widetilde{K}_p$ be the compositum of the $\mathbb{Z}_p$-extensions of K. We have:

4.14.1 Proposition. *Assume that* $\mathrm{tor}_{\mathbb{Z}_p}\Big(\bigoplus_{v|p} U_v^1 \big/ \mathrm{adh}_p(E'^{\mathrm{ord}})\Big) = 1$, *which is equivalent to the following two conditions (see 4.2.4):*

- $\bigoplus_{v|p} \mu_p(K_v) = i_p(\mu_p(K))$,
- $\mathbb{Z}_p \log_p(E'^{\mathrm{ord}})$ *is a direct summand in* $\bigoplus_{v|p} \log(U_v^1)$.

Then there exists a (nonunique) abelian extension F containing $H_{\mathrm{ta}}^{\mathrm{res}}(p)$, such that $\overline{K}^{\mathrm{ab}}(p)$ is the direct compositum of $\widetilde{K}_p$ with F over $\widetilde{K}_p \cap H^{\mathrm{ord}}(p)$. We then have $\mathrm{Gal}(F/H^{\mathrm{ord}}(p)) \simeq \prod_v (F_v^\times)_p$, and the extension $F/H_{\mathrm{ta}}^{\mathrm{res}}(p)$ is p-ramified with Galois group isomorphic to $\mathcal{E}^{\mathrm{ord}}$. □

4.14.2 Corollary. *If, in addition, the ordinary p-class group of K is trivial, $\overline{K}^{\mathrm{ab}}(p)$ is the direct compositum of $\widetilde{K}_p$ with F over K, and we have* $\mathrm{Gal}(F/K) \simeq \prod_v (F_v^\times)_p$. □

By 4.1.9, such a decomposition is always possible over K if we omit the inclusion $H_{\mathrm{ta}}^{\mathrm{res}}(p) \subseteq F$.

4.14.3 Remarks. (i) The reflection theorem implies that if we introduce $K' := K(\mu_p)$, we have, assuming the Leopoldt conjecture for p (see 4.2.2):

$$\mathrm{rk}_p(\mathcal{T}_p^{\mathrm{ord}}) = \mathrm{rk}_\omega(\mathcal{C}\ell_{K'}^{Pl'_p \mathrm{res}}) + \sum_{v|p} \delta_v - \delta,$$

where $Pl'_p := Pl_{K',p}$, which links the nonvanishing of $\mathcal{T}_p^{\mathrm{ord}}$ to that of the ω-component of the p-group of Pl'_p-ideal classes of the field K', where ω is the Teichmüller character of $\mathrm{Gal}(K'/K)$, $\delta_v := 1$ or 0 according as K_v contains μ_p or not, $\delta := 1$ or 0 according as $\omega = 1$ or not (so that $\sum_{v|p} \delta_v - \delta = 0$ for almost all p). But this point of view does not give any indication on the problem since the field K' depends on p.

(ii) From the analytic point of view, still (to simplify) in the case of a real quadratic field K, if $L_p(s, \chi)$ denotes the p-adic L function of the nontrivial character of K, then the residue formula at $s = 1$ yields, for an odd prime p unramified in $K/\mathbb{Q}$ and not dividing the class number of K (see [c, Wa, Ch. 5, § 6] or 2.6.5):

$$|\mathcal{T}_p^{\mathrm{ord}}| \overset{p}{=} \frac{\log_p(\varepsilon)}{p} \overset{p}{=} L_p(1, \chi) \ ;$$

the problem under study is thus equivalent to the question of whether or not there exists an infinite number of p such that $L_p(1, \chi)$ is divisible by p.

(iii) It is easy to find quadratic fields K for which the number of solutions p to $\mathcal{T}_p^{\mathrm{ord}} \neq 1$ is arbitrarily large. Indeed, choose:

$$d := q^2 (p_1 \cdots p_s)^4 + 1, \ q \geq 1,$$

$p_1, \ldots, p_s$ distinct primes, and assume that d is squarefree (the existence of infinitely many such d is a classical and reasonable conjecture); then the fundamental unit is equal to $\varepsilon = q(p_1 \cdots p_s)^2 + \sqrt{d}$, and we have $\varepsilon^2 \equiv 1 \bmod (p_1 \cdots p_s)^2$. It is true that such fields give many small solutions, but the total density is certainly unchanged. □

g) Structural Properties of $\overline{G}^{\mathrm{ab}}$ — Divisibility of the Connected Component — Cyclic Embedding Criterion

This subsection deals with the global structure of $\overline{G}^{\mathrm{ab}} := \mathrm{Gal}(\overline{K}^{\mathrm{ab}}/K)$ which is best studied using classical idelic methods.

Assuming the Leopoldt conjecture for all p, if we refer to Figure 4.2′ we see that the subgroups of $\overline{G}^{\mathrm{ab}}$ corresponding to:

$$\prod_{v \in Pl_0} F_v^\times \times \{\pm 1\}^{r_1} \ \text{ and } \ \prod_{v \in Pl_0} \mathrm{tor}_{\mathbb{Z}}(U_v^1) = \bigoplus_{\substack{v \in Pl_0 \\ \text{irregular}}} \mu(K_v)^1,$$

whose intersection is isomorphic to $\mu(K)$, produce elements of finite order in $\overline{G}^{\mathrm{ab}}$, and it is useful to know if they are all obtained in this way and from which idèles. We are going to solve this question by standard idelic methods, and not by the above p-adic localization techniques since here a direct global reasoning is more natural and gives a clear understanding of the underlying local-global problem. The use of the divisibility properties of the connected component D in C makes the solution of this problem quite trivial (see 4.15.2). However, so as to review the elementary techniques (mostly the Schmidt–Chevalley Theorem 4.3) which explain these results and divisibilities, we are going to give a proof which is essentially a direct historical proof [h, We2, III], knowing that later Artin–Tate [d, AT, Ch. 6, § 5] will obtain it as a consequence of the second inequality connected with the existence theorem, and Weil [d, We1, Ch. XII, XIII] by a similar method.

Definitions. Recall the main definitions and notations concerning infinite class field theory (see II.3.7):

(i) The reciprocity homomorphism $\rho : J \longrightarrow \overline{G}^{\mathrm{ab}}$ sends $\boldsymbol{x} \in J$ to the family $(\rho_N(\boldsymbol{x}))_N \in \overline{G}^{\mathrm{ab}} = \varprojlim_N \mathrm{Gal}(M/K)$, where $\rho_N := \rho_{M/K}$ is the reciprocity map relative to the finite abelian extension M/K corresponding to N.

(ii) We then have $\overline{G}^{\mathrm{ab}} \simeq \varprojlim_N J/N \simeq J\Big/\bigcap_N N$, where N ranges in the set of open subgroups of J containing $K^\times$; to fix ideas, we choose the cofinal subset formed by the $N := N_{\mathfrak{m}} := K^\times U_{\mathfrak{m}}^{\mathrm{res}}$, $\mathfrak{m} \in \langle\, Pl_0 \,\rangle_{\mathbb{N}}$.

(iii) In the special case (which occurs for $p = 2$, $e \geq 3$, $K \cap \mathbb{Q}(\mu_{2^e}) = \mathbb{Q}(\zeta_n + \zeta_n^{-1})$, $2 \leq n < e$, and $Pl_2^{\mathrm{ns}} = \emptyset$), $\boldsymbol{s}$ denotes an idèle such that $i(2 + \zeta_n + \zeta_n^{-1})^{2^n} = \boldsymbol{s}^{2^{n+1}}$ (see II.6.3.4.3). □

With the above, we have:

4.15 Theorem (elements of finite order of $\overline{G}^{\mathrm{ab}}$). *Let p be a prime number and e an integer ≥ 1. The group ${}_{p^e}\overline{G}^{\mathrm{ab}}$ of elements of $\overline{G}^{\mathrm{ab}}$ which are killed by p^e is equal to the image of ${}_{p^e}J$ (resp. of ${}_{2^e}J \cup \boldsymbol{s}\,.\,{}_{2^e}J$ in the special case) under the reciprocity map ρ.*

Proof. Let $\sigma \in {}_{p^e}\overline{G}^{\mathrm{ab}}$; there exists $\boldsymbol{x} \in J$ such that $\sigma = \rho(\boldsymbol{x})$, and we have:

$$\boldsymbol{x}^{p^e} \in \bigcap_N N,$$

where we recall that we choose $N := K^\times U_{\mathfrak{m}}^{\mathrm{res}}$, which means that:

$$\boldsymbol{x}^{p^e} =: i(x_N)\,\boldsymbol{u}_{\mathfrak{m}}, \ \ x_N \in K^\times, \ \boldsymbol{u}_{\mathfrak{m}} \in U_{\mathfrak{m}}^{\mathrm{res}}$$

for all N.

Fix an $N_0 = K^\times U^{\rm res}_{\mathfrak{m}_0}$; then, for all $\mathfrak{m} \subseteq \mathfrak{m}_0$ (i.e., $\mathfrak{m}_0|\mathfrak{m}$), the above equality implies that:

$$i(x_N) = i(x_{N_0})\, \boldsymbol{u}_{\mathfrak{m}_0}\, \boldsymbol{u}_{\mathfrak{m}}^{-1} =: i(x_{N_0})\, \boldsymbol{v}_{\mathfrak{m}_0,N},\ \ \boldsymbol{v}_{\mathfrak{m}_0,N} \in U^{\rm res}_{\mathfrak{m}_0},$$

showing that $\boldsymbol{v}_{\mathfrak{m}_0,N} =: i(\varepsilon_{\mathfrak{m}_0,N})$, $\varepsilon_{\mathfrak{m}_0,N} \in E^{\rm res}_{\mathfrak{m}_0}$. If we assume N_0 chosen so that $E^{\rm res}_{\mathfrak{m}_0} \subseteq (E^{\rm ord})^{p^e}$ (using 4.3), and if we set $\varepsilon_{\mathfrak{m}_0,N} =: \varepsilon_N^{p^e}$, we obtain:

$$(\boldsymbol{x}\, i(\varepsilon_N)^{-1})^{p^e} = i(x_{N_0})\, \boldsymbol{u}_{\mathfrak{m}}$$

for all $\mathfrak{m} \subseteq \mathfrak{m}_0$. It follows that x_{N_0} is everywhere locally a p^eth power (for each finite place v we consider a suitable N, in other words an $\mathfrak{m}$ which is sufficiently large at v), and by II.6.3.3 we have (apart for the special case) $x_{N_0} =: y_{N_0}^{p^e}$, $y_{N_0} \in K^\times$, and if we set $t_N := \varepsilon_N\, y_{N_0}$, we obtain:

$$(\boldsymbol{x}\, i(t_N)^{-1})^{p^e} \in U^{\rm res}_{\mathfrak{m}} \ \text{ for all } \mathfrak{m} \subseteq \mathfrak{m}_0.$$

Let us look at what happens on the support of any sufficiently large $\mathfrak{m}$ (in particular we assume that this support contains Pl_p); outside of this support, the v-component of the left hand side is in $(U_v)^{p^e}$. For any place $v|p$ of K we have $U_v^{m_v} \subseteq (U_v^{m_v - h_v})^{p^e}$ (see II.1.4.4), where h_v is a constant which does not depend on m_v assumed to be sufficiently large; for $v \nmid p$ finite, we have $U_v^{m_v} = (U_v^{m_v})^{p^e}$, and for $v|\infty$, we have $U_v = (U_v)^{p^e}$. Hence we can write (for all sufficiently large $\mathfrak{m}$).

$$\boldsymbol{x} =: \boldsymbol{\zeta}_N\, i(t_N)\, \boldsymbol{u}_{\mathfrak{m}_1},$$

where $\boldsymbol{\zeta}_N \in {}_{p^e}J = \prod_v {}_{p^e}\mu(K_v)$, $\boldsymbol{u}_{\mathfrak{m}_1} \in U^{\rm res}_{\mathfrak{m}_1}$, and where the $N_1 := K^\times U^{\rm res}_{\mathfrak{m}_1}$ form a cofinal subset for the family of subgroups N since $p^h\mathfrak{m}_1 \subseteq \mathfrak{m} \subseteq \mathfrak{m}_1$ for a suitable constant h (this is the crucial point). By compactness of ${}_{p^e}J$, there exists a limit point $\boldsymbol{\zeta}$ of $(\boldsymbol{\zeta}_N)_N$ which is therefore such that:

$$\boldsymbol{x} =: \boldsymbol{\zeta}\, i(t_{N'})\, \boldsymbol{u}'_{\mathfrak{m}'_1},$$

where the N'_1 form a cofinal subset to the set of N_1 above. Hence $\sigma = \rho(\boldsymbol{\zeta})$, proving the theorem in the general case. In the special case we must add the solutions $\sigma = \rho(\boldsymbol{s} \,.\, \boldsymbol{\zeta})$, where we recall that $\boldsymbol{s}^2 \in {}_{2^{e-1}}J\, K^\times$, $\boldsymbol{s} \notin {}_{2^e}J\, K^\times$. □

The following exercise shows why the above techniques are relative to the divisibility of $D := c\ell\Big(\bigcap_{\mathfrak{m}} (K^\times U^{\rm res}_{\mathfrak{m}})\Big) = c\ell\Big(\bigcap_N N\Big)$ (see II.3.7.4).

4.15.1 Exercise. (i) Using similar computations, show that D is divisible.

(ii) Show that, in general, we have the exact sequence:

$$1 \longrightarrow {}_{p^e}\mu(K) \,.\, {}_{p^e}U_\infty \longrightarrow {}_{p^e}J \xrightarrow{\ \rho\ } {}_{p^e}\overline{G}^{\rm ab} \longrightarrow 1,$$

which becomes, in the special case:

$$1 \longrightarrow \{\pm 1\} \,.\, \langle 2 + \zeta_n + \zeta_n^{-1} \rangle \,.\, {}_{2^e}U_\infty \longrightarrow \langle \boldsymbol{s} \rangle \,.\, {}_{2^e}J \xrightarrow{\ \rho\ } {}_{2^e}\overline{G}^{\mathrm{ab}} \longrightarrow 1.$$

(iii) Compute $\mathrm{tor}(D)$.

Answer. (i) It is enough to show that for all prime p we have $D = D^p$. Let $\boldsymbol{z} \in \bigcap_N N$; applying to $\boldsymbol{z}$ the reasoning made at the beginning of the proof of the theorem (N_0 fixed, N arbitrary), we obtain:

$$\boldsymbol{z} = i(x_{N_0})\, i(\varepsilon_{\mathfrak{m}_0,N})\, \boldsymbol{u}_{\mathfrak{m}}\ ;$$

here we only assume that $\mathfrak{m}_0$ has been chosen such that $\varepsilon_{\mathfrak{m}_0,N} =: \varepsilon_N^p$ for all $\mathfrak{m} \subseteq \mathfrak{m}_0$, in which case:

$$\boldsymbol{z}' := \boldsymbol{z}\, i(x_{N_0})^{-1} = i(\varepsilon_N)^p\, \boldsymbol{u}_{\mathfrak{m}}$$

for all $\mathfrak{m} \subseteq \mathfrak{m}_0$. The idèle $\boldsymbol{z}'$ is constant and in the right hand side $\mathfrak{m} \subseteq \mathfrak{m}_0$ is arbitrary; it follows that (by choosing an appropriate N at each place):

$$\boldsymbol{z}' =: \boldsymbol{x}^p, \quad \boldsymbol{x} \in J\ ;$$

hence $\boldsymbol{x}^p$ is an element of $J^p \cap \bigcap_N N$. The reasoning of 4.15 for $e = 1$ yields:

$$\boldsymbol{x} =: \boldsymbol{\zeta}\, \boldsymbol{y}, \quad \boldsymbol{\zeta} \in {}_pJ,$$

with $\boldsymbol{y} \in \bigcap_{N_1'} N_1'$ for a cofinal subset of the family of the groups N. We thus have $\mathcal{c}\ell(\boldsymbol{z}) = \mathcal{c}\ell(\boldsymbol{z}') = \mathcal{c}\ell(\boldsymbol{x}^p) = \mathcal{c}\ell(\boldsymbol{y})^p$, with $\mathcal{c}\ell(\boldsymbol{y}) \in D$, proving that D is divisible.

(ii) Here it is more convenient to use reduced idèles. Let $\boldsymbol{\zeta} \in {}_{p^e}J_0$ such that $\rho(\boldsymbol{\zeta}) = 1$; then $\mathcal{c}\ell_0(\boldsymbol{\zeta}) \in D_0$, and for all $h \in \mathbb{N}$ there exists $\boldsymbol{\theta}_h \in J_0$ such that:

$$\mathcal{c}\ell_0(\boldsymbol{\zeta}) = \mathcal{c}\ell_0(\boldsymbol{\theta}_h^{p^h}).$$

It follows that the $\mathcal{c}\ell_0(\boldsymbol{\theta}_h)$ are of finite order; by II.6.3.4, (γ), (iii), we may assume in the general case that $\boldsymbol{\theta}_h$ is of finite order. Then for all $h \in \mathbb{N}$ there exists $\xi_h \in \mu(K^\times)$ such that $\boldsymbol{\zeta} = i_0(\xi_h)\, \boldsymbol{\theta}_h^{p^h}$, and there exists $\xi \in \mu(K^\times)$ for which $\boldsymbol{\zeta}\, i_0(\xi)^{-1} = \boldsymbol{\theta}_h^{p^h}$, for infinitely many h; this implies that:

$$\boldsymbol{\zeta} = i_0(\xi),$$

since the subgroup of p-divisible elements of J_0 is trivial.

In the special case, if $\rho(\boldsymbol{s}^\lambda\, \boldsymbol{\zeta}) = 1$, with $\lambda \in \mathbb{Z}$, $\boldsymbol{\zeta} \in {}_{2^e}J_0$, then $\rho(\boldsymbol{s}^{2\lambda}\, \boldsymbol{\zeta}^2) = 1$ and, since $\boldsymbol{s}^2 = \boldsymbol{\zeta}_0\, i_0(y_n)$, $\boldsymbol{\zeta}_0 \in {}_{2^n}J_0$, $y_n = 2 + \zeta_n + \zeta_n^{-1}$, we obtain (noting that here we must replace $\boldsymbol{\theta}_h$ by $\boldsymbol{s}^\mu\, \boldsymbol{\theta}_h$):

$$\boldsymbol{\zeta}_0^\lambda\, \boldsymbol{\zeta}^2 = \pm 1.$$

If λ is odd, we can write $\boldsymbol{\zeta}_0 = \pm\boldsymbol{\zeta}'^2$, $\boldsymbol{\zeta}' \in {}_{2^e}J_0$, thus $\boldsymbol{s}^2 = \pm\boldsymbol{\zeta}'^2\, i_0(y_n)$, which shows that $\pm y_n$ (local square everywhere) is a square in $K^\times$, a contradiction. Then λ is even and we obtain $\boldsymbol{s}^\lambda\, \boldsymbol{\zeta} = \pm i_0(y_n)^{\lambda/2}$ (use the previous situation for $\rho(\boldsymbol{s}^\lambda\, \boldsymbol{\zeta}) = \rho(\boldsymbol{\zeta}_0^{\lambda/2}\, \boldsymbol{\zeta}) = 1$ which yields $\boldsymbol{\zeta}_0^{\lambda/2}\, \boldsymbol{\zeta} = \pm 1$). In the special case, $\rho(\boldsymbol{s})$ is indeed an element of order 2^{n+1} which is not obtained by means of the evident elements of finite order of $\overline{G}^{\rm ab}{}_{(2)}$ (i.e., the images of idèles of finite order); but $\rho(\boldsymbol{s})^2$ is such an element.

(iii) It is sufficient to study ${}_{p^e}D$. Let $\mathcal{d}(\boldsymbol{x}) \in {}_{p^e}D$; we may suppose that $\boldsymbol{x} \in {}_{p^e}J$ (resp. $\langle\, \boldsymbol{s}\,\rangle_{2^e}J$). But $\rho(\boldsymbol{x}) = 1$, hence (by (ii)) $\mathcal{d}(\boldsymbol{x}) \in \mathcal{d}({}_{p^e}U_\infty)$ in all the cases. Hence ${\rm tor}(D) = \mathcal{d}({\rm tor}(U_\infty)) \simeq {\rm tor}(\mathbb{C}^\times)^{r_2}$.

Note. This immediately proves that, in a finite extension L/K, the kernel of the transfer map $C_K/D_K \longrightarrow C_L/D_L$ is $\mathcal{d}_K\Big(\bigoplus_{v\in Pl_\infty^{\rm rc}} \{\pm 1\}\Big)\,.\,D_K\Big/D_K \simeq \bigoplus_{v\in Pl_\infty^{\rm rc}} \{\pm 1\}$, where $Pl_\infty^{\rm rc}$ denotes the set of real places of K which are totally complexified in L. Indeed, since ${\rm N}\circ j = [L:K]$, an element of the kernel is of finite order and can be represented by $c \in {\rm tor}(C_K)$. Thus $j(c) \in {\rm tor}(D_L) = \mathcal{d}_L({\rm tor}(U_{L,\infty}))$. Write $c = \mathcal{d}_K(\boldsymbol{x})$, $\boldsymbol{x} \in J_K$, then $j(\boldsymbol{x}) = y\,\boldsymbol{u}$, $y \in L^\times$, $\boldsymbol{u} \in {\rm tor}(U_{L,\infty})$. By II.2.1, this yields $y \in j(K^\times)$, and $\boldsymbol{u} \in {\rm tor}(U_{L,\infty}) \cap j(J_K) = j\Big(\bigoplus_{v\in Pl_\infty^{\rm rc}} \{\pm 1\}\,.\,{\rm tor}(U_{K,\infty})\Big)$ giving the result. □

4.15.2 Remarks. (i) It trivially follows (since C/D is profinite) that D is the maximal divisible subgroup, in which case if this fact is known *a priori*, the formula $\overline{G}^{\rm ab} \simeq C/D$ yields ${}_{p^e}\overline{G}^{\rm ab} \simeq {}_{p^e}(C/D) = {}_{p^e}C\,D/D$, since D is a divisible group, whence the result 4.15 by II.6.3.4, (γ), (iii).

(ii) For the global description of $\overline{G}^{\rm ab}$ we could have used the profinite group $J_0\Big/\bigcap_{\mathfrak{m}}(K^\times U_{0,\mathfrak{m}}^{\rm res}) = J_0/{\rm adh}_0(K^\times)$, in other words C_0/D_0 which is especially simple since $D_0 = {\rm adh}_0(E^{\rm ord})/E^{\rm ord} \simeq (\widehat{\mathbb{Z}}/\mathbb{Z})^{r_1+r_2-1}$ (D_0 is uniquely divisible contrary to D). □

The result of 4.15 shows for instance the following.

4.15.3 Corollary. *Except perhaps in the special case, the maximal subgroup of* $\mathcal{T}_p^{\rm ord} := {\rm Gal}(H_p^{\rm ord}{}_{(p)}/\widetilde{K}_p)$ *which lifts to a finite subgroup of* $(\overline{G}^{\rm ab})_p$ *is the image of* $\bigoplus_{v|p} \mu_p(K_v)$*, the latter being (assuming the Leopoldt conjecture for* p*) isomorphic to* $\Big(\bigoplus_{v|p} \mu_p(K_v)\Big)\Big/ i_p(\mu_p(K))$. □

Note that this group is also the kernel of the log map modulo $\mathbb{Z}_p{\rm log}_p(E^{\rm ord})$ (not modulo $\mathbb{Q}_p{\rm log}_p(E^{\rm ord})$!) on the torsion group of $\Big(\bigoplus_{v|p} U_v^1\Big)\Big/{\rm adh}_p(E'^{\rm ord})$ (see 4.2.4), which is isomorphic to the torsion group of ${\rm Gal}(H_p^{\rm ord}{}_{(p)}/H^{\rm ord}{}_{(p)})$.

Beware that $\bigoplus_{v|p} \mu_p(K_v)$ can be identified with a subgroup of $(\overline{G}^{\mathrm{ab}})_p$ but not with a subgroup of $\mathcal{T}_p^{\mathrm{ord}}$ (which is reasonable since $\mu_p(K) \subset \mathcal{E}^{\mathrm{ord}} \simeq \mathrm{Gal}(\overline{K}^{\mathrm{ab}}/M_0)$ by Figure 4.2).

A classical application of Theorem 4.15 is the following embedding criterion (after [d, AT, Ch. 10, § 3]).

4.15.4 Theorem (cyclic embedding criterion). *Let L/K be a cyclic extension of number fields of degree p^r. Then there exists a cyclic extension M of degree p^{r+e} of K containing L if and only if for any place v of K we have:*

$$ {}_{p^e}\mu(K_v) \subset \mathrm{N}_{L_v/K_v}(L_v^\times), $$

with in addition in the special case the extra condition $\prod_v \Big(\frac{s_v\,,\,L/K}{v}\Big) = 1.$[25]

Proof. Denote by $\chi \in \overline{G}^{\mathrm{ab}\,*}$ a character of order p^r of $\overline{G}^{\mathrm{ab}}$ such that the orthogonal complement $\langle\,\chi\,\rangle^\perp$ (i.e., the kernel of χ) fixes L. The existence of M is equivalent to that of $\psi \in \overline{G}^{\mathrm{ab}\,*}$ such that $\chi = \psi^{p^e}$. Since a profinite abelian group is reflexive, we can say that $\chi \in \overline{G}^{\mathrm{ab}\,*\,p^e}$ if and only if:

$$ \begin{aligned} \mathrm{Gal}(\overline{K}^{\mathrm{ab}}/L) &\supseteq (\overline{G}^{\mathrm{ab}\,*\,p^e})^\perp \\ &= \{\sigma \in \overline{G}^{\mathrm{ab}},\ \varphi^{p^e}(\sigma) = \varphi(\sigma^{p^e}) = 1 \ \ \forall \varphi \in \overline{G}^{\mathrm{ab}\,*}\} = {}_{p^e}\overline{G}^{\mathrm{ab}}. \end{aligned} $$

Thus this is equivalent to ${}_{p^e}\overline{G}^{\mathrm{ab}} \subseteq \mathrm{Ker}(\chi)$, hence (apart from the special case) to ${}_{p^e}\mu(K_v) \subset \mathrm{Ker}(\rho_{L/K})$ for all v, which is again equivalent to:

$$ {}_{p^e}\mu(K_v) \subset \mathrm{Ker}(\rho_{L_v/K_v}) = \mathrm{N}_{L_v/K_v}(L_v^\times) $$

for all v, by definition of $\rho_{L/K}$ (or because "$N \cap K_v^\times = N_v$").

In the special case we add the condition $\rho_{L/K}(\boldsymbol{s}) = 1$. □

4.15.5 Remarks. In practice we have the following informations:

- If v is an unramified finite place in L/K, any unit of K_v is a norm in L_v/K_v, hence the corresponding conditions are satisfied.
- If v is a tame finite place with ramification index $p^{r_v} > 1$, a necessary condition for the existence of M is that p^{r_v+e} (the ramification index of v in M/K) divides $q_v - 1$; it is sufficient to have ${}_{p^e}\mu(K_v) \subset \mathrm{N}_{L_v/K_v}(L_v^\times)$ since, if ζ is a generator of μ_{q_v-1}, we have:

$$ \zeta = \mathrm{N}_{L_v^{\mathrm{nr}}/K_v}(u), \quad u \in L_v^{\mathrm{nr}\times}, $$

[25] The special case happens for $p = 2$, $e \geq 3$, $K \cap \mathbb{Q}(\mu_{2^e}) = \mathbb{Q}(\zeta_n + \zeta_n^{-1})$, $2 \leq n < e$, and $Pl_2^{\mathrm{ns}} = \emptyset$. For the definition of $\boldsymbol{s} =: (s_v)_v$, see II.6.3.4.3.

and we check that the norm in L_v/K_v of $u^{\frac{q_v-1}{p^{e+r_v}}}$, has order p^e.

• For $p = 2$ and $e \geq 1$, since -1 is not a norm in $\mathbb{C}/\mathbb{R}$, no real infinite place should complexify in L/K.

• Finally there remains to check the local norm conditions at the wild places, and possibly also the extra condition in the special case, for which the components s_v of the idèle $\boldsymbol{s}$ are in U_v if $v \nmid 2$. A necessary condition is that the elements of ${}_{p^e}\mu(K)$ should be local norms everywhere. If $|Pl_p| = 1$ and ${}_{p^e}\mu(K) \simeq {}_{p^e}\mu(K_{v_0})$ for the place $v_0|p$, then by the product formula we can omit the condition at v_0 (this is the case for $K = \mathbb{Q}$ and $p = 2$). □

4.15.6 Example. A quadratic field $\mathbb{Q}(\sqrt{d})$ can be embedded in a cyclic extension of degree 4 of $\mathbb{Q}$ if and only if $d = p_1 \cdots p_t$ or $2p_1 \cdots p_t$, where the p_i are prime numbers congruent to 1 modulo (4). The generalization to the cyclic case of degree 2^e is similar since the special case does not occur ($Pl_2^{\rm ns} = \{2\}$ for $\mathbb{Q}$). □

4.15.7 Remarks. (i) If we compare this result 4.15.4 with that of Exercise I.6.2.3, we see the power of class field theory compared to an elementary and purely algebraic argument, limited to the Kummer case. Indeed, if $\mu_{p^e} \subset K$ we have ${}_{p^e}\mu(K_v) = \langle\, i_v(\zeta)\,\rangle$ for any place v, where here ζ is a generator of μ_{p^e}; the class field theory condition is only that ζ must be a local norm everywhere (the special case not occurring here), which is equivalent, by the Hasse's norm theorem for the cyclic extension L/K, to ζ being a global norm, which is the algebraic criterion.

For instance, a quadratic field $K(\sqrt{d})$ can be embedded in a cyclic extension of degree 4 of K if and only if d is the sum of two squares in K (equivalent to $-1 = x^2 - d\,y^2$, $x, y \in K$), but this is not the *practical* criterion since local normic conditions need easy computations contrary to the search of x and y (use $K = \mathbb{Q}(\sqrt{2})$ or $K = \mathbb{Q}(\sqrt{-1})$ to understand this since $K = \mathbb{Q}$ is too specific).

(ii) We have also proved a new local-global principle, saying that L/K (cyclic of arbitrary degree) can be embedded in M/K cyclic of degree $[L : K]p^e$ if and only if (outside of the special case) this happens locally everywhere (i.e., for every place v of K, L_v/K_v can be embedded in $M(v)/K_v$, cyclic of degree $[L_v : K_v]p^e$). If $M/L/K$ exists, then $M(v)/L_v/K_v$ clearly exists except if $[M_v : L_v] < p^e$; but in this case v is totally split in L/K (use the cyclicity of M/K), hence $L_v = K_v$, and we can always find a cyclic extension $M(v)$ of K_v of degree p^e containing L_v. It remains to show that the above local embedding property holds if and only if ${}_{p^e}\mu(K_v) \subseteq \mathrm{N}_{L_v/K_v}(L_v^\times)$, which is immediate since the structure of $\overline{G}_v^{\rm ab} := \mathrm{Gal}(\overline{K}_v^{\rm ab}/K_v)$ is known (see II.1.8.1) and is in particular such that ${}_{p^e}\overline{G}_v^{\rm ab} = {}_{p^e}\mu(K_v)$.

In the special case there is an obstruction since it is necessary to add the global norm condition on $\boldsymbol{s}$ to be able to globalize. □

4.15.8 Corollary. *The extension L/K (cyclic of degree p^r) can be embedded in a cyclic p-extension of arbitrarily large degree if and only if we have $L \subset H_p^{\mathrm{ord}}(p)$ (i.e., L/K is p-ramified and noncomplexified) and if, for any place v which is wildly ramified in L/K, we have $\mu_p(K_v) \subset \mathrm{N}_{L_v/K_v}(L_v^\times)$ with, in the special case, the additional condition* $\prod_{v|2} \Big(\frac{s_v\,,\,L/K}{v}\Big) = 1.$[26]

Proof. Simply look at the conditions of the theorem for e arbitrarily large (use 4.15.5).

In the special case, since the components s_v of $\boldsymbol{s}$ for $v \nmid 2$ are v-adic units, for any subextension L of $H_2^{\mathrm{ord}}(2)$ (which is then unramified at $v \nmid 2$) we have:

$$\rho_{L/K}(\boldsymbol{s}) = \prod_{v|2} \Big(\frac{s_v\,,\,L/K}{v}\Big),$$

which allow us to make any effective computation since s_v is (in K_v) one of the numbers $1 + \zeta_n$, $\zeta_{n+1} + \zeta_{n+1}^{-1}$, $\sqrt{-1}\,(\zeta_{n+1} + \zeta_{n+1}^{-1})$. □

4.15.9 Remark. In the p-elementary Kummer case (i.e., $\mu_p =: \langle\, \zeta \,\rangle \subset K$ and L/K cyclic of degree p), we recover the characterization of the $x \in K^\times$ such that $K(\sqrt[p]{x}\,)$ can be embedded in a cyclic p-extension of arbitrarily large degree ([BP], [Ja7], [Ng1]). Indeed, this happens if and only if the following two conditions are satisfied, where $\zeta_v := i_v(\zeta)$ and where $(\bullet\,,\,\bullet)_v$ denotes the Hilbert symbol of order p:

(i) $(i_v(x)\,,\,\zeta_v)_v = 1$ in the regular case $v \nmid p$; in other words:

$$\mathrm{v}(x) \equiv 0 \bmod (p) \text{ for all } v \nmid p\,,$$

which is equivalent to $K(\sqrt[p]{x}\,) \subset H_p^{\mathrm{ord}}(p)$,

(ii) $(i_v(x)\,,\,\zeta_v)_v =: \Big(\frac{x,\,\zeta}{v}\Big) = 1$ for all $v|p$.

In the special case, we add the condition $\prod_{v|2} (i_v(x)\,,\,s_v)_v = 1.$ □

The set of $x \in K^\times$ satisfying the above conditions is called the hilbertian radical of K (in the Kummer case).

We will solve in Section 6 the much more difficult problem of the initial radical of $\widetilde{K}_p$ which is of course contained in the above hilbertian radical.

h) The Grunwald–Wang Theorem — Weak Deployment Theorem for Decomposition Groups

We will show how the two theorems of the above title are in some sense analogous results although it is the first one which is famous.

[26] In this case, the value of $n \geq 2$ which characterizes the data of the special case is defined by $K \cap \mathbb{Q}(\mu_{2^\infty}) =: \mathbb{Q}(\zeta_n + \zeta_n^{-1})$.

4.16 THE IDELIC CHARACTERS. The problem is the following: is it possible to find a global abelian extension L/K for which the completions L_v/K_v are specified, at least for a finite set Σ of places of K? Once again this is a local-global problem whose most precise version is expressed in terms of characters on the idèle group.

4.16.1 Notation. If A is a topological abelian group, we denote by A^* the topological group (for the topology of pointwise convergence) of characters of A which are *continuous of finite order*, and we will always implicitly assume that these conditions are satisfied (see I.5.7). □

The continuity is essential (see I.5.3 which constructs an uninteresting noncontinuous character of order 2 of the group $J_{\mathbb{Q}}$ of idèles of $\mathbb{Q}$), but it is too general in practice: for example, on $J_{\mathbb{Q}}$ we can define χ by $\chi(\ell) := -1$ for any prime number $\ell > 0$ (seen as an idèle with support $\{\ell\}$), χ being trivial on $U_{\mathbb{Q}}^{\text{ord}}$; we thus obtain a continuous character of order 2 which has also nothing to do with the abelian arithmetic on $\mathbb{Q}$. Furthermore, a character of A of finite order is trivial on any divisible subgroup of A, and in our idelic context we are very naturally going to work in the group of reduced idèles.

All this leads to the following definitions.

4.16.2 Definitions (idelic characters). (i) A local character (implicitly, at a place v) is by definition any element of $(K_v^\times)^*$ (resp. of $\{\pm 1\}^*$) for v finite (resp. real infinite), and a semi-local character, regarding to a finite set Σ of places, is by definition any element of:

$$\langle \Sigma \rangle^* \simeq \bigoplus_{v\in\Sigma_0} (K_v^\times)^* \bigoplus_{v\in\Sigma_\infty} \{\pm 1\}^*,$$

where we have set by abuse of notation:

$$\langle \Sigma \rangle := \bigoplus_{v\in\Sigma_0} K_v^\times \bigoplus_{v\in\Sigma_\infty} \{\pm 1\},$$

for $\Sigma =: \Sigma_0 \cup \Sigma_\infty$. These characters can be considered as elements of J_0^*.

To a class $\overline{\chi}_v$ of local characters (for the relation $\chi'_v \sim \chi_v$ if and only if χ'_v and χ_v generate the same group) we can associate with the *cyclic* extension of K_v which corresponds, by local class field theory, to $N_v := \text{Ker}(\chi_v)$ (this is independent of the choice of representative). This defines a canonical bijection between the set of classes of local characters (at v) on that of cyclic extensions of K_v.

(ii) A global character is by definition any character, continuous of finite order, of C_0, hence of $C_0/D_0 \simeq \overline{G}^{\text{ab}}$. To link the local and global aspects, we consider in fact any global character χ as a character on J_0, trivial on a subgroup of the form $K^\times U_{0,\mathfrak{m}}^{\text{res}}$ (we then have $K^\times U_{0,\mathfrak{m}}^{\text{res}} \subseteq N := \text{Ker}(\chi) \subseteq J_0$); this is equivalent to saying that χ is a continuous character of J_0 trivial on $K^\times$. We still have $\chi \in J_0^*$.

Similarly, to a class $\overline{\chi}$ of global characters we can associate with the cyclic extension L of K which corresponds to $N := \mathrm{Ker}(\chi)$ by global class field theory: if $\mathrm{Ker}(\chi)$ contains $K^\times U_{0,\mathfrak{m}}^{\mathrm{res}}$, we have $L \subseteq K(\mathfrak{m})^{\mathrm{res}}$. □

Thus, all the characters defined above (local, semi-local, global) are *particular* characters, continuous of finite order, of J_0.

Note. Mention that in 1957 Kubota began in [Kub1] the study of the structure of the dual $\overline{G}^{\mathrm{ab}\,*}$ of $\overline{G}^{\mathrm{ab}}$, study which is based on the Grunwald–Wang theorem and which lead to a characterization of this group in terms of its fundamental invariants called, following Kaplansky, the "Ulm invariants".

4.16.3 Remark. By restriction to the $K_v^\times$, $v \in Pl_0$, and to the $\{\pm 1\}$, $v \in Pl_\infty^{\mathrm{r}}$, a global character χ defines the local characters χ_v. If N is the kernel of χ in J_0, that of χ_v in $K_v^\times$ (resp. in $\{\pm 1\}$) is hence $N_v = N \cap K_v^\times$ (resp. $N \cap \{\pm 1\}$), and by II.3.3.1, if L/K is the cyclic extension corresponding to $\overline{\chi}$, the local extensions corresponding to the $\overline{\chi}_v$ are the completions L_v/K_v of L/K; therefore there is an infinity of nontrivial χ_v (density Theorem II.4.6) hence χ is never a semi-local character. □

However we may ask whether it is possible to specify the χ_v on a *finite* set Σ of places of K. If χ exists, the restriction χ_Σ of χ to $\langle\, \Sigma \,\rangle$ is such that (in J_0^*):

$$\chi_\Sigma = \prod_{v \in \Sigma} \chi_v,$$

which is semi-local. In other words, the problem is to know under what conditions a given semi-local character $\chi_{(\Sigma)} := \prod_{v \in \Sigma} \chi_{(v)} \in \langle\, \Sigma \,\rangle^*$ is the restriction to $\langle\, \Sigma \,\rangle$ of a global character χ (i.e., such that $\chi_\Sigma = \chi_{(\Sigma)}$).

The answer is given by the following result, in which the special case enters once again.

4.16.4 Theorem (Grunwald–Wang (1933-1950)). *Let Σ be a finite set of noncomplex places of K. Let p be a fixed prime number. For each $v \in \Sigma$, we are given a local character $\chi_{(v)}$ of order p^{e_v}, $e_v \geq 0$, and we set:*

$$e := \max(e_v)_{v \in \Sigma}.$$

(i) Apart from the special case, there exists a global character χ of order p^e whose restriction to $\langle\, \Sigma \,\rangle$ is $\prod_{v \in \Sigma} \chi_{(v)}$.

(ii) In the Σ-special case [27], *the result is true if and only if the condition:*

[27] which occurs when $p = 2$, $K \cap \mathbb{Q}(\mu_{2^e}) = \mathbb{Q}(\zeta_n + \zeta_n^{-1})$, $2 \leq n < e$, $Pl_2^{\mathrm{ns}} := \{v | 2,\ \mathrm{Gal}(K_v(\zeta_{n+1})/K_v) \simeq \mathbb{Z}/2\mathbb{Z} \times \mathbb{Z}/2\mathbb{Z}\} \subseteq \Sigma$ (see II.6.3.3, II.6.3.4.2, II.6.3.4.3).

$$\prod_{v \in Pl_2^{\rm ns}} \chi_{(v)}(i_v(y_n^{2^{e-1}})) = 1,$$

is satisfied, where $y_n := 2 + \zeta_n + \zeta_n^{-1}$.

(iii) If in the Σ*-special case the above condition is not satisfied, there exists a solution* χ *of order* 2^{e+1}.

Proof. The reciprocity map $\rho : J_0 \longrightarrow \overline{G}^{\rm ab}$ yields the homeomorphism:

$$J_0/K^\times J_0^{p^e} \xrightarrow{\ \rho_e\ } \overline{G}^{\rm ab}/(\overline{G}^{\rm ab})^{p^e}.$$

Indeed, we have $\overline{G}^{\rm ab} \simeq C_0/D_0$, hence $\overline{G}^{\rm ab}/(\overline{G}^{\rm ab})^{p^e} \simeq C_0/C_0^{p^e} D_0 = C_0/C_0^{p^e}$, since $D_0 = D_0^{p^e}$; but $C_0/C_0^{p^e} \simeq J_0/K^\times J_0^{p^e}$. To simplify, set $F_e := \overline{K}^{\rm ab}{}_{[p^e]}$ (the maximal abelian pro-p-extension with exponent p^e of K); we thus have:

$${\rm Gal}(F_e/K) \simeq J_0/K^\times J_0^{p^e}.$$

The decomposition group $D_v(F_e/K)$ corresponds, under this homeomorphism, to the image of $K_v^\times$ (resp. of $\{\pm 1\}$) in $J_0/K^\times J_0^{p^e}$, and the extension F_e^Σ (maximal Σ-split subextension of F_e) is fixed under $\langle\, D_v(F_e/K)\,\rangle_{v\in\Sigma}$ which corresponds to:

$$\langle\, \Sigma\,\rangle / \langle\, \Sigma\,\rangle \cap (K^\times J_0^{p^e}).$$

Note. This computation of $D_v(F_e/K)$ does not follow directly from II.3.3, (iii), but from 4.12.5, (iii) since the image of $\widehat{K_v^\times}/(\widehat{K_v^\times})^{p^n} \simeq K_v^\times/K_v^{\times p^n}$ is still that of $K_v^\times$ for all $n \geq e$; this comes from the fact that the $D_v(F_e/K)$ are finite (thus compact) groups (review II.3.8.1).

The extension F_e/F_e^Σ is finite (but F_e/K is infinite, which implies that we must be careful).

4.16.4.1 Lemma 1. *The map* res : ${\rm Gal}(F_e/K)^* \longrightarrow {\rm Gal}(F_e/F_e^\Sigma)^*$ *is surjective.*

Proof. This follows from general properties of continuous characters on profinite abelian groups, but here it is also instructive to see it in terms of extensions, thanks to the following diagram:

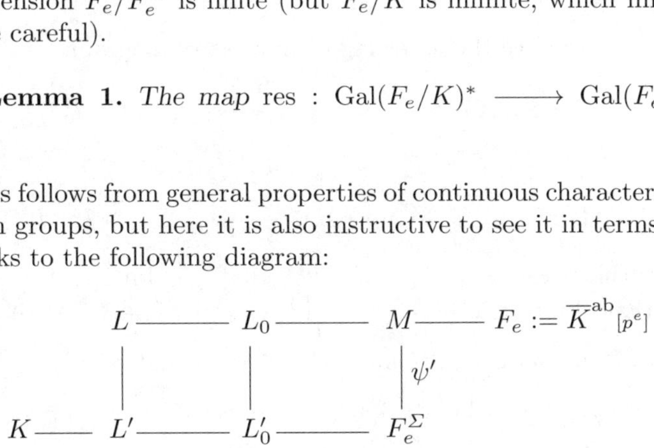

Fig. 4.3

Let ψ' be a character of $\mathrm{Gal}(F_e/F_e^{\Sigma})$; its kernel fixes a field M such that $F_e^{\Sigma} \subseteq M \subseteq F_e$. Set $M =: F_e^{\Sigma}(\theta)$, θ algebraic over F_e^{Σ} hence over K, then set $L_0 := K(\theta)$, which is a finite abelian extension of K such that M is the direct compositum of L_0 with F_e^{Σ} over $L'_0 := L_0 \cap F_e^{\Sigma}$. We may canonically identify $\mathrm{Gal}(M/F_e^{\Sigma})^*$ with $\mathrm{Gal}(L_0/L'_0)^*$. We are thus reduced to the finite case for which:

$$\mathrm{Gal}(L_0/K)^* \longrightarrow \mathrm{Gal}(L_0/L'_0)^*$$

is surjective; it follows that ψ', seen as a character of $\mathrm{Gal}(L_0/L'_0)$, is the restriction of a character ψ of $\mathrm{Gal}(L_0/K)$ which we lift to $\mathrm{Gal}(F_e/K)^*$; the character thus obtained (which we still denote by ψ) is a preimage of ψ'. □

4.16.4.2 Remark. The subfield L of L_0 fixed under the kernel of ψ is therefore such that M is the direct compositum over $L' := L \cap F_e^{\Sigma}$ of L with F_e^{Σ}. If ψ' has order p^e, it is clear that ψ has order p^e since F_e/K has exponent p^e; in this case we have $L' = K$, i.e., L/K cyclic of degree p^e. □

4.16.4.3 Lemma 2. *For any exact sequence of abelian topological groups $A \xrightarrow{f} B \xrightarrow{g} C \longrightarrow 1$, (with continuous maps) we have similarly the exact sequence $1 \longrightarrow C^* \xrightarrow{g^*} B^* \xrightarrow{f^*} A^*$, where f^*, g^* denote the usual dual maps of f, g.*

Proof. The injectivity of g^* is trivial because of the surjectivity of g.

Let $g^*(\psi_C) \in \mathrm{Im}(g^*)$; then we have $f^*(g^*(\psi_C)) = (g \circ f)^*(\psi_C) = 1$, hence $g^*(\psi_C) \in \mathrm{Ker}(f^*)$.

Now let $\psi_B \in \mathrm{Ker}(f^*)$; then $\psi_B(f(a)) = 1$ for all $a \in A$, thus ψ_B, which is trivial on $\mathrm{Im}(f) = \mathrm{Ker}(g)$, may be identified with a character of $B/\mathrm{Ker}(g) \simeq C$, giving $\psi_C \in C^*$ such that $g^*(\psi_C) = \psi_B$.

All the topological aspects are clear. □

By duality we have the following commutative diagram:

$$\begin{array}{ccccc}
1 \longrightarrow \mathrm{Gal}(F_e/K)^* & \xrightarrow{\rho_e^*} & J_0^* & \longrightarrow & (K^\times J_0^{p^e})^* \\
\downarrow \pi & & \downarrow & & \downarrow \\
1 \longrightarrow \mathrm{Gal}(F_e/F_e^{\Sigma})^* & \xrightarrow{\rho_e^*} & \langle \Sigma \rangle^* & \longrightarrow & (\langle \Sigma \rangle \cap (K^\times J_0^{p^e}))^*.
\end{array}$$

This shows that there exists $\chi \in J_0^*$ such that $\chi \in \mathrm{Im}(\rho_e^*)$ (i.e., χ is global) and such that (by restriction) $\chi_\Sigma = \prod_{v \in \Sigma} \chi_{(v)}$ given in $\langle \Sigma \rangle^*$, if and only if there exists a character $\psi' \in \mathrm{Gal}(F_e/F_e^{\Sigma})^*$ such that $\rho_e^*(\psi') = \prod_{v \in \Sigma} \chi_{(v)}$; we then have $\chi := \rho_e^*(\psi)$, where $\pi(\psi) = \psi'$. Since $\rho_e^*(\psi') = \prod_{v \in \Sigma} \chi_{(v)}$ has order p^e, ψ', hence also ψ and χ, have order p^e. The existence of χ is therefore equivalent to $\prod_{v \in \Sigma} \chi_{(v)}$ being trivial on:

$$\langle \Sigma \rangle \cap (K^\times J_0^{p^e}).$$

The following lemma shows that this last condition is generally satisfied.

4.16.4.4 Lemma 3. *We have* $\langle \Sigma \rangle \cap (K^\times J_0^{p^e}) = \langle \Sigma \rangle^{p^e}$, *except in the* Σ-*special case, where* $\langle \Sigma \rangle \cap (K^\times J_0^{2^e}) = \langle i_\Sigma(y_n^{2^{e-1}}) \rangle \langle \Sigma \rangle^{2^e}$, *in which* $\langle \Sigma \rangle^{2^e}$ *has index 2.*

Proof. Let $(x_v)_{v\in\Sigma} \in \langle \Sigma \rangle$ of the form $i_0(x)\,\boldsymbol{y}^{p^e}$, $x \in K^\times$, $\boldsymbol{y} \in J_0$. For $v \notin \Sigma$ we obtain:

$$1 = i_v(x)\, y_v^{p^e},$$

showing that x is locally a p^eth power on $Pl^{\mathrm{nc}} \backslash \Sigma$; thus in general $x = y^{p^e}$, $y \in K^\times$, with the possible additional solutions:

$$x = y_n^{2^{e-1}}\, y^{2^e},$$

in the Σ-special case (for which $Pl_2^{\mathrm{ns}} \subseteq \Sigma$). One inclusion ($\subseteq$) follows, and the other comes from the fact that in the Σ-special case:

$$i_\Sigma(y_n^{2^{e-1}}) = i_0(y_n^{2^{e-1}})\, i_{Pl^{\mathrm{nc}}\backslash\Sigma}(y_n^{2^{e-1}})^{-1} \in i_0(y_n^{2^{e-1}})\, J_0^{2^e},$$

since we know that for $v \notin Pl_2^{\mathrm{ns}}$ we have $i_v(y_n^{2^{e-1}}) \in K_v^{\times\, 2^e}$.[28] □

This proves the general case (points (i) and (ii)) of the theorem.

Let us now consider the particular case (of the special case), where:

$$\prod_{v \in Pl_2^{\mathrm{ns}}} \chi_{(v)}(i_v(y_n^{2^{e-1}})) = -1.$$

We again use the above commutative diagram for the exponent 2^{e+1}; the condition $\prod_{v \in Pl_2^{\mathrm{ns}}} \chi_{(v)}(i_v(y_n^{2^e})) = 1$ is then satisfied (the $\chi_{(v)}$ have order dividing 2^e) and ψ again exists in $\mathrm{Gal}(F_{e+1}/K)^*$ (it has order 2^{e+1} and $\psi' := \pi(\psi)$ has order 2^e). The global character $\chi := \rho_{e+1}^*(\psi)$ then has order 2^{e+1} and the corresponding cyclic extension L/K has degree 2^{e+1} (but the quadratic subextension L'/K is Σ-split), finishing the proof of the theorem. □

Note. Because of the descent that we have performed, it is not possible to control the ramification in L/K (see in V.3.3 a more effective but less general form of the Grunwald–Wang theorem).

4.16.5 Example. This is the simplest example, already considered in Artin–Tate. We take $K = \mathbb{Q}$, $p = 2$, $e = 3$, $\Sigma = \{v_0 := 2\}$. We are thus in the Σ-special case with:

[28] Indeed, $i_v(y_n^{2^{e-1}}) = (1+\zeta_n)^{2^e}$, $(\zeta_{n+1} + \zeta_{n+1}^{-1})^{2^e}$, or $(\sqrt{-1}\,(\zeta_{n+1} + \zeta_{n+1}^{-1}))^{2^e}$, in K_v, depending on the splitting of v in $K(\mu_{2^{n+1}})/K$.

$$\Sigma = Pl_2^{\mathrm{ns}} = \{v_0\},\ \ y_n^{2^{e-1}} = y_2^4 = 16.$$

We take for $\overline{\chi_{(v_0)}}$ the class corresponding to the unramified extension of degree 8 of $\mathbb{Q}_{v_0}$. The required condition:

$$\prod_{v \in Pl_2^{\mathrm{ns}}} \chi_{(v)}(i_v(y_n^{2^{e-1}})) = 1,$$

can be written here as $\chi_{(v_0)}(16) = 1$ and means that 16 must be a norm in the unramified extension of degree 8 of $\mathbb{Q}_{v_0}$, which is impossible since its norm group is $2^{8\mathbb{Z}} \oplus \mathbb{Z}_2^\times$.

This means that there do not exist global cyclic extensions $L/\mathbb{Q}$ of degree 8 in which 2 is totally inert. Let us prove again this surprising fact by a direct reasoning. Let $L/\mathbb{Q}$ be such an extension; for $v \neq v_0 := 2$ (including the place ∞), the extension $\mathbb{Q}_v(\mu_8)/\mathbb{Q}_v$ is cyclic of degree 1 or 2 (because it is unramified) and one of the numbers $1+\sqrt{-1}$, $\sqrt{-2}$, $\sqrt{2}$ is in $\mathbb{Q}_v$, so that:

$$16 = 2^4 = \ (1+\sqrt{-1})^8 \text{ or } (\sqrt{-2})^8 \text{ or } (\sqrt{2})^8 \in \mathbb{Q}_v^{\times\, 8},$$

showing that, for trivial reasons, 16 is a local norm at any place $v \neq v_0$ in $L/\mathbb{Q}$; but by the product formula, 16 is also a local norm at v_0 in $L/\mathbb{Q}$, contradiction. It is easily checked that $L := \mathbb{Q}(\mu_{17})$, cyclic of degree 16, gives the local extension $L_{v_0}/\mathbb{Q}_{v_0}$, unramified of degree 8 (the decomposition subfield of v_0 in $L/\mathbb{Q}$ being $\mathbb{Q}(\sqrt{17})$). □

4.16.6 Exercise. Show that, for any family $(n_v)_{v\in\Sigma}$ of integers such that n_v is the degree of a cyclic extension of K_v (i.e., $n_v = 1$ or 2 for $v \in \Sigma_\infty$, n_v arbitrary for $v \in \Sigma_0$), there exists a global cyclic extension L/K of degree g.c.d.$(n_v)_{v\in\Sigma}$ such that $[L_v : K_v] = n_v$ for all $v \in \Sigma$.

Answer. In the general case this is a trivial corollary to the above theorem. Since the required conditions are weaker, we can avoid the consequences of the Σ-special case by choosing the $\chi_{(v)}$ appropriately in the following way, in the spirit of [d, AT, Ch. 10, § 2, Lem. 8]. We omit the embeddings i_v and we localize the problem at $p = 2$. Suppose we are in the special case with $\prod_{v\in Pl_2^{\mathrm{ns}}} \chi_{(v)}(y_n^{2^{e-1}}) = -1$. Then there exists a $\chi_{(v_0)}$, $v_0 \in Pl_2^{\mathrm{ns}}$, of order 2^e such that $\chi_{(v_0)}(y_n^{2^{e-1}}) = -1$. By definition of Pl_2^{ns}, $L'_{v_0} := K_{v_0}(\zeta_{n+e} + \zeta_{n+e}^{-1})$ is a cyclic extension of degree 2^e of K_{v_0} and its quadratic subextension is $L''_{v_0} := K_{v_0}(\zeta_{n+1} + \zeta_{n+1}^{-1}) = K_{v_0}(\sqrt{y_n})$. Let $\chi'_{(v_0)}$ be a character of order 2^e of L'_{v_0} and $\chi''_{(v_0)} := {\chi'_{(v_0)}}^{2^{e-1}}$ that of L''_{v_0}. We have $\mathrm{N}_{L''_{v_0}/K_{v_0}}(\sqrt{y_n}) = -y_n$; thus, $\chi'_{(v_0)}(y_n^{2^{e-1}}) = \chi'_{(v_0)}((-y_n)^{2^{e-1}}) = {\chi'_{(v_0)}}^{2^{e-1}}(-y_n) = \chi''_{(v_0)}(-y_n) = 1$ since $-y_n = \mathrm{N}_{L''_{v_0}/K_v}(\sqrt{y_n})$ is in the norm group of L''_{v_0}. Therefore, replacing $\chi_{(v_0)}$ by $\chi'_{(v_0)}$, we get the result. □

This statement is much weaker than Grunwald–Wang Theorem 4.16.4; for example in the context of 4.16.5 it is easy to find cyclic extensions $L/\mathbb{Q}$ of degree 8 such that $L_{v_0}/\mathbb{Q}_{v_0}$ is of degree 8 (for $v_0 = 2$).

Any finite abelian p-extension L of K is contained in F_e for a suitable e, thus we have:

$$\mathrm{Gal}(\overline{K}^{\mathrm{ab}}{}_{(p)}/\overline{K}^{\mathrm{ab}}{}_{(p)}{}^{\Sigma}) := \varprojlim_{L} \mathrm{Gal}(L/L^{\Sigma}) = \varprojlim_{e\geq 1} \varprojlim_{L_e\subset F_e} \mathrm{Gal}(L_e/L_e^{\Sigma}).$$

Then, by going to the inverse limit on e for the expression of $\mathrm{Gal}(F_e/F_e^{\Sigma})$, we obtain the following important result of deployment of the decomposition groups of the elements of Σ.

4.16.7 Corollary (weak deployment theorem for decomposition groups). *For any finite set Σ of noncomplex places of K, we have:*

$$\mathrm{Gal}(\overline{K}^{\mathrm{ab}}{}_{(p)}/\overline{K}^{\mathrm{ab}}{}_{(p)}{}^{\Sigma}) = \bigoplus_{v\in\Sigma} D_v(\overline{K}^{\mathrm{ab}}{}_{(p)}/K),$$

where each $D_v(\overline{K}^{\mathrm{ab}}{}_{(p)}/K)$ is by 4.12.5 isomorphic to $(\widehat{K_v^\times})_p$.[29]

Proof. Indeed, this is clear in the general case since:

$$\begin{aligned}\mathrm{Gal}(\overline{K}^{\mathrm{ab}}{}_{(p)}/\overline{K}^{\mathrm{ab}}{}_{(p)}{}^{\Sigma}) &\simeq \varprojlim_{e\geq 1} (\langle\, \Sigma\,\rangle/\langle\, \Sigma\,\rangle^{p^e}) \\ &\simeq \bigoplus_{v\in\Sigma} \varprojlim_{e\geq 1} (\langle\, \{v\}\,\rangle/\langle\, \{v\}\,\rangle^{p^e}) = \bigoplus_{v\in\Sigma} (\widehat{K_v^\times})_p.\end{aligned}$$

In the Σ-special case, we have:

$$\mathrm{Gal}(\overline{K}^{\mathrm{ab}}{}_{(2)}/\overline{K}^{\mathrm{ab}}{}_{(2)}{}^{\Sigma}) \simeq \varprojlim_{e\geq 1} (\langle\, \Sigma\,\rangle/\langle\, i_\Sigma(y_n^{2^{e-1}})\,\rangle\langle\, \Sigma\,\rangle^{2^e}),$$

but the family of subgroups $\langle\, \Sigma\,\rangle^{2^e}$ is cofinal in that of the $\langle\, i_\Sigma(y_n^{2^{e-1}})\,\rangle\langle\, \Sigma\,\rangle^{2^e}$, so the result is the same. □

The above results obtained from the Grunwald–Wang theorem show that there is (apart in the Σ-special case) a weak deployment theorem for the decomposition groups in $\overline{K}^{\mathrm{ab}}{}_{[p^e]}/K$ which is an infinite extension with exponent p^e. This is stronger than 4.16.7.

4.16.8 Corollary. *If $\langle\, \Sigma\,\rangle/\langle\, \Sigma\,\rangle^{p^e} \simeq (\mathbb{Z}/p^e\mathbb{Z})^r$, $r \in \mathbb{N}$, then (apart from the Σ-special case), we can find a subextension F/K of $\overline{K}^{\mathrm{ab}}{}_{[p^e]}/K$ such that:*

[29] For v real infinite, we indeed have $\widehat{K_v^\times} = \widehat{\mathbb{R}^\times} \simeq \{\pm 1\}$.

$$\mathrm{Gal}(F/K) = \bigoplus_{v\in\Sigma} D_v(F/K) \simeq \langle\,\Sigma\,\rangle/\langle\,\Sigma\,\rangle^{p^e}.$$

The case $e = 1$ is thus always suitable. □

Note that the ramification of F/K cannot be controlled outside of Σ.

In addition, if an abelian extension L/K is such that the inertia groups are deployed in the (strong) sense of 1.4.3, this is not necessarily the case for the decomposition groups. For example, for $K = \mathbb{Q}$ and $L = \mathbb{Q}(\sqrt{5}, \sqrt{13})$, then $I_5(L/\mathbb{Q})$ and $I_{13}(L/\mathbb{Q})$ are deployed but not the decomposition groups; on the other hand, $F = \mathbb{Q}(\sqrt{5}, \sqrt{13}, \sqrt{17}, \sqrt{37})$ is such that:

$$\mathrm{Gal}(F/\mathbb{Q}) = D_5(F/\mathbb{Q}) \oplus D_{13}(F/\mathbb{Q}),$$

but it has been necessary to add some ramification.

4.16.9 Remark. Let Σ' be a finite set of noncomplex places of K containing Pl_p. Since $\mathrm{Gal}(\overline{K}^{\mathrm{ab}}(p)/\overline{K}^{\mathrm{ab}}(p)^{\Sigma'}) = \bigoplus_{v\in\Sigma'} D_v(\overline{K}^{\mathrm{ab}}(p)/K)$, and since $\overline{K}^{\mathrm{ab}}(p)^{\Sigma'} = H_{\mathrm{ta}}^{\mathrm{res}}(p)^{\Sigma'}$, under projection modulo $\mathrm{Gal}(\overline{K}^{\mathrm{ab}}(p)/H_{\mathrm{ta}}^{\mathrm{res}}(p)) = \bigoplus_{v|p} I_v(\overline{K}^{\mathrm{ab}}(p)/K)$, we obtain:

$$\mathrm{Gal}(H_{\mathrm{ta}}^{\mathrm{res}}(p)/H_{\mathrm{ta}}^{\mathrm{res}}(p)^{\Sigma'}) = \bigoplus_{v\in\Sigma'} D_v(H_{\mathrm{ta}}^{\mathrm{res}}(p)/K).$$

We thus obtain the weak deployment theorem for decomposition groups in $H_{\mathrm{ta}}^{\mathrm{res}}(p)/K$, for an arbitrary finite set Σ of (tame or not) noncomplex places of K (we have used $\Sigma' := \Sigma \cup Pl_p$ only to have a comfortable projection!). These decomposition groups are isomorphic to:

$$\mathbb{Z}_p, \quad \mathbb{Z}_p \times (F_v^\times)_p, \ \text{ or } \ (\{\pm 1\})_p,$$

according to whether $v \in Pl_p$, $v \in Pl_{\mathrm{ta}}$, or $v \in Pl_\infty^{\mathrm{r}}$ (compare with 4.5.4, 4.5.5). □

4.17 Roblot's Problem [Ro, Ch. II, § 2]. The above results on weak deployment of decomposition groups in the $\overline{K}^{\mathrm{ab}}[p^e]/K$ enable us to study some difficult questions such as the following (which can be considered as a generalization of the statement of Grunwald–Wang and can be studied in terms of local characters).

4.17.1 Question. Let $\Sigma \subset Pl_0$ and $S =: S_1 \cup S_2 \subset Pl^{\mathrm{nc}}$ (disjoint union) be two finite disjoint sets of places of K, and let $L \subset \overline{K}^{\mathrm{ab}\,S}$, L/K finite; does there exist a cyclic extension $M \subset \overline{K}^{\mathrm{ab}}$ of degree p over L, nonsplit on $\Sigma \cup S_2$ and split on S_1? □

Set $F_e := \overline{K}^{\mathrm{ab}}[p^e]$, $e \geq 1$ (in fact, everything will take place in $F_e^{S_1}/F_e^{S\cup\Sigma}$, where the decomposition of the places of $S_2 \cup \Sigma$ is deployed, apart from the special case, but we will not consider it). We can look for solutions $M =: LM_e$ split on $L_{[p^e]}$, with M_e contained in $F_e^{S_1}$, by increasing values of e.

For example, if $e = 1$ (in which case the Galois group is a vector space), the problem has a solution if and only if for all $v \in \Sigma$, the compositum of $L' := L_{[p]}F_1^{S\cup\Sigma}$ with $F_1^{S\cup\{v\}}$ is different from F_1^S (in other words, $|D_v(L_{[p]}/K)| < |D_v(F_1^S/K)|$, or $\mathrm{rk}_p(D_v(L_{[p]}/K)) < \mathrm{rk}_p(D_v(\overline{K}^{\mathrm{ab}}/K)) = \mathrm{rk}_p(K_v^\times))$, which can be read in L/K, the nonsplitting condition on S_2 being always satisfiable. Indeed, it is then easy to find M'/L' of degree p, $M' \subset F_1^{S_1}$, such that for all $v \in S_2 \cup \Sigma$, $M' \not\subset L'F_1^{S_1\cup\{v\}}$ (the details are left to the reader); any M of degree p over $L_{[p]}$ such that $ML' = M'$ is a solution.

For e sufficiently large, we will always have a solution M' on $L' := L_{[p^e]}F_e^{S\cup\Sigma}$ since each decomposition group $(\widehat{K_v^\times})_p$, $v \in \Sigma$, contains a subgroup isomorphic to $\mathbb{Z}_p$. Therefore we obtain M'/L' cyclic of degree p, corresponding to a character ψ' of order p of $\mathrm{Gal}(F_e^{S_1}/L')$, and we must then descend it to $L_{[p^e]}$, keeping under control the order as much as possible (this is the principle of Lemma 1 of the Grunwald–Wang theorem).

It is not always possible to descend M' to an extension of degree p over L, as is shown by the following example for the case $p = 2$ and arbitrary S.

4.17.2 Example. Let K be a totally real number field; we assume that for the prime numbers $\ell \in \{3, 7\}$ there exists a place v_ℓ of K above ℓ such that $[K_{v_\ell} : \mathbb{Q}_\ell]$ is odd; it follows that the 2-Sylow subgroup of $F_{v_\ell}^\times$ has order 2. Let $L := K(\sqrt{3}, \sqrt{7})$ and let Σ be the set of ramified places in L/K (it consists in places above 3 and 7, including v_3 and v_7, and possibly some places above 2). The extension L/K is biquadratic. If the quadratic extension M/L exists, $\mathrm{Gal}(M/K)$ is isomorphic to $(\mathbb{Z}/2\mathbb{Z})^3$ or to $\mathbb{Z}/4\mathbb{Z} \times \mathbb{Z}/2\mathbb{Z}$.

The first case is impossible since otherwise we would have:

$$\mathrm{Gal}(M_{v_7}/K_{v_7}) \simeq (\mathbb{Z}/2\mathbb{Z})^3$$

since $\mathrm{Gal}(L_{v_7}/K_{v_7})$ is biquadratic and M_{v_7}/L_{v_7} of degree 2, a contradiction (the 2-rank of this group is at most equal to 2 by II.1.8.2, (i)). In other words, the condition $\mathrm{rk}_2(D_v(L/K)) < \mathrm{rk}_2(K_v^\times)$ of the case $e = 1$ is not satisfied for the place $v = v_7$.

The second case is also impossible since none of the quadratic extensions:

$$K(\sqrt{3}),\quad K(\sqrt{7}),\quad K(\sqrt{21}),$$

of K, can be embedded in a cyclic extension F of degree 4 of K (by II.1.3.3 or 4.3.2, the inertia groups of v_3 and v_7 in an abelian 2-extension of K are at most of order 2; hence v_3 or v_7 would be totally ramified in F/K).

We stop here the technical part which depends on the context (without forgetting the possible special case).

4.18 Globalized Global Class Field Theory. We now give a formalism which takes directly into account the algebraic and topological aspects studied up to now: it is the doubly global one, given by Jaulent in [Ja7]. This formalism is adapted both to the localized description (i.e., of $\overline{K}^{\mathrm{ab}}{}_{(p)}/K$) and to the globalized description (i.e., of $\overline{K}^{\mathrm{ab}}/K$) of global class field theory over K, which explains the difficulty of starting from this to obtain the classical properties at a finite level, and is a justification for our inverse path which in any case is essential for the practical use of the theory. However, its statement a posteriori clarifies the role played by classical objects (in particular it is of reduced idelic type), it allows a natural study of abelian extensions with restricted ramification, and is probably the clearest local-global formulation (in particular because it is structurally based on the Schmidt–Chevalley theorem and expresses directly the Grunwald–Wang theorem in the form of weak deployment of decomposition groups).

4.18.1 Notations. Let K be a number field; consider the following objects:

• $\mathcal{J}_v := \widehat{K_v^\times}$, $v \in Pl^{\mathrm{nc}}$ (see II.1.7 and note that, for $v \in Pl_0$ then $\mathcal{J}_v = \pi_v^{\widehat{\mathbb{Z}}} \oplus \mathcal{U}_v$, where $\mathcal{U}_v = U_v$, and that for $v \in Pl_\infty^{\mathrm{r}}$ then $\mathcal{J}_v = \{\pm 1\}$ and $\mathcal{U}_v = 1$);

• $\mathcal{J} := \mathcal{J}_K := \prod_v' \mathcal{J}_v$ (idèle group), restricted product with respect to the family of unit groups $\mathcal{U}_v$;

• $\mathcal{U} := \mathcal{U}_K := \prod_v \mathcal{U}_v = \prod_{v \in Pl_0} U_v$ (group of unit idèles);

• $\mathcal{K}^\times := K^\times \otimes \widehat{\mathbb{Z}}$;

• $\overline{\jmath} := (\overline{\jmath}_v)_v : \mathcal{K}^\times \longrightarrow \mathcal{J}$, where $\overline{\jmath}_v$ is the extension by $\widehat{\mathbb{Z}}$-linearity of the canonical injection:

$$j_v : K^\times \otimes 1 \xrightarrow{\ i_v\ } K_v^\times \longrightarrow \mathcal{J}_v \ ;$$

• $\mathcal{C} := \mathcal{C}_K := \mathcal{J}/\overline{\jmath}(\mathcal{K}^\times)$ (idèle class group);

• $\rho : \mathcal{J} \longrightarrow \overline{G}^{\mathrm{ab}} := \mathrm{Gal}(\overline{K}^{\mathrm{ab}}/K)$ (global reciprocity map), defined for an idèle $\boldsymbol{x} =: (x_v)_v \in \mathcal{J}$ by:

$$\rho(\boldsymbol{x}) := \prod_v \rho_v(x_v),$$

where:

$$\rho_v : \mathcal{J}_v \longrightarrow \overline{G}_v^{\mathrm{ab}} := \mathrm{Gal}(\overline{K}_v^{\mathrm{ab}}/K_v) \xrightarrow{\ \simeq\ } D_v(\overline{K}^{\mathrm{ab}}/K),$$

is the local reciprocity map interpreted in $\overline{G}^{\mathrm{ab}}$ (see II.1.7, and 4.5 or 4.12.5). □

4.18.2 Remarks. (i) More precisely, if L/K is finite, consider the (finite) product:

$$\rho_{L/K}(\boldsymbol{x}) := \prod_v \rho_{v,L/K}(x_v),$$

where $\rho_{v,L/K}$ is the composition of ρ_v with the projection on $\mathrm{Gal}(L/K)$; then we check that $(\rho_{L/K}(\boldsymbol{x}))_L \in \varprojlim_L \mathrm{Gal}(L/K)$, thus defining $\rho(\boldsymbol{x})$.

(ii) The restricted product topology on $\mathcal{J}$ (with respect to the $\mathcal{U}_v$) is defined from that of the $\mathcal{J}_v$, recalling that a fundamental system of neighbourhoods of the unit element of $\mathcal{J}_v$ for $v \in Pl_0$ is formed by the:

$$\pi_v^{n\widehat{\mathbb{Z}}} \oplus U_v^m, \quad n \geq 1, \ m \geq 0.$$

Warning: the topology of $\mathcal{K}^\times$ is then not anymore that of the $\mathcal{K}^{\times n}$ (product of the topologies generally used for the p-completions of a group, as in 2.3.3), but that for which the open sub-$\widehat{\mathbb{Z}}$-modules $\mathcal{O}$ are characterized by the condition:

$$(\mathcal{E}^S : \mathcal{O} \cap \mathcal{E}^S) < \infty \quad \text{for all } S \subset Pl^{\mathrm{nc}}, \ |S| < \infty,$$

where $\mathcal{E}^S := E^S \otimes \widehat{\mathbb{Z}}$. □

4.18.3 Theorem. *We have the following results:*

(i) The map $\overline{\jmath}$ is continuous, the subgroup $\overline{\jmath}(\mathcal{K}^\times)$ is closed in $\mathcal{J}$, the quotient $\mathcal{J}/\overline{\jmath}(\mathcal{K}^\times)\mathcal{U}$ is isomorphic to the restricted class group of K, and $\mathcal{C}$ is compact.

(ii) The reciprocity map ρ has kernel equal to $\overline{\jmath}(\mathcal{K}^\times)$ and defines the homeomorphism:

$$\overline{G}^{\mathrm{ab}} \simeq \mathcal{C}.$$

(iii) The image of $\mathcal{J}_v$ (considered as a subgroup of $\mathcal{J}$) under ρ is the decomposition group of v in $\overline{K}^{\mathrm{ab}}/K$, and that of U_v^i, $i \geq 0$, is the ith higher ramification group (in upper numbering) of v in $\overline{K}^{\mathrm{ab}}/K$.

(iv) For any finite set Σ of noncomplex places, the image of $\bigoplus_{v\in\Sigma} \mathcal{J}_v$ under ρ is equal to $\bigoplus_{v\in\Sigma} D_v(\overline{K}^{\mathrm{ab}}/K)$ (weak deployment of the decomposition groups).

(v) The correspondence of infinite global class field theory can be expressed in terms of closed subgroups of $\mathcal{J}$ containing $\overline{\jmath}(\mathcal{K}^\times)$ (or of closed subgroups of $\mathcal{C}$). □

Everything can be justified from the classical results of the preceding chapters; all the statements that we have given can be given in this setting (see [Ja7] for additional material, as well as [GrJ, App.] for logarithmic aspects of the problem).

We can also write the localized version at p of the above (p-adic class field theory).

§5 Explicit Computations in Incomplete p-Ramification

We are now going to study $\mathrm{Gal}(H_T^S(p)/K)$ when T does not necessarily contain all the places above p, where we recall that $H_T^S(p)$ is the maximal S-split T-ramified abelian pro-p-extension of K.

In the preceding sections we have seen that all the $\mathbb{Z}_p$-rank problems rely on the knowledge of the T_p-adic rank $\mathrm{r}_{T_p}^{S_0}$ of the S_0-unit group or, equivalently, on that of the number $\widetilde{\mathrm{r}}_{T_p}^{S_0}$ of independent S_0-split T_p-ramified $\mathbb{Z}_p$-extensions since we have the relation:

$$\widetilde{\mathrm{r}}_{T_p}^{S_0} = \sum_{v \in T_p} [K_v : \mathbb{Q}_p] - \mathrm{r}_{T_p}^{S_0}.$$

When $T_p = Pl_p$, we have been able to prove theoretical results assuming the classical p-adic conjecture. When $T_p \neq Pl_p$, the computation of $\widetilde{\mathrm{r}}_{T_p}^{S_0}$ is more difficult because $\bigoplus_{v \in T_p} K_v$, and a fortiori $\mathcal{L}_{T_p}^{S_0} := \bigoplus_{v \in T_p} K_v \big/ \mathbb{Q}_p \log_{T_p}(E^{S_0})$, cannot be analyzed in terms of representations of the Galois group of the Galois closure of K. For instance this contains the case of the computation of the $\mathbb{Z}_p$-rank of the inertia or decomposition groups of places above p in $\overline{K}^{\mathrm{ab}}(p)/H_p^{\mathrm{ord}}(p)$ that we have done in Subsections (e) and (f) assuming Conjecture 4.12 since, for $T_p = Pl_p \backslash \{v\}$, $S_0 = \emptyset$, we conjecturally have:

$$\widetilde{\mathrm{r}}_{T_p} =: \widetilde{\mathrm{r}}_{p \backslash v} = \mathrm{d}_v + r_2 + 1 - [K_v : \mathbb{Q}_p],$$

where d_v is an accessible invariant (see 4.12.4).

To able to do this without assuming any p-adic conjecture, we are going to study the problem from a different, more numerical, point of view, using systematically the logarithm introduced in Section 2. This will enable us (in addition to the computation of $\widetilde{\mathrm{r}}_{T_p}^{S_0}$), to determine completely the decomposition of places in the compositum of the $\mathbb{Z}_p$-extensions of K when the usual data (class and unit groups) are known.

5.1 Notations. Recall the main notations already given in 2.1 and 2.2:

- T_p is a subset of Pl_p, S_0 is a finite set of finite places disjoint from T_p,
- $\widetilde{K}_{T_p}^{S_0}$ is the maximal T_p-ramified S_0-split subextension of the compositum $\widetilde{K}_p$ of the $\mathbb{Z}_p$-extensions of K,
- $\widetilde{K}_{T_p}^{S_0\,\mathrm{fr}}$ is the maximal $\mathbb{Z}_p$-free subextension of $\widetilde{K}_{T_p}^{S_0}$ (i.e., the compositum of the T_p-ramified S_0-split $\mathbb{Z}_p$-extensions of K),
- $\mathcal{Z}_{T_p}^{S_0} := \mathrm{Gal}(\widetilde{K}_{T_p}^{S_0\,\mathrm{fr}}/K) \simeq \mathbb{Z}_p \mathrm{Log}_{T_p}^{S_0}(I_{T_p}) \subset \bigoplus_{v \in T_p} K_v \big/ \mathbb{Q}_p \log_{T_p}(E^{S_0})$. □

By 2.5, (iii), if $v \nmid p$ is a finite place, the image in:

$$\mathbb{Z}_p\mathrm{Log}_p(I_p) \simeq \mathcal{Z}_p := \mathrm{Gal}(\widetilde{K}_p/K)$$

of the decomposition group $\widetilde{D}_v$ of v in $\widetilde{K}_p/K$ is equal to:

$$\mathrm{Log}_p(\widetilde{D}_v) = \mathbb{Z}_p\mathrm{Log}_p(\mathfrak{p}_v).$$

More generally, if S_0 is a finite set of tame finite places of K, we get that the group $\mathrm{Gal}(\widetilde{K}_p/\widetilde{K}_p^{S_0})$ is isomorphic to $\mathbb{Z}_p\mathrm{Log}_p(\langle\, S_0\,\rangle) := \sum_{v\in S_0} \mathbb{Z}_p\mathrm{Log}_p(\mathfrak{p}_v)$; we then have $\mathrm{Gal}(\widetilde{K}_p^{S_0}/K) \simeq \mathbb{Z}_p\mathrm{Log}_p(I_p)/\mathbb{Z}_p\mathrm{Log}_p(\langle\, S_0\,\rangle)$ whose $\mathbb{Z}_p$-torsion module yields $\mathrm{Gal}(\widetilde{K}_p^{S_0}/\widetilde{K}_p^{S_0\,\mathrm{fr}})$. Naturally, the inertia group in $\widetilde{K}_p/K$ of a tame place is trivial.

Assume now that $v|p$. In this case, the Artin map is not defined for $\mathfrak{p}_v$, and we must use the Hasse symbol II.3.1.2 which will bring us to Frobenius' computations (see II.4.4.3). Everything relies on the following general computation given in the context of incomplete p-ramification (i.e., which consists in working in $\widetilde{K}_\Sigma^{\mathrm{fr}}$, with $\Sigma \subseteq Pl_p$, instead of $\widetilde{K}_p$).

5.1.1 Lemma. *Let $\Sigma \subseteq Pl_p$ and let $v \in \Sigma$. Then for all $x \in K^\times$, the image of $\left(\frac{x\,,\,\widetilde{K}_\Sigma^{\mathrm{fr}}/K}{v}\right)$ in $\mathbb{Z}_p\mathrm{Log}_\Sigma(I_\Sigma) \simeq \mathcal{Z}_\Sigma$ is (by composition of* Log_Σ *with* Art^{-1}*):*

$$\mathrm{Log}_\Sigma\left(\frac{x\,,\,\widetilde{K}_\Sigma^{\mathrm{fr}}/K}{v}\right) = \mathrm{v}(x)\mathrm{Log}_\Sigma(\mathfrak{p}_v) - \log_v(x) \bmod \mathbb{Q}_p\log_\Sigma(E).$$

Note. In this expression, $\log_v(x) := \log(i_v(x)) \in K_v$ considered as a subgroup of $\bigoplus_{v'\in\Sigma} K_{v'}$, log is the Iwasawa logarithm, and $\mathrm{Log}_\Sigma(\mathfrak{p}_v) := \frac{1}{m}\log_\Sigma(\alpha) \bmod \mathbb{Q}_p\log_\Sigma(E)$, for any m such that $\mathfrak{p}_v^m = (\alpha)$, $\alpha \in K^\times$, and $\log_\Sigma = (\log_{v'})_{v'\in\Sigma}$.

Proof of the lemma. Set $\mathfrak{m}(n) = \prod_{v'\in\Sigma} \mathfrak{p}_{v'}^n$, $\mathfrak{m}_v(n) := \mathfrak{p}_v^n$, $n \in \mathbb{N}$, and let $\beta_n \in K^\times$ such that:

$$\beta_n x^{-1} \equiv 1 \bmod \mathfrak{m}_v(n),$$
$$\beta_n \equiv 1 \bmod \frac{\mathfrak{m}(n)}{\mathfrak{m}_v(n)}\ ;$$

by II.4.4.3, β_n is a v-associate of x for the computation of the Hasse's symbol $\left(\frac{x\,,\,K(\mathfrak{m}(n))^{\mathrm{ord}}/K}{v}\right)$; therefore we want to compute:

$$\left(\frac{x\,,\,\widetilde{K}_\Sigma^{\mathrm{fr}}/K}{v}\right) = \lim_n \left(\frac{x\,,\,K(\mathfrak{m}(n))^{\mathrm{ord}} \cap \widetilde{K}_\Sigma^{\mathrm{fr}}/K}{v}\right),$$

which is given by $\lim_n \left(\frac{K(\mathfrak{m}(n))^{\mathrm{ord}} \cap \widetilde{K}_\Sigma^{\mathrm{fr}}/K}{\mathfrak{b}_n}\right)^{-1}$ where we have set, for all sufficiently large n, $(\beta_n) =: \mathfrak{p}_v^{\mathrm{v}(x)}\mathfrak{b}_n$. The ideal $\mathfrak{b}_n$ is thus prime to Σ, but for the computation of $\mathrm{Log}_\Sigma(\mathfrak{b}_n)$ we must use an extension to $\langle \Sigma \rangle$ of Log_Σ, hence of $\log_\Sigma$, which must be continuous and respects the functional relation; we have chosen to base it on the Iwasawa logarithm, essentially characterized on $\mathbb{C}_p^\times$ by the relation $\log(p) = 0$. We then have:

$$\mathrm{Log}_\Sigma\left(\frac{x\,,\,\widetilde{K}^{\mathrm{fr}}_\Sigma/K}{v}\right) = -\lim_n\ \mathrm{Log}_\Sigma(\mathfrak{b}_n) = \mathrm{v}(x)\mathrm{Log}_\Sigma(\mathfrak{p}_v) - \lim_n\ \mathrm{Log}_\Sigma(\beta_n)\ ;$$

but $\mathrm{Log}_\Sigma(\beta_n) = \log_\Sigma(\beta_n) \bmod \mathbb{Q}_p\log_\Sigma(E)$ with $\log_\Sigma(\beta_n) = (\log_{v'}(\beta_n))_{v'\in\Sigma}$ where:

- $\lim\limits_n \log_v(\beta_n) = \log_v(x)$,
- $\lim\limits_n \log_{v'}(\beta_n) = 0$ for all $v' \in \Sigma$, $v' \neq v$,

proving the lemma. □

Note. To simplify certain formulas, we will often omit the expression "mod $\mathbb{Q}_p\log_\Sigma(E)$".

Let $\varpi_v \in K^\times$ of v-valuation 1. If $x \in K^\times$, we write $x =: \varpi_v^{\mathrm{v}(x)}u$, $u \in K^\times$ prime to v, and by 5.1.1, we obtain for all $v \in \Sigma$:

$$\begin{aligned}\mathrm{Log}_\Sigma\left(\frac{x\,,\,\widetilde{K}^{\mathrm{fr}}_\Sigma/K}{v}\right) &= \mathrm{v}(x)\mathrm{Log}_\Sigma(\mathfrak{p}_v) - \log_v(x)\\ &= \mathrm{v}(x)\mathrm{Log}_\Sigma(\mathfrak{p}_v) - \mathrm{v}(x)\log_v(\varpi_v) - \log_v(u)\\ &= \mathrm{v}(x)(\mathrm{Log}_\Sigma(\mathfrak{p}_v) - \log_v(\varpi_v)) - \log_v(u).\end{aligned}$$

We now use the fact that $D_v(\widetilde{K}^{\mathrm{fr}}_\Sigma/K) =: \widetilde{D}_v$ is the sub-$\mathbb{Z}_p$-module of $\mathrm{Gal}(\widetilde{K}^{\mathrm{fr}}_\Sigma/K)$ generated by the image of $K^\times$ under the Hasse symbol, and $I_v(\widetilde{K}^{\mathrm{fr}}_\Sigma/K) =: \widetilde{I}_v$ the one generated by the image of $K^\times_{\{v\}}$ (see II.3.1.3). To summarize all the above computations:

5.1.2 Proposition (inertia and decomposition groups in $\widetilde{K}^{\mathrm{fr}}_\Sigma/K$). *Let Σ be a subset of Pl_p.*

(i) For all $v \in \Sigma$, the image in $\mathbb{Z}_p\mathrm{Log}_\Sigma(I_\Sigma) \simeq \mathrm{Gal}(\widetilde{K}^{\mathrm{fr}}_\Sigma/K)$ of the inertia group $\widetilde{I}_v$ of v in $\widetilde{K}^{\mathrm{fr}}_\Sigma/K$ is equal to:

$$\mathrm{Log}_\Sigma(\widetilde{I}_v) = \log(U_v) \bmod \mathbb{Q}_p\log_\Sigma(E).$$

(ii) The image of the decomposition group $\widetilde{D}_v$ of $v \in \Sigma$ in $\widetilde{K}^{\mathrm{fr}}_\Sigma/K$ is the $\mathbb{Z}_p$-module generated by $\widetilde{I}_v$ and by the image of $\mathrm{Log}_\Sigma(\mathfrak{p}_v) - \log_v(\varpi_v)$, where ϖ_v is any element of $K^\times$ such that $\mathrm{v}(\varpi_v) = 1$:

$$\mathrm{Log}_\Sigma(\widetilde{D}_v) = \mathbb{Z}_p(\mathrm{Log}_\Sigma(\mathfrak{p}_v) - \log_v(\varpi_v)) + \log(U_v) \bmod \mathbb{Q}_p\log_\Sigma(E).$$

(iii) Finally, if $v \notin \Sigma$, we have:

$$\mathrm{Log}_\Sigma(\widetilde{D}_v) = \mathbb{Z}_p \mathrm{Log}_\Sigma(\mathfrak{p}_v). \qquad \square$$

5.1.3 Corollary. *If $\sigma \subseteq \Sigma$, the subgroup of $\mathrm{Gal}(\widetilde{K}_\Sigma^{\mathrm{fr}}/K)$ generated by the $\widetilde{I}_v$ for $v \in \Sigma \backslash \sigma$ is equal to $\mathrm{Gal}(\widetilde{K}_\Sigma^{\mathrm{fr}}/\widetilde{K}_\Sigma^{\mathrm{fr}} \cap \widetilde{K}_\sigma)$ and has image equal to:*

$$\langle\, \mathrm{Log}_\Sigma(\widetilde{I}_v)\,\rangle_{v \in \Sigma\backslash\sigma} = \bigoplus_{v \in \Sigma\backslash\sigma} \log(U_v) \bmod \mathbb{Q}_p \mathrm{log}_\Sigma(E).$$

We then have:

$$\mathrm{Gal}(\widetilde{K}_\Sigma^{\mathrm{fr}} \cap \widetilde{K}_\sigma/K) \simeq \mathbb{Z}_p\mathrm{Log}_\Sigma(I_\Sigma)\Big/\Big(\bigoplus_{v\in\Sigma\backslash\sigma} \log(U_v) \bmod \mathbb{Q}_p \mathrm{Log}_\Sigma(E)\Big)$$

whose torsion $\mathbb{Z}_p$-module yields $\mathrm{Gal}(\widetilde{K}_\Sigma^{\mathrm{fr}} \cap \widetilde{K}_\sigma/\widetilde{K}_\sigma^{\mathrm{fr}})$. $\square$

Note. In the above expression we can replace $\log_v(\varpi_v)$ by $\log(\pi_v)$, where π_v is a uniformizer of K_v, but then $\log(\pi_v)$ is an element of $K_v \times \{0\} \times \cdots \times \{0\}$.

It is easy to see that the higher ramification groups of v (in upper numbering) in $\widetilde{K}_\Sigma^{\mathrm{fr}}/K$ for $i \geq 1$ correspond to the $\log(U_v^i) \bmod \mathbb{Q}_p\mathrm{log}_\Sigma(E)$ [Gr5].

Grouping all the above results and coming back to the classical case of complete p-ramification (i.e., $\Sigma = Pl_p$ and σ is an arbitrary subset of Pl_p denoted T_p), we obtain a result which allows us to compute the decomposition of places in $\widetilde{K}_p/K$. Then the context is that of Notations 5.1.

5.2 Theorem (decomposition law of places in $\widetilde{K}_p/K$). *The isomorphisms Log_p and Log_{T_p} (see 2.5) lead to the following interpretations, respectively in $\mathcal{L}_p$ (case (i)) and in $\mathcal{L}_{T_p}$ (case (ii)):*

(i) The Galois group of $\widetilde{K}_p/\widetilde{K}_{T_p}^{S_0}$ is isomorphic to:

$$\sum_{v\in S_{\mathrm{ta}}} \mathbb{Z}_p\mathrm{Log}_p(\mathfrak{p}_v) + \sum_{v\in S_p} \mathbb{Z}_p(\mathrm{Log}_p(\mathfrak{p}_v) - \log(\pi_v)) + \bigoplus_{v\in Pl_p\backslash T_p} \log(U_v),$$

where $S_{\mathrm{ta}} = S_0\backslash S_p$; hence the Galois group of $\widetilde{K}_{T_p}^{S_0}/K$ is isomorphic to:

$$\mathbb{Z}_p\mathrm{Log}_p(I_p)\Big/\sum_{v\in S_{\mathrm{ta}}} \mathbb{Z}_p\mathrm{Log}_p(\mathfrak{p}_v) + \sum_{v\in S_p} \mathbb{Z}_p(\mathrm{Log}_p(\mathfrak{p}_v) - \log(\pi_v)) + \bigoplus_{v\in Pl_p\backslash T_p} \log(U_v),$$

whose torsion $\mathbb{Z}_p$-module yields $\mathrm{Gal}(\widetilde{K}_{T_p}^{S_0}/\widetilde{K}_{T_p}^{S_0\,\mathrm{fr}})$.

(ii) The Galois group of $\widetilde{K}_{T_p}^{\mathrm{fr}\,S_0}/K$ is isomorphic to:

$$\mathbb{Z}_p\mathrm{Log}_{T_p}(I_{T_p})\Big/\sum_{v\in S_0} \mathbb{Z}_p\mathrm{Log}_{T_p}(\mathfrak{p}_v),$$

whose torsion $\mathbb{Z}_p$*-module yields* $\mathrm{Gal}(\widetilde{K}_{T_p}^{\mathrm{fr}\, S_0}/\widetilde{K}_{T_p}^{S_0\, \mathrm{fr}})$. □

Note. In these expressions, π_v is a uniformizer of K_v, log is the Iwasawa logarithm, $\mathrm{Log}_{T_p}(\mathfrak{p}_v) := \frac{1}{m}\log_{T_p}(\alpha) \bmod \mathbb{Q}_p\log_{T_p}(E)$ as soon as $\mathfrak{p}_v^m = (\alpha)$, $\alpha \in K^\times$, with $\log_{T_p} := (\log \circ i_v)_{v\in T_p}$. Note also that $\log(K_v^\times)$ is a subgroup of $K_v \times \{0\} \times \cdots \times \{0\}$.

5.2.1 Exercise. (i) Show that we have the exact sequence:

$$1 \longrightarrow \mathrm{Gal}(\widetilde{K}_p/\widetilde{K}_{T_p}^{\mathrm{fr}}) \xrightarrow{\mathrm{Log}_p} \mathbb{Z}_p\mathrm{Log}_p(I_p) \xrightarrow{\widetilde{\mathrm{pr}}_{T_p}} \mathbb{Z}_p\mathrm{Log}_{T_p}(I_{T_p}) \longrightarrow 0,$$

where $\widetilde{\mathrm{pr}}_{T_p}$ comes from the projection $\mathrm{pr}_{T_p} : \bigoplus_{v|p} K_v \longrightarrow \bigoplus_{v\in T_p} K_v$. From this, identify the group $\mathrm{Gal}(\widetilde{K}_{T_p}/\widetilde{K}_{T_p}^{\mathrm{fr}})$.

(ii) Use a similar method to identify the groups $\mathrm{Gal}(\widetilde{K}_{T_p}/\widetilde{K}_{T_p}^{\mathrm{fr}\, S_0})$ and $\mathrm{Gal}(\widetilde{K}_{T_p}^{S_0}/\widetilde{K}_{T_p}^{\mathrm{fr}\, S_0})$. □

5.2.2 Example. Consider $K = \mathbb{Q}(\sqrt{-15})$ and $p = 2$. We have two prime ideals $\mathfrak{p}$ and $\mathfrak{q}$ above 2, which we can specify by the following relations:

$$\mathfrak{p}^2 := \left(\tfrac{1-\sqrt{-15}}{2}\right), \quad \sqrt{-15} \equiv 25 \bmod \mathfrak{p}^6.$$

Here we have $\bigoplus_{v|2} K_v \simeq \mathbb{Q}_2 \times \mathbb{Q}_2$ and $\bigoplus_{v|2} U_v \simeq \mathbb{Z}_2^\times \times \mathbb{Z}_2^\times$. We do not refer to the "res" or "ord" senses since they coincide. Since $E = \{\pm 1\}$, we have $\mathbb{Q}_2\log_2(E) = 0$ and we can merge the notations Log and log. Finally, the class group of K has order 2 and is generated by the class of $\mathfrak{p}$ (or of $\mathfrak{q}$, or of a prime ideal $\mathfrak{l}$ above 3).

(i) Description of $\mathcal{A}_2 := \mathrm{Gal}(H_{2(2)}/K)$. By 2.6.1, we have:

$$\begin{aligned} |\mathcal{T}_2| &= |\mathrm{tor}_{\mathbb{Z}_2}(\mathbb{Z}_2^\times \times \mathbb{Z}_2^\times/\langle\,(-1,\,-1)\,\rangle)| \times \frac{2}{(\mathbb{Z}_2\mathrm{Log}_2(I_2) : \mathbb{Z}_2\mathrm{Log}_2(P_2))} \\ &= 2 \times \frac{2}{(\mathbb{Z}_2\mathrm{Log}_2(I_2) : 4\mathbb{Z}_2 \times 4\mathbb{Z}_2)}, \end{aligned}$$

since $\mathbb{Z}_2^\times = \langle -1 \rangle \oplus (1 + 4\mathbb{Z}_2)$ and that $\log(\mathbb{Z}_2^\times) = 4\mathbb{Z}_2$. Let us compute $\mathbb{Z}_2\mathrm{Log}_2(I_2)$; since the class group of K is generated by the class of a prime ideal $\mathfrak{l}$ above 3, we have $I_2 = \langle \mathfrak{l} \rangle\, P_2$, hence:

$$\mathbb{Z}_2\mathrm{Log}_2(I_2) = \mathbb{Z}_2\mathrm{Log}_2(\mathfrak{l}) + 4\mathbb{Z}_2 \times 4\mathbb{Z}_2 \;\;;$$

but, since 3 is ramified, we obtain $\mathrm{Log}_2(\mathfrak{l}) = \frac{1}{2}\mathrm{Log}_2(3) = (2u, 2u)$, $u \in \mathbb{Z}_2^\times$, hence:

$$\mathbb{Z}_2\mathrm{Log}_2(I_2) = (2,2)\mathbb{Z}_2 + 4\mathbb{Z}_2 \times 4\mathbb{Z}_2,$$

showing that:

$$(\mathbb{Z}_2\mathrm{Log}_2(I_2) : \mathbb{Z}_2\mathrm{Log}_2(P_2)) = 2, \quad |\mathcal{T}_2| = 2,$$

and that H is contained in $\widetilde{K}_2$ (we will see in 6.7 an example of the opposite situation).

Furthermore, it is clear that $\mathrm{Gal}(\widetilde{K}_2/K) \simeq \mathbb{Z}_2 \times \mathbb{Z}_2$.

(ii) Description of $\mathcal{A}_{\{\mathfrak{p}\}} = \mathrm{Gal}(H_{\{\mathfrak{p}\}(2)}/K)$ (here $H_{\{\mathfrak{p}\}(2)}$ is the maximal $\{\mathfrak{p}\}$-ramified abelian 2-extension of K). We have (using 2.6):

$$\begin{aligned} |\mathcal{T}_{\{\mathfrak{p}\}}| &= |\mathrm{tor}_{\mathbb{Z}_2}(\mathbb{Z}_2^\times/\langle -1 \rangle)| \times \frac{2}{(\mathbb{Z}_2\mathrm{Log}_{\{\mathfrak{p}\}}(I_{\{\mathfrak{p}\}}) : \mathbb{Z}_2\mathrm{Log}_{\{\mathfrak{p}\}}(P_{\{\mathfrak{p}\}}))} \\ &= \frac{2}{(\mathbb{Z}_2\mathrm{Log}_{\{\mathfrak{p}\}}(I_{\{\mathfrak{p}\}}) : \mathbb{Z}_2\mathrm{Log}_{\{\mathfrak{p}\}}(P_{\{\mathfrak{p}\}}))}. \end{aligned}$$

We have $\mathbb{Z}_2\mathrm{Log}_{\{\mathfrak{p}\}}(P_{\{\mathfrak{p}\}}) = 4\mathbb{Z}_2$, and $\mathrm{Log}_{\{\mathfrak{p}\}}(\mathfrak{l}) = \frac{1}{2}\mathrm{Log}_{\{\mathfrak{p}\}}(3) = 2u,\ u \in \mathbb{Z}_2^\times$, giving:

$$\mathbb{Z}_2\mathrm{Log}_{\{\mathfrak{p}\}}(I_{\{\mathfrak{p}\}}) = 2\mathbb{Z}_2 + 4\mathbb{Z}_2 = 2\mathbb{Z}_2,$$

which thus yields:

$$|\mathcal{T}_{\{\mathfrak{p}\}}| = 1,$$

and shows that $\widetilde{K}^{\mathrm{fr}}_{\{\mathfrak{p}\}}$ again contains H. It also follows that:

$$H_{\{\mathfrak{p}\}(2)} = \widetilde{K}_{\{\mathfrak{p}\}} = \widetilde{K}^{\mathrm{fr}}_{\{\mathfrak{p}\}}.$$

We could also compute $\mathbb{Z}_2\mathrm{Log}_{\{\mathfrak{p}\}}(I_{\{\mathfrak{p}\}})$ by using $\mathfrak{q}$ (which belongs to $I_{\{\mathfrak{p}\}}$).

We have:

$$\mathcal{A}_{\{\mathfrak{p}\}} = \mathrm{Gal}(\widetilde{K}^{\mathrm{fr}}_{\{\mathfrak{p}\}}/K) \simeq \mathbb{Z}_2\mathrm{Log}_{\{\mathfrak{p}\}}(I_{\{\mathfrak{p}\}}) \simeq \mathbb{Z}_2,$$

which means that there exists a unique $\{\mathfrak{p}\}$-ramified $\mathbb{Z}_2$-extension (equal to $\widetilde{K}^{\mathrm{fr}}_{\{\mathfrak{p}\}}$).

As indicated in 5.2, (i), we can compute:

$$\mathrm{Gal}(\widetilde{K}_2 \cap H_{\{\mathfrak{p}\}}/K) = \mathrm{Gal}(\widetilde{K}_{\{\mathfrak{p}\}}/K) \simeq ((2,2)\mathbb{Z}_2 + 4\mathbb{Z}_2 \times 4\mathbb{Z}_2)/\{0\} \times 4\mathbb{Z}_2,$$

whose torsion group gives once again $\mathcal{T}_{\{\mathfrak{p}\}} = \mathrm{Gal}(\widetilde{K}_{\{\mathfrak{p}\}}/\widetilde{K}^{\mathrm{fr}}_{\{\mathfrak{p}\}})$ which is trivial, and we find once more that $\mathrm{Gal}(\widetilde{K}_{\{\mathfrak{p}\}}/K) \simeq \mathbb{Z}_2$.

Let us now compute $\mathrm{Gal}(\widetilde{K}^{\{\mathfrak{q}\}}_{\{\mathfrak{p}\}}/K)$. We still use the second formula of 5.2, (i), which here can be written:

$$\mathrm{Gal}(\widetilde{K}^{\{\mathfrak{q}\}}_{\{\mathfrak{p}\}}/K) \simeq (2,\, 2)\mathbb{Z}_2 + 4\mathbb{Z}_2 \times 4\mathbb{Z}_2 \,/\, \mathbb{Z}_2(\log_2(\mathfrak{q}) - \log(\pi_{\mathfrak{q}})) + \{0\} \times 4\mathbb{Z}_2.$$

Since 2 is split, $\pi_{\mathfrak{q}} = 2$ is suitable, hence $\log(\pi_{\mathfrak{q}}) = 0$. We have:

$$\mathfrak{q}^2 = \left(\tfrac{1+\sqrt{-15}}{2}\right),$$

which yields:

$$\log_{\mathfrak{p}}(\mathfrak{q}) = \tfrac{1}{2}\log_{\mathfrak{p}}\left(\tfrac{1+\sqrt{-15}}{2}\right) \equiv 2 \bmod (4), \quad \text{and } \log_{\mathfrak{q}}(\mathfrak{q}) = -\log_{\mathfrak{q}}(\mathfrak{p}).$$

We thus have:

$$\begin{aligned}\mathbb{Z}_2(\log_2(\mathfrak{q}) - \log(\pi_{\mathfrak{q}})) + \{0\} \times 4\mathbb{Z}_2 &= (2,\, -2)\mathbb{Z}_2 + \{0\} \times 4\mathbb{Z}_2 \\ &= (2,\, 2)\mathbb{Z}_2 + 4\mathbb{Z}_2 \times 4\mathbb{Z}_2\ ;\end{aligned}$$

it follows that $\widetilde{K}_{\{\mathfrak{p}\}}^{\{\mathfrak{q}\}} = K$ (the Frobenius of $\mathfrak{q}$ in $\mathcal{A}_{\{\mathfrak{p}\}}$ is a topological generator).

The Hilbert class field H is in fact the intersection of $\widetilde{K}_{\{\mathfrak{p}\}}$ with $\widetilde{K}_{\{\mathfrak{q}\}}$, and since $\mathfrak{q}$ is not split in $\widetilde{K}_{\{\mathfrak{p}\}}$, it is necessary that $\mathfrak{q}$ be inert in H, which is the case since $\mathfrak{q}$ is not principal; since here $H = \mathbb{Q}(\sqrt{-3},\, \sqrt{5})$ (genus theory), any elementary verification is possible.

(iii) Decomposition of tame places. We know that the decomposition group $\widetilde{D}_v$ of an odd place v in $\widetilde{K}_2/K$ is given by $\mathbb{Z}_2\mathrm{Log}(\mathfrak{p}_v)$ in:

$$\mathbb{Z}_2\mathrm{Log}(I_2) = (2,\, 2)\mathbb{Z}_2 + 4\mathbb{Z}_2 \times 4\mathbb{Z}_2.$$

For example, for $\mathfrak{p}_v = (1 + 2\sqrt{-15})$ (a prime ideal above 61), we have:

$$\begin{aligned}\mathrm{Log}(\mathfrak{p}_v) &= (\log_{\mathfrak{p}}(1 + 2\sqrt{-15})\,,\, \log_{\mathfrak{q}}(1 + 2\sqrt{-15})) \\ &= (\log(1 + 2\sqrt{-15})\,,\, \log(1 - 2\sqrt{-15})),\end{aligned}$$

which yields (knowing that $\sqrt{-15} \equiv 25 \bmod \mathfrak{p}^6$):

$$\mathrm{Log}(\mathfrak{p}_v) = (4u',\, 16u''),\ \ u',\ u'' \in \mathbb{Z}_2^{\times}.$$

Any concrete study is then possible; for instance we deduce that the Frobenius of v in $\widetilde{K}_2/K$ is such that its restriction to $\widetilde{K}_{\{\mathfrak{p}\}}$ (resp. to $\widetilde{K}_{\{\mathfrak{q}\}}$) is a topological generator of $\mathcal{A}_{\{\mathfrak{p}\}}^2$ (resp. of $\mathcal{A}_{\{\mathfrak{q}\}}^8$) (in particular it fixes H, but it cannot be otherwise since $\mathfrak{p}_v$ is principal, hence split in H/K). □

§6 Initial Radical of the $\mathbb{Z}_p$-Extensions

Let K be a number field containing the group μ_p of pth roots of unity, for a fixed prime number p. We denote by $W_p \subset K^\times/K^{\times p}$ the radical of the maximal p-elementary extension $H_p^{\mathrm{res}}[p]$ of K contained in H_p^{res} (i.e., p-ramified). It is easy to see that:

$$W_p = \{xK^{\times p} \in K^\times/K^{\times p},\ (x) \in I^p\langle\, Pl_p\,\rangle\}$$

by simply saying that $H_p^{\mathrm{res}}[p]/K$ is p-ramified (see I.6.3). The corresponding radical of $H_p^{\mathrm{ord}}[p]$ (maximal noncomplexified subextension of $H_p^{\mathrm{res}}[p]$) is, with an abuse of notation, $W_{p,\mathrm{pos}} := W_p \cap K_{\mathrm{pos}}^{\times}$ (case $p = 2$).

6.1 Lemma. *We have the following exact sequence giving W_p:*

$$1 \longrightarrow E^{Pl_p\,\mathrm{ord}}/(E^{Pl_p\,\mathrm{ord}})^p \longrightarrow W_p \xrightarrow{\;c\;} {}_p\mathcal{C}\ell^{Pl_p\,\mathrm{ord}} \longrightarrow 1,$$

where the map c is defined as follows. If $xK^{\times p} \in W_p$, we have $(x) =: \mathfrak{a}^p\mathfrak{a}_p$, $\mathfrak{a} \in I$, $\mathfrak{a}_p \in \langle\, Pl_p \,\rangle$, and we set $c(xK^{\times p}) := c\ell^{Pl_p\,\mathrm{ord}}(\mathfrak{a})$ which is a Pl_p-classe (in the ordinary sense) killed by p. □

Note. This is the exact sequence used in the proof of II.5.4.1, (i) for $T = \emptyset$, $S_0 = Pl_p$, and for the ordinary sense. We have $W_p = Y^{Pl_p\,\mathrm{ord}}/K^{\times p}$.

6.2 Remark. The maximal elementary subextension $\widetilde{K}_{p[p]} := \widetilde{K}_p \cap H_p^{\mathrm{res}}{}_{[p]}$ of $\widetilde{K}_p$ being Pl_∞^{r}-split, one could think that, for $p = 2$, we should directly work with the set $W_{2,\mathrm{pos}}$ formed by the totally positive elements of W_2; however the reflection phenomena imply that the corresponding exact sequence (in the restricted sense) does not exist anymore (if $x = y^2$ with $y \in K^\times$, then y is not necessarily totally positive and c is not defined; see V.2.3.3, (ii)). But it is then easy to go from W_2 to $W_{2,\mathrm{pos}}$. □

We are going to determine the radical of $\widetilde{K}_{p[p]}$. This problem has been studied for the first time in [Car], [CarK], and has been solved in 1985 in [Gr4]; it is related to the easier similar problem that we have solved in 4.15.8, 4.15.9. We will use the following diagram:

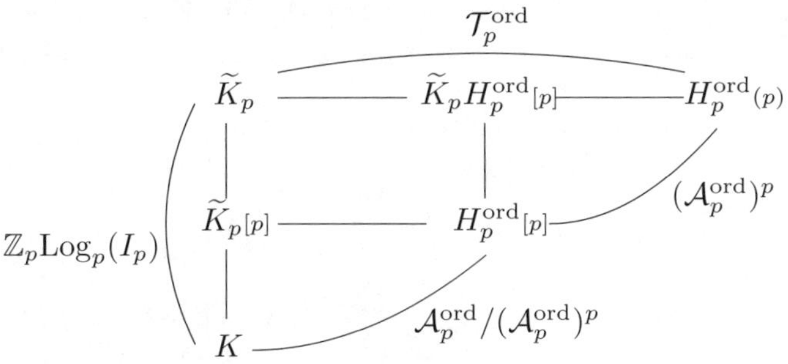

Fig. 6.1

Let $\mathfrak{f}$ be the conductor of $H_p^{\mathrm{ord}}{}_{[p]}$ (or a multiple in $\langle\, Pl_p \,\rangle_{\mathbb{N}}$); by definition, $H_p^{\mathrm{ord}}{}_{[p]} \subseteq K(\mathfrak{f})^{\mathrm{ord}}$ is fixed under $(I_p/P_{p,\mathfrak{f}})^p$ and its Artin group is equal to:

$$A = I_p^p\, P_{p,\mathfrak{f}}.$$

Since $H_p^{\mathrm{ord}}{}_{[p]}$ is also the subfield of $H_p^{\mathrm{ord}}{}_{(p)}$ fixed under $(\mathcal{A}_p^{\mathrm{ord}})^p$, we have by 2.5.1 with the notations of 2.2:

$$\mathrm{Gal}(\widetilde{K}_p/\widetilde{K}_{p[p]}) \simeq \mathbb{Z}_p\mathrm{Log}_p(A) = \mathrm{Log}_p \circ \mathrm{Art}^{-1}((\mathcal{A}_p^{\mathrm{ord}})^p) = p\mathbb{Z}_p\mathrm{Log}_p(I_p).$$

6.3 Proposition. *In the exact sequence:*

$$1 \longrightarrow \widetilde{A}/A \longrightarrow I_p/A \xrightarrow{\mathrm{Log}_p} \mathbb{Z}_p\mathrm{Log}_p(I_p)/p\mathbb{Z}_p\mathrm{Log}_p(I_p) \longrightarrow 0,$$

$\widetilde{A} := \{\mathfrak{a} \in I_p,\ \mathrm{Log}_p(\mathfrak{a}) \in p\mathbb{Z}_p\mathrm{Log}_p(I_p)\}$ *is the Artin group of* $\widetilde{K}_{p[p]}$.

Proof. If $\mathfrak{a} \in I_p$ is such that $\mathrm{Log}_p(\mathfrak{a}) \in p\mathbb{Z}_p\mathrm{Log}_p(I_p)$, there exist $\sigma \in \mathcal{A}_p^{\mathrm{ord}}$ and $\tau \in \mathcal{T}_p^{\mathrm{ord}}$ such that $\mathrm{Art}(\mathfrak{a}) = \tau\,\sigma^p$, hence $\mathrm{Art}(\mathfrak{a}) \in \mathrm{Gal}(H_p^{\mathrm{ord}}{}_{(p)}/\widetilde{K}_{p[p]})$ since the group generated by $\mathcal{T}_p^{\mathrm{ord}}$ and $(\mathcal{A}_p^{\mathrm{ord}})^p$ is exactly $\mathrm{Gal}(H_p^{\mathrm{ord}}{}_{(p)}/\widetilde{K}_{p[p]})$; hence $\mathfrak{a} \in A_{\widetilde{K}_p[p]/K}$. The inverse inclusion being clear by the characterization of $\mathrm{Gal}(\widetilde{K}_p/\widetilde{K}_{p[p]})$, the proposition follows. □

To come back to radicals, it is useful to introduce the pth power residue symbol studied in II.7.4; we simply recall what we will need.

6.4 Definition. Let K be a number field containing μ_p. Let $x \in K^\times$ and $\mathfrak{b}$ be an ideal of K prime to p and to the prime ideals of K ramified in $K(\sqrt[p]{x})/K$; we set:

$$\left(\frac{K(\sqrt[p]{x})/K}{\mathfrak{b}}\right)\sqrt[p]{x} =: \left(\frac{x}{\mathfrak{b}}\right)\sqrt[p]{x}. \qquad \square$$

The symbol $\left(\frac{x}{\mathfrak{b}}\right) \in \mu_p$ will allow us to translate in Kummer theory terms the action of the Artin symbol of $\mathfrak{b}$; in particular $\left(\frac{x}{\mathfrak{b}}\right) = 1$ if and only if $\mathfrak{b}$ is in the Artin group of $K(\sqrt[p]{x})/K$.[30]

From the exact sequence of 6.3, we set $\widetilde{A} =: \langle\, \mathfrak{b}_1, \ldots, \mathfrak{b}_t \,\rangle\, I_p^p\, P_{p,\mathfrak{f}}$, so that:

$$\mathrm{Gal}(H_p^{\mathrm{ord}}{}_{[p]}/\widetilde{K}_{p[p]}) = \Big\langle \left(\frac{H_p^{\mathrm{ord}}{}_{[p]}/K}{\mathfrak{b}_i}\right) \Big\rangle_{i=1,\ldots,t},$$

and we consider the map $\varphi : W_{p,\mathrm{pos}} \longrightarrow \mu_p^t$, which sends $xK^{\times p} \in W_{p,\mathrm{pos}}$ to the family of symbols $(\left(\frac{x}{\mathfrak{b}_i}\right))_{i=1,\ldots,t}$. We have $xK^{\times p} \in \widetilde{W}_p$ if and only if $K(\sqrt[p]{x}) \subseteq \widetilde{K}_{p[p]}$, hence if and only if $\sqrt[p]{x}$ is fixed under all the $\left(\frac{H_p^{\mathrm{ord}}{}_{[p]}/K}{\mathfrak{b}_i}\right)$, $i = 1, \ldots, t$, which characterizes the kernel of φ and gives the final result 6.6 below.

6.5 Notations. Let $W_{p,\mathrm{pos}} := \{xK^{\times p},\ (x) \in I^p\,\langle\, Pl_p \,\rangle\} \cap K^\times_{\mathrm{pos}}$ be the radical of the elementary abelian p-extension $H_p^{\mathrm{ord}}{}_{[p]}$ of K, unramified outside p and noncomplexified (see 6.2), and let $\mathfrak{f}$ be its conductor or the multiple $\prod_{v|p} \mathfrak{p}_v^{pe_v+1}$,

[30] The Kummer extensions considered below will be at most ramified at p hence it will be enough for $\mathfrak{b}$ to be prime to p.

where e_v is the ramification index of p in $K/\mathbb{Q}(\mu_p)$ (see II.1.6.3). Let $\mathfrak{b}_1, \ldots, \mathfrak{b}_t$ be ideals of K (prime to p) such that:

$$\langle \mathfrak{b}_1, \ldots, \mathfrak{b}_t \rangle I_p^p P_{p,\mathfrak{f}} = \{\mathfrak{a} \in I_p, \ \mathrm{Log}_p(\mathfrak{a}) \in p\mathbb{Z}_p\mathrm{Log}_p(I_p)\},$$

where $\mathrm{Log}_p(\mathfrak{a}) := \frac{1}{m}\mathrm{log}_p(\alpha) \bmod \mathbb{Q}_p\mathrm{log}_p(E)$, when $\mathfrak{a}^m =: (\alpha)$, $\alpha \in K^\times$. □

6.6 Theorem (initial radical of $\widetilde{K}_p$ [Gr4] (1985)). *Let K be a number field containing the group μ_p of pth roots of unity. Then the radical $\widetilde{W}_p$ of $\widetilde{K}_{p[p]}$ (i.e., the initial radical of the compositum of the $\mathbb{Z}_p$-extensions of K) is equal to the kernel of the map:*

$$\varphi : W_{p,\mathrm{pos}} \longrightarrow \mu_p^t$$

which sends $xK^{\times p} \in W_{p,\mathrm{pos}}$ to $\left(\left(\frac{x}{\mathfrak{b}_i}\right)\right)_{i=1,\ldots,t}$. □

6.7 Example. Let $K = \mathbb{Q}(\sqrt{-161}) = \mathbb{Q}(\sqrt{-7.23})$ and let $p = 2$ (here restricted and ordinary sense coincide). We have:

$$\mathcal{C}\ell = \langle\, c\ell(\mathfrak{l}_3) \,\rangle \oplus \langle\, c\ell(\mathfrak{l}_7) \,\rangle \simeq \mathbb{Z}/8\mathbb{Z} \times \mathbb{Z}/2\mathbb{Z},$$

where $\mathfrak{l}_3|3$ (split) and $\mathfrak{l}_7|7$ (ramified) in $K/\mathbb{Q}$. We check that $\mathfrak{f} = 4\mathfrak{p}$, where $\mathfrak{p}$ denotes the prime ideal of K above 2 (ramified in $K/\mathbb{Q}$ and corresponding to the unique place v_0 of K above 2).

A computation yields:

$$\mathbb{Z}_2\mathrm{Log}_2(P_2) = \mathbb{Z}_2\mathrm{Log}_2(1+\mathfrak{p}) = 4\mathbb{Z}_2 \oplus 2(1+\sqrt{-161})\mathbb{Z}_2,$$

so we obtain:

$$\begin{aligned}\mathrm{Log}_2(\mathfrak{l}_3) &= \tfrac{1}{8}\,\mathrm{log}_2(80+\sqrt{-161}) \\ &\equiv 2 - 2\sqrt{-161} \bmod 8\mathbb{Z}_2 \oplus 4(1+\sqrt{-161})\mathbb{Z}_2\ ;\end{aligned}$$

$$\mathrm{Log}_2(\mathfrak{l}_7) = \tfrac{1}{2}\,\mathrm{log}_2(7) \equiv 4 \bmod 8\mathbb{Z}_2 \oplus 4(1+\sqrt{-161})\mathbb{Z}_2.$$

We already deduce from these computations that:

$$\begin{aligned}\mathbb{Z}_2\mathrm{Log}_2(I_2) &= \mathrm{Log}_2(\langle\, \mathfrak{l}_3, \mathfrak{l}_7 \,\rangle) + \mathbb{Z}_2\mathrm{Log}_2(P_2) \\ &= \langle\, 2 - 2\sqrt{-161}, 4 \,\rangle + (8\mathbb{Z}_2 \oplus 4(1+\sqrt{-161})\mathbb{Z}_2) \\ &= \mathbb{Z}_2\mathrm{Log}_2(P_2),\end{aligned}$$

which yields (see 2.6.1, (i)) $\widetilde{K}_2 \cap H = K$ (the Hilbert class field of K is linearly disjoint from the compositum of the $\mathbb{Z}_2$-extensions of K). We then have $|\mathcal{T}_2| = 2 \times |\mathcal{C}\ell| = 32$ since $K_{v_0} = \mathbb{Q}_2(\sqrt{-1})$ and $\mathrm{tor}_{\mathbb{Z}_2}(U_{v_0}/\langle -1 \rangle) \simeq \mu_4/\mu_2 \simeq \mathbb{Z}/2\mathbb{Z}$.

Let us now consider the radical W_2. We have $E^{Pl_2} = \langle -1,\ 2 \rangle$ since the prime ideal $\mathfrak{p}$ above 2 is not principal. Since $\mathfrak{p}\,\mathfrak{l}_3^4 = (1 - \sqrt{-161})$, we have:

$$ {}_2\mathcal{C}\ell^{Pl_2} = \langle\, c\ell^{Pl_2}(\mathfrak{l}_3^2),\ c\ell^{Pl_2}(\mathfrak{l}_7)\,\rangle \simeq \mathbb{Z}/2\mathbb{Z} \times \mathbb{Z}/2\mathbb{Z}, $$

so we finally obtain:

$$ W_{2,\mathrm{pos}} = W_2 = \langle\, -1,\ 2,\ 7,\ 1 - \sqrt{-161}\,\rangle. $$

In particular $\widetilde{W}_2$ is a strict subgroup of W_2 (their respective $\mathbb{F}_2$-dimensions are 2 and 4).

To generate I_2 modulo $A = I_2^2\, P_{2,4\mathfrak{p}}$, we must simply know the generalized class group $\mathcal{C}\ell_{4\mathfrak{p}}$ (for this, we can use I.4.5 to be reduced to the usual class group). We will prove that:

$$ I_2 = \langle\, \mathfrak{l}_3,\ \mathfrak{l}_7,\ (2 + \sqrt{-161}),\ (2 - \sqrt{-161})\,\rangle P_{2,4\mathfrak{p}}\ ; $$

indeed, by assumption we have $I_2 = \langle\, \mathfrak{l}_3, \mathfrak{l}_7\,\rangle P_2$, and it is sufficient to represent $P_2 = P_{2,\mathfrak{p}}$ modulo $P_{2,4\mathfrak{p}}$. The exact sequence (a particular case of I.4.5, (i)):

$$ 1 \longrightarrow \langle\, -1\,\rangle \longrightarrow U^1_{v_0}/U^5_{v_0} \longrightarrow P_{2,\mathfrak{p}}/P_{2,4\mathfrak{p}} \longrightarrow 1 $$

shows that $P_{2,\mathfrak{p}}/P_{2,4\mathfrak{p}}$ has order 8 and exponent 2 (we have $U^1_{v_0} = \langle\, \sqrt{-1}\,\rangle \oplus U^3_{v_0}$). We then check that:

$$ P_{2,\mathfrak{p}} = \langle\, (\sqrt{-161}),\ (2 + \sqrt{-161}),\ (2 - \sqrt{-161})\,\rangle P_{2,4\mathfrak{p}}. $$

Since $\mathfrak{l}_3^8 = (80 + \sqrt{-161})$, with $(80 + \sqrt{-161}) \in (\sqrt{-161})P_{2,4\mathfrak{p}}$, we have proved our claim.

Set:

$$ \mathfrak{a} =: \mathfrak{l}_3^x\, \mathfrak{l}_7^y\, (2 + \sqrt{-161})^z\, (2 - \sqrt{-161})^t,\ \ x,\ y,\ z,\ t \in \mathbb{Z}/2\mathbb{Z}\ ; $$

the elements of $\widetilde{A}$ modulo A are characterized by the relation:

$$ x\,\mathrm{Log}_2(\mathfrak{l}_3) + y\,\mathrm{Log}_2(\mathfrak{l}_7) + z\,\mathrm{Log}_2(2 + \sqrt{-161}) + t\,\mathrm{Log}_2(2 - \sqrt{-161}) \in 8\mathbb{Z}_2 \oplus 4(1 + \sqrt{-161})\mathbb{Z}_2\ , $$

hence by:

$$ x(1 - \sqrt{-161}) + 2y + z(1 + \sqrt{-161}) + t(1 - \sqrt{-161}) \in 4\mathbb{Z}_2 \oplus 2(1 + \sqrt{-161})\mathbb{Z}_2\ , $$

which is equivalent to $x + y + t = 0$ and $y + z = 0$, and yields:

$$ \{(1,\ 0,\ 0,\ 1),\ (1,\ 1,\ 1,\ 0)\}, $$

as basis of solutions; it follows that:

$$\widetilde{A} = \langle\, \mathfrak{b}_1,\ \mathfrak{b}_2\, \rangle\, A, \text{ with } \mathfrak{b}_1 := (3)\, \mathfrak{l}'_5\, \mathfrak{l}'_{11},\ \mathfrak{b}_2 := \mathfrak{l}_7\, \mathfrak{l}_5\, \mathfrak{l}_{11},$$

where $\mathfrak{l}'$ denotes the conjugate of $\mathfrak{l}$, where we have set:

$$(2 + \sqrt{-161}) := \mathfrak{l}_3\, \mathfrak{l}_5\, \mathfrak{l}_{11} \text{ and } (1 - \sqrt{-161}) = \mathfrak{p}\, \mathfrak{l}_3^4.$$

The symbol computations (i.e., of Frobenius') are elementary and yield the following (with $\pi := 1 - \sqrt{-161}\,$):

- $\left(\frac{-1}{\mathfrak{l}_3}\right) = -1,\ \left(\frac{7}{\mathfrak{l}_3}\right) = +1,\ \left(\frac{\pi}{\mathfrak{l}_3}\right) = +1,$
- $\left(\frac{-1}{\mathfrak{l}'_3}\right) = -1,\ \left(\frac{7}{\mathfrak{l}'_3}\right) = +1,\ \left(\frac{\pi}{\mathfrak{l}'_3}\right) = -1,$
- $\left(\frac{-1}{\mathfrak{l}_7}\right) = -1,\ \left(\frac{7}{\mathfrak{l}_7}\right) = -1,\ \left(\frac{\pi}{\mathfrak{l}_7}\right) = +1,$
- $\left(\frac{-1}{\mathfrak{l}_5}\right) = +1,\ \left(\frac{7}{\mathfrak{l}_5}\right) = -1,\ \left(\frac{\pi}{\mathfrak{l}_5}\right) = -1,$
- $\left(\frac{-1}{\mathfrak{l}'_5}\right) = +1,\ \left(\frac{7}{\mathfrak{l}'_5}\right) = -1,\ \left(\frac{\pi}{\mathfrak{l}'_5}\right) = +1,$
- $\left(\frac{-1}{\mathfrak{l}_{11}}\right) = -1,\ \left(\frac{7}{\mathfrak{l}_{11}}\right) = -1,\ \left(\frac{\pi}{\mathfrak{l}_{11}}\right) = +1,$
- $\left(\frac{-1}{\mathfrak{l}'_{11}}\right) = -1,\ \left(\frac{7}{\mathfrak{l}'_{11}}\right) = -1,\ \left(\frac{\pi}{\mathfrak{l}'_{11}}\right) = -1.$

which yields:

- $\left(\frac{-1}{\mathfrak{b}_1}\right) = -1,\ \left(\frac{7}{\mathfrak{b}_1}\right) = +1,\ \left(\frac{\pi}{\mathfrak{b}_1}\right) = +1,$
- $\left(\frac{-1}{\mathfrak{b}_2}\right) = +1,\ \left(\frac{7}{\mathfrak{b}_2}\right) = -1,\ \left(\frac{\pi}{\mathfrak{b}_2}\right) = -1.$

Since 2 is an element of $\widetilde{W}_2$ (first floor of the cyclotomic $\mathbb{Z}_2$-extension of K), we can omit the symbol computations at 2. By 6.6, we finally obtain:

$$\widetilde{W}_2 = \langle\, 2,\ 7\,(1 - \sqrt{-161})\,\rangle K^{\times 2}. \qquad \square$$

6.7.1 Remarks. (i) Incidentally, we note that the radical of the maximal unramified elementary abelian 2-extension is $\langle\, -1,\, 7\rangle K^{\times 2}$ since $-1 \equiv 7 \equiv (\sqrt{-161}\,)^2 \bmod (4)$, but this information is insufficient to determine $\widetilde{W}_2$.

(ii) If we use 4.15.9 to see what are the elements:

$$x \in W_2 = \langle\, -1,\ 2,\ 7,\ (1 - \sqrt{-161})\,\rangle$$

for which $K(\sqrt{x})$ is cyclically embeddable with arbitrarily large degree (power of 2), we must compute the Hilbert symbols $(x\,,\, \sqrt{-1}\,)_{v_0}$ for the place $v_0|2$ in K since $K_{v_0} = \mathbb{Q}_2(\mu_4)$. But -1 and 7 are squares in K_{v_0}, 2 is a norm in the extension $\mathbb{Q}_2(\mu_8)/\mathbb{Q}_2(\mu_4)$ (this is automatic since $\sqrt{2}$ is in the cyclotomic $\mathbb{Z}_2$-extension) and:

$$(1-\sqrt{-161}\,,\,\sqrt{-1}\,)_{v_0} = (1-\sqrt{-161}\,,\,\sqrt{-161}\,)_{v_0} = 1.$$

We are here in the special case, with $s_{v_0} = 1+\sqrt{-1}$, and we check that $(x\,,\,1+\sqrt{-1}\,)_{v_0} = 1$ for all $x \in W_2$. Hence any element of W_2 is solution, and this embeddability criterion, much weaker than that of $\mathbb{Z}_2$-embeddability, is of no help at all.

(iii) The place v_0 is totally ramified in $\widetilde{K}_2/K$ (since v_0 is unique and $H \cap \widetilde{K}_2 = K$) and has in H/K a residue degree equal to 2 since the class of $\mathfrak{p}$ has order 2; v_0 being unique, its inertia group in $H_{2(2)}/K$ is equal to $\mathrm{Gal}(H_{2(2)}/H) \simeq \mathbb{Z}_2^2 \times \mathbb{Z}/2\mathbb{Z}$. It easily follows that the maximal locally free subextension L of $\overline{K}^{\mathrm{ab}}{}_{(2)}/K$ (i.e., such that $\mathrm{Gal}(L_v/K_v)$ is free for any place v) is equal to the compositum $H_0\widetilde{K}_2$, where H_0 is the decomposition subfield of v_0 in H/K. □

6.7.2 Exercise. Let $K = \mathbb{Q}(\sqrt{-\ell}\,)$, ℓ prime, $\ell \equiv 7 \bmod 8$; we put $-\ell = x^2 - 2y^2$, with x odd. We consider:

$$a = -\Big(\frac{x}{\ell}\Big)(x+\sqrt{-\ell}), \quad \text{if } \ell \equiv 7 \bmod\ 16,$$

$$a = +\Big(\frac{x}{\ell}\Big)(x+\sqrt{-\ell}), \quad \text{if } \ell \equiv 15 \bmod\ 16.$$

Check that $K(\sqrt{a})$ is contained in the compositum of the $\mathbb{Z}_2$-extensions of K (the sign of x does not matter). □

For analogous examples, concerning imaginary quadratic fields and $p = 2$, the reader can refer to [Gr4] and also to [Gr3] for similar questions.

If the Kummer hypothesis is not satisfied, we can solve the problem in $K' := K(\mu_p)$; then the Galois descent of the maximal elementary subextension of the compositum of the $\mathbb{Z}_p$-extensions of K is classical: if $\widetilde{W}'_p$ is the radical of $\widetilde{K}'_{p[p]}$, that of $K'\,\widetilde{K}_{p[p]}$ is the ω-component of $\widetilde{W}'_p$, where ω is the Teichmüller character.

§7 The Logarithmic Class Group

Let K be a number field and let p be a fixed prime number. We begin with the following exercise describing $\mathrm{Gal}(H_p^{\mathrm{lc}}/K\mathbb{Q}^{\mathrm{cycl}}{}_{(p)})$ by means of our usual correspondence of class field theory for pro-p-extensions, where $\mathbb{Q}^{\mathrm{cycl}}{}_{(p)}$ is the cyclotomic $\mathbb{Z}_p$-extension of $\mathbb{Q}$, and H_p^{lc} the maximal abelian locally cyclotomic pro-p-extension of K (i.e., the subfield of $H_p^{\mathrm{ord}}{}_{(p)}$ fixed by the relative decomposition groups of $v|p$, in $H_p^{\mathrm{ord}}{}_{(p)}/K\mathbb{Q}^{\mathrm{cycl}}{}_{(p)}$, since all the tame places split in this extension by 4.8.2).

7.1 Exercise (computation of $\mathrm{Gal}(H_p^{\mathrm{lc}}/K\mathbb{Q}^{\mathrm{cycl}}{}_{(p)})$). Let K/k be a relative extension.

(i) Show that in the isomorphism $\mathrm{Gal}(H_p^{\mathrm{ord}}{}_{(p)}/K) \simeq \mathcal{I}_p/\mathcal{P}_{p,\infty}$ of 2.4, we have $\mathrm{Gal}(H_p^{\mathrm{ord}}{}_{(p)}/K\widetilde{k}_p) \simeq \{\mathfrak{a} \in \mathcal{I}_p,\ \mathrm{Log}_{k,p}(\mathrm{N}_{K/k}(\mathfrak{a})) = 0\}/\mathcal{P}_{p,\infty}$.

(ii) Consider the case $k = \mathbb{Q}$ for which $K\widetilde{k}_p$ is the cyclotomic $\mathbb{Z}_p$-extension of K and deduce a way of computing of $\mathrm{Gal}(H_p^{\mathrm{lc}}/K\mathbb{Q}^{\mathrm{cycl}}{}_{(p)})$ which we will compare below with the logarithmic class group of K (see 4.13 and use 2.8.2 to compute the $D_v(H_p^{\mathrm{ord}}{}_{(p)}/K) \cap \mathrm{Gal}(H_p^{\mathrm{ord}}{}_{(p)}/K\mathbb{Q}^{\mathrm{cycl}}{}_{(p)})$ for $v|p$).

Answer. Point (i) is evident. For (ii), we check that the decomposition group of $v|p$ in $H_p^{\mathrm{ord}}{}_{(p)}/K\mathbb{Q}^{\mathrm{cycl}}{}_{(p)}$ is the image in $\mathcal{I}_p/\mathcal{P}_{p,\infty}$ of:

$$\mathcal{D}_v := \{(x_{\infty_{p\backslash v}})\,\mathfrak{p}_v^{-\mathrm{v}(x_{\infty_{p\backslash v}})},\ \mathrm{Log}_{\mathbb{Q},p}(\mathrm{N}_{K/\mathbb{Q}}((x_{\infty_{p\backslash v}})\mathfrak{p}_v^{-\mathrm{v}(x_{\infty_{p\backslash v}})})) = 0\},$$

where $x_{\infty_{p\backslash v}}$ ranges in the set of elements of $\mathcal{K}^\times := K^\times \otimes \mathbb{Z}_p$ such that $\bar{i}_{p\backslash v}(x_{\infty_{p\backslash v}}) = 1$. Note that $\mathrm{Log}_{\mathbb{Q},p}(\mathrm{N}_{K/\mathbb{Q}}(\mathfrak{p}_v)) = 0$ since here $\mathrm{Log}_{\mathbb{Q},p} = \log \circ\, \bar{i}_p$ on $\mathcal{Q}^\times$, where log is the Iwasawa logarithm for which $\log(p) = 0$. Moreover, it is clear that $\bar{i}_p(\mathrm{N}_{K/\mathbb{Q}}(x_{\infty_{p\backslash v}})) = \mathrm{N}_{K_v/\mathbb{Q}_p}(\bar{i}_v(x_{\infty_{p\backslash v}}))$. We merge the notations $\mathrm{Log}_{\mathbb{Q},p}$ and log. Therefore, we have:

$$\mathcal{D}_v := \{(x_{\infty_{p\backslash v}})\,\mathfrak{p}_v^{-\mathrm{v}(x_{\infty_{p\backslash v}})},\ \log(\mathrm{N}_{K_v/\mathbb{Q}_p}(\bar{i}_v(x_{\infty_{p\backslash v}}))) = 0\},$$

and

$$\mathrm{Gal}(H_p^{\mathrm{lc}}/K\mathbb{Q}^{\mathrm{cycl}}{}_{(p)}) \simeq \{\mathfrak{a} \in \mathcal{I}_p,\ \log(\mathrm{N}_{K/\mathbb{Q}}(\mathfrak{a})) = 0\}/\langle\, \mathcal{D}_v \,\rangle_{v|p}. \qquad \square$$

7.2 Definitions (degrees, logarithmic valuations [Ja5, Ja6]). (i) (degrees). Let v be a finite place.

If $v|p$, we call degree of v any element of $\mathbb{Z}_p$, denoted $\deg(v)$, such that $\deg(v)\,\mathbb{Z}_p = \log(\mathrm{N}_{K_v/\mathbb{Q}_p}(K_v^\times))$.

If $v \nmid p$, we put $\deg(v) := \log(\mathrm{N}\mathfrak{p}_v)$, where N is the absolute norm.[31]

(ii) (logarithmic p-adic absolute values). Let v be any place. We put, for all $x \in K^\times$:

- $|x|_v^\sim := (-1)^{\mathrm{v}(x)}$, if $v|\infty$,
- $|x|_v^\sim := (\mathrm{N}\mathfrak{p}_v)^{-\mathrm{v}(x)}$, if $v \nmid p$,
- $|x|_v^\sim := \mathrm{N}_{K_v/\mathbb{Q}_p}(i_v(x))\,(\mathrm{N}\mathfrak{p}_v)^{-\mathrm{v}(x)}$, if $v|p$.

These maps take their values in $\mathbb{Z}_p^\times$.

(iii) (logarithmic p-adic valuations). We define, for v finite:

$$\widetilde{\mathrm{v}} := -\frac{1}{\deg(v)} \log \circ\, |\bullet|_v^\sim,$$

with values in $\mathbb{Z}_p$. $\square$

[31] Any generator of $\mathbb{Z}_p \log(\mathrm{N}\mathfrak{p}_v)$ would be suitable for the definition.

7.3 Remarks. (i) The norm group of $\mathbb{Q}_p^{\mathrm{cycl}}(p)$ in $\widehat{\mathbb{Q}_p^\times}$ (infinite local class field theory!) is $p^{\mathbb{Z}_p} \oplus \mu(\mathbb{Q}_p)$ by II.1.8.3, (i) (exactly the kernel of log); therefore, the norm group of $K_v \cap \mathbb{Q}_p^{\mathrm{cycl}}(p)$, $v|p$, is:

$$\widetilde{N}_v = (p^{\mathbb{Z}_p} \oplus \mu(\mathbb{Q}_p)) \,.\, \mathrm{adh}(\mathrm{N}_{K_v/\mathbb{Q}_p}(K_v^\times)),$$

giving $\log(\mathrm{N}_{K_v/\mathbb{Q}_p}(K_v^\times)) = \log(\widetilde{N}_v)$. If $[K_v \cap \mathbb{Q}_p^{\mathrm{cycl}}(p) : \mathbb{Q}_p] = p^{n_0}$, we then have $\widetilde{N}_v = p^{\mathbb{Z}_p} \oplus \mu(\mathbb{Q}_p) \oplus (1 + p^{n_0+1+\delta_{2,p}}\mathbb{Z}_p)$, thus $\log(\mathrm{N}_{K_v/\mathbb{Q}_p}(K_v^\times)) = p^{n_0+1+\delta_{2,p}}\mathbb{Z}_p$, where $\delta_{2,p} = 1$ or 0 according as $p = 2$ or not.

(ii) The maps $\widetilde{\mathrm{v}}$ depend on the choice of the degrees contrary to the case of the logarithmic absolute values $|\bullet|_v^\sim$. □

7.4 Proposition (product formula). *We have* $\prod_v |x|_v^\sim = 1$ *for all* $x \in K^\times$.

Proof. We have $\prod_{v \in Pl_0} |x|_v^\sim = \prod_{v \in Pl_0} (\mathrm{N}\mathfrak{p}_v)^{-\mathrm{v}(x)} \prod_{v|p} \mathrm{N}_{K_v/\mathbb{Q}_p}(i_v(x))$. The first term is $\pm\mathrm{N}_{K/\mathbb{Q}}(x)^{-1} \in \mathbb{Q}_{\mathrm{pos}}^\times$. By the formula of local norms II.2.2, the second one is $\mathrm{N}_{K/\mathbb{Q}}(x)$. Since $\prod_{v|\infty} (-1)^{\mathrm{v}(x)}$ is the sign of $\mathrm{N}_{K/\mathbb{Q}}(x)$, the result follows. □

We will not need the absolute values $|\bullet|_v^\sim$ for $v|\infty$ and we note, once for all, that $\prod_{v \in Pl_0} |x|_v^\sim = \pm 1$.

The absolute values $|\bullet|_v^\sim$ can be extended to $\mathcal{K}^\times$, with values in $(\mathbb{Z}_p^\times)_p = 1 + p\mathbb{Z}_p$, and the corresponding maps $\widetilde{\mathrm{v}}$ on $\mathcal{K}^\times$ take their values in $\mathbb{Z}_p$. The product formula 7.4 is still valid on $\mathcal{K}^\times$.

7.5 Notations. (i) We consider $\mathcal{I} := I \otimes \mathbb{Z}_p$, $\mathcal{P} := P \otimes \mathbb{Z}_p$, $\mathcal{K}^\times := K^\times \otimes \mathbb{Z}_p$, and the corresponding subgroups $\mathcal{I}_p$, $\mathcal{P}_p$, $\mathcal{K}_p^\times$, of prime to p elements.

(ii) We extend the degree function by putting $\deg(\mathfrak{a}) := \sum_{v \in Pl_0} \mathrm{v}(\mathfrak{a})\deg(v)$ for all $\mathfrak{a} \in \mathcal{I}$, where $\mathrm{v}(\mathfrak{a})$ is the natural extension of the ordinary valuation v. For instance, if $\mathfrak{a} \in \mathcal{I}_p$, then $\deg(\mathfrak{a}) = \log(\mathrm{N}\mathfrak{a})$ if, for $v \nmid p$, we have choosen $\deg(v) = \log(\mathrm{N}\mathfrak{p}_v)$.

(iii) We put $\widetilde{\mathcal{I}} := \{\mathfrak{a} \in \mathcal{I},\ \deg(\mathfrak{a}) = 0\}$, and we consider the map $\widetilde{\mathrm{div}} : \mathcal{K}^\times \longrightarrow \mathcal{I}$ which associates with $x \in \mathcal{K}^\times$ the ideal:

$$\widetilde{\mathrm{div}}(x) := \prod_{v \in Pl_0} \mathfrak{p}_v^{\widetilde{\mathrm{v}}(x)}.$$ □

7.6 Proposition. *The* $\mathbb{Z}_p$*-module* $\widetilde{\mathcal{P}} := \widetilde{\mathrm{div}}(\mathcal{K}^\times)$ *is contained in* $\widetilde{\mathcal{I}}$.

Proof. We have $\sum_{v \in Pl_0} \widetilde{\mathrm{v}}(x)\deg(v) = -\sum_{v \in Pl_0} \log(|x|_v^\sim) = -\log\Big(\prod_{v \in Pl_0} |x|_v^\sim\Big) = 0$, by the product formula. □

7.6.1 Definition (the logarithmic class group of K [Ja5, Ja6]). We put:

$$\widetilde{\mathcal{C}\ell}_K := \widetilde{\mathcal{I}}/\widetilde{\mathcal{P}},$$

and call this $\mathbb{Z}_p$-module, the logarithmic class group of K. □

7.6.2 Remarque. Two systems of degrees give isomorphic logarithmic class groups: indeed, use the map $(\mathfrak{p}_v)_v \longmapsto (\mathfrak{p}_v^{u_v})_v$ on the free $\mathbb{Z}_p$-module $\mathcal{I}$, where $u_v \in \mathbb{Z}_p^\times$ is the quotient of the two degrees at v. □

7.7 Proposition. *We have* $\widetilde{\mathcal{C}\ell} \simeq \widetilde{\mathcal{I}}_p/\widetilde{\mathcal{P}}_p$, *where* $\widetilde{\mathcal{I}}_p := \widetilde{\mathcal{I}} \cap \mathcal{I}_p$, $\widetilde{\mathcal{P}}_p := \widetilde{\mathcal{P}} \cap \mathcal{I}_p$.

Proof. Let $\mathfrak{a} =: \mathfrak{a}_p\,\mathfrak{b} \in \widetilde{\mathcal{I}}$ with $\mathfrak{a}_p =: \prod_{v|p} \mathfrak{p}_v^{n_v}$, $n_v \in \mathbb{Z}_p$, $\mathfrak{b} \in \mathcal{I}_p$. To represent $\mathfrak{a}$ in $\widetilde{\mathcal{I}}_p$, it is necessary to find $x \in \mathcal{K}^\times$ such that $\widetilde{\mathrm{div}}(x)\,\mathfrak{a}_p \in \mathcal{I}_p$. This is equivalent to solve $\widetilde{\mathrm{v}}(x) = -n_v$ for all $v|p$, therefore to solve:

$$\log(\mathrm{N}_{K_v/\mathbb{Q}_p}(\bar{i}_v(x))) = n_v \deg(v) \ \text{ for all } v|p$$

since $\log(\mathrm{N}\mathfrak{p}_v) = 0$. By the Definition 7.2, (i), for $v|p$ there exist $x_v \in K_v^\times$ such that $\log(\mathrm{N}_{K_v/\mathbb{Q}_p}(x_v)) = n_v \deg(v)$. The result follows by taking for x a preimage in $\mathcal{K}^\times$ of $(x_v)_v$ for the map $\bar{i}_p : \mathcal{K}^\times \longrightarrow \bigoplus_{v|p}(\pi_v^{\mathbb{Z}_p} \oplus U_v^1)$ which is surjective (indeed, $\bar{i}_p(K^\times)$ is dense and $\langle\, \bar{i}_p(K^\times)\,\rangle_{\mathbb{Z}_p}$ is closed in the $\mathbb{Z}_p$-module of finite type $\bigoplus_{v|p}(\pi_v^{\mathbb{Z}_p} \oplus U_v^1) = \bigoplus_{v|p} \widehat{K_v^\times}$). □

7.8 Theorem. *For any number field* K, *we have* $\widetilde{\mathcal{C}\ell}_K \simeq \mathrm{Gal}(H_p^{\mathrm{lc}}/K\mathbb{Q}^{\mathrm{cycl}}{}_{(p)})$.

Proof. We now use the result of Exercise 7.1, (ii). It will be sufficient to prove that $\widetilde{\mathcal{P}}_p = \langle\, \mathcal{D}_v \,\rangle_{v|p}$ since $\{\mathfrak{a} \in \mathcal{I}_p,\ \log(\mathrm{N}\mathfrak{a}) = 0\}$ is precisely $\widetilde{\mathcal{I}}_p$. We check without any difficulties that:

$$\begin{aligned}
\widetilde{\mathcal{P}}_p &= \{\widetilde{\mathrm{div}}(x),\ x \in \mathcal{K}^\times\} \cap \mathcal{I}_p \\
&= \Big\{ \prod_{v|p} \mathfrak{p}_v^{-\frac{1}{\deg(v)}\log(\mathrm{N}_{K_v/\mathbb{Q}_p}(\bar{i}_v(x)))} \prod_{v\nmid p} \mathfrak{p}_v^{\mathrm{v}(x)},\ \ x \in \mathcal{K}^\times \Big\} \cap \mathcal{I}_p \\
&= \Big\{ \prod_{v\nmid p} \mathfrak{p}_v^{\mathrm{v}(x)},\ \ x \in \mathcal{K}^\times,\ \log(\mathrm{N}_{K_v/\mathbb{Q}_p}(\bar{i}_v(x))) = 0 \text{ for all } v|p \Big\} \\
&= \Big\{ (x) \,.\, \prod_{v|p} \mathfrak{p}_v^{-\mathrm{v}(x)},\ \ x \in \mathcal{K}^\times,\ \log(\mathrm{N}_{K_v/\mathbb{Q}_p}(\bar{i}_v(x))) = 0 \text{ for all } v|p \Big\}.
\end{aligned}$$

Thus, the inclusion $\langle\, \mathcal{D}_v \,\rangle_{v|p} \subseteq \widetilde{\mathcal{P}}_p$ is clear. For the other, taking $x \in \mathcal{K}^\times$, we can write $x = \prod_{v|p} x_{\infty_{p\backslash v}}$, with $\bar{i}_{p\backslash v}(x_{\infty_{p\backslash v}}) = 1$ as usual, so that $\mathrm{N}_{K_v/\mathbb{Q}_p}(\bar{i}_v(x)) = \mathrm{N}_{K_v/\mathbb{Q}_p}(\bar{i}_v(x_{\infty_{p\backslash v}}))$ for all $v|p$. Finally, this yields:

$$(x) \,.\, \prod_{v|p} \mathfrak{p}_v^{-\mathrm{v}(x)} = \prod_{v|p} \Big((x_{\infty_{p\setminus v}}) \mathfrak{p}_v^{-\mathrm{v}(x_{\infty_{p\setminus v}})} \Big),$$

with $\log(\mathrm{N}_{K_v/\mathbb{Q}_p}(\bar{i}_v(x_{\infty_{p\setminus v}}))) = 0$ for all $v|p$, giving the result. □

Note. In the various works of Jaulent–Soriano, the computations are done in the context of the p-adic class field theory 4.18. But it seems that, for practical computations, the interpretation in $\mathcal{I}_p/\mathcal{P}_{p,\infty} \simeq \mathrm{Gal}(H_p^{\mathrm{ord}}{}_{(p)}/K)$, which eliminates all the tame decomposition groups, is more convenient.

Thus, the Gross conjecture is equivalent to the finiteness of $\widetilde{\mathcal{C}\ell}$.

With the above techniques, the reader can look at the developpement of the theory by Jaulent–Soriano who have given a large study of the logarithmic class group in the spirit of that of the classical class group.

The final step for us will be to prove the relation of $\widetilde{\mathcal{C}\ell}_K$ with the Hilbert kernel $\mathrm{WK}_2(K)$ under Kummer assumptions. For this, we give the following explicit formula.

7.9 Theorem [Ja5, Th. 1]. *Let K be a number field containing μ_{2p}, and let ζ be a generator of μ_p. Then, for all finite place v of K and all $x \in K_v^\times$ we have, in terms of local Hilbert symbols: $(\zeta, x)_v = \zeta^{-\frac{1}{m_v}\log(|x|_v^{\sim})}$ (see 7.2, (ii)), where $m_v = |\mu(K_v)|$.*

Proof. Recall that $m_v = (q_v - 1)m_v^1$, where $q_v = \mathrm{N}\mathfrak{p}_v$, and where m_v^1 is a power of the residue characteristic of v (see II.7.1.3, (i)).

(i) (case $v \nmid p$). By the formula II.7.1.5, we have $(\zeta, x)_v^{\mathrm{reg}} := (\zeta, x)_v^{m_v^1} = \zeta^{\mathrm{v}(x)}$. Since $\widetilde{\mathrm{v}} = \mathrm{v}$ for $\deg(v) = \log(q_v)$, we have $\mathrm{v}(x) = -\frac{1}{\log(q_v)}\log(|x|_v^{\sim})$; but $\frac{1}{q_v-1}\log(q_v) = \frac{1}{q_v-1}\log(1 + q_v - 1) \equiv 1 \bmod (p)$ since $q_v \equiv 1 \bmod (p)$ if $p \neq 2$ and $q_v \equiv 1 \bmod (4)$ if $p = 2$. Thus $(\zeta, x)_v^{m_v^1} = \zeta^{-\frac{1}{q_v-1}\log(|x|_v^{\sim})}$, and finally we get the result since m_v^1 is prime to p.

(ii) (case $v|p$). We introduce the global cyclotomic field $C = \mathbb{Q}(\mu_{m_v^1})$ for which $C_u = \mathbb{Q}_p(\mu_{m_v^1}) \subseteq K_v$ for the unique place $u|p$ of C (thus, C_u is the maximal p-cyclotomic field contained in K_v). We then have $m_v = (q_v - 1)m_v^1$ and, in C_u, $m_u = (q_u - 1)m_u^1$ with $m_u^1 = m_v^1$.

By the formula II.7.1.1, (iv) in the extension K_v/C_u, since $\frac{m_v}{m_u} = \frac{q_v-1}{q_u-1}$, we have $(\zeta, x)_v^{q_v-1} =: (\zeta, y)_u^{q_u-1}$ with $y := \mathrm{N}_{K_v/C_u}(x)$.

We compute $(\zeta, y)_u^{q_u-1}$ for any $y \in C^\times$, using the product formula II.7.3.1 in C for Hilbert symbols of order $m = m_v^1$, and using (i):

$$\Big(\frac{\zeta,\, y}{u}\Big)^{q_u-1} = \Big(\frac{\zeta,\, y}{u}\Big)^{\frac{m_u}{m}} = \prod_{u' \neq u} \Big(\frac{\zeta,\, y}{u'}\Big)^{-\frac{m_{u'}}{m}} = \zeta^{\sum_{u'\neq u} \frac{1}{m}\log(|y|_{u'}^{\sim})}.$$

The product formula (for logarithmic p-adic valuations in C) yields:

$$\sum_{u' \neq u} \tfrac{1}{m} \log(|y|^{\sim}_{u'}) = -\tfrac{1}{m} \log(|y|^{\sim}_{u}).$$

By density, the formula is valid for $y := \mathrm{N}_{K_v/C_u}(x)$, and we check that:

$$-\tfrac{1}{m} \log(|y|^{\sim}_{u}) = -\tfrac{1}{m} \log(|x|^{\sim}_{v}).$$

Since $q_v - 1$ is prime to p, and since $(q_v - 1)m = m_v$, we have as expected $(\zeta, x)_v = \zeta^{-\frac{1}{m_v} \log(|x|^{\sim}_{v})}$. □

7.9.1 Lemma. *For any finite place* v, *we can choose* $\deg(v) = m_v$.

Proof. If $v|p$, $\log(\mathrm{N}_{K_v/\mathbb{Q}_p}(K_v^{\times})) = \log(\widetilde{N}_v)$ by 7.3, where $\widetilde{N}_v$ is the norm group of $K_v \cap \mathbb{Q}_p^{\mathrm{cycl}}(p)$. Since the maximal p-cyclotomic subfield of K_v is $\mathbb{Q}_p(\mu_{m_v^1})$, with $m_v^1 \equiv 0 \bmod (p)$ if $p \neq 2$ (resp. $m_v^1 \equiv 0 \bmod\ (4)$ if $p = 2$), $\log(\widetilde{N}_v) = m_v^1 \mathbb{Z}_p = m_v \mathbb{Z}_p$. The case $v \nmid p$ is clear. □

7.9.2 Corollary. *With these choices we have, for all finite place* v *of* K *and all* $x \in K^{\times}$, $(\zeta, x)_v = \zeta^{\widetilde{v}(x)}$, *and* $(\frac{\zeta,\, x}{v}) = \zeta^{-\frac{1}{m} \log(|x|^{\sim}_{v})}$ *in terms of Hilbert symbols of order* $m = |\mu(K)|$ *of* K. □

The exact sequence II.7.6 for the global Hilbert symbol is:

$$1 \longrightarrow \mathrm{WK}_2(K) \longrightarrow \mathrm{K}_2(K) \xrightarrow{\ h\ } \bigoplus_{v \in Pl^{\mathrm{nc}}}{}' \mu(K_v) \longrightarrow 1,$$

where $\bigoplus_{v \in Pl^{\mathrm{nc}}}{}' \mu(K_v) := \{(\xi_v)_v,\ \prod_v \xi_v^{\frac{m_v}{m}} = 1\}$, with $m = |\mu(K)|$. If we write the above exact sequence $1 \to B \longrightarrow A \longrightarrow C \to 1$, we obtain the standard exact sequence:

$$1 \to {}_pB \longrightarrow {}_pA \longrightarrow {}_pC \xrightarrow{\ \delta\ } B/B^p \longrightarrow A/A^p \longrightarrow C/C^p \to 1,$$

in which, if $c \in {}_pC$, then $\delta(c) = a^p\, B^p$ where $a \in A$ is a preimage of c. From the results of Tate [Ta2, § 6], we have $A/A^p \simeq C/C^p$ and ${}_pA = \{\{\zeta, x\},\ x \in K^{\times}\}$. Thus we obtain the exact sequence:

$$1 \to {}_p\mathrm{WK}_2(K) \longrightarrow {}_p\mathrm{K}_2(K) \xrightarrow{\ h\ } \bigoplus_{v \in Pl^{\mathrm{nc}}}{}' {}_p\mu(K_v) \longrightarrow \mathrm{WK}_2(K)/\mathrm{WK}_2(K)^p \to 1.$$

7.10 Proposition. *The map which sends* $\zeta \otimes \mathfrak{a}$, $\mathfrak{a} \in \widetilde{\mathcal{I}}$, *to* $(\zeta^{\mathrm{v}(\mathfrak{a})})_v$, *induces the isomorphism:*

$${}_p\mu(K) \otimes \widetilde{\mathcal{I}} \longrightarrow \bigoplus_{v \in Pl^{\mathrm{nc}}}{}' {}_p\mu(K_v).$$

Proof. With the above choice 7.9.1 of the degree of a place, we see that the map takes values in $\bigoplus_{v \in Pl^{\mathrm{nc}}}{}' {}_p\mu(K_v)$. Let $(\zeta^{n_v})_v$ such that $\sum_v n_v \frac{m_v}{m} \equiv 0 \bmod (p)$

(product formula). Let Σ be the support of $(n_v)_v$. By the density Theorem II.4.6, there exists $v_0 \notin \Sigma$, $v_0 \nmid p$, which does not split in $K(\sqrt[p]{\mu(K)})/K$, in other words such that $\frac{m_{v_0}}{m} \in \mathbb{Z}_p^\times$. Therefore, $n'_{v_0} := -\frac{m}{w_{v_0}} \sum_v n_v \frac{m_v}{m} \in p\mathbb{Z}_p$. Now consider $\mathfrak{a} := \mathfrak{p}_{v_0}^{n'_{v_0}} \cdot \prod_v \mathfrak{p}_v^{n_v}$; then $\deg(\mathfrak{a}) = 0$, and $\zeta \otimes \mathfrak{a}$ is a preimage of the familly $(\zeta^{n_v})_v$. This proves the surjectivity, the injectivity being immediate. □

We obtain the diagram of exact sequences:

$$\begin{array}{ccccccc}
{}_p\mathrm{K}_2(K) & \xrightarrow{\ h\ } & \bigoplus'_{v \in Pl^{\mathrm{nc}}} {}_p\mu(K_v) & \xrightarrow{\ \delta\ } & \mathrm{WK}_2(K)/\mathrm{WK}_2(K)^p & \longrightarrow & 1 \\
\uparrow & & \uparrow & & \uparrow & & \\
{}_p\mu(K) \otimes \mathcal{K}^\times & \xrightarrow{1 \otimes (\widetilde{\mathrm{v}})_v} & {}_p\mu(K) \otimes \widetilde{\mathcal{I}} & \longrightarrow & {}_p\mu(K) \otimes \widetilde{\mathcal{C}\ell} & \longrightarrow & 1
\end{array}$$

in which we identify ${}_p\mu(K) \otimes \mathcal{K}^\times$ with ${}_p\mu(K) \otimes K^\times$ and where the map ${}_p\mu(K) \otimes \mathcal{K}^\times \longrightarrow {}_p\mathrm{K}_2(K)$ is given by $\zeta \otimes x \longmapsto \{\zeta, x\}$.

The Corollary 7.9.2 means that the square on the left hand side is exact, proving the following.

7.11 Theorem (Jaulent [Ja5] (1992/1994)). *Let K be a number field containing μ_{2p}. Then we have a canonical isomorphism:*

$$\mathrm{WK}_2(K)/\mathrm{WK}_2(K)^p \simeq {}_p\mu(K) \otimes \widetilde{\mathcal{C}\ell}_K,$$

where $\widetilde{\mathcal{C}\ell}_K = \{\mathfrak{a} \in \mathcal{I}_p,\ \log(\mathrm{N}\mathfrak{a}) = 0\}/\langle\, \mathcal{D}_v \,\rangle_{v|p}$ (see 7.1, 7.7). □

We now have a way of computing $\mathrm{rk}_p(\mathrm{WK}_2(K))$, which is especially interesting when the p-part of the Hilbert kernel does not coincide with that of the regular kernel (see II.7.6.1). The details are left to the reader.

IV. Invariant Class Groups in p-Ramification Genus Theory

If the arithmetical invariants of K are known, in other words if class field theory over K is explicit, the situation for a finite extension L is a priori completely different, and one usually studies the corresponding invariants of L using several means. This chapter explains the two classical approaches: invariant classes formulas and genus theory.

Concerning the first point, we have already mentioned such a formula for class groups in II.6.2.3, II.6.2.4 but, although it is useful in practice, this formula does not involve the fundamental invariants, $\mathcal{A}_{L,p}^{\mathrm{ord}}$ and $\mathcal{T}_{L,p}^{\mathrm{ord}}$, [1] of class field theory, it is known only for cyclic extensions L/K, and in any case is covered by genus theory which does not even assume that L/K is Galois. Indeed, when L/K is any finite Galois extension with Galois group G, we will see, in Subsection 3, that the subgroups $(\mathcal{A}_{L,p}^{\mathrm{ord}})^G$ and $(\mathcal{T}_{L,p}^{\mathrm{ord}})^G$ are explicit (assuming the Leopoldt conjecture for p).

Before that, we will see, in Subsection 2, that the capitulation phenomena mentioned in principal ideal Theorem II.5.8.3 (tame case) never happen for the class groups which, under the Artin map, give the groups $\mathcal{A}_{K,p}^{\mathrm{ord}}$ and $\mathcal{T}_{K,p}^{\mathrm{ord}}$ (once again here the Leopoldt conjecture is necessary); in other words the transfer map $\mathcal{A}_{K,p}^{\mathrm{ord}} \longrightarrow \mathcal{A}_{L,p}^{\mathrm{ord}}$ is *injective* in any extension L of the base field K. Note that this injectivity is not true anymore if we do not take the ordinary sense, but then the kernel has a canonical value (Proposition 2.3). On the other hand, in a completely general setting the kernel of the transfer map $\mathcal{A}_{K,T}^{S} \longrightarrow \mathcal{A}_{L,T'}^{S'}$ is not known: for instance, for $T = \emptyset$ and $S = Pl_{\infty}^{\mathrm{r}}$, the transfer map corresponds to the map $\mathcal{C}\ell_K^{\mathrm{ord}} \longrightarrow \mathcal{C}\ell_L^{\mathrm{ord}}$ of extension of ideal classes, which depends on the (rather unpredictable) cohomology of the unit group since clearly $\mathrm{Ker}(\mathcal{C}\ell_K^{\mathrm{ord}} \longrightarrow \mathcal{C}\ell_L^{\mathrm{ord}})$ is canonically isomorphic to a subgroup of $H^1(G, E_L^{\mathrm{ord}})$.

Concerning genera theory, since the classical literature is incomplete on the subject [2], we will explain it, in Section 4, in complete generality (i.e., with ramification and decomposition, relative to an extension L/K which is not assumed to be Galois). It is a typical use of the correspondence of class field

[1] Galois group of the maximal p-ramified noncomplexified abelian pro-p-extension of L and its torsion group.

[2] [I] for the general unramified case, [Ja2] for the Galois case with T-ramification and S-decomposition.

theory which has many applications, and which sheds new light on the local-global normic phenomena already met in Chapter II, Section 2, in particular the results of II.2.6.

§1 Reduction to the Case of p-Ramification

Let L/K be a finite Galois extension and let $G := \mathrm{Gal}(L/K)$. By III.4.1.7 applied to L, we have the exact sequence of G-modules (assuming the Leopoldt conjecture for the prime number p):

$$1 \longrightarrow \prod_{w \nmid p} (F_w^\times)_p \longrightarrow (\overline{G}_L^{\mathrm{ab}})_p \longrightarrow \mathcal{A}_{L,p}^{\mathrm{ord}} \longrightarrow 1,$$

where $\overline{G}_L^{\mathrm{ab}} := \mathrm{Gal}(\overline{L}^{\mathrm{ab}}/L)$, $\mathcal{A}_{L,p}^{\mathrm{ord}} := \mathrm{Gal}(H_{L,p}^{\mathrm{ord}}(p)/L) \simeq \mathcal{T}_{L,p}^{\mathrm{ord}} \times \mathbb{Z}_p^{r_2(L)+1}$, $H_{L,p}^{\mathrm{ord}}(p)$ being the maximal p-ramified noncomplexified abelian pro-p-extension of L. The cohomology of the G-module:

$$\mathrm{Gal}(\overline{L}^{\mathrm{ab}}(p)/H_{L,p}^{\mathrm{ord}}(p)) \simeq \prod_{w \nmid p} (F_w^\times)_p$$

is known (because of the Shapiro's lemma, we are reduced to the case of the D_{w_0}-modules $(F_{w_0}^\times)_p$, $w_0 | v$, v finite, $v \nmid p$, whose cohomology may be nontrivial only when v ramifies). Thus, there is not much point in computing $(\overline{G}_L^{\mathrm{ab}})_p^G$, which would give a weaker information on $(\overline{G}_L^{\mathrm{ab}})_p$. The same holds for the groups of the form $\mathcal{A}_{L,T'}^{\mathrm{res}}$, T' being the set of places of L above the elements of a set T of finite places of K containing Pl_p, since we have a similar exact sequence:

$$1 \longrightarrow \prod_{w \in T'_{\mathrm{ta}} \cup Pl_{L,\infty}^{\mathrm{r}}} (F_w^\times)_p \longrightarrow \mathcal{A}_{L,T'}^{\mathrm{res}} \longrightarrow \mathcal{A}_{L,p}^{\mathrm{ord}} \longrightarrow 1.$$

On the other hand, the arithmetically difficult part is the group $\mathcal{A}_{L,p}^{\mathrm{ord}}$ (i.e., the case $T = Pl_p$) and in particular its torsion group $\mathcal{T}_{L,p}^{\mathrm{ord}}$ whose knowledge is sufficient for that of the groups $\mathcal{T}_{L,T'}^{\mathrm{ord}}$, $T \supseteq Pl_p$ (see III.4.1.5). We will thus limit our study to this case and we will use the following simplified notations.

1.1 Notations. (i) Let k be a number field and let p be a fixed prime number; we denote by:

- $\mathcal{A}_k^{\mathrm{ord}}$, the Galois group of the maximal p-ramified noncomplexified abelian pro-p-extension of k,
- $\mathcal{T}_k^{\mathrm{ord}}$, the $\mathbb{Z}_p$-torsion subgroup of $\mathcal{A}_k^{\mathrm{ord}}$,
- $\mathcal{E}_k^{\mathrm{ord}} := E_k^{\mathrm{ord}} \otimes \mathbb{Z}_p$,

- $k_p^\times := k_{Pl_p}^\times$, $\kappa_p^\times := k_p^\times \otimes \mathbb{Z}_p$,
- $I_{k,p} := I_{k,Pl_p}$, $\mathcal{I}_{k,p} := I_{k,p} \otimes \mathbb{Z}_p$,
- $\mathcal{P}_{k,\infty} := \{(x),\ x \in \kappa_p^\times,\ \bar{i}_{k,p}(x) = 1\}$ (see III.2.4.1),

where $\bar{i}_{k,p}$ is the canonical embedding $\kappa_p^\times \longrightarrow \bigoplus_{u|p} U_u^1$.

(ii) If L/K is a finite Galois extension and $\mathfrak{l}$ is a prime ideal of K, we denote by $e_{\mathfrak{l},p}$ the p-part of the ramification index of $\mathfrak{l}$ in L/K, and we consider the ideal groups:

- $R_{K,\rm ta} := \langle\, \mathfrak{l} \nmid p,\ e_{\mathfrak{l},p} > 1 \,\rangle$, $\mathcal{R}_{K,\rm ta} := R_{K,\rm ta} \otimes \mathbb{Z}_p$,
- $R_{L,\rm ta} := \langle\, \prod_{\mathfrak{L}|\mathfrak{l}} \mathfrak{L},\ \mathfrak{l} \nmid p,\ e_{\mathfrak{l},p} > 1 \,\rangle$, $\mathcal{R}_{L,\rm ta} := R_{L,\rm ta} \otimes \mathbb{Z}_p$. □

Note. In these notations, the lower index "ta" is not referring to the ramification in L/K but to that of the p-ramified extensions under consideration.

1.2 Remark. We will use cohomology groups of the form:

$$H^r(G,\, A \otimes \mathbb{Z}_p),\quad r \geq 0,$$

where A is a $\mathbb{Z}[G]$-module (G finite acting trivially on $\mathbb{Z}_p$). But, by flatness of $\mathbb{Z}_p$, we have the canonical isomorphisms:

$$H^r(G,\, A \otimes \mathbb{Z}_p) \simeq H^r(G,\, A) \otimes \mathbb{Z}_p.$$

Indeed, if $C^r(G,A)$ (resp. $B^r(G,A)$) is the group of r-cocycles (resp. r-coboundaries), it is easily checked, since C^r is the kernel of a homomorphism, that we have $C^r(G,\, A \otimes \mathbb{Z}_p) \simeq C^r(G,\, A) \otimes \mathbb{Z}_p$, and similarly for the group of coboundaries as image of a homomorphism. It follows that, to prove for instance that $H^r(G,\, A \otimes \mathbb{Z}_p) = 1$, it is sufficient to prove that we have:

$$C^r(G,\, A) \otimes 1 \subseteq B^r(G,\, A) \otimes \mathbb{Z}_p,$$

since $H^r(G,\, A) \otimes \mathbb{Z}_p \simeq C^r(G,\, A) \otimes \mathbb{Z}_p / B^r(G,\, A) \otimes \mathbb{Z}_p$. □

§2 Injectivity of the Transfer Map $\mathcal{A}_K^{\rm ord} \longrightarrow \mathcal{A}_L^{\rm ord}$

Let L/K be a finite extension of number fields. By III.2.4 and using the Notations 1.1, we have:

$$\mathcal{A}_K^{\rm ord} \simeq \mathcal{I}_{K,p}/\mathcal{P}_{K,\infty},\quad \mathcal{A}_L^{\rm ord} \simeq \mathcal{I}_{L,p}/\mathcal{P}_{L,\infty}.$$

In the map extension of ideals from K to L:

$$j_{L/K} =: j\, :\, \mathcal{I}_K \longrightarrow \mathcal{I}_L,$$

we trivially have $j(\mathcal{P}_{K,\infty}) \subset \mathcal{P}_{L,\infty}$, which defines the map, still denoted j by abuse of notation:

$$j : \mathcal{A}_K^{\mathrm{ord}} \longrightarrow \mathcal{A}_L^{\mathrm{ord}},$$

which sends $\mathrm{Art}_K(\mathfrak{a}) \in \mathcal{A}_K^{\mathrm{ord}}$, $\mathfrak{a} \in \mathcal{I}_{K,p}$, to $\mathrm{Art}_L(j(\mathfrak{a})) \in \mathcal{A}_L^{\mathrm{ord}}$, where Art_K, Art_L are the Artin maps corresponding to K and L, with values in $\mathcal{A}_K^{\mathrm{ord}}$ and $\mathcal{A}_L^{\mathrm{ord}}$ respectively, and which is nothing else than the transfer map defined from the finite case (see II.4.5, (v)). The first important result, which conditions all further ones, is the following (with Notations 1.1).

2.1 Theorem (injectivity of the transfer map). *Let L/K be a finite extension of number fields. Assuming the Leopoldt conjecture for p in the Galois closure of L over K, the map $j_{L/K} : \mathcal{A}_K^{\mathrm{ord}} \longrightarrow \mathcal{A}_L^{\mathrm{ord}}$ is injective. In particular $\mathcal{T}_L^{\mathrm{ord}}$ contains a subgroup isomorphic to $\mathcal{T}_K^{\mathrm{ord}}$.*

Proof. For the proof of the theorem, we may assume that the extension L/K is Galois: indeed, the injectivity of $j_{L/K}$ (with L/K Galois) implies that of $j_{L'/K}$ for any subextension L'/K of L/K.

Let $\mathfrak{a} \in \mathcal{I}_{K,p}$ be such that $j(\mathfrak{a}) \in \mathcal{P}_{L,\infty}$. We thus have $j(\mathfrak{a}) =: (y)$ with $y \in \mathcal{L}^\times$ and $\bar{i}_{L,p}(y) = 1$ (see III.2.4.1). Since $j(\mathfrak{a})$ is invariant under $G := \mathrm{Gal}(L/K)$, we have $y^{1-s} =: \eta(s) \in \mathcal{E}_L^{\mathrm{ord}}$ for all $s \in G$; since $\bar{i}_{L,p}(\eta(s)) = 1$, the Leopoldt conjecture implies $\eta(s) = 1$ for all $s \in G$, hence $y \in \mathcal{K}^\times$ (using 1.2 for $i = 0$) and the condition $\bar{i}_{L,p}(y) = 1$ descends to $\bar{i}_{K,p}(y) = 1$; therefore $\mathfrak{a} = (y)$ (j is injective on $\mathcal{I}_{K,p}$), hence $\mathfrak{a} \in \mathcal{P}_{K,\infty}$. □

2.2 Remark. Let $S = S_0$ be a finite set of finite prime to p places of K, and let S'_0 be the set of places of L above those of S_0. In the same way we can prove the injectivity of the map $\mathcal{A}_K^{S_0\,\mathrm{ord}} \longrightarrow \mathcal{A}_L^{S'_0\,\mathrm{ord}}$ under the hypothesis that $E_L^{S'_0\,\mathrm{ord}}$ is monogeneous and assuming the p-adic conjecture; the proof is the same starting from the fact that $\{\varepsilon \in \mathcal{E}_L^{S'_0\,\mathrm{ord}},\ \bar{i}_{L,p}(\varepsilon) = 1\} = 1$, under the above assumptions. □

Now, let K be a number field together with sets of places T and S, with $Pl_p \subseteq T$ and $S_0 = \emptyset$ (i.e., $S = S_\infty$). Let L/K be a finite Galois extension, and let T' and S'_∞ be the sets of places of L above those of T and of S_∞. Since for $p \neq 2$ the maps $j_{L/K}$ are injective on $\mathcal{A}_{K,T}^{S_\infty} = \mathcal{A}_{K,T}^{\mathrm{ord}}$ (Theorem 2.1), we have only to study the case $p = 2$ in the following result which will be of some importance in the Appendix.

2.3 Proposition. *Assuming the above hypothesis (especially $Pl_2 \subseteq T$) and the Leopoldt conjecture in L for $p = 2$, the transfer map:*

$$j_{L/K} =: j : \mathcal{A}_{K,T}^{S_\infty} \longrightarrow \mathcal{A}_{L,T'}^{S'_\infty}$$

has kernel equal to:

$$\bigoplus_{v \in \Delta_\infty^{\mathrm{c}}} D_v(H_T^{S_\infty}(2)/K) \simeq (\mathbb{Z}/2\mathbb{Z})^{|\Delta_\infty^{\mathrm{c}}|},$$

where $\Delta_\infty^{\mathrm{c}} := \Delta_\infty^{\mathrm{c}}(L/K)$ is the set of places of $\Delta_\infty := Pl_\infty^{\mathrm{r}} \backslash S_\infty$ complexified in L/K.

Proof. We have, with usual notations (see III.2.4, III.2.4.1):

$$\mathcal{A}_{K,T}^{S_\infty} \simeq \mathcal{I}_{K,T}/\mathcal{P}_{K,T,\infty,\Delta_\infty}, \quad \mathcal{A}_{L,T'}^{S'_\infty} \simeq \mathcal{I}_{L,T'}/\mathcal{P}_{L,T',\infty,\Delta'_\infty}.$$

Note that the sets S'_∞, Δ'_∞ must be replaced by the corresponding subsets of real infinite places of L.

If we use once again the proof of Theorem 2.1 to characterize the $\mathrm{Art}(\mathfrak{a}) \in \mathrm{Ker}(j)$, we obtain similarly:

$$j(\mathfrak{a}) =: (y), \quad y \in \mathcal{L}_{T',\Delta'_\infty}^\times \cap \mathcal{K}^\times, \quad \bar{\imath}_{L,T'}(y) = 1\ ;$$

although this implies $\bar{\imath}_{K,T}(y) = 1$, we only obtain that $\mathfrak{a} \in \mathcal{P}_{K,T,\infty,\Delta_\infty \backslash \Delta_\infty^{\mathrm{c}}}$ since the positivity of y on Δ'_∞ in L does not tell us anything about its signature on $\Delta_\infty^{\mathrm{c}}$ in K. The kernel of j is thus equal to:

$$\mathrm{Art}(\mathcal{P}_{K,T,\infty,\Delta_\infty \backslash \Delta_\infty^{\mathrm{c}}}) \simeq \mathcal{P}_{K,T,\infty,\Delta_\infty \backslash \Delta_\infty^{\mathrm{c}}}/\mathcal{P}_{K,T,\infty,\Delta_\infty}.$$

This Galois group is, by the correspondence of infinite 2-class field theory III.2.8:

$$\mathrm{Gal}(H_T^{S_\infty}(2)/H_T^{S_\infty \cup \Delta_\infty^{\mathrm{c}}}(2)),$$

which is equal to:

$$\bigoplus_{v \in \Delta_\infty^{\mathrm{c}}} D_v(H_T^{S_\infty}(2)/K) \simeq (\mathbb{Z}/2\mathbb{Z})^{|\Delta_\infty^{\mathrm{c}}|},$$

by the deployment Theorem III.4.1 (we have $Pl_2 \subseteq T$ and the Leopoldt conjecture in K for $p = 2$). □

2.4 Remark. These computations prove again that the kernel of the transfer map $\overline{G}_K^{\mathrm{ab}} \longrightarrow \overline{G}_L^{\mathrm{ab}}$ is exactly $\bigoplus_{v \in Pl_\infty^{\mathrm{rc}}} D_v(\overline{K}^{\mathrm{ab}}/K) \simeq (\mathbb{Z}/2\mathbb{Z})^{|Pl_\infty^{\mathrm{rc}}|}$, where Pl_∞^{rc} is the set of real places of K which are (totally) complexified in L. But the direct study of the map $C_K/D_K \longrightarrow C_L/D_L$ gives the result, in the non-Galois case, without the use of the Leopoldt conjecture (see the proof of III.4.15.1, (iii), Note, or Appendix, Section 1, (b)). □

§3 Determination of $(\mathcal{A}_L^{\mathrm{ord}})^G$ and $(\mathcal{T}_L^{\mathrm{ord}})^G$ — p-Rational Fields

Assumptions. We are still in the context of Section 1 and Notations 1.1 for a finite Galois extension L/K with Galois group G, and we assume that L satisfies the Leopoldt conjecture for p. □

a) Invariant Classes Formulas

We start with a few remarks:

(i) The above computations (the proof of Theorem 2.1) have shown the equality $\mathcal{P}_{L,\infty}^G = j(\mathcal{P}_{K,\infty})$.

(ii) The exact sequence:

$$1 \longrightarrow \mathcal{P}_{L,\infty} \longrightarrow \mathcal{I}_{L,p} \longrightarrow \mathcal{A}_L^{\mathrm{ord}} \longrightarrow 1,$$

yields:

$$1 \longrightarrow \mathcal{P}_{L,\infty}^G \longrightarrow \mathcal{I}_{L,p}^G \longrightarrow (\mathcal{A}_L^{\mathrm{ord}})^G \longrightarrow H^1(G, \mathcal{P}_{L,\infty}),$$

which can also be written:

$$1 \longrightarrow j(\mathcal{P}_{K,\infty}) \longrightarrow \mathcal{R}_{L,\mathrm{ta}} \,.\, j(\mathcal{I}_{K,p}) \longrightarrow (\mathcal{A}_L^{\mathrm{ord}})^G \longrightarrow H^1(G, \mathcal{P}_{L,\infty}),$$

because of (i) and the fact that $\mathcal{I}_{L,p}^G = \mathcal{R}_{L,\mathrm{ta}} \,.\, j(\mathcal{I}_{K,p})$ (see 1.1, (ii)).

Note. In fact, we have $I_{L,p}^G = \langle \prod_{\mathfrak{L} | \mathfrak{l}} \mathfrak{L},\ \mathfrak{l} \nmid p,\ e_{\mathfrak{l}} > 1 \rangle_{\mathbb{Z}} \,.\, j(I_{K,p})$, but, after tensoring by $\mathbb{Z}_p$, the primes $\mathfrak{l}$ for which $e_{\mathfrak{l}} \not\equiv 0 \bmod (p)$ do not enter in the result.

However, we have the following lemma.

3.1 Lemma. *Under the above assumptions, we have:*

$$H^1(G, \mathcal{P}_{L,\infty}) = 1.$$

Proof. If $\mathfrak{b}(s)$ is a 1-cocycle of $\mathcal{P}_{L,\infty}$, by setting $\mathfrak{b}(s) =: (y(s)),\ y(s) \in \mathcal{L}^\times$ such that $\bar{i}_{L,p}(y(s)) = 1$, this defines a 2-cocycle $\eta(s, s')$ of $\mathcal{E}_L^{\mathrm{ord}}$, because of the relation:

$$\frac{y(ss')}{s\, y(s') \,.\, y(s)} =: \eta(s, s') \ ;$$

therefore, we have $\eta(s, s') = 1$ under the Leopoldt conjecture, so $y(s)$ is a 1-cocycle of $\mathcal{L}^\times$. By 1.2 and the Hilbert–Speiser–Noether Theorem 90 (i.e., $H^1(G, L^\times) = 1$), it follows that it is a 1-coboundary of $\mathcal{L}^\times$, hence there exists $u \in \mathcal{L}^\times$ such that:

$$y(s) = u^{1-s} \ \text{ for all } s \in G.$$

Let us check that we can choose u prime to p. By flatness, we can reduce, for any n, to the case where $y(s)$ is a 1-cocycle of $L^\times$ congruent to 1 modulo (p^n) (see 2.4.1); the corresponding u is then given by:

$$u := \frac{1}{|G|} \sum_{s' \in G} y(s') \equiv 1 \bmod \left(\tfrac{p^n}{|G|}\right),$$

which is prime to p for n sufficiently large. Since $\bar{i}_{L,p}(y(s)) = 1$ for all $s \in G$, we have:

$$\bar{i}_{L,p}(u) \in \Big(\bigoplus_{w|p} U_w\Big)^G = j'\Big(\bigoplus_{v|p} U_v\Big)$$

(for the idelic embedding j'); finally $\bar{i}_{K,p}$ is surjective, hence there exists $a \in \mathcal{K}_p^\times$ such that:

$$\bar{i}_{L,p}(u) = j'(\bar{i}_{K,p}(a)) = \bar{i}_{L,p}(j(a)),$$

which yields $u =: j(a)\, v$, where $v \in \mathcal{L}_p^\times$ is such that $\bar{i}_{L,p}(v) = 1$. We have thus obtained:

$$\mathfrak{b}(s) = (y(s)) = (v)^{1-s} \text{ for all } s \in G, \text{ with } (v) \in \mathcal{P}_{L,\infty}. \qquad \square$$

Remark. In 2.1, 2.3, and 3.1, starting from III.2.4.1, we may have used more systematically the properties of the infinitesimals of K and L since the Leopoldt conjecture means here that $\mathcal{P}_{K,\infty} \simeq \mathcal{K}_\infty^\times$ and $\mathcal{P}_{L,\infty} \simeq \mathcal{L}_\infty^\times$ (canonically), and since classical properties (chinese remainder theorem, Theorem 90, ...) are valid with infinitesimals. This is left to the reader. □

This gives (with Notations 1.1) the following result.

3.2 Theorem (invariant classes subgroup [Gr1] (1982)). *Let L/K be a finite Galois extension of number fields with Galois group G, and let p be a prime number; we assume that L satisfies the Leopoldt conjecture for p. Then we have the following invariant classes formula:*

$$(\mathcal{A}_L^{\rm ord})^G = {\rm Art}_L(\mathcal{R}_{L,{\rm ta}})\,.\,j(\mathcal{A}_K^{\rm ord}),$$

where ${\rm Art}_L$ is the Artin map for L with values in $\mathcal{A}_L^{\rm ord}$, and $j := j_{L/K}$ is the transfer map (i.e., the extension of classes).

Proof. Let us show that the exact sequence:

$$1 \longrightarrow j(\mathcal{P}_{K,\infty}) \longrightarrow \mathcal{R}_{L,{\rm ta}}\,.\,j(\mathcal{I}_{K,p}) \longrightarrow (\mathcal{A}_L^{\rm ord})^G \longrightarrow 1,$$

which comes from the vanishing of $H^1(G, \mathcal{P}_{L,\infty})$ (Lemma 3.1), can indeed be interpreted in this way. We already have $j(\mathcal{I}_{K,p})/j(\mathcal{P}_{K,\infty}) \simeq j(\mathcal{A}_K^{\rm ord})$ and, by definition, ${\rm Art}_L(\mathcal{R}_{L,{\rm ta}}) \simeq \mathcal{R}_{L,{\rm ta}}/\mathcal{R}_{L,{\rm ta}} \cap \mathcal{P}_{L,\infty}$; but we have $\mathcal{R}_{L,{\rm ta}} \subset \mathcal{I}_{L,p}^G$ and $\mathcal{R}_{L,{\rm ta}} \cap \mathcal{P}_{L,\infty} = \mathcal{R}_{L,{\rm ta}} \cap j(\mathcal{P}_{K,\infty})$ since $\mathcal{P}_{L,\infty}^G = j(\mathcal{P}_{K,\infty})$. □

We can make more precise the relation giving $(\mathcal{A}_L^{\rm ord})^G$ (in particular concerning the $\mathbb{Z}_p$-module ${\rm Art}_L(\mathcal{R}_{L,{\rm ta}})$) in the following way, and with the same assumptions and notations as in the theorem.

3.2.1 Proposition. *We have the following relations:*

(i) $\mathrm{Art}_L(\mathcal{R}_{L,\mathrm{ta}}) \cap j(\mathcal{A}_K^{\mathrm{ord}}) = j(\mathrm{Art}_K(\mathcal{R}_{K,\mathrm{ta}}))$,

(ii) $\mathrm{Art}_L(\mathcal{R}_{L,\mathrm{ta}})/j(\mathrm{Art}_K(\mathcal{R}_{K,\mathrm{ta}})) \simeq (R_{L,\mathrm{ta}}/j(R_{K,\mathrm{ta}}))_p \simeq \bigoplus_{\mathfrak{l} \nmid p} \mathbb{Z}/e_{\mathfrak{l},p}\mathbb{Z}$.

Hence, we have:

$$(\mathcal{A}_L^{\mathrm{ord}})^G/j(\mathcal{A}_K^{\mathrm{ord}}) \simeq \bigoplus_{\mathfrak{l} \nmid p} \mathbb{Z}/e_{\mathfrak{l},p}\mathbb{Z}.$$

Proof. If $\mathfrak{b} \in \mathcal{R}_{L,\mathrm{ta}}$ is of the form $j(\mathfrak{a})\mathfrak{b}_\infty$ with $\mathfrak{a} \in \mathcal{I}_{K,p}$, $\mathfrak{b}_\infty \in \mathcal{P}_{L,\infty}$, we have $\mathfrak{b}_\infty \in (\mathcal{P}_{L,\infty})^G = j(\mathcal{P}_{K,\infty})$ and in particular $\mathfrak{b} \in j(\mathcal{I}_{K,p})$, which implies that $\mathfrak{b} \in j(\mathcal{R}_{K,\mathrm{ta}})$ (indeed, $R_{L,\mathrm{ta}} \cap j(I_{K,p}) = j(R_{K,\mathrm{ta}})$). This proves (i).

Furthermore, consider the canonical (surjective) map:

$$(R_{L,\mathrm{ta}}/j(R_{K,\mathrm{ta}}))_p \xrightarrow{\mathrm{Art}_L} \mathrm{Art}_L(\mathcal{R}_{L,\mathrm{ta}})/j(\mathrm{Art}_K(\mathcal{R}_{K,\mathrm{ta}}))\ ;$$

the above computations show that this map is injective, proving (ii). □

We can now start the study of the $\mathbb{Z}_p$-torsion in the Galois extension L/K. Recall that, for a number field k, the function $\mathrm{Log}_{k,p}$ composed with Art^{-1} takes its values in:

$$\mathcal{L}_{k,p} := \bigoplus_{u \mid p} k_u \Big/ \mathbb{Q}_p \log_{k,p}(E_k),$$

and yields the exact sequence (Theorem III.2.5):

$$1 \longrightarrow \mathcal{T}_k^{\mathrm{ord}} \longrightarrow \mathcal{A}_k^{\mathrm{ord}} \longrightarrow \mathbb{Z}_p \mathrm{Log}_{k,p}(I_{k,p}) \longrightarrow 0.$$

3.2.2 Lemma. *We have the exact sequence:*

$$1 \longrightarrow (\mathcal{T}_L^{\mathrm{ord}})^G/j(\mathcal{T}_K^{\mathrm{ord}}) \longrightarrow (\mathcal{A}_L^{\mathrm{ord}})^G/j(\mathcal{A}_K^{\mathrm{ord}}) \longrightarrow \mathrm{Log}_{L,p}((\mathcal{A}_L^{\mathrm{ord}})^G)/\mathrm{Log}_{L,p}(j(\mathcal{A}_K^{\mathrm{ord}})) \longrightarrow 0.$$

Proof. The surjectivity being clear, let σ_L be an element of $(\mathcal{A}_L^{\mathrm{ord}})^G$ such that:

$$\mathrm{Log}_{L,p}(\sigma_L) = \mathrm{Log}_{L,p}(j(\sigma_K)),\ \sigma_K \in \mathcal{A}_K^{\mathrm{ord}}.$$

We then have $\sigma_L = \tau\, j(\sigma_K)$, with $\tau \in \mathcal{T}_L^{\mathrm{ord}}$; hence $\tau \in (\mathcal{T}_L^{\mathrm{ord}})^G$. Finally we have $(\mathcal{T}_L^{\mathrm{ord}})^G \cap j(\mathcal{A}_K^{\mathrm{ord}}) = j(\mathcal{T}_K^{\mathrm{ord}})$ since $\mathcal{T}_L^{\mathrm{ord}}$ is finite and j is injective, thus giving the kernel. □

Furthermore, by Theorem 3.2, we have:

$$\mathrm{Log}_{L,p}((\mathcal{A}_L^{\mathrm{ord}})^G) = \mathrm{Log}_{L,p}(\mathrm{Art}_L(\mathcal{R}_{L,\mathrm{ta}})) + \mathrm{Log}_{L,p}(j(\mathcal{A}_K^{\mathrm{ord}})).$$

Let us now check that the canonical map:

$$\mathcal{L}_{K,p} := \bigoplus_{v \mid p} K_v \Big/ \mathbb{Q}_p \log_{K,p}(E_K) \longrightarrow \mathcal{L}_{L,p} := \bigoplus_{w \mid p} L_w \Big/ \mathbb{Q}_p \log_{L,p}(E_L)$$

(coming from the injective map $j' : \bigoplus_{v|p} K_v \longrightarrow \bigoplus_{w|p} L_w$), is injective. Indeed, if $j'(x) \in \mathbb{Q}_p\mathrm{log}_{L,p}(E_L)$, we have, setting $e_G := \frac{1}{|G|} \sum_{s\in G} s =: \frac{1}{|G|}\,\nu_{L/K}$:

$$\begin{aligned} j'(x) \in (\mathbb{Q}_p\mathrm{log}_{L,p}(E_L))^G &= e_G\mathbb{Q}_p\mathrm{log}_{L,p}(E_L) \\ &= \mathbb{Q}_p\mathrm{log}_{L,p}(\nu_{L/K}(E_L)) \subseteq j'(\mathbb{Q}_p\mathrm{log}_{K,p}(E_K)), \end{aligned}$$

since $\mathrm{log}_{L,p}$ is a homomorphism of G-modules. But j' is injective on $\bigoplus_{v|p} K_v$, proving the result. By abuse of notation, we will again denote by j this injection of $\mathcal{L}_{K,p}$ in $\mathcal{L}_{L,p}$.

On the other hand, we have, with the usual abuse of notation "Log $=$ Log $\circ$ Art^{-1}":

$$\begin{aligned} \mathrm{Log}_{L,p}(\mathrm{Art}_L(\mathcal{R}_{L,\mathrm{ta}})) &= \Big\langle \mathrm{Log}_{L,p}\Big(\prod_{\mathfrak{L}|\mathfrak{l}} \mathfrak{L}\Big) \Big\rangle \\ &= \Big\langle \frac{1}{e_{\mathfrak{l},p}}\mathrm{Log}_{L,p}(j(\mathfrak{l})) \Big\rangle = \Big\langle \frac{1}{e_{\mathfrak{l},p}} j(\mathrm{Log}_{K,p}(\mathfrak{l})) \Big\rangle \end{aligned}$$

since $j(\mathfrak{l}) = \Big(\prod_{\mathfrak{L}|\mathfrak{l}} \mathfrak{L}\Big)^{e_{\mathfrak{l}}}$, where only the primes $\mathfrak{l} \nmid p$ ramified in L/K enter in these computations, and:

$$\mathrm{Log}_{L,p}(j(\mathcal{A}_K^{\mathrm{ord}})) = j(\mathrm{Log}_{K,p}(\mathcal{A}_K^{\mathrm{ord}})) = j(\mathbb{Z}_p\mathrm{Log}_{K,p}(I_{K,p})).$$

Hence, because of the injectivity of j:

$$\begin{aligned} &\mathrm{Log}_{L,p}((\mathcal{A}_L^{\mathrm{ord}})^G)/\mathrm{Log}_{L,p}(j(\mathcal{A}_K^{\mathrm{ord}})) \simeq \\ &\qquad \Big(\Big\langle \frac{1}{e_{\mathfrak{l},p}}\mathrm{Log}_{K,p}(\mathfrak{l}) \Big\rangle + \mathbb{Z}_p\mathrm{Log}_{K,p}(I_{K,p})\Big)\Big/\mathbb{Z}_p\mathrm{Log}_{K,p}(I_{K,p})\ ; \end{aligned}$$

we can replace $e_{\mathfrak{l}}$ by $e_{\mathfrak{l},p}$ since we are concerned with $\mathbb{Z}_p$-modules.

Since we have by Proposition 3.2.1, (ii):

$$((\mathcal{A}_L^{\mathrm{ord}})^G : j(\mathcal{A}_K^{\mathrm{ord}})) = (R_{L,\mathrm{ta}} : j(R_{K,\mathrm{ta}})) = \prod_{\mathfrak{l}\nmid p} e_{\mathfrak{l},p},$$

we obtain, from the exact sequence 3.2.2, the following result of [Gr1; Gr2], proved again by different methods in [Ja2], [MoNg] (see Notations 1.1).

3.3 Theorem (invariant classes formula for the torsion group (1982)). *Let L/K be a finite Galois extension of number fields and $G := \mathrm{Gal}(L/K)$. Let p be a prime number; we assume that L satisfies the Leopoldt conjecture for p. We then have the following formula:*

$$|(\mathcal{T}_L^{\mathrm{ord}})^G| = |\mathcal{T}_K^{\mathrm{ord}}| \times \frac{\prod_{\mathfrak{l}\nmid p} e_{\mathfrak{l},p}}{\Big(\sum_{\mathfrak{l}\nmid p} \frac{1}{e_{\mathfrak{l},p}}\mathbb{Z}_p\mathrm{Log}_p(\mathfrak{l}) + \mathbb{Z}_p\mathrm{Log}_p(I_{K,p}) : \mathbb{Z}_p\mathrm{Log}_p(I_{K,p})\Big)},$$

where $e_{\mathfrak{l},p}$ is the p-part of the ramification index of $\mathfrak{l}$ in L/K, and where Log_p *is the logarithm function* $I_{K,p} \longrightarrow \bigoplus_{v|p} K_v \big/ \mathbb{Q}_p \mathrm{log}_{K,p}(E_K)$ *defined in III.2.2.* □

3.3.1 Exercise (totally real case). Let K be a totally real number field and let L/K be Galois with Galois group G; we assume that L satisfies the Leopoldt conjecture for p.

(i) Let A and B be subgroups of finite index of I_p such that $B \subseteq A$; show that:

$$(\mathbb{Z}_p \mathrm{Log}_p(A) : \mathbb{Z}_p \mathrm{Log}_p(B)) = (\mathbb{Z}_p \mathrm{log}(\mathrm{N}A) : \mathbb{Z}_p \mathrm{log}(\mathrm{N}B))$$

(where N is the absolute norm on K).

(ii) Show that $\mathbb{Z}_p \mathrm{log}(\mathrm{N}I_p) = qp^{n_0}\mathbb{Z}_p$, where $q = p$ (resp. 4) if $p \neq 2$ (resp. $p = 2$) and $p^{n_0} = [K \cap \mathbb{Q}^{\mathrm{cycl}}(p) : \mathbb{Q}]$.

(iii) Deduce that the formula of Theorem 3.3 can be written here:

$$|(\mathcal{T}_L^{\mathrm{ord}})^G| = |\mathcal{T}_K^{\mathrm{ord}}| \times p^h, \quad h := \rho - r + \sum_{\mathfrak{l} \nmid p} \varepsilon_{\mathfrak{l}},$$

with the following data:

- $p^{\varepsilon_{\mathfrak{l}}}$:= the p-part of the ramification index $e_{\mathfrak{l}}$ of $\mathfrak{l}$ in L/K,
- p^r := the p-part of $[L : K]$,
- $\rho := \min(n_0 + r\,; \ldots, \nu_{\mathfrak{l}} + \varphi_{\mathfrak{l}} + \gamma_{\mathfrak{l}}, \ldots) - n_0$, where:
  $p^{\nu_{\mathfrak{l}}}$:= the p-part of $q^{-1}\mathrm{log}(\ell)$, where $\mathfrak{l} \cap \mathbb{Z} =: \ell\mathbb{Z}$,
  $p^{\varphi_{\mathfrak{l}}}$:= the p-part of the residue degree of ℓ in $L/\mathbb{Q}$,
  $p^{\gamma_{\mathfrak{l}}}$:= the p-part of $|Pl_{L,\mathfrak{l}}|$.

Answer. (i) Assuming the Leopoldt conjecture, the kernel of the trace:

$$\mathrm{Tr}_{K/\mathbb{Q}} : \bigoplus_{v|p} K_v \longrightarrow \mathbb{Q}_p$$

is exactly equal to $\mathbb{Q}_p \mathrm{log}_p(E^{\mathrm{ord}})$; the first result follows by definition of Log_p.

(ii) By III.2.6.1, (i), and III.2.6.4, we have:

$$(\mathbb{Z}_p \mathrm{Log}_p(I_p) : \mathbb{Z}_p \mathrm{Log}_p(P_p)) = p^{-n_0}(q\mathbb{Z}_p : \mathrm{log}(\mathrm{N}_{K/\mathbb{Q}}(U_K))),$$

where $U_K := \bigoplus_{v|p} U_v$, which we can write here as follows (using (i)):

$$(\mathbb{Z}_p \mathrm{log}(\mathrm{N}I_p) : \mathbb{Z}_p \mathrm{log}(\mathrm{N}P_p)) = p^{-n_0}(q\mathbb{Z}_p : \mathrm{log}(\mathrm{N}_{K/\mathbb{Q}}(U_K)))\ ;$$

but $\mathbb{Z}_p \mathrm{log}(\mathrm{N}P_p) = \mathrm{log}(\mathrm{N}_{K/\mathbb{Q}}(U_K)))$ and the only possibility is:

$$\mathbb{Z}_p \mathrm{log}(\mathrm{N}I_p) = qp^{n_0}\mathbb{Z}_p.$$

Other (more natural) proof (see III.2.8, III.2.8.1). The usual identity:

$$\left(\frac{\widetilde{K}_p/K}{\mathfrak{a}}\right)\Big|_{\mathbb{Q}^{\rm cycl}(p)} = \left(\frac{\mathbb{Q}^{\rm cycl}(p)/\mathbb{Q}}{\mathrm{N}\mathfrak{a}}\right)$$

for all $\mathfrak{a} \in I_p$ means that $\mathrm{Gal}(\mathbb{Q}^{\rm cycl}(p)/K \cap \mathbb{Q}^{\rm cycl}(p))$ corresponds to $\mathbb{Z}_p\log(\mathrm{N}I_p)$ under the isomorphism:

$$\mathrm{Gal}(\mathbb{Q}^{\rm cycl}(p)/\mathbb{Q}) \simeq \mathbb{Z}_p\log(I_{\mathbb{Q},p}) = q\mathbb{Z}_p$$

since $\mathrm{Log}_{\mathbb{Q},p} = \log$; the result again follows.

(iii) It remains to compute the index:

$$\Big(\sum_{\mathfrak{l}\nmid p} \frac{1}{e_{\mathfrak{l},p}}\mathbb{Z}_p\mathrm{Log}_p(\mathfrak{l}) + \mathbb{Z}_p\mathrm{Log}_p(I_p) : \mathbb{Z}_p\mathrm{Log}_p(I_p)\Big),$$

hence (using (i)) the index $\Big(\sum_{\mathfrak{l}\nmid p} \frac{1}{e_{\mathfrak{l},p}}\mathbb{Z}_p\log(\mathrm{N}\mathfrak{l}) + qp^{n_0}\mathbb{Z}_p : qp^{n_0}\mathbb{Z}_p\Big)$; but:

$$\mathrm{N}\mathfrak{l} = \ell^{f_\mathfrak{l}(K/\mathbb{Q})}, \quad \frac{1}{e_{\mathfrak{l},p}}\log(\mathrm{N}\mathfrak{l}) = \frac{1}{e_{\mathfrak{l},p}}\, f_\mathfrak{l}(K/\mathbb{Q})\,.\,\log(\ell),$$

and we have $[L : K] = e_\mathfrak{l}\,.\,f_\mathfrak{l}(L/K)\,.\,|Pl_{L,\mathfrak{l}}|$, hence:

$$\frac{1}{e_{\mathfrak{l},p}}\log(\mathrm{N}\mathfrak{l}) = \frac{1}{[L:K]}\, f_\mathfrak{l}(L/\mathbb{Q})\,.\,|Pl_{L,\mathfrak{l}}|\,.\,\log(\ell),$$

and the formula easily follows. □

b) *p*-Primitive Ramification — *p*-Rationality

We go back to a general base field K. In 1986 we introduced in [Gr6, III, 1, 2], the following notion, valid without any hypothesis.

3.4 Definition (p-primitive ramification). A p-primitively ramified p-extension is a p-extension L/K such that:

$$\Big(\sum_{\mathfrak{l}\nmid p} \frac{1}{e_{\mathfrak{l},p}}\mathbb{Z}_p\mathrm{Log}_p(\mathfrak{l}) + \mathbb{Z}_p\mathrm{Log}_p(I_{K,p}) : \mathbb{Z}_p\mathrm{Log}_p(I_{K,p})\Big) = \prod_{\mathfrak{l}\nmid p} e_{\mathfrak{l},p}$$

(i.e., $(\mathcal{T}_L^{\rm ord})^G = j(\mathcal{T}_K^{\rm ord})$, using 3.3 when one assumes the Leopoldt conjecture for p in L). □

3.4.1 Proposition. *The above definition is equivalent to the fact that* $\sum_{\mathfrak{l}\nmid p} \mathbb{Z}_p\mathrm{Log}_p(\mathfrak{l})$ *is a pure sub-*$\mathbb{Z}_p$*-module of* $\mathbb{Z}_p\mathrm{Log}_p(I_{K,p})$ *(in other words, a direct summand in the latter), and that its dimension is equal to the number of tamely ramified places* $\mathfrak{l}$ *in* L/K *(with* $e_{\mathfrak{l},p} \neq 1$*).*

Proof. Indeed, consider the $\mathbb{Z}_p$-lattices:

$$A := \mathbb{Z}_p \mathrm{Log}_p(I_{K,p}), \quad B := \sum_{\mathfrak{l} \nmid p} \mathbb{Z}_p \mathrm{Log}_p(\mathfrak{l}), \quad C := \sum_{\mathfrak{l} \nmid p} \mathbb{Z}_p \frac{1}{e_{\mathfrak{l},p}} \mathrm{Log}_p(\mathfrak{l}).$$

We have $B \subseteq A \cap C \subseteq C$, showing that:

$$(A + C : A) = (C : A \cap C) = \prod_{\mathfrak{l} \nmid p} e_{\mathfrak{l},p}$$

if and only if:

$$B = \bigoplus_{\mathfrak{l} \nmid p} \mathbb{Z}_p \mathrm{Log}_p(\mathfrak{l}) \ \text{ and } \ A \cap C = B.$$

It is then immediate to show that this happens if and only if B is a pure sub-$\mathbb{Z}_p$-module of maximal rank of A. □

This leads to the following intrinsic definition.

3.4.2 Definition (p-primitive sets). Any finite set Σ of prime ideals $\mathfrak{l} \nmid p$ of K such that:

$$\mathbb{Z}_p \mathrm{Log}_p(I_{K,p}) = \bigoplus_{\mathfrak{l} \in \Sigma} \mathbb{Z}_p \mathrm{Log}_p(\mathfrak{l}) \bigoplus \mathcal{Z},$$

for some $\mathbb{Z}_p$-module $\mathcal{Z}$, is called a p-primitive set of K. □

3.4.3 Remarks. (i) By III.2.5, (iii), this means that the Frobenius':

$$\left(\frac{\widetilde{K}_p/K}{\mathfrak{l}}\right), \quad \mathfrak{l} \in \Sigma,$$

satisfy the following two conditions:

- $\mathrm{Gal}(\widetilde{K}_p/\widetilde{K}_p^{\Sigma}) = \bigoplus_{\mathfrak{l} \in \Sigma} \left\langle \left(\frac{\widetilde{K}_p/K}{\mathfrak{l}}\right) \right\rangle_{\mathbb{Z}_p}$,
- $\mathrm{Gal}(\widetilde{K}_p/\widetilde{K}_p^{\Sigma})$ is a direct summand in $\mathrm{Gal}(\widetilde{K}_p/K)$.

(ii) Since $\mathbb{Z}_p \mathrm{Log}_p(I_{K,p})$ has a $\mathbb{Z}_p$-rank equal to $\dim_{\mathbb{Q}_p}(\mathcal{L}_{K,p}) = r_2 + 1$ (assuming the Leopoldt conjecture for p), we deduce that the p-primitive sets have in this case at most $r_2 + 1$ elements. There exists an infinite number of p-primitive sets by the density Theorem II.4.6. The first example of such a maximal p-primitive set in a nontrivial case (i.e., K not totally real) was given in [Gr6, III, 2, Rem., v]: it is $K = \mathbb{Q}(\sqrt{-23})$ for $p = 3$ and for the set of the two prime ideals of K above 13.

(iii) These p-primitive sets have been found useful for the description of certain families of extensions having simpler arithmetical properties (which is clear since we thus minimize the invariant $\mathcal{T}_L^{\mathrm{ord}}$ for certain p-extensions of K).[3]

[3] In [JaS] and [Ng3], similar notions have been used for the computation of irregular Hilbert symbols from regular symbols.

In particular, we can even kill more arithmetical properties by asking that (relative to a fixed p and for finite p-extensions L/K, L satisfying the Leopoldt conjecture for p):

$$\mathcal{T}_L^{\mathrm{ord}} = 1,$$

which by 3.3, using the fixed point theorem for p-groups, is equivalent to the following two conditions:

- $\mathcal{T}_K^{\mathrm{ord}} = 1$,
- L/K is p-primitively ramified. □

We thus see that the base field cannot be arbitrary.

3.4.4 Definition (p-rational fields ([MoNg], [GrJ])). The number field K is said to be p-rational if it satisfies the Leopoldt conjecture for p and if $\mathcal{T}_K^{\mathrm{ord}} = 1$. □

3.4.5 Remarks. (i) By III.2.6.1, (ii$_2$), using III.4.2.4, III.4.2.5, and assuming the Leopoldt conjecture, the p-rationality of K is equivalent to the following three conditions:

- $(\mathcal{C}\ell^{\mathrm{ord}})_p \simeq \mathbb{Z}_p\mathrm{Log}_p(I_p)\Big/\bigoplus_{v|p}\log(U_v^1) \bmod \mathbb{Q}_p\log_p(E)$,

which can be nontrivial (this point is equivalent to the fact that the p-Hilbert class field is contained in the compositum of the $\mathbb{Z}_p$-extensions of K),

- $\bigoplus_{v|p}\mu_p(K_v) = i_p(\mu_p(K))$,

which means that $(\mathrm{R}_2^{\mathrm{ord}}(K))_p = (\mathrm{WK}_2(K))_p$ (see II.7.6.1),

- $\mathbb{Z}_p\log_p(E'^{\mathrm{ord}})$ is a direct summand in $\bigoplus_{v|p}\log(U_v^1)$,

which expresses the minimality of the p-adic regulator.

(ii) If K is p-rational, then all its subfields are p-rational (from the injectivity of j). By III.4.1.11, $\mathbb{Q}$ is p-rational for all p, which ensures that this notion is not empty!

(iii) For a p-rational field, we have (noncanonically):

$$\mathrm{Gal}(\overline{K}^{\mathrm{ab}}(p)/K) \simeq \prod_{v\nmid p}(F_v^\times)_p \times \mathbb{Z}_p^{r_2+1}.$$ □

From Exercise 3.3.1 above, we easily deduce:

3.4.6 Proposition. *Let K be a totally real number field satisfying the Leopoldt conjecture for p. If L is a p-extension of K, satisfying the Leopoldt conjecture, then L is p-rational if and only if the following two conditions are satisfied:*

- $\mathcal{T}_K^{\mathrm{ord}} = 1$ *(i.e., K is p-rational);*

• *either L/K is p-ramified, or there exists a unique tame place $\mathfrak{l}$ of K such that L/K is $Pl_p \cup \{\mathfrak{l}\}$-ramified, such that $\nu_{\mathfrak{l}} \leq n_0$, and with a residue degree in $K/K \cap \mathbb{Q}^{\mathrm{cycl}}(p)$ which is prime to p.* □

3.5 Theorem (after [GrJ]). *For a number field K, each of the following properties is equivalent to the p-rationality of K:*

(i) $\mathcal{A}^{\mathrm{ord}} := \mathrm{Gal}(H_p^{\mathrm{ord}}(p)/K) \simeq \mathbb{Z}_p^{r_2+1}$,

(ii) the Galois group $\mathcal{G}^{\mathrm{ord}}$ of the maximal p-ramified noncomplexified pro-p-extension of K is a free pro-p-group on r_2+1 generators, which is equivalent to the following four conditions:

- *K satisfies the Leopoldt conjecture for p,*
- *$(\mathcal{C}\ell^{\mathrm{ord}})_p \simeq \mathbb{Z}_p\mathrm{Log}_p(I_p) \Big/ \bigoplus_{v|p} \log(U_v^1) \bmod \mathbb{Q}_p\log_p(E)$,*
- *$\bigoplus_{v|p} \mu_p(K_v) = i_p(\mu_p(K))$,*
- *$\mathbb{Z}_p\log_p(E'^{\mathrm{ord}})$ is a direct summand in $\bigoplus_{v|p} \log(U_v^1)$,*

(iii) we have the following alternative:

- *either $\mu_p \subset K$, $|Pl_p| = 1$, and $(\mathcal{C}\ell^{Pl_p\mathrm{res}})_p = 1$,*

or

- *$\mu_p \not\subset K$, no element of Pl_p is totally split in $K' := K(\mu_p)$, and the ω-component of $\mathcal{C}\ell_{K'}^{Pl_{K',p}}$ is trivial.*

Proof. Recall that ω denotes the character of the action of $\mathrm{Gal}(K'/K)$ on μ_p ($\omega = 1$ if and only if $\mu_p \subset K$). Using reflection Theorem II.5.4.5 in K' with $T' = Pl'_p$, $S' = Pl'^{\mathrm{r}}_\infty$, $\chi = \omega$ (which gives $\chi^* = 1$), we have:

$$\mathrm{rk}_p(\mathcal{A}^{\mathrm{ord}}) = \mathrm{rk}_\omega\left(\mathcal{C}\ell_{K'}^{Pl'_p\mathrm{res}}\right) + \rho_\omega(Pl'_p,\, Pl'^{\mathrm{r}}_\infty).$$

If $\omega = 1$, we obtain:

$$\rho_\omega = r_2 + r_1 + |Pl_p| - \delta_{2,p}\, r_1 = r_2 + |Pl_p|$$

since $r_1 = 0$ if $p \neq 2$. If $\omega \neq 1$ (which implies that $p \neq 2$), we obtain:

$$\rho_\omega = r_2 + 1 + |\{v|p,\ D_v(K'/K) = 1\}|.$$

It follows that each of the conditions (i) and (iii) is equivalent to $\mathrm{rk}_p(\mathcal{A}^{\mathrm{ord}}) = r_2 + 1$, which is easily seen to be equivalent to the p-rationality of K.

If K is p-rational, then $|H^2(\mathcal{G}^{\mathrm{ord}}, \mathbb{Z}/p\mathbb{Z})| = |\mathcal{T}^{\mathrm{ord}}| = 1$ (see the Section 2 of the Appendix), giving (ii). But (ii) implies (i). □

Note. Movahhedi has shown in 1988 in his thesis [Mo, § 3] that a p-extension L of K is p-rational if and only if K is p-rational and if L/K is p-primitively ramified (hence we obtain a going up theorem for the Leopoldt conjecture in this particular case). This result has also been proved in [GrJ] by p-adic class field theory, where the case of p-regular p-extensions is also dealt with.

3.5.1 Examples (p-rational p-extensions of $\mathbb{Q}$). Let p be a prime number and let L be a p-extension of $\mathbb{Q}$. Then, from Proposition 3.4.6, a necessary and sufficient condition for L to be p-rational is that $L/\mathbb{Q}$ satisfies one of the following two conditions (assuming the Leopoldt conjecture for p):

(i) $L/\mathbb{Q}$ is unramified outside of p;

(ii) $L/\mathbb{Q}$ is unramified outside of $\{p, \ell\}$, where $\ell \neq p$ is a prime number such that $\log(\ell)$ generates $\log(\mathbb{Z}_p^\times)$ (which means that $\ell \not\equiv \pm 1 \bmod (8)$ for $p = 2$ and $\ell^{p-1} \not\equiv 1 \bmod (p^2)$ if $p \neq 2$).

For example (after [Gr6, III]), for $p = 2$, the 2-rational *abelian* 2-extensions of $\mathbb{Q}$ are the subfields of the fields:

$$\mathbb{Q}(\mu_{2^\infty})\mathbb{Q}(\sqrt{-\ell}), \ \ -\ell \equiv 5 \bmod (8),$$

or of the fields:

$$\mathbb{Q}(\mu_{2^\infty})\mathbb{Q}\Big(\sqrt{\sqrt{\ell}\,\tfrac{a-\sqrt{\ell}}{2}}\,\Big), \ \ \ell = a^2 + 4b^2 \equiv 5 \bmod (8).$$

For $p = 3$, the 3-rational *abelian* 3-extensions of $\mathbb{Q}$ are the subfields of the fields:

$$\mathbb{Q}^{\mathrm{cycl}}{}_{(3)}\, k_\ell, \ \ \ell \equiv 4, \ 7 \bmod (9),$$

where $\mathbb{Q}^{\mathrm{cycl}}{}_{(3)}$ is the cyclotomic $\mathbb{Z}_3$-extension of $\mathbb{Q}$ and k_ℓ the cyclic cubic field of conductor ℓ. □

3.5.2 Remark. For $p = 2$ and 3, the discussions in II.7.8 imply that the above p-extensions L are also the p-regular abelian p-extensions of $\mathbb{Q}$, i.e., those for which the p-part of $\mathrm{R}_2^{\mathrm{ord}}(L)$ is trivial. Recall that this comes from the fact that the notions of p-regularity and p-rationality coincide as soon as L contains the maximal real subfield of $\mathbb{Q}(\mu_p)$, and that 2 and 3 are the only values of p which always satisfy these conditions, the reflection arguments doing the rest of the work since $\omega^{-1} = \omega$ (see also [Gr6, II, th. 1]). In other words, for the abelian extensions L above, we also have that the p-part of $\mathrm{WK}_2(L)$ is trivial (no exotic symbols of order 2 (resp. 3)).

Of course there are p-rational nonabelian p-extensions L of $\mathbb{Q}$ for which the same conclusion holds. □

We refer to the literature, such as [Gr6], [GrJ], [Ja7], [JaS], [Mo], [MoNg], [RØ], for complementary results about these particular families of fields.

§4 Genus Theory with Ramification and Decomposition

This aspect of class field theory has its origin in Gauss's theory of binary quadratic forms with coefficients in $\mathbb{Z}$ [e, Ko3, Ch. 1, § 1] or, equivalently, in that of quadratic fields over $\mathbb{Q}$ (see Exercise 4.2.10). The context of reduction of quadratic forms, although very rich, is not generalizable (it is in fact closely related to the theory of continued fractions in quadratic fields, which is specific); on the contrary, the interpretation in terms of ideal classes of Gauss's theory is generalizable, and this is the point of view that we will now use.

More precisely, let L be an arbitrary finite extension of K. Let $\mathfrak{a}'$ be an ideal of L; its Artin symbol in a given abelian extension M of L depends on class field theory over L (a priori unknown). But the Artin symbol of $\mathrm{N}_{L/K}(\mathfrak{a}')$, being known in any abelian extension of K, is known in M^{ab}/K (maximal abelian extension of K in M); thus, because of the norm lifting Theorem II.4.7.2, the Artin symbol of $\mathrm{N}_{L/K}(\mathfrak{a}')$ in M^{ab}/K yields that of $\mathfrak{a}'$ in LM^{ab}/L, and one can hope that this is a nontrivial extension. This will lead to the general definition given by Fröhlich in [Fr3] of the genus field in the above context $M/L/K$.

In other words, genus theory in L/K is class field theory (over K) "restricted" to the subgroup $\mathrm{N}_{L/K}(I_L)$ of I_K. When $K = \mathbb{Q}$ and L/K is abelian (case of Gauss's quadratic genus theory), the extension $LM^{\mathrm{ab}}/\mathbb{Q}$ is abelian and, as we know, all the theory may be expressed by means of Dirichlet characters.

Classically, the natural context has been that of the Hilbert class fields H_K and H_L (in the ordinary or restricted sense) for an extension L/K of number fields, the genus field being the maximal subextension $H_{L/K}$ of H_L which is equal to the compositum of L with an abelian extension of K. Thus $H_{L/K}$ is for instance equal to the compositum of L with H_L^{ab} (the maximal abelian subextension of H_L in H_L/K), according to the following diagram:

$$\begin{array}{ccccccc}
L & \text{———} & LH_K & \text{———} & H_{L/K} & \text{———} & H_L \\
| & & | & & | & & \\
L^{\mathrm{ab}} & \text{———} & L^{\mathrm{ab}}H_K & \text{———} & H_L^{\mathrm{ab}} & & \\
| & & | & & & & \\
L \cap H_K & \text{———} & H_K & & & & \\
| & & & & & & \\
K & & & & & &
\end{array}$$

Fig. 4.1

The objective is then to describe $H_{L/K}/L$, which gives partial information on H_L/L. The literature on this subject is so large that we can only give some classical references such as [h, Ha2], [Fr2, Fr3], [Fu1], [I], [Le1].

The introduction of sets T and S has also been considered by several authors (Jaulent [Ja2], Federer [Fe]), the genus field being then the field $H^S_{L/K,T}$, compositum of L with the maximal abelian subextension $H^{S'\,\mathrm{ab}}_{L,T'}$ of

$H_{L,T'}^{S'}$ in $H_{L,T'}^{S'}/K$, which gives a diagram analogous to the one above (by II.1.2.5, the nonramification and the S-decomposition are carried over and hence we have $LH_{K,T}^{S} \subseteq H_{L/K,T}^{S}$); this general setting has been developed in [Ja2, Ch. II, 2.4, Ch. III, 2.1] for Galois extensions, thanks to p-adic class field theory and the formalism of infinitesimals III.2.4.1, III.2.4.2.

Finally, we note a synthetic approach to genus theory (for number fields and function fields) in [AJ].

In the case $T \neq \emptyset$, the extensions $H_{K,T}^{S}$, $H_{L,T'}^{S'}$, and $H_{L/K,T}^{S}$ may be infinite, but we will see that $H_{L,T'}^{S'\,\mathrm{ab}}$ is always a finite extension of $H_{K,T}^{S}$, so that we will be able to give a completely general result on the number of genera:

$$g_{L/K,T}^{S} := [H_{L/K,T}^{S} : LH_{K,T}^{S}].$$

We will be in the general context of a finite extension L/K of number fields which is not necessarily Galois.

a) Computation of the Number of Genera — Examples

Let K be a number field together with sets of places T and S, and let L/K be an arbitrary finite extension; denote by T' and S' the sets of places of L above those of T and S. We fix a prime number p.

4.1 Notations. (i) We consider the usual notations:

$$T_p := T \cap Pl_p,\ T_{\mathrm{ta}} := T \backslash T_p,\ H_{K,T}^{S}(p),\ H_{L,T'}^{S'}(p),\ H_{L/K,T}^{S}(p)\ ;$$

in particular $H_{L/K,T}^{S}(p)$ denotes the p-genus field of L/K (relative to the given sets T and S), in other words the maximal pro-p-subextension of $H_{L,T'}^{S'}/L$, compositum of L with an abelian pro-p-extension of K[4].

(ii) We consider moduli of L of the following form:

$$\mathfrak{m}' = \prod_{w \in T'_p} \mathfrak{p}_w^{m'} \,.\, \mathfrak{m}'_{\mathrm{ta}},\ m' \geq 0, \ \ \text{with } \mathfrak{m}'_{\mathrm{ta}} = \prod_{w \in T'_{\mathrm{ta}}} \mathfrak{p}_w,$$

and we say that $\mathfrak{m}'$ is large enough if m' is large enough.

(iii) We denote by $L(\mathfrak{m}')^{S'}$ the corresponding S'-split ray class field over L; we thus have, by III.1.3.2, $\bigcup_{m'} L(\mathfrak{m}')^{S'}(p) = H_{L,T'}^{S'}(p)$. □

Note. For $T_{\mathrm{ta}} \neq \emptyset$, the choice of moduli implies that some computations are valid only for the corresponding p-subextensions and their norm groups.

Let us now use the results on the idelic correspondence of class field theory (II.3.5 and II.5.1.4). Over K, the field $L(\mathfrak{m}')^{S'\,\mathrm{ab}}$ corresponds to:

[4] which we can take equal to $H_{L,T'}^{S'\,\mathrm{ab}}(p)$ as we have said in the introduction to this section.

$$K^\times \, \mathrm{N}_{L(\mathfrak{m}')^{S'}/K}(J_{L(\mathfrak{m}')^{S'}}) \,;$$

over L, the field $L(\mathfrak{m}')^{S'}$ corresponds to:

$$L^\times \, \mathrm{N}_{L(\mathfrak{m}')^{S'}/L}(J_{L(\mathfrak{m}')^{S'}}) = L^\times \, U^{S'}_{L,\mathfrak{m}'}$$

(by definition of ray class fields). It follows that we can write:

$$\begin{aligned} K^\times \, \mathrm{N}_{L(\mathfrak{m}')^{S'}/K}(J_{L(\mathfrak{m}')^{S'}}) &= K^\times \, \mathrm{N}_{L/K} \circ \mathrm{N}_{L(\mathfrak{m}')^{S'}/L}(J_{L(\mathfrak{m}')^{S'}}) \\ &= K^\times \, \mathrm{N}_{L/K}(L^\times \, . \, \mathrm{N}_{L(\mathfrak{m}')^{S'}/L}(J_{L(\mathfrak{m}')^{S'}})) \end{aligned}$$

(adding $L^\times$ does not change anything since $\mathrm{N}_{L/K}(L^\times) \subset K^\times$); hence:

$$K^\times \, \mathrm{N}_{L(\mathfrak{m}')^{S'}/K}(J_{L(\mathfrak{m}')^{S'}}) = K^\times \, \mathrm{N}_{L/K}(L^\times \, U^{S'}_{L,\mathfrak{m}'}) = K^\times \, \mathrm{N}_{L/K}(U^{S'}_{L,\mathfrak{m}'})$$

(using the two expressions for the group corresponding to $L(\mathfrak{m}')^{S'}$ over L). The idèle group corresponding to $L(\mathfrak{m}')^{S'\,\mathrm{ab}}$ over K is therefore:

$$K^\times \, \mathrm{N}_{L/K}(U^{S'}_{L,\mathfrak{m}'}).$$

By the norm lifting theorem, we obtain the following partial result, which is already quite useful.

4.1.1 Proposition. *The field $L\,L(\mathfrak{m}')^{S'\,\mathrm{ab}}$ corresponds over L to the group:*

$$\mathrm{N}^{-1}_{L/K}(K^\times \, \mathrm{N}_{L/K}(U^{S'}_{L,\mathfrak{m}'})) = \mathrm{N}^{-1}_{L/K}(K^\times) \, . \, U^{S'}_{L,\mathfrak{m}'}.$$

For $T = \emptyset$, the genus field $H^S_{L/K}$ corresponds over L to the group:

$$\mathrm{N}^{-1}_{L/K}(K^\times \, \mathrm{N}_{L/K}(U^{S'}_{L})) = \mathrm{N}^{-1}_{L/K}(K^\times) \, . \, U^{S'}_{L}. \qquad \square$$

To simplify notations, set temporarily:

$$H := H^S_{K,T}, \; H' := H^{S'}_{L,T'}, \; M' := L(\mathfrak{m}')^{S'\,\mathrm{ab}}, \; M := H \cap M',$$

and consider the following diagram:

$$\begin{array}{ccccc} H & \text{———} & H\,M' & \text{———} & H'^{\mathrm{ab}} \\ | & & | & & \\ M & \text{———} & M' & & \end{array}$$

Fig. 4.2

in which the fields $M_{(p)}$ and $M'_{(p)}$ tend respectively to $H_{(p)}$ and to $H'^{\text{ab}}_{(p)}$ with regard to $\mathfrak{m}'$.

The intersection M is equal to $K(\mathfrak{m})^S \cap M'$ for all $\mathfrak{m} = \prod_{v\in T_p} \mathfrak{p}_v^m \prod_{v\in T_{\text{ta}}} \mathfrak{p}_v$, sufficiently large with respect to $\mathfrak{m}'$ (for instance $\mathfrak{m}$ multiple of the conductor of M), and therefore it corresponds over K to:

$$K^\times U^S_{K,\mathfrak{m}} \,.\, K^\times \mathrm{N}_{L/K}(U^{S'}_{L,\mathfrak{m}'}) = K^\times \langle\, S\,\rangle\, U^{\text{res}}_{K,\mathfrak{m}}\, \mathrm{N}_{L/K}(U^{\text{res}}_{L,\mathfrak{m}'})$$

since $\mathrm{N}_{L/K}(\langle\, S'\,\rangle) \subseteq \langle\, S\,\rangle$. We first simplify the term:

$$V := U^{\text{res}}_{K,\mathfrak{m}}\, \mathrm{N}_{L/K}(U^{\text{res}}_{L,\mathfrak{m}'}).$$

Since we have:

$$U^{\text{res}}_{K,\mathfrak{m}} = \bigoplus_{v\in T_p} U_v^m \bigoplus_{v\in T_{\text{ta}}} U_v^1 \,.\, \prod_{v\notin T} U_v,$$

$$\mathrm{N}_{L/K}(U^{\text{res}}_{L,\mathfrak{m}'}) = \bigoplus_{v\in T_p} \mathrm{N}_{L/K}\Big(\bigoplus_{w|v} U_w^{m'}\Big) \bigoplus_{v\in T_{\text{ta}}} \mathrm{N}_{L/K}\Big(\bigoplus_{w|v} U_w^1\Big) \,.\, \prod_{v\notin T} \mathrm{N}_{L/K}\Big(\bigoplus_{w|v} U_w\Big),$$

and that, by assumption, we can assume (for m sufficiently large):

$$\bigoplus_{v\in T_p} U_v^m \subseteq \bigoplus_{v\in T_p} \mathrm{N}_{L/K}\Big(\bigoplus_{w|v} U_w^{m'}\Big),$$

we obtain $V = \bigoplus_{v\in T_p} \mathrm{N}_{L/K}\Big(\bigoplus_{w|v} U_w^{m'}\Big) \bigoplus_{v\in T_{\text{ta}}} U_v^1 \,.\, \prod_{v\notin T} U_v$ which can be written $V =: W \bigoplus_{v\in T_{\text{ta}}} U_v^1 \,.\, \prod_{v\notin T} U_v$, where $W := \bigoplus_{v\in T_p} \mathrm{N}_{L/K}\Big(\bigoplus_{w|v} U_w^{m'}\Big)$ is a neighbourhood of 1 in $\bigoplus_{v\in T_p} U_v$, arbitrarily small. By considering the corresponding p-extensions, we thus have:

$$\begin{aligned}
[M'_{(p)} : M_{(p)}] &= (K^\times \langle\, S\,\rangle V : K^\times \mathrm{N}_{L/K}(U^{S'}_{L,\mathfrak{m}'}))_p \\
&= \frac{(\langle\, S\,\rangle V : \mathrm{N}_{L/K}(U^{S'}_{L,\mathfrak{m}'}))_p}{(K^\times \cap (\langle\, S\,\rangle V) : K^\times \cap \mathrm{N}_{L/K}(U^{S'}_{L,\mathfrak{m}'}))_p}
\end{aligned}$$

which is a divisor of the numerator equal to:

$$\prod_{v\in S} \Big(K_v^\times : \mathrm{N}_{L/K}\Big(\bigoplus_{w|v} L_w^\times\Big)\Big)_p \times \prod_{v\notin T\cup S} \Big(U_v : \mathrm{N}_{L/K}\Big(\bigoplus_{w|v} U_w\Big)\Big)_p$$

(for p-parts, the tame ramification indices are trivial, and on T_p we obtain the index $(W : W) = 1$).

We again introduce the local fields (see II.1.5.3, II.2.6.3):

$$L_v^{\text{ab}} := \bigcap_{w|v} L_w^{\text{ab}},\ \hat{L}_v^{\text{ab}} := \langle\, L_w^{\text{ab}}\,\rangle_{w|v},\ \check{L}_v^{\text{ab}} \subseteq \hat{L}_v^{\text{ab}}.$$

Denote by e_v^{ab} and f_v^{ab} the ramification index and the residue degree of L_v^{ab}/K_v, and by $\check{e}_v^{\mathrm{ab}}$ the ramification index of $\check{L}_v^{\mathrm{ab}}/K_v$. The finite degree $[M'(p) : M(p)]$ is nondecreasing (with respect to $\mathfrak{m}'$) and is a divisor of:

$$\prod_{v\in S}\Big(K_v^{\times} : \mathrm{N}_{L/K}\Big(\bigoplus_{w|v} L_w^{\times}\Big)\Big)_p \times \prod_{v\notin T\cup S}\Big(U_v : \mathrm{N}_{L/K}\Big(\bigoplus_{w|v} U_w\Big)\Big)_p = \prod_{v\in S} e_{v,p}^{\mathrm{ab}} f_{v,p}^{\mathrm{ab}} \times \prod_{v\notin T\cup S} \check{e}_{v,p}^{\mathrm{ab}}.$$

It follows that for all sufficiently large $\mathfrak{m}'$, we have $H(p)M'(p) = H'^{\mathrm{ab}}(p)$. Indeed, there exists $\mathfrak{m}'_0$ such that $H(p)M'(p) = H(p)M'_0(p)$ (where $M'_0(p)$ corresponds to $\mathfrak{m}'_0$) for all $\mathfrak{m}'$ multiple of $\mathfrak{m}'_0$; since any x of $H'^{\mathrm{ab}}(p)$ belongs to a suitable $M'(p)$, our claim follows. Hence the expression that we have obtained above for $[M'(p) : M(p)]$ gives in fact $[H'^{\mathrm{ab}}(p) : H(p)]$ for all sufficiently large $\mathfrak{m}'$.

Furthermore, we have:

$$K^{\times} \cap (\langle S\rangle V) = \{\varepsilon \in E^S_{\mathfrak{m}_{\mathrm{ta}}},\ i_{T_p}(\varepsilon) \in W\},$$
$$K^{\times} \cap \mathrm{N}_{L/K}(U^{S'}_{L,\mathfrak{m}'}) = \{\varepsilon \in E^S_{\mathfrak{m}_{\mathrm{ta}}},\ i_{T_p}(\varepsilon) \in W\} \cap \mathrm{N}_{L/K}(U_L^{S'}),$$

and by the above, the index:

$$(\{\varepsilon \in E^S_{\mathfrak{m}_{\mathrm{ta}}},\ i_{T_p}(\varepsilon) \in W\} : \{\varepsilon \in E^S_{\mathfrak{m}_{\mathrm{ta}}},\ i_{T_p}(\varepsilon) \in W\} \cap \mathrm{N}_{L/K}(U_L^{S'}))_p$$

is nonincreasing and therefore constant for sufficiently large $\mathfrak{m}'$. If $T_p = \emptyset$ (tame genus theory), this index is always equal to:

$$(E^S_{\mathfrak{m}} : E^S_{\mathfrak{m}} \cap \mathrm{N}_{L/K}(U_L^{S'}))_p,$$

where $\mathfrak{m} := \prod_{v\in T} \mathfrak{p}_v$. When $T_p \neq \emptyset$, we introduce $\mathcal{E}^S_T \subseteq \mathcal{E}^S := E^S \otimes \mathbb{Z}_p$ defined as the kernel of the canonical map (see III.1.6.6):

$$\bar{i}_{T_p} : \mathcal{E}^S_{\mathfrak{m}_{\mathrm{ta}}} \longrightarrow \bigoplus_{v\in T_p} U_v^1,$$

or that of:

$$\bar{i}_T : \mathcal{E}^S \longrightarrow \bigoplus_{v\in T_p} U_v^1 \bigoplus_{v\in T_{\mathrm{ta}}} (F_v^{\times})_p\ ;$$

it is indeed easy to check that, for all sufficiently large $\mathfrak{m}'$ we have:

$$(\{\varepsilon \in E^S_{\mathfrak{m}_{\mathrm{ta}}},\ i_{T_p}(\varepsilon) \in W\} : \{\varepsilon \in E^S_{\mathfrak{m}_{\mathrm{ta}}},\ i_{T_p}(\varepsilon) \in W\} \cap \mathrm{N}_{L/K}(U_L^{S'}))_p = (\mathcal{E}^S_T : \mathcal{E}^S_T \cap \mathrm{N}_{L/K}(U_L^{S'})).$$

4.1.2 Remarks. (i) The abuse of notation $\mathcal{E}^S_T \cap \mathrm{N}_{L/K}(U_L^{S'})$ means in fact $\mathcal{E}^S_T \cap \mathrm{N}_{L/K}(\widehat{U_L^{S'}})$, where $\widehat{U_L^{S'}} := \prod_{w\in Pl'\setminus S'} (U_w)_p \prod_{w\in S'} \widehat{L_w^{\times}}$ is the profinite p-completion of $U_L^{S'}$. In practice, however, we do the following. We lift $\mathcal{E}^S_T/\mathcal{E}^S_T \cap \mathcal{E}^{S\,[L:K]_p}$ into $E^S_T/E^S_T \cap E^{S\,[L:K]_p}$, where:

$$E_T^S := \{\varepsilon \in E^S,\ \ \varepsilon \otimes 1 \,.\, \mathcal{E}^{S\,[L:K]_p} \in \mathcal{E}_T^S / \mathcal{E}_T^S \cap \mathcal{E}^{S\,[L:K]_p}\}\ ;$$

then $(\mathcal{E}_T^S : \mathcal{E}_T^S \cap \mathrm{N}_{L/K}(\widehat{U_L^{S'}})) = (E_T^S : E_T^S \cap \mathrm{N}_{L/K}(U_L^{S'}))_p$.

For $\varepsilon \in \mathcal{E}_T^S$ the local norm conditions are trivially satisfied on T.

(ii) Note that:

$$\mathrm{N}_{L/K}(U_L^{S'}) = \Big\{(x_v)_v \in J_K,\\ x_v \in \mathrm{N}_{L/K}\Big(\bigoplus_{w|v} U_w\Big)\ \ \forall v \notin S,\ \ x_v \in \mathrm{N}_{L/K}\Big(\bigoplus_{w|v} L_w^\times\Big)\ \ \forall v \in S\Big\},$$

and that $K^\times \cap \mathrm{N}_{L/K}(U_L^{S'})$ is the subgroup of the elements of $K^\times$ which are local norms everywhere in L/K and in addition norms of local units outside of S, the expression "norm of local units at v" meaning belonging to $\mathrm{N}_{L/K}\Big(\bigoplus_{w|v} U_w\Big)$. Recall that in the case where L/K is not Galois, if $x \in K^\times$ is a local norm at v such that $\mathrm{v}(x) = 0$, then x is not necessarily a norm of local units (Remark 4.2.1 below). On the contrary, in the Galois case we have the double simplification $\mathcal{E}_T^S \cap \mathrm{N}_{L/K}(U_L^{S'}) = \mathcal{E}_T^S \cap \mathrm{N}_{L/K}(J_L)$. □

We thus have obtained the following general result where K is a number field given together with sets of places T and S, L is a finite extension of K, T' and S' are the sets of places of L above those of T and S, and where, for a prime number p, $H_T^S(p)$ is the maximal T-ramified S-split abelian pro-p-extension of K.

4.2 Theorem (expression for the number of genera). *Let $H_{L/K,T}^S(p)$ be the p-genus field relative to T and S, in other words the maximal T'-ramified S'-split pro-p-extension of L, compositum of L with an abelian extension of K, and let $g_{L/K,T}^S(p) := [H_{L/K,T}^S(p) : LH_T^S(p)]$ be the number of p-genera relative to T and S.*

(i) We then have:

$$g_{L/K,T}^S(p) = \frac{\displaystyle\prod_{v\in S} e_{v,p}^{\mathrm{ab}}\, f_{v,p}^{\mathrm{ab}} \times \prod_{v\notin T\cup S} \check{e}_{v,p}^{\mathrm{ab}}}{[L^{\mathrm{ab}} : L^{\mathrm{ab}} \cap H_T^S]_p \times (\mathcal{E}_T^S : \mathcal{E}_T^S \cap \mathrm{N}_{L/K}(U_L^{S'}))},$$

where e_v^{ab}, f_v^{ab} are the ramification index and the residue degree of L_v^{ab}/K_v, $\check{e}_v^{\mathrm{ab}}$ is the ramification index of $\check{L}_v^{\mathrm{ab}}/K_v$ (see II.2.6.3), and where:

$$\mathcal{E}_T^S := \{\varepsilon \in \mathcal{E}^S := E^S \otimes \mathbb{Z}_p,\ \ \bar{i}_T(\varepsilon) = 1\}.$$

(ii) In addition, if L/K is Galois, we have:

$$g_{L/K,T}^S(p) = \frac{\displaystyle\prod_{v\in S} e_{v,p}^{\mathrm{ab}}\, f_{v,p}^{\mathrm{ab}} \times \prod_{v\notin T\cup S} e_{v,p}^{\mathrm{ab}}}{[L^{\mathrm{ab}} : L^{\mathrm{ab}} \cap H_T^S]_p \times (\mathcal{E}_T^S : \mathcal{E}_T^S \cap \mathrm{N}_{L/K}(J_L))}. \qquad \square$$

Note. In the case $T_p = \emptyset$, we replace $\mathcal{E}_T^S$ by $E_{\mathfrak{m}}^S$ in all the above formulas, with $\mathfrak{m} = \prod_{v \in T} \mathfrak{p}_v$.

4.2.1 Remark. The important point in practice is the computation of:

$$\mathcal{E}_T^S \cap \mathrm{N}_{L/K}(U_L^{S'}) \;;$$

but this computation is purely local since if $\varepsilon \in \mathcal{E}_T^S$ (which is in general represented by an element of E^S modulo this norm group), we have $\varepsilon \in \mathrm{N}_{L/K}(U_L^{S'})$ if and only if (see II.2.6.7):

$$(i_v(\varepsilon),\, L_v^{\mathrm{ab}}/K_v) = 1 \text{ for all } v \in S,$$
$$(i_v(\varepsilon),\, \check{L}_v^{\mathrm{ab}}/K_v) = 1 \text{ for all } v \notin T \cup S.$$

These local symbol computations must be done only if $i_v(\varepsilon)$ is not a local unit or if the local extension under consideration is ramified. In the Galois case, the condition is:

$$(i_v(\varepsilon),\, L_v^{\mathrm{ab}}/K_v) = 1 \text{ for all } v \notin T. \qquad \square$$

4.2.2 Corollary (usual genus theory: $T = \emptyset$). *(i) Globalizing, we have:*

$$g_{L/K}^S = \frac{\prod_{v \in S} e_v^{\mathrm{ab}}\, f_v^{\mathrm{ab}} \times \prod_{v \notin S} \check{e}_v^{\mathrm{ab}}}{[L^{\mathrm{ab}} : L^{\mathrm{ab}} \cap H^S] \times (E^S : E^S \cap \mathrm{N}_{L/K}(U_L^{S'}))},$$

$$[H_{L/K}^S : L] = \frac{|\mathcal{C}\ell^S| \times \prod_{v \in S} e_v^{\mathrm{ab}}\, f_v^{\mathrm{ab}} \times \prod_{v \notin S} \check{e}_v^{\mathrm{ab}}}{[L^{\mathrm{ab}} : K] \times (E^S : E^S \cap \mathrm{N}_{L/K}(U_L^{S'}))}.$$

(ii) If in addition L/K is Galois, we have:

$$g_{L/K}^S = \frac{\prod_{v \in S} e_v^{\mathrm{ab}}\, f_v^{\mathrm{ab}} \times \prod_{v \notin S} e_v^{\mathrm{ab}}}{[L^{\mathrm{ab}} : L^{\mathrm{ab}} \cap H^S] \times (E^S : E^S \cap \mathrm{N}_{L/K}(J_L))},$$

$$[H_{L/K}^S : L] = \frac{|\mathcal{C}\ell^S| \times \prod_{v \in S} e_v^{\mathrm{ab}}\, f_v^{\mathrm{ab}} \times \prod_{v \notin S} e_v^{\mathrm{ab}}}{[L^{\mathrm{ab}} : K] \times (E^S : E^S \cap \mathrm{N}_{L/K}(J_L))}. \qquad \square$$

4.2.3 Corollary. *(i) We have the following formulas:*

$$g_{L/K}^{\mathrm{res}} = \frac{\prod_{v \in Pl_0} \check{e}_v^{\mathrm{ab}}}{[L^{\mathrm{ab}} : L^{\mathrm{ab}} \cap H^{\mathrm{res}}] \times (E^{\mathrm{res}} : E^{\mathrm{res}} \cap \mathrm{N}_{L/K}(U_L^{\mathrm{res}}))},$$

$$[H_{L/K}^{\mathrm{res}} : L] = \frac{|\mathcal{C}\ell^{\mathrm{res}}| \times \prod_{v \in Pl_0} \check{e}_v^{\mathrm{ab}}}{[L^{\mathrm{ab}} : K] \times (E^{\mathrm{res}} : E^{\mathrm{res}} \cap \mathrm{N}_{L/K}(U_L^{\mathrm{res}}))},$$

$$g_{L/K}^{\rm ord} = \frac{2^{r_1^{\rm c}} \times \prod_{v \in Pl_0} \check e_v^{\rm ab}}{[L^{\rm ab} : L^{\rm ab} \cap H^{\rm ord}] \times (E^{\rm ord} : E^{\rm ord} \cap {\rm N}_{L/K}(U_L^{\rm ord}))},$$

$$[H_{L/K}^{\rm ord} : L] = \frac{|\mathcal{C}\!\ell^{\rm ord}| \times 2^{r_1^{\rm c}} \times \prod_{v \in Pl_0} \check e_v^{\rm ab}}{[L^{\rm ab} : K] \times (E^{\rm ord} : E^{\rm ord} \cap {\rm N}_{L/K}(U_L^{\rm ord}))},$$

where $r_1^{\rm c}$ is the number of real infinite places v of K for which all the completions L_w for $w|v$ are equal to $\mathbb{C}$ (i.e., $f_v^{\rm ab} = 2$).

(ii) If L/K is Galois, $\check e_v^{\rm ab} = e_v^{\rm ab}$, $r_1^{\rm c}$ is the number of real infinite places of K complexified in L, and we can replace ${\rm N}_{L/K}(U_L^{\rm res})$, ${\rm N}_{L/K}(U_L^{\rm ord})$ by ${\rm N}_{L/K}(J_L)$. □

4.2.4 Remark. The formula giving $[H_{L/K}^S : L]$ (for $T = \emptyset$) is similar to that giving the number of invariant classes when the extension L/K is cyclic; this is absolutely normal since it is easily checked that if L/K is Galois with cyclic Galois group $G =: \langle \sigma \rangle$, the genus field $H_{L/K}^S$ is the subfield of $H_L^{S'}$ fixed under $(\mathcal{C}\!\ell_L^{S'})^{1-\sigma}$. Since $(\mathcal{C}\!\ell_L^{S'} : (\mathcal{C}\!\ell_L^{S'})^{1-\sigma}) = |(\mathcal{C}\!\ell_L^{S'})^G|$, we indeed have the claimed equality. □

4.2.5 Exercise. For the numerical data of Exercise II.2.6.6, compute the number of genera (in the restricted and ordinary sense) of $L/\mathbb{Q}$.

Answer. We know that $|\mathcal{C}\!\ell^{\rm res}| = |\mathcal{C}\!\ell^{\rm ord}| = 1$. The roots of $X^4 + 14X^2 - 19$ being:

$$\pm\sqrt{-7 \pm 2\sqrt{17}}\,,$$

at infinity L has two real completions and one complex completion; therefore $L/\mathbb{Q}$ is not Galois, $L^{\rm ab} = \mathbb{Q}(\sqrt{17})$, and $r_1^{\rm c} = 0$.

The ramified places in $L/\mathbb{Q}$ are 17, 19, and 2. For $v = 17$, $\check e_v^{\rm ab} = e_v^{\rm ab} = 2$ since $L_w^{\rm ab} = \mathbb{Q}_{17}(\sqrt{17}, \sqrt{-7})$ is unique. For $v = 19$, two of the completions are $\mathbb{Q}_{19}$, the third is $\mathbb{Q}_{19}(\sqrt{-19})$, which immediately yields $\check e_v^{\rm ab} = 1$. For $v = 2$, we know that $\check e_v^{\rm ab} = 2$.

We already obtain:

$$g_{L/\mathbb{Q}}^{\rm res} = [H_{L/\mathbb{Q}}^{\rm res} : L] = 2,$$

$$g_{L/\mathbb{Q}}^{\rm ord} = [H_{L/\mathbb{Q}}^{\rm ord} : L] = \frac{2}{(\langle -1 \rangle \ : \ \langle -1 \rangle \cap {\rm N}_{L/\mathbb{Q}}(U_L^{\rm ord}))}.$$

The computations of II.2.6.6 show that -1 is not in ${\rm N}_{L/\mathbb{Q}}(U_L^{\rm ord})$ because of the place 2 for which ${\rm N}_{L/\mathbb{Q}}\Big(\bigoplus_{v|2} U_w\Big) = 1 + 4\mathbb{Z}_2$, although -1 is a local norm everywhere in $L/\mathbb{Q}$; hence we have $[H_{L/\mathbb{Q}}^{\rm ord} : L] = 1$.

As a complement, we can check that -1 is indeed a local norm everywhere in $L/\mathbb{Q}$, the verifications being needed only for $v \in \{\infty, 17, 19, 2\}$.

The extension $L_v^{\rm ab}/\mathbb{Q}_v$ is trivial for these places, except for $v = 17$, where $L_v^{\rm ab}/\mathbb{Q}_v = \mathbb{Q}_{17}(\sqrt{17}, \sqrt{-7})/\mathbb{Q}_{17}$; in this case, we can compute the norm group of $L_v^{\rm ab}/\mathbb{Q}_v$ as an intersection, using II.1.6.5. Note that we cannot use the product formula since $L/\mathbb{Q}$ is not abelian.

On this example we see that $g_{L/\mathbb{Q}}^{\rm res}$ is different from $g_{L^{\rm ab}/\mathbb{Q}}^{\rm res}$, the latter being trivial. □

4.2.6 Exercise (genus theory for pure cubic fields). Consider:

$$L = \mathbb{Q}\Big(\sqrt[3]{3^{n_0}\, p_1^{n_1} \cdots p_t^{n_t}}\,\Big),$$

$t \geq 0$, $n_0 \in \{0, 1\}$, $n_i \in \{1, 2\}$ for $1 \leq i \leq t$, where the p_i are distinct prime numbers different from 3. Compute the number of genera in the restricted sense. Find the the genus field.

Answer. We apply Corollary 4.2.3, which yields here:

$$g_{L/\mathbb{Q}}^{\rm res} = [H_{L/\mathbb{Q}}^{\rm res} : L] = \prod_v \check{e}_v^{\rm ab},$$

where v ranges in set of finite places of $\mathbb{Q}$. We refer to II.2.6.3 for the following computation of the $\check{e}_v^{\rm ab}$. It is clear that if $v \notin \{3, p_1, \ldots, p_t\}$, we have $\check{e}_v^{\rm ab} = 1$. If $v = p_i$, we have $e_w = 3$, hence $\hat{L}_v^{\rm ab} = \check{L}_v^{\rm ab} = L_w^{\rm ab}$ for the unique completion L_w of L above v. Kummer theory shows that $L_w/\mathbb{Q}_v$ is abelian if and only if $\mu_3 \subset \mathbb{Q}_v$; hence $\hat{L}_v^{\rm ab} = L_w$ or $\mathbb{Q}_v$ according as $\mu_3 \subset \mathbb{Q}_v$ or not. Therefore, we obtain:

$$\check{e}_v^{\rm ab} = 3 \ \text{ if } p_i \equiv 1 \bmod (3),$$
$$\check{e}_v^{\rm ab} = 1 \ \text{ if } p_i \not\equiv 1 \bmod (3).$$

If $v = 3$ and if $e_w = 3$, we have similarly $\check{e}_v^{\rm ab} = 1$; if there are two completions with $e_{w_1} = 2$ (for which necessarily $L_{w_1} = \mathbb{Q}_v(\mu_3)$), $e_{w_2} = 1$, we check that we again have $\check{e}_v^{\rm ab} = 1$. Hence:

$$[H_{L/\mathbb{Q}}^{\rm res} : L] = 3^\gamma,$$

where γ is the number of p_i congruent to 1 modulo (3).

For $p_i \equiv 1 \bmod (3)$, there exists a unique cyclic cubic extension L_i of conductor p_i (class field theory over $\mathbb{Q}$); by considering the completions and using local class field theory on $\mathbb{Q}_{p_i}$ (with II.1.8.2, (i)), we check that LL_i/L is unramified. It follows that $H_{L/\mathbb{Q}}^{\rm res}$ is the compositum of the L_i, and we even have:

$$\mathrm{Gal}(H_{L/\mathbb{Q}}^{\rm res}/L) \simeq (\mathbb{Z}/3\mathbb{Z})^\gamma.$$ □

4.2.7 Remarks. (i) In the above example we can have $\gamma = 0$ and $|\mathcal{C}\ell_L^{\rm res}| \equiv 0 \bmod (3)$ (for example if $L = \mathbb{Q}(\sqrt[3]{20}\,)$); this comes from the fact that $L/\mathbb{Q}$ is not Galois.

(ii) It is easy to show that if L/K is a *cyclic* p-extension, we have $(\mathcal{C}\ell_L^{\rm res})_p = 1$ if and only if $H^{\rm res}(p) \subseteq L$ and $g_{L/K}^{\rm res}(p) = 1$. In the case of an arbitrary p-extension, the criterion is the same except that we must replace the number of p-genera by the number of p-central classes (see 4.7.4 for $S = \emptyset$). For instance, the only *cyclic* p-extensions $L/\mathbb{Q}$ for which $(\mathcal{C}\ell_L^{\rm res})_p = 1$ are the following:

- for $p \neq 2$, the subfields of the cyclotomic $\mathbb{Z}_p$-extension of $\mathbb{Q}$,
- for $p = 2$, the cyclic subextensions of $\mathbb{Q}(\mu_{2^\infty})$,
- the p-subextensions of the fields $\mathbb{Q}(\mu_\ell)$, ℓ prime, $\ell \equiv 1 \bmod (p)$.

Although noncyclic, the 2-extensions $\mathbb{Q}(\mu_{2^n})$ for $n \geq 3$ have a trivial 2-class group (use genus theory in the cyclic extension $\mathbb{Q}(\mu_{2^n})/\mathbb{Q}(\mu_4)$). □

4.2.8 Exercise (case of ray class fields). Let $\mathfrak{m}$ be a modulus of K with support T, and let $L := K(\mathfrak{m})^{\rm res}$. What is the number of genera in the restricted sense of L/K? Deduce a simple expression for $(E^{\rm res} : E^{\rm res} \cap {\rm N}(J_L))$.

Answer. The genus field $M := H_{L/K}^{\rm res}$ is an abelian extension of K, unramified over L; by II.4.2.2, its conductor $\mathfrak{f}$ is characterized by the condition:

$$U_{\mathfrak{f}}^{\rm res} \subseteq {\rm N}_{M/K}(U_M^{\rm res}),\ \ \mathfrak{f} \text{ minimal} \ \ ;$$

since M/L is unramified, ${\rm N}_{M/L}(U_M^{\rm res}) = U_L^{\rm res}$ and we have equivalently:

$$U_{\mathfrak{f}}^{\rm res} \subseteq {\rm N}_{L/K}(U_L^{\rm res}),\ \ \mathfrak{f} \text{ minimal},$$

which characterizes the conductor of L. It follows, by II.5.1.1, that $L = K(\mathfrak{f})^{\rm res}$ and $M = L$. The number of genera of a ray class field is therefore equal to 1.

Using II.5.2.2, (i), and formula I.4.5.5, (i) for $S = \emptyset$, formula 4.2.3 implies the relation:

$$(E^{\rm res} : E^{\rm res} \cap {\rm N}(J_L)) = \frac{(E^{\rm res} : E_{\mathfrak{m}}^{\rm res})}{\prod_{v \in T} (E_{\frac{\mathfrak{m}}{\mathfrak{m}_v}}^{\rm res} : E_{\mathfrak{m}}^{\rm res})},$$

which measures the deployment defect of the inertia groups in L/K, in the sense of III.1.4.3, and which gives a method for the explicit computation of $(E^{\rm res} : E^{\rm res} \cap {\rm N}(J_L))$ (only for a ray class field). We have $E^{\rm res} \subseteq {\rm N}(J_L)$ if and only if there is deployment, and the opposite normic situation $E^{\rm res} \cap {\rm N}(J_L) = E_{\mathfrak{m}}^{\rm res}$ holds if and only if we have the equalities $E_{\frac{\mathfrak{m}}{\mathfrak{m}_v}}^{\rm res} = E_{\mathfrak{m}}^{\rm res}$ for each $v \in T$ (which is not incompatible with deployment if $E^{\rm res} = E_{\mathfrak{m}}^{\rm res}$).

This does not mean that ray class fields have a trivial class group! In fact, the computation of the number of genera with respect to the base field does not give the best information on this subject. We illustrate this with an example used in [Cor1].

Example. Let $m := 4\,p_1 \cdots p_t$, $t \geq 1$, where the p_i are distinct prime numbers congruent to 1 modulo (4), and let $L := \mathbb{Q}(\mu_m)$; thus, we have $g_{L/\mathbb{Q}}^{\rm res} = 1$. Consider $K' := \mathbb{Q}(\sqrt{-1}) \subset L$. We easily obtain:

$$g^{\rm res}_{L/K'} = \tfrac{1}{4} \prod_{i=1}^{t} (p_i - 1),$$

by noting that the p_i are split in $K'/\mathbb{Q}$ and using II.3.4.3 to show that $(\mu_4 : \mu_4 \cap \mathrm{N}_{L/K'}(J_L)) = 4$ (reduce to the norm group of a single $\mathbb{Q}_{p_i}(\mu_{p_i})$). □

This kind of trick, using a relative base field K' for which the ramified places (in L/K) are very split in K'/K, enables us to give examples of class groups of large rank, and is used for the problem of infinite class fields towers (see [Mar1], [Scho] and in general the papers of Hajir and Maire on this subject; we have detailed such examples in II.5.9.4, (iii), and Exercises II.5.9.5, II.5.9.6, II.5.9.7).

4.2.9 Exercise (genus theory for quadratic fields). Let $L = \mathbb{Q}(\sqrt{d})$ be a quadratic field, with d a squarefree natural integer. Write d in the form:

$$d =: s\, 2^{\delta} \,.\, \prod_{i=1}^{t} s_i\, p_i, \quad t \geq 0,$$

where the p_i are the odd prime divisors of d, $\delta \in \{0, 1\}$, $s_i := (-1)^{\frac{p_i - 1}{2}}$, $s \in \{-1, 1\}$ then being determined in a unique way.

(i) Show that the genus field in the restricted sense of $L/\mathbb{Q}$ is:

$$H^{\rm res}_{L/\mathbb{Q}} = \mathbb{Q}(\sqrt{s 2^{\delta}}; \sqrt{s_1 p_1}, \ldots, \sqrt{s_t p_t}).$$

From this, compute $H^{\rm ord}_{L/\mathbb{Q}}$.

(ii) Show that for $T = \{2\}$, the corresponding 2-genus field (in the restricted sense) is $\mathbb{Q}(\mu_{2^\infty})(\sqrt{p_1}, \ldots, \sqrt{p_t})$.

Answer. (i) The extension $M := \mathbb{Q}(\sqrt{s 2^{\delta}}; \sqrt{s_1 p_1}, \ldots, \sqrt{s_t p_t})$ is defined so that the ramification index in $M/\mathbb{Q}$ of an arbitrary prime number p is equal to that of p in $\mathbb{Q}(\sqrt{d})/\mathbb{Q}$. Therefore this extension is unramified over L, abelian over $\mathbb{Q}$, and its degree (over $\mathbb{Q}$) is equal to 2^{t+1} or to 2^t according to whether $s 2^{\delta} \neq 1$ or not; it is therefore contained in $H^{\rm res}_{L/\mathbb{Q}}$. Here, Corollary 4.2.3 yields:

$$[H^{\rm res}_{L/\mathbb{Q}} : \mathbb{Q}] = 2 \times \frac{\prod_v e_v}{2} = \prod_v e_v = e_2 \times 2^t,$$

which is equal to 2^{t+1} or to 2^t according to whether 2 is ramified or not in $L/\mathbb{Q}$, which is also equivalent to the fact that $s 2^{\delta} \neq 1$ or not. It follows that $M = H^{\rm res}_{L/\mathbb{Q}}$.

It is clear that $H^{\rm ord}_{L/\mathbb{Q}} = H^{\rm res}_{L/\mathbb{Q}}$ for $d < 0$, or for $d > 0$ and $H^{\rm res}_{L/\mathbb{Q}}$ real (i.e., $s_i = 1$ for all i), otherwise $H^{\rm ord}_{L/\mathbb{Q}}$ is the maximal real subfield of $H^{\rm res}_{L/\mathbb{Q}}$; however, it is more instructive to obtain this from the formula giving $[H^{\rm ord}_{L/\mathbb{Q}} :$

$\mathbb{Q}]$, which will show incidentally that the restricted sense is more canonical than the ordinary sense.

Thus, we have:

$$[H^{\mathrm{ord}}_{L/\mathbb{Q}} : \mathbb{Q}] = \frac{2^{r_1^{\mathrm{c}}} \times [H^{\mathrm{res}}_{L/\mathbb{Q}} : \mathbb{Q}]}{(\langle -1 \rangle : \langle -1 \rangle \cap \mathrm{N}(J_L))}.$$

If $d < 0$, we have $r_1^{\mathrm{c}} = 1$ but $-1 \notin \mathrm{N}(J_L)$ since -1 is not a norm in $\mathbb{C}/\mathbb{R}$, and $H^{\mathrm{ord}}_{L/\mathbb{Q}} = H^{\mathrm{res}}_{L/\mathbb{Q}}$ (clearly, since in this case $H^{\mathrm{res}}_{L/\mathbb{Q}}$ is noncomplexified over L). If $d > 0$, we have $r_1^{\mathrm{c}} = 0$ and $-1 \in \mathrm{N}(J_L)$ if and only if -1 is a norm at each p_i, $i = 1, \ldots, t$ (if 2 is ramified, the product formula implies that -1 is also a norm at 2). By Exercise II.1.6.5, if an odd prime p is ramified in $L/\mathbb{Q}$ (i.e., if $p|d$), then -1 is in the norm group of $\mathbb{Q}_p(\sqrt{d})/\mathbb{Q}_p$ if and only if $p \equiv 1 \bmod (4)$; this can also be seen by computing the regular Hilbert symbol $\left(\frac{-1\,,\,d}{p}\right)$. Therefore, we still have $H^{\mathrm{ord}}_{L/\mathbb{Q}} = H^{\mathrm{res}}_{L/\mathbb{Q}}$ (for $d > 0$), if and only if $s_1 = \cdots = s_t = 1$, hence $s = 1$ ($H^{\mathrm{res}}_{L/\mathbb{Q}}$ is noncomplexified over L).

If -1 is not a norm, then $H^{\mathrm{ord}}_{L/\mathbb{Q}}$ is the maximal real subfield of $H^{\mathrm{res}}_{L/\mathbb{Q}}$ (complexified over L since one of the s_i or s is equal to -1).

Result (i) is thus proved.

(ii) In this case, the number of 2-genera $g^{\mathrm{res}}_{L/\mathbb{Q},T^{(2)}}$ is given by:

$$\frac{\prod\limits_{v \neq 2} e_v}{[L : L \cap H_T^{\mathrm{res}}]} = \frac{2^t}{[L : L \cap H_T^{\mathrm{res}}]}$$

since $\mathcal{E}_T^{\mathrm{res}} = 1$ here. We have $H_T^{\mathrm{res}} = \mathbb{Q}(\mu_{2^\infty})$, the maximal 2-ramified extension of $\mathbb{Q}$. In the case $t = 0$ ($L = \mathbb{Q}(\sqrt{-1}),\ \mathbb{Q}(\sqrt{2})$, or $\mathbb{Q}(\sqrt{-2})$), the genus field is $\mathbb{Q}(\mu_{2^\infty})$, so we may assume that $t \geq 1$; then $L \cap H_T^{\mathrm{res}} = \mathbb{Q}$, the number of genera is equal to 2^{t-1} and it is easy to see that, by taking the compositum of the field $H^{\mathrm{res}}_{L/\mathbb{Q}}(2)$ with $\mathbb{Q}(\mu_{2^\infty})$, we obtained the desired field, equal to:

$$\mathbb{Q}(\mu_{2^\infty})(\sqrt{s2^\delta}; \sqrt{s_1 p_1}, \ldots, \sqrt{s_t p_t}) = \mathbb{Q}(\mu_{2^\infty})(\sqrt{p_1}, \ldots, \sqrt{p_t})$$

since $\sqrt{-1}$, $\sqrt{2}$, $\sqrt{-2}$ are in $\mathbb{Q}(\mu_{2^\infty})$. □

In [Le1], one can find the generalization of the principle of part (i) of this exercise to the case of an arbitrary abelian extension of $\mathbb{Q}$, and note that reasoning by duality (introducing the group of Dirichlet characters of L) gives a convenient description of the genus field.

4.2.10 Exercise (representation of prime numbers by integral quadratic forms). We keep the assumptions and notations of the above exercise. In addition, we assume that the restricted Hilbert class field of L is equal to its restricted genus field or, equivalently, that the restricted class number of L is equal to the number of invariant classes (equal to the number of genera). To check this in practice, because of Lemma II.6.1.2, it is necessary and sufficient that the restricted class number of L is equal to 2^{r-1}, where r is the number of ramified primes in $L/\mathbb{Q}$, or, if we know the structure of this class group, that it is 2-elementary (i.e., $(\mathcal{C}\ell_L^{\rm res})^2 = 1$) since in $L/\mathbb{Q}$, a class c is invariant if and only if $c^2 = 1$.

(α) Let $p > 0$ be a prime number. Show that the following two properties are equivalent:

(i) There exist $x,\ y \in \mathbb{Z}$ such that:

$$p = x^2 - dy^2 \text{ if } d \equiv 2,\ 3 \bmod (4), \quad p = \tfrac{1}{4}\,(x^2 - dy^2) \text{ if } d \equiv 1 \bmod (4)\ ;$$

(ii) we have one of the following:

- $\left(\dfrac{s2^\delta}{p}\right) = \left(\dfrac{s_1p_1}{p}\right) = \cdots = \left(\dfrac{s_tp_t}{p}\right) = 1$, if p is unramified in $L/\mathbb{Q}$,
- $\left(\dfrac{s_1p_1}{2}\right) = \cdots = \left(\dfrac{s_tp_t}{2}\right) = 1$ [5], if $p = 2$ is ramified,
- $\left(\dfrac{s2^\delta}{p_i}\right) = \left(\dfrac{s_1p_1}{p_i}\right) = \cdots = \left(\dfrac{s_{i-1}p_{i-1}}{p_i}\right) = \left(\dfrac{s_{i+1}p_{i+1}}{p_i}\right) = \cdots = \left(\dfrac{s_tp_t}{p_i}\right) = 1,$

if $p = p_i$.

(β) Describe in detail the case $d = 15$.

(γ) By giving an example, show that the above equivalence may be false if $H_L^{\rm res} \neq H_{L/\mathbb{Q}}^{\rm res}$.

Answer. (α) Condition (i) (p norm of a totally positive integer of L) is equivalent to the following fact: the residue degree of p in $L/\mathbb{Q}$ is equal to 1 and a prime ideal $\mathfrak{p}$ above p is principal in the restricted sense. Class field theory tells us that this is equivalent to p having residue degree equal to 1 in $H_L^{\rm res}/\mathbb{Q}$ (see II.5.2). Since $H_L^{\rm res} = H_{L/\mathbb{Q}}^{\rm res}$ is given in Exercise 4.2.9, the equivalence with (ii) is clear, at least in the unramified case (which means that p is totally split in $H_{L/\mathbb{Q}}^{\rm res}/\mathbb{Q}$). In the ramified case, we express the fact that the completion at p of the extension $H_{L/\mathbb{Q}}^{\rm res}/\mathbb{Q}$ is quadratic (hence equal to $\mathbb{Q}_2(\sqrt{s2^\delta}\,)/\mathbb{Q}_2$ for $p = 2$, to $\mathbb{Q}_{p_i}(\sqrt{s_ip_i}\,)/\mathbb{Q}_{p_i}$, for $p = p_i$). This proves (α).

Remarks. (i) If $H_L^{\rm res} = L$ (i.e., L is principal in the restricted sense), the number of genera is equal to 1 and we have the relation $e_2\, 2^t = 2$; hence d must be of the form -1, 2, -2 or $(-1)^{\frac{\ell-1}{2}}\ell$ with ℓ an odd prime, and the condition is equivalent to the fact that p is not inert in $\mathbb{Q}(\sqrt{d}\,)$. For $d = \ell \equiv 1 \bmod (4)$, formulas II.6.1.2, II.6.2.3, or those for the number of genera, show that $(\mathcal{C}\ell_L^{\rm res})_2 = (\mathcal{C}\ell_L^{\rm ord})_2 = 1$; hence $E_L^{\rm ord}$ represents every signature and it follows that the norm of the fundamental unit is equal to -1.

(ii) It is conjectured that there exists an infinite number of primes $\ell \equiv 1 \bmod (4)$ such that the real quadratic field $L = \mathbb{Q}(\sqrt{\ell})$ is principal. It is

[5] These symbols are such that $(\frac{s_ip_i}{2}) = 1$ if and only if $s_ip_i \equiv 1 \bmod (8)$.

known (Baker–Stark (1966/1967)) that the only imaginary quadratic fields which are principal are the fields $\mathbb{Q}(\sqrt{d}\,)$ for $d = -1, -2, -3, -7, -11, -19, -43, -67, -163$, and it is conjectured that there are exactly 65 discriminants of imaginary quadratic fields for which $\mathcal{C}\ell^2 = 1$ (they are given in [a, BŠa, Tab. 5]; for an overview of all these problems, we refer for instance to [Lou]). □

(β) In the case $d = 15$ ($\mathcal{C}\ell_L^{\rm res} \simeq \mathbb{Z}/2\mathbb{Z} \times \mathbb{Z}/2\mathbb{Z}$), we have $H_{L/\mathbb{Q}}^{\rm res} = \mathbb{Q}(\sqrt{-1}, \sqrt{-3}, \sqrt{5})$ and we obtain the condition (for $p \neq 2,\ 3,\ 5$):

$$\left(\frac{-1}{p}\right) = \left(\frac{-3}{p}\right) = \left(\frac{5}{p}\right) = 1$$

which yields $p \equiv 1$ or -11 modulo (60); for instance, for $p = 61$ we indeed have condition (i) ($61 = 11^2 - 15 \times 2^2$).

By (ii), the integers 2, 3, and 5 are not of the form $x^2 - 15y^2$.

For $p = 11$, we find that $-11 = 2^2 - 15 \times 1^2$, and since the fundamental unit of $\mathbb{Q}(\sqrt{15}\,)$ is equal to $4 + \sqrt{15}$ and is totally positive, 11 cannot be representable by the form $x^2 - 15y^2$, in accordance to the above criterion, although 11 is split in $L/\mathbb{Q}$; however, the prime ideals above 11 are principal in the ordinary sense.

(γ) The field $L = \mathbb{Q}(\sqrt{34}\,)$ is such that $\mathcal{C}\ell_L^{\rm res} \simeq \mathbb{Z}/4\mathbb{Z}$, $\mathcal{C}\ell_L^{\rm ord} \simeq \mathbb{Z}/2\mathbb{Z}$. The genus field is $\mathbb{Q}(\sqrt{2}\,, \sqrt{17}\,)$ and condition (ii) is here (for $p \neq 2,\ 17$):

$$\left(\frac{2}{p}\right) = \left(\frac{17}{p}\right) = 1.$$

The prime number $p = 127$ satisfies this condition but is not representable by $x^2 - 34y^2$ since we have $-127 = 3^2 - 34.2^2$ and the fundamental unit $\varepsilon = 35 + 6\sqrt{34}$ of L is totally positive. Note that 17 also yields a counterexample since $-17 = 17^2 - 34.3^2$, but not 2 ($2 = 6^2 - 34.1^2$). □

In this context, the quadratic symbols $\left(\frac{s2^\delta}{p}\right)$, $\left(\frac{s_i p_i}{p}\right)$ determine the genus of the prime number p (for example unramified), the principal genus corresponding to their triviality.

4.2.11 Exercise. Consider the case $K = \mathbb{Q}$, $L/\mathbb{Q}$ arbitrary, $T = S = \emptyset$. Give the number of genera in this case by computing explicitly the $\check{e}_p^{\rm ab}$, for all primes p, from the fact that the abelian extensions of $\mathbb{Q}_p$ are known (they are the subfields of the cyclotomic extensions of $\mathbb{Q}_p$, as we have seen in II.1.8.3). We refer to [I] for this result. □

4.2.12 Exercise. Show that the global norm conductor $\mathfrak{f}_{L/K}^{\rm gen\,S}$ of the genus field $H_{L/K}^S$ over K (equal to that of the extension $H_L^{S'\,\rm ab}/K$) is a divisor of $\prod_{v\in S} \mathfrak{f}_{L_v^{\rm ab}/K_v} \prod_{v\notin S} \mathfrak{f}_{\check{L}_v^{\rm ab}/K_v}$.

Answer. We keep the notations that we have used in all of Section 4. We are here in the case $T = \emptyset$. The idèle group corresponding to $H_L^{S'\,\mathrm{ab}} = L_{(1)}{}^{S'\,\mathrm{ab}}$ over K is $K^\times \mathrm{N}_{L/K}(U_L^{S'})$, and the desired conductor is the smallest modulus $\mathfrak{f}$ of K such that $U_\mathfrak{f}^{\mathrm{res}} \subseteq K^\times \mathrm{N}_{L/K}(U_L^{S'})$. This inclusion is equivalent to $U_\mathfrak{f}^{\mathrm{res}} \subseteq E^S \mathrm{N}_{L/K}(U_L^{S'})$.

If we only ask for the inclusion $U_\mathfrak{m}^{\mathrm{res}} \subseteq \mathrm{N}_{L/K}(U_L^{S'})$, we obtain a multiple $\mathfrak{m}$ of the conductor. This is equivalent to the following inclusions:

$$U_v^{\mathrm{v}(\mathfrak{m})} \subseteq \mathrm{N}_{L/K}\Big(\bigoplus_{w|v} L_w^\times\Big) \ \text{ if } v \in S,$$

$$U_v^{\mathrm{v}(\mathfrak{m})} \subseteq \mathrm{N}_{L/K}\Big(\bigoplus_{w|v} U_w\Big) \ \text{ if } v \notin S.$$

But, by II.2.6.3, if we denote by $\mathfrak{m}_v := \mathfrak{p}_v^{\mathrm{v}(\mathfrak{m})}$ the v-component of $\mathfrak{m}$, we obtain respectively $\mathfrak{m}_v = \mathfrak{f}_{L_v^{\mathrm{ab}}/K_v}$ if $v \in S$, $\mathfrak{m}_v = \mathfrak{f}_{\breve{L}_v^{\mathrm{ab}}/K_v}$ if $v \notin S$. The result follows.

If L/K is abelian, $\mathfrak{f}_{L/K}^{\mathrm{gen}\,S}$ is independent of S and coincides with the norm conductor of L/K (this is clear since $H_{L/K}^S$ is abelian over K and unramified over L), and in this case we see that the multiple obtained above again gives $\mathfrak{f}_{L/K}$.

For $K = \mathbb{Q}$, $L/\mathbb{Q}$ arbitrary, and $S = \emptyset$, since $E^{\mathrm{res}} = 1$, the conductor of $H_{L/\mathbb{Q}}^{\mathrm{res}}$ is given by $\mathfrak{f}_{L/\mathbb{Q}}^{\mathrm{gen\,res}} = \prod_v \mathfrak{f}_{\breve{L}_v^{\mathrm{ab}}/\mathbb{Q}_v}$. In the case of Exercise 4.2.5, we easily see that this conductor is equal to (4×17), which shows that the genus field is equal to $L(\sqrt{-1})$. □

b) The Genus Exact Sequence

The computation of the number of genera will enable us to show the existence of a fundamental exact sequence, which entirely summarizes genus theory for an extension L/K not assumed to be Galois.

The analog of Figure 4.1, relative to the sets T and S where, to simplify, we set $H := H_{K,T}^S$, $H' := H_{L,T'}^{S'}$, gives the following diagram:

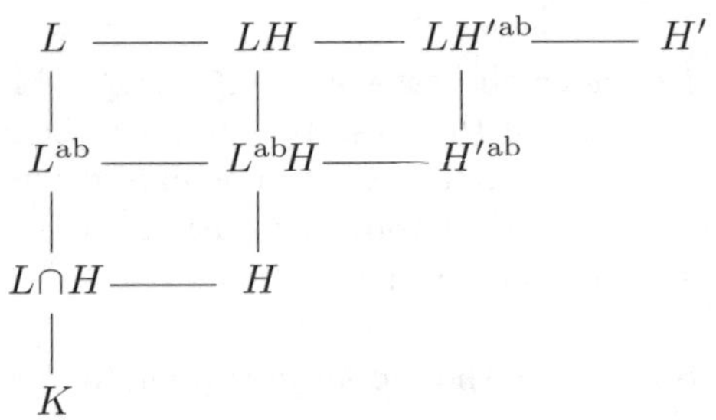

Fig. 4.3

where the genus field $H^S_{L/K,T}$ is given by LH'^{ab}.

Let $v \in Pl$ and let $w \in Pl_{L,v}$. Since H'/L is abelian, the corresponding completion above w can be denoted H'_w/L_w since the completions $H'_{w'}$ for $w'|w$ are independent of the choice of $w'|w$. As usual we set, relative to the extensions L/K and H'/K:

$$L_v^{\mathrm{ab}} := \bigcap_{w|v} L_w^{\mathrm{ab}},\ H_v'^{\mathrm{ab}} := \bigcap_{w|v} H_w'^{\mathrm{ab}},\ \hat{L}_v^{\mathrm{ab}} := \langle\, L_w^{\mathrm{ab}} \,\rangle_{w|v},\ \hat{H}_v'^{\mathrm{ab}} := \langle\, H_w'^{\mathrm{ab}} \,\rangle_{w|v},$$

and we also introduce the subfields $\check{L}_v^{\mathrm{ab}}$ and $\check{H}_v'^{\mathrm{ab}}$, of $\hat{L}_v^{\mathrm{ab}}$ and $\hat{H}_v'^{\mathrm{ab}}$ respectively. We will implicitly use the inclusions:

$$(L^{\mathrm{ab}})_v \subseteq L_v^{\mathrm{ab}} \subseteq \check{L}_v^{\mathrm{ab}},$$

and similarly with H'.

Since H'/L is S'-split, if $v \in S$ we have $H_v'^{\mathrm{ab}} = L_v^{\mathrm{ab}}$, showing that the groups:

$$G_v^{\mathrm{ab}}(H'/K) := \mathrm{Gal}(H_v'^{\mathrm{ab}}/K_v) \text{ and } G_v^{\mathrm{ab}}(L/K) := \mathrm{Gal}(L_v^{\mathrm{ab}}/K_v)$$

coincide. Furthermore, if $v \notin T \cup S$, then, the inertia groups:

$$\check{G}_v^{\mathrm{ab}\,0}(H'/K) := (\mathrm{Gal}(\check{H}_v'^{\mathrm{ab}}/K_v))^0 \text{ and } \check{G}_v^{\mathrm{ab}\,0}(L/K) := (\mathrm{Gal}(\check{L}_v^{\mathrm{ab}}/K_v))^0$$

are canonically isomorphic. To prove this, the simplest is to note that, since H'/L is unramified at v, by local class field theory, and with evident notations, $U_w = \mathrm{N}_{H'_w/L_w}(U'_w)$, hence:

$$U_v/\langle\, \mathrm{N}_{L_w/K_v}(U_w) \,\rangle_{w|v} = U_v/\langle\, \mathrm{N}_{H'_w/K_v}(U'_w) \,\rangle_{w|v},$$

which by II.2.6.3 gives the claimed isomorphism.

4.3 Definitions. (i) We denote by h the canonical isomorphism:

$$\bigoplus_{v\in S} G_v^{\mathrm{ab}}(L/K) \bigoplus_{v\notin T\cup S} \check{G}_v^{\mathrm{ab}\,0}(L/K) \longrightarrow \bigoplus_{v\in S} G_v^{\mathrm{ab}}(H'/K) \bigoplus_{v\notin T\cup S} \check{G}_v^{\mathrm{ab}\,0}(H'/K).$$

(ii) We define a canonical map:

$$\nu : E^S \longrightarrow \bigoplus_{v\in S} G_v^{\mathrm{ab}}(L/K) \bigoplus_{v\notin T\cup S} \check{G}_v^{\mathrm{ab}\,0}(L/K),$$

sending $\varepsilon \in E^S$ to $((i_v(\varepsilon)\,,\, L_v^{\mathrm{ab}}/K_v)_{v\in S}\,;\ (i_v(\varepsilon)\,,\, \check{L}_v^{\mathrm{ab}}/K_v)_{v\notin T\cup S})$. [6]

(iii) We also consider the map:

$$\pi : \bigoplus_{v\in S} G_v^{\mathrm{ab}}(L/K) \bigoplus_{v\notin T\cup S} \check{G}_v^{\mathrm{ab}\,0}(L/K) \longrightarrow \mathrm{Gal}(H'^{\mathrm{ab}}/H),$$

[6] If ε is an S-unit of K, the local norm residue symbols $(i_v(\varepsilon)\,,\, L_v^{\mathrm{ab}}/K_v)$ for $v \in S$ (resp. $(i_v(\varepsilon)\,,\, \check{L}_v^{\mathrm{ab}}/K_v)$ for $v \notin T \cup S$) have values in $G_v^{\mathrm{ab}}(L/K)$ (resp. in $\check{G}_v^{\mathrm{ab}\,0}(L/K)$).

sending $(\sigma_v)_{v\notin T}$ in the left hand side to the product (for $v \notin T$) of the canonical images of the $h(\sigma_v)_{|(H'^{\mathrm{ab}})_v}$ in $\mathrm{Gal}(H'^{\mathrm{ab}}/K)$ (the result does indeed belong to the subgroup $\mathrm{Gal}(H'^{\mathrm{ab}}/H)$ since H/K is T-ramified and S-split). □

Note. By abuse of notation we denote by $h(\sigma_v)_{|H'^{\mathrm{ab}}}$ the image of $h(\sigma_v)_{|(H'^{\mathrm{ab}})_v}$ in $\mathrm{Gal}(H'^{\mathrm{ab}}/K)$. The definition of π therefore becomes: $\pi((\sigma_v)_{v\notin T}) = \prod_{v\notin T} h(\sigma_v)_{|H'^{\mathrm{ab}}}$.

Notations. Recall that K is a number field given together with sets of places T and S, L is a finite extension of K, T' and S' are the sets of places of L above those of T and S. For the prime number p, put $T_p := T \cap Pl_p$, and $\mathcal{E}^S_T := \mathrm{Ker}\Big(\mathcal{E}^S := E^S \otimes \mathbb{Z}_p \longrightarrow \bigoplus_{v\in T_p} U^1_v \bigoplus_{v\in T\backslash T_p} (F^\times_v)_p\Big)$. If $T_p = \emptyset$, we replace $\mathcal{E}^S_T$ by $E^S_{\mathfrak{m}}$ where $\mathfrak{m} := \prod_{v\in T} \mathfrak{p}_v$.

The field H^S_T (resp. $H^{S'}_{L,T'}$) is the maximal T-ramified S-split (resp. T'-ramified S'-split) abelian extension of K (resp. of L), and $H^{S'\,\mathrm{ab}}_{L,T'}$ is the maximal subextension of $H^{S'}_{L,T'}$ which is abelian over K.

Finally, $G^{\mathrm{ab}}_v(L/K) := \mathrm{Gal}(L^{\mathrm{ab}}_v/K_v)$, and $G^{\mathrm{ab}\,0}_v(L/K)$ (resp. $\check{G}^{\mathrm{ab}\,0}_v(L/K)$) are the inertia groups of L^{ab}_v/K_v (resp. $\check{L}^{\mathrm{ab}}_v/K_v$). □

4.4 Theorem (genus exact sequence). *We have the exact sequence of finite p-groups:*

$$1 \longrightarrow \mathcal{E}^S_T/\mathcal{E}^S_T \cap \mathrm{N}_{L/K}(U^{S'}_L) \xrightarrow{\ \nu\ } \bigoplus_{v\in S} (G^{\mathrm{ab}}_v(L/K))_p \bigoplus_{v\notin T\cup S} (\check{G}^{\mathrm{ab}\,0}_v(L/K))_p \xrightarrow{\ \pi\ } (\mathrm{Gal}(H^{S'\,\mathrm{ab}}_{L,T'}/H^S_T))_p \longrightarrow 1,$$

which can be written in the Galois case:

$$1 \longrightarrow \mathcal{E}^S_T/\mathcal{E}^S_T \cap \mathrm{N}_{L/K}(J_L) \xrightarrow{\ \nu\ } \bigoplus_{v\in S} (G^{\mathrm{ab}}_v(L/K))_p \bigoplus_{v\notin T\cup S} (G^{\mathrm{ab}\,0}_v(L/K))_p \xrightarrow{\ \pi\ } (\mathrm{Gal}(H^{S'\,\mathrm{ab}}_{L,T'}/H^S_T))_p \longrightarrow 1. \ \square$$

When $T = \emptyset$, globalizing we obtain:

4.4.1 Corollary (classical genus theory). *We have the exact sequence:*

$$1 \longrightarrow E^S/E^S \cap \mathrm{N}_{L/K}(U^{S'}_L) \xrightarrow{\ \ \nu\ \ } \bigoplus_{v\in S} G^{\mathrm{ab}}_v(L/K) \bigoplus_{v\notin S} \check{G}^{\mathrm{ab}\,0}_v(L/K) \xrightarrow{\ \ \pi\ \ } \mathrm{Gal}(H^{S'\,\mathrm{ab}}_L/H^S) \longrightarrow 1,$$

which can be written in the Galois case:

$$1 \longrightarrow E^S/E^S \cap \mathrm{N}_{L/K}(J_L) \xrightarrow{\ \ \nu\ \ } \bigoplus_{v\in S} G^{\mathrm{ab}}_v(L/K) \bigoplus_{v\notin S} G^{\mathrm{ab}\,0}_v(L/K) \xrightarrow{\ \ \pi\ \ } \mathrm{Gal}(H^{S'\,\mathrm{ab}}_L/H^S) \longrightarrow 1. \ \square$$

4.4.2 Corollary. *If L/K is Galois, we have the exact sequences:*

$$1 \longrightarrow E^{\mathrm{res}}/E^{\mathrm{res}} \cap \mathrm{N}_{L/K}(J_L) \xrightarrow{\ \nu\ }$$
$$\bigoplus_{v \in Pl_0} G_v^{\mathrm{ab}\,0}(L/K) \xrightarrow{\ \pi\ } \mathrm{Gal}(H_L^{\mathrm{res\,ab}}/H^{\mathrm{res}}) \longrightarrow 1,$$
$$1 \longrightarrow E^{\mathrm{ord}}/E^{\mathrm{ord}} \cap \mathrm{N}_{L/K}(J_L) \xrightarrow{\ \nu\ }$$
$$\bigoplus_{v \in Pl_\infty^{\mathrm{r}}} G_v^{\mathrm{ab}}(L/K) \bigoplus_{v \in Pl_0} G_v^{\mathrm{ab}\,0}(L/K) \xrightarrow{\ \pi\ } \mathrm{Gal}(H_L^{\mathrm{ord\,ab}}/H^{\mathrm{ord}}) \longrightarrow 1. \ \square$$

Proof of the theorem. Let $\varepsilon \in \mathcal{E}_T^S$; if we have:

$$((\bar{i}_v(\varepsilon)\,,\, L_v^{\mathrm{ab}}/K_v)_{v \in S}\,;\ (\bar{i}_v(\varepsilon)\,,\, \check{L}_v^{\mathrm{ab}}/K_v)_{v \notin T \cup S}) = 1$$

in $\bigoplus_{v \in S} (G_v^{\mathrm{ab}})_p \bigoplus_{v \notin T \cup S} (\check{G}_v^{\mathrm{ab}\,0})_p$, this means that ε is a local norm on S, and is a norm of local units on $Pl\backslash(T \cup S)$ in L/K; but by definition of $\mathcal{E}_T^S$, ε is evidently a norm of local units on T, hence finally on $Pl\backslash S$; this proves the injectivity.

The surjectivity of π follows from the fact that the subgroup generated by the decomposition groups $D_v(H'^{\mathrm{ab}}/K)$ for $v \in S$ and the inertia groups $I_v(H'^{\mathrm{ab}}/K)$ for $v \notin T \cup S$ (which are respectively images under π of the $G_v^{\mathrm{ab}}(L/K)$ for $v \in S$ and of the $\check{G}_v^{\mathrm{ab}\,0}(L/K)$ for $v \notin T \cup S$) fixes the maximal T-ramified S-split subextension of H'^{ab}/K, which is equal to H by definition.

The composite $\pi \circ \nu$ is such that, for all $\varepsilon \in \mathcal{E}_T^S$:

$$\begin{aligned}\pi \circ \nu(\varepsilon) &= \prod_{v \in S} (\bar{i}_v(\varepsilon)\,,\, H_v'^{\mathrm{ab}}/K_v)_{|H'^{\mathrm{ab}}} \prod_{v \notin T \cup S} (\bar{i}_v(\varepsilon)\,,\, \check{H}_v'^{\mathrm{ab}}/K_v)_{|H'^{\mathrm{ab}}} \\ &= \prod_v \left(\frac{\varepsilon\,,\, H'^{\mathrm{ab}}/K}{v}\right)_p\end{aligned}$$

which is equal to 1 by the product formula for Hasse symbols (here for their p-parts) in the abelian extension H'^{ab}/K (since these symbols are equal to 1 on T). The resulting inclusion $\mathrm{Im}(\nu) \subseteq \mathrm{Ker}(\pi)$ yields:

$$\begin{aligned}(\mathrm{Ker}(\pi) : \mathrm{Im}(\nu)) &= |\mathrm{Ker}(\pi)| \times \frac{1}{|\mathrm{Im}(\nu)|} \\ &= \frac{\prod_{v \in S} |(G_v^{\mathrm{ab}})_p| \prod_{v \notin T \cup S} |(\check{G}_v^{\mathrm{ab}\,0})_p|}{[H'^{\mathrm{ab}} : H]_p} \times \frac{1}{(\mathcal{E}_T^S : \mathcal{E}_T^S \cap \mathrm{N}_{L/K}(U_L^{S'}))},\end{aligned}$$

which is indeed equal to 1 by the formula of Theorem 4.2, proving the exactness of the sequence. $\square$

From this, in the case of complete p-ramification, we obtain another deployment theorem [Ja2, Ch. II, Cor. 2.45]. We assume the p-adic conjecture in K (i.e., the Leopoldt conjecture for p if $S_0 = \emptyset$).

4.4.3 Corollary. *If T contains Pl_p and if E^S is monogeneous, we have:*

$$\mathrm{Gal}(H_{L,T'}^{S'\,\mathrm{ab}}(p)/H_T^S(p)) \simeq \bigoplus_{v\in S}(G_v^{\mathrm{ab}}(L/K))_p \bigoplus_{v\notin T\cup S}(\check{G}_v^{\mathrm{ab}\,0}(L/K))_p.$$

For $T = Pl_p$, $S = \emptyset$ or $S = Pl_\infty^{\mathrm{r}}$, we obtain:

$$\mathrm{Gal}(H_{L,p}^{\mathrm{res\,ab}}(p)/H_p^{\mathrm{res}}(p)) \simeq \bigoplus_{v\in Pl_{\mathrm{ta}}}(\check{G}_v^{\mathrm{ab}\,0}(L/K))_p,$$

$$\mathrm{Gal}(H_{L,p}^{\mathrm{ord\,ab}}(p)/H_p^{\mathrm{ord}}(p)) \simeq (\mathbb{Z}/2\mathbb{Z})_p^{r_1^{\mathrm{c}}} \bigoplus_{v\in Pl_{\mathrm{ta}}}(\check{G}_v^{\mathrm{ab}\,0}(L/K))_p,$$

where r_1^{c} is the number of real places of K which are totally complexified in L.

Proof. The assumptions imply that $\mathcal{E}_T^S = 1$ (see III.3.6.2, (vi)), proving the result. □

We come back to the general situation of Figure 4.3, Definition 4.3, and Notations of 4.4.

4.5 Proposition (another way of writing the genus exact sequence). *Set*[7]*:*

$$\Omega_T^S(L/K) := \Big\{(\sigma_v)_{v\notin T} \in \bigoplus_{v\in S}(G_v^{\mathrm{ab}}(L/K))_p \bigoplus_{v\notin T\cup S}(\check{G}_v^{\mathrm{ab}\,0}(L/K))_p,\ \prod_{v\notin T}\sigma_v\big|_{L^{\mathrm{ab}}} = 1\Big\}.$$

Then, the genus exact sequence can be written:

$$1 \longrightarrow \mathcal{E}_T^S/\mathcal{E}_T^S \cap \mathrm{N}_{L/K}(U_L^{S'}) \xrightarrow{\ \nu\ } \Omega_T^S(L/K) \xrightarrow{\ \pi\ } (\mathrm{Gal}(H_{L,T'}^{S'\,\mathrm{ab}}/L^{\mathrm{ab}}\,H_T^S))_p \longrightarrow 1.$$

Note. If L/K is Galois, we can replace $\check{G}_v^{\mathrm{ab}\,0}$ by $G_v^{\mathrm{ab}\,0}$, $\mathrm{N}_{L/K}(U_L^{S'})$ by $\mathrm{N}_{L/K}(J_L)$. If moreover $T_p = \emptyset$, we replace $\mathcal{E}_T^S$ by $E_{\mathfrak{m}}^S$ where $\mathfrak{m} := \prod_{v\in T}\mathfrak{p}_v$.

Proof of the proposition. Under the projection:

$$\mathrm{Gal}(H'^{\mathrm{ab}}/K) \longrightarrow \mathrm{Gal}(L^{\mathrm{ab}}/K),$$

the image of $\pi((\sigma_v)_{v\notin T}) := \prod_{v\notin T} h(\sigma_v)\big|_{H'^{\mathrm{ab}}}$ is equal to $\prod_{v\notin T}(\sigma_v)\big|_{L^{\mathrm{ab}}}$, hence $\pi(\Omega_T^S(L/K)) \subseteq (\mathrm{Gal}(H'^{\mathrm{ab}}/L^{\mathrm{ab}}\,H))_p$.

The kernel of the restriction of π to $\Omega_T^S(L/K)$ is still equal to the image of ν since for all $\varepsilon \in \mathcal{E}_T^S$ we have the relation:

$$\prod_v \Big(\bar{i}_v(\varepsilon)\,,\,(L^{\mathrm{ab}})_v/K_v\Big)\Big|_{L^{\mathrm{ab}}} = \prod_v \left(\frac{\varepsilon\,,\,L^{\mathrm{ab}}/K}{v}\right)_p = 1$$

showing that:

[7] were we recall that $\sigma_v\big|_{L^{\mathrm{ab}}}$ denotes the canonical image of $\sigma_v\big|_{(L^{\mathrm{ab}})_v}$ in G^{ab}.

$$\nu(\mathcal{E}_T^S) \subseteq \Omega_T^S(L/K).$$

The image of the restriction of π to $\Omega_T^S(L/K)$ is $(\mathrm{Gal}(H'^{\mathrm{ab}}/L^{\mathrm{ab}}\, H))_p$ since, by definition of H, $L \cap H$ is the maximal T-ramified S-split abelian subextension of L, and we even have the following evident exact sequence:

$$1 \longrightarrow \Omega_T^S(L/K) \longrightarrow$$
$$\bigoplus_{v\in S}(G_v^{\mathrm{ab}}(L/K))_p \bigoplus_{v\notin T\cup S}(\check{G}_v^{\mathrm{ab}\,0}(L/K))_p \xrightarrow{\ \pi'\ } (\mathrm{Gal}(L^{\mathrm{ab}}/L\cap H))_p \longrightarrow 1,$$

where π' is the similar map for L^{ab}/K, which proves the desired conclusion by looking at orders. □

Note. When L/K is abelian, we can replace $G_v^{\mathrm{ab}}(L/K)$ and $\check{G}_v^{\mathrm{ab}\,0}(L/K)$ by their canonical images in G, which are $D_v(L/K)$ and $I_v(L/K)$ respectively.

4.5.1 Corollary (abelian case, tame genus theory). *When L/K is an abelian p-extension and $T_p = \emptyset$, we have:*

$$\Omega_T^S(L/K) \simeq \Big\{(\tau_v)_v \in \bigoplus_{v\in S} D_v(L/K) \bigoplus_{v\notin T\cup S} I_v(L/K),\ \ \prod_{v\notin T} \tau_v = 1\Big\},$$

and the exact sequence:

$$1 \longrightarrow E_{\mathfrak{m}}^S/E_{\mathfrak{m}}^S \cap \mathrm{N}_{L/K}(J_L) \xrightarrow{\ \nu\ } \Omega_T^S(L/K) \xrightarrow{\ \pi\ } (\mathrm{Gal}(H_{L/K,T}^S/L\,H_T^S))_p \longrightarrow 1,$$

where $\mathfrak{m} := \prod_{v\in T} \mathfrak{p}_v$ and where ν associates with $\varepsilon \in E_{\mathfrak{m}}^S$ the family of Hasse symbols $\left(\frac{\varepsilon,\, L/K}{v}\right)$, $v \notin T$. Therefore, this yields:

$$\mathrm{rk}_p(\mathcal{C}\ell_{L,T'}^{S'}) \geq \mathrm{rk}_p(\Omega_T^S(L/K)) - \mathrm{rk}_p(E_{\mathfrak{m}}^S/E_{\mathfrak{m}}^S \cap \mathrm{N}_{L/K}(J_L)).$$

Proof. To obtain the inequality, write that $1 \longrightarrow B \longrightarrow A \longrightarrow C \longrightarrow 1$ implies $\mathrm{rk}_p(A) \leq \mathrm{rk}_p(B) + \mathrm{rk}_p(C)$, and use the trivial inequality:

$$\mathrm{rk}_p(\mathrm{Gal}(H_{L/K,T}^S/L\,H_T^S)) \leq \mathrm{rk}_p(\mathrm{Gal}(H_{L,T'}^{S'}/L)) = \mathrm{rk}_p(\mathcal{C}\ell_{L,T'}^{S'}).$$ □

Example. When L/K is cyclic of degree p, with $L \not\subseteq H^{\mathrm{res}}$ (resp. $L \not\subseteq H^{\mathrm{ord}}$), we recover the classical lower bounds:

$$\begin{aligned}\mathrm{rk}_p(\mathcal{C}\ell_L^{\mathrm{res}}) &\geq t - 1 - \mathrm{rk}_p(E^{\mathrm{res}}/E^{\mathrm{res}} \cap \mathrm{N}_{L/K}(J_L))\\ &\geq t - 1 - \mathrm{rk}_p(E^{\mathrm{res}}),\\ \mathrm{rk}_p(\mathcal{C}\ell_L^{\mathrm{ord}}) &\geq t + r_1^{\mathrm{c}} - 1 - \mathrm{rk}_p(E^{\mathrm{ord}}/E^{\mathrm{ord}} \cap \mathrm{N}_{L/K}(J_L))\\ &\geq t + r_1^{\mathrm{c}} - 1 - \mathrm{rk}_p(E^{\mathrm{ord}}),\end{aligned}$$

respectively, where t is the number of finite places of K ramified in L/K and r_1^c the number of real places of K complexified in L/K. $\square$

Finally, coming back to the case of an arbitrary extension L/K, we easily obtain from the above way of writing the genus exact sequence, a similar lower bound for $\mathrm{rk}_p(\mathcal{C}\ell_{L,T'}^{S'})$, which for $T = S = \emptyset$ proves the following result.

4.5.2 Corollary. *Let $L^{\mathrm{nr}} := H^{\mathrm{res}} \cap L^{\mathrm{ab}}$ be the maximal unramified subextension of L^{ab}. For any prime number p we have:*

$$\mathrm{rk}_p(\mathcal{C}\ell_L^{\mathrm{res}}) \geq \max\big(\mathrm{rk}_p(\mathrm{Gal}(H^{\mathrm{res}}/L^{\mathrm{nr}})),\\ \mathrm{rk}_p(\Omega(L/K)) - \mathrm{rk}_p(E^{\mathrm{res}}/E^{\mathrm{res}} \cap \mathrm{N}_{L/K}(U_L^{\mathrm{res}}))\big).$$

Proof. We first write that:

$$\mathrm{rk}_p(\mathcal{C}\ell_L^{\mathrm{res}}) \geq \max\big(\mathrm{rk}_p(\mathrm{Gal}(H^{\mathrm{res}}/L^{\mathrm{nr}}))\ ,\ \mathrm{rk}_p(\mathrm{Gal}(H_{L/K}^{\mathrm{res}}/L\,H^{\mathrm{res}}))\big).$$

We then consider the exact sequence of 4.5 for $T = S = \emptyset$:

$$1 \longrightarrow E^{\mathrm{res}}/E^{\mathrm{res}} \cap \mathrm{N}_{L/K}(U_L^{\mathrm{res}}) \longrightarrow \Omega(L/K) \longrightarrow \mathrm{Gal}(H_{L/K}^{\mathrm{res}}/L\,H^{\mathrm{res}}) \longrightarrow 1,$$

with here $\Omega(L/K) = \Big\{(\sigma_v)_v \in \bigoplus_v \check{G}_v^{\mathrm{ab}\,0}(L/K),\ \ \prod_v \sigma_v|_{L^{\mathrm{ab}}} = 1\Big\}$, where we recall that $\check{G}_v^{\mathrm{ab}\,0}(L/K)$ is the inertia group of $\check{L}_v^{\mathrm{ab}}/K_v$. $\square$

4.5.3 Remarks. (i) A lower bound for $\mathrm{rk}_p(\mathrm{Gal}(H^{\mathrm{res}}/L^{\mathrm{nr}}))$ is given by $\mathrm{rk}_p(\mathcal{C}\ell^{\mathrm{res}}) - \mathrm{rk}_p(\mathrm{Gal}(L^{\mathrm{nr}}/K))$, and a lower bound for $\mathrm{rk}_p(\Omega(L/K))$ is given by $\sum_v \mathrm{rk}_p(\check{G}_v^{\mathrm{ab}\,0}(L/K)) - \mathrm{rk}_p(\mathrm{Gal}(L^{\mathrm{ab}}/L^{\mathrm{nr}}))$.

(ii) An upper bound for $\mathrm{rk}_p(E^{\mathrm{res}}/E^{\mathrm{res}} \cap \mathrm{N}_{L/K}(U_L^{\mathrm{res}}))$ is $\mathrm{rk}_p(E^{\mathrm{res}})$.

(iii) The lower bound 4.5.2 is useful in the non-Galois case since the groups G_v^{ab}, $\check{G}_v^{\mathrm{ab}\,0}$ yield more information than the corresponding groups in L^{ab}/K; for instance, in the context of Exercise 4.2.6, it gives again $\mathrm{rk}_3(\mathcal{C}\ell_L) \geq \gamma$. $\square$

4.5.4 Example. We take $K = \mathbb{Q}$ and $p = 2$. Let q, ℓ be two distinct prime numbers congruent to 1 modulo (8), let k be a cyclic field of degree 4, of conductor $q\ell$, containing $\mathbb{Q}(\sqrt{q\ell})$, and let $L = k\,\mathbb{Q}(\sqrt{q})$. We easily have:

$$I_q(L/\mathbb{Q}) = \mathrm{Gal}(L/\mathbb{Q}(\sqrt{\ell})),\ \ I_\ell(L/\mathbb{Q}) = \mathrm{Gal}(L/\mathbb{Q}(\sqrt{q})),$$

hence we obtain that $\Omega(L/\mathbb{Q})$ has order 2. Since $E^{\mathrm{res}} = 1$, we thus have:

$$\mathrm{rk}_2(\mathcal{C}\ell_L^{\mathrm{res}}) \geq 1.$$

The brutal lower bound $\mathrm{rk}_2(\mathcal{C}\ell_L^{\mathrm{res}}) \geq \sum_v \mathrm{rk}_2(I_v(L/\mathbb{Q})) - \mathrm{rk}_2(\mathrm{Gal}(L/L^{\mathrm{nr}}))$ yields here $\mathrm{rk}_2(\mathcal{C}\ell_L^{\mathrm{res}}) \geq 0$. We thus see that it is important to take into account the product formula. $\square$

4.5.5 Remark. The usual genus exact sequence (case $T = \emptyset$) allows to prove a slightly stronger form of the result II.3.4.4′ on the converse of the product formula for the generalized residue symbol. Indeed, for a fixed family

$$(\sigma_v)_v \in \bigoplus_v G_v^{\mathrm{ab}}(L/K),$$

such that $\prod_v \sigma_v|_{L^{\mathrm{ab}}}$ is the identity in G^{ab}, we can choose S sufficiently large, containing the support of this family and the places which are ramified in L/K (or at least those for which $\check{G}_v^{\mathrm{ab}\,0} \neq 1$), and such that $H^S_{L/K} = LH^S$, so that the conditions $\prod_v \sigma_v|_{L^{\mathrm{ab}}} = 1$ and $\prod_v h(\sigma_v)|_{H'^{\mathrm{ab}}} = 1$ will be equivalent; we thus obtain a solution $x \in E^S$. The condition $H^S_{L/K} = LH^S$ is absolutely necessary as it is shown by the following example.

Example (after [d, CF, Exer. 2.16]). Consider $K = \mathbb{Q}$, $L = \mathbb{Q}(\sqrt{-14})$ and set $G := \mathrm{Gal}(L/\mathbb{Q}) =: \langle\, s \,\rangle$. We want to have:

$$\sigma_2 = s, \ \ \sigma_7 = 1, \ \ \sigma_\infty = s, \ \ \sigma_v = 1 \ \text{ for } v \neq 2, \ 7, \ \infty.$$

We have $H_{L/\mathbb{Q}} = \mathbb{Q}(\sqrt{2}, \sqrt{-7})$ which is of degree 2 over L. Take $S = \{2, \ 7, \ \infty\}$. We have $LH^S = L$ since $H = \mathbb{Q}$, hence $H^S_{L/\mathbb{Q}} = H_{L/\mathbb{Q}}$: indeed, the places of L above those of S are totally split in $H_{L/\mathbb{Q}}/L$ (immediate by studying $\mathbb{Q}(\sqrt{2}, \sqrt{-7})/\mathbb{Q}$). Hence all the conditions are satisfied except for $H^S_{L/\mathbb{Q}} = LH^S$.

Finally, since $L = \mathbb{Q}(\sqrt{-14})$, the problem is equivalent (in terms of Hilbert symbols) to solving with $x \in E^S$:

$$\Big(\frac{-14, \, x}{2}\Big) = -1, \ \ \Big(\frac{-14, \, x}{7}\Big) = 1, \ \ \Big(\frac{-14, \, x}{\infty}\Big) = -1, \ \ \Big(\frac{-14, \, x}{v}\Big) = 1,$$

for $v \neq 2, \ 7, \ \infty$. But the S-unit group of $\mathbb{Q}$ is $E^S = \langle\, -1, 2, 7 \,\rangle$, and a computation easily yields:

$$\Big(\frac{-14, \, -1}{2}\Big) = \Big(\frac{-14, \, 2}{2}\Big) = \Big(\frac{-14, \, 7}{2}\Big) = 1,$$

so that a solution with an S-unit is impossible. □

4.5.6 Exercise (after an example of [Mai1]). Consider $L = \mathbb{Q}(\sqrt{23}, \sqrt{-m})$ with $m = 2 \times 5 \times 7 \times 41$. For a subfield K of L, the inequality 4.5.1 yields, for $T = \emptyset$, $S = \emptyset$ or $S = Pl^{\mathrm{r}}_\infty$:

$$\mathrm{rk}_2(\mathcal{C}\!\ell_L) \geq \mathrm{rk}_2(\Omega^S(L/K)) - \mathrm{rk}_2(E^S/E^S \cap \mathrm{N}_{L/K}(J_L)).$$

Compute this lower bound in the following cases.

(i) $K = \mathbb{Q}$, $S = \emptyset$, $S = \{\infty\}$.

(ii) $K = \mathbb{Q}(\sqrt{-23m})$, $S = \emptyset$.
(iii) $K = \mathbb{Q}(\sqrt{-m})$, $S = \emptyset$.
(iv) $K = \mathbb{Q}(\sqrt{23})$, $S = \emptyset$, $S = \{\infty, \infty'\}$.

Answer. In case (i), the lower bound is 4 for each S. In cases (ii) and (iii), this lower bound is 0 and 1 respectively. In case (iv), $S = \emptyset$, one finds $\mathrm{rk}_2(\mathcal{C}\ell_L) \geq 5$; with $S = \{\infty, \infty'\}$, one finds $\mathrm{rk}_2(\mathcal{C}\ell_L) \geq 6$: using evident notations, $\Omega^{\mathrm{ord}}(L/K)$ is a subspace of index 2 of $D_\infty \oplus D_{\infty'} \oplus I_{v_2} \oplus I_{v_5} \oplus I_{v_7} \oplus I_{v_7'} \oplus I_{v_{41}} \oplus I_{v_{41}'} \simeq (\mathbb{Z}/2\mathbb{Z})^8$, $E^{\mathrm{ord}} = \langle\, -1,\, \varepsilon\,\rangle$ with $\varepsilon = 24 + 5\sqrt{23}$, and we check that -1 is not a local norm, that ε is a local norm (compute the quadratic Hilbert's symbols $(\varepsilon, -m)_v$; for instance, since $\sqrt{23} \equiv 3 \bmod \mathfrak{p}_{v_7}$, we have $(\varepsilon, -m)_{v_7} = (24 + 5 \times 3, -m)_7 = 1$ in $\mathbb{F}_7^\times$).

Note that the use of PARI yields $(\mathcal{C}\ell_L)_2 \simeq \mathbb{Z}/4\mathbb{Z} \times (\mathbb{Z}/2\mathbb{Z})^5$. □

c) Central Classes — Knot Groups

Once genus theory is known, it is useful to develop the theory of central classes, at least for the case $T = \emptyset$, the central class field being, when L/K is Galois with Galois group G, the largest subextension $C^S_{L/K}$ of $H^{S'}_L/L$, Galois over K, such that $\mathrm{Gal}(C^S_{L/K}/L)$ is contained in the center of $\mathrm{Gal}(C^S_{L/K}/K)$. [8]

Using II.3.3, (vi), or II.4.5, (vi), it is immediately checked that the central class field is the subfield of $H^{S'}_L$ corresponding to $L^\times . I_G J_L . U^{S'}_L$ in idelic terms, or to $I_G \mathcal{C}\ell^{S'}_L$ in terms of class groups (where I_G is the augmentation ideal of G); it contains the genus field $H^S_{L/K}$, and hence gives a larger subextension of $H^{S'}_L/L$ which is however less explicit because of the existence of a knot group at the level of units in the expression of $[C^S_{L/K} : L]$: More precisely, we will see in 4.7 that for any Galois extension L/K:

$$[C^S_{L/K} : H^S_{L/K}] = \frac{(K^\times \cap \mathrm{N}_{L/K}(J_L) : \mathrm{N}_{L/K}(L^\times))}{(E^S \cap \mathrm{N}_{L/K}(J_L) : E^S \cap \mathrm{N}_{L/K}(L^\times))},$$

which involves two knot groups. The first one (the defect group of the Hasse norm principle) has been mentioned in II.6.2.6, II.6.2.7 and may be considered as a Galois invariant, contrary to the second one which is more arithmetical in nature.

These aspects can already be found in pioneering work of Scholz [Scholz], and have been the object of many papers such as [Fr1; Fr3], [Fu2], [Go], and in particular [Jeh], where one can find a historical survey of these questions which (cohomologically) involve the Schur multiplier group of G (or its dual). [9] Moreover, in [Jeh] one can find refinements of the Grunwald–Wang

[8] As usual, $H^{S'}_L$ is the S'-split Hilbert class field of L, where S' is the set of places of L above those of S.

[9] [d, CF, Ch. VIII, § 2.4], [e, Ko3, Ch. 2, § 7.5], [Ra].

theorem to construct abelian extensions with given Galois group and given number knot $(K^\times \cap \mathrm{N}_{L/K}(J_L) : \mathrm{N}_{L/K}(L^\times))$.

The case of central classes with a modulus has been studied by Shirai in [Shi], where the corresponding class fields towers are also studied; this point of view, with controlled wild ramification, has then been developed in [HM3]. For a number of extensions of the theory of central classes, see [Miy4].

As for genus theory, it is interesting to be able to study extensions L/K which are not necessarily Galois; for this Jaulent has given the following definition ([Ja2; Ja7] and [AJ]).

4.6 Definition. Let S be a finite set of noncomplex places of K. If L is an arbitrary finite extension of K, the central class field of L/K (relative to S) is by definition the abelian extension of L which, by class field theory, corresponds to the subgroup:

$$L^\times \,.\, {}_{\mathrm{N}}J_L \,.\, U_L^{S'}$$

of J_L, where S' is the set of places of L above those of S, where $\mathrm{N} := \mathrm{N}_{L/K}$, and where ${}_{\mathrm{N}}J_L$ is the kernel of N. □

4.6.1 Proposition. *Let L/K be a finite Galois extension with Galois group G. Then:*

$${}_{\mathrm{N}}J_L \,.\, U_L^{\mathrm{res}} = I_G J_L \,.\, U_L^{\mathrm{res}},$$

where I_G is the augmentation ideal of G.

Proof. We have, by considering the map which sends $\boldsymbol{y} =: (y_w)_w \in J_L$ to the pair formed by the ideal $\prod_{w \in Pl_{L,0}} \mathfrak{p}_w^{\mathrm{w}(y_w)}$ and the $r_1(L)$-uple $((-1)^{\mathrm{w}(y_w)})_{w \in Pl_{L,\infty}^{\mathrm{r}}}$, the canonical G-module isomorphism:

$$J_L/U_L^{\mathrm{res}} \simeq I_L \times \{\pm 1\}^{r_1(L)} \;;$$

it is easily checked, by considering separately the two factors of the right hand side which are induced G-modules, that:

$${}_{\nu}(J_L/U_L^{\mathrm{res}}) = I_G(J_L/U_L^{\mathrm{res}}),$$

where $\nu := \nu_{L/K}$ is the algebraic norm. Using the arithmetic norm with values in J_K/U_K^{res}, this can also be written ${}_{\mathrm{N}}(J_L/U_L^{\mathrm{res}}) = I_G(J_L/U_L^{\mathrm{res}})$ since the map:

$$j := j_{L/K} : I_K \times \{\pm 1\}^{r_1(K)} \longrightarrow I_L \times \{\pm 1\}^{r_1(L)},$$

is injective on the subgroup $\mathrm{N}(I_L) \times \mathrm{N}(\{\pm 1\}^{r_1(L)}) = \mathrm{N}(I_L) \times \{\pm 1\}^{r_1^{\mathrm{nc}}(K)}$, where $r_1^{\mathrm{nc}}(K)$ is the number of real places at infinity of K noncomplexified in L, and $\nu = j \circ \mathrm{N}$. Therefore we obtain, since $\mathrm{N}(J_L) \cap U_K^{\mathrm{res}} = \mathrm{N}(U_L^{\mathrm{res}})$:

$$_{\mathrm{N}}J_L \,.\, U_L^{\mathrm{res}} = I_G J_L \,.\, U_L^{\mathrm{res}},$$

and a fortiori:

$$_{\mathrm{N}}J_L \,.\, U_L^{S'} = I_G J_L \,.\, U_L^{S'}. \qquad \square$$

The given definition is thus legitimate and incidentally proves an interesting relation.

4.6.2 Definition (knot group). For any finite extension L/K, we put:

$$\mathcal{K} := \mathcal{K}(L/K) := K^\times \cap \mathrm{N}_{L/K}(J_L)/\mathrm{N}_{L/K}(L^\times).$$

This group is called the knot group corresponding to L/K. The number $|\mathcal{K}|$ is called the knot number. $\square$

We can now give the number $[C_{L/K}^S : L]$ of central classes; this computation will use the number of genera, and more precisely the formulas 4.2.2 giving $[H_{L/K}^S : L]$. The notations are those of 4.2.

4.7 Theorem (number of central classes). *If L is an arbitrary finite extension of K, the number of central classes is given by the following formula:*

$$\begin{aligned}[C_{L/K}^S : L] &= \frac{|\mathcal{C}\ell^S| \times \prod\limits_{v\in S} e_v^{\mathrm{ab}}\, f_v^{\mathrm{ab}} \times \prod\limits_{v\notin S} \breve{e}_v^{\mathrm{ab}}}{[L^{\mathrm{ab}} : K] \times (E^S : E^S \cap \mathrm{N}_{L/K}(U_L^{S'}) \cap \mathrm{N}_{L/K}(L^\times))} |\mathcal{K}| \\ &= [H_{L/K}^S : L] \times \frac{|\mathcal{K}|}{(E^S \cap \mathrm{N}_{L/K}(U_L^{S'}) : E^S \cap \mathrm{N}_{L/K}(U_L^{S'}) \cap \mathrm{N}_{L/K}(L^\times))}.\end{aligned}$$

If L/K is Galois, the formula becomes:

$$\begin{aligned}[C_{L/K}^S : L] &= \frac{|\mathcal{C}\ell^S| \times \prod\limits_{v\in S} e_v^{\mathrm{ab}}\, f_v^{\mathrm{ab}} \times \prod\limits_{v\notin S} e_v^{\mathrm{ab}}}{[L^{\mathrm{ab}} : K] \times (E^S : E^S \cap \mathrm{N}_{L/K}(L^\times))} |\mathcal{K}| \\ &= [H_{L/K}^S : L] \times \frac{|\mathcal{K}|}{(E^S \cap \mathrm{N}_{L/K}(J_L) : E^S \cap \mathrm{N}_{L/K}(L^\times))}.\end{aligned}$$

Proof. To simplify, we omit the usual embeddings, and we recall that $\mathrm{N} := \mathrm{N}_{L/K}$. Set:

$$\begin{aligned} N^{\mathrm{c}} &:= \{\boldsymbol{y} \in J_L,\ \mathrm{N}(\boldsymbol{y}) \in \mathrm{N}(U_L^{S'} L^\times)\} = L^\times \,.\, {}_{\mathrm{N}}J_L \,.\, U_L^{S'}, \\ N^{\mathrm{g}} &:= \{\boldsymbol{y} \in J_L,\ \mathrm{N}(\boldsymbol{y}) \in K^\times \,.\, \mathrm{N}(U_L^{S'})\}, \end{aligned}$$

which correspond respectively to $C_{L/K}^S$ and to $H_{L/K}^S$ on L (by 4.1.1 for the genus field). We thus have:

$$\mathrm{Gal}(C_{L/K}^S/H_{L/K}^S) \simeq N^{\mathrm{g}}/N^{\mathrm{c}}.$$

Consider the map:

$$f : N^{\mathrm{g}} \longrightarrow K^\times \cap \mathrm{N}(J_L)/(E^S \cap \mathrm{N}(U_L^{S'})) \,.\, \mathrm{N}(L^\times),$$

defined in the following way. For $\boldsymbol{y} \in N^{\mathrm{g}}$, we have $\mathrm{N}(\boldsymbol{y}) =: x\,\mathrm{N}(\boldsymbol{u})$, with $x \in K^\times$, $\boldsymbol{u} \in U_L^{S'}$ (we then have $x \in K^\times \cap \mathrm{N}(J_L)$); we denote by $f(\boldsymbol{y})$ the class of x modulo $(E^S \cap \mathrm{N}(U_L^{S'}))\,.\,\mathrm{N}(L^\times)$. This map is well defined since, if $\mathrm{N}(\boldsymbol{y}) =: x'\,\mathrm{N}(\boldsymbol{u}')$ with $x' \in K^\times$, $\boldsymbol{u}' \in U_L^{S'}$, we obtain $x\,x'^{-1} = \mathrm{N}(\boldsymbol{u}'\,\boldsymbol{u}^{-1})$, hence $x\,x'^{-1} \in E^S \cap \mathrm{N}(U_L^{S'})$.

Let $x \in K^\times$ be such that $x =: \mathrm{N}(\boldsymbol{y})$, $\boldsymbol{y} \in J_L$; we then have $\boldsymbol{y} \in N^{\mathrm{g}}$, showing that f is surjective.

The inclusion $N^{\mathrm{c}} \subseteq \mathrm{Ker}(f)$ being clear, let $\boldsymbol{y} \in \mathrm{Ker}(f)$; we have:

$$\mathrm{N}(\boldsymbol{y}) =: \varepsilon\,\mathrm{N}(y)\,\mathrm{N}(\boldsymbol{u}),$$

with $\varepsilon \in E^S \cap \mathrm{N}(U_L^{S'})$, $y \in L^\times$, $\boldsymbol{u} \in U_L^{S'}$. It follows that $\mathrm{N}(\boldsymbol{y}) \in K^\times\,.\,\mathrm{N}(U_L^{S'})$, hence $\boldsymbol{y} \in N^{\mathrm{c}}$. We thus have the isomorphism:

$$\begin{aligned}\mathrm{Gal}(C_{L/K}^S/H_{L/K}^S) &\simeq K^\times \cap \mathrm{N}(J_L)/(E^S \cap \mathrm{N}(U_L^{S'}))\,.\,\mathrm{N}(L^\times) \simeq \\ &(K^\times \cap \mathrm{N}(J_L)/\mathrm{N}(L^\times)) \,/\, (E^S \cap \mathrm{N}(U_L^{S'})/E^S \cap \mathrm{N}(U_L^{S'}) \cap \mathrm{N}(L^\times)).\end{aligned}$$

We can write:

$$[C_{L/K}^S : H_{L/K}^S] = \frac{(K^\times \cap \mathrm{N}(J_L) : \mathrm{N}(L^\times))}{(E^S \cap \mathrm{N}(U_L^{S'}) : E^S \cap \mathrm{N}(U_L^{S'}) \cap \mathrm{N}(L^\times))},$$

and the theorem follows using 4.2.2.

Finally, recall that in the Galois case we can replace $\mathrm{N}_{L/K}(U_L^{S'})$ by $\mathrm{N}_{L/K}(J_L)$, hence $E^S \cap \mathrm{N}_{L/K}(U_L^{S'}) \cap \mathrm{N}_{L/K}(L^\times)$ by $E^S \cap \mathrm{N}_{L/K}(L^\times)$. □

4.7.1 Remarks. (i) In the non-Galois case, $E^S \cap \mathrm{N}_{L/K}(U_L^{S'}) \cap \mathrm{N}_{L/K}(L^\times)$ is not necessarily equal to $E^S \cap \mathrm{N}_{L/K}(L^\times)$. For $K = \mathbb{Q}$, $L = \mathbb{Q}(\sqrt{-7 + 2\sqrt{17}})$ (see Exercises II.2.6.6, 4.2.5), and $S = \{\infty\}$, we have $E^{\mathrm{ord}} = \langle\,-1\,\rangle$ and we check that $-1 = \mathrm{N}_{L/K}\Big(\frac{3+\sqrt{17}}{2} \times \frac{1+\sqrt{-7+2\sqrt{17}}}{2}\Big) \in \mathrm{N}_{L/K}(L^\times)$, but we know that $-1 \notin \mathrm{N}_{L/K}(U_L^{\mathrm{ord}})$.

(ii) When L/K is Galois, the group $\mathcal{K}$ is a relatively accessible Galois invariant; this is not true anymore for $E^S \cap \mathrm{N}_{L/K}(J_L)/E^S \cap \mathrm{N}_{L/K}(L^\times)$. □

4.7.2 Examples (after [Go]). (i) (example due to Scholz). Consider (for $K = \mathbb{Q}$) the case of $L = \mathbb{Q}(\sqrt{13}, \sqrt{17})$, given in II.6.2.7, (iii), and for which $|\mathcal{K}| = 2$ by Razar's criterion. It is clear that L is the genus field of $M = \mathbb{Q}(\sqrt{221})$ in the ordinary sense; since $\mathcal{C}\ell_M^{\mathrm{ord}} \simeq \mathbb{Z}/2\mathbb{Z}$, we have $\mathcal{C}\ell_L^{\mathrm{ord}} = 1$ (invariant class formula in L/M, for instance), hence $C_{L/\mathbb{Q}}^{\mathrm{ord}} = H_{L/\mathbb{Q}}^{\mathrm{ord}} = L$, showing that the knot group of the units:

$$E^{\mathrm{ord}} \cap \mathrm{N}_{L/\mathbb{Q}}(J_L)/E^{\mathrm{ord}} \cap \mathrm{N}_{L/\mathbb{Q}}(L^\times),$$

has order 2.

(ii) (example due to Hasse). If $L = \mathbb{Q}(\sqrt{13}, \sqrt{-3})$, we check by Razar's criterion that the Hasse norm principle is also not true; but:

$$E^{\mathrm{ord}} \cap \mathrm{N}_{L/\mathbb{Q}}(J_L)/E^{\mathrm{ord}} \cap \mathrm{N}_{L/\mathbb{Q}}(L^\times) = 1,$$

for a trivial reason (-1 is not a local norm at the place at infinity), which yields $C_{L/\mathbb{Q}} \neq H_{L/\mathbb{Q}}$ (here, ordinary and restricted senses coincide). More precisely, $H_{L/\mathbb{Q}} = L$ and $C_{L/\mathbb{Q}} = H_M$ (the Hilbert class field of $M := \mathbb{Q}(\sqrt{-39})$ which is a cyclic extension of degree 4 of M). □

4.7.3 Corollary. *In the case L/K Galois, we have the formulas:*

$$[C^{\mathrm{res}}_{L/K} : L] = \frac{|\mathcal{C}\!\ell^{\mathrm{res}}| \times \prod_{v \in Pl_0} e_v^{\mathrm{ab}}}{[L^{\mathrm{ab}} : K] \times (E^{\mathrm{res}} : E^{\mathrm{res}} \cap \mathrm{N}_{L/K}(L^\times))}\,|\mathcal{K}|,$$

$$[C^{\mathrm{ord}}_{L/K} : L] = \frac{|\mathcal{C}\!\ell^{\mathrm{ord}}| \times 2^{r_1^{\mathrm{c}}} \times \prod_{v \in Pl_0} e_v^{\mathrm{ab}}}{[L^{\mathrm{ab}} : K] \times (E^{\mathrm{ord}} : E^{\mathrm{ord}} \cap \mathrm{N}_{L/K}(L^\times))}\,|\mathcal{K}|,$$

where r_1^{c} is the number of real places of K complexified in L/K. □

We give in the following proposition an example of the use of the number of central classes for the study of p-class groups in a p-extension.

4.7.4 Proposition (p-triviality criterion in a p-extension). *Let L/K be a p-extension with Galois group G, let S be a finite set of noncomplex places of K, and S' the set of places of L above those of S. Then $(\mathcal{C}\!\ell_L^{S'})_p = 1$ if and only if the following three conditions are satisfied:*

(i) $H^S{}_{(p)} \subseteq L$,

(ii) $(E^S : E^S \cap \mathrm{N}_{L/K}(J_L)) = \dfrac{\prod_{v \in S} e_v^{\mathrm{ab}} f_v^{\mathrm{ab}} \times \prod_{v \notin S} e_v^{\mathrm{ab}}}{[L^{\mathrm{ab}} : H^S{}_{(p)}]}$,

(iii) $|\mathcal{K}| = (E^S \cap \mathrm{N}_{L/K}(J_L) : E^S \cap \mathrm{N}_{L/K}(L^\times))$.

Proof. Since G is a p-group, we have $(\mathcal{C}\!\ell_L^{S'})_p = 1$ if and only if:

$$(\mathcal{C}\!\ell_L^{S'})_p = I_G(\mathcal{C}\!\ell_L^{S'})_p,$$

in other words $C^S_{L/K}{}_{(p)} = L$. Indeed, from [d, Se2, Ch. IX, § 1, Cor. to Th. 2] applied for instance to $\mathbb{F}_p[G]$, we deduce that there exists a power of I_G contained in $p\mathbb{Z}[G]$. We deduce the three conditions by using Theorem 4.7, and Formula 4.2.2, (ii) for the number of genera in the Galois case. □

Note. Since, by II.2.4.1, (ii), $H_0(G, \mathcal{C}\ell_L^{S'})$ and $H^0(G, (\mathcal{C}\ell_L^{S'})^*)$ have same order, we obtain $(\mathcal{C}\ell_L^{S'} : I_G(\mathcal{C}\ell_L^{S'}))_p = |(\mathcal{C}\ell_L^{S'})_p^{*G}|$, then $|C^S_{L/K(p)}| = |(\mathcal{C}\ell_L^{S'})_p^{*G}|$; thus we recover the result by using the classical fixed point theorem for finite p-groups.

4.7.5 Remark. Condition (iii) is empty when G is cyclic (Hasse principle), or when $\mathcal{K} = 1$ (see Razar's criterion); on the contrary it becomes nontrivial in the other cases so that, in practice, there does not exist an easy numerical criterion for the triviality of the p-class group in a noncyclic p-extension.

In the particular case $K = \mathbb{Q}$, $S = \emptyset$, condition (i) is empty, (ii) is easy to check, and condition (iii) is equivalent to $\mathcal{K} = 1$; this implies that for $K = \mathbb{Q}$ and in the restricted sense, the above problem is reduced to that of the Hasse principle. □

4.8 A Normic Criterion for p-Rationality or for p-Regularity (after [BGr]). Suppose $\mu_p \subset K$ and let L/K be a p-extension with Galois group G. We know that in this case the two notions of p-rationality and p-regularity coincide (see II.7.8 and III.4.2.2). Then using the condition (iii) of Theorem 3.5, we deduce that L is p-rational if and only if:

- p does not split in $L/\mathbb{Q}$,
- $(\mathcal{C}\ell_L^{Pl'_p\mathrm{res}})_p = 1$.

If p does not split, $\mathcal{K}_{L/K} = 1$ (apply II.6.2.7, (i) with $D_{w_0}(L/K) = G$ for the single place $w_0|v_0|p$ in $L/K/\mathbb{Q}$). Then, using the p-triviality criterion, the above is equivalent to:

- p does not split in $L/\mathbb{Q}$,
- $H^{Pl_p\mathrm{res}}(p) \subseteq L$,
- $(E^{Pl_p\mathrm{res}} : E^{Pl_p\mathrm{res}} \cap \mathrm{N}_{L/K}(J_L)) = \dfrac{e^{\mathrm{ab}}_{v_0} f^{\mathrm{ab}}_{v_0}}{[L^{\mathrm{ab}} : H^{Pl_p\mathrm{res}}(p)]} \times \prod_{v \nmid p} e_v^{\mathrm{ab}}$.

But if p is not split in $L/\mathbb{Q}$ and if $H^{Pl_p\mathrm{res}}(p)$ (the Pl_p-split Hilbert class field) is contained in L, we trivially have $H^{Pl_p\mathrm{res}}(p) = K$ by class field theory, and $[L^{\mathrm{ab}} : K] = e^{\mathrm{ab}}_{v_0} f^{\mathrm{ab}}_{v_0}$ since $D_{w_0}(L/K) = G$. Then the p-rationality (or p-regularity) of L is characterized, when K contains μ_p, in the following way.

4.8.1 Proposition. *Let L be a p-extension of $K \supset \mu_p$. Then L is p-rational (or p-regular) if and only if the following three conditions are satisfied:*

(i) p does not split in $L/\mathbb{Q}$,
(ii) $(\mathcal{C}\ell^{Pl_p\mathrm{res}})_p = 1$,
(iii) $(E^{Pl_p\mathrm{res}} : E^{Pl_p\mathrm{res}} \cap \mathrm{N}_{L/K}(J_L)) = \prod_{v \nmid p} e_v^{\mathrm{ab}}$,

where $e_v^{\mathrm{ab}} := e(L_v^{\mathrm{ab}}/K_v)$, noting that $E^{Pl_p\mathrm{res}} =: \langle\, \pi \,\rangle \oplus E^{\mathrm{res}}$ with $(\pi) = \mathfrak{p}_{v_0}^n$ for the least integer $n > 0$ such that $\mathfrak{p}_{v_0}^n$ is principal in the restricted sense. □

Since, by 3.4.3, (iii), L is p-rational if and only if:

- K is p-rational,

- L/K is p-primitively ramified,

we obtain, in the Kummer case, the following.

4.8.2 Proposition. *Let K be a p-rational field containing μ_p. A p-extension L/K is p-primitively ramified if and only if the following two conditions are satisfied:*

(i) p does not split in L/K,

(ii) $(E^{Pl_p \mathrm{res}} : E^{Pl_p \mathrm{res}} \cap \mathrm{N}_{L/K}(J_L)) = \prod_{v \nmid p} e_v^{\mathrm{ab}}$. □

We leave the context of p-rational or p-regular fields, and complete this subsection with the following splitting theorem. Let K be a number field together with sets of places T and S.

4.9 Theorem (from questions studied by Herz, Wyman, Gold, Cornell–Rosen, Oriat, Bond, ...). *Let L/K be a cyclic extension of degree d, with Galois group $G =: \langle\, s \,\rangle$. Let $H := H_T^S$ be the maximal T-ramified S-split abelian extension of K and $H' := H_{L,T'}^{S'}$ the analogous extension over L. We assume that $L \cap H = K$. Then $H'/L/K$ is split (i.e., H' is the direct compositum over K of L with an extension F).*

Proof. Set:

$$A := \mathrm{Gal}(H/K), \quad A' := \mathrm{Gal}(H'/L), \quad \Gamma := \mathrm{Gal}(H'/K).$$

Let U be an open subgroup of A and M the subfield of H fixed under U.

Since $L \cap H = K$, there exists a finite place v of K, unramified in H'/K, such that $s_M := \left(\frac{LM/K}{v}\right)$ is a lift of s in $\mathrm{Gal}(LM/M)$. Let $\sigma_M := \left(\frac{H'/K}{w'}\right)$, where $w'|w|v$ in $H'/L/K$; since $\left(\frac{L/K}{v}\right) = s$, w is unique and has residue degree equal to d in L/K. By construction (using the fact that H'/L is abelian when writing the Frobenius'), we have (see II.1.2.1.3):

$$\sigma_M^d = \left(\frac{H'/K}{w'}\right)^d = \left(\frac{H'/L}{w}\right) = \left(\frac{H'/L}{\mathfrak{p}_w}\right) = \left(\frac{H'/L}{(\mathfrak{p}_v)}\right),$$

and since $\left(\frac{H/K}{\mathfrak{p}_v}\right) \in U$, we have finally:

$$\sigma_M^d \in j(U),$$

where $j := j_{L/K}$ is the transfer map $A \to A'$ which is continuous (use the characterization of A and A' given in III.2.4). Taking a limit point σ of the σ_M in Γ, we obtain:

$$\sigma^d = 1.$$

The field F fixed under $\langle\, \sigma \,\rangle$ is a solution; it even contains H. □

It is easily checked that if we do not assume that $L \cap H = K$, then the subextension M' of H' fixed under $j(A)$ is split over K.

If $T = S = \emptyset$, we recover the Hilbert class field situation (without any limiting process since we can take $U = 1$).

For additional material (which can be generalized with ramification and decomposition), and for the relations with genus theory and the theory of central classes, see [CorR].

For other classical applications of the theory of central classes, we refer to [Fr3], where in particular the case of $K = \mathbb{Q}$ is treated in detail.

4.10 Exercise. Let L/K be Galois with Galois group G. Show that:

$$\mathcal{K} := K^\times \cap \mathrm{N}_{L/K}(J_L)/\mathrm{N}_{L/K}(L^\times) \simeq {}_{\mathrm{N}}C_L/\mathcal{C}\ell_L({}_{\mathrm{N}}J_L)$$

(Hint: if $x = \mathrm{N}(\boldsymbol{y})$ with evident notations, use the image of $\mathcal{C}\ell_L(\boldsymbol{y}) \in C_L$). Deduce that:

$$\mathcal{K} \simeq H^{-1}(G, C_L)/\mathrm{Im}(H^{-1}(G, J_L)).$$

Consider the formalism of II.3.2 and prove Proposition II.6.2.6. □

V. Cyclic Extensions with Prescribed Ramification

In this chapter we give an approach to the study of ramification in $\overline{K}^{\rm ab}{}_{[p^e]}/K$, the maximal pro-$p$-subextension of $\overline{K}^{\rm ab}/K$ with exponent p^e, in particular through the study of the ramification possibilities for cyclic extensions of degree p^e of K. We will apply these results to the case of the maximal tamely ramified abelian extension $H_{\rm ta}^{\rm res}/K$ whose structure is always complicated as soon as the invariants $\mathcal{C}\ell^{\rm res}$ or $E^{\rm res}$ are nontrivial. Concerning this, we will have to make an assumption on the group $(\mathcal{C}\ell^{\rm res})_p$ when $e \geq 2$, but the case $e = 1$ can be solved without any assumption.

Let $S = S_0 \cup S_\infty$ be a fixed finite set of noncomplex places of K. If we are given $e \geq 1$ and a finite set T disjoint from S_0, and formed of finite places of K, we may ask whether or not there exists a cyclic extension L of K of degree p^e which is S-split and such that, for example:

$$e_v(L/K) = p^e \text{ for all } v \in T, \quad e_v(L/K) = 1 \text{ for all } v \notin T,$$

such an extension being said to be T-totally ramified and S-split. With the method that we are going to describe, it would be possible to study any similar problem dealing with prescribed ramification indices.

In practice, the answer is contained, by class field theory, in the numerical study of the generalized S-class groups $\mathcal{C}\ell_{\mathfrak{m}}^S$ since, for $\mathfrak{m} = \prod_{v \in T} \mathfrak{p}_v^{m_v}$ sufficiently large, the p-ray class field $K(\mathfrak{m})^S(p)$ contains all solutions relative to e, S, and T. This is possible using PARI/KANT and the effective techniques of [CoDO], [j, Coh2]. We can first use the exact sequence (see III.1.4 for $t = \delta_\infty = \emptyset$):

$$1 \longrightarrow (E^S/E^S_{\mathfrak{m}})_p \longrightarrow \bigoplus_{v \in T} (U_v/U_v^{m_v})_p \longrightarrow \mathrm{Gal}(K(\mathfrak{m})^S(p)/H^S(p)) \longrightarrow 1,$$

which enables us to solve the problem *on the S-split p-Hilbert class field* (we will call this the weak form of the problem under study). Since this exact sequence does not involve the class group, it is not sufficient for the solution of the problem over K.

In any case, from the point of view of generalized class groups, the computations must be redone as soon as T is modified, and it does not seem possible to *characterize* the sets T which are suitable for the given problem.

We are going to construct a governing field (depending only on the data p, e, S) which will allow us to characterize solutions T by using simple arithmetical conditions; this method uses the Kummer duality of Chapter I, Section 6, which is the origin of the reflection theorem, and can be considered as another application of the reflection process.

The notion of governing field has been used by Stevenhagen in [Ste], where one can find a study of this type of problem based on the notion of group extension, which is therefore of a cohomological nature [e, Ko3, Ch. 3, § 3]. Similarly, in [Neu3], Neukirch treats in depth the embedding problem whose obstructions include the use of such governing fields. On the contrary, the method that we suggest is elementary thanks to a precise use of class field theory as developed in the preceding chapters. However, although it gives a good description of these extensions (particularly adapted to the use of Čebotarev's theorem), it does not construct them, and for that it is necessary to use [j, Coh2], [DaP], [Fi] among others.

Before that, and to show what a naïve approach can give, let us start by a numerical example which can also be used as an illustration of the computational techniques in ray class fields.

§1 Study of an Example

We take $K = \mathbb{Q}(\sqrt{10})$ and $p = 2$. It is easily checked that:

$$E^{\text{ord}} = \langle -1,\ \varepsilon \rangle,\ E^{\text{res}} = \langle \varepsilon^2 \rangle,$$

where $\varepsilon = 3 + \sqrt{10}$, and that by III.1.1.5 $\mathcal{C}\ell^{\text{ord}} = \mathcal{C}\ell^{\text{res}}$, of order 2, is generated by the class of a prime ideal $\mathfrak{l}_3$ above 3. Here, genus theory (see IV.4.2.9) easily gives:

$$H^{\text{ord}} = H^{\text{res}} =: H = \mathbb{Q}(\sqrt{2},\ \sqrt{5}).$$

We consider the set T formed by the prime ideals (7) and $\mathfrak{l} := (9 + 2\sqrt{10})$ (above 41) of K, with respective residue degrees 2 and 1 in $K/\mathbb{Q}$; we must thus consider $\mathfrak{m} = (7)\,\mathfrak{l}$ (tame ramification). We will mainly use $S = Pl_\infty^{\text{r}}$ (formed by the two places at infinity of K) and $S = \emptyset$.

We must first determine the groups:

$$E_{(7)}^{\text{ord}},\ E_{(7)}^{\text{res}},\ E_{\mathfrak{l}}^{\text{ord}},\ E_{\mathfrak{l}}^{\text{res}},\ E_{\mathfrak{m}}^{\text{ord}},\ E_{\mathfrak{m}}^{\text{res}},$$

by noting that $|F_{(7)}^\times| = 3 \times 16$ and $|F_{\mathfrak{l}}^\times| = 5 \times 8$, giving here:

$$\bigoplus_{v \in T} (F_v^\times)_2 \simeq \mathbb{Z}/16\mathbb{Z} \oplus \mathbb{Z}/8\mathbb{Z}.$$

We have $\varepsilon^8 \equiv -1 \bmod (7)$, which shows that $i_{(7)}(\varepsilon)$ is a generator of the group $(F_{(7)}^\times)_2$; we also note that in E^{ord} we have:

$$-\varepsilon^8 \equiv 1 \bmod (7).$$

Furthermore, we have $\varepsilon^8 \equiv 37 \bmod \mathfrak{l}$ and $\varepsilon^{20} \equiv -1 \bmod \mathfrak{l}$, which shows that $i_{\mathfrak{l}}(\varepsilon)$ is a generator of $F_{\mathfrak{l}}^\times$ (or that $i_{\mathfrak{l}}(\varepsilon^5)$ is a generator of $(F_{\mathfrak{l}}^\times)_2$) and that in E^{ord} we have:

$$-\varepsilon^{20} \equiv 1 \bmod \mathfrak{l}.$$

Thus, we see that:

$$E_{(7)}^{\text{ord}} = \langle -\varepsilon^8 \rangle,\ E_{\mathfrak{l}}^{\text{ord}} = \langle -\varepsilon^{20} \rangle,\ E_{\mathfrak{m}}^{\text{ord}} = \langle\ \varepsilon^{80} \rangle,$$

$$E_{(7)}^{\text{res}} = \langle\ \varepsilon^{16} \rangle,\ E_{\mathfrak{l}}^{\text{res}} = \langle\ \varepsilon^{40} \rangle,\ E_{\mathfrak{m}}^{\text{res}} = \langle\ \varepsilon^{80} \rangle.$$

For explicit computations, set:

$$\alpha := i_{(7)}(\varepsilon^5),\ \text{of order 16 in } F_{(7)}^\times,$$
$$\beta := i_{\mathfrak{l}}(\varepsilon^5),\ \text{of order 8 in } F_{\mathfrak{l}}^\times.$$

By the embeddings $i_{(7)}$ and $i_{\mathfrak{l}}$ in the residue fields (noting that $\alpha^8 = -1,\ \beta^4 = -1$), we have, for the 2-Sylow subgroups:

$$i_{(7)}(E^{\text{ord}})_2 = \langle -1,\ \alpha \rangle = \langle \alpha \rangle,\ \ i_{(7)}(E^{\text{res}})_2 = \langle \alpha^2 \rangle,$$
$$i_{\mathfrak{l}}(E^{\text{ord}})_2 = \langle -1,\ \beta \rangle = \langle \beta \rangle,\ \ i_{\mathfrak{l}}(E^{\text{res}})_2 = \langle \beta^2 \rangle.$$

We thus have:

$$(F_{(7)}^\times)_2 \oplus (F_{\mathfrak{l}}^\times)_2 \simeq \langle (\alpha, 1),\ (1, \beta) \rangle$$

and, under this isomorphism, we have:

$$i_T(E^{\text{ord}})_2 = \langle (-1,-1),\ (\alpha,\beta) \rangle,\ \ i_T(E^{\text{res}})_2 = \langle (\alpha,\beta)^2 \rangle.$$

Therefore, with the usual formula of I.4.5.6 we obtain:

- $\mathrm{Gal}(K_{(7)}{}^{\text{ord}}(2)/H) \simeq \langle \alpha \rangle / \langle \alpha \rangle = 1$,
- $\mathrm{Gal}(K_{(7)}{}^{\text{res}}(2)/H) \simeq \langle \alpha \rangle / \langle \alpha^2 \rangle \simeq \mathbb{Z}/2\mathbb{Z}$,
- $\mathrm{Gal}(K_{(\mathfrak{l})}{}^{\text{ord}}(2)/H) \simeq \langle \beta \rangle / \langle \beta \rangle = 1$,
- $\mathrm{Gal}(K_{(\mathfrak{l})}{}^{\text{res}}(2)/H) \simeq \langle \beta \rangle / \langle \beta^2 \rangle \simeq \mathbb{Z}/2\mathbb{Z}$,
- $\mathrm{Gal}(K_{(\mathfrak{m})}{}^{\text{ord}}(2)/H) \simeq \langle (\alpha,1),\ (1,\beta) \rangle / \langle (-1,-1),\ (\alpha,\beta) \rangle$
  $\simeq \langle (1,\beta) \rangle / \langle (-1,-1),\ (\alpha,\beta) \rangle \cap \langle (1,\beta) \rangle \simeq \mathbb{Z}/4\mathbb{Z}$,

since $(1,\beta)^4 = (1,-1) = (-1,-1)(\alpha,\beta)^8$ and $(1,\beta)^2 \notin \langle (-1,-1),\ (\alpha,\beta) \rangle$,

- $\mathrm{Gal}(K_{(\mathfrak{m})}{}^{\text{res}}(2)/H) \simeq \langle (\alpha,1),\ (1,\beta) \rangle / \langle (\alpha,\beta)^2 \rangle$
  $= \langle (1,\beta),\ (\alpha,\beta) \rangle / \langle (\alpha,\beta)^2 \rangle$;

to compute this last group, denote by q the map sending an element to its image in the quotient group; then $q(\alpha,\beta)$ has order 2, $q(1,\beta)$ has order 8, and we clearly have $\langle q(1,\beta) \rangle \cap \langle q(\alpha,\beta) \rangle = 1$, hence:

- $\mathrm{Gal}(K_{(\mathfrak{m})}{}^{\text{res}}(2)/H) = \langle q(1,\beta) \rangle \oplus \langle q(\alpha,\beta) \rangle \simeq \mathbb{Z}/8\mathbb{Z} \times \mathbb{Z}/2\mathbb{Z}$.

In $K_{(\mathfrak{m})}{}^{\text{res}}(2)/H$, the inertia groups of (7) and $\mathfrak{l}$ are respectively equal to:

$$\langle\, q(\alpha,1)\,\rangle \simeq \mathbb{Z}/8\mathbb{Z}, \text{ and } \langle\, q(1,\beta)\,\rangle \simeq \mathbb{Z}/8\mathbb{Z},$$

and by II.5.2.2, (i), they fix respectively $K_{(7)}^{\mathrm{res}}(2)$ and $K_{(\mathfrak{l})}^{\mathrm{res}}(2)$.

Since $\mathrm{Gal}(K_{(\mathfrak{m})}^{\mathrm{res}}(2)/K_{(\mathfrak{m})}^{\mathrm{ord}}(2))$ has order 4, and is noncyclic since it is generated by the decomposition groups of the two places at infinity of K, we have the following diagram.

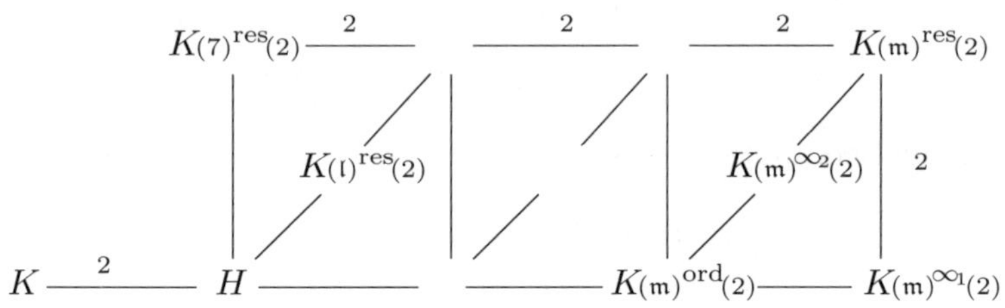

Fig. 1.1

This being said, it is not easy to deduce the structure of $\mathrm{Gal}(K_{(\mathfrak{m})}^{\mathrm{res}}(2)/K)$ and in particular to say if it is possible to descend to K all or part of $K_{(\mathfrak{m})}^{\mathrm{res}}(2)/H$. For this, we would need to study in detail the structure of the 2-Sylow subgroup of $\mathcal{C}\ell_{\mathfrak{m}}^{\mathrm{res}}$, which has order 32, and this would be rather painful although it could be done, but we are looking for a systematic method. Nonetheless, the above study shows at least that there does not exist any $\{(7)\}$-totally ramified Pl_∞^{r}-split quadratic extension of K, or a $\{\mathfrak{l}\}$-totally ramified Pl_∞^{r}-split quadratic extension of K.

However, we will see at the end of this chapter (Example 3.1) that $K_{(7)}^{\mathrm{res}}(2)/H$ and $K_{(\mathfrak{l})}^{\mathrm{res}}(2)/H$ can be descended to K into nontotally real $\{(7)\}$ and $\{\mathfrak{l}\}$-totally ramified quadratic extensions. We will also see that $K_{(\mathfrak{m})}^{\mathrm{ord}}(2)/H$ can also be descended, in other words that there exists a cyclic extension of degree 4 of K which is $\{(7), \mathfrak{l}\}$-totally ramified and totally real. The reader can verify this claim by showing that $K_{(\mathfrak{m})}^{\mathrm{ord}}(2)$ contains the biquadratic field:

$$K\Big(\sqrt{5},\ \sqrt{7(9+2\sqrt{10})}\,\Big),$$

proving that:

$$\mathrm{Gal}(K_{(\mathfrak{m})}^{\mathrm{ord}}(2)/K) \simeq \mathbb{Z}/4\mathbb{Z} \times \mathbb{Z}/2\mathbb{Z}.$$

§2 Construction of a Governing Field

Fix a prime power p^e, $e \geq 1$, and a finite set $S = S_0 \cup S_\infty$ of noncomplex places of K, the idea being that we are going to vary $T = T_p \cup T_{\mathrm{ta}}$ as a nonempty finite set disjoint from S_0 such that the condition $\mathrm{N}\mathfrak{p}_v \equiv 1 \bmod (p^e)$ for all $v \in T_{\mathrm{ta}}$ (trivially necessary to have $e_v(L/K) = p^e$) is satisfied (see II.1.3.3).

We recall that H_T^S is the maximal T-ramified S-split abelian extension of K and H^S the S-split Hilbert class field.

2.1 Notations and Basic Tools. (i) Let $\mathfrak{m} = \prod_{v\in T} \mathfrak{p}_v^{m_v}$ be a sufficiently large mudulus so that we have $U_v^{m_v} \subseteq (U_v)^{p^e}$ for all $v \in T$ ($m_v = 1$ is sufficient when $v \in T_{\rm ta}$). We check that for all S, $K(\mathfrak{m})^S(p)$ contains the maximal subextension of H_T^S/H^S with exponent dividing p^e, which is clear since this subextension is fixed under the image of $\bigoplus_{v\in T}(U_v)^{p^e}$ under the reciprocity map. We set:

$$A := \mathrm{Gal}(H_T^S/K), \quad B := \mathrm{Gal}(H_T^S/H^S).$$

(ii) Consider the exact sequence (see III.1.1 for $t = \delta_\infty = \emptyset$):

$$1 \longrightarrow E^S/E^S_{\mathfrak{m}} \xrightarrow{\ i_T\ } \bigoplus_{v\in T} U_v/U_v^{m_v} \xrightarrow{\ \rho\ } \mathrm{Gal}(K(\mathfrak{m})^S/H^S) \longrightarrow 1,$$

where ρ denotes the reciprocity map, which we write in the form:

$$1 \longrightarrow i_T(E^S) \longrightarrow \bigoplus_{v\in T} U_v/U_v^{m_v} \xrightarrow{\ \rho\ } \mathrm{Gal}(K(\mathfrak{m})^S/H^S) \longrightarrow 1,$$

where i_T denotes, by abuse of notation, the map sending $\varepsilon \in E^S$ to the family $(i_v(\varepsilon) \bmod U_v^{m_v})_{v\in T}$. We deduce the exact sequences (for $e \geq 1$):

$$1 \longrightarrow i_{T,e}(E^S) \longrightarrow \bigoplus_{v\in T} U_v/(U_v)^{p^e} \xrightarrow{\ \rho_e\ } B/B^{p^e} \longrightarrow 1,$$

where $i_{T,e}$ is the composite map:

$$i_{T,e} : E^S \xrightarrow{\ i_T\ } \bigoplus_{v\in T} U_v/U_v^{m_v} \longrightarrow \bigoplus_{v\in T} U_v/(U_v)^{p^e}.$$

We have $i_{T,e} =: (i_{v,e})_{v\in T}$ with an evident meaning.

(iii) If G is a finite abelian group, we denote by G^* the dual group $\mathrm{Hom}(G, \mathbb{C}_1^\times)$ (see I.5.7), and for any group homomorphism $h : G \longrightarrow G'$, we denote the dual map by $h^* : G'^* \longrightarrow G^*$, defined by $h^*(\chi')(\sigma) := \chi'(h(\sigma))$ for all $\chi' \in G'^*$ and all $\sigma \in G$.

2.2 Kummer Duality (review of (Ch. I; § 6, (a))). Let $K_e := K(\mu_{p^e})$ and let $Q_e := K_e(\sqrt[p^e]{X}\,)$, where X is a subgroup of $K_e^\times$ containing $K_e^{\times p^e}$ and such that $W := X/K_e^{\times p^e}$ is finite. In what follows, X will be of the form $YK_e^{\times p^e}$, where Y is a subgroup of $K^\times$ containing $K^{\times p^e}$; because of this, one should not mistake the quotients $W = Y/Y \cap K_e^{\times p^e}$ (the radical for Q_e/K_e) and $Y/K^{\times p^e}$, because of the possibility of the global exceptional case, for $p = 2$ and $e \geq 2$, studied in II.6.3.2.

We set $G_e := \mathrm{Gal}(Q_e/K_e)$. Then the map $G_e \longrightarrow W^*$, sending $\sigma \in G_e$ to $\chi_\sigma \in W^*$ defined by:

$$\chi_\sigma(\overline{\alpha}) := (\sqrt[p^e]{\overline{\alpha}}\,)^{\sigma-1} \quad \text{for all } \overline{\alpha} \in W,$$

is a canonical isomorphism (see I.6.2.2). The inverse map is defined as follows. If $\chi \in W^*$, there exists $\overline{\beta} \in W$ such that $W = \langle \overline{\beta} \rangle \operatorname{Ker}(\chi)$ and the image of χ is the unique generator σ_χ of $\operatorname{Gal}(Q_e/K_e(\sqrt[p^e]{\operatorname{Ker}(\chi)}\,))$ such that:

$$\left(\sqrt[p^e]{\overline{\beta}}\,\right)^{\sigma_\chi - 1} = \chi(\overline{\beta}).$$

a) Solution to the Cyclic Case of Degree p

In this subsection, we solve the case $e = 1$, which represents a special step for the general case, and which is in practice the most commonly used; in addition, this case has a complete solution, without any special assumption. We use the notations of 2.1, (i), where T is nonempty.

2.3 THE FUNDAMENTAL DIAGRAM. Let M/H^S be the (p-elementary) subextension of H^S_T/H^S fixed under $C := B \cap A^p$, where $A := \operatorname{Gal}(H^S_T/K)$. We have the following diagram:

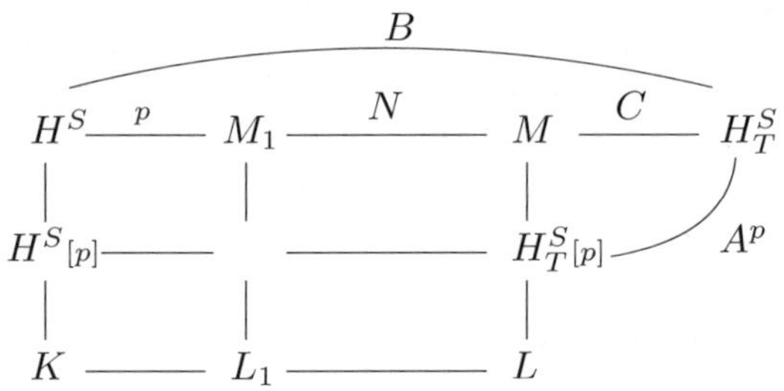

Fig. 1.2

in which $H^S_T[p]$ is the subfield of H^S_T fixed under A^p, $H^S[p] := H^S \cap H^S_T[p]$ is fixed under $B\,A^p$; hence M is the direct compositum of H^S with $H^S_T[p]$ over $H^S[p]$, and since $\operatorname{Gal}(H^S_T[p]/K)$ is an $\mathbb{F}_p$-vector space, there exists $L \subseteq H^S_T[p]$ such that $H^S_T[p]$ is the direct compositum of L with $H^S[p]$, hence such that M is the direct compositum of L with H^S over K; note that $H^S_T[p]/K$ (resp. $H^S[p]/K$) is indeed the maximal p-elementary subextension of H^S_T/K (resp. of H^S/K).

In the diagram, we have implicitly assumed that $H^S_T[p]$ is different from $H^S[p]$, otherwise the problem is trivially impossible (we will see in 2.4.3 what this case means).

2.3.1 Lemma. *A necessary and sufficient condition for the existence of a T-totally ramified S-split cyclic extension L_1/K of degree p, is that there exists a hyperplane N of B/C such that:*

$$B/C = N + I_v C/C \text{ for all } v \in T,$$

where $I_v \subseteq B$ is the inertia group of v in H_T^S/K. [1]

Proof. If L_1 exists, it is clear that $N := \mathrm{Gal}(M/M_1)$, where $M_1 := L_1 H^S$, is a hyperplane of B/C since L_1 is linearly disjoint from H^S/K for $T \neq \emptyset$, hence $[M_1 : H^S] = p$; furthermore, for each $v \in T$, $I_v C/C$ is the inertia group of v in M/H^S, and this group is not contained in N since L_1/K, hence also M_1/H^S, is totally ramified at v. It follows that $N + I_v C/C = B/C$.

Conversely, if there exists a hyperplane N of B/C such that $B/C = N + I_v C/C$ for all $v \in T$, then the subextension M_1/H^S fixed under N must be T-totally ramified since $I_v C/C$ is the inertia group of v in M/H^S and is not contained in N.

Since M is the direct compositum of H^S with L over K, M_1/H^S can be descended to an extension $L_1 \subseteq L$ of K, cyclic of degree p, and T-totally ramified since M_1/L_1 is unramified (L_1/K is S-split by construction). □

We are going to translate 2.3.1 in terms of numerical invariants of the field K; for this, we prove a fundamental exact sequence which is a reformulation of the computations that we performed to prove I.4.5, (ii). Let us consider $Y_T^S := \{\alpha \in K_T^{\times p} K_{T,\Delta_\infty}^\times,\ (\alpha) = \mathfrak{a}^p\, \mathfrak{a}_{S_0},\ \mathfrak{a} \in I_T,\ \mathfrak{a}_{S_0} \in \langle\, S_0 \,\rangle\}$, with $\Delta_\infty := Pl_\infty^{\mathrm{r}} \backslash S_\infty$, and the embedding $i_{T,1}$ defined in 2.1, (ii).

2.3.2 Lemma. *We have the exact sequence:*

$$1 \longrightarrow i_{T,1}(Y_T^S) \longrightarrow \bigoplus_{v \in T} U_v/(U_v)^p \xrightarrow{\ \widetilde{\rho}_1\ } B/C \longrightarrow 1,$$

where $\widetilde{\rho}_1$ is the composed (surjective) map:

$$\bigoplus_{v \in T} U_v/(U_v)^p \xrightarrow{\ \rho_1\ } B/B^p \longrightarrow B/C.$$

Proof. Indeed, with the condition of 2.1, (i) for the choice of $\mathfrak{m}$, we have the following isomorphism of class field theory:

$$A/A^p \simeq \mathcal{C}\ell_{\mathfrak{m}}^S/(\mathcal{C}\ell_{\mathfrak{m}}^S)^p \simeq I_T/P_{T,\mathfrak{m},\mathrm{pos}}\langle\, S \,\rangle I_T^p,$$

in which the image of B is given by that of $P_{T,\mathrm{pos}}\langle\, S \,\rangle$ (see II.5.1.4), thus giving:

$$\begin{aligned} B/C &\simeq P_{T,\mathrm{pos}}\langle\, S \,\rangle P_{T,\mathfrak{m},\mathrm{pos}}\langle\, S \,\rangle I_T^p / P_{T,\mathfrak{m},\mathrm{pos}}\langle\, S \,\rangle I_T^p \\ &\simeq P_{T,\Delta_\infty}\,.\langle\, S_0 \,\rangle I_T^p / P_{T,\mathfrak{m},\Delta_\infty}\,.\langle\, S_0 \,\rangle I_T^p. \end{aligned}$$

Furthermore, we have the exact sequence (see the proof of I.4.5, (ii)):

$$\begin{aligned} 1 \longrightarrow Y_T^S/Y_{T,\mathfrak{m}}^S \longrightarrow K_T^{\times p} K_{T,\Delta_\infty}^\times / K_T^{\times p} K_{T,\mathfrak{m},\Delta_\infty}^\times \longrightarrow \\ P_{T,\Delta_\infty}\,.\langle\, S_0 \,\rangle I_T^p / P_{T,\mathfrak{m},\Delta_\infty}\,.\langle\, S_0 \,\rangle I_T^p \longrightarrow 1, \end{aligned}$$

[1] The sum is direct for the tame places, but not necessarily for the wild places since their inertia group may not be cyclic.

where:

$$Y^S_{T,\mathfrak{m}} := Y^S_T \cap (K_T^{\times p} K^\times_{T,\mathfrak{m},\Delta_\infty}),$$

in which we replace $K_T^{\times p} K^\times_{T,\Delta_\infty} / K_T^{\times p} K^\times_{T,\mathfrak{m},\Delta_\infty}$ by:

$$\bigoplus_{v\in T} U_v/(U_v)^p U_v^{m_v} = \bigoplus_{v\in T} U_v/(U_v)^p$$

since $U_v^{m_v} \subseteq (U_v)^p$ by assumption, and $Y^S_T/Y^S_{T,\mathfrak{m}}$ by $i_{T,1}(Y^S_T)$. □

2.3.3 Remarks. (i) For the sequel, it is important to note that $K^{\times p} Y^S_T$ does not depend on T since we have (using I.5.1.2 for an inclusion):

$$K^{\times p} Y^S_T = Y^S := \{\alpha \in K^{\times p} K^\times_{\Delta_\infty},\ (\alpha) = \mathfrak{a}^p\, \mathfrak{a}_{S_0},\ \mathfrak{a} \in I,\ \mathfrak{a}_{S_0} \in \langle\, S_0 \,\rangle\}.$$

(ii) Note that Y^S contains the group E^S of S-units and that we have the exact sequence:

$$1 \longrightarrow E^S K^{\times p}/K^{\times p} \longrightarrow Y^S/K^{\times p} \longrightarrow {}_p\mathcal{C}\ell^S/c\ell^S(P) \longrightarrow 1$$

(send $\alpha \in K^{\times p} K^\times_{\Delta_\infty}$, such that $(\alpha) = \mathfrak{a}^p\, \mathfrak{a}_{S_0}$, to the class of $\mathfrak{a}$ modulo $c\ell^S(P)$). It shows that:

$$Y^S = E^S K^{\times p}$$

if and only if ${}_p\mathcal{C}\ell^S = c\ell^S(P)$. For $p \neq 2$ we have ${}_p\mathcal{C}\ell^S = c\ell^S(P)$ if and only if $(\mathcal{C}\ell^{S_0\,\mathrm{ord}})_p = 1$, and for $p = 2$, ${}_2\mathcal{C}\ell^S = c\ell^S(P)$ if and only if:

$${}_2\mathcal{C}\ell^S = P/P_{\Delta_\infty}(E^{S_0\,\mathrm{ord}}) \simeq K^\times/K^\times_{\Delta_\infty} E^{S_0\,\mathrm{ord}},$$

which is equivalent to:

$$|{}_2\mathcal{C}\ell^S| = \frac{2^{|\Delta_\infty|}}{|\mathrm{sgn}_{\Delta_\infty}(E^{S_0\,\mathrm{ord}})|}.$$

Note that for $p = 2$, the condition $(\mathcal{C}\ell^{S_0\,\mathrm{ord}})_p = 1$ is only sufficient. □

2.3.4 Lemma. *There exists a hyperplane N of B/C such that $B/C = N + I_v C/C$ for all $v \in T$, if and only if there exist nontrivial characters $\varphi_{1,v} \in (U_v/(U_v)^p)^*$, for $v \in T$, such that $\varphi_1 := \prod_{v\in T} \varphi_{1,v}$ is trivial on $i_{T,1}(Y^S_T)$.*

Proof. We will use the dual exact sequence to that of 2.3.2:

$$1 \longrightarrow (B/C)^* \xrightarrow{\ \widetilde{\rho}_1^*\ } \Big(\bigoplus_{v\in T} U_v/(U_v)^p\Big)^* \longrightarrow (i_{T,1}(Y^S_T))^* \longrightarrow 1.$$

If we note, by the characterization of inertia groups, that:

$$\widetilde{\rho}_1(U_v/(U_v)^p) = I_v C/C \text{ for all } v \in T,$$

we have the following equivalent statements:

(i) there exists a hyperplane N of B/C such that:

$$B/C = N + I_v C/C \text{ for all } v \in T,$$

hence such that:

$$B/C = N + \widetilde{\rho}_1(U_v/(U_v)^p) \text{ for all } v \in T\ ;$$

(ii) there exists $\psi_1 \in (B/C)^*$ such that:

$$\psi_1(\widetilde{\rho}_1(U_v/(U_v)^p)) \neq 1 \text{ for all } v \in T,$$

hence such that, by duality:

$$\widetilde{\rho}_1^*(\psi_1)(U_v/(U_v)^p) \neq 1 \text{ for all } v \in T\ ;$$

(iii) there exists $\varphi_1 =: \prod_{v \in T} \varphi_{1,v}$, with $\varphi_{1,v} \in (U_v/(U_v)^p)^*$, such that:

$$\varphi_1 \in \mathrm{Im}(\widetilde{\rho}_1^*), \text{ and } \varphi_{1,v} \neq 1 \text{ for all } v \in T\ ;$$

(iv) there exists $\varphi_1 =: \prod_{v \in T} \varphi_{1,v}$, $\varphi_{1,v} \neq 1$ for all $v \in T$, such that:

$$i_{T,1}(Y_T^S) \subseteq \mathrm{Ker}(\varphi_1)\ ;$$

in other words:

(v) there exist $\varphi_{1,v} \in (U_v/(U_v)^p)^*$, $\varphi_{1,v} \neq 1$ for all $v \in T$, such that:

$$\prod_{v \in T} \varphi_{1,v}(y_v) = 1$$

for all $(y_v)_{v \in T} := i_{T,1}(y) = (i_{v,1}(y))_{v \in T}$ such that $y \in Y_T^S$. □

This finishes the direct part of our study.

We now consider the field:

$$Q_{1(S)} := K_1(\sqrt[p]{Y^S}\,), \text{ where } K_1 := K(\mu_p).$$

Its radical is therefore $Y^S K_1^{\times p}/K_1^{\times p} \simeq Y^S/K^{\times p}$ since for pth powers the global exceptional case does not happen. Kummer theory shows that $Q_{1(S)}/K_1$ is $Pl_p \cup \langle S_0 \rangle$-ramified and Δ_∞-split (see I.6.3).

Then, let T be given, nonempty, disjoint from S_0, satisfying $\mathrm{N}\mathfrak{p}_v \equiv 1 \bmod (p)$ for all $v \in T_{\mathrm{ta}}$. By 2.3.3, (i), we also have $Q_{1(S)} = K_1\Big(\sqrt[p]{Y_T^S}\,\Big)$. For each place $v \in T$, consider:

$$V_{v,1}^S := \{y \in Y^S,\ i_v(y) \in K_v^{\times p}\} \text{ and } \delta_{1,v}^S := \mathrm{Gal}\Big(Q_{1(S)}\Big/K_1\Big(\sqrt[p]{V_{v,1}^S}\,\Big)\Big)$$

(noting that we have $V_{v,1}^S = K^{\times p}\{y \in Y_T^S,\ i_{v,1}(y) = 1\}$). It is clear that $K_1\left(\sqrt[p]{V_{v,1}^S}\right)$ is contained in the decomposition subfield of any place v_1 of K_1 above v in $Q_1(S)/K_1$; but since there is no local exceptional case either (i.e., $K_v^\times \cap K_{1,v_1}^{\times p} = K_v^{\times p}$), we have equality, hence $\delta_{1,v}^S$ is exactly the decomposition group of an arbitrary place of K_1 above v (this group only depends on v). This remark on the decomposition groups is mainly interesting for the wild places since, by assumption, the tame places are totally split in K_1/K and the result is trivial. To determine in practice the groups $V_{v,1}^S$, it is sufficient to use Kummer theory for the case $e = 1$. Finally, we have:

$$\delta_{1,v}^S \simeq Y^S/V_{1,v}^S.$$

2.3.5 Lemma. *The condition:*

(v) there exist $\varphi_{1,v} \in (U_v/(U_v)^p)^*$, $\varphi_{1,v} \neq 1$ for all $v \in T$, such that $\prod_{v\in T} \varphi_{1,v}(y_v) = 1$ for all $(y_v)_{v\in T} := i_{T,1}(y) = (i_{v,1}(y))_{v\in T}$ such that $y \in Y_T^S$, in the proof of Lemma 2.3.4, is equivalent to:

(v′) there exist elements $\sigma_{1,v} \in \delta_{1,v}^S$, $v \in T$, $\sigma_{1,v}$ of order p (in other words nontrivial) when $\delta_{1,v}^S \simeq U_v/(U_v)^p$, such that $\prod_{v\in T} \sigma_{1,v} = 1$.

Proof. For this, for each $v \in T$ we use the exact sequence:

$$1 \longrightarrow V_{v,1}^S/K^{\times p} \longrightarrow Y_T^S/K_T^{\times p} \xrightarrow{i_{v,1}} i_{v,1}(Y_T^S) \subseteq U_v/(U_v)^p \longrightarrow 1,$$

and the dual exact sequence:

$$1 \longrightarrow (i_{v,1}(Y_T^S))^* \longrightarrow (Y_T^S/K_T^{\times p})^* \longrightarrow (V_{v,1}^S/K^{\times p})^* \longrightarrow 1.$$

(v) $\Rightarrow$ (v′): we send each $\varphi_{1,v}$ to its restriction to $i_{v,1}(Y_T^S)$, whose image under $i_{v,1}^*$ yields a character $\chi_{1,v} \in (Y^S/K^{\times p})^* \simeq (Y_T^S/K_T^{\times p})^*$ defined by:

$$\chi_{1,v}(y) := \varphi_{1,v}(i_{v,1}(y)) \text{ for all } y \in Y_T^S.$$

By Kummer duality, to $\chi_{1,v}$ corresponds $\sigma_{1,v} \in \delta_{1,v}^S$ since $\mathrm{Ker}(\chi_{1,v})$ contains $V_{v,1}^S/K^{\times p}$ (by the above exact sequence). We thus have, by our assumption on φ_1:

$$\prod_{v\in T} \chi_{1,v}(y) = \prod_{v\in T} \varphi_{1,v}(i_{v,1}(y)) = 1 \text{ for all } y \in Y_T^S\ ;$$

this can be written $\prod_{v\in T} \chi_{1,v} = 1$, hence $\prod_{v\in T} \sigma_{1,v} = 1$.

When $\delta_{1,v}^S \simeq U_v/(U_v)^p$, the above exact sequence shows that $i_{v,1}(Y_T^S) = U_v/(U_v)^p$, in which case $\sigma_{1,v} = 1$ would mean that $\chi_{1,v} = 1$, hence that $\varphi_{1,v} = 1$, a contradiction.

(v′) $\Rightarrow$ (v): to each $\sigma_{1,v} \in \delta_{1,v}^S$ corresponds:

$$\chi_{1,v} \in (Y^S/K^{\times p})^* \simeq (Y_T^S/K_T^{\times p})^*,$$

trivial on $V_{v,1}^S/K^{\times p}$, and the relation $\prod_{v\in T} \sigma_{1,v} = 1$ yields $\prod_{v\in T} \chi_{1,v} = 1$. Since $\chi_{1,v}$ is trivial on $V_{v,1}^S/K^{\times p}$, it comes from a character of $i_{v,1}(Y_T^S) \subseteq U_v/(U_v)^p$ which can be extended to a character $\varphi_{1,v} \in (U_v/(U_v)^p)^*$ which we can choose to be nontrivial if $i_{v,1}(Y_T^S)$ is strictly contained in $U_v/(U_v)^p$, and we will have the relation:

$$\Big(\prod_{v\in T} \varphi_{1,v}\Big)(i_{T,1}(y)) = \prod_{v\in T} \varphi_{1,v}(i_{v,1}(y)) = \prod_{v\in T} \chi_{1,v}(i_{v,1}(y)) = \prod_{v\in T} \chi_{1,v}(y) = 1$$

for all $y \in Y_T^S$.

Finally, $i_{v,1}(Y_T^S) = U_v/(U_v)^p$ means that $\delta_{1,v}^S \simeq U_v/(U_v)^p$, hence that $\varphi_{1,v} = \chi_{1,v}$ which is nontrivial by assumption. □

2.3.6 Remark. For the computations, we have chosen representatives α (of the elements of $Y^S/K^{\times p}$) in Y_T^S, but in fine the condition that we have found is about the extension $Q_{1(S)}/K_1$ which does not depend on T. □

We have obtained the following result for which we recall the main notations: $S = S_0 \cup S_\infty$ is a finite set of noncomplex places, $Y^S := \{\alpha \in K^{\times p} K_{\Delta_\infty}^\times,\ (\alpha) = \mathfrak{a}^p \mathfrak{a}_{S_0},\ \mathfrak{a} \in I,\ \mathfrak{a}_{S_0} \in \langle S_0 \rangle\}$, where $\Delta_\infty := Pl_\infty^{\mathrm{r}} \setminus S_\infty$, $Q_{1(S)} := K_1(\sqrt[p]{Y^S})$ with $K_1 := K(\mu_p)$, and $\delta_{1,v}^S$ is the decomposition group of the place v in $Q_{1(S)}/K_1$.

2.4 Theorem (cyclic case of degree p). *Let T be a nonempty finite set of finite places of K such that $T \cap S_0 = \emptyset$ and such that $\mathrm{N}\mathfrak{p}_v \equiv 1 \bmod (p)$ for all $v \in T_{\mathrm{ta}}$. Then there exists a T-totally ramified and S-split cyclic extension of degree p of K if and only if there exist $\sigma_{1,v} \in \delta_{1,v}^S$, $v \in T$, with $\sigma_{1,v}$ of order p when $\delta_{1,v}^S \simeq U_v/(U_v)^p$, such that:*

$$\prod_{v\in T} \sigma_{1,v} = 1. \qquad \square$$

2.4.1 Remarks. (i) Such an extension always exists if no place $v \in T$ satisfies the condition $\delta_{1,v}^S \simeq U_v/(U_v)^p$.

(ii) If $v \in T_p$, we have:

$$U_v/(U_v)^p = U_v^1/(U_v^1)^p \simeq (\mathbb{Z}/p\mathbb{Z})^{[K_v : \mathbb{Q}_p]+\delta_v},$$

where $\delta_v := 1$ or 0 according to whether or not K_v contains μ_p. We know that the condition $\delta_{1,v}^S \simeq U_v/(U_v)^p$ is equivalent to $i_{v,1}(Y_T^S) = U_v^1/(U_v^1)^p$, and the knowledge of the dimension of $i_{v,1}(Y_T^S)$ is sufficient to see if this condition is satisfied or not.

(iii) The wild places of T can ramify in $Q_{1(S)}/K_1$, contrary to the tame places of T.

(iv) The method used would easily allow us to compute the number of solutions for a given T. □

Let us now look at the case of tame places v. If $v \in T_{\rm ta}$, $\delta^S_{1,v}$ is a cyclic group (of order 1 or p) generated by the Frobenius of v, and since $U_v/(U_v)^p \simeq F_v^\times/F_v^{\times p}$ is cyclic of order p, the condition "$\sigma_{1,v}$ of order p when $\delta^S_{1,v} \simeq U_v/(U_v)^p$" can be written $\sigma_{1,v} =: \left(\frac{K_1(\sqrt[p]{Y^S})/K_1}{v_1}\right)^{a_v}$, $a_v \in (\mathbb{Z}/p\mathbb{Z})^\times$, where v_1 is any place of K_1 above v. In other words we have:

2.4.2 Corollary (tame case [GrM]). *If $T_p = \emptyset$, the necessary and sufficient condition of existence of a T-totally ramified S-split cyclic extension of degree p of K can be written: there exist $a_v \in (\mathbb{Z}/p\mathbb{Z})^\times$ such that:*

$$\prod_{v \in T} \left(\frac{K_1(\sqrt[p]{Y^S})/K_1}{v_1}\right)^{a_v} = 1.$$ □

In fact the above conditions and the product are only for the $v \in T$ whose Frobenius is nontrivial.

2.4.3 Remark. When $T \neq \emptyset$, the case where $H^S_T[p] = H^S[p]$ (i.e., $B/C = 1$), which makes the problem trivially impossible, is equivalent to $i_{T,1}(Y^S_T) = \bigoplus_{v\in T} U_v/(U_v)^p$ (see 2.3.2). But it is easily checked that the radical of the field $Q_1(S)^T$ (where, as usual, $Q_1(S)^T$ denotes the maximal T-split subextension of $Q_1(S)$ over K_1) is $\bigcap_{v\in T} V^S_{v,1} \cap K^\times_T$, the kernel of $i_{T,1}$ in Y^S_T, hence that:

$$\mathrm{Gal}(Q_1(S)/Q_1(S)^T) \simeq i_{T,1}(Y^S_T).$$

The case $B/C = 1$ is thus finally equivalent to:

$$\mathrm{Gal}(Q_1(S)/Q_1(S)^T) \simeq \bigoplus_{v\in T} U_v/(U_v)^p \simeq \bigoplus_{v\in T} \delta^S_{1,v}\ ;$$

it is clear that in this case the existence of $\sigma_{1,v} \neq 1$ with trivial product is impossible. □

2.4.4 Corollary (case of a single place). *If $T = \{v\}$, where v is a tame place, the problem has a solution if and only if $\left(\frac{Q_1(S)/K_1}{v_1}\right) = 1$ (i.e., the place v is totally split in $Q_1(S)/K$ or, equivalently, $i_v(Y^S_{\{v\}}) \subset (U_v)^p$) [Cor2].*

If v is a wild place, the problem has a solution if and only if $\delta^S_{1,v}$ is not isomorphic to $U_v/(U_v)^p \simeq U^1_v/(U^1_v)^p$ or, equivalently, if the canonical map $i_v : Y^S_{\{v\}} \longrightarrow U_v/(U_v)^p$ is not surjective. □

2.4.5 Corollary. *The field K has, for any nonempty set of finite places T (such that $\mathrm{N}\mathfrak{p}_v \equiv 1 \bmod (p)$ for all $v \in T_{\rm ta}$), a T-totally ramified (complexified* *or not) cyclic extension of degree p, if and only if the following conditions are satisfied:*

(i) (case $p \neq 2$). The field K is equal to $\mathbb{Q}$ or to an imaginary quadratic field different from $\mathbb{Q}(\mu_3)$ if $p = 3$, and whose p-class group is trivial.

(ii) (case $p = 2$). The field K is totally real and its restricted 2-class group is trivial (i.e., its ordinary 2-class group is trivial and $E^{\mathrm{res}} = (E^{\mathrm{ord}})^2$).

If $K = \mathbb{Q}(\mu_3)$ and $p = 3$, the only exceptions are for $T = T_{\mathrm{ta}}$ containing a single place v with $\mathrm{N}\mathfrak{p}_v \not\equiv 1 \bmod (9)$.

Proof. By 2.4.4 (for $S = \emptyset$) and the surjectivity of the Artin map, the necessary and sufficient condition is $Q_1 = K_1$, hence $Y^{\mathrm{res}} \subseteq K_1^{\times p}$ and finally $Y^{\mathrm{res}} \subseteq K^{\times p}$ by II.6.3.2. But we have the exact sequence (see 2.3.3, (ii) for $S = \emptyset$):

$$1 \longrightarrow E^{\mathrm{res}}/E^{\mathrm{res}} \cap K^{\times p} \longrightarrow Y^{\mathrm{res}}/K^{\times p} \longrightarrow {}_p\mathcal{C}\ell^{\mathrm{res}}/c\ell^{\mathrm{res}}(P) \longrightarrow 1.$$

Thus in every case, this is equivalent to:

$${}_p\mathcal{C}\ell^{\mathrm{res}} = c\ell^{\mathrm{res}}(P) \text{ and } E^{\mathrm{res}} = E^{\mathrm{res}} \cap K^{\times p}.$$

If $p \neq 2$, this yields the triviality of $(\mathcal{C}\ell^{\mathrm{ord}})_p$ and the equality $E^{\mathrm{res}} = (E^{\mathrm{res}})^p$, hence $r_1 + r_2 = 1$ and $\mu_p(K) = 1$, giving (i). The converse is immediate.

In the case $p = 3$, we have $K = K_1 = \mathbb{Q}(\mu_3)$, $Y^{\mathrm{res}} = \mu_3 K^{\times 3}$, and the governing field is $Q_1 = K(\sqrt[3]{\mu_3}) = \mathbb{Q}(\mu_9)$. But the Frobenius of a tame place v in $\mathbb{Q}(\mu_9)/\mathbb{Q}(\mu_3)$ is trivial if and only if $\mathrm{N}\mathfrak{p}_v \equiv 1 \bmod (9)$ since $\mathrm{N}\mathfrak{p}_v \equiv 1 \bmod (3)$. The result easily follows since if T contains the place above 3, there is always a solution.

If $p = 2$ we obtain ${}_2\mathcal{C}\ell^{\mathrm{res}} = c\ell^{\mathrm{res}}(P)$ and $E^{\mathrm{res}} = (E^{\mathrm{ord}})^2$. We have $(E^{\mathrm{ord}} : E^{\mathrm{res}}) = |\mathrm{sgn}(E^{\mathrm{ord}})| \leq 2^{r_1}$ and $(E^{\mathrm{ord}} : (E^{\mathrm{ord}})^2) = 2^{r_1+r_2}$, hence $r_2 = 0$ (K is totally real, in which case the relation $(E^{\mathrm{ord}} : E^{\mathrm{res}}) = 2^{r_1}$ yields $c\ell^{\mathrm{res}}(P) = 1$), hence (ii). The converse is again easy.

Examples. (i) In the case $p = 2$, if K is a real quadratic field, genus theory implies that K is equal to $\mathbb{Q}(\sqrt{2})$ or to $\mathbb{Q}(\sqrt{\ell})$ for a prime $\ell \equiv 1 \bmod (4)$ (see IV.4.2.9, (i)).

(ii) Still for $p = 2$ and any K, but for the ordinary sense, the analogous condition $Y^{\mathrm{ord}} \subset K^{\times 2}$ is not possible since in the exact sequence:

$$1 \longrightarrow E^{\mathrm{ord}}/(E^{\mathrm{ord}})^2 \longrightarrow Y^{\mathrm{ord}}/K^{\times 2} \longrightarrow {}_2\mathcal{C}\ell^{\mathrm{ord}} \longrightarrow 1,$$

we have $E^{\mathrm{ord}}/(E^{\mathrm{ord}})^2 \neq 1$ since $\mu_2(K) \neq 1$. Therefore, there does not exist any number field such that for any $T \neq \emptyset$ there is a T-totally ramified noncomplexified quadratic extension (for instance $T = \{3\}$ for $K = \mathbb{Q}$ or $T = \{(1 + \sqrt{-2})\}$ for $K = \mathbb{Q}(\sqrt{-2})$).[2] □

[2] In these examples, the governing field is $K(\sqrt{-1})$, then the counterexamples $T = \{v\}$, v tame, are given by the places v for which $q_v \not\equiv 1 \bmod (4)$.

2.4.6 Corollary. *Let $\delta_{1,p}^S$ be the subgroup of* $\mathrm{Gal}(Q_{1(S)}/K_1)$ *generated by the groups $\delta_{1,v}^S$ for $v|p$. Then, for a nonempty $T_{\mathrm{ta}} \subset Pl_{\mathrm{ta}}$, there exists a cyclic extension of degree p which is $T_{\mathrm{ta}} \cup Pl_p$-ramified, totally ramified at the places $v \in T_{\mathrm{ta}}$, and S-split, if and only if there exist $a_v \in (\mathbb{Z}/p\mathbb{Z})^\times$, $v \in T_{\mathrm{ta}}$, such that:*

$$\prod_{v \in T_{\mathrm{ta}}} \left(\frac{Q_{1(S)}/K_1}{v_1}\right)^{a_v} \in \delta_{1,p}^S.$$

Proof. If we assume the existence of a_v satisfying the above relation, we have:

$$\prod_{v \in T_{\mathrm{ta}}} \left(\frac{Q_{1(S)}/K_1}{v_1}\right)^{a_v} = \prod_{v|p} \sigma'_{1,v}$$

with $\sigma'_{1,v} \in \delta_{1,v}^S$; if we take $T_p := \{v|p,\ \sigma'_{1,v} \neq 1\}$ and set $\sigma_{1,v} := \sigma'^{-1}_{1,v}$ for all $v \in T_p$, we obtain the result by applying the theorem to $T := T_{\mathrm{ta}} \cup T_p$. The converse is trivial. □

It is clear that, if $\delta_{1,p}^S = \mathrm{Gal}(Q_{1(S)}/K_1)$, the problem of the existence of a cyclic extension of degree p which is $T_{\mathrm{ta}} \cup Pl_p$-ramified ($T_{\mathrm{ta}} \neq \emptyset$), totally ramified at the places $v \in T_{\mathrm{ta}}$ and S-split, always has a solution. Note also that there does not exist any cyclic extension of degree p, ramified in at least one place above p, unramified outside p and S-split, if and only if:

$$\delta_{1,p}^S := \langle\, \delta_{1,v}^S \,\rangle_{v|p} \simeq \bigoplus_{v|p} U_v^1/(U_v^1)^p$$

(use 2.4.3 with $T = Pl_p$).

2.4.7 Corollary. *Let t be a nonempty finite set of places of K satisfying $\mathrm{N}\mathfrak{p}_v \equiv 1 \bmod (p)$ for all $v \in t_{\mathrm{ta}}$, and disjoint from S_0. Then there exists an infinity of tame places v' for which there exists an S-split cyclic extension of degree p of K which is t or $t \cup \{v'\}$-totally ramified.*

Proof. In $Q_{1(S)}/K_1$, for an arbitrary choice of $\sigma_{1,v} \in \delta_{1,v}^S$, $v \in t$, $\sigma_{1,v}$ of order p when $\delta_{1,v}^S \simeq U_v/(U_v)^p$, set $\prod_{v \in t} \sigma_{1,v} =: \tau$; if $\tau \neq 1$, by the Čebotarev theorem we can find v' such that $\left(\frac{Q_1(S)/K}{v'_1}\right) = \tau^{-1}$, and $T := t \cup \{v'\}$ satisfies the criterion of Theorem 2.4. □

2.4.8 Remark. More generally, we see that as soon as T *strictly* contains a set of tame places whose Frobenius' generate $\mathrm{Gal}(Q_{1(S)}/K_1)$, there exists a cyclic extension of degree p of K, T-ramified (not necessarily totally), and S-split. □

b) Minimal Ramification Sets

As suggested by the work of Martinet on the search for small discriminants [Mar2] [3], we can set the following definition.

2.5 Definition (minimal ramification). Let K be a number field, S a fixed finite set of noncomplex places of K, and p a prime number. We say that $T \subset Pl_0$ ($T \neq \emptyset$, $T \cap S_0 = \emptyset$, $\mathrm{N}\mathfrak{p}_v \equiv 1 \bmod (p)$ for all $v \in T_{\mathrm{ta}}$) is a minimal ramification set of K (relative to S and p) if there exists at least one cyclic extension of degree p of K which is T-totally ramified and S-split, and if for all $t \subset T$, $t \neq T$, $t \neq \emptyset$, this is not true for t. □

If T (minimal) contains wild places, the solutions which are T-totally ramified can have conductors, hence relative discriminants, which differ for wild places. The notion of an extension (cyclic of degree p) with minimal relative discriminant (in some sense which can be made more precise) is finer but can be handled from the above notion which is more intrinsic.

If $|T| = 1$, the minimality condition is equivalent only to the existence and refers to Corollary 2.4.4. As soon as $|T| \geq 2$, if T contains a place v for which $\delta^S_{1,v}$ is not isomorphic to $U_v/(U_v)^p$, T is not minimal (indeed, there exists an S-split $\{v\}$-totally ramified extension since we can take $\sigma_{1,v} = 1$ by 2.4). Thus, we implicitly assume that the sets T that we consider have at least two elements and are such that $\delta^S_{1,v} \simeq U_v/(U_v)^p$ for all $v \in T$. If $v \in T_{\mathrm{ta}}$, this means that the Frobenius of v in $Q_{1(S)}/K_1$ has order p and, if $v \in T_p$, the condition means that $i_{v,1}(Y^S_T) = U^1_v/(U^1_v)^p$ (which is more difficult to satisfy).

In the context of Definition 2.5, using the notations of 2.4, we thus have:

2.6 Theorem. *Let T be a finite set of finite places of K,* [4] *such that for all $v \in T$ the decomposition group $\delta^S_{1,v}$ of v in $Q_{1(S)}/K_1$ is isomorphic to $U_v/(U_v)^p$. We assume that $T_{\mathrm{ta}} \neq \emptyset$. Then T is a minimal ramification set if and only if the $\mathbb{F}_p$-vector space of families $(\sigma_{1,v})_{v\in T}$, $\sigma_{1,v} \in \delta^S_{1,v}$, such that $\prod_{v\in T} \sigma_{1,v} = 1$, has dimension 1 and is generated by a family $(\tau_{1,v})_{v\in T}$, $\tau_{1,v} \in \delta^S_{1,v}$, such that $\tau_{1,v} \neq 1$ for all $v \in T$.*

Proof. If T is minimal, there exists a family $(\sigma_{1,v})_{v\in T}$, $\sigma_{1,v} \in \delta^S_{1,v}$, $\sigma_{1,v} \neq 1$ for all $v \in T$, such that $\prod_{v\in T} \sigma_{1,v} = 1$; if we had a second independent family, since $\delta^S_{1,v}$ is cyclic of order p for tame places (we have assumed that $T_{\mathrm{ta}} \neq \emptyset$), we would obtain, by linear combination, a solution with support $t \subset T$, $t \neq \emptyset$, a contradiction. The converse is clear. □

[3] and "On some 2-class fields" (private communication).

[4] $|T| \geq 2$, $T \cap S_0 = \emptyset$, $\mathrm{N}\mathfrak{p}_v \equiv 1 \bmod (p)$ for all $v \in T_{\mathrm{ta}}$.

2.6.1 Remarks. (i) If $T \subseteq Pl_p$ and if none of the $\delta^S_{1,v}$ is cyclic (otherwise we can use the above argument), there can be several independent solutions with support T.

(ii) We have, for arbitrary T, the exact sequence:

$$1 \longrightarrow \mathrm{Ker}(\pi) \longrightarrow \bigoplus_{v \in T} \delta^S_{1,v} \xrightarrow{\ \pi\ } \mathrm{Gal}(Q_{1(S)}/K_1),$$

where $\pi((\sigma_{1,v})_v) := \prod_{v \in T} \sigma_{1,v}$, showing that:

$$\sum_{v \in T} \mathrm{rk}_p(\delta^S_{1,v}) \leq \mathrm{rk}_p(\mathrm{Gal}(Q_{1(S)}/K_1)) + \mathrm{rk}_p(\mathrm{Ker}(\pi)).$$

A necessary condition for minimality is thus that we have (when $T_{\mathrm{ta}} \neq \emptyset$):

$$\sum_{v \in T} \mathrm{rk}_p(U_v) \leq \mathrm{rk}_p(Y^S/K^{\times p}) + 1,$$

which yields an upper bound for $|T|$; if $T_p = \emptyset$, we simply obtain:

$$|T| \leq \mathrm{rk}_p(Y^S/K^{\times p}) + 1. \qquad \square$$

2.6.2 Corollary (tame case). *When $T_p = \emptyset$, the minimality condition is equivalent to the fact that the Frobenius' of the places of T are nontrivial and satisfy a "unique" relation of the form:*

$$\prod_{v \in T} \left(\frac{Q_{1(S)}/K_1}{v}\right)^{a_v} = 1,$$

$a_v \in (\mathbb{Z}/p\mathbb{Z})^\times$ for all $v \in T$. $\square$

2.6.3 Example. Let $K = \mathbb{Q}(\sqrt{5}\,)$, $S = Pl^{\mathrm{r}}_\infty$ and $p = 2$. The governing field is therefore:

$$Q_{1(S)} =: Q_{1(\infty)} = K(\sqrt{-1}\,, \sqrt{\varepsilon}\,), \ \ \varepsilon := \frac{1+\sqrt{5}}{2}.$$

This is a biquadratic extension of K totally ramified at 2. Now consider the set $T = \{\mathfrak{l}_5, \mathfrak{l}_{11}, \mathfrak{l}_{19}\}$, where $\mathfrak{l}_5 := (\sqrt{5}\,)$, $\mathfrak{l}_{11} := (4+\sqrt{5}\,)$, $\mathfrak{l}_{19} := (1-2\sqrt{5}\,)$; we easily obtain, in terms of quadratic residue symbols in the residue fields:

$$\left(\frac{-1}{\mathfrak{l}_5}\right) = +1, \ \left(\frac{-1}{\mathfrak{l}_{11}}\right) = -1, \ \left(\frac{-1}{\mathfrak{l}_{19}}\right) = -1,$$
$$\left(\frac{\varepsilon}{\mathfrak{l}_5}\right) = -1, \ \left(\frac{\varepsilon}{\mathfrak{l}_{11}}\right) = +1, \ \left(\frac{\varepsilon}{\mathfrak{l}_{19}}\right) = -1.$$

This shows that:

- $\delta_{\mathfrak{l}_5} = \mathrm{Gal}(Q_{1(\infty)}/K(\sqrt{-1}\,))$,
- $\delta_{\mathfrak{l}_{11}} = \mathrm{Gal}(Q_{1(\infty)}/K(\sqrt{\varepsilon}\,))$,
- $\delta_{\mathfrak{l}_{19}} = \mathrm{Gal}(Q_{1(\infty)}/K(\sqrt{-\varepsilon}\,))$,

so it follows that T is a minimal set.

But we can check that $T = \{\mathfrak{l}_5,\ \mathfrak{l}_{11},\ \overline{\mathfrak{l}}_{19} := (1+2\sqrt{5}\,)\}$ is not minimal (indeed, $t = \{\mathfrak{l}_{11},\ \overline{\mathfrak{l}}_{19}\}$ is minimal since $\delta_{\overline{\mathfrak{l}}_{19}} = \delta_{\mathfrak{l}_{11}}$). We see on this last example that the choice of conjugate of the places is crucial. □

2.6.4 Exercise. Using Kummer theory, check that there exists a minimal $\{\mathfrak{l}_5,\ \mathfrak{l}_{11},\ \mathfrak{l}_{19}\}$-totally ramified and totally real quadratic extension L/K.

Answer. Consider:

$$\alpha := \sqrt{5}\,(4+\sqrt{5}\,)(2\sqrt{5}\,-1) = 35+6\sqrt{5}\ ;$$

it is clear that α is totally positive and that $L := K(\sqrt{\alpha}\,)$ is $\{\mathfrak{l}_5,\ \mathfrak{l}_{11},\ \mathfrak{l}_{19}\}$-totally ramified if it is unramified at 2; but we have:

$$\alpha \equiv -1+2\sqrt{5} = 1-4\frac{1-\sqrt{5}}{2} \equiv 1 \bmod (4).$$

It remains to check that there do not exist any quadratic extension $\{\mathfrak{l}_5\}$, $\{\mathfrak{l}_{11}\}$, $\{\mathfrak{l}_{19}\}$, $\{\mathfrak{l}_5,\ \mathfrak{l}_{11}\}$, $\{\mathfrak{l}_5,\ \mathfrak{l}_{19}\}$, $\{\mathfrak{l}_{11},\ \mathfrak{l}_{19}\}$-totally ramified and totally real; this is done with the radical:

$$W_{\mathrm{pos}} := \langle\, \varepsilon\sqrt{5}\,,\ 4+\sqrt{5}\,,\ \varepsilon(2\sqrt{5}\,-1)\,\rangle K^{\times 2}.$$

We must then check that the six corresponding quadratic extensions are ramified at 2 by using the Kummer congruences; this is a little painful numerically (numbers *not* congruent to a square modulo 4), but easy.

We can also check that the extension $\{\mathfrak{l}_{11},\ \overline{\mathfrak{l}}_{19}\}$-totally ramified and totally real is $K(\sqrt{\beta}\,)$ with:

$$\beta := \frac{59+23\sqrt{5}}{2} \equiv \Big(\frac{1-\sqrt{5}}{2}\Big)^2 \bmod (4).$$ □

c) Approach to the Cyclic Case of Degree p^e

Assumptions. We now implicitly assume that $e \geq 2$ and that the sets T which will be considered are nonempty, disjoint from S, and such that $\mathrm{N}\mathfrak{p}_v \equiv 1 \bmod (p^e)$ for all $v \in T_{\mathrm{ta}}$. We assume that the p-Sylow subgroup of $\mathrm{Gal}(H^S/K) \simeq \mathcal{C}\ell^S$ is p-elementary. □

We then have the following preliminary result (refer to Figure 1.2).

2.7 Lemma. *With the above assumptions, considering a set T, there exists a T-totally ramified S-split cyclic extension L_e of K of degree p^e if and only if there exists a T-totally ramified S-split cyclic extension L_1 of K of degree p such that $M_1 := L_1 H^S$ is contained in a cyclic subextension M_e/H^S of degree p^e of H_T^S/H^S.*

Proof. One direction being clear, let M_e/H^S be cyclic of degree p^e containing $M_1 = L_1 H^S$, where L_1/K is T-totally ramified S-split of degree p; it follows that M_1/H^S is also T-totally ramified, which implies this property for M_e/H^S since it is cyclic of prime power degree.

Set $\Gamma_e := \mathrm{Gal}(M_e{}_{(p)}/K)$ and $\gamma_e := \mathrm{Gal}(M_e{}_{(p)}/H^S{}_{(p)})$. Let $\sigma_i \in \Gamma_e \backslash \gamma_e$ such that $\mathrm{Gal}(H^S{}_{(p)}/K) = \bigoplus_i \langle \sigma_i \gamma_e \rangle$. Since $\mathrm{Gal}(M_1{}_{(p)}/K)$ is an $\mathbb{F}_p$-vector space, we have $\sigma_i^p|_{M_1(p)} = 1$, hence $\sigma_i^p \in \gamma_e^p$ since $[M_1{}_{(p)} : H^S{}_{(p)}] = p$. We thus have $\sigma_i = \tau_i\, s_i$, $\tau_i \in \gamma_e$, $s_i^p = 1$, $s_i \notin \gamma_e$; we then check that:

$$\Gamma_e = \gamma_e \bigoplus_i \langle\, s_i \,\rangle.$$

Therefore, there exists a cyclic extension L_e of K of degree p^e, linearly disjoint from H^S/K, such that $L_e H^S = M_e$; since M_e/L_e is unramified, L_e/K is T-totally ramified. □

The existence of M_e/H^S (cyclic of degree p^e) such that its subextension M_1/H^S of degree p is T-totally ramified is not sufficient (this corresponds to weak solutions); the fact that M_1/H^S *comes from* an extension L_1/K is the key point which uses the invariant Y^S.

However, when $(\mathcal{C}\ell^S)_p$ is not killed by p, the lemma can be false: choose $\Gamma_e = \langle\, \sigma_1 \,\rangle \oplus \langle\, \sigma_2 \,\rangle$, with σ_1 of order 8, σ_2 of order 2, and $\gamma_e = \langle\, \sigma \,\rangle$, where $\sigma = \sigma_1^2\, \sigma_2$.

2.7.1 Example. Consider $K = \mathbb{Q}(\sqrt{-14})$, $T = \{(11)\}$, $S = \emptyset$, $p = 2$. By I.4.5.6, the 2-Sylow subgroup of $\mathcal{C}\ell_T = \mathcal{C}\ell_{(11)}$ has order 16 (we have $\mathcal{C}\ell \simeq \mathbb{Z}/4\mathbb{Z}$ and $\mathrm{N}(11) - 1 = 120$). On the other hand, we check that if $\mathfrak{p}_3$ is a prime ideal above 3, $c\ell(\mathfrak{p}_3)$ and $c\ell_{(11)}(\mathfrak{p}_3)$ have respective orders 4 and 8. We thus have $V_T = \langle\, -1,\, 5 + 2\sqrt{-14}\,\rangle K_T^{\times 2}$ since the residual images modulo (11) of -1 and $5 + 2\sqrt{-14}$ are squares, and $\mathrm{rk}_2(\mathcal{C}\ell_{(11)}) = 2$ by I.4.6. Hence $(\mathcal{C}\ell_{(11)})_2 \simeq \mathbb{Z}/8\mathbb{Z} \times \mathbb{Z}/2\mathbb{Z}$.

There are still two possible cases, according to whether or not the Hilbert class field is fixed under a subgroup isomorphic to $\mathbb{Z}/2\mathbb{Z} \times \mathbb{Z}/2\mathbb{Z}$ or to $\mathbb{Z}/4\mathbb{Z}$ (only the second case yields a counterexample). But, by III.1.1.1, we have:

$$\mathrm{Gal}(K_{(11)(2)}/H) \simeq (U_{(11)}/\langle\, -1\,\rangle)_2 \simeq (F_{(11)}^\times/\langle\, \overline{-1}\,\rangle)_2,$$

which is a cyclic group of order 4.

This gives an illustration of the phenomenon: there exists an extension M_2 of H in $K_{(11)}$ which is cyclic of degree 4, (11)-totally ramified, whose quadratic subextension comes from a quadratic extension L_1 of K (it is $K(\sqrt{-11})$), but M_2/H is not split over K. □

2.7.2 Lemma. *The existence of L_e is equivalent to the existence of characters $\varphi_{e,v} \in (U_v/(U_v)^{p^e})^*$ for $v \in T$, of order p^e, such that $\varphi_e := \prod_{v \in T} \varphi_{e,v}$ is trivial on $i_{T,e}(E^S(Y_T^S)^{p^{e-1}})$.*

Proof. Consider the following commutative diagram, obtained from the exact sequences recalled in 2.1, (ii):

$$\begin{array}{ccccccccc} 1 & \longrightarrow & (B/B^{p^e})^* & \xrightarrow{\rho_e^*} & \bigoplus_{v \in T} (U_v/(U_v)^{p^e})^* & \longrightarrow & (i_{T,e}(E^S))^* & \longrightarrow & 1 \\ & & \uparrow & & \uparrow & & \uparrow & & \\ 1 & \longrightarrow & (B/B^p)^* & \xrightarrow{\rho_1^*} & \bigoplus_{v \in T} (U_v/(U_v)^p)^* & \longrightarrow & (i_{T,1}(E^S))^* & \longrightarrow & 1. \end{array}$$

The first condition for the existence of L_e is that of L_1 (first condition of Lemma 2.7), hence that of $\varphi_1 \in \bigoplus_{v \in T} (U_v/(U_v)^p)^*$ satisfying the following conditions, where $C := B \cap A^p$ (characterization (iii) in the proof of 2.3.4):

$$\varphi_1 =: \widetilde{\rho}_1^*(\psi_1), \ \psi_1 \in (B/C)^*,$$
$$\varphi_1 = \prod_{v \in T} \varphi_{1,v}, \ \varphi_{1,v} \neq 1 \text{ for all } v \in T,$$

where we recall that $\widetilde{\rho}_1^*$ is the composite of the two injective maps:

$$(B/C)^* \longrightarrow (B/B^p)^* \xrightarrow{\rho_1^*} \bigoplus_{v \in T} (U_v/(U_v)^p)^* \ ;$$

by definition, the kernel N of ψ_1 fixes the field M_1, and the existence of M_e (second condition of Lemma 2.7) is therefore equivalent to the additional condition on the image ψ_1' of ψ_1 in $(B/B^{p^e})^*$:

$$\psi_1' =: \psi_e^{p^{e-1}},$$

for $\psi_e \in (B/B^{p^e})^*$, whose kernel defines M_e. We thus have:

$$\rho_e^*(\psi_1') = \rho_e^*(\psi_e)^{p^{e-1}},$$

hence, considering by abuse of notation φ_1 as an element of $(B/B^{p^e})^*$, in which case φ_1 can also be written $\rho_e^*(\psi_1')$, we get:

$$\varphi_1 = \rho_e^*(\psi_e)^{p^{e-1}} =: \varphi_e^{p^{e-1}},$$

where we have set:

$$\varphi_e := \rho_e^*(\psi_e) \in \bigoplus_{v\in T}(U_v/(U_v)^{p^e})^*.$$

We have obtained the following equivalent conditions to the existence of L_e:

(i) there exists $\varphi_e \in \mathrm{Im}(\rho_e^*)$ such that:

$$\varphi_1 = \varphi_e^{p^{e-1}} \in \mathrm{Im}(\widetilde{\rho}_1^*) \text{ and } \varphi_1 =: \prod_{v\in T}\varphi_{1,v},$$

$\varphi_{1,v}$ of order p for all $v \in T$;

(ii) there exists $\varphi_e =: \prod_{v\in T}\varphi_{e,v}$, $\varphi_{e,v}$ of order p^e for all $v \in T$ (it is clear that we then have $\varphi_{e,v}^{p^{e-1}} = \varphi_{1,v}$), such that:

$$i_{T,e}(E^S) \subseteq \mathrm{Ker}(\varphi_e) \text{ and } i_{T,1}(Y_T^S) \subseteq \mathrm{Ker}(\varphi_e^{p^{e-1}}),$$

or, equivalently, $i_{T,e}(E^S) \subseteq \mathrm{Ker}(\varphi_e)$ and $i_{T,e}((Y_T^S)^{p^{e-1}}) \subseteq \mathrm{Ker}(\varphi_e)$; in other words:

(iii) there exist $\varphi_{e,v} \in (U_v/(U_v)^{p^e})^*$, $\varphi_{e,v}$ of order p^e for all $v \in T$, satisfying the following condition:

$$\prod_{v\in T}\varphi_{e,v}(y_v) = 1$$

for all $(y_v)_{v\in T} := i_{T,e}(y) = (i_{v,e}(y))_{v\in T}$, such that $y \in E^S(Y_T^S)^{p^{e-1}}$. □

This finishes the proof of the lemma and indicates what is the governing field to be considered.

2.7.3 Notations. We set independently of any T:

$$Q_{e(S)} := K_e\Big(\sqrt[p^e]{E^S(Y^S)^{p^{e-1}}}\Big),$$

with $K_e := K(\mu_{p^e})$, $Y^S := \{\alpha \in K^{\times p}K^{\times}_{\Delta_\infty},\ (\alpha) = \mathfrak{a}^p\mathfrak{a}_{S_0},\ \mathfrak{a} \in I,\ \mathfrak{a}_{S_0} \in \langle\, S_0\,\rangle\}$, where $\Delta_\infty := Pl^{\mathrm{r}}_\infty \backslash S_\infty$, and for any finite place v of K:

$$\delta_{e,v}^S := \mathrm{Gal}\Big(Q_{e(S)}\big/K_e\Big(\sqrt[p^e]{V_{v,e}^S}\Big)\Big),$$

where :

$$V_{v,e}^S := \{y \in E^S(Y^S)^{p^{e-1}},\ i_v(y) \in K_v^{\times p^e}\}.$$ □

2.7.4 Remark. We have, for any T disjoint from S_0 (see 2.3.3, (i)):

$$E^S(Y^S)^{p^{e-1}} = K^{\times p^e}E^S(Y_T^S)^{p^{e-1}},$$
$$V_{v,e}^S = K^{\times p^e}\{y \in E^S(Y_T^S)^{p^{e-1}},\ i_{v,e}(y) = 1\},\quad v \in T.$$ □

To simplify, set temporarily:

$$E^S(Y^S)^{p^{e-1}} =: R^S_e, \quad E^S(Y^S_T)^{p^{e-1}} =: R^S_{e,T} \;;$$

we have, for any finite place v, the exact sequence:

$$1 \longrightarrow (R^S_e K_e^{\times p^e} / V^S_{v,e} K_e^{\times p^e})^* \longrightarrow (R^S_e / V^S_{v,e})^* \longrightarrow (R^S_e \cap K_e^{\times p^e} / V^S_{v,e} \cap K_e^{\times p^e})^* \longrightarrow 1,$$

showing that the interpretation by Kummer duality is possible for a set T only if we have:

$$R^S_e \cap K_e^{\times p^e} = V^S_{v,e} \cap K_e^{\times p^e}$$

for all $v \in T$.

2.7.5 Remark (exceptional case). We know that $R^S_e \cap K_e^{\times p^e} = K^{\times p^e}$, except if $p = 2,\ e \geq 2,\ K \cap \mathbb{Q}(\mu_{2^e}) = \mathbb{Q}(\zeta_n + \zeta_n^{-1}),\ 2 \leq n \leq e$, and if:

$$x := (-1)^{2^{e-n}} (2 + \zeta_n + \zeta_n^{-1})^{2^{e-1}} \in R^S_e$$

(exceptional case studied in II.6.3.2, for which $x = (1 + \zeta_n)^{2^e}$ in K_e).

If v is tame and if $x \in R^S_e$, then $x = (1 + \zeta_n)^{2^e}$ in K_e and, since v is totally split in K_e/K, we have $\zeta_n \in K_v$, hence $i_v(x) \in K_v^{\times 2^e}$, so that we automatically have $x \in V^S_{v,e}$.

We are thus led to assume that, if $x \in R^S_e$, then $x \in V^S_{v,e}$ for all $v \in T_2$. It is easily checked that this is the case if and only if $Pl_2^{\mathrm{ns}} \cap T_2 = \emptyset$ (see II.6.3.4.2). In the case $e > n$, we obtain that for all $v \in T_2$, K_v must contain one of the three numbers:

$$1 + \zeta_n, \ \zeta_{n+1} + \zeta_{n+1}^{-1}, \ \sqrt{-1}(\zeta_{n+1} + \zeta_{n+1}^{-1}).$$

In the case $n = e$, we have:

$$(1 + \zeta_n)^{2^n} = x, \ \ (\zeta_{n+1} + \zeta_{n+1}^{-1})^{2^n} = (\sqrt{-1}(\zeta_{n+1} + \zeta_{n+1}^{-1}))^{2^n} = -x,$$

and in this case, it is $1 + \zeta_n$ which must be in K_v for all $v \in T_2$. □

We still have the isomorphisms (Notation 2.7.3):

$$\delta^S_{e,v} \simeq (R^S_e / V^S_{v,e})^*.$$

For $p = 2$, if the local exceptional case occurs for a place $v \in T_2$, the decomposition group of v in $Q_{e(S)}/K_e$ can be of index 2 in $\delta^S_{e,v}$.

As for the case $e = 1$, for each $v \in T$ we send $\varphi_{e,v}$ to $\chi_{e,v} \in (R^S_{e,T})^*$, then to $\sigma_{e,v} \in \delta^S_{e,v}$ by Kummer duality, and we check that we have $\prod_{v \in T} \sigma_{e,v} = 1$. However the converse is not as simple, and we must explain precisely how to choose the $\sigma_{e,v}$ such that $\chi_{e,v}$, which may be identified with a character $\varphi'_{e,v}$ of $i_{v,e}(R^S_{e,T})$, can be extended to a character $\varphi_{e,v}$ of order p^e of $U_v/(U_v)^{p^e}$. For this we observe that $i_{v,e}(R^S_{e,T}) = i_{v,e}(R^S_{e,\{v\}})$ depends only on v and not on the whole set T.

2.8 Definition (admissibility). We will say that, for $v \in Pl_0$, $\sigma_{e,v} \in \delta^S_{e,v}$ is admissible if $\chi_{e,v}$, seen as a character of $i_{v,e}(R^S_{e,\{v\}})$, can be extended to a character $\varphi_{e,v}$ of order p^e of $U_v/(U_v)^{p^e}$. □

Set temporarily (for $e \geq 1$):

$$\delta := \delta^S_{e,v}, \quad \overline{U} := U_v/(U_v)^{p^e}, \quad \overline{W} := i_{v,e}(R^S_{e,\{v\}}).$$

We have the exact sequence:

$$1 \longrightarrow V^S_{v,e}/K^{\times p^e} \longrightarrow R^S_e/K^{\times p^e} \longrightarrow \overline{W} \longrightarrow 1,$$

showing that:

$$\delta \simeq \overline{W}^{\,*}.$$

We give now practical conditions for the characterization of admissibility.

2.8.1 Lemma 1. *If $\overline{U}/\overline{W}$ has exponent p^e, the admissibility condition is empty.*

Proof. The exact sequence:

$$1 \longrightarrow (\overline{U}/\overline{W})^* \longrightarrow \overline{U}^{\,*} \longrightarrow \overline{W}^{\,*} \longrightarrow 1$$

shows that any $\varphi' \in \overline{W}^{\,*}$ can be lifted to an element of order p^e. □

This deals with the particular case $\delta = 1$ (i.e., $\overline{W} = 1$).

Thus we assume now that $\overline{U}^{p^{e-1}} \subseteq \overline{W}$ (we then have $\delta \neq 1$). We already have the following general principle.

2.8.2 Lemma 2. *If $\overline{U}^{p^{e-1}} \subseteq \overline{W}$ and if $\overline{U} \simeq (\mathbb{Z}/p^e\mathbb{Z})^r$, $r \geq 1$, then $\sigma \in \delta$ is admissible if and only if $\sigma \notin \delta^p$.*

Proof. Indeed, if $\varphi \in \overline{U}^{\,*}$ has order p^e and if $\varphi' =: \psi'^p$ in $\overline{W}^{\,*}$, we obtain $\varphi =: \psi^p\theta$ in $\overline{U}^{\,*}$, with $\theta \in (\overline{U}/\overline{W})^*$; but such a φ has order p^{e-1} (since $\theta^{p^{e-1}} = 1$), a contradiction. Conversely, let $\varphi' \in \overline{W}^{\,*}$, $\varphi' \notin (\overline{W}^{\,*})^p$; if a lift φ does not have order p^e, this means that $\varphi =: \psi^p$, $\psi \in \overline{U}^{\,*}$, hence that $\varphi' = \psi'^p$ in $\overline{W}^{\,*}$, a contradiction. □

Note that this deals with the tame case for which $\overline{U} \simeq \mathbb{Z}/p^e\mathbb{Z}$: indeed, $\overline{U}/\overline{W}$ has exponent p^e means that $\overline{W} = 1$, hence $\delta = 1$ (the Frobenius is trivial); if $\delta \neq 1$, σ must be a power prime to p of the Frobenius. We also recover the formulation of the case $e = 1$ since in that case $\delta \simeq \overline{U}$ means that we are in case of Lemma 2, and $\sigma \notin \delta^p$ is equivalent to $\sigma \neq 1$ or to σ of order p.

It remains to look at the case $v|p$ for which we have $U_v^1 = \mu_p(K_v) \oplus U_1$, where $\mu_p(K_v)$ has order p^h with $0 < h < e$, and where U_1 is a free $\mathbb{Z}_p$-module of rank $r \geq 1$. We thus have:

$$\overline{U} =: \overline{U}_0 \oplus \overline{U}_1 \simeq \mathbb{Z}/p^h\mathbb{Z} \times (\mathbb{Z}/p^e\mathbb{Z})^r.$$

Let φ_0 be a generator of $\overline{U}_0^{\,*}$. Thus we denote by $\varphi'_0 \in \overline{W}^{\,*}$ the restriction of φ_0 to $\overline{W}$, and by χ_0 this same φ'_0 seen as a character of $R^S_{e,\{v\}}$. We then have:

2.8.3 Lemma 3. *If $\overline{U}^{p^{e-1}} \subseteq \overline{W}$ and if $\overline{U} \simeq \mathbb{Z}/p^h\mathbb{Z} \times (\mathbb{Z}/p^e\mathbb{Z})^r$, $r \geq 1$, $0 < h < e$, then $\sigma \in \delta$ is admissible if and only if $\sigma \notin \langle\, \tau \,\rangle\, \delta^p$, where τ corresponds to χ_0 defined above.*

Proof. Indeed, if $\varphi \in \overline{U}^{\,*}$ has order p^e, the relation $\varphi' \in \langle\, \varphi'_0 \,\rangle (\overline{W}^{\,*})^p$ yields $\varphi =: \varphi_0^\lambda \psi^p \theta$ which does not have order p^e. Conversely, if $\varphi' \notin \langle\, \varphi'_0 \,\rangle (\overline{W}^{\,*})^p$, we have $\varphi =: \varphi_0^\lambda \psi$ with $\psi \in \overline{U}_1^{\,*}$ of order p^e (otherwise, since $\overline{U}_1 \simeq (\mathbb{Z}/p^e\mathbb{Z})^r$, we would have $\psi \in (\overline{U}_1^{\,*})^p$, a contradiction). □

In practice, we determine the $\tau_{e,v} \in \delta^S_{e,v}$ for $v|p$ if necessary, but the admissibility condition is usually $\sigma_{e,v} \notin (\delta^S_{e,v})^p$.

2.8.4 Remark. Is it possible that $\langle\, \tau \,\rangle\, \delta^p = \delta$, in other words that there do not exist any admissible element in case of Lemma 3 ? The condition $\langle\, \tau \,\rangle\, \delta^p = \delta$ is equivalent to $\delta = \langle\, \tau \,\rangle$ (i.e., $\overline{W}^{\,*} = \langle\, \varphi'_0 \,\rangle$); the condition $\overline{U}^{p^{e-1}} \subseteq \overline{W}$ shows that the $\mathbb{Z}_p$-rank of U_1 is equal to 1, so that $U_v = \mathbb{Z}_p^\times$; hence $p = 2$ (otherwise $h = 0$) and $\overline{U} = \langle -\overline{1} \rangle \oplus \langle\, \overline{5} \,\rangle$. The character φ_0 has order 2 ($\varphi_0(-\overline{1}) = -1$, $\varphi_0(\overline{5}) = 1$), and similarly φ'_0 has order 2 (since here we have $\delta \neq 1$) and we have $\overline{W} = \langle\, \overline{w_0} \,\rangle$ of order 2; hence $\overline{w_0} \in \{-\overline{1},\ -\overline{5}^{2^{e-1}},\ \overline{5}^{2^{e-1}}\}$. The cases $\overline{w_0} = -\overline{1},\ -\overline{5}^{2^{e-1}}$ are not possible since $\overline{W}$ does not contain $\overline{U}^{2^{e-1}}$, and the case $\overline{w_0} = \overline{5}^{2^{e-1}}$ is not possible since it would give $\varphi'_0 = 1$. Thus this situation does not occur and we deduce that, in every case, there always exists at least one admissible element in δ. □

2.8.5 Example. For $K = \mathbb{Q}(\sqrt{141})$, $S = \emptyset$, $e = 2$, we have $Q_2 = K_2(\sqrt[4]{\varepsilon}\,)$ with $\varepsilon = 95 + 8\sqrt{141}$, and we check that $\delta = \mathrm{Gal}(Q_2/K_2(\sqrt{\varepsilon}\,)) = \langle \tau \rangle$; but $\overline{U}/\overline{W}$ has exponent 4 and any element $\sigma \in \delta$ is admissible. □

We thus obtain the following theorem using Notations 2.7.3 and Definition 2.8, where $S = S_0 \cup S_\infty$ is a finite set of noncomplex places of K.

2.9 Theorem (general case). *We assume that the p-Sylow subgroup of $\mathcal{C}\ell^S$ is killed by p. Let T be a nonempty finite set of finite places of K such that $T \cap S_0 = \emptyset$ and $\mathrm{N}\mathfrak{p}_v \equiv 1 \bmod (p^e)$ for all $v \in T_{\rm ta}$. We assume that the condition of 2.7.5 is satisfied in the exceptional case.*[5] *Then there exists a T-totally ramified S-split cyclic extension of degree p^e of K if and only if there exist $\sigma_{e,v} \in \delta^S_{e,v}$, $\sigma_{e,v}$ admissible for all $v \in T$, such that:*

$$\prod_{v \in T} \sigma_{e,v} = 1. \qquad \square$$

As in the case $e = 1$, when $T_p = \emptyset$ the exceptional case does not occur; thus we can write the above condition only in terms of Frobenius'.

2.9.1 Corollary (tame case). *If $T_p = \emptyset$, the necessary and sufficient condition for the existence of a T-totally ramified S-split cyclic extension of degree p^e of K is that there exist $a_v \in (\mathbb{Z}/p^e\mathbb{Z})^\times$, $v \in T$, such that:*

$$\prod_{v \in T} \left(\frac{K_e(\sqrt[p^e]{E^S (Y^S)^{p^{e-1}}})/K_e}{v_e} \right)^{a_v} = 1,$$

where for each v, v_e is a place of K_e above v. $\square$

Note. For $e = 1$, $K_e\Big(\sqrt[p^e]{E^S\,(Y^S)^{p^{e-1}}}\Big) = K_1(\sqrt[p]{E^S\,Y^S}) = K_1(\sqrt[p]{Y^S}) = Q_{1(S)}$, showing that the notations are compatible. In the case $e = 1$, Y^S does not play any special role with respect to E^S, which also explains the need for the two steps $e = 1$ and $e \geq 2$.

From Corollary 2.4.5, which shows incidentally that the class group is not anymore an obstruction, we easily obtain the following result by translating the additional condition $E^{\rm res} = 1$.

2.9.2 Corollary. *A number field K has, for any integer $e \geq 1$, for any nonempty set T of finite places (such that $\mathrm{N}\mathfrak{p}_v \equiv 1 \bmod (p^e)$ for all $v \in T_{\rm ta}$), a T-totally ramified cyclic extension of degree p^e (complexified or not) if and only if K is equal to $\mathbb{Q}$, or if $p \neq 2$ and K is an imaginary quadratic field, different from $\mathbb{Q}(\mu_3)$ if $p = 3$, whose p-class group is trivial.*

If $K = \mathbb{Q}(\mu_3)$ and $p = 3$, the exceptions are for $T = T_{\rm ta}$ containing a single place v with $\mathrm{N}\mathfrak{p}_v \not\equiv 1 \bmod (3^{e+1})$. $\square$

[5] If $p = 2$, $e \geq 2$, $K \cap \mathbb{Q}(\mu_{2^e}) = \mathbb{Q}(\zeta_n + \zeta_n^{-1})$ for some n such that $2 \leq n \leq e$, and if $x := (-1)^{2^{e-n}}(2 + \zeta_n + \zeta_n^{-1})^{2^{e-1}} \in E^S(Y^S)^{p^{e-1}}$, then we assume that $Pl_2^{\rm ns} \cap T_2 = \emptyset$ and that $1 + \zeta_n \in K_v$ for all $v \in T_2$ if $e = n$.

By the fundamental diagram III.4.4.1, this exactly corresponds to the case where $\overline{K}^{\rm ab}(p)$ is the direct compositum of $H_p^{\rm res}(p)$ with $H_{\rm ta}^{\rm res}(p)$ over K, case for which (the Leopoldt conjecture here being empty !) we have:

$$\mathrm{Gal}(H_{\rm ta}^{\rm res}(p)/K) \simeq \prod_{v \in Pl_{\rm ta}} (F_v^\times)_p \ \text{ and } \ \mathrm{Gal}(H_p^{\rm res}(p)/K) \simeq \bigoplus_{v|p} U_v^1 .$$

This quite predictable fact confirms that it is indeed $E^{\rm res}$ and $\mathcal{C}\ell^{\rm res}$ which are obstructions to a simple description of $H_{\rm ta}^{\rm res}/K$; but $H_{\rm ta}^{\rm res}[p]/K$ can be studied thanks to the invariant $Y^{\rm res}$ (which depends on p).

Using proofs similar to the case $e = 1$, we can prove a number of corollaries, such as the following (but with the assumption $(\mathcal{C}\ell^S)_p^p = 1$).

2.9.3 Corollary. *Let t be a nonempty finite set of places of K satisfying $\mathrm{N}\mathfrak{p}_v \equiv 1 \bmod (p^e)$ for all $v \in t_{\rm ta}$ and disjoint from S_0. We assume that we are not in the exceptional case (see 2.7.5). Then there exists an infinity of tame places v' for which there exists an S-split cyclic extension of degree p^e of K which is t or $t \cup \{v'\}$-totally ramified.* □

We give now some examples for $p = 2$ showing that the exceptional case is not so “exceptional” in the practice.

2.9.4 Examples (illustrations of the exceptional case). (i) Let $K = \mathbb{Q}$, $p = 2$, $e = 2$, $S = \{2, \infty\}$. We have $Y^S = E^S \mathbb{Q}^{\times 2} = \langle -1, 2 \rangle \mathbb{Q}^{\times 2}$ hence, setting $R^S := E^S (Y^S)^2$:

$$R^S = \langle -1, 2 \rangle \mathbb{Q}^{\times 4},$$

and, since $K_2 = \mathbb{Q}(\sqrt{-1}\,)$:

$$Q_{2(S)} = \mathbb{Q}(\sqrt{-1}\,)(\sqrt[4]{-1}, \ \sqrt[4]{2}\,) = \mathbb{Q}(\sqrt{-1}\,)(\sqrt[4]{2}\,)$$

since $\mathbb{Q}(\sqrt{-1}, \ \sqrt{2}\,)$ contains $\sqrt[4]{-1}$. The extension $Q_{2(S)}/K_2$ is here a cyclic extension of degree 4. This confirms that we are in the exceptional case; in fact, we have:

$$R^S/\mathbb{Q}^{\times 4} \simeq \langle -1, 2 \rangle / \langle 2^4 \rangle \simeq \mathbb{Z}/2\mathbb{Z} \times \mathbb{Z}/4\mathbb{Z},$$
$$R^S K_2^{\times 4}/K_2^{\times 4} \simeq \langle -1, 2 \rangle / \langle -4, 2^4 \rangle = \langle -1, 2 \rangle / \langle -4 \rangle \simeq \mathbb{Z}/4\mathbb{Z},$$

since $-4 = (1 + \sqrt{-1}\,)^4$.

If $T = \{\ell\}$, $\ell \equiv 1 \bmod (4)$, there exists an ℓ-totally ramified 2-split real cyclic extension of degree 4 of $\mathbb{Q}$ if and only if ℓ is totally split in $\mathbb{Q}(\sqrt{-1}\,)(\sqrt[4]{2}\,)/\mathbb{Q}(\sqrt{-1}\,)$, then in $\mathbb{Q}(\sqrt{-1}, \ \sqrt[4]{2}\,)/\mathbb{Q}$, hence if and only if:

$$2^{\frac{\ell-1}{4}} \equiv 1 \bmod (\ell).$$

The first example is for $\ell = 73$. Note that any solution ℓ is totally split in $\mathbb{Q}(\sqrt{-1}\,)(\sqrt[4]{2}\,)/\mathbb{Q}$ which contains $\mathbb{Q}(\mu_8)$, and that we must have:

$$\ell \equiv 1 \bmod (8).$$

For $S = \{2\}$, the governing field is the same, hence all solutions (ℓ-totally ramified, 2-split, cyclic of degree 4) will be real.

If $T = \{\ell_1, \ldots, \ell_r\}$, $\ell_i \equiv 1 \bmod (4)$ for $1 \leq i \leq r$, we must have the relation:

$$\prod_{i=1}^{r} \left(\frac{\mathbb{Q}(\sqrt{-1})(\sqrt[4]{2})/\mathbb{Q}}{\ell_i} \right)^{a_i} = 1, \quad a_i \in \{\pm 1\}.$$

(ii) Let $K = \mathbb{Q}(\sqrt{3})$, $p = 2$, $e = 2$, $S = \{\infty\}$. Since $\mathcal{C}\ell^S = \mathcal{C}\ell^{\text{ord}} = 1$, we have $Y^S = E^{\text{ord}} K^{\times 2} = \langle -1, \varepsilon \rangle K^{\times 2}$, where $\varepsilon := 2 + \sqrt{3}$. It follows that:

$$R^{\text{ord}} := E^S (Y^S)^2 = \langle -1, \varepsilon \rangle K^{\times 4}.$$

Here, one could think that because of the nature of the radical, we are not concerned with the exceptional case; but this is not sure because, 2 being ramified in $K/\mathbb{Q}$, we have, for the place v_0 of K above 2, $\mathrm{v}_0(-4) = 4$, so that we can find $u \in K^\times$ such that $-4 = \eta u^4$ with $\eta \in K^\times$ prime to v_0 and $\mathrm{v}_0(u) = 1$. Thus, we must see whether or not $-4 \in R^{\text{ord}}$; but we have the relation:

$$-4 = -\varepsilon^{-2} (1 + \sqrt{3})^4,$$

showing that in fact we are indeed in the exceptional case, with:

$$-\varepsilon^2 = \left(\frac{1 + \sqrt{3}}{1 + \sqrt{-1}} \right)^4$$

in $K_2 = K(\sqrt{-1})$. We again have:

$$\begin{aligned} R^{\text{ord}} / K^{\times 4} &\simeq \mathbb{Z}/2\mathbb{Z} \times \mathbb{Z}/4\mathbb{Z}, \\ R^{\text{ord}} K_2^{\times 4} / K_2^{\times 4} &\simeq \langle -1, \varepsilon \rangle / \langle -\varepsilon^2, \varepsilon^4 \rangle = \langle -1, \varepsilon \rangle / \langle -\varepsilon^2 \rangle \simeq \mathbb{Z}/4\mathbb{Z}. \end{aligned}$$

The extension $Q_{2(\infty)}/K_2$ is therefore $K_2(\sqrt[4]{\varepsilon})/K_2$, cyclic of degree 4. □

d) Solution to the Weak Form

To finish, we return to the weak form of the problem that we are studying, in other words to the existence of an abelian extension L_e of K, which is cyclic of degree p^e over H^S, T-totally ramified over H^S and S-split over K. In this study, we do not need any assumption on $\mathcal{C}\ell^S$.

We can restart the reasoning from the exact sequence of 2.1, (ii) which yields the dual exact sequence:

$$1 \longrightarrow (B/B^{p^e})^* \xrightarrow{\rho_e^*} \bigoplus_{v \in T} (U_v/(U_v)^{p^e})^* \longrightarrow (i_{T,e}(E^S))^* \longrightarrow 1.$$

It is clear that such an extension L_e/H^S is identified by a character $\psi_e \in (B/B^{p^e})^*$ of order p^e, which must satisfy the following condition: if $\rho_e^*(\psi_e) =$

$\varphi_e = \prod_{v \in T} \varphi_{e,v}$ in $\bigoplus_{v \in T} (U_v/(U_v)^{p^e})^*$, then $\varphi_{e,v}$ has order p^e for all $v \in T$. Indeed, this is the total ramification condition that can be obtained directly by writing that $\rho_e(U_v)B^{p^e}/B^{p^e}$ (the inertia group of v in the maximal subextension of H_T^S/H^S with exponent dividing p^e) is projected modulo $\mathrm{Ker}(\psi_e)$ on a group of order p^e, hence $\psi_e(\rho_e(U_v)) \simeq \mathbb{Z}/p^e\mathbb{Z}$, so that $\rho_e^*(\psi_e)(U_v) \simeq \mathbb{Z}/p^e\mathbb{Z}$, proving our assertion.

By translating the fact that $\varphi_e \in \mathrm{Im}(\rho_e^*)$, we obtain that L_e exists if and only if there exist $\varphi_{e,v} \in (U_v/(U_v)^{p^e})^*$, $\varphi_{e,v}$ of order p^e for all $v \in T$, satisfying the following condition:

$$\prod_{v \in T} \varphi_{e,v}(\varepsilon_v) = 1$$

for all $(\varepsilon_v)_{v \in T} := i_{T,e}(\varepsilon) = (i_{v,e}(\varepsilon))_{v \in T}$, such that $\varepsilon \in E^S$.

The rest of the reasoning is then known (by Kummer duality whenever possible), this time with the radical defined by E^S, over K_e, still with the same remarks concerning the exceptional case for $p = 2$ when $T_2 \neq \emptyset$.

2.10 Notations. Let $S = S_0 \cup S_\infty$ be a finite set of noncomplex places of K and let $e \geq 1$ be fixed; we set:

$$Q'_e(S) := K_e(\sqrt[p^e]{E^S}),$$

where $K_e := K(\mu_{p^e})$, and for any finite place v of K:

$$\delta'^S_{e,v} := \mathrm{Gal}\Big(Q'_e(S) \Big/ K_e\Big(\sqrt[p^e]{V'^S_{v,e}}\Big)\Big),$$

where $V'^S_{v,e} := \{\varepsilon \in E^S,\ i_v(\varepsilon) \in (U_v)^{p^e}\}$. □

We thus obtain the following result, using Definition 2.8 and the above notations; recall that H^S is the S-split Hilbert class field of K.

2.11 Theorem (weak form). *Let T be a nonempty finite set of finite places of K such that $T \cap S_0 = \emptyset$ and $\mathrm{N}\mathfrak{p}_v \equiv 1 \bmod (p^e)$ for all $v \in T_{\mathrm{ta}}$. We assume that the condition of 2.7.5 is satisfied in the exceptional case.*[6] *Then there exists a cyclic extension of degree p^e of H^S, abelian over K, T-totally ramified over H^S, S-split over K, if and only if there exist $\sigma'_{e,v} \in \delta'^S_{e,v}$, $\sigma'_{e,v}$ admissible for all $v \in T$, such that:*

$$\prod_{v \in T} \sigma'_{e,v} = 1.$$ □

2.11.1 Remarks. (i) Recall that for $e = 1$, the admissibility condition is, for a finite place v:

[6] If $p = 2$, $e \geq 2$, $K \cap \mathbb{Q}(\mu_{2^e}) = \mathbb{Q}(\zeta_n + \zeta_n^{-1})$ for some n such that $2 \leq n \leq e$, and if $x := (-1)^{2^{e-n}}(2 + \zeta_n + \zeta_n^{-1})^{2^{e-1}} \in E^S(Y^S)^{p^{e-1}}$, then we assume that $Pl_2^{\mathrm{ns}} \cap T_2 = \emptyset$ and that $1 + \zeta_n \in K_v$ for all $v \in T_2$ if $e = n$.

$$\sigma'_{1,v} \neq 1 \text{ when } \delta'^{S}_{1,v} \simeq U_v/(U_v)^p,$$

and that for a tame place (e arbitrary) it means that $\sigma'_{e,v}$ is of the form:

$$\left(\frac{K_e(\sqrt[p^e]{E^S})/K_e}{v_e}\right)^{a'_v}, \ a'_v \in (\mathbb{Z}/p^e\mathbb{Z})^\times,$$

for any place v_e of K_e above v.

(ii) As usual, when $T_p = \emptyset$, the groups $\delta'^{S}_{e,v}$, $v \in T$, are the decomposition groups in $Q'_{e(S)}/K_e$ of a place v_e of K_e above v, and the condition of the above theorem is equivalent to the existence of $a'_v \in (\mathbb{Z}/p^e\mathbb{Z})^\times$ for $v \in T$, such that:

$$\prod_{v \in T} \left(\frac{K_e(\sqrt[p^e]{E^S})/K_e}{v_e}\right)^{a'_v} = 1. \qquad \square$$

We leave to the reader the task of deducing from the theorem several applications. For example that which consists in characterizing weak solutions which are not split over K (in other words, which do not correspond to a strong solution); in the tame case, the extensions which are solutions (in the weak sense) are in one-to-one correspondence with solutions $(\sigma'_{e,v})_{v\in T}$ seen modulo the equivalence relation $(\sigma'_{e,v})_{v\in T} \sim (\sigma''_{e,v})_{v\in T}$ if and only if there exists $c \in (\mathbb{Z}/p^e\mathbb{Z})^\times$ such that $\sigma''_{e,v} = \sigma'^{c}_{e,v}$ for all $v \in T$.

§3 Conclusion and Perspectives

So as to see the practical interest of this dualization of the problem using a governing field, we again consider the example studied in Section 1 in the tame case.

3.1 Example. Thus, consider $K = \mathbb{Q}(\sqrt{10})$ for $p = 2$. We have:

$$E^{\mathrm{ord}} = \langle -1,\ \varepsilon \rangle, \ \ E^{\mathrm{res}} = \langle \varepsilon^2 \rangle,$$

where $\varepsilon = 3 + \sqrt{10}$. For $S = Pl^{\mathrm{r}}_\infty$ and $\emptyset$, we have respectively:

$$Y^{\mathrm{ord}} = \langle -1,\ \varepsilon,\ \eta \rangle K^{\times 2}, \ \ Y^{\mathrm{res}} = \langle \varepsilon\eta \rangle K^{\times 2},$$

where $\eta = 1 + \sqrt{10}$ and $\varepsilon\eta = 13 + 4\sqrt{10}$: indeed, $\mathcal{C}\ell^{\mathrm{ord}} = \mathcal{C}\ell^{\mathrm{res}}$ is generated by the class of a prime ideal $\mathfrak{l}_3$ above 3 whose square is the principal ideal (η).

(i) (case $e = 1$). Let us start with the case $S = \emptyset$. The governing field is:

$$Q_1 := K_1(\sqrt{Y^{\mathrm{res}}}) = K\left(\sqrt{13 + 4\sqrt{10}}\right);$$

by Kummer theory, Q_1/K is unramified (it is at most 2-ramified, but we have $13+4\sqrt{10} = 1+4(3+\sqrt{10}) \equiv 1 \bmod (4)$) hence we obtain that here $Q_1 = H$, the Hilbert class field (in the restricted or the ordinary sense), and that the prime ideal of K above 2 is inert in H/K; this can be seen either by Kummer theory or by the decomposition law of places in $K_{(1)}^{\mathrm{res}} =: H$ since this ideal is not principal in K (indeed, $\mathfrak{l}_2\mathfrak{l}_3 = (2+\sqrt{10})$ and $\mathfrak{l}_3$ is not principal).

By 2.4.6, the case where we allow wild ramification always has a solution, then we will only treat the tame case. A subset $T \subset Pl_{\mathrm{ta}}$ is solution if and only if:

$$\prod_{v\in T} \left(\frac{Q_1/K}{v}\right) = 1$$

(but the fields L_1 which are solutions may be complex).

For instance, by 2.4.4, if $T =: \{v\}$, there exists a $\{v\}$-totally ramified quadratic extension of K if and only if v is split in H/K, hence if and only if $\mathfrak{p}_v$ is principal in K.

Let us resume the analysis of this case which allows us to make several observations. Indeed, we always have the field $\mathbb{Q}\Big(\sqrt{(-1)^{\frac{\ell-1}{2}}\ell}\,\Big)$, ℓ (odd) being the residue characteristic of v, which yields the solution $K\Big(\sqrt{(-1)^{\frac{\ell-1}{2}}\ell}\,\Big)$ if we have $K\Big(\sqrt{(-1)^{\frac{\ell-1}{2}}\ell}\,\Big) \neq H$ and if ℓ is not split in $K/\mathbb{Q}$ (otherwise $|T| = 2$); but ℓ not split is equivalent to:

$$\left(\frac{10}{\ell}\right) = -1 \text{ for } \ell \neq 5, \text{ or } \ell = 5 \text{ (ramification).}$$

The condition $\left(\frac{10}{\ell}\right) = -1$ ($\ell \neq 5$) is equivalent to $\mathfrak{p}_v = (\ell)$ (inertia) which is indeed principal. There remains the case of the prime ideal $\mathfrak{l}_5$ above 5 ramified in $K/\mathbb{Q}$; we can check that it is not principal (we have $\mathfrak{l}_5\mathfrak{l}_3 = (5+\sqrt{10})$ with $\mathfrak{l}_3$ not principal); thus, we must have $\mathbb{Q}(\sqrt{5}) \subseteq H$ (otherwise we would have a contradiction), which is indeed the case since we have already noted in Section 1 that H is the genus field, equal to $\mathbb{Q}(\sqrt{2},\ \sqrt{5})$.

Of course the use of $Q_1 = H$ remains nontrivial, even for $|T| = 1$, since there exist principal prime ideals $\mathfrak{l}$, split in $K/\mathbb{Q}$ (those above 41 for example), which give $\{\mathfrak{l}\}$-totally ramified quadratic extensions of K.

Now, if $S = Pl_\infty^{\mathrm{r}}$ (i.e., we want totally real solutions), the governing field is then:

$$\begin{aligned} Q_{1(\infty)} := K_1(\sqrt{Y^{\mathrm{ord}}}\,) &= K(\sqrt{-1},\ \sqrt{\varepsilon},\ \sqrt{\eta}) \\ &= H(\sqrt{-1},\ \sqrt{\varepsilon}) = K(\sqrt{5},\ \sqrt{-1},\ \sqrt{\varepsilon}), \end{aligned}$$

and the condition:

$$\prod_{v\in T} \left(\frac{Q_{1(\infty)}/K}{v}\right) = 1,$$

is stronger.

For instance, for $T = \{v\}$, the necessary and sufficient condition is that the following conditions be satisfied (in terms of quadratic residue symbols in $F_v^\times$, $v \nmid 5$):

$$\left(\frac{5}{\mathfrak{p}_v}\right) = \left(\frac{-1}{\mathfrak{p}_v}\right) = \left(\frac{\varepsilon}{\mathfrak{p}_v}\right) = 1.$$

If v is split in $K/\mathbb{Q}$, the above conditions can be written:

$$\ell \equiv \pm 1 \bmod (5), \quad \ell \equiv 1 \bmod (4), \quad \text{and} \quad \left(\frac{3+a}{(\ell)}\right) = 1,$$

where ℓ is still the residue characteristic of v and where the rational integer a is such that $\sqrt{10} \equiv a \bmod \mathfrak{p}_v$. If v is inert in $K/\mathbb{Q}$, the residue field of v is $\mathbb{F}_{\ell^2}$, in which $\mathbb{F}_\ell^\times = (\mathbb{F}_{\ell^2}^\times)^{\ell+1} \subseteq \mathbb{F}_{\ell^2}^{\times 2}$; the conditions are then simply $(\frac{\varepsilon}{(\ell)}) = 1$; but since ε is of norm -1, we have $\varepsilon^{1+\ell} \equiv -1 \bmod (\ell)$, hence:

$$\left(\frac{\varepsilon}{(\ell)}\right) \equiv \varepsilon^{\frac{\ell^2-1}{2}} \equiv \varepsilon^{(\ell+1)\frac{\ell-1}{2}} \equiv (-1)^{\frac{\ell-1}{2}} \bmod (\ell),$$

so that we obtain a solution in the inert case if and only if $\ell \equiv 1 \bmod (4)$. The ramified case $(v|5)$ does not give any solution since $\varepsilon \equiv 3 \bmod \mathfrak{l}_5$. We thus recover the fact that neither (7) nor $\mathfrak{l} = (9+2\sqrt{10})$ give a solution which is a real field: indeed, for (7) we have $7 \not\equiv 1 \bmod (4)$, and for $\mathfrak{l}$, although we have $41 \equiv \pm 1 \bmod (5)$ and $41 \equiv 1 \bmod (4)$, we have however (with $a = 16$):

$$\left(\frac{19}{41}\right) = -1.$$

Let us now look at the case where T is formed of two tame places. If $T =: \{u, v\}$, there exists a T-totally ramified totally real quadratic extension of K if and only if u and v are such that:

$$\left(\frac{Q_1(\infty)/K}{u}\right) = \left(\frac{Q_1(\infty)/K}{v}\right).$$

We check, by restricting the action of the Frobenius' to each of the elements $\sqrt{5}$, $\sqrt{-1}$, $\sqrt{\varepsilon}$, that this is equivalent to the following three conditions:

$$\left(\frac{5}{\mathfrak{p}_u}\right) = \left(\frac{5}{\mathfrak{p}_v}\right), \ \mathrm{N}\mathfrak{p}_u \equiv \mathrm{N}\mathfrak{p}_v \bmod (4), \quad \text{and} \quad \left(\frac{\varepsilon}{\mathfrak{p}_u}\right) = \left(\frac{\varepsilon}{\mathfrak{p}_v}\right),$$

for which we can perform the computations as in the case $|T| = 1$ for the quadratic residue symbols. It immediately follows from these computations and the preceding ones that there exists a $\{(7), \mathfrak{l}\}$-totally ramified totally real quadratic extension of K since:

$$\left(\frac{5}{(7)}\right) = \left(\frac{5}{\mathfrak{l}}\right) = 1, \ \ \mathrm{N}(7) \equiv \mathrm{N}\mathfrak{l} \bmod (4), \ \ \left(\frac{\varepsilon}{(7)}\right) = \left(\frac{\varepsilon}{\mathfrak{l}}\right) = -1.$$

(ii) (case $e = 2$). Still with $K = \mathbb{Q}(\sqrt{10})$ and $S = Pl_\infty^{\mathrm{r}}$, but now for $e = 2$, the governing field is:

$$Q_{2(\infty)} = K_2\Big(\sqrt[4]{E^{\mathrm{ord}}(Y^{\mathrm{ord}})^2}\,\Big)$$
$$= K_2(\sqrt[4]{-1},\ \sqrt[4]{\varepsilon},\ \sqrt{\eta}) = K_2(\sqrt{5},\ \sqrt[4]{-1},\ \sqrt[4]{\varepsilon}),$$

and since $K_2(\sqrt{5}) = K_2(\sqrt{2})$ and that $\sqrt{2} \in \mathbb{Q}(\sqrt[4]{-1})$, we obtain:

$$Q_{2(\infty)} = K_2\Big(\sqrt[4]{-1},\ \sqrt[4]{\varepsilon}\Big) = Q'_{2(\infty)},$$

showing that in this case any weak solution is split over K. We note that $Q_{2(\infty)}/K_2$ has degree 8, while $\langle -1,\ \varepsilon,\ \eta^2 \rangle K^{\times 4}/K^{\times 4}$ has order 16; we are thus in the exceptional case, here without any consequence (tame case).

For $T = \{u,\, v\}$ such that $\mathrm{N}\mathfrak{p}_u \equiv \mathrm{N}\mathfrak{p}_v \equiv 1 \bmod (4)$, the condition on the Frobenius' is the existence of odd integers a and b such that:

$$\left(\frac{Q_{2(\infty)}/K_2}{u_2}\right)^a \left(\frac{Q_{2(\infty)}/K_2}{v_2}\right)^b = 1,$$

which can also be written:

$$\left(\frac{Q_{2(\infty)}/K_2}{u_2}\right) = \left(\frac{Q_{2(\infty)}/K_2}{v_2}\right)^{\pm 1};$$

for example, for $T = \{(7),\, \mathfrak{l}\}$ we check that:

$$\left(\frac{K_2(\sqrt[4]{-1})/K_2}{u_2}\right) = 1, \quad \left(\frac{K_2(\sqrt[4]{\varepsilon})/K_2}{u_2}\right) = \sigma \text{ of order } 4,$$
$$\left(\frac{K_2(\sqrt[4]{-1})/K_2}{v_2}\right) = 1, \quad \left(\frac{K_2(\sqrt[4]{\varepsilon})/K_2}{v_2}\right) = \tau \text{ of order } 4,$$

for $u_2|7$ and $v_2|\mathfrak{l}$; the condition is thus simply $\sigma = \tau^{\pm 1}$, which is indeed the case since σ and τ have order 4. Thus, we have shown the existence of a $\{(7),\, \mathfrak{l}\}$-totally ramified, totally real cyclic quartic extension of K.

For arbitrary (tame) $u,\ v$, there exists a T-totally ramified totally real cyclic extension of K of degree 4 if and only if u and v satisfy simultaneously to the following two conditions:

(α) $\mathrm{N}\mathfrak{p}_u \equiv \mathrm{N}\mathfrak{p}_v \equiv 1 \bmod (4)$,

(β) $\left(\frac{K_2(\sqrt[4]{E^{\mathrm{ord}}})/K_2}{u_2}\right) = \left(\frac{K_2(\sqrt[4]{E^{\mathrm{ord}}})/K_2}{v_2}\right)^{\pm 1}$,

with $u_2|u = u_1,\ v_2|v = v_1$. Condition ($\beta$) implies (assuming condition (α)):

$$\mathrm{N}\mathfrak{p}_u \equiv \mathrm{N}\mathfrak{p}_v \bmod (8) \ \text{ and } \left(\frac{\varepsilon}{\mathfrak{p}_u}\right)_4 = \left(\frac{\varepsilon}{\mathfrak{p}_v}\right)_4^{\pm 1},$$

where $\left(\frac{\bullet}{\mathfrak{p}_u}\right)_4$, $\left(\frac{\bullet}{\mathfrak{p}_v}\right)_4$ are 4th power residue symbols modulo $\mathfrak{p}_u$ and $\mathfrak{p}_v$. It follows that there exists a T-totally ramified totally real cyclic extension of degree 4 of K if and only if:

$$\mathrm{N}\mathfrak{p}_u \equiv \mathrm{N}\mathfrak{p}_v \equiv 1,\ 5 \bmod (8) \ \text{ and } \ \left(\frac{\varepsilon}{\mathfrak{p}_u}\right)_4 = \left(\frac{\varepsilon}{\mathfrak{p}_v}\right)_4^{\pm 1}. \qquad \square$$

3.2 Remark. The above study has shown that we have already succeeded since, for e and S fixed (and still under the assumption that $(\mathcal{C}\ell^S)_p^p = 1$ when $e \geq 2$), the governing field $Q_{e(S)}$ gives us the possibility to consider any set of places T. But we can go even further by better using class field theory for $Q_{e(S)}/K_e$. Let us simply give a glimpse of this question. Since the governing field:

$$Q_{e(S)} = K_e\Big(\sqrt[p^e]{E^S\,(Y^S)^{p^{e-1}}} \Big)$$

is $Pl_p \cup S_0$-ramified and Δ_∞-split over K_e, it is easy to compute its conductor $\mathfrak{f}_e^S$ (or some multiple), so this enables us to reduce the problem of finding solutions T to computations in the generalized class group $\mathcal{C}\ell_{K_e,\mathfrak{f}_e^S}^{\Delta_\infty}$ of K_e which is given once and for all as soon as S and e are fixed. More precisely, restricting to the tame case to simplify, from any abstract solution:

$$\prod_{i=1}^{r} \sigma_i^{a_i} = 1,\ \ \sigma_i \in \mathrm{Gal}(Q_{e(S)}/K_e),\ \ a_i \in (\mathbb{Z}/p^e\mathbb{Z})^\times,\ \ r \geq 1,$$

the Čebotarev theorem gives an infinity of solutions $T = \{v^1, \ldots, v^r\}$ by asking that:

$$\left(\frac{Q_{e(S)}/K}{v_e^i}\right) = \sigma_i,\ \ i = 1, \ldots, r.$$

We can also widen this point of view in the context of 2.11 where the Hilbert class field H^S comes into play (search for weak solutions). $\square$

The results of this chapter must be considered as essentially *constructive* for many reasonings of class field theory; in particular we see clearly the possibility of computing densities of abelian extensions ordered by their relative discriminant.

Using a global idelic method, we have proved the Grunwald–Wang Theorem III.4.16.4 which in particular gives us the possibility to ask for given ramifications (total or not). Unfortunately, this type of result is an existence theorem and does not give any information outside the set of places under consideration. See also the results of [Neu3] already mentioned, which use the point of view of the embedding problem to obtain effective forms of the Grunwald–Wang theorem.

To finish, we propose an exercise which relies on the techniques of this chapter and which gives such explicit constructions, around which many variants can be obtained.

3.3 Exercise (effective form of the Grunwald–Wang theorem). Let K be a number field such that the p-Sylow subgroup of $\mathcal{C}\ell^{\mathrm{res}}$ is p-elementary. Let T be a nonempty finite set of finite places of K, $(e_v)_{v\in T}$ a family of nonnegative integers, and $e := \max_{v\in T}(e_v)$; we assume that $\mathrm{N}\mathfrak{p}_v \equiv 1 \bmod (p^{e_v})$ for all $v \in T_{\mathrm{ta}}$. If $p = 2$ we assume that we are not in the exceptional case.

Let $t \subseteq T$.

(i) We assume here that $t = \{v_0\}$ with $e_{v_0} \geq 1$. Show that there exists an extension $L^{(t)}$ of K, cyclic of degree $p^{e_{v_0}}$, $(T\backslash t)$-split, such that $[L^{(t)}_{v_0} : K_{v_0}] = p^{e_{v_0}}$.

(ii) For $t \subset T$ such that $e_v \geq 1$ for all $v \in t$, we assume that we have constructed an extension $L^{(t)}$ of K, cyclic of degree $p^{\max_{v\in t}(e_v)}$, $(T\backslash t)$-split, such that $[L^{(t)}_v : K_v] = p^{e_v}$ for all $v \in t$. Let $u \in T\backslash t$, and let M (linearly disjoint from $L^{(t)}/K$), be an extension of K, cyclic of degree p^{e_u}, $(T\backslash\{u\})$-split, and such that $[M_u : K_u] = p^{e_u}$ (see (i)). Show that, in the compositum $L^{(t)}M$, there exists a subextension $L^{(t\cup\{u\})}$ of K, cyclic of degree $p^{\max_{v\in t\cup\{u\}}(e_v)}$, $(T\backslash(t\cup\{u\}))$-split, such that $[L^{(t\cup\{u\})}_v : K_v] = p^{e_v}$ for all $v \in t \cup \{u\}$.

(iii) Conclude.

Answer. (i) We apply Corollary 2.9.3 to t and to $S = T\backslash t$ (we obtain the case $|t| = 1$ for an induction).

(ii) We realize M/K as an extension $\{u\}$ or $\{u, u'\}$-totally ramified, $(T\backslash \{u\})$-split of degree p^{e_u}, for a suitable u'.

If χ (resp. ψ) is a character of order $p^{\max_{v\in t}(e_v)}$ (resp. of order p^{e_u}) corresponding to $L^{(t)}$ (resp. to M), the extension L fixed under the kernel of $\chi\psi$ is cyclic of degree $p^{\max_{v\in t\cup\{u\}}(e_v)}$; since we have $LM = LL^{(t)} = ML^{(t)}$, it is clear that we have $[L_v : K_v] = [L^{(t)}_v : K_v] = p^{e_v}$ for all $v \in t$ (M/K is $(T\backslash\{u\})$-split hence t-split) and $[L_u : K_u] = [M_u : K_u] = p^{e_u}$ ($L^{(t)}/K$ is $(T\backslash t)$-split hence $\{u\}$-split). Thus, $L^{(t\cup\{u\})} = L$ is a solution to the problem and it is $(T\backslash(t \cup \{u\}))$-split as can be immediately checked.

(iii) The induction stops when $t = \{v \in T,\ e_v \geq 1\}$, which yields an extension L of K, cyclic of degree p^e, satisfying $[L_v : K_v] = p^{e_v}$ for all $v \in t$, and $(T \setminus t)$-split (for the last e_v equal to 0, we obtain that L/K is split for these places, so that the S-splitting is in fact implicitly treated in the above). □

Appendix
Arithmetical Interpretation of $H^2(\mathcal{G}_T^S, \mathbb{Z}/p^e\mathbb{Z})$

Let p be a fixed prime number and let K be a number field together with sets of places T and $S = S_0 \cup S_\infty$. We denote by:

$$\overline{H}_T^S(p)$$

the maximal T-ramified S-split pro-p-extension of K. By definition its maximal abelian subextension is $H_T^S(p)$, and if we set:

$$\mathcal{G}_T^S := \mathrm{Gal}(\overline{H}_T^S(p)/K),$$

its abelianization is:

$$\mathcal{G}_T^{S\,\mathrm{ab}} =: \mathcal{A}_T^S = \mathrm{Gal}(H_T^S(p)/K),$$

studied in Chapter III, Sections 2 and 4, and Chapter IV, Sections 1, 2, and 3. The extension $\overline{H}_T^S(p)$ is even the p-tower of the successive maximal T-ramified S-split abelian pro-p-extensions.

The study of $\mathcal{G}_T^S$ by generators and relations goes back (at least for $S_0 = \emptyset$) to a paper of Šafarevič [Ša] from 1964 and has been largely extended in the works that we have already mentioned. [1]

The dimension of $H^1(\mathcal{G}_T^S, \mathbb{Z}/p\mathbb{Z})$ (Šafarevič's formula given in II.5.4.1) is computable in complete generality with the usual invariants of class field theory; it gives the minimal number of generators of $\mathcal{G}_T^S$.

On the contrary, the dimension of $H^2(\mathcal{G}_T^S, \mathbb{Z}/p\mathbb{Z})$, which gives the minimal number of relations between these generators (in the corresponding free pro-p-group), is easily obtained only when T contains the places dividing p and $S_0 = \emptyset$ (see [e, Ko3, Ch. 3, § 2.6] where numerical examples are given); for instance, recall that for $T = S_0 = \emptyset$ and $S_\infty = Pl_\infty^{\mathrm{r}}$, $\overline{H}_T^S(p) =: \overline{H}^{\mathrm{ord}}(p)$ is the p-Hilbert class fields tower, which can be finite or infinite, and that the structure of $\mathcal{G}^{\mathrm{ord}}$ is a difficult problem (Ch. II; § 5, (f)). We will see later what can be said about it.

To understand the arithmetical obstructions to a generalization (to arbitrary T and S), we suggest here a methodology whose arguments use class

[1] [g, Hab; Ko4; NSW, Ch. VIII, § 7; Se3, Ch. I, § 4], [Neum2]; see also the survey by Yamagishi in [i, Miy0].

field theory as much as possible, and which leads to the following two sets of results:

(i) We give, when $S_0 = \emptyset$ and $Pl_p \subseteq T$, an arithmetical interpretation of $H^2(\mathcal{G}_T^{S_\infty}, \mathbb{Z}/p^e\mathbb{Z})$, $e \geq 1$ (and even of $H^2(\mathcal{G}_T^{S_\infty}, \mathbb{Z}_p)$) in terms of the invariant $\mathcal{T}_T^{S_\infty}$ which is very well understood (including numerically) thanks to the study which has been done in the preceding chapters (Theorem 2.2). We essentially recover in an unified way the interpretation obtained by cohomological means (i.e., in the context of Poitou–Tate duality) by Haberland in [g, Hab] and by Nguyen Quang Do in [Ng1; Ng2], which we will occasionally mention.

(ii) We obtain a new general result about the number of relations for the case of incomplete p-ramification with decomposition, which contains everything which is known up to now (including Šafarevič's results) (Theorems 3.6 and 3.7).

§1 A General Approach by Class Field Theory

For the moment, we do not make any assumptions on T and S.

1.1 Notations. (i) To simplify notations, we set:

$$\overline{H}_{K,T}^{S}{}^{(p)} =: \overline{H}, \quad H_{K,T}^{S}{}^{(p)} =: H, \quad \overline{H}_{L,T'}^{S'}{}^{(p)} =: \overline{H}', \quad H_{L,T'}^{S'}{}^{(p)} =: H',$$

$$\mathcal{G}_{K,T}^S =: \mathcal{G}, \quad \mathcal{A}_{K,T}^S =: \mathcal{A}, \quad \mathcal{T}_{K,T}^S =: \mathcal{T}, \quad \mathcal{G}_{L,T'}^{S'} =: \mathcal{G}', \quad \mathcal{A}_{L,T'}^{S'} =: \mathcal{A}',$$

where T' and S' are the sets of places above those of T and S in a finite extension L of K. We will have to consider the situation obtained by replacing T by:

$$T_\sharp := T \cup Pl_p$$

(complete p-ramification), and $S = S_0 \cup S_\infty$ by:

$$S_\sharp := S_\infty$$

(no finite decomposition); in this case, we add the index $\sharp$ to the above notations:

$$\overline{H}_\sharp,\ H_\sharp,\ \overline{H}'_\sharp,\ H'_\sharp,\ \mathcal{G}_\sharp,\ \mathcal{A}_\sharp,\ \mathcal{T}_\sharp,\ \mathcal{G}'_\sharp,\ \mathcal{A}'_\sharp.$$

(ii) In this Appendix, G will always denote a finite p-group, and the cohomology groups $H^r(G, A)$, A discrete with trivial action of G, will denote Tate's modified groups for $r \in \mathbb{Z}$;[2] they are written additively, althought they are isomorphic to multiplicative groups. □

We start from the following fact:

[2] [d, CF, Ch. IV, § 6], [g, NSW, Ch. I].

1.2 Lemma. *For any fixed $e \geq 1$, we have:*

$$H^2(\mathcal{G}, \mathbb{Z}/p^e\mathbb{Z}) = \varinjlim_{\mathcal{H}} \operatorname{Inf}(H^2(\mathcal{G}/\mathcal{H}, \mathbb{Z}/p^e\mathbb{Z})),$$

where $\mathcal{H}$ ranges in the set of open (i.e., closed of finite index) normal subgroups of $\mathcal{G}$, and Inf *is the inflation map.*

Proof. See [g, Se3, Ch. I, § 2.1] or [g, NSW, Ch. I, § 5]. □

A necessary and sufficient condition for the finiteness of $H^2(\mathcal{G}, \mathbb{Z}/p^e\mathbb{Z})$ is that $\operatorname{Inf}(H^2(\mathcal{G}/\mathcal{H}, \mathbb{Z}/p^e\mathbb{Z}))$ stabilizes with respect to $\mathcal{H}$, which will be indeed the case (Proposition 3.5). [3]

For any $\mathcal{H}$ as above, let:

$$G := \mathcal{G}/\mathcal{H}.$$

We will determine:

$$\operatorname{Inf}(H^2(G, \mathbb{Z}/p^e\mathbb{Z})),$$

by computing separately (respectively in Subsections (a) and (b)):

$$\operatorname{Ker}_{e,G} := \operatorname{Ker}(H^2(G, \mathbb{Z}/p^e\mathbb{Z}) \xrightarrow{\operatorname{Inf}} H^2(\mathcal{G}, \mathbb{Z}/p^e\mathbb{Z})) \ \text{ and } \ H^2(G, \mathbb{Z}/p^e\mathbb{Z}).$$

In this study, we can (and we will have to) assume that G is arbitrarily large.

Note. If A is a finite G-module, we denote by A^* the dual G-module of A; this notation, defined in I.5.7, is compatible with those used elsewhere.

a) Study of $\operatorname{Ker}(H^2(G, \mathbb{Z}/p^e\mathbb{Z}) \xrightarrow{\operatorname{Inf}} H^2(\mathcal{G}_T^S, \mathbb{Z}/p^e\mathbb{Z}))$

The Hochschild–Serre exact sequence can be written in this context:

$$1 \longrightarrow H^1(G, \mathbb{Z}/p^e\mathbb{Z}) \xrightarrow{\operatorname{Inf}} H^1(\mathcal{G}, \mathbb{Z}/p^e\mathbb{Z}) \xrightarrow{\operatorname{Res}} H^1(\mathcal{H}, \mathbb{Z}/p^e\mathbb{Z})^G \xrightarrow{\operatorname{Tra}} \operatorname{Ker}_{e,G} \longrightarrow 1,$$

where the transgression Tra and the G-module action on $H^1(\mathcal{H}, \mathbb{Z}/p^e\mathbb{Z})$ (which is still the usual one for H^1 groups) are recalled in [e, Ko3, Ch. 2, § 3.6] (see also [g, NSW, Ch. I, § 6]). We obtain:

$$H^1(G, \mathbb{Z}/p^e\mathbb{Z}) = (G^{\mathrm{ab}}/G^{\mathrm{ab}\,p^e})^*,$$
$$H^1(\mathcal{G}, \mathbb{Z}/p^e\mathbb{Z}) = (\mathcal{A}/\mathcal{A}^{p^e})^* \ ;$$

we can assume that G^{ab} is large enough so that the above two groups are isomorphic, i.e., we ask that:

[3] We thank a referee for his remark about the fact that $H^2(\mathcal{G}, \mathbb{Z}/p^e\mathbb{Z})$ is, a priori, not finite.

$$\mathcal{H}\,[\mathcal{G},\mathcal{G}] \subseteq \mathcal{G}^{p^e}\,[\mathcal{G},\mathcal{G}],$$

which is possible (with $\mathcal{H}$ of finite index) since, by class field theory, $\mathcal{A} = \mathcal{G}/[\mathcal{G},\mathcal{G}]$ is a $\mathbb{Z}_p$-module of finite type. Therefore, we have:

$$\mathrm{Ker}_{e,\,G} \simeq H^1(\mathcal{H}, \mathbb{Z}/p^e\mathbb{Z})^G.$$

Let L be the field fixed under $\mathcal{H}$. Since $\overline{H}/L$ is T'-ramified, S'-split, the analogous extension $\overline{H}'$ of L contains $\overline{H}$, and since $\overline{H}'/K$ is Galois (by maximality), T-ramified and S-split, we have $\overline{H}' = \overline{H}$. Thus $\mathcal{H} = \mathrm{Gal}(\overline{H}'/L) = \mathcal{G}'$, hence $\mathcal{H}^{\mathrm{ab}} = \mathcal{A}'$, so that:

$$\mathrm{Ker}_{e,\,G} \simeq (\mathcal{A}'/\mathcal{A}'^{p^e})^{*\,G},$$

hence, by duality [4], we can write:

$$(\mathrm{Ker}_{e,\,G})^* \simeq (\mathcal{A}'/\mathcal{A}'^{p^e})_G := (\mathcal{A}'/\mathcal{A}'^{p^e})/I_G(\mathcal{A}'/\mathcal{A}'^{p^e}) = \mathcal{A}'/\mathcal{A}'^{p^e}.\,I_G\mathcal{A}',$$

where I_G is the augmentation ideal of G. Therefore, we have obtained:

1.3 Proposition. *For any sufficiently large G, we have:*

$$\mathrm{Ker}_{e,\,G} \simeq (\mathcal{A}'/\mathcal{A}'^{p^e}.\,I_G\mathcal{A}')^*. \qquad \square$$

b) Study of $H^2(G, \mathbb{Z}/p^e\mathbb{Z})$ — The Schur Multiplier

We start from the exact sequence:

$$0 \longrightarrow \mathbb{Z} \xrightarrow{\;p^e\;} \mathbb{Z} \longrightarrow \mathbb{Z}/p^e\mathbb{Z} \longrightarrow 0,$$

which yields:

$$H^{-3}(G,\mathbb{Z}) \xrightarrow{\;p^e\;} H^{-3}(G,\mathbb{Z}) \longrightarrow H^{-3}(G,\mathbb{Z}/p^e\mathbb{Z}) \longrightarrow H^{-2}(G,\mathbb{Z}) \xrightarrow{\;p^e\;} H^{-2}(G,\mathbb{Z}),$$

which can be written:

$$1 \longrightarrow H^{-3}(G,\mathbb{Z})/p^e H^{-3}(G,\mathbb{Z}) \longrightarrow H^{-3}(G,\mathbb{Z}/p^e\mathbb{Z}) \longrightarrow {}_{p^e}H^{-2}(G,\mathbb{Z}) \longrightarrow 1.$$

Since the group G is finite, we have:

$${}_{p^e}H^{-2}(G,\mathbb{Z}) := {}_{p^e}H_1(G,\mathbb{Z}) \simeq {}_{p^e}G^{\mathrm{ab}}$$

[4] See the Definition 1.2.2 and the proof of Proposition 3.1.1 of [g, NSW, Ch. I, § 2; Ch. III, § 1] or use the definition of φ^s for a character φ and $s \in G$.

[d, CF, Ch. IV, § 3, Prop. 1]; furthermore, we know that:

$$H^{-3}(G, \mathbb{Z}/p^e\mathbb{Z}) := H_2(G, \mathbb{Z}/p^e\mathbb{Z}),$$

is the dual of $H^2(G, (\mathbb{Z}/p^e\mathbb{Z})^*)$ (Proposition 3.1.1 of [g, NSW, Ch. III, § 1] already mentioned). We thus obtain the exact sequence:

$$1 \longrightarrow H^{-3}(G, \mathbb{Z})/p^e H^{-3}(G, \mathbb{Z}) \longrightarrow H^2(G, (\mathbb{Z}/p^e\mathbb{Z})^*)^* \longrightarrow {}_{p^e}G^{\mathrm{ab}} \longrightarrow 1.$$

We are reduced to the study of the Schur multiplier $H^{-3}(G, \mathbb{Z})$ of G. [5] For this we use Tate's fundamental result on cohomological global class field theory mentioned in II.3.2, which says that:

$$H^{-3}(G, \mathbb{Z}) \simeq H^{-1}(G, C'),$$

where $C' := C_L$ is the idèle class group of L [d, CF, Ch. VII, § 11.3]. [6]

By introducing the connected component D' of the unit element of C', we obtain the exact sequence of G-modules:

$$1 \longrightarrow D' \longrightarrow C' \xrightarrow{\ \rho_L\ } \mathrm{Gal}(\overline{L}^{\mathrm{ab}}/L) \longrightarrow 1,$$

which yields:

$$\begin{aligned} H^{-1}(G, D') \longrightarrow H^{-1}(G, C') &\longrightarrow H^{-1}(G, \mathrm{Gal}(\overline{L}^{\mathrm{ab}}/L)) \\ &\longrightarrow H^0(G, D') \xrightarrow{\ f\ } H^0(G, C'). \end{aligned}$$

By [d, AT, Ch. 9, § 2], for all $k \in \mathbb{Z}$ we have:

$$H^{2k-1}(G, D') = 1, \quad H^{2k}(G, D') \simeq (\mathbb{Z}/2\mathbb{Z})^{r_1^{\mathrm{c}}},$$

where r_1^{c} is the number of real places of K which are complexified in L/K (equal to 0 for $p \neq 2$, or for $p = 2$ and $S_\infty = Pl_\infty^{\mathrm{r}}$ since L/K is S_∞-split), but we need some more details.

The following study (until Proposition 1.4) is more complicated because of the case $p = 2$, which is often excluded in the literature (or treated under assumptions which imply $r_1^{\mathrm{c}} = 0$).

Computation of the Cohomology Group $H^{-1}(G, D')$. Using III.4.15.1, (iii), a direct computation of $H^{-1}(G, D') := {}_\nu D'/I_G D'$, where $\nu := \nu_{L/K}$ is the algebraic norm, becomes easy.

Indeed, let $d' \in {}_\nu D'$, then $1 = d'^\nu \equiv d'^{|G|} \bmod I_G D' = I_G D'^{|G|}$, and d' is represented by $\delta' \in {}_{|G|}D' \simeq {}_{|G|}U'_\infty$ with $\delta'^\nu = 1$. By Shapiro's lemma, the

[5] This study is analogous to that of Roquette in [d, CF, Ch. IX] on class fields towers.

[6] To ease notations, in this Appendix, $(H^r)_{r\in\mathbb{Z}}$ always denotes Tate's modified cohomology.

computation of $H^{-1}(G, \mathrm{tor}(U'_\infty))$ is reduced to that of $H^{-1}(\langle c\rangle, \mathrm{tor}(\mathbb{C}^\times)) = \mathrm{tor}(\mathbb{C}^\times)/(\mathrm{tor}(\mathbb{C}^\times))^2 = 1$ for a complex conjugation c corresponding to a complexified real place. Thus $\delta' \in I_G D'$ giving the triviality of $H^{-1}(G, D')$.

COMPUTATION OF $H^0(G, C')$ AND $H^0(G, D')$ — IDENTIFICATION OF f. Denote by $\Delta_\infty^{\mathrm{c}}(L/K) =: \Delta_\infty^{\mathrm{c}} \subseteq \Delta_\infty := Pl_\infty^{\mathrm{r}} \backslash S_\infty$ the subset of the r_1^{c} places v which are complexified in L/K.

We have, since the algebraic and arithmetical norms $\nu_{L/K}$, $\mathrm{N}_{L/K}$ satisfy $\nu_{L/K} = j_{L/K} \circ \mathrm{N}_{L/K}$, with $j_{L/K}$ injective on C:

$$H^0(G, C') = C'^G/\nu_{L/K}(C') \simeq C/\mathrm{N}_{L/K}(C'),$$

which can be identified with G^{ab} by the reciprocity map $\rho_{L/K}$.

It is easily checked, from III.4.15.1, (ii), that:

$$H^0(G, D') = D'^G/\nu_{L/K}(D') \simeq \{c \in C,\ j_{L/K}(c) \in D'\}/D$$

since $\mathrm{N}_{L/K}(D') = D$. Then:

$$H^0(G, D') \simeq \prod_{v \in \Delta_\infty^{\mathrm{c}}} \{\pm 1\},$$

as a subgroup of $C/D \simeq \mathrm{Gal}(\overline{K}^{\mathrm{ab}}/K)$, can be identified with the *direct product* of the decomposition groups of the $v \in \Delta_\infty^{\mathrm{c}}$ in $\overline{K}^{\mathrm{ab}}/K$, which we interpret as being (deployment Theorem III.4.1):

$$\bigoplus_{v \in \Delta_\infty^{\mathrm{c}}} D_v(\overline{K}^{\mathrm{ab}}/K) \subset \mathrm{Gal}(\overline{K}^{\mathrm{ab}}/K)$$

(this defect which only happens in the case $p = 2$ comes from the fact that $-1 \in \mathbb{R}^\times$ becomes divisible in $\mathbb{C}^\times$; hence $j_{L/K}\Big(\prod_{v \in \Delta_\infty^{\mathrm{c}}} \{\pm 1\}\Big)$ represents $D'^G/\nu_{L/K}(D')$ in J_L). Thus, the map f can be considered as the map sending:

$$(\sigma_v)_v \in \bigoplus_{v \in \Delta_\infty^{\mathrm{c}}} D_v(\overline{K}^{\mathrm{ab}}/K)$$

to the product of the restrictions of the σ_v in L^{ab}/K (which happens non-trivially only for $p = 2$ since G is a p-group).

Thus, in every case we obtain the exact sequence:

$$1 \longrightarrow H^{-1}(G, C') \longrightarrow H^{-1}(G, \mathrm{Gal}(\overline{L}^{\mathrm{ab}}/L)) \longrightarrow \bigoplus_{v \in \Delta_\infty^{\mathrm{c}}} D_v(\overline{K}^{\mathrm{ab}}/K) \xrightarrow{\ f\ } G^{\mathrm{ab}},$$

which we write in the form:

$$1 \longrightarrow H^{-1}(G, C') \longrightarrow H^{-1}(G, \mathrm{Gal}(\overline{L}^{\mathrm{ab}}/L)) \longrightarrow \mathrm{Ker}(f) \longrightarrow 1, \qquad (0)$$

where:

$$\mathrm{Ker}(f) = \Big(\bigoplus_{v\in\Delta^{\mathrm{c}}_\infty} D_v(\overline{K}^{\mathrm{ab}}/K)\Big) \bigcap \mathrm{Gal}(\overline{K}^{\mathrm{ab}}/L^{\mathrm{ab}}).$$

We may assume that $\Delta^{\mathrm{c}}_\infty(L/K) = \Delta^{\mathrm{c}}_\infty(\overline{H}/K)$: for this, it is sufficient that L contain elements $\theta \in \overline{H}$ having suitable complex conjugates. Therefore, $\mathrm{Ker}(f)$ stabilizes.

Computation of the Cohomology Group $H^{-1}(G, \mathrm{Gal}(\overline{L}^{\mathrm{ab}}/L))$. To compute $H^{-1}(G, C')$, we will first simplify the profinite group:

$$H^{-1}(G, \mathrm{Gal}(\overline{L}^{\mathrm{ab}}/L)).$$

Since G is a finite p-group and that we are interested in:

$$H^{-1}(G, C')/p^e H^{-1}(G, C'),$$

we may replace $\mathrm{Gal}(\overline{L}^{\mathrm{ab}}/L)$ by $\mathrm{Gal}(\overline{L}^{\mathrm{ab}}{}_{(p)}/L)$. The general deployment Theorem III.4.1 (for $\mathrm{Gal}(\overline{L}^{\mathrm{ab}}{}_{(p)}/L)$, assuming the Leopoldt conjecture for p in $\overline{H}$) can then be expressed by the exact sequence:

$$1 \longrightarrow \mathcal{F}'_\sharp := \prod_{w\in(Pl'_{\mathrm{ta}}\setminus T'_{\mathrm{ta}})\cup S'_\infty} (F_w^\times)_p \longrightarrow \mathrm{Gal}(\overline{L}^{\mathrm{ab}}{}_{(p)}/L) \longrightarrow \mathcal{A}'_\sharp \longrightarrow 1,$$

where we recall that $\mathcal{A}'_\sharp := \mathrm{Gal}(H'_\sharp/L)$ (see 1.1, (i)); this yields:

$$H^{-1}(G, \mathcal{F}'_\sharp) \longrightarrow H^{-1}(G, \mathrm{Gal}(\overline{L}^{\mathrm{ab}}{}_{(p)}/L)) \longrightarrow H^{-1}(G, \mathcal{A}'_\sharp) \longrightarrow H^0(G, \mathcal{F}'_\sharp).$$

A group $H^r(G, \mathcal{F}'_\sharp)$, $r \geq 0$, is equal to the product of the groups:

$$H^r\Big(G, \bigoplus_{w|v}(F_w^\times)_p\Big) = H^r(D_{w_0}, (F_{w_0}^\times)_p),\ \ v \in (Pl_{\mathrm{ta}} \setminus T_{\mathrm{ta}}) \cup S_\infty,\ \ w_0|v,$$

using Shapiro's lemma. Since this product is only over places v not dividing p, unramified and noncomplexified in L/K, we deduce that the cohomology of $\mathcal{F}'_\sharp$ (thus its homology) is trivial. Hence:

$$H^{-1}(G, \mathrm{Gal}(\overline{L}^{\mathrm{ab}}{}_{(p)}/L)) \simeq H^{-1}(G, \mathcal{A}'_\sharp) \simeq {}_\nu\mathcal{A}'_\sharp/I_G\mathcal{A}'_\sharp,$$

where ${}_\nu\mathcal{A}'_\sharp$ is the kernel of the map $\nu_{L/K} =: \nu$ (the *algebraic* norm in L/K), and I_G is the augmentation ideal of G.

Conclusion: An Expression of the Schur Multiplier $H^{-3}(G, \mathbb{Z})$. Consider the kernel ${}_{\mathrm{N}}\mathcal{A}'_\sharp$ of the arithmetic norm $\mathrm{N}_{L/K} =: \mathrm{N}$, which can be identified by class field theory with the restriction of automorphisms $\mathcal{A}'_\sharp \longrightarrow \mathcal{A}_\sharp$, according to the following diagram (see II.5.7.1, II.5.7.2):

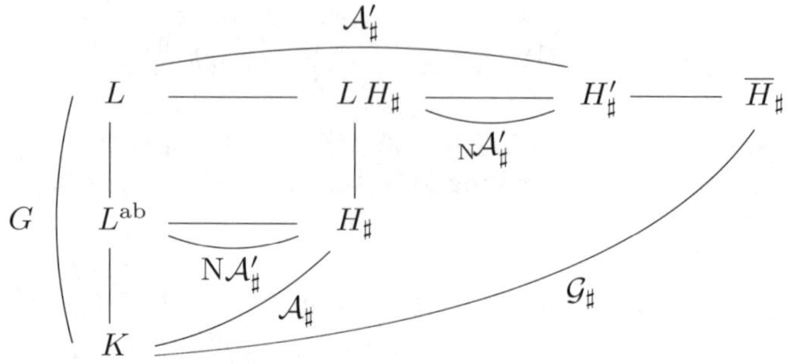

Fig. A.1

Assuming the Leopoldt conjecture for p, we know by IV.2.3 that the transfer map:

$$j_{L/K} =: j : \mathcal{A}_\sharp \longrightarrow \mathcal{A}'_\sharp$$

has kernel equal to:

$$\bigoplus_{v\in\Delta^{\mathrm{c}}_\infty} D_v(H_\sharp/K),$$

where $\Delta^{\mathrm{c}}_\infty := \Delta^{\mathrm{c}}_\infty(L/K) = \Delta^{\mathrm{c}}_\infty(\overline{H}/K)$ (for sufficiently large G). Therefore, we have the isomorphism:

$$_\nu\mathcal{A}'_\sharp/_{\mathrm{N}}\mathcal{A}'_\sharp \simeq \Big(\bigoplus_{v\in\Delta^{\mathrm{c}}_\infty} D_v(H_\sharp/K)\Big) \bigcap \mathrm{Gal}(H_\sharp/L^{\mathrm{ab}})$$

(send $\sigma' \in {}_\nu\mathcal{A}'_\sharp$ to $\mathrm{N}\sigma'$ which is an element of the kernel of j since $\nu = j \circ \mathrm{N}$; if $\sigma \in \mathrm{Ker}(j) \cap \mathrm{N}\mathcal{A}'_\sharp$, then $\sigma = \mathrm{N}\sigma'$ with $\sigma' \in \mathcal{A}'_\sharp$ and $\nu(\sigma') = 1$, giving the surjectivity).

The group on the right hand side is the kernel of the map (similar to f):

$$\bigoplus_{v\in\Delta^{\mathrm{c}}_\infty} D_v(H_\sharp/K) \longrightarrow G^{\mathrm{ab}}.$$

The above quotient is thus isomorphic to $\mathrm{Ker}(f)$ since, by the deployment Theorem III.4.1 (for $\mathrm{Gal}(\overline{K}^{\mathrm{ab}}{}_{(p)}/K)$), the projection:

$$\bigoplus_{v\in\Delta^{\mathrm{c}}_\infty} D_v(\overline{K}^{\mathrm{ab}}/K) \longrightarrow \bigoplus_{v\in\Delta^{\mathrm{c}}_\infty} D_v(H_\sharp/K),$$

is an isomorphism, and the exact sequence (0) which is now:

$$1 \longrightarrow H^{-1}(G, C') \longrightarrow {}_\nu\mathcal{A}'_\sharp/I_G\mathcal{A}'_\sharp \xrightarrow{\ \mathrm{N}\ } \mathrm{Ker}(j) \cap \mathrm{Gal}(H_\sharp/L^{\mathrm{ab}}) \longrightarrow 1,$$

expresses the isomorphisms:

$$H^{-3}(G, \mathbb{Z}) \simeq H^{-1}(G, C') \simeq {}_{\mathrm{N}}\mathcal{A}'_\sharp/I_G\mathcal{A}'_\sharp.$$

Hence, for all sufficiently large G we have for any p:

$$H^{-3}(G,\mathbb{Z})/p^e H^{-3}(G,\mathbb{Z}) \simeq {}_{\mathrm{N}}\mathcal{A}'_\sharp/({}_{\mathrm{N}}\mathcal{A}'_\sharp)^{p^e}.\, I_G\mathcal{A}'_\sharp.$$

Finally, we have obtained, using Notations 1.1:

1.4 Proposition (fundamental exact sequences). *Assuming the Leopoldt conjecture for p in $\overline{H}$, we have the exact sequences:*

$$1 \longrightarrow {}_{\mathrm{N}}\mathcal{A}'_\sharp/({}_{\mathrm{N}}\mathcal{A}'_\sharp)^{p^e}.\, I_G\mathcal{A}'_\sharp \longrightarrow H^2(G,(\mathbb{Z}/p^e\mathbb{Z})^*)^* \xrightarrow{\ \delta\ } {}_{p^e}G^{\mathrm{ab}} \longrightarrow 1, \quad (1)$$

$$1 \longrightarrow (\mathrm{Inf}(H^2(G,\mathbb{Z}/p^e\mathbb{Z})))^* \longrightarrow H^2(G,\mathbb{Z}/p^e\mathbb{Z})^* \longrightarrow \mathcal{A}'/\mathcal{A}'^{p^e}.I_G\mathcal{A}' \longrightarrow 1, \quad (2)$$

for any sufficiently large quotient G of $\mathcal{G}$. □

The exact sequence (2) is the one which defines $\mathrm{Ker}_{e,\,G}$ in the dual form (Proposition 1.3) (note that there we have come back to the group $\mathcal{A}'$).

In the next subsection, we will study more closely the group $\mathcal{A}'/\mathcal{A}'^{p^e}.\, I_G\mathcal{A}'$.

c) A Class Field Theory Formula for $|\mathrm{Inf}(H^2(G,\mathbb{Z}/p^e\mathbb{Z}))|$

Relative to the extension $\overline{H}/L/K$, we have the following diagram, analogous to the preceding one (except that here, in an evident sense, L and L^{ab} tend respectively to $\overline{H}$ and H, that p-ramification is incomplete, and that there is finite decomposition):

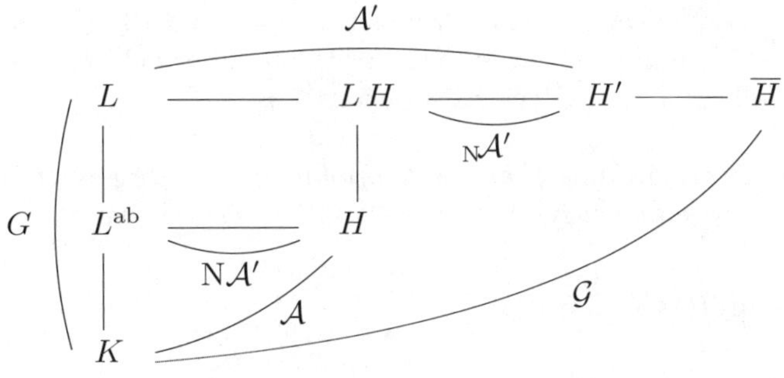

Fig. A.2

Recall that $\mathrm{N} := \mathrm{N}_{L/K}$. We can always assume that G is such that:

$$\mathrm{N}\mathcal{A}' \subseteq \mathcal{A}^{p^n},$$

for $n \geq e$ such that $\mathcal{A}^{p^n}$ does not have any $\mathbb{Z}_p$-torsion (it is sufficient that p^n kills $\mathcal{T}$), so that $\mathrm{N}\mathcal{A}'$ is free with finite index in $\mathcal{A}$. Let us show that in the canonical exact sequence:

$$1 \longrightarrow {}_{\mathrm{N}}\mathcal{A}'/{}_{\mathrm{N}}\mathcal{A}' \cap (\mathcal{A}'^{p^e}.\, I_G\mathcal{A}') \longrightarrow \mathcal{A}'/\mathcal{A}'^{p^e}.\, I_G\mathcal{A}' \xrightarrow{\ \mathrm{N}\ } \mathrm{N}\mathcal{A}'/\mathrm{N}\mathcal{A}'^{p^e} \longrightarrow 1,$$

we have:

$${}_{\mathrm{N}}\mathcal{A}' \cap (\mathcal{A}'^{p^e}.\, I_G\mathcal{A}') = ({}_{\mathrm{N}}\mathcal{A}')^{p^e}.\, I_G\mathcal{A}'.$$

If $\sigma =: \sigma_1^{p^e}\sigma_2$, with $\sigma \in {}_{\mathrm{N}}\mathcal{A}'$, $\sigma_1 \in \mathcal{A}'$, $\sigma_2 \in I_G\mathcal{A}'$, we obtain $\mathrm{N}\sigma_1^{p^e} = 1$, hence $\mathrm{N}\sigma_1 \in \mathcal{T}$, so that $\sigma_1 \in {}_{\mathrm{N}}\mathcal{A}'$ since $\mathrm{N}\mathcal{A}'$ is free. Now, it is easily checked that there exists a *canonical* isomorphism:

$${}_{p^e}G^{\mathrm{ab}}/\pi({}_{p^e}\mathcal{T}) \longrightarrow \mathrm{N}\mathcal{A}'/\mathrm{N}\mathcal{A}'^{p^e},$$

where π is the projection $\mathcal{A} \longrightarrow G^{\mathrm{ab}}$: use the fact that $G^{\mathrm{ab}} \simeq \mathcal{A}/\mathrm{N}\mathcal{A}'$, with $\mathrm{N}\mathcal{A}' \subseteq \mathcal{A}^{p^n}$ for a large $n \geq e$; then, associate with $\sigma \in \mathcal{A}$, such that $\sigma^{p^e} \in \mathrm{N}\mathcal{A}'$, the image of σ^{p^e} in $\mathrm{N}\mathcal{A}'/\mathrm{N}\mathcal{A}'^{p^e}$. Therefore, we have obtained, using Notations 1.1:

1.5 Proposition. *For any sufficiently large finite quotient G of $\mathcal{G}$ we have the exact sequence:*

$$1 \longrightarrow {}_{\mathrm{N}}\mathcal{A}'/({}_{\mathrm{N}}\mathcal{A}')^{p^e}.\, I_G\mathcal{A}' \longrightarrow \mathcal{A}'/\mathcal{A}'^{p^e}.\, I_G\mathcal{A}' \longrightarrow {}_{p^e}G^{\mathrm{ab}}/\pi({}_{p^e}\mathcal{T}) \longrightarrow 1. \qquad (3)$$

where π is the projection $\mathcal{A} \longrightarrow G^{\mathrm{ab}}$. □

Note. A fortiori ${}_{\mathrm{N}}\mathcal{A}'$ is not necessarily equal to ${}_{\nu}\mathcal{A}'$ in incomplete p-ramification (since the transfer map j may not be injective even in the case $p \neq 2$) and ${}_{\mathrm{N}}\mathcal{A}'/I_G\mathcal{A}'$ is therefore not ${}_{\nu}\mathcal{A}'/I_G\mathcal{A}' := H^{-1}(G, \mathcal{A}')$, a phenomenon that we have already seen.

The exact sequences (1), (2) (Proposition 1.4), and (3) (Proposition 1.5) imply the first general result as follows, for the number field K given together with sets of places T and S (with Notations 1.1).

1.6 Proposition. *Assume that the Leopoldt conjecture is satisfied for p in $\overline{H}$. Then for any sufficiently large finite quotient G of $\mathcal{G}$ we have:*

$$|(\mathrm{Inf}(H^2(G, \mathbb{Z}/p^e\mathbb{Z})))^*| = \frac{|{}_{\mathrm{N}}\mathcal{A}'_\sharp/({}_{\mathrm{N}}\mathcal{A}'_\sharp)^{p^e}.\, I_G\mathcal{A}'_\sharp|}{|{}_{\mathrm{N}}\mathcal{A}'/({}_{\mathrm{N}}\mathcal{A}')^{p^e}.\, I_G\mathcal{A}'|} \times |{}_{p^e}\mathcal{T}|. \qquad \square$$

§2 Complete p-Ramification Without Finite Decomposition

We are going to give a new proof of the corresponding classical results using the above computations.

Assumptions. We assume in this section that T contains Pl_p and that $S_0 = \emptyset$. We will see that the subsets T_{ta} (formed by the tame places of T)

and S_∞ are only involved in a simple way. We still assume that the Leopoldt conjecture for p is satisfied in $\overline{H}$. □

The formula of 1.6 that we have proved above yields, since $\mathcal{A}' = \mathcal{A}'_\sharp$ and since $\mathrm{Inf}(H^2(G, \mathbb{Z}/p^e\mathbb{Z}))$ stabilizes, $|H^2(\mathcal{G}, \mathbb{Z}/p^e\mathbb{Z})^*| = |{}_{p^e}\mathcal{T}|$ for all $e \geq 1$. It is immediate to exhibit a canonical isomorphism:

$$H^2(\mathcal{G}, (\mathbb{Z}/p^e\mathbb{Z})^*)^* \longrightarrow {}_{p^e}\mathcal{T},$$

since the exact sequences (1), (2), (3) are equivalent to a commutative diagram giving the result, but this will be done later in complete generality (use Theorem 3.7 in which $\mathcal{E}_T^S = 1$ because of the Leopoldt conjecture, or show directly that the restriction of δ to $H^2(\mathcal{G}, (\mathbb{Z}/p^e\mathbb{Z})^*)^*$ is injective).

2.1 Cohomological Variant of (3). To relate this with what is known in the case $Pl_p \subseteq T$, $S_0 = \emptyset$, Nguyen Quang Do has noted that we have, by duality, the isomorphism:

$$H^{-3}(G, \mathbb{Z})/p^e H^{-3}(G, \mathbb{Z}) \simeq {}_{p^e}H^2(G, \mathbb{Q}_p/\mathbb{Z}_p)^*,$$

and that we can prove, without any assumption, the following general exact sequence:

$$\begin{aligned}1 \longrightarrow {}_{p^e}H^2(\mathcal{G}, \mathbb{Q}_p/\mathbb{Z}_p)^* &\longrightarrow {}_{p^e}H^2(G, \mathbb{Q}_p/\mathbb{Z}_p)^* \\ &\longrightarrow (\mathrm{Ker}_{e,\,G})^* \longrightarrow {}_{p^e}G^{\mathrm{ab}}/\pi({}_{p^e}\mathcal{T}) \longrightarrow 1,\end{aligned}$$

where π is the projection $\mathcal{A} \longrightarrow G^{\mathrm{ab}}$.

Hence, reorganizing the above arguments, we obtain the exact sequence (for sufficiently large G):

$$\begin{aligned}1 \longrightarrow {}_{p^e}H^2(\mathcal{G}, \mathbb{Q}_p/\mathbb{Z}_p)^* &\longrightarrow {}_{\mathrm{N}}\mathcal{A}'/({}_{\mathrm{N}}\mathcal{A}')^{p^e} . I_G\mathcal{A}' \\ &\longrightarrow \mathcal{A}'/\mathcal{A}'^{p^e} . I_G\mathcal{A}' \longrightarrow {}_{p^e}G^{\mathrm{ab}}/\pi({}_{p^e}\mathcal{T}) \longrightarrow 1. \quad (3')\end{aligned}$$

The Leopoldt conjecture for p in $\overline{H}$ then classically implies that we have $H^2(\mathcal{G}, \mathbb{Q}_p/\mathbb{Z}_p) = 1$ (after [g, Hab; NSW, Ch. X, § 3], [Ng1]), and thus we recover (3) which, together with (1) and (2), which remain the crucial points of the reasoning, gives again Proposition 1.6. Moreover, (1) and (2), independently of (3), need the Leopoldt conjecture.

This allows us to conclude on the desired interpretation under the assumptions of complete p-ramification without finite decomposition, but with arbitrary infinite decomposition, which the literature on the subject usually does not mention.

Recall that $\overline{H}_{T}^{S}(p)$ is the maximal T-ramified S-split pro-p-extension of K, $\mathcal{G}_T^S := \mathrm{Gal}(\overline{H}_{T}^{S}(p)/K)$, and $\mathcal{T}_T^S$ is the torsion group of the abelianization of $\mathcal{G}_T^S$.

2.2 Theorem. *Assume that T contains Pl_p, that $S_0 = \emptyset$, and that the Leopoldt conjecture is satisfied for p in $\overline{H}_T^{S_\infty}(p)$.*

(i) We have the following rank formulas:

$$\mathrm{rk}_p(H^2(\mathcal{G}_T^{S_\infty}, \mathbb{Z}/p\mathbb{Z})) = \mathrm{rk}_p(\mathcal{T}_T^{S_\infty}) = \mathrm{rk}_p(V_T^{S_\infty}/K_T^{\times p}) + \sum_{v \in T} \delta_v - \delta + \delta_{2,p}|\Delta_\infty|,$$

$$\mathrm{rk}_p(H^2(\mathcal{G}_T^{S_\infty}, \mathbb{Z}/p\mathbb{Z})) - \mathrm{rk}_p(H^1(\mathcal{G}_T^{S_\infty}, \mathbb{Z}/p\mathbb{Z})) = -(r_2 + 1),$$

where $V_T^{S_\infty} := \{\alpha \in K_T^{\times p} K_{T,\Delta_\infty}^\times,\ (\alpha) = \mathfrak{a}^p,\ \mathfrak{a} \in I_T,\ i_v(\alpha) \in K_v^{\times p}\ \ \forall v \in T\}$.

(ii) We have:

$$H^2(\mathcal{G}_T^{S_\infty}, \mathbb{Z}_p) := \varprojlim_{e \geq 1} H^2(\mathcal{G}_T^{S_\infty}, \mathbb{Z}/p^e\mathbb{Z}) \simeq (\mathcal{T}_T^{S_\infty})^*.$$

Proof. (i) For $e = 1$ then $H^2(\mathcal{G}_T^{S_\infty}, \mathbb{Z}/p\mathbb{Z})$ and ${}_p\mathcal{T}_T^{S_\infty}$ are $\mathbb{F}_p$-vector spaces whose common dimension is given by Formula III.4.2 for $S_0 = \emptyset$ and assuming the Leopoldt conjecture for p, proving the first equality.

Using the expression I.4.6, (i) of $\mathrm{rk}_p(H^1(\mathcal{G}_T^{S_\infty}, \mathbb{Z}/p\mathbb{Z}))$, we obtain the second one.

(ii) Since $H^2(\mathcal{G}_T^{S_\infty}, \mathbb{Z}/p^e\mathbb{Z})$ is constant for e sufficiently large, the isomorphisms:

$$H^2(\mathcal{G}_T^{S_\infty}, \mathbb{Z}/p^e\mathbb{Z}) \simeq (\mathcal{T}_T^{S_\infty})^*,$$

enable us to go to the limit and to obtain the interpretation of $H^2(\mathcal{G}_T^{S_\infty}, \mathbb{Z}_p)$. □

We thus obtain the necessary and sufficient conditions for the vanishing of $H^2(\mathcal{G}_T^{\mathrm{ord}}, \mathbb{Z}/p\mathbb{Z})$ stated in III.4.2.5 to characterize the cases where $\mathcal{G}_T^{\mathrm{ord}}$ (with $Pl_p \subseteq T$) is a free pro-p-group ($\mathcal{G}_T^{S_\infty}$ for $p = 2$ and $S_\infty \neq Pl_\infty^{\mathrm{r}}$ never being one).

2.2.1 Corollary. *The group $\mathcal{G}_T^{\mathrm{ord}}$, $T \supseteq Pl_p$, is a free pro-p-group if and only if the following four conditions are satisfied:*

- *$T_{\mathrm{ta}} = \emptyset$ (i.e., $T = Pl_p$),*
- *$\displaystyle\bigoplus_{v|p} \mu_p(K_v) = i_p(\mu_p(K))$,*
- *$\mathbb{Z}_p \log_p(E'^{\mathrm{ord}})$ is a direct summand in $\displaystyle\bigoplus_{v|p} \log(U_v^1)$,*
- *the ordinary p-Hilbert class field is contained in the compositum of the $\mathbb{Z}_p$-extensions of K.*

In that case, it is pro-p-free on $r_2 + 1$ generators. □

We also deduce from this theorem, and from Theorem III.4.1.5, a formula which refers to the classical arithmetic invariants (classes and units) of the base field K.

2.2.2 Corollary. *When $T \supseteq Pl_p$, we obtain:*

$$|H^2(\mathcal{G}_T^{S_\infty}, \mathbb{Z}_p)| = |\mathcal{T}_p^{\mathrm{ord}}| \times \prod_{v \in T_{\mathrm{ta}} \cup \Delta_\infty} |(F_v^\times)_p|$$
$$= |\mathcal{T}_p^{\mathrm{ord}}| \times \prod_{v \in T_{\mathrm{ta}}} (q_v - 1)_p \times (2^{|\Delta_\infty|})_p,$$

where $\Delta_\infty := Pl_\infty^{\mathrm{r}} \backslash S_\infty$ *and where* $|\mathcal{T}_p^{\mathrm{ord}}|$ *is given by III.2.6.1, (ii*$_2$*).* □

§3 The General Case — Infinitesimal Knot Groups

From now on we do not make any assumption on T and S, but we still assume that the Leopoldt conjecture is satisfied for p in $\overline{H}$. Recall that we have set $T_\sharp := T \cup Pl_p$, $S_\sharp := S_\infty$, and that the invariants indexed by $\sharp$ are relative to $T_\sharp$ and $S_\sharp$.

We then only have the formula proved in Proposition 1.6, where $\mathrm{N} := \mathrm{N}_{L/K}$, $G = \mathrm{Gal}(L/K)$:

$$|(\mathrm{Inf}(H^2(G, \mathbb{Z}/p^e\mathbb{Z})))^*| = \frac{|{}_{\mathrm{N}}\mathcal{A}'_\sharp/({}_{\mathrm{N}}\mathcal{A}'_\sharp)^{p^e} . I_G \mathcal{A}'_\sharp|}{|{}_{\mathrm{N}}\mathcal{A}'/({}_{\mathrm{N}}\mathcal{A}')^{p^e} . I_G \mathcal{A}'|} \times |{}_{p^e}\mathcal{T}|,$$

which comes from the exact sequences (1), (2), and (3).

Consider the following general diagram:

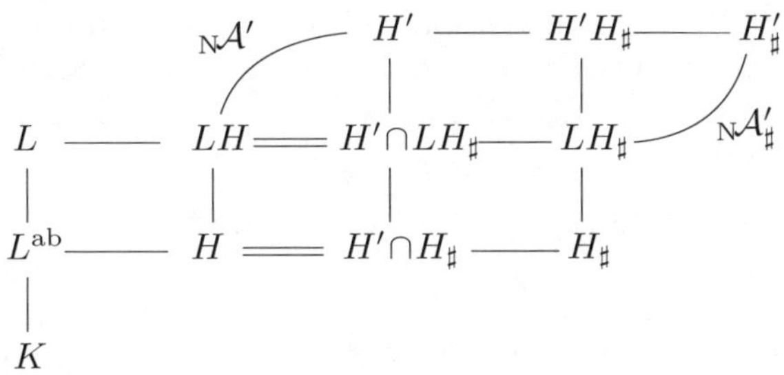

Fig. A.3

We have (see Fig. A.1 and A.2):

$${}_{\mathrm{N}}\mathcal{A}'_\sharp = \mathrm{Gal}(H'_\sharp/LH_\sharp), \quad {}_{\mathrm{N}}\mathcal{A}' = \mathrm{Gal}(H'/LH),$$

and $H' \cap LH_\sharp = LH$ (indeed, H'/K is T-ramified and S-split, and $H' \cap H_\sharp = H' \cap LH_\sharp \cap H_\sharp$ is a T-ramified S-split abelian extension of K, hence equal to H). The restriction $\mathcal{A}'_\sharp \longrightarrow \mathcal{A}'$ thus gives on ${}_{\mathrm{N}}\mathcal{A}'_\sharp$ the exact sequence:

$$1 \longrightarrow X_e \longrightarrow {}_{\mathrm{N}}\mathcal{A}'_\sharp/({}_{\mathrm{N}}\mathcal{A}'_\sharp)^{p^e} . I_G \mathcal{A}'_\sharp \longrightarrow {}_{\mathrm{N}}\mathcal{A}'/({}_{\mathrm{N}}\mathcal{A}')^{p^e} . I_G \mathcal{A}' \longrightarrow 1. \quad (4)$$

a) Infinitesimal Computations

We are going to see that this kernel X_e is an infinitesimal knot group of units which is quite difficult to understand and which also involves complete p-ramification.

We will use the following additional simplified notations (see III.1.6.6 and III.2.4.1).

3.1 Notations. (i) For arbitrary sets T and S, we put:[7]

- $\mathcal{E}_\infty := \mathcal{E}_T^S := \{\varepsilon \in E^S \otimes \mathbb{Z}_p,\ \bar{i}_T(\varepsilon) = 1\}$,
- $\mathcal{K}_\infty^\times := \{x \in K_{T_\sharp, \Delta_\infty}^\times \otimes \mathbb{Z}_p,\ \bar{i}_T(x) = 1\}$,
- $\mathcal{L}_\infty^\times := \{y \in L_{T'_\sharp, \Delta'_\infty}^\times \otimes \mathbb{Z}_p,\ \bar{i}'_{T'}(y) = 1\}$,
- $\mathcal{P}_\infty := \{(x),\ x \in \mathcal{K}_\infty^\times\}$, $\mathcal{P}'_\infty := \{(y),\ y \in \mathcal{L}_\infty^\times\}$,
- $\mathcal{I} := I_{T_\sharp} \otimes \mathbb{Z}_p$, $\mathcal{I}' := I_{L, T'_\sharp} \otimes \mathbb{Z}_p$.

Relative to the extension L/K (T-ramified and S-split), we set (with the usual abuse of notation):

- $\mathcal{N}_\infty := \mathcal{K}_\infty^\times \cap \mathrm{N}_{L/K}(J_L)$.

This defines the group of Δ_∞-positive T-infinitesimal elements of K, prime to p, which are local norms everywhere in L/K.[8]

(ii) In the case of complete ramification without finite decomposition (i.e., in the context of $T \cup Pl_p$-ramification and S_∞-decomposition), we systematically add the index $\sharp$ to the above notations (the symbol ∞, relative to $T_\sharp$, is then not the same and is denoted $\sharp\infty$). We thus obtain the following objects:

- $\mathcal{E}_{\sharp, \sharp\infty} := \mathcal{E}_{T_\sharp}^{S_\infty} := \{\varepsilon \in E^{S_\infty} \otimes \mathbb{Z}_p,\ \bar{i}_{T_\sharp}(\varepsilon) = 1\}$,

which is trivial assuming the Leopoldt conjecture for p,

- $\mathcal{K}_{\sharp\infty}^\times := \{x \in K_{T_\sharp, \Delta_\infty}^\times \otimes \mathbb{Z}_p,\ \bar{i}_{T_\sharp}(x) = 1\}$,
- $\mathcal{L}_{\sharp\infty}^\times := \{y \in L_{T'_\sharp, \Delta'_\infty}^\times \otimes \mathbb{Z}_p,\ \bar{i}'_{T'_\sharp}(y) = 1\}$,
- $\mathcal{P}_{\sharp\infty} := \{(x),\ x \in \mathcal{K}_{\sharp\infty}^\times\}$, $\mathcal{P}'_{\sharp\infty} := \{(y),\ y \in \mathcal{L}_{\sharp\infty}^\times\}$,
- $\mathcal{A}_\sharp \simeq \mathcal{I}/\mathcal{P}_{\sharp\infty}$, $\mathcal{A}'_\sharp \simeq \mathcal{I}'/\mathcal{P}'_{\sharp\infty}$,
- $\mathcal{N}_{\sharp\infty} := \mathcal{K}_{\sharp\infty}^\times \cap \mathrm{N}_{L/K}(J_L)$. □

We then have, by 3.1, (i), $\mathcal{A} \simeq \mathcal{I}/\mathcal{P}_\infty \,.\, \mathcal{S}_0$ and $\mathcal{A}' \simeq \mathcal{I}'/\mathcal{P}'_\infty \,.\, \mathcal{S}'_0$, where $\mathcal{S}_0 := \langle\, S_0 \,\rangle \otimes \mathbb{Z}_p$ and $\mathcal{S}'_0 := \langle\, S'_0 \,\rangle \otimes \mathbb{Z}_p$.

3.2 Proposition. *We have the canonical isomorphisms:*

[7] Here, the use in some notations of $T_\sharp$ instead of T is unimportant but will allow us to represent classes by ideals prime to T and p, which will be essential later.

[8] The elements of $\mathcal{K}_\infty^\times$ are trivially local norms on $T \cup S$ and on Δ_∞.

$$ {}_{\mathrm{N}}\mathcal{A}'/I_G\mathcal{A}' \simeq \mathcal{N}_\infty/\mathrm{N}\mathcal{L}_\infty^\times \,.\, \mathcal{E}_\infty, \quad {}_{\mathrm{N}}\mathcal{A}'_\sharp/I_G\mathcal{A}'_\sharp \simeq \mathcal{N}_{\sharp\infty}/\mathrm{N}\mathcal{L}_{\sharp\infty}^\times. $$

Proof. Let $\sigma' \in \mathcal{A}'$ be such that $\mathrm{N}\sigma' = 1$. We have $\sigma' =: \mathrm{Art}'(\mathfrak{a}')$, $\mathfrak{a}' \in \mathcal{I}'$, where $\mathrm{Art}' : \mathcal{I}' \longrightarrow \mathcal{A}'$ is the Artin map, and the ideal $\mathfrak{a}'$ satisfies:

$$ \mathrm{N}\mathfrak{a}' =: (x_\infty)\,\mathfrak{a}_{S_0}, \quad x_\infty \in \mathcal{K}_\infty^\times, \quad \mathfrak{a}_{S_0} \in \mathcal{S}_0. $$

Since $\mathfrak{a}'$ is defined modulo $\mathcal{P}'_\infty \,.\, \mathcal{S}'_0$, it is clear that x_∞ is defined modulo $\mathrm{N}\mathcal{L}_\infty^\times \,.\, \mathcal{E}_\infty$. Finally, (x_∞) can be written:

$$ (x_\infty) = \mathfrak{a}_{S_0}^{-1}\,\mathrm{N}\mathfrak{a}' = \mathrm{N}(\mathfrak{a}'^{-1}_{S_0}\,\mathfrak{a}'), \quad \mathfrak{a}'_{S_0} \in \mathcal{S}'_0 $$

since L/K is S_0-split; it follows that x_∞, which is already a local norm on $T \cup S \cup \Delta_\infty$, is a local norm everywhere in L/K (i.e., $x_\infty \in \mathcal{N}_\infty$). We send σ' to the class of x_∞ modulo $\mathrm{N}\mathcal{L}_\infty^\times \,.\, \mathcal{E}_\infty$.

If $x_\infty = \mathrm{N}y_\infty \,.\, \varepsilon_\infty$, $y_\infty \in \mathcal{L}_\infty^\times$, $\varepsilon_\infty \in \mathcal{E}_\infty$, we obtain $\mathrm{N}\mathfrak{a}' =: \mathrm{N}(y_\infty) \,.\, \mathfrak{b}_{S_0}$, $\mathfrak{b}_{S_0} \in \mathcal{S}_0$, but $\mathfrak{b}_{S_0} = \mathrm{N}\mathfrak{b}'_{S_0}$, $\mathfrak{b}'_{S_0} \in \mathcal{S}'_0$. Since $\mathcal{I}'$ is an induced G-module, we have ${}_{\mathrm{N}}\mathcal{I}' = I_G\mathcal{I}'$, and:

$$ \mathfrak{a}' = (y_\infty) \,.\, \mathfrak{b}'_{S_0} \,.\, \mathfrak{c}', \quad \mathfrak{c}' \in I_G\mathcal{I}', $$

hence:

$$ \sigma' = \mathrm{Art}'((y_\infty) \,.\, \mathfrak{b}'_{S_0} \,.\, \mathfrak{c}') \in I_G\mathcal{A}'. $$

We have thus proved the injectivity of the given map.

Its surjectivity comes from the fact that, in $\mathcal{N}_\infty$ any x_∞ is such that $(x_\infty) = \mathrm{N}\mathfrak{a}'$, $\mathfrak{a}' \in \mathcal{I}'$: indeed, the ideal (x_∞) is prime to T and, since L/K is T-ramified, the condition that x_∞ is a local norm at $v \notin T$ implies that the v-valuation (here in $\mathbb{Z}_p$) of x_∞ is a multiple of the residue degree of v in L/K (see II.1.4.3); this yields the preimage $\sigma' := \mathrm{Art}'(\mathfrak{a}')$.

Similarly, we have:

$$ {}_{\mathrm{N}}\mathcal{A}'_\sharp/I_G\mathcal{A}'_\sharp \simeq \mathcal{N}_{\sharp\infty}/\mathrm{N}\mathcal{L}_{\sharp\infty}^\times \,.\, \mathcal{E}_{\sharp,\sharp\infty} = \mathcal{N}_{\sharp\infty}/\mathrm{N}\mathcal{L}_{\sharp\infty}^\times. \qquad \square $$

The surjectivity of ${}_{\mathrm{N}}\mathcal{A}'_\sharp \longrightarrow {}_{\mathrm{N}}\mathcal{A}'$ is equivalent to the equality:

$$ \mathcal{N}_\infty = \mathcal{N}_{\sharp\infty} \,.\, \mathrm{N}\mathcal{L}_\infty^\times \,.\, \mathcal{E}_\infty. $$

The surjection ${}_{\mathrm{N}}\mathcal{A}'_\sharp/({}_{\mathrm{N}}\mathcal{A}'_\sharp)^{p^e} \,.\, I_G\mathcal{A}'_\sharp \longrightarrow {}_{\mathrm{N}}\mathcal{A}'/({}_{\mathrm{N}}\mathcal{A}')^{p^e} \,.\, I_G\mathcal{A}'$ in the exact sequence (4) gives rise to the surjection:

$$ \mathcal{N}_{\sharp\infty}/\mathcal{N}_{\sharp\infty}^{p^e} \,.\, \mathrm{N}\mathcal{L}_{\sharp\infty}^\times \longrightarrow \mathcal{N}_\infty/\mathcal{N}_\infty^{p^e} \,.\, \mathrm{N}\mathcal{L}_\infty^\times \,.\, \mathcal{E}_\infty \simeq \mathcal{N}_{\sharp\infty}/\mathcal{N}_{\sharp\infty}^{p^e} \,.\, ((\mathrm{N}\mathcal{L}_\infty^\times \,.\, \mathcal{E}_\infty) \cap \mathcal{N}_{\sharp\infty}), $$

which yields:

3.2.1 Proposition. *The kernel X_e in (4) is isomorphic to:*

$$\mathcal{N}_{\sharp\infty}^{p^e} . ((\mathrm{N}\mathcal{L}_\infty^\times . \mathcal{E}_\infty) \cap \mathcal{N}_{\sharp\infty}) / \mathcal{N}_{\sharp\infty}^{p^e} . \mathrm{N}\mathcal{L}_{\sharp\infty}^\times . \qquad \square$$

We can already obtain from this expression an extension of the results of Section 2 in the monogeneous case for complete p-ramification.

3.2.2 Corollary. *If T contains Pl_p and if E^S is monogeneous then, assuming the p-adic conjecture, we have $X_1 = 1$, in other words:*

$$\mathrm{rk}_p(H^2(\mathcal{G}_T^S, \mathbb{Z}/p\mathbb{Z})) = \mathrm{rk}_p(V_T^S / K_T^{\times p}) + \sum_{v \in T} \delta_v - \delta + \delta_{2,p} |\Delta_\infty| . \qquad \square$$

However, we are going to deduce a number of new consequences.

b) Infinitesimal Knot Groups — The Number of Relations — A Generalization of Šafarevič's Results

In the case $S_0 = \emptyset$ and T arbitrary, Šafarevič in his 1964 paper [Ša] [9] has given an upper bound for $\mathrm{rk}_p(H^2(\mathcal{G}_T^{S_\infty}, \mathbb{Z}/p\mathbb{Z}))$ which is, with our notations:

$$\mathrm{rk}_p(H^2(\mathcal{G}_T^{S_\infty}, \mathbb{Z}/p\mathbb{Z})) \leq \mathrm{rk}_p(V_T^{S_\infty} / K_T^{\times p}) + \sum_{v \in T} \delta_v - \delta + \delta_{2,p} |\Delta_\infty|,$$

except for $T_p = \{v \in T_{\mathrm{ta}},\ \mathrm{N}\mathfrak{p}_v \equiv 1 \bmod (p)\} = \Delta_\infty = \emptyset$ and $\delta = 1$, where we have:

$$\mathrm{rk}_p(H^2(\mathcal{G}^{\mathrm{ord}}, \mathbb{Z}/p\mathbb{Z})) \leq \mathrm{rk}_p(V^{\mathrm{ord}} / K^{\times p})$$

(in other words, in this extreme case we gain one dimension with respect to the general expression).

Let us show how to generalize this result from the above expression 3.2.1 for X_e which we must first simplify. Thus, let T and S be arbitrary. We need the following two lemmas, in which we set $\Delta_p := Pl_p \backslash T_p$, and we use the infinitesimal chinese remainder theorem which says that we have, for instance in L, the exact sequence:

$$1 \longrightarrow \mathcal{L}_{\sharp\infty}^\times \longrightarrow \mathcal{L}_\infty^\times \xrightarrow{\overline{i}'_{\Delta'_p}} \bigoplus_{w \in \Delta'_p} U_w^1 \longrightarrow 1.$$

3.3 Lemma 1. *Let $\varepsilon_\infty \in \mathcal{E}_\infty$. Then there exists $y_\infty \in \mathcal{L}_\infty^\times$ such that $\mathrm{N}y_\infty . \varepsilon_\infty \in \mathcal{N}_{\sharp\infty}$.*

Proof. Since L/K is Galois and unramified in Δ_p, the norm:

[9] See also [e, Ko3, Ch. 3, § 2, Th. 3.76], [g, NSW, Ch. VIII, § 7, Cor. 8.7.5], [Mai2], and [Mai4].

$$\mathrm{N} : \bigoplus_{w \in \Delta'_p} U^1_w \longrightarrow \bigoplus_{v \in \Delta_p} U^1_v$$

is surjective. By approximation, there exists $y_\infty \in \mathcal{L}^\times_\infty$ such that:

$$\mathrm{N}(\overline{i}'_{\Delta'_p}(y_\infty)) = \overline{i}_{\Delta_p}(\varepsilon_\infty)^{-1},$$

which can also be written as $\overline{i}_{\Delta_p}(\mathrm{N}y_\infty) = \overline{i}_{\Delta_p}(\varepsilon_\infty)^{-1}$, which yields:

$$\overline{i}_{\Delta_p}(\mathrm{N}y_\infty \,.\, \varepsilon_\infty) = 1 \;;$$

the result follows since ε_∞ is a local norm everywhere (look separately at the cases $v \in T$, $v \in \Delta_p$, $v \in S$, and $v \notin T_\sharp \cup S$). □

3.4 Lemma 2. *We have* $\mathrm{N}\mathcal{L}^\times_\infty \cap \mathcal{N}_{\sharp\infty} = \mathrm{N}\mathcal{L}^\times_{\sharp\infty}$.

Proof. Let $y_\infty \in \mathcal{L}^\times_\infty$ be such that $\mathrm{N}y_\infty \in \mathcal{N}_{\sharp\infty}$. Therefore $\overline{i}_{\Delta_p}(\mathrm{N}y_\infty) = 1$, in other words $\mathrm{N}(\overline{i}'_{\Delta'_p}(y_\infty)) = 1$; then, since L/K is unramified in Δ_p, we easily obtain (Shapiro's lemma, which is reduced here to the cyclic case where $H^{-1} = H^1$):

$$H^{-1}\Big(G, \bigoplus_{w \in \Delta'_p} U^1_w\Big) = \bigoplus_{v \in \Delta_p} H^{-1}(D_{w_0}, U^1_{w_0}) = 1,$$

hence $\overline{i}'_{\Delta'_p}(y_\infty) \in I_G\Big(\bigoplus_{w \in \Delta'_p} U^1_w\Big)$. By approximation, we can find $y^*_\infty \in I_G\mathcal{L}^\times_\infty$ such that:

$$\overline{i}'_{\Delta'_p}(y^*_\infty) = \overline{i}'_{\Delta'_p}(y_\infty)^{-1},$$

in which case:

$$y_{\sharp\infty} := y_\infty\, y^*_\infty \in \mathcal{L}^\times_{\sharp\infty}$$

is such that $\mathrm{N}y_\infty = \mathrm{N}y_{\sharp\infty}$. □

Thus, let $\varepsilon_\infty \in \mathcal{E}_\infty$; there exists $y_\infty \in \mathcal{L}^\times_\infty$ such that $x := \mathrm{N}y_\infty \,.\, \varepsilon_\infty \in \mathcal{N}_{\sharp\infty}$ (Lemma 1); we send ε_∞ to the image of x in X_e. If $x' := \mathrm{N}y'_\infty \,.\, \varepsilon_\infty \in \mathcal{N}_{\sharp\infty}$, $x'\, x^{-1} = \mathrm{N}(y'_\infty \,.\, y^{-1}_\infty) \in \mathrm{N}\mathcal{L}^\times_\infty \cap \mathcal{N}_{\sharp\infty}$, hence $x' = x\,\mathrm{N}(y_{\sharp\infty})$ (Lemma 2) which has the same image in X_e. Finally, if $x = \mathrm{N}y_\infty \,.\, \varepsilon_\infty \in \mathcal{N}_{\sharp\infty}$, it is clear that ε_∞ is a preimage of x for the above map $\mathcal{E}_\infty \longrightarrow X_e$ which is thus surjective.

3.5 Proposition. *Assuming the Leopoldt conjecture for p in $\overline{H}$ we have, in complete generality (for any sufficiently large G), the exact sequence:*

$$1 \longrightarrow \mathcal{E}_\infty \cap (\mathcal{N}^{p^e}_{\sharp\infty} . \mathrm{N}\mathcal{L}^\times_\infty) \longrightarrow \mathcal{E}_\infty \longrightarrow X_e \longrightarrow 1.$$

Moreover, the group X_e stabilizes with respect to L (or to G), and for any sufficiently large G we obtain:

$$H^2(\mathcal{G},\mathbb{Z}/p^e\mathbb{Z}) = \mathrm{Inf}(H^2(G,\mathbb{Z}/p^e\mathbb{Z})).$$

Proof. The computation of the kernel is immediate. We know from exact sequence (4) that the exponent of X_e divides p^e; then, being a nondecreasing quotient of $\mathcal{E}_\infty$ which is a $\mathbb{Z}_p$-module of finite type, X_e stabilizes. The final result follows. □

3.5.1 Remark. Since $\kappa_\infty := (\mathcal{E}_\infty : \mathcal{E}_\infty \cap (\mathcal{N}_{\sharp\infty}^p\,.\,\mathrm{N}\mathcal{L}_\infty^\times))$ is a nondecreasing function with respect to L (or to G), any computation of some value of κ_∞ for a given L which does not satisfy the condition G large enough (for instance a subfield of H), yields an effective lower bound for $\mathrm{rk}_p(H^2(\mathcal{G}_T^S,\mathbb{Z}/p\mathbb{Z}))$. □

To summarize, we have obtained the following result (case $e=1$) for which we return to the standard notations that we recall: $\overline{H}_{T}^{S}(p)$ is the maximal T-ramified S-split pro-p-extension of K, $\mathcal{G}_T^S := \mathrm{Gal}(\overline{H}_{T}^{S}(p)/K)$, and $\mathcal{T}_T^S$ is the torsion group of the abelianization of $\mathcal{G}_T^S$. We use the rank formula of $\mathcal{T}_T^S$ given in III.4.2, in which enter:

$$\begin{aligned} V_T^S &:= \{\alpha \in K_T^{\times p}K_{T,\Delta_\infty}^\times,\ (\alpha) = \mathfrak{a}^p\mathfrak{a}_{S_0}, \\ &\qquad \mathfrak{a} \in I_T,\ \mathfrak{a}_{S_0} \in \langle\, S_0\,\rangle,\ \ i_v(\alpha) \in K_v^{\times p}\ \ \forall v \in T\}, \\ \mathrm{r}_{T_p}^{S_0} &:= \mathrm{rk}_{\mathbb{Z}_p}(\mathrm{adh}_{T_p}(E'^{S_0})) = \dim_{\mathbb{Q}_p}(\mathbb{Q}_p\mathrm{log}_{T_p}(E^{S_0})) \end{aligned}$$

(the T_p-adic rank of E^{S_0} defined in III.1.6.2), $\delta_v := 1$ or 0 according as K_v contains μ_p or not, $\delta := 1$ or 0 according as K contains μ_p or not, where $\delta_{2,p}$ is the Kronecker symbol, and $\Delta_\infty := Pl_\infty^{\mathrm{r}} \backslash S_\infty$.

We do not make any assumption on T and S, except $T \cap S_0 = \emptyset$.

3.6 Theorem (general expression for the number of relations). *We assume that the Leopoldt conjecture is satisfied for p in $\overline{H}_{T}^{S}(p)$. Then, for any sufficiently large finite Galois subextension L/K of $\overline{H}_{T}^{S}(p)/K$, we have:*

$$\begin{aligned} \mathrm{rk}_p(H^2(\mathcal{G}_T^S,\mathbb{Z}/p\mathbb{Z})) &= \mathrm{rk}_p(\mathcal{E}_T^S/\mathcal{E}_T^S \cap (\mathcal{N}_{\sharp\infty}^p\,.\,\mathrm{N}\mathcal{L}_\infty^\times)) + \mathrm{rk}_p(\mathcal{T}_T^S) \\ &= \mathrm{rk}_p(\mathcal{E}_T^S/\mathcal{E}_T^S \cap (\mathcal{N}_{\sharp\infty}^p\,.\,\mathrm{N}\mathcal{L}_\infty^\times)) + \mathrm{rk}_p(V_T^S/K_T^{\times p}) \\ &\qquad + \sum_{v\in T}\delta_v - \delta - (r_1+r_2-1+|S_0| - \mathrm{r}_{T_p}^{S_0}) + \delta_{2,p}|\Delta_\infty|, \end{aligned}$$

where:

$$\begin{aligned} \mathcal{E}_T^S &:= \{\varepsilon \in E^S \otimes \mathbb{Z}_p,\ \bar{i}_T(\varepsilon) = 1\}, \\ \mathcal{N}_{\sharp\infty} &:= \{x \in K_{T\cup Pl_p,\Delta_\infty}^\times \otimes \mathbb{Z}_p,\ \bar{i}_{T\cup Pl_p}(x) = 1\} \cap \mathrm{N}_{L/K}(J_L), \\ \mathcal{L}_\infty^\times &:= \{y \in L_{T'\cup Pl'_p,\Delta'_\infty}^\times \otimes \mathbb{Z}_p,\ \bar{i}'_{T'}(y) = 1\}. \end{aligned}$$

□

Note that in this statement, only $X_1 = \mathcal{E}_T^S/\mathcal{E}_T^S \cap (\mathcal{N}_{\sharp\infty}^p \,.\, \mathrm{N}\mathcal{L}_\infty^\times)$ depends on the extension L/K and that it is the maximum of the $\mathrm{rk}_p(X_1)$ which occurs.

3.7 Theorem. *In the same general context of the above Theorem 3.6, we have, for L/K sufficiently large, the exact sequence :*

$$1 \longrightarrow \mathcal{E}_T^S/\mathcal{E}_T^S \cap (\mathcal{N}_{\sharp\infty}^{p^e} \,.\, \mathrm{N}\mathcal{L}_\infty^\times) \longrightarrow H^2(\mathcal{G}_T^S, (\mathbb{Z}/p^e\mathbb{Z})^*)^* \longrightarrow {}_{p^e}\mathcal{T}_T^S \longrightarrow 1.$$

Proof. Consider the exact sequences (1), (2), (3) of Section 1, the exact sequence (4) of Section 3 which defines X_e, and Proposition 3.5. Let π be the projection $\mathcal{A} \longrightarrow G^{\mathrm{ab}}$ whose kernel is $\mathrm{N}\mathcal{A}'$ (see Fig. A.2 in Section 1). Then the exactness of the following diagram for any $e \geq 1$:

$$\begin{array}{ccccccc}
 & {}_{\mathrm{N}}\mathcal{A}'_\sharp/({}_{\mathrm{N}}\mathcal{A}'_\sharp)^{p^e} \,.\, I_G\mathcal{A}'_\sharp & \longrightarrow & H^2(G, (\mathbb{Z}/p^e\mathbb{Z})^*)^* & \xrightarrow{\delta} & {}_{p^e}G^{\mathrm{ab}} & \longrightarrow 1 \\
 & \downarrow & & \downarrow & & \downarrow & \\
1 \longrightarrow & {}_{\mathrm{N}}\mathcal{A}'/({}_{\mathrm{N}}\mathcal{A}')^{p^e} \,.\, I_G\mathcal{A}' & \longrightarrow & \mathcal{A}'/\mathcal{A}'^{p^e} \,.\, I_G\mathcal{A}' & \longrightarrow & {}_{p^e}G^{\mathrm{ab}}/\pi({}_{p^e}\mathcal{T}) &
\end{array}$$

is a tedious verification using the original definition of the maps. Finally, we apply the snake lemma to it. □

When $\mathcal{E}_T^S = 1$ (which is the case if $Pl_p \subseteq T$ and $S_0 = \emptyset$ under the Leopoldt conjecture for p), then $H^2(\mathcal{G}_T^S, (\mathbb{Z}/p^e\mathbb{Z})^*)^* \simeq {}_{p^e}\mathcal{T}_T^S$.

When $\mathcal{G}_T^S$ is finite, the left hand term of the exact sequence can be simplified (Theorem 3.8).

3.7.1 Corollary. *The existence of a surjection of the form $\mathcal{E}_T^S \longrightarrow X_e$, where the exponent of X_e divides p^e, implies the upper bound:*

$$|X_e| \leq (\mathcal{E}_T^S : (\mathcal{E}_T^S)^{p^e}),$$

and (for $e = 1$) yields:

$$\mathrm{rk}_p(X_1) \leq \mathrm{rk}_p(\mathcal{E}_T^S).$$

□

Let us stay with the case $e = 1$; by definition of $\mathcal{E}_T^S$ we have the exact sequence:

$$1 \longrightarrow \mathcal{E}_T^S \otimes \mathbb{Q}_p \longrightarrow E^S \otimes \mathbb{Q}_p \longrightarrow \mathbb{Q}_p\mathrm{log}_{T_p}(E^S) = \mathbb{Q}_p\mathrm{log}_{T_p}(E^{S_0\,\mathrm{ord}}) \longrightarrow 1,$$

hence:

$$\begin{aligned}
\mathrm{rk}_p(\mathcal{E}_T^S) &= \mathrm{rk}_{\mathbb{Z}_p}(\mathcal{E}_T^S) + \mathrm{rk}_p(\mathrm{tor}(\mathcal{E}_T^S)) \\
&= r_1 + r_2 - 1 + |S_0| - \mathrm{r}_{T_p}^{S_0} + \mathrm{rk}_p(\mathrm{tor}(\mathcal{E}_T^S)).
\end{aligned}$$

If $\xi \in \mathrm{tor}(\mathcal{E}_T^S)$, we have $\xi \in \mathrm{tor}(E^S \otimes \mathbb{Z}_p) = \mathrm{tor}(E^S) \otimes \mathbb{Z}_p$ (by flatness); but $\mathrm{tor}(E^S) \otimes \mathbb{Z}_p \simeq \mu_p(K) \cap K_{\Delta\infty}^\times$ which is trivial, except if $\mu_p(K) \neq 1$ (i.e., $\delta = 1$)

and $\Delta_\infty = \emptyset$ (when $p = 2$ and $r_1 \neq 0$). But in this case $\mu_p(K) \otimes 1 \subseteq \mathcal{E}_T^{S_0\,\mathrm{ord}}$ if and only if:

$$T_p = \{v \in T_{\mathrm{ta}},\ \mathrm{N}\mathfrak{p}_v \equiv 1 \bmod (p)\} = \emptyset.$$

We recover (for $S_0 = \emptyset$) Šafarevič's inequality (including the particular case), which is thus valid for arbitrary S.

3.7.2 Corollary. *We have:*

$$\begin{aligned}\mathrm{rk}_p(H^2(\mathcal{G}_T^S, \mathbb{Z}/p\mathbb{Z})) &= \mathrm{rk}_p(V_T^S/K_T^{\times p}) + \sum_{v\in T} \delta_v - \delta + \delta_{2,p}|\Delta_\infty| \\ &\quad + \mathrm{rk}_p(\mathrm{tor}(\mathcal{E}_T^S)) - \mathrm{rk}_p(\mathcal{E}_T^S \cap (\mathcal{N}_{\sharp\infty}^p \,.\, \mathrm{N}\mathcal{L}_\infty^\times)/(\mathcal{E}_T^S)^p), \\ &\leq \mathrm{rk}_p(V_T^S/K_T^{\times p}) + \sum_{v\in T} \delta_v - \delta + \delta_{2,p}|\Delta_\infty|,\end{aligned}$$

except in the case where $\mathcal{E}_T^S = E^{S_0\,\mathrm{ord}} \otimes \mathbb{Z}_p$ and $\mu_p(K) \neq 1$, where we have:

$$\mathrm{rk}_p(H^2(\mathcal{G}^{S_0\,\mathrm{ord}}, \mathbb{Z}/p\mathbb{Z})) \leq \mathrm{rk}_p(V^{S_0\,\mathrm{ord}}/K^{\times p}). \qquad \square$$

Note. As usual, the tame finite places v for which $(F_v^\times)_p = 1$ do not occur, and it is preferable to assume that T does not contain any; in that case we have $\sum_{v\in T} \delta_v = \sum_{v\in T_p} \delta_v + |T_{\mathrm{ta}}|$.

c) Finite Generalized p-Class Fields Towers

This Subsection (c) generalizes some results of Šafarevič [Ša], followed by Kisilevsky–Labute [KL] for the problem of class fields towers, and Maire [Mai1, Mai2] for the tame case, these results being obtained by cohomological methods on the idèle groups.

For simplicity, we first consider the context of the p-Hilbert class fields towers (in the ordinary sense, for instance). In this case, we must take $T = \emptyset$, $S = S_\infty = Pl_\infty^{\mathrm{r}}$, which yields (the notion of an infinitesimal element for T being trivial here):

$$\mathrm{rk}_p(H^2(\mathcal{G}^{\mathrm{ord}}, \mathbb{Z}/p\mathbb{Z})) = \mathrm{rk}_p(X_1) + \mathrm{rk}_p(\mathcal{C}\ell^{\mathrm{ord}}),$$

with the exact sequence (Proposition 3.5):

$$1 \longrightarrow \mathcal{E} \cap (\mathcal{N}_{\sharp\infty}^p \,.\, \mathrm{N}\mathcal{L}^\times) \longrightarrow \mathcal{E} \longrightarrow X_1 \longrightarrow 1,$$

still for any sufficiently large G (i.e., the finite extension L is sufficiently large in the p-Hilbert tower), where:

- $\mathcal{E} := E^{\mathrm{ord}} \otimes \mathbb{Z}_p$,
- $\mathcal{K}^\times := K_p^\times \otimes \mathbb{Z}_p$, $\ \mathcal{L}^\times := L_p^\times \otimes \mathbb{Z}_p$,

- $\mathcal{N}_{\sharp\infty} := \mathcal{K}^{\times}_{\sharp\infty} \cap \mathrm{N}(J_L)$,

with:

- $\mathcal{K}^{\times}_{\sharp\infty} := \{x \in K_p^{\times} \otimes \mathbb{Z}_p,\ \bar{i}_p(x) = 1\}$.

Let $\varepsilon \in \mathcal{E}$ be of the form:

$$\varepsilon =: a^p_{\sharp\infty} \,.\, \mathrm{N}y, \quad a_{\sharp\infty} \in \mathcal{N}_{\sharp\infty}, \quad y \in \mathcal{L}^{\times} \ ;$$

we have:

$$(a_{\sharp\infty}) =: \mathrm{N}\mathfrak{a}', \quad \mathfrak{a}' \in \mathcal{I}' := I_{L,p} \otimes \mathbb{Z}_p,$$

so we obtain from the above equality $(1) = \mathrm{N}\mathfrak{a}'^p \,.\, \mathrm{N}(y)$, which yields:

$$\mathfrak{a}'^p \,.\, (y) \in I_G \,\mathcal{I}'.$$

If the p-Hilbert tower $\overline{H}^{\mathrm{ord}}{}_{(p)}$ is *finite*, we can then take for L this tower, so $\mathcal{I}'$ is formed of principal ideals. Set $\mathfrak{a}' =: (y')$, which implies that $a_{\sharp\infty} =: \mathrm{N}y' \,.\, \eta$, $\eta \in \mathcal{E}$, and the existence of $\varepsilon' \in \mathcal{E}' := E_L^{\mathrm{ord}} \otimes \mathbb{Z}_p$ such that:

$$y'^p \,.\, y =: \omega \,.\, \varepsilon', \quad \omega \in I_G \mathcal{L}^{\times} \ \ ;$$

we then have $\mathrm{N}y'^p \,.\, \mathrm{N}y = \mathrm{N}\varepsilon'$, which implies:

$$\varepsilon = a^p_{\sharp\infty} \,.\, \mathrm{N}y'^{-p} \,.\, \mathrm{N}\varepsilon' = \eta^p \,.\, \mathrm{N}\varepsilon',$$

showing the inclusion $\mathcal{E} \cap (\mathcal{N}^p_{\sharp\infty} \,.\, \mathrm{N}\mathcal{L}^{\times}) \subseteq \mathcal{E}^p \,.\, \mathrm{N}\mathcal{E}'$, hence the equality:

$$\mathcal{E} \cap (\mathcal{N}^p_{\sharp\infty} \,.\, \mathrm{N}\mathcal{L}^{\times}) = \mathcal{E}^p \,.\, \mathrm{N}\mathcal{E}'$$

(the other inclusion using the fact that $X_1^p = 1$), so finally:

$$X_1 \simeq E^{\mathrm{ord}} / (E^{\mathrm{ord}})^p \,.\, \mathrm{N}(E_L^{\mathrm{ord}}),$$

which yields the equality:

$$\mathrm{rk}_p(H^2(\mathcal{G}^{\mathrm{ord}}, \mathbb{Z}/p\mathbb{Z})) = \mathrm{rk}_p(E^{\mathrm{ord}}/\mathrm{N}(E_L^{\mathrm{ord}})) + \mathrm{rk}_p(\mathcal{C}\ell^{\mathrm{ord}}),$$

for the p-tower L (assumed to be finite) of the p-Hilbert class fields of K. This is a classical result, usually proved in a different way.

The computation of X_e is identical; then, taking p^e divisible by $[L : K]$, we obtain the exact sequence (Theorem 3.7):

$$1 \longrightarrow E^{\mathrm{ord}}/\mathrm{N}(E_L^{\mathrm{ord}}) \longrightarrow H^2(\mathcal{G}^{\mathrm{ord}}, \mathbb{Z}_p)^* \longrightarrow (\mathcal{C}\ell^{\mathrm{ord}})_p \longrightarrow 1.$$

These results can be largely generalized. We come back to the general case for T and S and we still assume the Leopoldt conjecture for p in $\overline{H}^S_{T(p)}$ (the p-tower of the maximal T-ramified S-split abelian pro-p-extensions of K).

Recall that $\mathcal{A}_T^S := (\mathcal{C}\ell_T^S)_p$ is the abelianization of the group $\mathcal{G}_T^S$, that $\mathcal{A}_T^S \simeq \mathcal{T}_T^S \times \mathbb{Z}_p^{\widetilde{\mathrm{r}}_{T_p}^{S_0}}$, with $\widetilde{\mathrm{r}}_{T_p}^{S_0} = \sum_{v\in T_p} [K_v : \mathbb{Q}_p] - \mathrm{r}_{T_p}^{S_0}$ (see III.1.6.3), and finally, that $\mathcal{E}_T^S := \{\varepsilon \in E^S \otimes \mathbb{Z}_p,\ \bar{\imath}_T(\varepsilon) = 1\}$.

Note. We will have to suppose that $\mathcal{G}_T^S$ is finite; in this case $\mathcal{A}_T^S$ is also finite and, to put emphasis on the analogy with the above case of Hilbert class fields, we preferably denote $\mathcal{A}_T^S$ by $(\mathcal{C}\ell_T^S)_p$.

3.8 Theorem. *Let K be given together with sets of places T and S. If the p-tower $\overline{H}_{T}^{S}{}_{(p)} =: L$ is finite, we have $X_e \simeq \mathcal{E}_T^S/(\mathcal{E}_T^S)^{p^e}.\mathrm{N}(\mathcal{E}_{L,T'}^{S'})$, which yields the rank formula:*

$$\begin{aligned}\mathrm{rk}_p(H^2(\mathcal{G}_T^S, \mathbb{Z}/p\mathbb{Z})) &= \mathrm{rk}_p(\mathcal{E}_T^S/\mathrm{N}(\mathcal{E}_{L,T'}^{S'})) + \mathrm{rk}_p(\mathcal{T}_T^S) \\ &= \mathrm{rk}_p(\mathcal{E}_T^S/\mathrm{N}(\mathcal{E}_{L,T'}^{S'})) + \mathrm{rk}_p(V_T^S/K_T^{\times p}) + \sum_{v\in T}\delta_v - \delta \\ &\quad - \Big(r_1 + r_2 - 1 + |S_0| - \sum_{v\in T_p}[K_v : \mathbb{Q}_p]\Big) + \delta_{2,p}|\Delta_\infty|,\end{aligned}$$

and the exact sequence:

$$1 \longrightarrow \mathcal{E}_T^S/\mathrm{N}(\mathcal{E}_{L,T'}^{S'}) \longrightarrow H^2(\mathcal{G}_T^S, \mathbb{Z}_p)^* \longrightarrow \mathcal{T}_T^S = (\mathcal{C}\ell_T^S)_p \longrightarrow 1.$$

Proof. The finiteness of $\mathcal{G}_T^S$ implies that of $(\mathcal{C}\ell_T^S)_p$ which coincides with $\mathcal{T}_T^S$. Moreover $\widetilde{\mathrm{r}}_{T_p}^{S_0} = 0$ and this yields $\mathrm{r}_{T_p}^{S_0} = \sum_{v\in T_p}[K_v : \mathbb{Q}_p]$ in the usual rank formula for $\mathcal{T}_T^S$.

We then copy the proof of the case of the p-Hilbert class fields towers, noting that being principal in L means belonging to $\mathcal{P}_{L,T',\infty,\Delta'_\infty}\cdot\mathcal{S}'_0$. □

We deduce the following result of which some particular cases are known (Folk, Lemmermeyer, Maire, Schmithals).

3.8.1 Corollary. *If $(\mathcal{C}\ell_T^S)_p := \mathcal{G}_T^{S\,\mathrm{ab}}$ is a cyclic group, then $\mathcal{G}_T^S = (\mathcal{C}\ell_T^S)_p = \mathcal{T}_T^S$ and $\mathcal{E}_T^S = \mathrm{N}_{L/K}(\mathcal{E}_{L,T'}^{S'})$ for $L := \overline{H}_{T}^{S}{}_{(p)} = H_T^S{}_{(p)}$.*

Proof. To simplify, put $K' := H_T^S{}_{(p)}$, $G := \mathrm{Gal}(K'/K)$, $\Gamma := \mathrm{Gal}(K''/K)$, and $H := \mathrm{Gal}(K''/K')$, where K'' is the second floor of the tower. Let s be a generator of G. Since $H = [\Gamma, \Gamma]$ is abelian and G cyclic, it is an elementary exercise to prove that:

- H^{s-1} is a normal subgroup of Γ,
- Γ/H^{s-1} is abelian.

Then $H \subseteq H^{s-1}$, proving that $H = 1$ since G and H are p-groups. Thus the tower is finite and is reduced to $K' = H_T^S{}_{(p)}$, $H^2(\mathcal{G}_T^S, \mathbb{Z}/p\mathbb{Z}) = 1$ or $\mathbb{Z}/p\mathbb{Z}$

depending on whether or not $\mathcal{G}_T^S = \mathcal{T}_T^S$ is a trivial cyclic group, and the result follows by the p-rank formula of the theorem. □

For example, if $\mathcal{C}\ell^{\mathrm{ord}}$ is cyclic, any unit of K is the norm of a unit of the Hilbert class field H^{ord}: the globalisation of this normic property is easy, but note that the Hilbert tower is not necessarily equal to H^{ord} (e.g., $K = \mathbb{Q}(\sqrt{5 \times 257}\,)$, where $\mathcal{C}\ell^{\mathrm{ord}} \simeq \mathbb{Z}/2\mathbb{Z}$ and where the class number of H^{ord} is equal to 3).

3.8.2 Corollary. *If the extension* $\overline{H}{}^S_{T(p)}/K$ *is finite, we have:*

$$\mathrm{rk}_p(\mathcal{C}\ell_T^S) < 2 + 2\sqrt{\mathrm{rk}_p(\mathcal{E}_T^S) + 1}.$$

Note. If $T_p = \emptyset$, we replace $\mathcal{E}_T^S$ by $E_{\mathfrak{m}}^S$ with $\mathfrak{m} := \prod_{v \in T} \mathfrak{p}_v$.

Proof of the corollary. Using the rank formula of 3.8, we have:

$$\mathrm{rk}_p(H^2(\mathcal{G}_T^S, \mathbb{Z}/p\mathbb{Z})) = \mathrm{rk}_p(\mathcal{E}_T^S/\mathrm{N}_{L/K}(\mathcal{E}_{L,T'}^{S'})) + \mathrm{rk}_p(\mathcal{C}\ell_T^S).$$

Applying the Gaschütz–Vinberg Theorem II.5.9.1, we easily get the result. □

3.8.3 Remark. Recall that if $(\mathcal{C}\ell_T^S)_p$ is finite, we have:

$$\mathrm{rk}_p(\mathcal{E}_T^S) = r_1 + r_2 - 1 + |S_0| - \sum_{v \in T_p} [K_v : \mathbb{Q}_p] + \mathrm{rk}_p(\mathrm{tor}(\mathcal{E}_T^S)),$$

where $\mathrm{tor}(\mathcal{E}_T^S) = 1$ except if $\mathcal{E}_T^S = \mathcal{E}^{S_0\,\mathrm{ord}}$ and $\mu_p(K) \neq 1$, in which case the above rank is $r_1 + r_2 + |S_0|$. □

d) A Lower Bound for $\mathrm{rk}_p(H^2(\mathcal{G}_T^S, \mathbb{Z}/p\mathbb{Z}))$ — Conclusion

Since $|X_e|$ is greater or equal to 1, we obtain (for arbitrary T and S):

$$|H^2(\mathcal{G}_T^S, \mathbb{Z}/p^e\mathbb{Z})| \geq |{}_{p^e}\mathcal{T}_T^S|.$$

If $e = 1$ we have, in terms of $\mathbb{F}_p$-dimensions:

$$\mathrm{rk}_p(H^2(\mathcal{G}_T^S, \mathbb{Z}/p\mathbb{Z})) \geq \mathrm{rk}_p(\mathcal{T}_T^S),$$

which yields, using III.4.2:

$$\mathrm{rk}_p(H^2(\mathcal{G}_T^S, \mathbb{Z}/p\mathbb{Z})) \geq \mathrm{rk}_p(V_T^S/K_T^{\times p}) + \sum_{v \in T} \delta_v - \delta$$
$$- (r_1 + r_2 - 1 + |S_0| - \mathrm{r}_{T_p}^{S_0}) + \delta_{2,p}|\Delta_\infty|.$$

We know (Šafarevič's formula) that:

$$\mathrm{rk}_p(H^1(\mathcal{G}_T^S, \mathbb{Z}/p\mathbb{Z})) = \mathrm{rk}_p(V_T^S/K_T^{\times p}) + \sum_{v\in T_p} [K_v : \mathbb{Q}_p] + \sum_{v\in T} \delta_v - \delta - (r_1 + r_2 - 1 + |S_0|) + \delta_{2,p}|\Delta_\infty|,$$

which yields:

$$\mathrm{rk}_p(H^2(\mathcal{G}_T^S, \mathbb{Z}/p\mathbb{Z})) - \mathrm{rk}_p(H^1(\mathcal{G}_T^S, \mathbb{Z}/p\mathbb{Z})) \geq - \sum_{v\in T_p} [K_v : \mathbb{Q}_p] + \mathrm{r}_{T_p}^{S_0} = -\widetilde{\mathrm{r}}_{T_p}^{S_0}.$$

Then by 3.7.2 we have:

3.9 Proposition. *In the case where T and S are arbitrary, we obtain the following general upper and lower bounds:*

$$- \sum_{v\in T_p} [K_v : \mathbb{Q}_p] + \mathrm{r}_{T_p}^{S_0} \leq \mathrm{rk}_p(H^2(\mathcal{G}_T^S, \mathbb{Z}/p\mathbb{Z})) - \mathrm{rk}_p(H^1(\mathcal{G}_T^S, \mathbb{Z}/p\mathbb{Z})) \leq - \sum_{v\in T_p} [K_v : \mathbb{Q}_p] + r_1 + r_2 - 1 + |S_0|,$$

except for $T_p = \{v \in T_{\mathrm{ta}},\ \mathrm{N}\mathfrak{p}_v \equiv 1 \bmod (p)\} = \Delta_\infty = \emptyset$ and $\mu_p(K) \neq 1$, where we have:

$$0 \leq \mathrm{rk}_p(H^2(\mathcal{G}^{S_0\,\mathrm{ord}}, \mathbb{Z}/p\mathbb{Z})) - \mathrm{rk}_p(H^1(\mathcal{G}^{S_0\,\mathrm{ord}}, \mathbb{Z}/p\mathbb{Z})) \leq r_1 + r_2 + |S_0|. \quad \square$$

3.9.1 Remarks. (i) Assume that $T_p \neq \emptyset$. When:

$$\mathrm{r}_{T_p}^{S_0} = r_1 + r_2 - 1 + |S_0|$$

($\mathrm{r}_{T_p}^{S_0}$ is equal to the $\mathbb{Z}$-rank of $E^{S_0\,\mathrm{ord}}$, which is equivalent to $\mathcal{E}_T^S = 1$), we obtain the following equality (proved independently in [Mai4]):

$$\mathrm{rk}_p(H^2(\mathcal{G}_T^S, \mathbb{Z}/p\mathbb{Z})) - \mathrm{rk}_p(H^1(\mathcal{G}_T^S, \mathbb{Z}/p\mathbb{Z})) = r_1 + r_2 - 1 + |S_0| - \sum_{v\in T_p} [K_v : \mathbb{Q}_p],$$

and in addition:

$$\mathrm{rk}_p(H^2(\mathcal{G}_T^S, \mathbb{Z}/p\mathbb{Z})) = \mathrm{rk}_p(\mathcal{T}_T^S).$$

Since we know a simple expression for $|\mathcal{T}_T^S|$ (see III.2.6, (ii)), if we ask that $\mathcal{T}_T^S = 1$, we can easily find examples where $\mathcal{G}_T^S$ is pro-p-free on:

$$\sum_{v\in T_p} [K_v : \mathbb{Q}_p] - (r_1 + r_2 - 1 + |S_0|)$$

generators.

(ii) Note that for $T_p = Pl_p$, $S_0 = \emptyset$ (complete p-ramification without finite decomposition), the lower and upper bounds given above are optimal because of the results of Section 2.

(iii) For $T_p = S_0 = \Delta_\infty = \emptyset$, $r_1 + r_2 - 1 = 0$ (i.e., K is equal to $\mathbb{Q}$ or to an imaginary quadratic field), and $\{v \in T_{\rm ta},\ {\rm N}\mathfrak{p}_v \equiv 1 \bmod (p)\} \neq \emptyset$ or $\mu_p(K) = 1$, we obtain from Theorem 3.6 and Corollary 3.7.2:

$$\begin{aligned}\mathrm{rk}_p(H^1(\mathcal{G}^{\rm ord}_{T_{\rm ta}}, \mathbb{Z}/p\mathbb{Z})) &= \mathrm{rk}_p(H^2(\mathcal{G}^{\rm ord}_{T_{\rm ta}}, \mathbb{Z}/p\mathbb{Z}))\\ &= \mathrm{rk}_p(\mathcal{C}\ell^{\rm ord}_{T_{\rm ta}})\\ &= \mathrm{rk}_p(V^{\rm ord}_{T_{\rm ta}}/K^{\times p}_{T_{\rm ta}}) + \sum_{v \in T_{\rm ta}} \delta_v - \delta. \qquad \square\end{aligned}$$

3.10 CONCLUSION. Apart from the assumptions of Section 2, the method used in this Appendix gives the possibility to go further than the classical points of view. Indeed, it is easily seen that at least the strong p-adic conjecture III.4.12 occurs (for instance, the extensions $H_\sharp/H$ and $H'_\sharp/H'$ depend on inertia groups of wild places and the groups $\mathcal{E}^S_T$ are generalizations of the $\mathcal{E}^{\rm ord}_{p\backslash v}$ which are the $I_v(\overline{K}^{\rm ab}{}_{(p)}/H^{\rm ord}_{p}{}_{(p)})$), which (in part) explains some cohomological difficulties of the standard methods which do not take in account these p-adic questions in a natural way, in other words which do not use all informations on the group $\overline{G}^{\rm ab}$ and its p-adic properties.

In addition, classical cohomology may be partially inadapted, in particular for $p = 2$ according to the places above 2 and ∞, because the arithmetic norm does not have any existence in the sole context of the G-module structure; this is an additional argument in favor of the above arithmetic approach.

Of course, knot groups of the form:

$$\mathcal{E}^S_T / \mathcal{E}^S_T \cap (\mathcal{N}^{p^e}_{\sharp\infty} \,.\, \mathrm{N}\mathcal{L}^\times_\infty),$$

are in general not computable, although we have reduced the problem to a *finite* (but quite large) subextension L/K of $\overline{H}^S_{T(p)}/K$.

In another direction, we have obtained:

$$|H^2(\mathcal{G}^S_T, \mathbb{Z}/p^e\mathbb{Z})^*| = (\mathcal{E}^S_T : \mathcal{E}^S_T \cap (\mathcal{N}^{p^e}_{\sharp\infty} \,.\, \mathrm{N}\mathcal{L}^\times_\infty)) \times |{}_{p^e}\mathcal{T}^S_T|$$

for all $e \geq 1$. Then it is easily checked that $|H^2(\mathcal{G}^S_T, \mathbb{Z}/p^e\mathbb{Z})^*|$ is a nondecreasing function with respect to e (note that the choice of the sufficiently large subextension L depends on e); is it bounded or not ? It is not difficult to deduce, from Theorem 3.7, an exact sequence:

$$1 \longrightarrow (\mathcal{T}^S_T)^* \longrightarrow H^2(\mathcal{G}^S_T, \mathbb{Z}_p) \longrightarrow (\mathcal{F}^S_T)^* \longrightarrow 1,$$

where $\mathcal{F}^S_T$ is a quotient $\mathbb{Z}_p$-module of $\mathcal{E}^S_T$.

Bibliography

Books on number fields and class field theory

a) Directly accessible general books:

[BŠa] Borevič, Z.I. and Šafarevič, I.R., Number Theory, Academic Press, New York 1966; French translation: Théorie des nombres, Gauthier-Villars 1967.

[Co1] Cohn, H., A classical introduction to algebraic numbers and class fields, Universitext, Springer-Verlag 1978, second edition 1988.

[D] Descombes, R., Eléments de théorie des nombres, P.U.F., Dunod 1986.

[He] Hecke, E., Lectures on the theory of algebraic numbers (translated from German), Graduate Texts in Math. 77, Springer-Verlag 1981.

[IR] Ireland, K. and Rosen, M., A classical introduction to modern number theory, Graduate Texts in Math. 84, Springer-Verlag 1982, second edition 1990.

[Ko1] Koch, H., Introduction to classical mathematics I. From the quadratic reciprocity law to the uniformization theorem (translated from German), Mathematics and its applications 70, Kluwer Academic Publishers 1991.

[Lg] Long, R.L., Algebraic number theory, Pure and Applied Mathematics 41, Marcel Dekker 1977.

[Ma] Marcus, D.A., Number fields, Universitext, Springer-Verlag 1977.

[Ri1] Ribenboim, P., Arithmétique des corps, Coll. Méthodes, Hermann 1970.

[Ri2] Ribenboim, P., Classical theory of algebraic numbers, Universitext, Springer-Verlag 2001.

[Sam] Samuel, P., Théorie des nombres, Coll. Méthodes, Hermann 1967; English translation: Algebraic theory of numbers, Houghton Mifflin Co., Boston 1970.

[Se1] Serre, J-P., Cours d'arithmétique, P.U.F., Dunod 1970; English translation: A course in arithmetic, Graduate Texts in Math. 7, Springer-Verlag 1973.

[ST] Stewart, I.N. and Tall, D.O., Algebraic number theory, Chapman and Hall, London 1979.

b) Elementary p-adic analysis:

[A] Amice, Y., Les nombres p-adiques, Le Mathématicien 14, P.U.F., Dunod 1975.

[Gou] Gouvêa, F., p-adic numbers, Universitext, Springer-Verlag 1991.

[K] Koblitz, N., p-adic numbers, p-adic analysis and zêta-functions, Graduate Texts in Math. 58, Springer-Verlag, second edition 1984.

[Ri3] Ribenboim, P., The theory of classical valuations, Springer Monographs in Math., Springer-Verlag 1999.

[Rob] Robert, A., A course in p-adic analysis, Graduate Texts in Math. 198, Springer-Verlag 2000.

c) More complete basic books on algebraic number theory:

[Ca] Cassels, J.W.S., Local fields, London Mathematical Society Student Texts 3, Cambridge Univ. Press 1986.

[FT] Fröhlich, A. and Taylor, M., Algebraic number theory, Cambridge Studies in Adv. Math. 27, Cambridge Univ. Press 1991.

[Ha1] Hasse, H., Number theory (English translation of "Zahlentheorie" 1969), Grundlehren 229, Springer-Verlag 1979.

[Jan] Januz, G., Algebraic number fields, Academic Press, New York 1973, second edition Graduate Studies in Mathematics 7, American Math. Soc., Providence, Rhode Island 1996.

[Ko2] Koch, H., Number theory. Algebraic numbers and functions (translated from German), Graduate Studies in Mathematics 24, American Math. Soc., Providence, Rhode Island 2000.

[Nar1] Narkiewicz, W., Elementary and analytic theory of algebraic numbers, Warszawa, P.W.N., second edition Springer-Verlag 1990.

[Neu1] Neukirch, J., Algebraic number theory (English translation of "Algebraische Zahlentheorie" 1992), Grundlehren 322, Springer-Verlag 1999.

[Wa] Washington, L.C., Introduction to cyclotomic fields, Graduate Texts in Math. 83, Springer-Verlag, third edition 1999.

[WsE] Weiss, E., Algebraic number theory, Mc. Graw-Hill, New York 1963.

d) Classical reference books on class field theory:

[AT] Artin, E. and Tate, J., Class field theory, Benjamin, New York, Amsterdam 1968, second edition, Advanced Book Classics, Addison-Wesley Publ. Comp., Advanced Book Program, Redwood City, California 1990.

[CF] Cassels, J.W.S. and Fröhlich, A. (Eds.), Algebraic number theory, Academic Press, New York 1967.

[Iy1] Iyanaga, S. (Ed.), The theory of numbers, North-Holland Publ. Comp. 1975 (English translation of: Iwanami Shoten Publ., Tokyo 1969).

[Lang1] Lang, S., Algebraic Number Theory, Addison Wesley Publ. Comp. 1970, second edition Graduate Texts in Math. 110, Springer-Verlag 1994.

[Se2] Serre, J-P., Corps locaux, Hermann 1962, seconde édition 1968; English translation: Local fields, Graduate Texts in Math. 67, Springer-Verlag 1979.

[We1] Weil, A., Basic number theory, Grundlehren 144, Springer-Verlag 1967, third edition 1974.

e) Synthesis of algebraic number theory and of the cohomological approach to class field theory — Geometrical aspects of number theory (usually without proofs):

[Ko3] Koch, H. (Parshin, A.N. and Šafarevič, I.R., Eds.), Number theory II, Encycl. of Math. Sci., vol. 62, Springer-Verlag 1992; Algebraic Number Theory, second edition 1997.

[MP] Manin, Y.I. and Panchishkin, A.A. (Parshin, A.N. and Šafarevič, I.R., Eds.), Number theory I, Encycl. of Math. Sci., vol. 49, Springer-Verlag 1995.

f) Additional material — Other approaches:

[Art1] Artin, E., Algebraic numbers and algebraic functions, lectures notes by Adamson, I., Gordon and Breach, New York 1967.

[Co2] Cohn, H., Introduction to the construction of class fields, Corrected reprint of the 1985 original, Dover Publ. Inc., New York 1994.

[Haz] Hazewinkel, M., Local class field theory is easy, Adv. in Math. 18 (1975), 148–181.

[Iw1] Iwasawa, K., Local class field theory, Oxford University press, New York; Clarendon Press, Oxford 1986.

[Lang2] Lang, S., Cyclotomic fields I and II, Springer-Verlag, combined second edition 1990.

[Lem] Lemmermeyer, F., Reciprocity laws from Euler to Eisenstein, Springer Monographs in Math., Springer-Verlag 2000.

[Neu2] Neukirch, J., Class field theory, Grundlehren 280, Springer-Verlag 1986.

[Neum1] Neumann, O., Two proofs of the Kronecker-Weber theorem "according to Kronecker and Weber", J. reine angew. Math. 323 (1981), 105–126.

g) Galois cohomology — Pro-p-extensions of number fields:

[Hab] Haberland, K., Galois cohomology of algebraic number fields, V.E.B. Deutscher Verlag der Wissenschaften 1978.

[Ko4] Koch, H., Galoissche Theorie der p-Erweiterungen, V.E.B. Deutscher Verlag der Wissenschaften 1970; English translation: Lemmermeyer, F., Springer Monographs in Mathematics, Springer-Verlag 2002.

[NSW] Neukirch, J., Schmidt, A. and Wingberg, K., Galois cohomology of number fields, Grundlehren 323, Springer-Verlag 2000.

[P] Poitou, G., Cohomologie galoisienne des modules finis, Séminaire de l'Institut de Mathématiques de Lille, Travaux et Recherches Mathématiques, Vol. XIII, Dunod, Paris 1967.

[Se3] Serre, J-P., Cohomologie galoisienne, Lect. Notes in Math. 5, Springer-Verlag, fifth edition 1991; English translation: Galois cohomology, Springer-Verlag 1997; corrected second printing, Springer Monographs in Mathematics, Springer-Verlag 2002.

h) Historical books (important steps in class field theory, surveys):

[Art2] Artin, E., Beweis des allgemeinen Reziprozitätsgesetzes, Abh. Math. Semin. Univ. Hamburg 5 (1927), 353–363 (Collected Papers 1965).

[Che1] Chevalley, C., Sur la théorie du corps de classes dans les corps finis et les corps locaux (Thèse), Jour. of the Faculty of Sciences Tokyo (1933), 365–476.

[Che2] Chevalley, C., La théorie du corps de classes, Ann. of Math. II, 41 (1940), 394–418.

[Che3] Chevalley, C., Class field theory, Nagoya University, Nagoya 1954.

[Ha2] Hasse, H., Bericht über neuere Untersuchungen und Probleme aus der Theorie der algebraischen Zahlkörper I, Ia, II, Physica Verlag, Würzburg 1965.

[Her] Herbrand, M.J., Le développement moderne de la théorie des corps algébriques, corps de classes et lois de réciprocité, Mémorial des Sciences Mathématiques, fasc. 75, Acad. Sci. Paris 1936.

[Hil] Hilbert, D., Die Theorie der algebraischer Zahlkörper (Zahlbericht), Jahresber. der Deutsch. Math. Ver. 4 (1897), 177–546; French translation: Levy, A. et Got, T., Ann. Fac. Sci. Univ. Toulouse 3, 1 (1909), 2 (1910), 3 (1911), reissued by Gabay, Paris 1991; English translation: Adamson, I., with an Introduction by Lemmermeyer, F. and Schappacher, N., Springer-Verlag 2000.

[HN] Hochschild, G. and Nakayama, T., Cohomology in class field theory, Ann. of Math. 55 (1952), 348–366.

[Iy2] Iyanaga, S., Collected Papers, Iwanami Shoten Publ., Tokyo 1994.

[Tak] Takagi, T., Über eine Theorie des relativ-abelschen Zahlkörpers, J. Coll. Sci. Imp. Univ. Tokyo 41, 9 (1920), 1–133, In: The Collected Papers, Iwanami Shoten Publ., Tokyo, second edition 1973, Springer-Verlag (1990), 73–167.

[We2] Weil, A., Sur la théorie du corps de classes, J. Math. Soc. Japan 3, 1 (1951), 1–35.

i) HISTORICAL SURVEYS OF CLASS FIELD THEORY (texts containing a large number of bibliographical material):

[F1] Frei, G., Heinrich Weber and the emergence of class field theory, In: The history of modern mathematics (Rowe, D.E. and Mc Cleary, J., Eds.), Academic Press (1989), 425–450.

[F2] Frei, G., The reciprocity law from Euler to Eisenstein, In: The intersection of History and Mathematics (Sasaki, Ch., Sugiura, M. and Dauben, J.W., Eds), Birkhäuser-Verlag, Basel (1994), 67-88.

[Gar] Garbanati, D., Class field theory summarized, Rocky Mountain J. of Math. 11, 2 (1981), 195–225.

[Ha3] Hasse, H., Class field theory (Notes rédigées par Frei, G.), Coll. Math. 11, Université Laval 1973.

[Miy0] Miyake, K. (Ed.), Class field theory — Its centenary and prospect, Advanced Studies in Pure Mathematics 30, Math. Soc. Japan, Tokyo 2001.

[Miy1] Miyake, K., The etablishment of the Takagi-Artin class field theory, In: The intersection of History and Mathematics (Sasaki, Ch., Sugiura, M. and Dauben, J.W., Eds), Birkhäuser-Verlag, Basel (1994), 109–128.

[Nar2] Narkiewicz, W., Global class field theory (Hazewinkel, M., Ed.), Handbook of algebra, Vol.1, Elsevier Sc. B.V., Amsterdam, North-Holland (1996), 365–393.

[Tau] Taussky, O., Some noncommutative methods in algebraic number theory, In: A Century of Mathematics in America, part II, American Math. Soc., Providence, Rhode Island (1988), 493–511.

j) ALGORITHMIC ASPECTS OF ALGEBRAIC NUMBER THEORY:

[Coh1] Cohen, H., A course in computational algebraic number theory, Graduate Texts in Math. 138, Springer-Verlag 1993, fourth corrected printing 2000.

[Coh2] Cohen, H., Advanced topics in computational number theory, Graduate Texts in Math. 193, Springer-Verlag 2000.

[PZ] Pohst, M. and Zassenhaus, H., Algorithmic algebraic number theory, third edition, Cambridge Univ. Press, Cambridge 1993.

Additional references

[AF] Amice, Y. et Fresnel, J., Fonctions zêta p-adiques des corps de nombres abéliens réels, Acta Arith. 20 (1972), 353–384.

[AJ] Anglès, B. et Jaulent, J-F., Théorie des genres des corps globaux, Manuscripta Math. 101, 4 (2000), 513–532.

[BGr] Berger, R.I. and Gras, G., Regular fields: normic criteria in p-extensions, Publ. Math. Fac. Sci. Besançon (Théorie des Nombres), Année 1991/1992.

[BP] Bertrandias, F. et Payan, J-J., Γ-extensions et invariants cyclotomiques, Ann. Sci. Ec. Norm. Sup. 4, 5 (1972), 517–548.

[Br] Brumer, A., Galois groups of extensions of algebraic number fields with given ramification, Michigan Math. J. 13 (1966), 33–40.

[BS] Buchmann, J. and Sands, J.W., An algorithm for testing Leopoldt's conjecture, Jour. Number Theory 27 (1987), 92–105.

[Car] Carrol, J.E., On determining the quadratic subfields of $\mathbb{Z}_2$-extensions of complex quadratic fields, Compositio Math. 30 (1975), 259–271.

[CarK] Carrol, J.E. and Kisilevsky, H., Initial layers of $\mathbb{Z}_\ell$-extensions of complex quadratic fields, Compositio Math. 32 (1976), 157–168.

[ChaW] Chase, S.U. and Waterhouse, W.C., Moore's theorem on uniqueness of reciprocity laws, Invent. Math. 16 (1972), 267–270.

[Che4] Chevalley, C., Généralisation de la théorie du corps de classes pour les extensions infinies, Jour. Math. Pures Appl. 15 (1936), 359–371.

[Che5] Chevalley, C., Deux théorèmes d'arithmétique, J. Math. Soc. Japan 3, 1 (1951), 36–44.

[Coa] Coates, J., p-adic L-functions and Iwasawa's theory, Proc. of Durham Symposium 1975, New York-London (1977), 269–353.

[CoDO] Cohen, H., Diaz y Diaz, F. and Olivier, M., Computing ray class groups, conductors and discriminants, Math. Comput. 67 (1998), 773–795.

[CohL] Cohen, H. and Lenstra, H.W., Heuristics on class groups of number fields, Number theory, Noordwijkerhout 1983, Lect. Notes in Math. 1068, Springer-Verlag (1984), 33–62.

[CohM] Cohen, H. et Martinet, J., Etudes heuristiques des groupes de classes des corps de nombres, J. reine angew. Math. 404 (1990), 39–76.

[CohR] Cohen, H. and Roblot, X-F., Computing the Hilbert class field of real quadratic fields, Math. Comput. 69 (2000), 1229–1244.

[CoHu] Conner, P.E. and Hurrelbrink, J., Class number parity, Series in Pures Mathematics 8, Singapore, World Scientific 1988.

[Col] Colmez, P., Résidu en $s = 1$ des fonctions zêta p-adiques, Invent. Math. 91 (1988), 371–389.

[Cor1] Cornell, G., On the construction of relative genus field, Transactions of the American Math. Soc. 271, 2 (1982), 501–511.

[Cor2] Cornell, G., The structure of the ray class group, In: Algebraic Number Theory, RIMS, Kokyuroku (1987).

[CorR] Cornell, G. and Rosen, M., A note on the splitting of the Hilbert class field, Jour. Number Theory 28 (1988), 152–158.

[DaP] Daberkow, M. and Pohst, M., On the computation of Hilbert class fields, Jour. Number Theory 69 (1998), 213–230.

[Deu] Deuring, M., Die Klassenkörper der complexen Multiplikation, Teubner-Verlag, Stuttgart 1958.

[DS] Diaz y Diaz, F. et Soriano, F., Approche algorithmique du groupe des classes logarithmiques, Jour. Number Theory 76 (1999), 1–15.

[EKW] Emsalem, M., Kisilevsky, H.K. et Wales, D.B., Indépendance linéaire sur $\overline{\mathbb{Q}}$ de logarithmes de nombres algébriques et rang p-adique du groupe des unités d'un corps de nombres, Jour. Number Theory 19 (1981), 181–198.

[Em] Emsalem, M., Rang p-adique de groupes de S-unités d'un corps de nombres, C.R. Acad. Sci. Paris 297, Série I (1983), 225–228.

[Fe] Federer, L.J., Genera theory for S-class groups, Houston J. Math. 12, 4 (1986), 497–502.

[FG] Federer, L.J. and Gross, B.N., Regulators and Iwasawa modules, Invent. Math. 62 (1981), 443-457.

[Fi] Fieker, C., Computing class fields via the Artin map, Math. Comput. 70, 235 (2001), 1293–1303.

[Fl] Fleckinger, V., Une interprétation de la conjecture de Leopoldt, C.R. Acad. Sci. Paris 302, Série I (1986), 607–610.

[Fr1] Fröhlich, A., On fields of class two, Proc. London Math. Soc. III, 4 (1954), 235–256.

[Fr2] Fröhlich, A., The genus field and genus number in algebraic number fields, Mathematika 6 (1959), 40–46, 142–146.

[Fr3] Fröhlich, A., Central extensions, Galois groups and ideal class groups of number fields, Contemporary Mathematics 24, Amer. Math. Soc. 1983.

[Fu1] Furuta, Y., The genus field and genus number in algebraic number fields, Nagoya Math. J. 29 (1967), 281–285.

[Fu2] Furuta, Y., A norm residue map for central extensions of an algebraic number field, Nagoya Math. J. 93 (1984), 61–69.

[Fu3] Furuta, Y., The notion of restricted idèles with an application to some extension fields, Nagoya Math. J. 27 (1966), 121–132; J. Math. Soc. Japan 18 (1966), 247–252.

[Ga] Garland, H., A finitness theorem for K_2 of a number field, Ann. of Math. 94 (1971), 534–548.

[Go] Gold, R., The principal genus and Hasse's norm theorem, Indiana Univ. Math. J. 26, 1 (1977), 183–189.

[Gr1] Gras, G., Groupe de Galois de la p-extension abélienne p-ramifiée maximale d'un corps de nombres, J. reine angew. Math. 333 (1982), 86–132.

[Gr2] Gras, G., Logarithme p-adique et groupes de Galois, J. reine angew. Math. 343 (1983), 64–80.

[Gr3] Gras, G., Sur les $\mathbb{Z}_2$-extensions d'un corps quadratique imaginaire, Ann. Inst. Fourier 33, 4 (1983), 1–18.

[Gr4] Gras, G., Plongements kummériens dans les $\mathbb{Z}_p$-extensions, Compositio Math. 55 (1985), 383–396.

[Gr5] Gras, G., Decomposition and inertia groups in $\mathbb{Z}_p$-extensions, Tokyo J. of Math. 9, 1 (1986), 41–51.

[Gr6] Gras, G., Remarks on K_2 of number fields, Jour. Number Theory 23 (1986), 322–335.

[Gr7] Gras, G., Théorie des genres analytique des fonctions L p-adiques des corps totalement réels, Invent. Math. 86 (1986), 1–17.

[Gr8] Gras, G., Classes généralisées invariantes, J. Math. Soc. Japan 46, 3 (1994), 467–476.

[Gr9] Gras, G.,Principalisation d'idéaux par extensions absolument abéliennes, Jour. Number Theory 62 (1997), 403–421.

[Gr10] Gras, G., Théorèmes de réflexion, J. Théorie des Nombres de Bordeaux 10, 2 (1998), 399–499.

[GrJ] Gras, G. et Jaulent, J-F., Sur les corps de nombres réguliers, Math. Z. 202 (1989), 343–365.

[GrM] Gras, G. et Munnier, A., Extensions cycliques T-totalement ramifiées, Publ. Math. Fac. Sci. Besançon (Théorie des Nombres), Années 1996/97–1997/98.

[Grt] Greither, C., Class groups of abelian fields, and the main conjecture, Ann. Inst. Fourier 42 (1992), 449–499.

[GW] Gruenberg, K.W. and Weiss, A., Capitulation and transfer kernels, J. Théorie des Nombres de Bordeaux 12, 1 (2000), 219–226.

[Haj] Hajir, F., On the growth of p-class groups in p-class fields towers, Jour. Alg. 188 (1997), 256–271.

[HM1] Hajir, F. and Maire, C., Tamely ramified towers and discriminant bounds for number fields, Compositio Math. 128 (2001), 35–53; Tamely ramified towers and discriminant bounds for number fields II, Jour. of Symbolic Computation 33 (2002), 425–423.

[HM2] Hajir, F. and Maire, C., Asymptotically good towers of global fields, Proceedings of the European Congress of Mathematics, Bacelona 2000, Progress in Math. 202 (2002), 207–218.

[HM3] Hajir, F. and Maire, C., Extensions of number fields with wild ramification of bounded depth, Intern. Math. Research Notices 13 (2002), 667–696.

[HM4] Hajir, F. and Maire, C., Unramified subextensions of ray class fields towers, Jour. of Algebra 249 (2002), 528–543

[Hu] Hubbard, D., The non-existence of certain free pro-p-extensions and capitulation in a family of dihedral extensions of $\mathbb{Q}$, Thesis, University of Washington (1996).

[I] Ishida, M., The genus fields of algebraic number fields, Lect. Notes in Math. 555, Springer-Verlag 1976.

[Iw2] Iwasawa, K., On p-adic L-functions, Ann. of Math. II, 89 (1969), 198–205.

[Iw3] Iwasawa, K., On $\mathbb{Z}_\ell$-extensions of algebraic number fields, Ann. of Math. 98 (1973), 246–326.

[Iw4] Iwasawa, K., On the μ-invariant of $\mathbb{Z}_\ell$-extensions, Conf. on Number Theory, Algebraic Geometry, and Commutative Algebra (in honor of Y. Akizuki), Tokyo, Kinokuniya (1977), 1–11.

[J] Jannsen, U., Iwasawa modules up to isomorphism, Advanced Studies in Pure Mathematics 17 (1989), 171–207.

[Ja1] Jaulent, J-F., Introduction au K_2 des corps de nombres, Publ. Math. Fac. Sci. Besançon (Théorie des Nombres), Années 1981/82–1982/83.

[Ja2] Jaulent, J-F., L'arithmétique des ℓ-extensions (Thèse d'Etat, Université de Franche-Comté, Besançon), Publ. Math. Fac. Sci. Besançon (Théorie des Nombres), Années 1984/85–1985/86.

[Ja3] Jaulent, J-F., Sur l'indépendance ℓ-adique de nombres algébriques, Jour. Number Theory 20 (1985), 149–158.

[Ja4] Jaulent, J-F., Sur les conjectures de Leopoldt et de Gross, Astérisque 147/148 (1987), 107–120.

[Ja5] Jaulent, J-F., Sur le noyau sauvage des corps de nombres, Acta Arith. 67 (1994), 335–348.

[Ja6] Jaulent, J-F., Classes logarithmiques des corps de nombres, J. Théorie des Nombres de Bordeaux 6 (1995), 301–325.

[Ja7] Jaulent, J-F., Théorie ℓ-adique globale du corps de classes, J. Théorie des Nombres de Bordeaux 10, 2 (1998), 355–397.

[JaMai] Jaulent, J-F. et Maire, C., A propos de la tour localement cyclotomique d'un corps de nombres, Abh. Math. Semin. Univ. Hamburg 70 (2000), 239–250.

[JaMi] Jaulent, J-F. and Michel, A., On certain étale K-groups and their logarithmic interpretations, preprint 2001.

[JaN] Jaulent, J-F. et Nguyen Quang Do, T., Corps p-rationnels, corps p-réguliers et ramification restreinte, J. Théorie des Nombres de Bordeaux 5 (1993), 343–363.

[JaS] Jaulent, J-F. et Sauzet, O., Pro-ℓ-extensions de corps de nombres $\mathfrak{l}$-rationnels, Jour. Number Theory 65 (1997), 240–267.

[JaSor] Jaulent, J-F. et Soriano, F., Sur le noyau sauvage des corps de nombres et le groupe des classes logarithmiques, Math. Z., 238, 2 (2001), 335–354.

[Jeh] Jehne, W., On knots in algebraic number theory, J. reine angew. Math. 311/312 (1979), 215–254.

[Keu] Keune, F., On the structure of the K_2 of the ring of integers in a number field, K-Theory 2 (1989), 625–645.

[KL] Kisilevsky, H. and Labute, J., On a sufficient condition for the p-class tower of a CM-field to be infinite, In: Proceedings of Int. Conf. Théorie des nombres (De Konink, J-M. and Levesque, C., Eds.), Laval, Québec (1987), de Gruyter, Berlin 1989.

[KM1] Kolster, M. and Movahhedi, A., Bi-quadratic number fields with trivial 2-primary Hilbert kernels, preprint 2000.

[KM2] Kolster, M. and Movahhedi, A., Galois co-descent for étale wild kernels and capitulation, Ann. Inst. Fourier 50, 1 (2000), 35–65.

[Kol] Kolster, M., An idelic approach to the wild kernel, Invent. Math. 103 (1991), 9–24.

[Koly] Kolyvagin, V.A., Euler systems, The Grothendieck Festschrift, Vol. II, P.M. 87, Birkhäuser, Boston (1990), 435–483.

[Kub1] Kubota, T., Galois group of the maximal abelian extension of an algebraic number field, Nagoya Math. J. 12 (1957), 177–189.

[Kub2] Kubota, T., Geometry of numbers and class field theory, Japan. J. Math. 13, 2 (1987), 235–275; Japan. J. Math. 26, 2 (2000), 407.

[Kub3] Kubota, T., The foundation of class field theory based on the principles of space diagrams (japanese), Sūgaku 44, 1 (1992), 1–12.

[KubO] Kubota, T. and Oka, S., On the deduction of the class field theory from the general reciprocity of powers residues, Nagoya Math. J. 160 (2000), 135–142.

[Kur] Kurihara, M., On the ideal class group of the maximal real subfield of a number field with all roots of unity, J. Eur. Math. Soc. 1, 1 (1999), 35–49.

[La] Lannuzel, A., Sur les extensions pro-p-libres d'un corps de nombres, Thèse, Université de Franche-Comté, Besançon (1997).

[LaNg] Lannuzel, A. et Nguyen Quang Do, T., Conjectures de Greenberg et extensions pro-p-libres d'un corps de nombres, Manuscripta Math. 102 (2000), 187–209.

[Lau] Laurent, M., Rang p-adique d'unités et action de groupes, J. reine angew. Math. 399 (1989), 81–108.

[Le1] Leopoldt, H.W., Zur Geschlechtertheorie in abelschen Zahlkörper, J. Math. Soc. Japan 3 (1951), 45–51.

[Le2] Leopoldt, H.W., Zur Struktur der ℓ-Klassengruppe galoisscher Zahlkörper, J. reine angew. Math. 199 (1958), 165–174.

[Le3] Leopoldt, H.W., Zur Arithmetik in abelschen Zahlkörpern, J. reine angew. Math. 209 (1962), 54–71.

[Lou] Louboutin, S., The nonquadratic imaginary cyclic fields of 2-power degrees with class numbers equal to their genus class number, Proc. of the A.M.S. 127 (1999), 355–361.

[Mai1] Maire, C., Extensions T-ramifiées modérées, S-décomposées, Thèse, Université de Franche-Comté, Besançon (1995).

[Mai2] Maire, C., Complément à un résultat de Šafarevič, Math. Nachr. 198 (1999), 149–168.

[Mai3] Maire, C., On infinite unramified extensions, Pacific Jour. of Math. 192 (2000), 135–142.

[Mai4] Maire, C., On the $\mathbb{Z}_\ell$-rank of abelian extensions with restricted ramification, Jour. Number Theory 92 (2002), 376–404.

[Mar1] Martinet, J., Tours de corps de classes et estimations de discriminants, Invent. Math. 44 (1978), 65–73.

[Mar2] Martinet, J., Méthodes géométriques dans la recherche des petits discriminants, Sém. Théorie des Nombres, Paris (1983/84), Birkhäuser, Bâle (1985), 147–179.

[Mat] Matsumura, N., On the class field tower of an imaginary quadratic number field, Mem. Fac. Sci. Kyushu University 31 (1977), 165–171.

[MaW] Mazur, B. and Wiles, A., Class fields of abelian extensions of $\mathbb{Q}$, Invent. Math. 76 (1984), 179–330.

[Mi] Miki, H., On the maximal abelian ℓ-extension of a finite algebraic number field with given ramification, Nagoya Math. J. 70 (1978), 183–202.

[Mil] Milnor, J., Introduction to algebraic K-theory, Ann. Math. Stud. 72, Princeton Univ. Press 1971.

[Miy2] Miyake, K., On the structure of the idèle group of an algebraic number field, Nagoya Math. J. 80 (1980), 117–127; Tôhoku Math. J. 34 (1982), 101–112.

[Miy3] Miyake, K., Algebraic investigations of Hilbert's theorem 94, the principal ideal theorem and the capitulation problem, Exp. Math. 7 (1989), 289–346.

[Miy4] Miyake, K., On central extensions, In: Proceedings of Int. Conf. Théorie des nombres (De Konink, J-M. and Levesque, C., Eds.), Laval, Québec (1987), de Gruyter, Berlin 1989.

[Mo] Movahhedi, A., Sur les p-extensions des corps p-rationnels, Math. Nachr. 149 (1990), 163–176.

[MoNg] Movahhedi, A. et Nguyen Quang Do, T., Sur l'arithmétique des corps de nombres p-rationnels, Sém. Théorie des Nombres, Paris (1987/89), P.M. 81, Birkhäuser, Boston (1990), 155–200.

[Neu3] Neukirch, J., Über das Einbettungsproblem der algebraischen Zahlentheorie, Invent. Math. 21 (1973), 59–116.

[Neum2] Neumann, O., On p-closed algebraic number fields with restricted ramification, Izv. Akad. Nauk USSR, ser. Math. 39, 2 (1975), 259–271 (English translation: Math. USSR, Izv. 9, 243–254).

[Ng1] Nguyen Quang Do, T., Sur la $\mathbb{Z}_p$-torsion de certains modules galoisiens, Ann. Inst. Fourier 36, 2 (1986), 27–46.

[Ng2] Nguyen Quang Do, T., Une étude cohomologique de la partie 2-primaire de $K_2\mathcal{O}$, K-Theory 3 (1990), 523–542.

[Ng3] Nguyen Quang Do, T., Lois de réciprocité primitives, Manuscripta Math. 72 (1991), 307–324.

[Pon] Pontrjagin, L.S., Topological groups, Gordon and Breach 1966, third edition 1973.

[Ra] Razar, J., Central and genus class fields and the Hasse norm theorem, Compositio Math. 35 (1977), 281–298.

[Ro] Roblot, X-F., Algorithmes de factorisation dans les extensions relatives et applications de la conjecture de Stark à la construction des corps de classes de rayon, Thèse, Université de Bordeaux I, Bordeaux (1997).

[RØ] Rognes, J. and Østvær, P.A., Two-primary algebraic K-theory of two-regular number fields, Math. Z. 233 (2000), 251–263.

[Roy1] Roy, D. (Gouvêa, F., Ed.), On the v-adic independance of algebraic numbers, Advances in number theory, Proc. 3e conf. théorie des nombres, Queen's Univ., Kingston, Canada 1991, Clarendon Press, Oxford (1993), 441–451.

[Roy2] Roy, D., Matrices whose coefficients are linear forms in logarithms, Jour. Number Theory 41 (1992), 22–47.

[Ru1] Rubin, K., Stark units and Kolyvagin's "Euler systems", J. reine angew. Math. 425 (1992), 141–154.

[Ru2] Rubin, K., Euler systems, Ann. Math. Stud. 67, Princeton Univ. Press 2000.

[Ša] Šafarevič, I.R., Extensions with given points of ramification, Publ. Math. Inst. Hautes Etudes Sci. 18 (1964), 71–95 (A.M.S. Transl., Ser. 2, 59 (1966), 128–149).

[San] Sands, J., Kummer's and Iwasawa's version of Leopoldt's conjecture, Can. Math. Bull. 31, 3 (1988), 338–346.

[SchFK] Schmidt, F.K., Zur Klassenkörpertheorie im Kleinen, J. reine angew. Math. 162 (1930), 155–168.

[Schm] Schmithals, B., Konstruktion imaginärquadratischer Körper mit unendlichem Klassenkörperturm, Arch. Math. 34 (1980), 307–312.

[Schn] Schneider, P., Über gewisse Galoiscohomologie gruppen, Math. Z. 168 (1979), 181–205.

[Scho] Schoof, R., Infinite class field towers of quadratic fields, J. reine angew. Math. 372 (1986), 209–220.

[Scholz] Scholz, A., Totale Normenreste, die keine Normen sind, als Erzeuger nicht abelscher Körpererweiterungen, J. reine angew. Math. 172 (1936), 100–107; J. reine angew. Math. 182 (1940), 217–234.

[Se4] Serre, J-P., Représentations linéaires des groupes finis, Coll. Méthodes, Hermann, Paris, troisième édition 1978; English translation: Linear representations of finite groups, Graduate Texts in Math. 42, Springer-Verlag 1977.

[Se5] Serre, J-P., Groupes algébriques et corps de classes, Hermann, Paris, seconde édition 1975, reprint of the second edition 1984.

[Se6] Serre, J-P., Sur le résidu de la fonction zêta p-adique d'un corps de nombres, C.R. Acad. Sci. Paris, 287, Série I (1978), 183–188.

[Shi] Shirai, S., On the central class fields mod $\mathfrak{m}$ of Galois extensions of an algebraic number field, Nagoya Math. J. 71 (1978), 61–85.

[Sil] Silvester, J., Introduction to algebraic K-theory, Chapman and Hall 1981.

[Sim] Simon, D., Solving norm equations in relative number fields using S-units, Math. Comput. 71, 239 (2002), 1287–1305.

[Sor] Soriano-Gafiuk, F., Sur le noyau hilbertien d'un corps de nombres, C.R. Acad. Sci. Paris 330, Série I (2000), 863–866.

[St] Stern, L., Criterion for the equality of norm groups of idèle groups of algebraic number fields, Jour. Number Theory 62 (1997), 338–352.

[Ste] Stevenhagen, P., Ray class groups and governing fields, Thesis, University of Amsterdam, Amsterdam (1988).

[Su] Suzuki, H., A generalization of Hilbert's theorem 94, Nagoya Math. J. 121 (1991), 161–169.

[Ta1] Tate, J., Symbols in arithmetic, Actes Congrès International de Mathématiques 1970, Tome 1, 201–211.

[Ta2] Tate, J., Relations between K_2 and galois cohomology, Invent. Math. 36 (1976), 257–274.

[TBS] Tate, J., Les conjectures de Stark sur les fonctions L d'Artin en $s = 0$ (Notes rédigées par Bernardi, D. et Schappacher, N.), P.M. 47, Birkhäuser, Boston 1984.

[Wal1] Waldschmidt, M., Transcendance et exponentielles en plusieurs variables, Invent. Math. 63 (1981), 97–127.

[Wal2] Waldschmidt, M., A lower bound for the p-adic rank of the units of an algebraic number field, Topics in classical number theory, Budapest 1981, Coll. Math. Soc. Jànos Bolyai 34, North Holland (1984), 97–127.

[Wal3] Waldschmidt, M., Diophantine approximation on linear algebraic groups. Transcendence properties of the exponential function in several variables, Grundlehren 326, Springer-Verlag 2000.

[WsA] Weiss, A., Multiplicative module structure, Fields Institute Monographs, A.M.S. 1996.

[Wy] Wyman, B.F., What is a reciprocity law?, Amer. Math. Monthly, 79 (1972), 571–586.

[Y] Yamagishi, M., A note on free pro-p-extensions of algebraic number fields, J. Théorie des Nombres de Bordeaux 5 (1993), 165–178; Manuscripta Math. 91 (1996), 231–233.

Index of Notations[10]

places — valuations — absolute values — moduli:

p	prime number
K	number field (base field)
Pl	set of places of K
Pl_0, Pl_∞	set of finite, infinite, places of K
Pl_p	set of places of K dividing p
Pl_{ta}	$Pl_0 \backslash Pl_p$ (finite tame places of K)
Pl_∞^{r}, Pl_∞^{c}	set of real, complex, infinite places of K
Pl^{nc}	$Pl \backslash Pl_\infty^{\mathrm{c}}$ (set of noncomplex places of K)
r_1, r_2	$\lvert Pl_\infty^{\mathrm{r}} \rvert$, $\lvert Pl_\infty^{\mathrm{c}} \rvert$ (signature of K)
v	place (finite or infinite) of K
ℓ	residue characteristic of the finite place v
v	valuation corresponding to v
$\lvert \bullet \rvert_v$	absolute value corresponding to v
$\mathfrak{p}_v$	prime ideal corresponding to the finite place v
$\mathfrak{p}$, $\mathrm{N}\mathfrak{p}$	prime ideal, absolute norm of $\mathfrak{p}$
T	finite subset of Pl_0 (corresponding to ramification)
S	finite subset of Pl^{nc} (corresponding to decomposition)
T_p, S_p	$T \cap Pl_p$, $S \cap Pl_p$ (wild part of T, S)
T_{ta}	$T \backslash T_p$ (tame part of T)
S_0, S_∞	$S \cap Pl_0$, $S \cap Pl_\infty^{\mathrm{r}}$
$\mathfrak{m}$	$\prod_{v \in T} \mathfrak{p}_v^{m_v}$, $m_v \geq 0$ (modulus of K)
$\mathfrak{m}_p$	$\prod_{v \in T_p} \mathfrak{p}_v^{m_v}$ (wild part of $\mathfrak{m}$)
$\mathfrak{m}_{\mathrm{ta}}$	$\prod_{v \in T_{\mathrm{ta}}} \mathfrak{p}_v$ (tame part of $\mathfrak{m}$)
$\varphi(\mathfrak{m})$	$\prod_{v \in T,\, m_v \geq 1} (\mathrm{N}\mathfrak{p}_v^{m_v - 1}(\mathrm{N}\mathfrak{p}_v - 1))$ (Euler φ-function)

[10] The notations are not put in alphabetic order but are, roughly speaking, classified according to the main notions and subjects gradually encountered.

ray groups — ideal groups — class groups:

$K_T^\times$	group of $x \in K^\times$ prime to T
$K_{T,\mathfrak{m},\Delta_\infty}^\times$	group of $x \in K^\times$, prime to T, congruent to 1 modulo $\mathfrak{m} \in \langle T \rangle_\mathbb{N}$, and positive at the places of $\Delta_\infty \subseteq Pl_\infty^\mathrm{r}$
$K_{T,\mathfrak{m},\mathrm{pos}}^\times$	$K_{T,\mathfrak{m},Pl_\infty^\mathrm{r}}^\times$ (group of totally positive elements of $K_{T,\mathfrak{m}}^\times$)
P_T	$\{(x),\ x \in K_T^\times\}$
$P_{T,\mathfrak{m},\Delta_\infty}$	$\{(x),\ x \in K_{T,\mathfrak{m},\Delta_\infty}^\times\}$
$P_{T,\mathfrak{m},\mathrm{pos}}$	$P_{T,\mathfrak{m},Pl_\infty^\mathrm{r}} := \{(x),\ x \in K_{T,\mathfrak{m},\mathrm{pos}}^\times\}$
$P_{T,\mathfrak{m},\mathrm{pos}}\langle S \rangle$	$P_{T,\mathfrak{m},\Delta_\infty} . \langle S_0 \rangle$ where $\langle S_0 \rangle := \langle \mathfrak{p}_v,\ v \in S_0 \rangle_\mathbb{Z}$ and $\Delta_\infty := Pl_\infty^\mathrm{r} \setminus S_\infty$
I	group of fractional ideals of K
I_T	subgroup of ideals of K prime to T
$\mathcal{C}\ell_\mathfrak{m}^S$	$I_T/P_{T,\mathfrak{m},\mathrm{pos}}\langle S \rangle := I_T/P_{T,\mathfrak{m},\Delta_\infty} . \langle S_0 \rangle$ (group of S-ray classes modulo $\mathfrak{m}$)
$c\ell_\mathfrak{m}^S$	canonical map $I_T \longrightarrow \mathcal{C}\ell_\mathfrak{m}^S$
$\mathcal{C}\ell^\mathrm{res}$	$\mathcal{C}\ell$ (restricted (or narrow) class group)
$\mathcal{C}\ell^\mathrm{ord}$	$\mathcal{C}\ell^{Pl_\infty^\mathrm{r}}$ (ordinary class group)
$\mathcal{C}\ell^{S_0\,\mathrm{res}}$	$\mathcal{C}\ell^{S_0}$ (restricted S_0-class group)
$\mathcal{C}\ell^{S_0\,\mathrm{ord}}$	$\mathcal{C}\ell^{S_0 \cup Pl_\infty^\mathrm{r}}$ (ordinary S_0-class group)
$\mathcal{C}\ell_T^S$	$\varprojlim_{\mathfrak{m} \in \langle T \rangle_\mathbb{N}} \mathcal{C}\ell_\mathfrak{m}^S$
$\widetilde{\mathcal{C}\ell}$	logarithmic p-class group

global units — S-units — infinitesimal units — pseudo-units:

$E_\mathfrak{m}^S$	group of S-units of K congruent to 1 modulo $\mathfrak{m}$
E^res	E (group of units in the restricted sense, i.e., totally positive)
E^ord	$E^{Pl_\infty^\mathrm{r}}$ (group of units in the ordinary sense)
$E^{S_0\mathrm{res}}$	E^{S_0} (group of S_0-units in the restricted sense)
$E^{S_0\mathrm{ord}}$	$E^{S_0 \cup Pl_\infty^\mathrm{r}}$ (group of S_0-units in the ordinary sense)
E'^S	$\Big\{\varepsilon \in E^S,\ i_T(\varepsilon) \in \bigoplus_{v \in T_p} U_v^1 \bigoplus_{v \in T_\mathrm{ta}} (F_v^\times)_p\Big\}$ (T-principal S-units)
$\mathrm{adh}_T(E'^S)$	closure ("adherence") in $\bigoplus_{v \in T_p} U_v^1 \bigoplus_{v \in T_\mathrm{ta}} (F_v^\times)_p$ of $i_T(E'^S)$
$\mathcal{E}_\mathfrak{m}^S,\ \mathcal{E}_T^S$	$E_\mathfrak{m}^S \otimes \mathbb{Z}_p,\ \mathrm{Ker}\Big(\mathcal{E}^S \longrightarrow \bigoplus_{v \in T_p} U_v^1 \bigoplus_{v \in T_\mathrm{ta}} (F_v^\times)_p\Big)$
$Y_{T,\mathfrak{m}}^S$	$\{y \in K_T^{\times p} K_{T,\mathfrak{m},\Delta_\infty}^\times,\ (y) = \mathfrak{a}^p \mathfrak{a}_{S_0},\ \mathfrak{a} \in I_T,\ \mathfrak{a}_{S_0} \in \langle S_0 \rangle\}$
V_T^S	$\{\alpha \in Y_T^S := Y_{T,(1)}^S,\ i_v(\alpha) \in K_v^{\times p}\ \ \forall v \in T\}$

global extensions — Galois groups — p-subextensions:

$L/K,\ \mathrm{N}_{L/K}$	arbitrary finite extension, norm in L/K
L^ab	maximal abelian extension of K in L

$L^{\mathrm{ab}\,S}$	maximal S-split extension of K in L^{ab}
G	Galois group of a Galois extension L/K
G^{ab}	$\mathrm{Gal}(L^{\mathrm{ab}}/K)$ (abelianization of G when it exists)
$G^{\mathrm{ab}\,S}$	Galois group of $L^{\mathrm{ab}\,S}/K$
M/K	arbitrary abelian extension
$M(p)$	maximal p-extension of K in M
$M[p^e]$	maximal extension of K in M of exponent dividing p^e
M^{nc}	maximal noncomplexified extension of K in M

completions of an extension — local Galois groups — local data:

$Pl_{L,v}$	set of places of L above $v \in Pl$
$w\|v$	place of L above v
K_v	completion of K at the place v
L_w, N_{L_w/K_v}	completion of L at $w\|v$, norm in L_w/K_v
L_w^{ab}	maximal abelian extension of K_v in L_w
L_w^{nr}	inertia field of L_w/K_v (we have $L_w^{\mathrm{nr}} \subseteq L_w^{\mathrm{ab}}$)
G_w	Galois group of a Galois extension L_w/K_v
G_w^0	$\mathrm{Gal}(L_w/L_w^{\mathrm{nr}})$ (inertia group of a Galois extension L_w/K_v)
G_w^{ab}	$\mathrm{Gal}(L_w^{\mathrm{ab}}/K_v)$ ($G_w/[G_w, G_w]$ in the Galois case)
$G_w^{\mathrm{ab}\,0}$	$\mathrm{Gal}(L_w^{\mathrm{ab}}/L_w^{\mathrm{nr}})$ ($G_w^0/[G_w, G_w]$ in the Galois case)
e_w, e_w^{ab}	ramification index of L_w/K_v, L_w^{ab}/K_v
$f_w = f_w^{\mathrm{ab}}$	residue degree of L_w/K_v or of L_w^{ab}/K_v
L_v^{ab}	$\bigcap_{w\|v} L_w^{\mathrm{ab}}$
G_v^{ab}	$\mathrm{Gal}(L_v^{\mathrm{ab}}/K_v)$
$G_v^{\mathrm{ab}\,0}$	inertia group of L_v^{ab}/K_v
e_v^{ab}, f_v^{ab}	ramification index, residue degree, of L_v^{ab}/K_v
π_v	uniformizer of K_v (if $K_v = \mathbb{R}$, $\pi_v \in -\mathbb{R}^{\times +}$)
$U_v =: U_v^0$	unit group of K_v (if $K_v = \mathbb{R}, \mathbb{C}$, $U_v := \mathbb{R}^{\times +}, \mathbb{C}^\times$)
U_v^i	subgroups $1 + (\pi_v^i)$ of U_v, $i \geq 1$ (if $K_v = \mathbb{R}, U_v^i := \mathbb{R}^{\times +}$)
F_v	residue field of K at v (if $K_v = \mathbb{R}$, $F_v := \mathbb{R}$)
q_v	cardinality of F_v, $v \in Pl_0$
$(F_v^\times)_p$	p-Sylow subgroup of $F_v^\times$ (if $F_v = \mathbb{R}$, $(F_v^\times)_p := (\{\pm 1\})_p$)
$\widehat{\mathbb{Z}}$, $\widehat{K_v^\times}$	profinite completion of $\mathbb{Z}$, $K_v^\times$

roots of unity (global, local):

μ_n	group of nth roots of unity
$\mu(K)$, $\mu(K_v)$	group of roots of unity of K, K_v
$\mu_p(K)$, $\mu_p(K_v)$	p-Sylow subgroup of $\mu(K)$, $\mu(K_v)$

algebraic and abelian closures (global, local):

$\overline{K},\ \overline{K}_v$	algebraic closure of K, K_v
$\overline{K}^{\mathrm{ab}},\ \overline{K}_v^{\mathrm{ab}}$	abelian closure of K, K_v, in $\overline{K}$, $\overline{K}_v$
$\overline{K}_v^{\mathrm{nr}}$	maximal unramified ("non ramifiée") extension of K_v in $\overline{K}_v$
$\overline{G},\ \overline{G}_v$	Galois group of $\overline{K}/K$, $\overline{K}_v/K_v$
$\overline{G}^{\mathrm{ab}},\ \overline{G}_v^{\mathrm{ab}}$	abelianization of $\overline{G}$, $\overline{G}_v$ ($\mathrm{Gal}(\overline{K}^{\mathrm{ab}}/K)$, $\mathrm{Gal}(\overline{K}_v^{\mathrm{ab}}/K_v)$)

global inertia and decomposition:

D_w	decomposition group of w in a Galois extension L/K
L^{D_w}	decomposition field of w in a Galois extension L/K
I_w	inertia group of w in a Galois extension L/K
L^{I_w}	inertia field of w in a Galois extension L/K
$D_w^{\mathrm{ab}},\ I_w^{\mathrm{ab}}$	abelianization of D_w, image of I_w in D_w^{ab}

idèle groups — idelic embeddings — idèle class groups:

$J,\ J_0$	group of idèles, reduced idèles, of K
$J_{T,\mathfrak{m}}$	group of idèles prime to T, congruent to 1 modulo $\mathfrak{m}$
$J_{T,\mathfrak{m},\Delta_\infty}$	group of idèles prime to T, congruent to 1 modulo $\mathfrak{m}$, and positive at the places of $\Delta_\infty \subseteq Pl_\infty^{\mathrm{r}}$
$J_{T,\mathfrak{m},\mathrm{pos}}$	$J_{T,\mathfrak{m},Pl_\infty^{\mathrm{r}}}$ (group of idèles prime to T, congruent to 1 modulo $\mathfrak{m}$, and totally positive)
$U_{\mathfrak{m}}^S$	group of S-unit idèles congruent to 1 modulo $\mathfrak{m}$
U^{res}	U (unit idèles in the restricted sense)
U^{ord}	$U^{Pl_\infty^{\mathrm{r}}}$ (unit idèles in the ordinary sense)
$\langle\, S\,\rangle$	$\bigoplus_{v\in S} K_v^\times := \{(x_v)_v \in J,\ x_v = 1\ \ \forall v \notin S\}$ (subgroup of J of idèles with support contained in S)
$\boldsymbol{x}\ \boldsymbol{y},\ \ \boldsymbol{u}$	idèles of K, unit idèle
$i_v,\ i_w$	embedding of K in K_v, L in L_w
sgn_v	sign function: $K^\times \longrightarrow K_v^\times/K_v^{\times 2} \simeq \{\pm 1\}$ ($v \in Pl_\infty^{\mathrm{r}}$)
sgn	$(\mathrm{sgn}_v)_{v\in Pl_\infty^{\mathrm{r}}}$ (signature homomorphism of K)
i_{T,Δ_∞}	$(i_v)_{v\in T\cup\Delta_\infty}$
$i,\ i_0$	$(i_v)_{v\in Pl} : K^\times \longrightarrow J$, $(i_v)_{v\in Pl_0} : K^\times \longrightarrow J_0$
$\overline{i}_v$	embedding $K_{\{v\}}^\times \otimes \mathbb{Z}_p \longrightarrow (U_v)_p$
$C,\ C_0$	idèle class group, reduced idèle class group
$\mathcal{cl},\ \mathcal{cl}_0$	canonical map $J \to C$, $J_0 \to C_0$
$D,\ D_0$	connected component of the unit element of C, C_0

conductors — Frobenius' — normic symbols (global, local):

$R \subset Pl_0$	set of places of K ramified in $L^{\rm ab}/K$
$\mathfrak{f}_{L/K} =: \mathfrak{f}$	norm (or Artin) conductor of L/K (equal to $\mathfrak{f}_{L^{\rm ab}/K}$)
$\mathfrak{f}_v(L^{\rm ab}/K)$	local v-conductor of $L^{\rm ab}/K$ (equal to $\mathfrak{f}_{(L^{\rm ab})_v/K_v}$)
$\left(\frac{L^{\rm ab}/K}{v}\right)$	Frobenius automorphism of $v \notin R$ in $L^{\rm ab}/K$
$\left(\frac{L^{\rm ab}/K}{\mathfrak{a}}\right)$	Artin symbol of $\mathfrak{a}$ in $L^{\rm ab}/K$ ($\mathfrak{a}$ prime to R)
(L_w/K_v)	Frobenius automorphism of L_w/K_v (unramified)
$(x, L_w/K_v)$	local norm residue symbol of $x \in K_v^\times$ (or local reciprocity map) in L_w/K_v (values in $G_w^{\rm ab}$)
$\left(\frac{x\,,\,L/K}{v}\right)$	norm residue symbol (or Hasse symbol) of $x \in K^\times$ at v in L/K (values in $G^{\rm ab}$)
$\left(\frac{x\,,\,y}{v}\right)$, $(x\,,\,y)_v$	Hilbert symbol on K, on K_v
$\mathrm{WK}_2(K)$	Hilbert (= wild) kernel in $\mathrm{K}_2(K)$
$\mathrm{R}_2(K)$	regular kernel in $\mathrm{K}_2(K)$

global reciprocity maps — ray fields — restricted ramification:

$\rho_{L/K}$, $\rho_{L/K}^S$	global reciprocity map for L/K (values in $G^{\rm ab}$, $G^{{\rm ab}\,S}$)
$\alpha_{L/K}$, $\alpha_{L/K}^S$	Artin map for L/K (values in $G^{\rm ab}$, $G^{{\rm ab}\,S}$)
$A_{L/K,T}$, $A_{L/K}$	Artin group in I_T, Artin group, for L/K
$K(\mathfrak{m})^S$	S-split ray class field modulo $\mathfrak{m}$
$K(\mathfrak{m})^{\rm res}$	$K(\mathfrak{m})$ (ray class field modulo $\mathfrak{m}$ in the restricted sense)
$K(\mathfrak{m})^{\rm ord}$	$K(\mathfrak{m})^{Pl_\infty^{\rm r}}$ (ray class field modulo $\mathfrak{m}$ in the ordinary sense (or noncomplexified))
H_T^S	maximal T-ramified S-split abelian extension of K
$H_{\rm ta}^{\rm res}$	$H_{\rm ta}$ (maximal tamely ramified abelian extension of K)
$H^{\rm res}$	H (restricted (or narrow) Hilbert class field of K, i.e., maximal unramified (at finite places) abelian extension of K)
$H^{\rm ord}$	$H^{Pl_\infty^{\rm r}}$ (ordinary Hilbert class field of K, i.e., maximal unramified noncomplexified abelian extension of K)
$H_{L/K,T}^S$	genus field (relative to T and S) for L/K
$g_{L/K,T}^S$	number of genera (relative to T and S) for L/K

pro-p-extensions — p-adic logarithms — p-adic ranks:

$\mathcal{G}_T^S$	Galois group of the maximal T-ramified S-split pro-p-extension $\overline{H}_T^S(p)$ of K
$\mathcal{A}_T^S$, $\mathcal{B}_T^S$	$\mathcal{G}_T^{S\,{\rm ab}} = \mathrm{Gal}(H_T^S(p)/K)$, $\mathrm{Gal}(H_T^S(p)/H^S(p))$
$\mathcal{T}_T^S$	torsion subgroup of $\mathcal{A}_T^S$
$\widetilde{K}_p$	compositum of the $\mathbb{Z}_p$-extensions of K

$\widetilde{K}_{T_p}^{S_0}$	maximal T_p-ramified S_0-split subextension of $\widetilde{K}_p$
$\widetilde{K}_{T_p}^{S_0\,\mathrm{fr}}$	compositum of the T_p-ramified S_0-split $\mathbb{Z}_p$-extensions of K (maximal free subextension of $\widetilde{K}_{T_p}^{S_0}$)
$\mathbb{Q}^{\mathrm{cycl}}(p)$	cyclotomic $\mathbb{Z}_p$-extension of $\mathbb{Q}$
$\log$, $\log_v$	Iwasawa logarithm on $\mathbb{C}_p^\times$, $\log \circ i_v : K^\times \to K_v$
$\log_{T_p}$	$(\log_v)_{v \in T_p}$
$\mathcal{L}_{T_p}^{S_0}$	$\mathbb{Q}_p$-vector space $\Big(\bigoplus_{v \in T_p} K_v\Big) \Big/ \mathbb{Q}_p \log_{T_p}(E^{S_0})$
$\mathcal{I}_T$	$I_T \otimes \mathbb{Z}_p$
$\mathcal{P}_{T,\infty,\Delta_\infty}$	$\bigcap_n (P_{T,\mathfrak{m}(n),\Delta_\infty} \otimes \mathbb{Z}_p)$ with $\mathfrak{m}(n) := \prod_{v \in T_p} \mathfrak{p}_v^n \prod_{v \in T_{\mathrm{ta}}} \mathfrak{p}_v$
$\mathrm{Log}_{T_p}^{S_0}$	logarithm on $\mathcal{I}_T$ (values in $\mathcal{L}_{T_p}^{S_0}$)
$\mathcal{Z}_{T_p}^{S_0}$	$\mathrm{Gal}(\widetilde{K}_{T_p}^{S_0\,\mathrm{fr}}/K)$
$\mathrm{r}_{T_p}^{S_0}$, $\mathrm{r}_p^{S_0}$	T_p-adic rank of E^{S_0}, $\mathrm{r}_{Pl_p}^{S_0}$ (p-adic rank of E^{S_0})
$\widetilde{\mathrm{r}}_{T_p}^{S_0}$	$\mathbb{Z}_p$-rank of $\mathcal{Z}_{T_p}^{S_0}$
$\widetilde{\mathrm{r}}_p$	$\widetilde{\mathrm{r}}_{Pl_p}$ (number of independent $\mathbb{Z}_p$-extensions of K)
$\mathcal{O}_p$	$\mathcal{O}_{Pl_p}$ for the object $\mathcal{O}$

algebraic notions — characters and representations:

$\mathrm{rk}_p(A)$	$\mathbb{F}_p$-dimension of A/A^p for a $\mathbb{Z}$-module A
$(A)_p$	$A/\{\alpha \in A,\ \alpha$ of order prime to $p\}$ or p-Sylow subgroup of A (A commutative finite or profinite)
a_p	p-part of the natural integer a
${}_{p^e}A$	$\{a \in A,\ a^{p^e} = 1\}$
$A \otimes \mathbb{Z}_p$	$A \otimes_{\mathbb{Z}} \mathbb{Z}_p$ (p-completion of the $\mathbb{Z}$-module A)
$\mathrm{tor}_\Lambda(A)$	torsion sub-Λ-module of the Λ-module A
${}_{\mathrm{N}}A$, ${}_{\nu}A$	kernel in A of the arithmetic norm N, the algebraic norm ν, in an extension L/K
$A^\times$	group of invertible elements of the ring A
A^*	dual of the (finite or profinite) abelian group A
N, Γ	Galois closure of K over $\mathbb{Q}$, Galois group of $N/\mathbb{Q}$
e_H	$\frac{1}{\lvert H \rvert} \sum_{t \in H} t$, for a subgroup H of Γ
ψ	$\mathbb{C}_p$-irreducible character of N (i.e., of Γ)
V_ψ	absolutely irreducible representation with character ψ
Ψ_N	set of $\mathbb{C}_p$-irreducible characters of N

General Index

abelian closure
  global, 163, 274, 279, 291, 323
  local, 86, 90, 115, 298
absolute value, 9, 14
approximation theorem, 35, 94, 211, 286, 456, 457
Artin
  conductor, 133, 144, 145
  group, 131, 145, 163, 256, 349
  map, 131, 138, 146, 162, 164, 222, 242, 343, 361, 364, 367, 419, 455
  reciprocity law, 113, 132
  symbol, 132, 137, 138, 205, 350

central classes, 398, 400
Chevalley's formula, 180
chinese remainder theorem, 35, 94, 211, 286, 456, 457
class (ideal), 141
class field
  T-ramified, S-split, 152
  general, 143, 221, 235, 236, 398, 405, 433
  tower, 170, 172, 175, 176, 441, 445, 460, 462
class field theory
  cohomological, 108, 445
  global infinite, 121, 257, 340
  Jaulent's global, 340
  local infinite, 86
class group
  ideal, 37, 43, 45, 47, 164, 178, 221, 223, 283, 398, 402
  idèle, 28, 45, 113, 340
  logarithmic, 215, 319, 355, 357
  ordinary sense, 8, 38, 223
  restricted sense, 8, 38, 223
cohomology, 96, 97, 108, 182, 287, 321, 361–363, 442, 447, 458, 465
completion
  of a number field, 13, 15, 21
  of an extension, 20
  profinite, 87, 229
  semi-local, 331
complexification
  extension, 112, 139, 143, 146, 161, 167, 223, 241, 275, 280, 296, 419
  real place, 7, 24, 30, 140, 161, 365, 445
conductor
  computation, 81, 126, 127
  congruence group, 164
  global, 118, 126, 133, 144, 231, 389
  local, 80, 83
  nonabelian, 128
  theorem, 127

congruence group, 164
conjecture
  Gross's p-adic, 215, 318, 358
  Leopoldt, 237, 253, 258, 262, 271, 274
  p-adic, 258–260, 262, 271, 275
  strong p-adic, 305, 313
connected component, 30, 32, 34, 50, 122, 234, 248, 282, 297, 325, 445
cyclotomic field
  global, 7, 137, 161, 168, 232, 233
  local, 90, 115, 389

decomposition (group, field), 17, 19, 39, 66, 69, 71, 119, 140, 151, 224, 226, 287, 289, 299, 302, 304, 317, 333, 337, 341, 344
decomposition law
  general, 112, 140, 147, 163
  Kummer, 59
  wild place, 300, 345
  $\mathbb{Z}_p$-extension, 237, 258, 316, 342, 345
decomposition of a place in an extension, 10
density theorem, 141
deployment theorem, 232, 275, 277, 278, 289, 337, 338, 341, 365, 385, 393, 446, 448
Dirichlet theorem, 268, 282
Dirichlet–Herbrand theorem, 24, 157, 261, 265, 307, 308
divisible (element, group), 31, 32, 49, 124, 245, 297, 325, 327, 331, 446
duality, 53, 56, 57, 97, 182, 287, 331, 332, 334, 411, 414, 432, 443, 444, 449

embedding
  cyclic, 328
  for a p-completion, 238, 301
  idelic, 28, 41, 48, 235, 264
  Kummer extension, 351
  of a number field, 12, 13, 15, 18, 66, 91
exact sequences (of global class field theory), 4, 5, 236, 239, 250, 279, 280, 294
exceptional case, 183, 190, 191, 411, 415, 427, 430, 431, 433, 437, 439
existence theorem
  global, 118, 141, 160
  local, 77

free pro-p-extension, 73, 234, 241, 257, 280, 284, 285, 300, 374, 452, 464
Frobenius automorphism
  global, 39, 70, 111, 131, 148, 204, 250, 343, 372, 404
  local, 68, 69, 75, 107
fundamental class, 109, 121

genera (number of), 381, 383, 385, 389
generalized class group, 37, 40, 130, 157, 164, 407, 438
genus
  exact sequence, 390, 394
  field, 376–378, 435
  theory, 375, 390, 408, 419
governing field, 410, 438
Gross's conjecture, 215, 318, 358

Grunwald–Wang theorem, 332, 335, 439

Hasse principle
- defect, 181, 398, 400, 402, 405
- general, 176
- normic, 179, 180, 329, 402
- powers, 183, 191
- quadratic form, 192, 202

Hasse symbol
- computation, 135, 137
- general, 105, 106, 113, 116, 135, 207, 343, 395

Hasse–Minkowski theorem, 192, 202

H_T^S, H_p^{ord}, $H_{\mathrm{ta}}^{\mathrm{res}}$, 152, 221, 236, 240, 243, 250, 293, 300, 441

Hensel's lemma, 22

higher ramification group, 72, 129, 140

Hilbert
- class field, 62, 134, 143, 146, 148, 152, 169, 221, 223, 235, 376, 405, 433
- kernel, 213, 318, 360
- tower, 170, 172, 175, 176, 441, 460

Hilbert symbol
- global, 200, 201, 210, 358
- global computation, 207
- local, 85, 195, 200, 217
- quadratic, 192
- regular, 198, 200

idèle class (of finite order), 184, 324

idèle class group, 28, 45, 113, 340

idèle group
- Galois action, 91, 94
- general, 28
- globalized, 340
- reduced, 33, 48, 331
- topology, 31, 33, 49
- units, 30, 340

inertia (group, field), 66, 69, 71, 119, 140, 150, 224, 287, 296, 302, 344

infinitesimal number, 247, 286, 355, 366, 454, 456, 458

invariant classes, 178, 180, 365, 369

K_2 (fundamental diagram), 210

K_2 of a field, 199, 217, 359

knot group, 182, 398, 400, 401, 405, 454, 465

Kummer
- duality, 57, 156, 411
- theory, 54, 58, 59, 81, 85, 160, 184, 348, 384, 423

Leopoldt conjecture, 237, 253, 258, 262, 271, 274

local norm group (definition), 98

logarithm, 242, 284, 343, 355, 370

Martinet's constants, 171, 176

modulus, 23

monogeneous, 27, 259, 262, 268, 275, 308, 322, 364, 394, 456

Moore's theorem, 210, 214, 359

norm
- local-global relation, 92
- of class group, 164
- of local elements, 104
- of local units, 104

norm group
- Galois action, 120
- global, 118, 120, 127, 142, 146, 160, 329, 398, 402, 405

local, 77, 78, 80, 83, 85, 88, 98, 100, 106, 115, 137, 254, 381, 454
local-global relation, 111
norm invariants (local), 100
norm lifting theorem
global, 119, 142
local, 79
norm principle defect, 181, 398, 400, 402, 405
norm residue symbol, 74–76, 107, 115, 195, 196, 343, 382, 391, 397

ordinary class group, 38, 223
ordinary sense, 8, 223
Ostrowski theorem, 9

p-adic conjecture, 258–260, 262, 271, 275, 305, 313
p-adic rank, 237, 258, 266, 268, 304, 316, 342
p-adic zeta function, 255
p-completion, 228, 238, 244, 245, 301, 341
place (of a number field), 7, 9, 11, 14
power residue symbol, 204, 350
p-primitive (set, extension), 371–374
p-rank
class group, 40, 42, 44, 152, 246, 386, 396
cohomology group, 452
torsion group, 282, 452, 458
p-rational field, 373, 375, 403
p-regular field, 216, 375, 403
principal ideal theorem, 168
principal idèle, 28
product formula, 112, 116, 135, 179, 201, 202, 206, 209, 212, 219, 336, 356, 358, 387, 393, 395, 397

quadratic
field, 83, 144, 223, 329, 386
form, 180, 192, 202, 387
reciprocity law, 113, 217

radical
initial, 330, 349, 351
Kummer, 54, 58, 153, 156, 215, 411, 415, 418, 423, 433
ramification (global)
p-primitive, 371, 373, 374
minimal, 421
tame, 73, 77, 81, 89, 91, 118, 140, 150, 152, 168, 173, 221, 230, 289, 348, 418, 430
wild, 59, 82, 230, 240, 300, 345
ray class field
general, 143, 146, 147, 161, 168, 276, 299, 377, 385, 407
ramification index, 150
residue degree, 140, 150
ray class group, 37
ray group, 23
reciprocity law
Artin, 113, 132
nth, 204, 206
quadratic, 113, 217
reciprocity map
global, 105, 108, 110, 122, 146, 148, 222, 244, 295, 324, 340, 411, 446
local, 74, 88, 109, 115

reflection theorem, 44, 152, 157, 214, 216, 283, 323, 374, 408
regular kernel, 213, 216
residue field, 21
restricted class group, 38, 223
restricted sense, 8, 223

Šafarevič's formula, 44, 152, 160, 441
Schmidt–Chevalley theorem, 124, 288, 340
S-class group, 37, 40, 130, 157, 164, 407, 438
Shapiro's lemma, 96, 362, 445, 447, 457
signature, 12
special case, 183, 184, 190, 324, 328, 332, 335, 354
Spiegelungssatz, 156, 158, 159
splitting of a class field, 404
S-unit group
- global, 23, 24, 235, 261, 271, 397, 414
- idelic, 30

symbol
- Artin, 132, 137, 138, 205, 350
- Hasse, 105, 106, 113, 116, 135, 207, 343, 395
- Hilbert, 85, 195, 200, 201, 210, 217, 358
- norm residue, 74–76, 107, 115, 195, 196, 343, 358, 382, 391, 397
- on a field, 199
- power residue, 204, 350

Takagi group, 164
tame kernel (= regular kernel), 213
tame ramification, 73, 77, 81, 89, 91, 118, 140, 150, 152, 168, 173, 221, 230, 289, 348, 418, 430
topological group, 30, 50, 51, 53
topology (idèle group), 31, 33, 49
torsion group ($\mathcal{T}_T^S$, $\mathcal{T}_p^{\mathrm{ord}}$), 241, 250, 253, 280, 282, 319, 345, 362, 369, 451, 458
transfer map, 74, 75, 106, 139, 168, 327, 361, 364, 367, 404, 448, 450

unit group
- global, 23, 180, 223, 261
- idelic, 30, 340
- local, 21, 230
- restricted sense, 8, 24

unramified extension
- global, 62, 139, 143, 148, 176, 223, 302
- local, 68, 75, 77

valuation, 12, 355
v-associate, 136, 194, 208
v-conductor, 80, 81

wild kernel (= Hilbert kernel), 213, 318, 360
wild ramification, 59, 82, 230, 240, 300, 345

$\mathbb{Z}_p$-extension
- cyclotomic, 253, 267, 318, 354
- decomposition law, 237, 258, 316, 342, 345
- general, 234, 240, 280, 342
- initial radical, 349, 351
- local, 90

$\mathbb{Z}_p$-rank (Galois group), 237, 258, 266, 268, 269, 304, 313, 316, 342

Springer **M***onographs in* **M***athematics*

This series publishes advanced monographs giving well-written presentations of the "state-of-the-art" in fields of mathematical research that have acquired the maturity needed for such a treatment. They are sufficiently self-contained to be accessible to more than just the intimate specialists of the subject, and sufficiently comprehensive to remain valuable references for many years. Besides the current state of knowledge in its field, an SMM volume should also describe its relevance to and interaction with neighbouring fields of mathematics, and give pointers to future directions of research.

Abhyankar, S.S. **Resolution of Singularities of Embedded Algebraic Surfaces** 2nd enlarged ed. 1998
Andrievskii, V.V.; Blatt, H.-P. **Discrepancy of Signed Measures and Polynomial Approximation** 2002
Armitage, D.H.; Gardiner, S.J. **Classical Potential Theory** 2001
Arnold, L. **Random Dynamical Systems** corr. 2nd printing 2003 (1st ed. 1998)
Aubin, T. **Some Nonlinear Problems in Riemannian Geometry** 1998
Bang-Jensen, J.; Gutin, G. **Digraphs** 2001
Baues, H.-J. **Combinatorial Foundation of Homology and Homotopy** 1999
Brown, K.S. **Buildings** 3rd printing 2000 (1st ed. 1998)
Cherry, W.; Ye, Z. **Nevanlinna's Theory of Value Distribution** 2001
Ching, W.K. **Iterative Methods for Queuing and Manufacturing Systems** 2001
Crabb, M.C.; James, I.M. **Fibrewise Homotopy Theory** 1998
Dineen, S. **Complex Analysis on Infinite Dimensional Spaces** 1999
Elstrodt, J.; Grunewald, F. Mennicke, J. **Groups Acting on Hyperbolic Space** 1998
Fadell, E.R.; Husseini, S.Y. **Geometry and Topology of Configuration Spaces** 2001
Fedorov, Y.N.; Kozlov, V.V. **A Memoir on Integrable Systems** 2001
Flenner, H.; O'Carroll, L. Vogel, W. **Joins and Intersections** 1999
Gelfand, S.I.; Manin, Y.I. **Methods of Homological Algebra** 2nd ed. 2002
Griess, R.L.Jr. **Twelve Sporadic Groups** 1998
Gras, G. **Class Field Theory** 2003
Ivrii, V. **Microlocal Analysis and Precise Spectral Asymptotics** 1998
Jech, T. **Set Theory** (3rd revised edition 2002)
Jorgenson, J.; Lang, S. **Spherical Inversion on SLn (R)** 2001
Khoshnevisan, D. **Multiparameter Processes** 2002
Koch, H. **Galois Theory of *p*-Extensions** 2002
Kozlov, V.; Maz'ya, V. **Differential Equations with Operator Coefficients** 1999
Landsman, N.P. **Mathematical Topics between Classical & Quantum Mechanics** 1998
Lebedev, L.P.; Vorovich, I.I. **Functional Analysis in Mechanics** 2002
Lemmermeyer, F. **Reciprocity Laws: From Euler to Eisenstein** 2000
Malle, G.; Matzat, B.H. **Inverse Galois Theory** 1999
Mardesic, S. **Strong Shape and Homology** 2000
Murdock, J. **Normal Forms and Unfoldings for Local Dynamical Systems** 2002
Narkiewicz, W. **The Development of Prime Number Theory** 2000
Parker, C.; Rowley, P. **Symplectic Amalgams** 2002
Prestel, A.; Delzell, C.N. **Positive Polynomials** 2001
Puig, L. **Blocks of Finite Groups** 2002
Ranicki, A. **High-dimensional Knot Theory** 1998
Ribenboim, P. **The Theory of Classical Valuations** 1999
Rowe, E.G.P. **Geometrical Physics in Minkowski Spacetime** 2001
Rudyak, Y.B. **On Thom Spectra, Orientability and Cobordism** 1998
Ryan, R.A. **Introduction to Tensor Products of Banach Spaces** 2002
Saranen, J.; Vainikko, G. **Periodic Integral and Pseudodifferential Equations with Numerical Approximation** 2002
Schneider, P. **Nonarchimedean Functional Analysis** 2002
Serre, J-P. **Complex Semisimple Lie Algebras** 2001 (reprint of first ed. 1987)
Serre, J-P. **Galois Cohomology** corr. 2nd printing 2002 (1st ed. 1997)

Serre, J-P. **Local Algebra** 2000
Smirnov, E. **Hausdorff Spectra in Functional Analysis** 2002
Springer, T.A.; Veldkamp, F.D. **Octonions, Jordan Algebras, and Exceptional Groups** 2000
Sznitman, A.-S. **Brownian Motion, Obstacles and Random Media** 1998
Tits, J.; Weiss, R.M. **Moufang Polygons** 2002
Uchiyama, A. **Hardy Spaces on the Euclidean Space** 2001
Üstünel, A.-S.; Zakai, M. **Transformation of Measure on Wiener Space** 2000
Yang, Y. **Solitons in Field Theory and Nonlinear Analysis** 2001